Birkhäuser

Trends in Mathematics

Trends in Mathematics is a series devoted to the publication of volumes arising from conferences and lecture series focusing on a particular topic from any area of mathematics. Its aim is to make current developments available to the community as rapidly as possible without compromise to quality and to archive these for reference.

Proposals for volumes can be submitted using the Online Book Project Submission Form at our website www.birkhauser-science.com.

Material submitted for publication must be screened and prepared as follows:

All contributions should undergo a reviewing process similar to that carried out by journals and be checked for correct use of language which, as a rule, is English. Articles without proofs, or which do not contain any significantly new results, should be rejected. High quality survey papers, however, are welcome.

We expect the organizers to deliver manuscripts in a form that is essentially ready for direct reproduction. Any version of TeX is acceptable, but the entire collection of files must be in one particular dialect of TeX and unified according to simple instructions available from Birkhäuser.

Furthermore, in order to guarantee the timely appearance of the proceedings it is essential that the final version of the entire material be submitted no later than one year after the conference.

More information about this series at http://www.springer.com/series/4961

Valérie Berthé • Michel Rigo
Editors

Sequences, Groups, and Number Theory

Birkhäuser

Editors
Valérie Berthé
IRIF, Université Paris Diderot
Paris, France

Michel Rigo
Department of Mathematics
University of Liège
Liège, Belgium

ISSN 2297-0215 ISSN 2297-024X (electronic)
Trends in Mathematics
ISBN 978-3-030-09874-2 ISBN 978-3-319-69152-7 (eBook)
https://doi.org/10.1007/978-3-319-69152-7

Mathematics Subject Classification (2010): 68R15, 20F10, 20M35, 11K16, 11B85, 03D40, 03B25, 03D05

Printed on acid-free paper

This book is published under the imprint Birkhäuser, www.birkhauser-science.com by the registered company Springer International Publishing AG part of Springer Nature.
The registered company address is: Gewerbestrasse 11, 6330 Cham, Switzerland

Preface

This collaborative volume aims at presenting and developing recent trends at the interface between the study of *sequences*, *groups*, and *number theory*, as the title may suggest. It is inspired by the celebrated Lothaire series [385–387] and animated by the same spirit as in the books [78, 79]. Among the various topics developed in this volume, let us quote the notions of automatic and regular sequences, of normality, of amenability of groups, but also of tilings and multidimensional subshifts, as striking examples of such bridges. These topics are handled with a viewpoint combining mathematics and theoretical computer science. On the one hand, some of the newest results in these areas have been selected for this volume and benefit from a synthetic exposition. On the other hand, emphasis on the connections existing between the main topics of the book is sought.

This book is primarily intended for graduate students or research mathematicians and computer scientists interested in combinatorics on words, automatic and regular sequences, numeration systems, normal numbers, automata theory, group theory, automaton groups, amenable groups, number theory and arithmetics, formal language theory and discrete dynamical systems, symbolic dynamics, but also tilings. We hope that some of the chapters can serve as a useful material for lecturing at a master or graduate level.

Let us succinctly sketch the contents of this contributed volume. The book can roughly be divided into four general blocks. The first block which is made of Chapters 2, 3, and 4 pertains to number theory and focuses on sequences. The second one made of Chapters 5 and 6 is devoted to word combinatorics. The third block, made of Chapters 7 and 8, focuses on normal numbers and provides two viewpoints on normal numbers, namely, a computer scientist and a dynamical perspective. The last block is concerned with group theory with Chapters 9, 10, and 11. Note that short abstracts of each chapter can be found below and at the beginning of each chapter.

Number theory is one of the main frames underpinning this book. One can represent a real number by an infinite word, for instance, by considering its development in an integer base. One can also code a set of natural integers by its characteristic sequence considered as an infinite word over the alphabet $\{0, 1\}$. Connections between number theory and the study of sequences are therefore natural.

Automatic and **regular sequences** provide rich and widely studied classes of sets, numbers, or functions, illustrating remarkably well these connections. Automatic sequences correspond to the most basic objects in terms of Chomsky–Schützenberger hierarchy, namely, regular languages, i.e., languages accepted by finite automata, and they allow the definition of "simple sets" of numbers by recognizing sets of representations in a given numeration system. Similarly, the notion of a regular sequence extends the concept of automatic sequence to sequences taking infinitely many values. For more on automatic and regular sequences, see the monograph [14]. This hierarchy can be revisited in terms of sequences, numbers, and functions, such as developed in Chapter 2 with the study of Mahler functions. Chapter 2 focuses in particular on the algebraic, analytic, and Diophantine properties of Mahler functions, by highlighting the rational-transcendental dichotomy. The question of the number theoretic properties of real numbers whose expansions are highly structured is also developed in Chapter 2, whereas Chapter 4 looks at applications of the theory of polynomial identities for automatic and regular sequences: a characterization of regular sequences is provided in terms of the so-called shuffle and power properties, stated in the context of noncommutative rational series by Berstel and Reutenauer in [77, Chap. 3].

Recently several authors have proposed formal methods to obtain automatic proofs (in the sense of automated theorem proving) about properties of k-automatic sequences and, more generally, for particular families of morphic words (see, for instance, [248, 540]). Indeed, some properties of interest for automatic sequences are expressible by a first-order (FO) formula of $\langle \mathbb{N}, +, V_b \rangle$, where V_b is a base-dependent predicate. One can derive the decidability of $\langle \mathbb{N}, +, V_b \rangle$ from the decidability of Presburger arithmetic $\langle \mathbb{N}, + \rangle$ together with the Büchi–Bruyère theorem [112, 114, 115] (see Theorem 3.3.4), and one thus can algorithmically decide, for instance, whether or not a given automatic sequence contains a repetition like a square or a cube. Chapter 3 deals with these connections involving formal logic, decision problems, automaticity, regularity, and numeration. It also shows how problems linked with the enumeration of combinatorial objects associated with automatic sequences give rise to regular sequences. Questions on repetitions and avoidance in words are also considered in Chapter 5.

Note that **logic** is a crucial notion in the present context that goes through several chapters, in particular in the context of tilings. Indeed, the domino problem asks for the existence of an algorithm deciding whether a finite set of Wang tiles may tile the plane. This problem was formulated on $\mathbb{Z}^2$ by Wang [580] in 1961 in order to study a fragment of first-order (FO) logic, as recalled in Chapter 9. As another example, the set theory notion of filters and ultrafilters is also considered in Chapters 6 and 11.

Similarly, **machines** issued from theoretical computer science are ubiquitous in this book and occur under various forms by providing hierarchies and measures of computational complexity or generating and constructing devices: let us quote Büchi automata in Chapter 3, transducers in Chapter 7, Mealy automata in

Chapter 10, Turing machines in Chapter 9, but also cellular automata in Chapters 10 and 11.

The study of **normal numbers** is another representative illustration of the numerous connections existing between numbers, sequences, computability, dynamical systems, probability, arithmetics, and algebra. Several (fast) effective constructions of normal numbers and of absolutely normal numbers are provided in Chapter 7, with the speed of convergence to normality being also considered. The analysis of the computational complexity is obtained by counting the number of mathematical operations required to output the first k digits of the expansion of the computed number in a designated base. Moreover, in Chapter 7, normality is also expressed in terms of non-compressibility by a bounded-to-one nondeterministic transducer, according to [56], together with further generalizations. The degree of randomness in the expansions is investigated in Chapter 8 under a dynamical viewpoint. Normality is discussed both from a topological and from a measure-theoretic viewpoint, by stressing the fact that normality corresponds to genericity for an invariant measure of maximal entropy. The symbolic dynamical systems considered in this chapter are assumed to fulfill the specification property and thus to have a unique measure of maximal entropy. Constructions by concatenations of words are then provided.

Dynamical systems and **computation** are known to have closed connections. Indeed, numerous computational models benefit from being viewed as a dynamical system, whereas the ability to encode computations in various dynamical systems provides information on their complexity and predictability. As an illustration of this interplay, Hochman's breakthrough [294, 295] on sofic multidimensional subshifts has shown how computability theory is needed for their understanding (see also [79, Chap. 9]). The main techniques in this framework rely on the use of Turing machines that can be encoded by tilings. These topics will be developed in Chapter 9 where the interactions between symbolic dynamics, substitution tilings, computability, and group combinatorics are stressed: subshifts are considered here both as computational models and discrete models for dynamical systems, and the decidability of the Domino Problem is reinterpreted as a group property. Different notions of effectiveness in subshifts defined over groups are also discussed. Recall that the emptiness problem for subshifts of finite type is equivalent to the domino problem, that is, the problem of tiling the plane with Wang tiles. Note that Wang tilings are also considered as a tool for the understanding of the behavior of automaton (semi)groups, such as developed in Chapter 10, where Wang tilesets are associated with complete and deterministic Mealy automata.

Combinatorics on words deals with problems that can be stated in a noncommutative monoid such as construction and properties of infinite words, k-automatic and k-regular sequences, unavoidable regularities or patterns, factorization and colorings, etc. Word combinatorics is a quite recent subject in discrete mathematics. One can trace it back to the early twentieth century with the works of Thue, dealing with repetition-free words, and then in the 1930s with Morse and Hedlund, with their fundamental work on symbolic dynamics. The expansion of this research topic is mostly due to Schützenberger in France and Novikov and Adjan in former Russia.

Several important problems in combinatorics on words (e.g., pattern avoidance of various forms such as squares, cubes, fractional, abelian powers, binomial powers, etc.) are explored and often solved by developing algorithmic methods. In particular, Chapter 5 is devoted to existence (or nonexistence) results for infinite words avoiding a repetition, a pattern, or even a formula. The size of a language of words avoiding some repetition pattern is also considered. Morphic words play a crucial role in this context. This approach complements the formal methods of Chapter 3. Another recurrent topic in combinatorics deals with factorization and coloring problems such as in Ramsey theory. Chapter 6 is about monochromatic factorizations of nonperiodic words and presents striking connections with topological compactification, Hindman's finite sums theorem, partition regularity of IP sets (an IP set is a set of natural numbers which contains all finite sums of some infinite set), and the Milliken–Taylor theorem.

Groups are ubiquitous in this book. They provide dynamical systems such as developed in Chapter 9, which deals with decidability problems (domino problem) involving subshifts of finite type on a finitely generated group. Group actions are considered in Chapter 11 with the notion of **self-similarity**, a notion which occurs in geometry, algebra, holomorphic dynamics, and computer science. The common language is the one of finite automata such as developed in Chapter 10. Self-similarity is interpreted in group theory as a group that contains permuted copies of itself as a group. Constructions using finite automata have allowed the developments of spectacular and unexpected group zoologies. They have permitted the proof of existence of finitely generated groups with intermediate growth using the automaton group of Grigorchuk or with nonuniform exponential growth, etc. Two classes of (semi)groups are considered in Chapter 10, namely, automatic and automata groups. Growth issues are also naturally considered in Chapters 10 and 11. The notion of **amenability** for group actions, whose study is thoroughly developed in Chapter 11, is again a striking example of the interaction between combinatorics, dynamics, group theory, functional analysis, probability, etc. Amenability for a group G acting on a set X can be formulated in terms of the existence of a G-invariant mean on subsets of X.

Let us conclude our brief presentation of the book with **numeration systems**. In a generic way, a numeration system allows the expansion of numbers as words over an alphabet of digits. A numeration system usually is either defined by an algorithm providing expansions or by an iterative process associated with a dynamical system. So again, words are demonstrating their representation power. Among the various questions related to the expansions of numbers, we have chosen to develop two focused viewpoints on β-numeration (i.e., numeration systems with non-integer bases), namely, connections with logic in Chapter 3 and normality issues in Chapter 8.

Parts of the material developed in this book were presented during the fourth CANT (Combinatorics, Automata and Number Theory) school that was organized at the Centre International de Rencontres Mathématiques (CIRM) from 28 November to 2 December 2016 in Marseille. We thank the CIRM for supporting this event.

We now give, by order of appearance, abstracts of every chapter. Chapter 1 is a general introduction where the main notions that will occur in this book are presented. The reader may skip this chapter in a first reading and use it as a reference if needed.

Chapter 2 by Michael Coons and Lukas Spiegelhofer
Number Theoretic Aspects of Regular Sequences

We present a survey of results concerning regular sequences and related objects. Regular sequences were defined in the early 1990s by Allouche and Shallit as a combinatorially, algebraically, and analytically interesting generalization of automatic sequences. In this chapter, after a historical introduction, we follow the development from automatic sequences to regular sequences, and their associated generating functions, to Mahler functions. We then examine size and growth properties of regular sequences. The last half of the chapter focuses on the algebraic, analytic, and Diophantine properties of Mahler functions. In particular, we survey the rational-transcendental dichotomies of Mahler functions, due to Bézivin, and of regular numbers, due to Bell, Bugeaud, and Coons.

Chapter 3 by Émilie Charlier
First-Order Logic and Numeration Systems

The Büchi–Bruyère theorem states that a subset of $\mathbb{N}^d$ is b-recognizable if and only if it is b-definable. This result is a powerful tool for showing that many properties of b-automatic sequences are decidable. Going a step further, first-order logic can be used to show that many enumeration problems of b-automatic sequences can be described by b-regular sequences. The latter sequences can be viewed as a generalization of b-automatic sequences to integer-valued sequences. These techniques were extended to two wider frameworks: U-recognizable subsets of $\mathbb{N}^d$ and β-recognizable subsets of $\mathbb{R}^d$. In the second case, real numbers are represented by infinite words, and hence, the notion of β-recognizability is defined by means of Büchi automata. Again, logic-based characterizations of U-recognizable (resp. β-recognizable) sets allow us to obtain various decidability results. The aim of this chapter is to present a survey of this very active research domain.

Chapter 4 by Jason Bell
Some Applications of Algebra to Automatic Sequences

We give an overview of the theory of rings satisfying a polynomial identity and use this to give a proof of a characterization due to Berstel and Reutenauer of automatic and regular sequences in terms of two properties, which we call the shuffle property and the power property. These properties show that if one views an automatic sequence f as a map on a free monoid on k-letters to a finite subset of a ring, then the values of f are closely related to values of f on related words obtained by permuting letters of the word. We use this characterization to give answers to three questions from Allouche and Shallit, two of which have not appeared in the literature. The final part of the chapter deals more closely with the shuffle property, and we view this as a generalization of regular sequences. We show that sequences with the shuffle property are closed under the process of taking sums, taking products; in addition we show that there is closure under a noncommutative product, which turns the collection of shuffled sequences into a noncommutative algebra. We show that this algebra is very large, in the sense that it contains a copy of a free associative algebra on countably many generators. We conclude by giving some open questions, which we hope will begin a more careful study of shuffled sequences.

Chapter 5 by Pascal Ochem, Michaël Rao, and Matthieu Rosenfeld Avoiding or Limiting Regularities in Words

It is commonly admitted that the origin of combinatorics on words goes back to the work of Axel Thue in the beginning of the twentieth century, with his results on repetition-free words. Thue showed that one can avoid cubes on infinite binary words and squares on ternary words. Up to now, a large part of the work on the theoretic part of combinatorics on words can be viewed as extensions or variations of Thue's work, that is, showing the existence (or nonexistence) of infinite words avoiding, or limiting, a repetition-like pattern. The goal of this chapter is to present the state of the art in the domain and also to present general techniques used to prove a positive or a negative result. Given a repetition pattern P and an alphabet, we want to know if an infinite word without P exists. If it exists, we are also interested in the size of the language of words avoiding P, that is, the growth rate of the language. Otherwise, we are interested in the minimum number of factor P that a word must contain. We talk about limitation of usual, fractional, abelian, and k-abelian repetitions and other generalizations such as patterns and formulas. The last sections are dedicated to the presentation of general techniques to prove the existence or the nonexistence of an infinite word with a given property.

Chapter 6 by Caïus Wojcik and Luca Zamboni Coloring Problems for Infinite Words

Given a finite coloring (or finite partition) of the free semigroup $\mathscr{A}^+$ over a set $\mathscr{A}$, we consider various types of monochromatic factorizations of right-sided infinite words $x \in \mathscr{A}^\omega$. In 2006, Brown asked the following question in the spirit of Ramsey theory: Given a nonperiodic infinite word $x = x_1x_2x_3\cdots$ with values in a set $\mathscr{A}$, does there exist a finite coloring $\varphi : \mathscr{A}^+ \to C$ relative to which x does not admit a φ-monochromatic factorization, i.e., a factorization of the form $x = u_1 u_2 u_3 \cdots$ with $\varphi(u_i) = \varphi(u_j)$ for all $i, j \geq 1$? We give an optimal affirmative answer to this question by showing that if $x = x_1 x_2 x_3 \cdots$ is an infinite nonperiodic word with values in a set $\mathscr{A}$, then there exists a 2-coloring $\varphi : \mathscr{A}^+ \to \{0, 1\}$ such that for any factorization $x = u_1 u_2 u_3 \cdots$, we have $\varphi(u_i) \neq \varphi(u_j)$ for some $i \neq j$. Some stronger versions of the usual notion of monochromatic factorization are also introduced and studied. We establish links, and in some cases equivalences, between the existence of these factorizations and fundamental results in Ramsey theory including the infinite Ramsey theorem, Hindman's finite sums theorem, partition regularity of IP sets, and the Milliken–Taylor theorem.

Chapter 7 by Verónica Becher and Olivier Carton
Normal Numbers and Computer Science

Émile Borel defined normality more than one hundred years ago to formalize the most basic form of randomness for real numbers. A number is normal to a given integer base if its expansion in that base is such that all blocks of digits of the same length occur in it with the same limiting frequency. This chapter is an introduction to the theory of normal numbers. We present five different equivalent formulations of normality, and we prove their equivalence in full detail. Four of the definitions are combinatorial, and one is in terms of finite automata, analogous to the characterization of Martin-Löf randomness in terms of Turing machines. All known examples of normal numbers have been obtained by constructions. We show three constructions of numbers that are normal to a given base and two constructions of numbers that are normal to all integer bases. We also prove Agafanov's theorem that establishes that a number is normal to a given base exactly when its expansion in that base is such that every subsequence selected by a finite automaton is also normal.

Chapter 8 by Manfred Madritsch
Normal Numbers and Symbolic Dynamics

The present chapter takes a dynamical viewpoint on normal numbers. Starting with a description of the link between dynamical systems and numeration systems, we present the concept of normal and non-normal numbers providing two different views on the dynamics of the system. Normal numbers are "normal" with respect to randomly chosen objects, whereas non-normal numbers and extreme variants thereof are examples of general objects, from a topological viewpoint. In the following sections, we present how to obtain maximal randomness as well as constructions of numbers with a given degree of chaos. Then we turn our attention to non-normal numbers. Since they are not completely random, we have to find a different measurement for analyzing their structure. The Hausdorff dimension will provide us with an interesting parameter in this context.

Chapter 9 by Nathalie Aubrun, Sebastián Barbieri, and Emmanuel Jeandel About the Domino Problem for Subshifts on Groups

From a classical point of view, the domino problem is the question of the existence of an algorithm which can decide whether a finite set of square tiles with colored edges can tile the plane, subject to the restriction that adjacent tiles share the same color along their adjacent edges. This question has already been settled in the negative by Berger in 1966; however, these tilings can be reinterpreted in dynamical terms using the formalism of subshifts of finite type, and hence, the same question can be formulated for arbitrary finitely generated groups. In this chapter, we present the state of the art concerning the domino problem in this extended framework. We also discuss different notions of effectiveness in subshifts defined over groups, that is, the ways in which these dynamical objects can be described through Turing machines.

Chapter 10 by Ines Klimann and Matthieu Picantin Automaton (Semi)groups: Wang Tilings and Schreier Tries

Groups and semigroups generated by Mealy automata were formally introduced in the early 1960s. They revealed their full potential over the years, by contributing to important conjectures in group theory. In the current chapter, we intend to provide various combinatorial and dynamical tools to tackle some decision problems all related to some extent to the growth of automaton (semi)groups. In the first part, we consider Wang tilings as a major tool in order to study and understand the behavior of automaton (semi)groups. There are various ways to associate a Wang tileset with a given complete and deterministic Mealy automaton and various ways to interpret the induced Wang tilings. We describe some of these fruitful combinations, as well as some promising research opportunities. In the second part, we detail some toggle switch between a classical notion from group theory—Schreier graphs—and some properties of an automaton group about its growth or the growth of its monogenic subgroups. We focus on polynomial activity automata and on reversible automata, which are somehow diametrically opposed families.

Chapter 11 by Laurent Bartholdi
Amenability Groups and *G*-Sets

This text surveys classical and recent results in the field of amenability of groups, from a combinatorial standpoint. It has served as the support of courses at the University of Göttingen and the École Normale Supérieure. The goals of the text are to be as self-contained as possible, so as to serve as a good introduction for newcomers to the field; to stress the use of combinatorial tools, in collaboration with functional analysis, probability, etc., with discrete groups in focus; to consider from the beginning the more general notion of amenable actions; and, lastly, to describe recent classes of examples and, in particular, groups acting on Cantor sets and topological full groups.

Paris, France
Liège, Belgium
August 2017

Valérie Berthé
Michel Rigo

Chapter 11 by Laurent Bartholdi
Amenability: Groups and G-Sets

[illegible]

[illegible]

Acknowledgments

The editors would like to express their gratitude to Julien Leroy and Narad Rampersad who were kind enough to read drafts of this book and who suggested many improvements. They would also like to thank the series editor whose constant support has been a precious help through all this project.

Contents

Contributors

Nathalie Aubrun LIP, ENS de Lyon, Lyon, France

Sebastián Barbieri LIP, ENS de Lyon, Lyon, France

Laurent Bartholdi École Normale Supérieure, Paris, France. Mathematical Institute, Georg-August University of Göttingen, Bunsenstrasse, Göttingen, Germany

Verónica Becher Departamento de Computación, Facultad de Ciencias Exactas y Naturales, Universidad de Buenos Aires. CONICET, Pabellón I, Ciudad Universitaria, Buenos Aires, Argentina

Jason Bell Department of Pure Mathematics, University of Waterloo, Waterloo, ON, Canada

Valérie Berthé IRIF, UMR 8243, CNRS & Université Paris Diderot, Paris Cedex 13, France

Olivier Carton IRIF, UMR 8243, CNRS & Université Paris Diderot, Paris Cedex 13, France

Émilie Charlier Department of Mathematics, University of Liège, Liège, Belgium

Michael Coons School of Mathematical and Physical Sciences, University of Newcastle, Callaghan, NSW, Australia

Emmanuel Jeandel LORIA, Campus Scientifique, Vandœuvre-Lès-Nancy, France

Ines Klimann IRIF, UMR 8243, CNRS & Univ. Paris Diderot, Paris Cedex 13, France

Manfred Madritsch Institut Élie Cartan de Lorraine, Universtité de Lorraine, Vandœuvre-Lès-Nancy Cedex, France

Pascal Ochem CNRS & LIRMM, Montpellier, Montpellier Cedex 5, France

Matthieu Picantin IRIF, UMR 8243, CNRS & Université Paris Diderot, Paris, France

Michaël Rao LIP, ENS de Lyon, Lyon, France

Michel Rigo Department of Mathematics, University of Liège, Liège, Belgium

Matthieu Rosenfeld LIP, ENS de Lyon, Lyon, France

Lukas Spiegelhofer Institut für Diskrete Mathematik und Geometrie, Technische Universität Wien, Wien, Austria

Caïus Wojcik Institut Camille Jordan, Université Lyon 1, CNRS UMR 5208, Villeurbanne Cedex, France

Luca Q. Zamboni Institut Camille Jordan, Université Lyon 1, CNRS UMR 5208, Villeurbanne Cedex, France

Chapter 1
General Framework

Valérie Berthé and Michel Rigo

Abstract This introductory chapter briefly presents some of the main notions that appear in the subsequent chapters of this book. We recap a few definitions and results from combinatorics on groups and words, formal language theory, morphic words, k-automatic and k-regular sequences, and dynamical systems. Our aim is not to be exhaustive. The reader can consult this chapter when studying other parts of this book.

1.1 Conventions

The set of nonnegative integers (respectively integers, rational numbers, real numbers, and complex numbers) is written $\mathbb{N}$ (respectively, $\mathbb{Z}$, $\mathbb{Q}$, $\mathbb{R}$, and $\mathbb{C}$). In particular, the set $\mathbb{N}$ is $\{0, 1, 2, \ldots\}$. We use the notation $[\![i, j]\!]$ for the set of integers $\{i, i+1, \ldots, j\}$. The *floor* of a real number x is $\lfloor x \rfloor = \sup\{z \in \mathbb{Z} \mid z \leq x\}$, whereas $\{x\} = x - \lfloor x \rfloor$ stands for the *fractional part* of x. Recall that $\lceil \cdot \rceil$ denotes the *ceiling function*, i.e., $\lceil x \rceil = \inf\{z \in \mathbb{Z} \mid z \geq x\}$. The characteristic sequence χ_X of a set $X \subset \mathbb{N}^d$ takes its values in $\{0, 1\}$ and satisfies $\chi_X(n) = 1$ if and only if $n \in X$.

Let us recall the notation about asymptotics. Let $f, g : \mathbb{R} \to \mathbb{R}$ be two functions. The definitions given below can also be applied to functions defined on another domain like $\mathbb{R}_{>a}$, $\mathbb{N}$ or $\mathbb{Z}$. We assume implicitly that the following notions are defined for $x \to +\infty$. We write $f \in \mathcal{O}(g)$, if there exist two constants x_0 and $C > 0$ such that, for all $x \geq x_0$, $|f(x)| \leq C|g(x)|$. We also write $f \ll g$ or $g \gg f$, or else $g \in \Omega(f)$. Note that we can write either $f \in \mathcal{O}(g)$ or $f = \mathcal{O}(g)$. Be aware that in the literature, authors sometimes give different meanings to the notation $\Omega(f)$. Here we consider a bound, for all large enough x, but there exist

---

V. Berthé (✉)
IRIF, UMR 8243, CNRS & Université Paris Diderot, Case 7014, F-75205 Paris Cedex 13, France
e-mail: berthe@irif.fr

M. Rigo
Department of Mathematics, University of Liège, Allée de la découverte 12 (B37), B-4000 Liège, Belgium
e-mail: M.Rigo@ulg.ac.be

V. Berthé, M. Rigo (eds.), *Sequences, Groups, and Number Theory*, Trends in Mathematics, https://doi.org/10.1007/978-3-319-69152-7_1

variants where the bound holds only for an increasing sequence $(x_n)_{n\geq 0}$ of reals, i.e., $\limsup_{x\to+\infty} |g(x)|/|f(x)| > 0$.

If g belongs to $\mathcal{O}(f) \cap \Omega(f)$, i.e., there exist constants x_0, C_1, C_2 with $C_1, C_2 > 0$ such that, for all $x \geq x_0$, $C_1|f(x)| \leq |g(x)| \leq C_2|f(x)|$, then we write $g \in \Theta(f)$. As an example, the function $x^2 + \sin 6x$ is in $\Theta(x^2)$ and $x^2|\sin(4x)|$ is in $\mathcal{O}(x^2)$ but not in $\Theta(x^2)$.

1.2 Algebraic Structures

We briefly recall the basic definitions of monoid, (semi)group, (semi)ring, field, ideal, vector space, and module.

Definition 1.2.1. Let $\mathbb{S}$ be a set equipped with a single binary operation

$$\star : \mathbb{S} \times \mathbb{S} \to \mathbb{S}.$$

It is convenient to call this operation a *multiplication* over $\mathbb{S}$, and the product of $x, y \in \mathbb{S}$ is usually denoted by xy.

If this multiplication is *associative*, i.e., for all $x, y, z \in \mathbb{S}$, $(xy)z = x(yz)$, then the algebraic structure given by the pair $(\mathbb{S}, \star)$ is a *semigroup*.

If, moreover, multiplication has an identity element, i.e., there exists some element $1 \in \mathbb{S}$ such that, for all $x \in \mathbb{S}$, $x1 = x = 1x$, then $(\mathbb{S}, \star)$ is a *monoid*.

In addition if every element $x \in \mathbb{S}$ has an *inverse*, i.e., there exists $y \in \mathbb{S}$ such that $xy = 1 = yx$, then $(\mathbb{S}, \star)$ is a *group*.

Definition 1.2.2. A *semiring* is a set R equipped with two binary operations $+$ and $\cdot$ such that

1. $(R, +)$ is a commutative monoid with identity element 0.
2. $(R, \cdot)$ is a monoid with identity element 1.
3. The product is distributive with respect to the sum.
4. For all $r \in R$, $0 \cdot r = 0 = r \cdot 0$.

If, moreover, $\cdot$ is commutative, then the semiring is said to be *commutative*. A *ring* is a semiring where $(R, +)$ is a commutative group. A *field* is a commutative ring where $(R, \cdot)$ is a group.

Definition 1.2.3. A (two-sided) *ideal* of a ring $(R, +, \cdot)$ is a nonempty subset I of R, such that $(I, +)$ is a subgroup of $(R, +)$ and for all $i \in I$ and all $r \in R$, $i \cdot r$ and $r \cdot i$ belong to I.

Definition 1.2.4. Let K be a field with identity element 1 for its multiplication. A *vector space* over K is a set V equipped with a binary operation $+ : V \times V \to V$ such that $(V, +)$ is a commutative group and a binary operation $\cdot : K \times V \to V$ such that, for all $k, \ell \in K$ and all $x, y \in V$,

1. $k \cdot (\ell \cdot x) = (k\ell) \cdot x$
2. $1 \cdot x = x$
3. $(k + \ell) \cdot x = k \cdot x + \ell \cdot x$
4. $k \cdot (x + y) = k \cdot x + k \cdot y$

A *K-module* is similarly defined but it is built over a ring K instead of a field.

We now consider natural notions specific to group and semigroup theory (see also Section 9.3.1 for further basic definitions on group theory and Chapter 11).

For a given property $\mathscr{P}$ of groups (abelian, free, nilpotent, soluble, ...), group G is called *virtually* $\mathscr{P}$ if G contains a finite-index subgroup satisfying property $\mathscr{P}$. See also Definition 9.3.36 and Section 9.3.4.1 for properties of virtually free groups such as the decidability for the word problem (Theorem 9.3.37).

Schreier graphs generalize Cayley graphs. Let G be a group generated by S and acting on a set X, the vertices of its *Schreier graph* (depending on S) are the elements of X, and there is an edge from x to y if y is the image of x under the action of some element of S. By considering the action of the group on itself by right multiplication, this graph coincides with its *Cayley graph*. See also Definition 10.3.1 and Section 11.3.

Let G be a finitely generated group with a generator system given by $S = \{g_1, \ldots, g_m\}$. The *length* of $g \in G$ (with respect to S) is the smallest integer ℓ such that g can be represented by a product of the form

$$g = g_{i_1}^{\pm 1} \cdots g_{i_\ell}^{\pm 1},$$

i.e., the length of the shortest decomposition of g. The *growth* of the group G (with respect to S) is the map

$$\gamma_S : \mathbb{N} \to \mathbb{N},\ n \mapsto \operatorname{Card}\{g \in G \mid d_S(g) \leq n\},$$

where $d_S(g)$ is the length of g with respect to S. This definition can be made independent of S by noticing that the growths corresponding to two generating sets are equivalent [409]. Note that a finite group has a bounded growth, an infinite abelian group has a polynomial growth, and a non-abelian free group has an exponential growth. The growth of a finitely generated group can also be seen as the growth of its Cayley graph: we count the vertices which are within distance n of the identity element. This notion is considered in Sections 10.3.4.1, 11.3.1, and 11.4.

1.3 Words

This section is intended to give basic definitions about words either finite or infinite. Words are ubiquitous when encoding a piece of information. As an example, a finite word over the alphabet of digits $\{0, \ldots, k-1\}$ can be seen as the k-ary expansion of an integer. On the other hand, an infinite word over $\{0, 1\}$ could be used as the characteristic sequence for a subset of $\mathbb{N}$. For material not covered here, see the classical Lothaire's textbooks on finite or infinite words and their properties are [385–387]. Also see Allouche and Shallit's book [14] about automatic sequences or, Queffélec's book [488] for a dynamical point of view. For a quick overview, the reader can have a look at the chapter [150] or the tutorial [75]. The book [504] is also intended to serve as introductory lecture notes on the subject.

1.3.1 Finite Words

An *alphabet* is a finite nonempty set. Its elements are called *symbols* or *letters*.

Definition 1.3.1. A (finite) *word* over Σ is a finite sequence of letters from Σ. The empty sequence is called the *empty word* and it is denoted by ε. The sets of all finite words (respectively, finite nonempty words) over Σ are denoted by Σ^* (respectively, Σ^+). A word $w = w_0 w_2 \cdots w_{n-1}$ where $w_i \in \Sigma, 0 \leq i < n$, can be seen as a function $w : \{0, 1, \ldots, n-1\} \to \Sigma$ in which $w(i) = w_i$ for all i. The empty word is the word whose domain is the empty set.

Let $u = u_0 \cdots u_{m-1}$ and $v = v_0 \cdots v_{n-1}$ be two words over Σ. The *concatenation* of u and v is the word $w = w_0 \cdots w_{m+n-1}$ defined by $w_i = u_i$ if $0 \leq i < m$, and $w_i = v_{i-m}$ otherwise. We write $u \cdot v$ or simply uv to express the concatenation of u and v. The concatenation (or catenation) of words is an associative operation, i.e., given three words u, v and w, $(uv)w = u(vw)$. Hence, parenthesis can be omitted. In particular, the set Σ^* (respectively, Σ^+) equipped with the concatenation product is a monoid (respectively, a semigroup).

The *length* of a word w, denoted by $|w|$, is the number of occurrences of the letters in w. In other words, if $w = w_0 w_2 \cdots w_{n-1}$ with $w_i \in \Sigma, 0 \leq i < n$, then $|w| = n$. In particular, the length of the empty word is zero. The set of words of length k (respectively, at most k) over Σ is denoted by Σ^k (respectively, $\Sigma^{\leq k}$). For $a \in \Sigma$ and $w \in \Sigma^*$, we write $|w|_a$ for the number of occurrences of a in w. Therefore, we have

$$|w| = \sum_{a \in \Sigma} |w|_a .$$

If u and v are two words over Σ such that $|u|_a = |v|_a$ for all $a \in \Sigma$, then u is obtained by permuting the letters of v: u and v are said to be *abelian equivalent*. These are anagrams.

A word u is a *factor* of a word v (respectively, a *prefix* or a *suffix*), if there exist words x and y such that $v = xuy$ (respectively, $v = uy$, or $v = xu$). A factor (respectively, a prefix or a suffix) u of a word v is called *proper* if $u \neq v$ and $u \neq \varepsilon$. Prefixes and suffixes are sometimes called initial and terminal factors. Thus, for example, if $w =$ `concatenation`, then `con` is a prefix, `ate` is a factor, and `nation` is a suffix of w. If $w = w_0 \cdots w_n$ and u is a factor of w such that $u = w_i \cdots w_{i+|u|-1}$, we say that u *occurs* in w at position i. For instance, in abbabaabbaab, the factor ab occurs at positions 0, 3, 6, 10. The set of factors of u (respectively, of prefixes of u) is denoted by $\mathrm{Fac}(u)$ (respectively, $\mathrm{Pref}(u)$).

The *mirror* (sometimes called *reversal*) of a word $u = u_0 \cdots u_{m-1}$ is the word $\tilde{u} = u_{m-1} \cdots u_0$. It can be defined inductively on the length of the word by $\tilde{\varepsilon} = \varepsilon$ and $\widetilde{au} = \tilde{u}a$ for $a \in \Sigma$ and $u \in \Sigma^*$. Notice that for $u, v \in \Sigma^*$, $\widetilde{uv} = \tilde{v}\tilde{u}$. A *palindrome* is a word u such that $\tilde{u} = u$. For instance, the palindromes of length at most 3 in $\{0, 1\}^*$ are $\epsilon, 0, 1, 00, 11, 000, 010, 101, 111$.

1.3.2 Infinite Words

Instead of considering finite sequences of elements belonging to an alphabet Σ, considering infinite sequences of elements in Σ is also relevant.

Definition 1.3.2. An (one-sided right) *infinite word* is a map from $\mathbb{N}$ to Σ. If $\mathbf{w}$ is an infinite word, we often write

$$\mathbf{w} = a_0 a_1 a_2 \cdots ,$$

where each $a_i \in \Sigma$. The set of all infinite words of Σ is denoted Σ^ω (one can also find the notation $\Sigma^{\mathbb{N}}$).

Example 1.3.3. Consider the infinite word $\mathbf{x} = x_0 x_1 x_2 \cdots$ where the letters $x_i \in \{0, \ldots, 9\}$ are given by the digits appearing in the usual decimal expansion of $\pi - 3$,

$$\pi - 3 = \sum_{i=0}^{+\infty} x_i \, 10^{-i-1},$$

i.e., $\mathbf{x} = 14159265358979323846264338327950288419 \cdots$ is an infinite word.

The notions of *factor*, *prefix*, or *suffix* introduced for finite words can be extended to infinite words. Factors and prefixes are finite words, but a suffix of an infinite word is also infinite. We still make use of the notation $\mathrm{Fac}(\mathbf{w})$ and $\mathrm{Pref}(\mathbf{w})$.

Definition 1.3.4. The *language* of the infinite word $\mathbf{x}$ is the set of all its factors. It is denoted by $\mathrm{Fac}(\mathbf{x})$. The set of factors of length n occurring in $\mathbf{x}$ is denoted by $\mathrm{Fac}_n(\mathbf{x})$.

Definition 1.3.5. The *complexity function*, or *factor complexity*, of an infinite word $\mathbf{x}$ maps $n \in \mathbb{N}$ onto the number $p_{\mathbf{x}}(n) = \mathrm{Card}(\mathrm{Fac}_n(\mathbf{x}))$ of distinct factors of length n occurring in $\mathbf{x}$.

Example 1.3.6. The Thue–Morse word $\mathbf{t} = t_0 t_1 t_2 \cdots$ (ubiquitous word encountered in combinatorics on words [18]) can be defined over $\{\mathtt{a}, \mathtt{b}\}$ by $t_n = \mathtt{a}$ if and only if there is an even number of ones in the base-2 expansion of $n \geq 0$. Otherwise stated, if the sum of base-2 digits of n is even. Thus a prefix of $\mathbf{t}$ is given

$$\mathtt{abbabaabbaababbabaababbaabbabaab} \cdots .$$

If we replace a with 1 and b with 0, then we get the characteristic sequence χ_E of the set of integers whose sum of base-2 digits is even. The factor complexity of the Thue–Morse word **t** is well known [107, 391]. See also [78, p. 225] where a chapter is devoted to the factor complexity of morphic words. We have

$$p_{\mathbf{t}}(n) = \begin{cases} 4n - 2 \cdot 2^m - 4, & \text{if } 2 \cdot 2^m < n \leq 3 \cdot 2^m; \\ 2n + 4 \cdot 2^m - 2, & \text{if } 3 \cdot 2^m < n \leq 4 \cdot 2^m. \end{cases}$$

Definition 1.3.7. A two-sided or *bi-infinite word* is a map from $\mathbb{Z}$ to Σ. The set of all bi-infinite words is denoted ${}^{\omega}\Sigma^{\omega}$ (one can also find the notation $\Sigma^{\mathbb{Z}}$).

Definition 1.3.8. An infinite word $\mathbf{x} = x_0 x_1 \cdots$ is *(purely) periodic* if there exists a finite word $u = u_0 \cdots u_{k-1} \neq \epsilon$ such that $x = u^{\omega}$, i.e., for all $n \geq 0$, we have $x_n = u_r$ where $n = dk + r$ with $r \in \{0, \ldots, k-1\}$. An infinite word x is *eventually periodic* (or *ultimately periodic*) if there exist two finite words $u, v \in \Sigma^*$, with $v \neq \epsilon$ such that $x = uvvv \cdots = uv^{\omega}$. Notice that purely periodic words are special cases of eventually periodic words. For any eventually periodic word x, there exist words u, v of shortest length such that $x = uv^{\omega}$, then the integer $|u|$ (respectively $|v|$) is referred to as the *preperiod* (respectively *period*) of x. An infinite word is said to be *nonperiodic* if it is not eventually periodic.

Let us mention the next result called Morse–Hedlund theorem.

Theorem 1.3.9. *Let* $\mathbf{w}$ *be an infinite word over a finite alphabet. The word* $\mathbf{w}$ *is eventually periodic if and only if there exists some integer* N *such that* $p_{\mathbf{w}}(N) \leq N$.

Among the nonperiodic words of low factor complexity, Sturmian words play a special role and have been extensively studied. An infinite word $\mathbf{x}$ is *Sturmian* if $p_{\mathbf{x}}(n) = n + 1$ for all $n \geq 0$. Note that Sturmian words are over a 2-letter alphabet. For general references, see [386, Chapter 2] or [487, Chapter 6]. They will be considered in Chapter 6.

Definition 1.3.10. An infinite word $\mathbf{x}$ is *recurrent* if all its factors occur infinitely often in $\mathbf{x}$. It is *uniformly recurrent* if it is recurrent and for every factor u of $\mathbf{x}$, for the infinite set

$$\left\{ i_1^{(u)} < i_2^{(u)} < i_3^{(u)} < \cdots \right\}$$

of positions where u occurs in $\mathbf{x}$, there exists a constant C_u such that, for all $j \geq 1$,

$$i_{j+1}^{(u)} - i_j^{(u)} \leq C_u .$$

Note that, by Furstenberg's theorem, for any infinite word $\mathbf{w}$, there is a uniformly recurrent word $\mathbf{r}$ over the same alphabet such that every finite factor of $\mathbf{r}$ is a factor of $\mathbf{w}$, i.e., $\mathrm{Fac}(\mathbf{r}) \subseteq \mathrm{Fac}(\mathbf{w})$ (see Theorem 4.4.9).

Let $\mathbf{x}$ be an infinite word, the function $R_{\mathbf{x}} : \mathrm{Fac}(\mathbf{x}) \to \mathbb{N} \cup \{\infty\}$ maps a factor u of $\mathbf{x}$ to the smallest k such that every factor of $\mathbf{x}$ of length k contains u, or ∞ if no such k exists. Otherwise stated, an infinite word $\mathbf{x}$ is uniformly recurrent, if for every factor u of $\mathbf{x}$, $R_{\mathbf{x}}$ is finite. The *recurrence function* maps $n \in \mathbb{N}$ to

$$R_{\mathbf{x}}(n) = \max_{u \in L_n(\mathbf{x})} R_{\mathbf{x}}(u) .$$

Otherwise stated, if $\mathbf{x}$ is uniformly recurrent, then for every factor of length n of $\mathbf{x}$, $R_{\mathbf{x}}(n)$ is finite and u occurs in all factors of length $R_{\mathbf{x}}(n)$ of $\mathbf{x}$.

Assume that Σ is totally ordered: $(\Sigma, <)$. Let $\mathbf{x}, \mathbf{y}$ be two infinite words over Σ. We say that $\mathbf{x}$ is *lexicographically less* than $\mathbf{y}$ if there exists N such that $x_i = y_i$ for all $i < N$ and $x_N < y_N$.

Definition 1.3.11. One can endow Σ^ω with a *distance* d defined as follows. Let $\mathbf{x}, \mathbf{y}$ be two infinite words over Σ. Let $\mathbf{x} \wedge \mathbf{y}$ denote the longest common prefix of $\mathbf{x}$ and $\mathbf{y}$. Then the distance d is given by

$$d(\mathbf{x}, \mathbf{y}) := \begin{cases} 0, & \text{if } \mathbf{x} = \mathbf{y}, \\ 2^{-|\mathbf{x} \wedge \mathbf{y}|}, & \text{otherwise.} \end{cases}$$

This notion of distance extends to $\Sigma^{\mathbb{Z}}$. Notice that the topology on Σ^ω is the product topology (of the discrete topology on Σ). The space Σ^ω is a compact *Cantor set*, that is, a totally disconnected compact space without isolated points. Since Σ^ω is a (complete) metric space, it is therefore relevant to speak of convergent sequences of infinite words. The sequence $(\mathbf{z}_n)_{n \geq 0}$ of infinite words over Σ *converges* to $\mathbf{x} \in \Sigma^\omega$, if for all $\epsilon > 0$, there exists $N \in \mathbb{N}$ such that, for all $n \geq N$, $d(\mathbf{z}_n, \mathbf{x}) < \epsilon$. To express the fact that a sequence of finite words $(w_n)_{n \geq 0}$ over Σ converges to an infinite word $\mathbf{y}$, it is assumed that Σ is extended with an extra letter $c \notin \Sigma$. Any finite word w_n is replaced with the infinite word $w_n ccc \cdots$, and if the sequence of infinite words $(w_n ccc \cdots)_{n \geq 0}$ converges to $\mathbf{y}$, then the sequence $(w_n)_{n \geq 0}$ is said to converge to $\mathbf{y}$.

Let $(u_n)_{n \geq 0}$ be a sequence of nonempty finite words. If we define, for all $\ell \geq 0$, the finite word v_ℓ as the concatenation $u_0 u_1 \cdots u_\ell$, then the sequence $(v_\ell)_{\ell \geq 0}$ of finite words converges to an infinite word. This latter word is said to be the concatenation of the elements in the infinite sequence of finite words $(u_n)_{n \geq 0}$. In particular, for a constant sequence $u_n = u$ for all $n \geq 0$, $v_\ell = u^{\ell+1}$ and the concatenation of an infinite number of copies of the finite word u is denoted by u^ω.

We have discussed the fact that a (finite) word u may appear as a factor of an infinite word $\mathbf{x}$. It may occur a finite number of times, infinitely often, or even in such a way that $R_{\mathbf{x}}(u)$ is finite. But we could also introduce the *frequency* of a factor u occurring in $\mathbf{x}$ as the following limit, if it exists,

$$\lim_{n \to +\infty} \frac{\operatorname{Card}\left(\{i \leq n - |u| \mid x_i \cdots x_{i+|u|-1} = u\}\right)}{n} .$$

For instance, for the infinite word $\mathbf{w} = 01\,0011\,0^4 1^4\,0^8\,1^8\,0^{16} 1^{16} \cdots$ where we have longer and longer blocks of consecutive zeroes followed by longer and longer blocks of ones. The frequencies of 0 and 1 do not exist. Frequency appears naturally in the definition of normal numbers given below. See also Theorem 1.6.10 about the frequency of symbols in automatic sequences and morphic words. Frequencies are also considered in Chapter 5 in the framework of repetitions, and in Chapter 7 and 8 in the framework of normality.

1.3.3 Number Representations

We refer the reader to Frougny's chapter [386] or to [227] for a general presentation of numeration systems. The book [503] can also serve as an introduction to the subject. We also mention the survey [36]. More details are also discussed in Section 3.2 of this book.

Let $k \geq 2$ be an integer. Let us recall how base-k expansion of integers may be computed. For any positive integer n, there exist $\ell \geq 0$ such that $k^\ell \leq n < k^{\ell+1}$ and unique coefficients $c_0, \ldots, c_\ell \in \{0, \ldots, k-1\}$ such that

$$n = \sum_{i=0}^{\ell} c_i\, k^i \text{ and } c_\ell \neq 0\,.$$

The coefficients $c_\ell, \ldots, c_0$ can be computed by successive Euclidean divisions. Set $n_0 := n$. We have $n_0 = c_\ell\, k^\ell + n_1$ with $n_1 < k^\ell$ and for $i = 1, \ldots, \ell$, $n_i = c_{\ell-i}\, k^{\ell-i} + n_{i+1}$ with $n_{i+1} < k^{\ell-i}$. The word $c_\ell \cdots c_0$ is said to be the *k-ary representation* or *k-ary expansion* of n (sometimes called *greedy* representation) and denoted by $\mathrm{rep}_p(n)$. If $d_\ell \cdots d_0$ is a word over an alphabet of digits included in $\mathbb{Z}$, we define

$$\mathrm{val}_k(d_\ell \cdots d_0) = \sum_{i=0}^{\ell} d_i\, k^i\,.$$

If one replaces the sequence $(k^n)_{n\geq 0}$ with an increasing sequence $(U_n)_{n\geq 0}$ of integer such that $U_0 = 1$, then a similar algorithm may be applied. The corresponding U-expansions are over the alphabet $\{0, \ldots, \sup\lceil \frac{U_{n+1}}{U_n} \rceil - 1\}$. One finds the general terminology *positional numeration system*. It is also possible to extend the procedure to represent real numbers. Let $x \in (0, 1)$. There exists a decomposition of the form

$$x = \sum_{i=1}^{+\infty} c_i\, k^{-i}$$

where $c_i \in \{0, \ldots, k-1\}$ for all $i \geq 1$. If we forbid sequences where $c_i = k-1$ for all large enough i, then the sequence $(c_i)_{i\geq 1}$ is unique. Given $x \in [0, 1)$, the algorithm in Table 1.1 provides the corresponding sequence $(c_i)_{i\geq 0}$ of digits.

Table 1.1 An algorithm for computing the base-k expansion of $x \in [0, 1)$.

```
i ← 0
y ← x
REPEAT FOREVER
    c_i ← ⌊ky⌋
    y ← {ky}
    INCREMENT i
END-REPEAT.
```

In this algorithm, we iterate a map from the interval $[0, 1)$ onto itself, i.e.,

$$T_k : [0, 1) \to [0, 1), y \mapsto \{ky\} \tag{1.1}$$

and the value taken by the image determines the next digit in the expansion. This yields a dynamical system such as discussed in Section 1.7. The interval $[0, 1)$ is thus split into k subintervals $[j/k, (j+1)/k)$, for $j = 0, \dots, k-1$. For all $i \geq 0$, if $T_k^i(x)$ belongs to the subinterval $[j/k, (j+1)/k)$, then the digit c_i occurring in $\mathrm{rep}_k(x)$ is equal to j. It is indeed natural to consider such subintervals. If y belongs to $[j/k, (j+1)/k)$, then ky has an integer part equal to j and the map T_k is continuous and increasing on every subinterval $[j/k, (j+1)/k)$. Note also that the range of T_k on any of these subintervals is $[0, 1)$. So applying T_k to a point in one of these subintervals can lead to a point belonging to any of these subintervals (later on, we shall introduce some other transformation, e.g., β-transformations, where a restriction appears on the intervals that can be reached). So to speak, the base-k expansion of x can be derived from the trajectory of x under T_k, i.e., from the sequence $(T_k^n(x))_{n\geq 0}$.

As an example, consider the base $k = 3$ and the expansion of $x = 3/10$. The point lies in the interval $[0, 1/3)$; thus the first digit of the expansion is 0. Then $T_3(3/10) = 9/10$ lies in the interval $[2/3, 1)$; thus the second digit is 2. If we apply again T_3, we get $T_3^2(3/10) = \{27/10\} = 7/10$, which belongs again to $[2/3, 1)$ giving the digit 2. Then $T_3^3(3/10) = 1/10$ giving the digit 0 and finally $T_3^4(3/10) = 3/10$. So $\mathrm{rep}_3(3/10) = (0220)^\omega$.

A natural generalization of base-k expansion (discussed in Section 3.6 and in Example 8.1.2) is to replace the base k with a real number $\beta > 1$. In particular, the transformation T_k will be replaced by the so-called β-transformation. Note that we shall be concerned with expansions of numbers in $[0, 1)$. If $x \geq 1$, then there exists a smallest d such that x/β^d belongs to $[0, 1)$. It is therefore enough[1] to concentrate on $[0, 1)$.

Definition 1.3.12 (β-Expansions). We will only represent real numbers in the interval $[0, 1)$. Let $\beta > 1$ be a real number. The representations discussed here

[1]If the β-expansion of x/β^d is $d_0 d_1 \cdots$, then using an extra decimal point, the expansion of x is conveniently written $d_0 \cdots d_{\ell-1} \bullet d_\ell d_{\ell+1} \cdots$. Note that the presentation in Chapter 1 is not entirely consistent with our present treatment if x belongs to $[0, 1/(\beta-1)] \setminus [0, 1)$.

are a direct generalization of the base-k expansions. Every real number $x \in [0, 1)$ can be written as a series

$$x = \sum_{i=0}^{+\infty} c_i \beta^{-i-1} \tag{1.2}$$

where c_i belong to $\{0, \lceil\beta\rceil - 1\}$. Note that if β is an integer, then $\lceil\beta\rceil - 1 = \beta - 1$. For integer base-$b$ expansions, a number may have more than one representation, namely, those ending with 0^ω or $(b-1)^\omega$. For a real base β, we obtain many more representations. Consider the Golden mean ϕ, which satisfies $\phi^2 - \phi - 1 = 0$, and thus

$$\frac{1}{\phi^n} = \frac{1}{\phi^{n+1}} + \frac{1}{\phi^{n+2}}, \quad \forall n \geq 0 .$$

As an example, the number $1/\phi$ has thus infinitely many representations as a power series with negative powers of ϕ and coefficients 0 and 1:

$$\frac{1}{\phi} = \frac{1}{\phi^2} + \frac{1}{\phi^3} = \frac{1}{\phi^2} + \frac{1}{\phi^4} + \frac{1}{\phi^5} = \frac{1}{\phi^2} + \frac{1}{\phi^4} + \frac{1}{\phi^6} + \frac{1}{\phi^7} = \cdots .$$

To get a canonical expansion for a real $x \in [0, 1)$, we just have to replace the integer base b with β and consider the so-called *β-transformation*

$$T_\beta : [0, 1) \to [0, 1),\ x \mapsto \{\beta x\}$$

in the algorithm from Table 1.1. For $i = 0, 1, \ldots$, the idea is to remove the largest integer multiple c_i of β^{-i-1} and then repeat the process with the remainder and the next negative power of β to get (1.2). Note that c_i is less than $\lceil\beta\rceil$ because of the greediness of the process. Otherwise, one could have removed a larger multiple of the power of β at a previous step. The corresponding infinite word $c_0c_1\cdots$ is called the *β-expansion* of x and is usually denoted by $\mathsf{d}_\beta(x)$. Any word $d_0d_1\cdots$ over a finite alphabet of nonnegative integers satisfying

$$x = \sum_{i=0}^{+\infty} d_i \beta^{-i-1}$$

is said to be a *β-representation* of x. Thus, the β-expansion of x is the lexicographically maximal word among the β-representations of x.

The greediness of the algorithm can be reformulated as follows.

Lemma 1.3.13. *A word $d_0d_1\cdots$ over $\{0, \ldots, \lceil\beta\rceil - 1\}$ is the β-expansion of a real number $x \in [0, 1)$ if and only if, for all $j \geq 0$,*

$$\sum_{i=j}^{+\infty} d_i\, \beta^{-i-1} < \beta^{-j}\,.$$

Proposition 1.3.14. *Let x, y be real numbers in $[0, 1)$. We have $x < y$ if and only if $\mathsf{d}_\beta(x)$ is lexicographically less than $\mathsf{d}_\beta(y)$.*

1.3.4 Normality

Now that number representations and the frequency of a factor have been introduced, we can define normal numbers.

A real number x is *simply normal* with respect to base $b \geq 2$ if in the base-b expansion of x (which is an infinite word over $\{0, \ldots, b-1\}$), the frequency of every digit $d \in \{0, 1, \ldots, b-1\}$ exists and is equal to $1/b$. Furthermore x is *normal* in base b if it is simply normal with respect to the bases $b, b^2, b^3, \ldots$. An equivalent definition is to say that for all $k \geq 1$ and every word $u = u_1 \ldots u_k \in \{0, 1, \ldots, b-1\}^k$, the frequency of u in the base-b expansion of x exists and is equal to $1/b^k$. A real number x is *absolutely normal* if x is normal to every integer base greater than or equal to 2.

Normality can also be expressed in terms of uniform distribution modulo 1 [578] (see Section 7.6 for corresponding definitions). Indeed, a real number x is normal to base b if and only if the sequence $(b^j x)_{j\geq 0}$ is uniformly distributed modulo 1.

These notions were introduced by Borel [99] and are discussed in Chapters 2, 7, and 8. In particular, constructions of normal numbers are provided in Sections 7.7 and 7.8. See also Theorem 7.4.1 (the so-called Hot Spot Lemma according to [101]) for a further convenient characterization of normality in terms of limsups instead of limits. For a dynamical viewpoint, see Section 8.2, where the definition of a normal number is transferred to symbolic dynamical systems, and constructions with concatenation of words for languages with specification are provided.

1.3.5 Repetitions in Words

In combinatorics on words, a question that naturally arises is to study the repetitions that should occur or may be avoided in words. See in particular Chapter 5 and Chapters 4 and 5 in [79].

Concatenating a word w with itself k times is abbreviated by w^k. In particular, $w^0 = \varepsilon$. Furthermore, for an integer m and a word $w = w_1 w_2 \cdots w_n$, where $w_i \in \Sigma$ for $1 \leq i \leq n$ (here it is convenient to start indexing with 1), the *rational power*

$$w^{m/n}$$

is $w^q w_1 w_2 \cdots w_r$, where $m = qn + r$ for $0 \le r < n$. For instance, we have

$$(\texttt{abbab})^{9/5} = \texttt{abbababba}\,.$$

Consider definitions that have to do with repetitions in words. A *square* is a nonempty word of the form xx, where $x \in \Sigma^*$. An example of a square in English is the word murmur with x equal to mur. An *overlap* is a word of the form $axaxa$, where $a \in \Sigma$ and $x \in \Sigma^*$. The word alfalfa is an example of an overlap in English with x equal to lf. It is obvious that every overlap has a square as prefix. For any positive integer $k \ge 2$, a *k-power* is a nonempty word of the form x^k. Thus a 2-power is a square, and a 3-power is a *cube*. A nonempty word that is not a k-power for any $k \ge 2$ is *primitive*.

Let us say a few words about avoidance (which is the topic of Chapter 5). It is an easy exercise to show that over a 2-letter alphabet, every word of a length of at least 4 contains a square. This raises several questions. Over a 3-letter alphabet, can we build longer words with no square as a factor? In particular, does there exist an infinite word with no square in it? Also over a 2-letter alphabet, if squares cannot be avoided, could we avoid cubes or even overlaps?

We say that a word w (finite or infinite) is *square-free* (or avoids squares) if no factor of w is a square. A finite or infinite word is *overlap-free* if it contains no factor that is an overlap. Thue [563] was the first to show the existence of an infinite overlap-free binary word. The Thue–Morse word (see Example 1.3.6) is overlap-free. See [79, Chapter 4] for more on avoidable repetitions and regularities in words. More generally, a (finite or infinite) word is *k-power-free* (or *avoids k-powers*) if none of its factors is a k-power. For instance, one can check that abbabaabbaab is overlap-free. (It is indeed a prefix of the Thue–Morse word). The goal of Chapter 5 is to present general techniques to prove positive or negative results about the appearance of a repetition pattern. The general question is to know whether an infinite word without a given pattern exists over an alphabet of a given size. Another question is to consider the growth function (in the sense of Definition 1.5.7) of the language of finite words avoiding a particular pattern.

Many variations on these topics exist. For instance, an abelian square is a word of the form uv where u and v are abelian equivalent. One can check that over a 3-letter alphabet, every long enough finite word contains an abelian square.

In Chapter 6, the addressed question is this: given a nonperiodic word $\mathbf{x} \in \Sigma^\omega$, does there exist a finite nonempty set C and a mapping $\varphi : \Sigma^+ \to C$ such that for each factorization $\mathbf{x} = u_1 u_2 u_3 \cdots$ there exist $i, j \ge 1$ such that $\varphi(u_i) \ne \varphi(u_j)$?

1.4 Morphisms

Infinite words of particular interest can be obtained by iterating morphisms of free monoids. They have many interesting combinatorial properties and can be generated by a simple mean.

Definition 1.4.1. A map $h : \Sigma^* \to \Delta^*$, where Σ and Δ are alphabets, is called a *morphism* if h satisfies $h(xy) = h(x)h(y)$ for all $x, y \in \Sigma^*$. In particular, we have $h(\varepsilon) = \varepsilon$. When $\Sigma = \Delta$, morphisms are also called *substitutions*.

A morphism may be specified by providing the values $h(a)$ for all $a \in \Sigma$. For example, we may define the morphism $t : \{0,1\}^* \to \{0,1\}^*$ by

$$\begin{aligned} 0 &\mapsto 01 \\ 1 &\mapsto 10. \end{aligned} \tag{1.3}$$

This morphism is often referred to as the *Thue–Morse morphism*. The domain Σ^* of a morphism h is easily extended to the set Σ^ω of (one-sided) infinite words. Let $h : \Sigma^* \to \Delta^*$ be a morphism and $\mathbf{x} = x_0x_1x_2\cdots$ be an infinite word over Σ. Simply consider the sequence of finite words $(h(x_0\cdots x_n))_{n\geq 0}$ of images of the prefixes of $\mathbf{x}$. The limit of this sequence is $h(\mathbf{x})$. In particular, if $h : \Sigma^* \to \Sigma^*$ and $\mathbf{x}$ is an infinite word such that $h(\mathbf{x}) = \mathbf{x}$, then $\mathbf{x}$ is said to be a *fixed point* of h.

A morphism $h : \Sigma^* \to \Sigma^*$ such that $h(a) = ax$ for some $a \in \Sigma$ and $x \in \Sigma^*$ with $h^i(x) \neq \varepsilon$ for all i is said to be *prolongable on a*. The Thue–Morse morphism t given by (1.3) is prolongable on 0 (and also on 1). The first few iterations of t are

$$\begin{aligned} t(0) &= 01 \\ t^2(0) &= 0110 \\ t^3(0) &= 01101001 \\ t^4(0) &= 0110100110010110 \\ &\vdots \end{aligned}$$

Since $|t(0)| = |t(1)| = 2$, we have $|t^n(0)| = 2^n$ for all $n \geq 0$. It is easy to prove that $t^n(0)$ is a proper prefix of $t^{n+1}(0)$, and thus the sequence $(t^n(0))_{n\geq 0}$ converges to an infinite word. So we get the fixed point of t

$$t^\omega(0) = 0110100110010110\cdots.$$

One can prove that the fixed point $t^\omega(0)$ is the *Thue–Morse word* introduced in Example 1.3.6.

More generally, if $h : \Sigma^* \to \Sigma^*$ is a morphism prolongable on a, we may then repeatedly iterate h to obtain the infinite *fixed point*

$$h^\omega(a) = a\,x\,h(x)\,h^2(x)\,h^3(x)\cdots.$$

This infinite word is said to be *purely morphic*.

The factor complexity of purely morphic word is well known. The next result was stated by Pansiot in [467] and then generalized in [468]. For a comprehensive presentation, see [78, Section 4.7]. Recall that the case of eventually periodic words is settled by Morse–Hedlund theorem.

Theorem 1.4.2. *Let* $\mathbf{w}$ *be a pure morphic word. If* $\mathbf{w}$ *is not eventually periodic, then its factor complexity* $p_{\mathbf{w}}$ *belongs to* $\Theta(n)$, $\Theta(n \log \log n)$, $\Theta(n \log n)$, *or* $\Theta(n^2)$.

Definition 1.4.3. A morphism h is *non-erasing* if $h(a) \neq \epsilon$ for all $a \in \Sigma$. Otherwise it is *erasing*. A morphism is *k-uniform* if $|h(a)| = k$ for all $a \in \Sigma$; it is *uniform* if it is k-uniform for some k. A 1-uniform morphism is often said to be a *letter-to-letter morphism* or a *coding*.

The Thue–Morse morphism t given in (1.3) is 2-uniform.

Example 1.4.4 (Fibonacci Word). Another significant example of a purely morphic word is the *Fibonacci word*. It is obtained from the non-uniform morphism defined over the alphabet $\{0, 1\}$ by $\sigma : 0 \mapsto 01, 1 \mapsto 0$,

$$\sigma^{\omega}(0) = (x_n)_{n \geq 0} = 0100101001001010010100100101001001010010100 \cdots .$$

It is a Sturmian word and can be obtained as follows. Let $\phi = (1 + \sqrt{5})/2$ be the Golden mean. For all $n \geq 1$, if $\lfloor (n+1)\phi \rfloor - \lfloor n\phi \rfloor = 2$, then $x_{n-1} = 0$; otherwise $x_{n-1} = 1$.

An infinite word $\mathbf{x}$ over Δ is *morphic* if there exists a purely morphic word $\mathbf{y}$ over Σ and a morphism $g : \Sigma^* \to \Delta^*$ such that $\mathbf{x} = g(\mathbf{y})$.

We can always restrict ourselves to non-erasing prolongable morphisms and codings. This result was already stated in [154]. J.-J. Pansiot also considered this result in [466]. For a proof, see [14]. An alternative short proof is given in [298]. This result is also discussed in detail in [134] and [146].

Theorem 1.4.5. *Let* $f : \Sigma^* \to \Sigma^*$ *be a (possibly erasing) morphism that is prolongable on a letter* $a \in \Sigma$. *Let* $g : \Sigma^* \to \Gamma^*$ *be a (possibly erasing) morphism. If the word* $g(f^{\omega}(a))$ *is infinite, there exists a non-erasing morphism* $h : \Delta^* \to \Delta^*$ *prolongable on a letter* $c \in \Delta$ *and a coding* $j : \Delta^* \to \Gamma^*$ *such that* $g(f^{\omega}(a)) = j(h^{\omega}(c))$.

1.5 Languages and Machines

Formal languages theory is mostly concerned with the study of the mathematical properties of sets of words. For a comprehensive exposition on regular (or rational) languages and automata theory, see, for instance, Sakarovitch's book [518]. For the connections with infinite words, see [476]. For an overview see the chapter [590]. Finally see [555], Hopcroft and Ullman's classic book [301], or its updated version [300] for general books on formal languages theory.

1.5.1 Languages of Finite Words

Let Σ be an alphabet. A subset L of Σ^* is said to be a *language*. Since a language is a *set* of words, we can apply all the usual set operations like union, intersection, or set difference: $\cup$, $\cap$, or $\setminus$. The concatenation of words can be extended to define an operation on languages. If L, M are languages, LM is the language of the words obtained by concatenation of a word in L and a word in M, i.e.,

$$LM = \{uv \mid u \in L, v \in M\}.$$

We can of course define the concatenation of a language with itself, so it permits us to introduce the power of a language. Let $n \in \mathbb{N}$, Σ be an alphabet, and $L \subseteq \Sigma^*$ be a language. The language L^n is the set of words obtained by concatenating n words in L. We set $L^0 := \{\epsilon\}$. In particular, we recall that Σ^n denotes the set of words of length n over Σ, i.e., concatenations of n letters in Σ. The *(Kleene) star* of the language L is defined as

$$L^* = \bigcup_{i \geq 0} L^i.$$

Otherwise stated, L^* contains the words that are obtained as the concatenation of an arbitrary number of words in L. Notice that the definition of Kleene star is compatible with the notation Σ^* introduced to denote the set of finite words over Σ. We also write $L^{\leq n}$ as a shorthand for

$$L^{\leq n} = \bigcup_{i=0}^{n} L^i.$$

Note that if the empty word belongs to L, then $L^{\leq n} = L^n$. We recall that $\Sigma^{\leq n}$ is the set of words over Σ of length at most n.

Example 1.5.1. Let $L = \{a, ab, aab\}$ and $M = \{a, ab, ba\}$ be two finite languages. We have

$$L^2 = \{aa, aab, aaab, aba, abab, abaab, aaba, aabab, aabaab\}$$

and

$$M^2 = \{aa, aab, aba, abab, abba, baa, baab, baba\}.$$

One can notice that $\operatorname{Card}(L^2) = (\operatorname{Card} L)^2$ but $\operatorname{Card}(M^2) < (\operatorname{Card} M)^2$. This is due to the fact that all words in L^2 have a unique factorization as concatenation of two elements in L, but this is not the case for M, where $(ab)a = a(ba)$. We can notice that

$$L^* = \{a\}^* \cup \{a^{i_1}ba^{i_2}b\cdots a^{i_n}ba^{i_{n+1}} \mid \forall n \geq 1, i_1,\ldots,i_n \geq 1,\ i_{n+1} \geq 0\}.$$

Since languages are sets of (finite) words, a language can be either *finite* or *infinite*. For instance, a language L differs from $\emptyset$ or $\{\epsilon\}$ if and only if the language L^* is infinite. Let L be a language, we set $L^+ = LL^*$. The mirror operation can also be extended from words to languages: $\tilde{L} = \{\tilde{u} \mid u \in L\}$.

Definition 1.5.2. A language is *prefix-closed* (respectively *suffix-closed*) if it contains all prefixes (respectively suffixes) of any of its elements. A language is *factorial* if it contains all factors of any of its elements.

Obviously, any factorial language is prefix-closed and suffix-closed. The converse does not hold. For instance, the language $\{a^n b \mid n > 0\}$ is suffix-closed but not factorial.

Example 1.5.3. Connected with the Thue–Morse word (see Example 1.3.6), the set of words over $\{0, 1\}$ containing an even number of ones is the language

$$\begin{aligned} E &= \{w \in \{0,1\}^* \mid |w|_1 \equiv 0 \pmod 2\} \\ &= \{\epsilon, 0, 00, 11, 000, 011, 101, 110, 0000, 0011, \ldots\}. \end{aligned}$$

This language is closed under mirror, i.e., $\tilde{L} = L$. Notice that the concatenation $E\{1\}E$ is the language of words containing an odd number of ones and $E \cup E\{1\}E = E(\{\epsilon\} \cup \{1\}E) = \{0,1\}^*$. Notice that E is neither prefix-closed, since $1001 \in E$ but $100 \notin E$, nor suffix-closed.

Definition 1.5.4. The set of factors of a language L is denoted as $\mathrm{Fac}(L)$, whereas the set of prefixes of a language L is denoted as $\mathrm{Pref}(L)$. The notation $w^{-1}L$ stands for $w^{-1}L = \{u \mid wu \in L\}$.

If a language L over Σ can be obtained by applying to some finite languages a finite number of operations of union, concatenation, and Kleene star, then this language is said to be a *regular language*. This generation process leads to *regular expressions* which are well-formed expressions used to describe how a regular language is built in terms of these operations.

Note that the *Chomsky–Schützenberger hierarchy* introduced in the theory of formal languages provides a classification depending on the machine needed to recognize an infinite language of finite words. From a computational perspective, the simplest languages are the regular languages. They are accepted (or recognized) by finite automata, and described by regular expressions. One then has context-free languages that are recognized by non-deterministic pushdown automata, context-sensitive languages recognized by linear-bounded non-deterministic Turing machines, and lastly, recursively enumerable languages recognized by Turing machines. See Section 2.1.2 for a similar hierarchy for Mahler functions and regular sequences.

From the definition of a regular language, the following result is immediate.

Theorem 1.5.5. *The class of regular languages over Σ is the smallest subset of 2^{Σ^*} (for inclusion) containing the languages $\emptyset$, $\{a\}$ for all $a \in \Sigma$ and closed under union, concatenation, and Kleene star.*

Example 1.5.6. For instance, the language L over $\{0, 1\}$ whose words do not contain the factor 11 is regular. It is called the *Golden mean shift*. This language can be described by the regular expression $L = \{0\}^*\{1\}\{0,01\}^* \cup \{0\}^*$. Otherwise stated, it is generated from the finite languages $\{0\}$, $\{0,01\}$, and $\{1\}$ by applying union, concatenation, and star operations. Its complement in Σ^* is also regular and is described by the regular expression $\Sigma^*\{11\}\Sigma^*$. The language E from Example 1.5.3 is also regular; we have the following regular expression $\{0\}^*(\{1\}\{0\}^*\{1\}\{0\}^*)^*$ describing E.

Definition 1.5.7. Let $L \subseteq \Sigma^*$ be a language over the alphabet Σ. The *growth function* of L is the map

$$g_L : \mathbb{N} \to \mathbb{N},\ n \mapsto \mathrm{Card}(L \cap \Sigma^n)\,.$$

In particular, $g_L(n) \leq (\mathrm{Card}\,\Sigma)^n$ for all $n \geq 0$. Note that the complexity function of an infinite word $\mathbf{x}$ (see Definition 1.3.5) is exactly the growth function of the language $\mathrm{Fac}(\mathbf{x})$ of $\mathbf{x}$.

1.5.2 Formal Series

Let R be a semiring (see Definition 1.2.2). We can consider a map m from Σ^* to R. This map can be represented as a formal series

$$S = \sum_{w \in \Sigma^*} m(w)\, w\,.$$

This means that the coefficient (S, w) of the series S for the word w is given by $m(w)$. The sets of those formal series is denoted by $R\langle\langle \Sigma^* \rangle\rangle$ and has a semiring structure for the two operations defined as follows:

$$(S + T, w) = (S, w) + (T, w)$$

and

$$(ST, w) = \sum_{uv=w} (S, u)(T, v)\,.$$

In particular, a finite word w of length n can be factored in $n + 1$ concatenation products. This means that the sum above is finite. When R is limited to the Boolean semiring $\mathbb{B}$, then $\mathbb{B}\langle\langle\Sigma^*\rangle\rangle$ is just the set of languages over Σ. As a prominent example, Mahler functions are studied in details in Chapter 2.

1.5.3 Codes

A subset $X \subset \Sigma^+$ is a *code* if every word in X^* has a unique factorization with factors in X, i.e.,

$$(x_1 \cdots x_m = y_1 \cdots y_n,\ x_1, \ldots, x_m, y_1, \ldots, y_n \in X) \Rightarrow (m = n \text{ and } x_i = y_i\ \forall i)\,.$$

As an example, the set $X = \{a, ab, ba\}$ is not a code because the word aba has two X-factorizations: $a(ba)$ and $(ab)a$. The language $\{a^i b \mid i \geq 0\}$ is clearly a code. For an introduction to codes, see Bruyère's chapter in [386].

Let X be a set of words where no word in X is a proper prefix of another word in X. Then X is said to be a *prefix code*. The terminology of code comes from the following proposition.

Proposition 1.5.8. *A subset $X \subset \Sigma^+$ is a code if and only if any morphism $f : \Gamma^* \to \Sigma^*$ induced by a one-to-one correspondence (i.e., bijection) from Γ to X is one to one (injective).*

The notion can be extended to deal with infinite words. A subset $X \subset \Sigma^+$ is an *ω-code* if every word in Σ^ω has at most one factorization with words in X. As an example, $X = \{a, ab, bb\}$ is a code but it is not an ω-code. The infinite word $abbb\cdots$ has two X-factorizations $(a, bb, bb, \ldots)$ and $(ab, bb, bb, \ldots)$.

1.5.4 Automata

As we shall briefly explain in this section, the regular languages are exactly the languages recognized by finite automata. We start with non-deterministic automata in Definition 1.5.9, then we present the deterministic ones in Definition 1.5.13. Finally, we introduce automata with output in Definition 1.5.17. The notions recalled here will be used in particular in Section 7.5 in connection with normality, and in Chapter 10 with the notion of Mealy automaton.

Definition 1.5.9. A *finite automaton* is a labeled graph given by a 5-tuple $\mathscr{A} = (Q, \Sigma, E, I, T)$ where Q is the (finite) *set of states*, $E \subseteq Q \times \Sigma^* \times Q$ is the finite set of *edges* defining the *transition relation*, $I \subseteq Q$ is the set of *initial states*, and T is the *set of terminal (or final) states*. A *path* in the automaton is a sequence

$$(q_0, u_0, q_1, u_1, \ldots, q_{k-1}, u_{k-1}, q_k)$$

such that, for all $i \in \{0, \ldots, k-1\}$, $(q_i, u_i, q_{i+1}) \in E$, $u_0 \cdots u_{k-1}$ is the *label* of the path. Such a path is *successful* if $q_0 \in I$ and $q_k \in T$. The language $L(\mathscr{A})$ *recognized* (or *accepted*) by $\mathscr{A}$ is the set of labels of all successful paths in $\mathscr{A}$.

Any finite automaton $\mathscr{A}$ gives a partition of Σ^* into $L(\mathscr{A})$ and $\Sigma^* \setminus L(\mathscr{A})$. When depicting an automaton, initial states are marked with an incoming arrow and terminal states are marked with an outgoing arrow. A transition like (q, u, r) is represented by a directed edge from q to r with label u, $q \xrightarrow{u} r$.

Example 1.5.10. In Figure 1.1 the automaton has two initial states p and r and three terminal states q, r, and s. For instance, the word ba is recognized by the automaton. There are two successful paths corresponding to the label ba: (p, b, q, a, s) and (p, b, p, a, s). For this latter path, we can write $p \xrightarrow{b} p \xrightarrow{a} s$. On the other hand, the word $baab$ is not recognized by the automaton.

Example 1.5.11. The automaton in Figure 1.2 recognizes exactly the language E of the words having an even number of 1 from Example 1.5.3.

Definition 1.5.12. Let $\mathscr{A} = (Q, \Sigma, E, I, T)$ be a finite automaton. A state $q \in Q$ is *accessible* (respectively *co-accessible*) if there exists a path from an initial state to q (respectively from q to some terminal state). If all states of $\mathscr{A}$ are both accessible and co-accessible, then $\mathscr{A}$ is said to be *trim*.

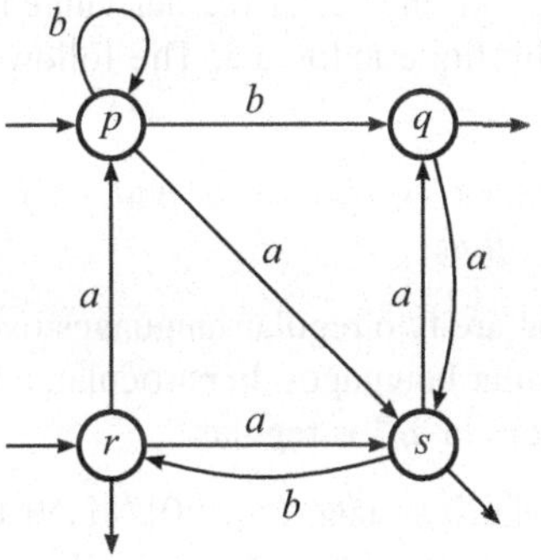

Fig. 1.1 A finite automaton.

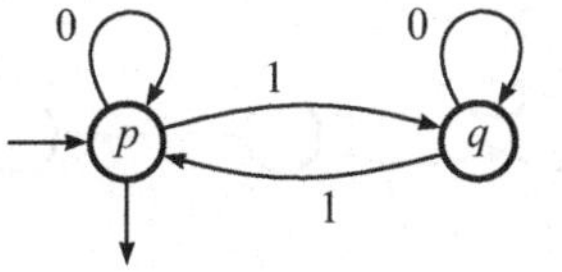

Fig. 1.2 An automaton recognizing words with an even number of 1.

Definition 1.5.13. A finite automaton $\mathscr{A} = (Q, \Sigma, E, I, T)$ is said to be *deterministic* (*DFA*) if it has only one initial state q_0, if E is a subset of $Q \times \Sigma \times Q$ and for each $(q, a) \in Q \times \Sigma$ there is at most one state $r \in Q$ such that $(q, a, r) \in E$. In that case, E defines a partial function $\delta_{\mathscr{A}} : Q \times \Sigma \to Q$ that is called the *transition function* of $\mathscr{A}$. The adjective *partial* means that the domain of $\delta_{\mathscr{A}}$ can be a strict subset of $Q \times \Sigma$. To express that the partial transition function is total, the DFA can be said to be *complete*. To get a total function, one can add to Q a new "sink state" s and, for all $(q, a) \in Q \times \Sigma$ such that $\delta_{\mathscr{A}}$ is not defined, set $\delta_{\mathscr{A}}(q, a) := s$. This operation does not alter the language recognized by $\mathscr{A}$. We can extend $\delta_{\mathscr{A}}$ to be defined on $Q \times \Sigma^*$ by $\delta_{\mathscr{A}}(q, \epsilon) = q$ and, for all $q \in Q$, $a \in \Sigma$, and $u \in \Sigma^*$, $\delta_{\mathscr{A}}(q, au) = \delta_{\mathscr{A}}(\delta_{\mathscr{A}}(q, a), u)$. Otherwise stated, the language recognized by $\mathscr{A}$ is $L(\mathscr{A}) = \{u \in \Sigma^* \mid \delta_{\mathscr{A}}(q_0, u) \in F\}$ where q_0 is the initial state of $\mathscr{A}$. If the automaton is deterministic, it is sometimes convenient to refer to the 5-tuple $\mathscr{A} = (Q, \Sigma, \delta_{\mathscr{A}}, I, T)$.

As explained by the following result, for languages of finite words, finite automata and deterministic finite automata recognize exactly the same languages. The following result is referred to as Rabin–Scott theorem [489].

Theorem 1.5.14. *If L is recognized by a finite automaton $\mathscr{A}$, there exists a DFA which can be effectively computed from $\mathscr{A}$ and recognizing the same language L.*

A proof and more details about classical results in automata theory can be found in textbooks like [300, 518] or [539]. For standard material in automata theory, we shall not refer again to these references below.

One important result is that the set of regular languages coincides with the set of languages recognized by finite automata. The following result is referred to as Kleene's theorem [349].

Theorem 1.5.15. *A language is regular if and only if it is recognized by a (deterministic) finite automaton.*

Observe that if L and M are two regular languages over Σ, then $L \cap M$, $L \cup M$, LM, and $L \setminus M$ are also regular languages. In particular, a language over Σ is regular if and only if its complement in Σ^* is regular.

Example 1.5.16. The regular language $L = \{0\}^*\{1\}\{0, 01\}^* \cup \{0\}^*$ introduced in Example 1.5.6 is recognized by the DFA depicted in Figure 1.3. Notice that the state s is a *sink*: a non-terminal state and all transitions remain in s.

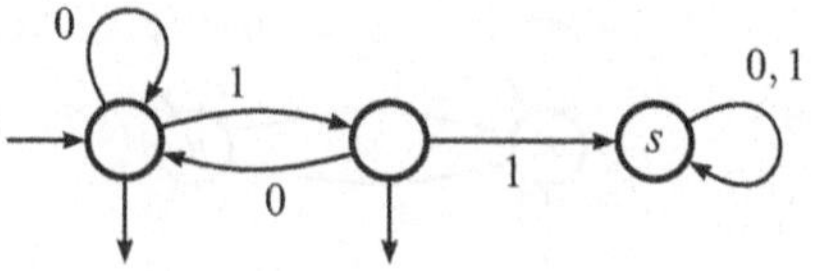

Fig. 1.3 A DFA accepting words without factor 11.

We introduce the notion of automaton with output (see also more generally Definition 7.5.1 for the notion of a transducer). It generalizes the classical DFA: if the output function takes at most two values, then it is a DFA. The extra output function will take care of the extra coding.

Definition 1.5.17. A *deterministic finite automaton with output* or DFAO for short is given by a 5-tuple $\mathcal{A} = (Q, q_0, A, \delta, \mu)$ where Q is a finite set of states, $q_0 \in Q$ is the initial state, $\delta : Q \times A \to Q$ is the transition function, and $\mu : Q \to B$ is the output map (where B is some finite set).

Finite automata accepting languages of infinite words are not presented here. Büchi automata (where an accepting run goes infinitely often through an accepting state) are introduced in Section 3.6.

1.6 Sequences and Machines

1.6.1 Automatic Sequences

We now consider how finite automata can be used to generate sequences with values in a finite alphabet, namely, we present the *automatic sequences*. As we shall soon see, they are particular morphic words and are deeply linked with the integer base-k numeration system. They were introduced by A. Cobham [156] under the name *uniform tag sequences*. Automatic sequences will appear in Chapters 2, 3, and 4. See in particular Section 2.2 for definitions, properties and examples, and connections with Mahler functions. We will recall that automatic sequences may be obtained as the image under a coding of the fixed point of a k-uniform morphism. Equivalently, for all $n \geq 0$, the nth symbol of such a sequence is the output of a deterministic finite automaton with output fed with the k-ary expansion of n.

Definition 1.6.1. Let $k \geq 2$. Consider an infinite word $\mathbf{w} = g(f^{\omega}(a))$ where $f : \Sigma^* \to \Sigma^*$ is a k-uniform morphism prolongable on a and $g : \Sigma^* \to \Gamma^*$ is a coding. We say that $\mathbf{w}$ is *k-automatic*.

Observe that $|f^n(a)| = k^n$ for all $n \geq 0$. We first consider the "internal sequence," i.e., the fixed point $\mathbf{x} = f^{\omega}(a) = x_0x_1x_2\cdots$. Let j such that $k \leq j < k^2$; then $j = kq + r$ with $1 \leq q < k$ and $0 \leq r < k$. The symbol x_j is the $(r+1)$st symbol occurring in $f(x_q)$. As depicted in Figure 1.4, this simply comes from one iteration of the k-uniform morphism.

We obtain the following result by induction on $m \geq 0$. Even though it is not surprising, it has an important consequence about how the word can be obtained.

Lemma 1.6.2. *Let j such that $k^m \leq j < k^{m+1}$, for some $m \geq 0$. Then $j = kq + r$ with $k^{m-1} \leq q < k^m$ and $0 \leq r < k$ and the symbol x_j is the $(r+1)$st symbol occurring in $f(x_q)$.*

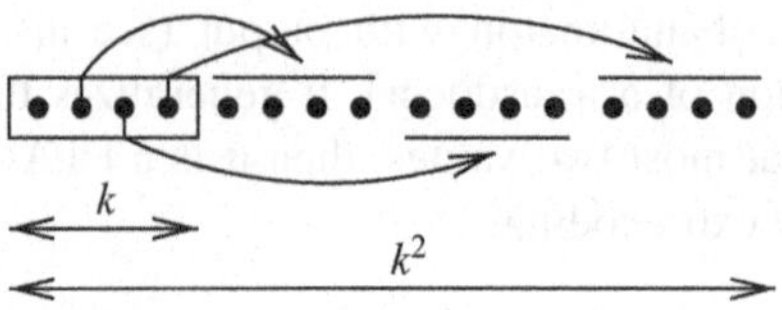

Fig. 1.4 Iterating a k-uniform morphism (with $k = 4$).

$$f : \begin{cases} \mathtt{a} \mapsto \mathtt{abc} \\ \mathtt{b} \mapsto \mathtt{cbc} \\ \mathtt{c} \mapsto \mathtt{bca} \end{cases}$$

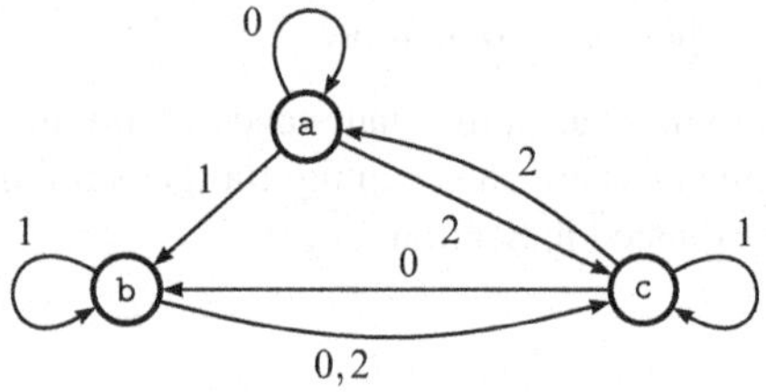

Fig. 1.5 A 3-uniform morphism and the associated automaton $\mathscr{A}_f$.

The quotient $\lfloor j/k \rfloor$ of the Euclidean division of j by k is denoted by j DIV k. So to speak, for any symbol x_j occurring in $\mathbf{x} = f^\omega(a)$, we can track its history: x_j has been produced by f from $x_{j \text{ DIV } k}$. The latter symbol appears itself in the image by f of $x_{(j \text{ DIV } k) \text{ DIV } k}$, and so on and so forth.

Note that if the base-k expansion of j is $\text{rep}_k(j) = c_i \cdots c_1 c_0$, then the base-$k$ expansion of j DIV k is $c_i \cdots c_1$. This simple observation permits one to easily track the past of a given symbol by considering the prefixes of $\text{rep}_k(j)$. Consider, for instance, the symbol $\mathbf{t}_{28}$ occurring in the Thue–Morse word:

$$\mathbf{t} = 0\underline{1}1\overline{0}100\underline{1}100101\overline{1}0100101100110\underline{1}001\cdots.$$

Since $\text{rep}_2(28) = 11100$, this symbol comes from $\mathbf{t}_{14}$ because $\text{rep}_2(14) = 1110$. Then $\mathbf{t}_{14}$ appears in the image of $\mathbf{t}_7$, itself appearing in the image of $\mathbf{t}_3$ and finally in the image of $\mathbf{t}_1$.

But Lemma 1.6.2 provides some extra knowledge. Let j such that $j = kq + r$ with $k^{m-1} \leq q < k^m$ and $0 \leq r < k$, for some $m \geq 0$. We have just explained how x_j comes from x_q. But the knowledge of x_q and r entirely determines x_j. It is thus time to explain where does the term of automatic sequence come from.

We can associate with a k-uniform morphism $f : \Sigma^* \to \Sigma^*$ and a letter $a \in \Sigma$, a DFA $\mathscr{A}_f = (\Sigma, a, [\![0, k-1]\!], \delta_f, \Sigma)$ where $\delta_f(b, i) = w_{b,i}$ if $f(b) = w_{b,0} \cdots w_{b,k-1}$. Note that the alphabet Σ is the set of states of this automaton.

Example 1.6.3. Consider the morphism f and the associated automaton depicted in Figure 1.5.

The next propositions explain the terminology of automatic sequences.

Proposition 1.6.4. *Let* $\mathbf{x} = f^{\omega}(a) = x_0x_1\cdots$ *with* f *a* k*-uniform morphism. With the above notation, for all* $j \geq 0$,

$$x_j = \delta_f(a, \text{rep}_k(j)).$$

Proof. This is a direct consequence of Lemma 1.6.2. □

The converse also holds.

Proposition 1.6.5. *Let* $(\Sigma, a, [\![0, k-1]\!], \delta, \Sigma)$ *be a DFA such that* $\delta(a, 0) = a$. *Then the word* $\mathbf{x} = x_0x_1x_2\cdots$ *defined by* $x_j = \delta(a, \text{rep}_k(j))$, *for all* $j \geq 0$, *is the fixed point of a* k*-uniform morphism* f *prolongable on* a *where* $f(b) = \delta(b, 0)\cdots\delta(b, k-1)$ *for all* $b \in \Sigma$.

Proof. This is again a direct consequence of Lemma 1.6.2. □

The reader will object that we have not taken into account that an extra coding can be applied to $\mathbf{x} = f(\mathbf{x})$. This does not require many changes. We simply have to make use of automata with output as stated below in Cobham's theorem on automatic sequences [156].

Theorem 1.6.6. *Let* $\mathbf{w} = w_0w_1w_2\cdots$ *be an infinite word over an alphabet* Γ. *It is of the form* $g(f^{\omega}(a))$ *where* $f : \Sigma^* \to \Sigma^*$ *is a* k*-uniform morphism prolongable on* $a \in \Sigma$ *and* $g : \Sigma^* \to \Gamma^*$ *is a coding if and only if there exists a DFAO*

$$(\Sigma, a, [\![0, k-1]\!], \delta, \mu : \Sigma \to \Gamma)$$

such that $\delta(a, 0) = a$ *and, for all* $j \geq 0$, $w_j = \mu(\delta(a, \text{rep}_k(j)))$.

Proof. Proceed as above and the coding g coincides with the output function μ. □

Example 1.6.7. From the morphism t given in (1.3) generating the Thue–Morse word, we derive the automaton depicted in Figure 1.2. Again considering 28, which is written 11100 in base 2, if we start from the initial state p and we read consecutively the symbols in $\text{rep}_2(28)$ from left to right, then we follow some path in the automaton, and the state q finally reached gives the symbol $\mathbf{t}_{28}$. The output function maps p to 0 and q to 1.

Example 1.6.8. Let us consider a more intricate example where a coding, and thus an output function, is used. The morphism f and the coding g are given in Figure 1.6. The corresponding automaton is represented on the right of the same figure. We have

$$f^{\omega}(\mathtt{a}) = \mathtt{acabaccaacababacacabaccaaccaacab}\cdots$$

and

$$g(f^{\omega}(\mathtt{a})) = 00010000000101000001000000000001\cdots.$$

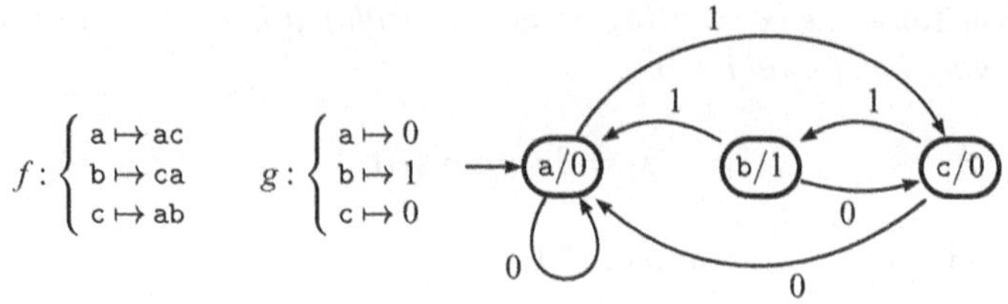

Fig. 1.6 A 2-uniform morphism, a coding and the corresponding DFAO.

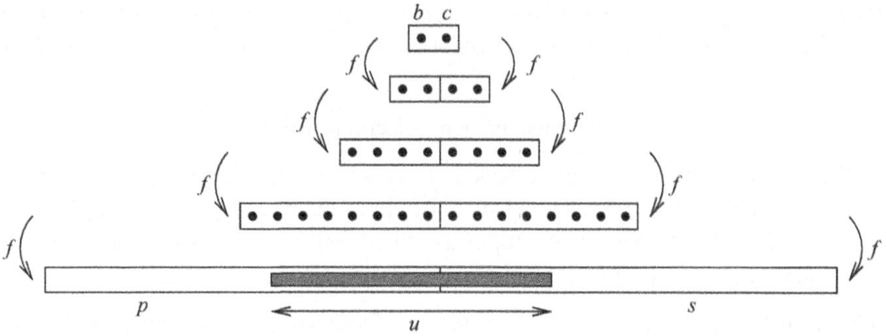

Fig. 1.7 Iterating a 2-uniform morphism.

Again, the jth symbol in $g(f^\omega(\mathrm{a}))$ can be readily obtained from $\mathrm{rep}_2(j)$ fed to the DFAO represented in Figure 1.6 where the states contain the information about the value of the output function.

Now we turn to the factors occurring in an automatic sequence $\mathbf{w} = g(\mathbf{x})$, where $\mathbf{x}$ is a fixed point of the k-uniform morphism $f : \Sigma^* \to \Sigma^*$. Let u be a factor of length n occurring in $\mathbf{x}$. There exists i such that $k^{i-1} \leq n < k^i$. Note that $|f^i(b)| = k^i$ for all $b \in \Sigma$. We consider the factorization of $\mathbf{x}$ into consecutive blocks of length k^i of the form $f^i(b)$. Therefore, the factor u either occurs inside some $f^i(b)$ or it overlaps two images, i.e., u occurs in $f^i(bc)$ for some letters $b, c \in \Sigma$. Actually, there exist two letters b and c such that $f^i(bc) = pus$ with $|p| < k^i$. This last condition tells us that u starts inside $f^i(b)$. Such a simple observation, where we look backwards at the images of the morphism, as suggested by Figure 1.7, is sometimes called a *desubstitution*. It provides us with an upper bound on the number of factors of length n that may occur in $\mathbf{x}$: the number of pairs of letters (b, c) is $(\mathrm{Card}\ \Sigma)^2$ and u should start in one of the k^i symbols of $f^i(b)$. Therefore, the number of factors of length n in $\mathbf{x}$ is at most

$$(\mathrm{Card}\ \Sigma)^2\, k^i \leq (\mathrm{Card}\ \Sigma)^2\, k\, n\,.$$

We can even replace $(\mathrm{Card}\ \Sigma)^2$ with $p_{\mathbf{x}}(2)$ because only the factors bc occurring in $\mathbf{x}$ give factors of the form $f^i(b)f^i(c)$ occurring in $\mathbf{x} = f^i(\mathbf{x})$. Since applying a coding

g cannot increase the number of factors, we get

$$\mathrm{Card}(\mathrm{Fac}(\mathbf{x}) \cap \Sigma^n) \geq \mathrm{Card}\{g(u) \mid u \in \mathrm{Fac}(\mathbf{x}) \cap \Sigma^n\},$$

and so we have obtained the following result.

Theorem 1.6.9. *Let* $\mathbf{w}$ *be a* k*-automatic sequence. Then* $p_{\mathbf{w}}(n)$ *is in* $\mathcal{O}(n)$.

A proof of the following result can be found in [14, Section 8.4].

Theorem 1.6.10. *If the frequency of a letter in a morphic sequence exists, then it is an algebraic number. If the frequency of a letter in an automatic sequence exists, then it is a rational number.*

To conclude this section, we present another characterization of k-automatic sequences. This is not the last one; in Chapter 3, Section 3.3, a logical characterization of k-automatic sequences will be discussed, whereas Chapter 4 will provide an algebraic characterization in terms of polynomial identities (see Corollary 4.5.3).

Definition 1.6.11. Let $k \geq 2$ be an integer. Given a sequence $s = (s(n))_{n\geq 0}$, we define a particular set of subsequences called the *k-kernel* of s

$$\mathrm{Ker}_k(s) := \left\{(s(k^\ell n + r))_{n\geq 0} \mid \ell \geq 0,\ 0 \leq r < k^\ell\right\}.$$

An equivalent definition of the k-kernel is to introduce k operators of *k-decimation* acting on the set of sequences and defined, for $r \in \{0, \ldots, k-1\}$, by

$$\rho_{k,r}((s(n))_{n\geq 0}) = (s(kn + r))_{n\geq 0}.$$

Thus $\mathrm{Ker}_k(s)$ is the set of sequences of the form

$$\rho_{k,r_1} \circ \cdots \circ \rho_{k,r_m}((s(n))_{n\geq 0}) \tag{1.4}$$

for all $m \geq 0$ and $r_1, \ldots, r_m \in \{0, \ldots, k-1\}$. These decimation operators are close to the Cartier operators discussed in Chapter 2. The following result appeared in Eilenberg's book [211]. Note that if a sequence t belongs to $\mathrm{Ker}_k(s)$, then $\rho_{k,r}(t)$ also belongs to $\mathrm{Ker}_k(s)$.

Theorem 1.6.12. *A sequence is* k*-automatic if and only if its* k*-kernel is finite.*

Example 1.6.13. The 2-kernel of the Thue–Morse sequence contains exactly two sequences (the sequence itself and its "complement"). Indeed, let $s_2(n)$ be the sum of digits of the binary expansion of n, we have

$$s_2(2n) = s_2(n), \quad s_2(2n + 1) = s_2(n) + 1. \tag{1.5}$$

1.6.2 Regular Sequences

We have seen that k-automatic sequences may be defined through the finiteness of their k-kernel (Theorem 1.6.12). This characterization is used to extend the notion to sequences taking infinitely many values. Allouche and Shallit considered sequences taking values in a ring R containing a commutative Noetherian ring R' (i.e., every ideal of R' is finitely generated). Examples of such rings R' are given by all finite rings, all principal ideal domains, and in particular $\mathbb{Z}$, the ring of polynomials with coefficients in a field, or all fields. We may consider linear combinations with coefficients in R' (R'-linear combinations) of sequences in $R^{\mathbb{N}}$. Endowed with pointwise addition and multiplication by an element in R', the set $R^{\mathbb{N}}$ has a R'-module structure: if $r = (r(n))_{n\geq 0}$ and $s = (s(n))_{n\geq 0}$ belong to $R^{\mathbb{N}}$ and α belongs to R', then, for all $n \in \mathbb{N}$,

$$(r + s)(n) = r(n) + s(n)$$

and

$$(\alpha \cdot r)(n) = \alpha \cdot r(n).$$

In this short section, we mainly consider sequences in $\mathbb{Z}^{\mathbb{N}}$, i.e., $R = R' = \mathbb{Z}$. We will encounter regular sequences in Chapters 2, 3, and 4 of this book. To have stand-alone chapters, these notions will also be repeated there. In Chapter 3 (see in particular Section 3.4.1), k-regularity will be extended to sequences taking values in a semiring.

Regular sequences appeared in [16]. Many examples are given in [15]. See also [14, Chapter 16] and the updated version of Berstel and Reutenauer's book [77] where a chapter is devoted to regular sequences and linked with rational series.

Let M be a R-module and a subset $X \subset M$. The *submodule generated* by X is the intersection of all submodules of M containing X. It is denoted by $\langle X\rangle$. It is the set of all finite R-linear combinations of elements in X. A module is *finitely generated* (over R) when it is generated by a finite set (i.e., it is the R-span of a finite set). One also says that the module is of *finite type* or even *finite over R*. Note that the finite set of generators is not necessarily a basis.

Definition 1.6.14. Let $k \geq 2$ be an integer. A sequence $s = (s(n))_{n\geq 0}$ taking integer values is *k-regular* if the $\mathbb{Z}$-module generated by its k-kernel $\langle \mathrm{Ker}_k(s)\rangle$ is finitely generated, i.e., there exists a finite number of sequences in $\mathbb{Z}^{\mathbb{N}}$

$$t_1 = (t_1(n))_{n\geq 0}, \ldots, t_\ell = (t_\ell(n))_{n\geq 0}$$

such that

$$\langle \mathrm{Ker}_k(s)\rangle = \langle t_1, \ldots, t_\ell\rangle.$$

In particular, every sequence in $\mathrm{Ker}_k(s)$ is a $\mathbb{Z}$-linear combination of the t_js. For all $i \geq 0$ and for all $r \in \{0, \ldots, k^i - 1\}$, there exist integers $c_{i,1}, \ldots, c_{i,\ell}$ such that

$$\forall n \geq 0, \quad s(k^i n + r) = \sum_{j=1}^{\ell} c_{i,j}\, t_j(n).$$

One can consider another point of view. A sequence is said to be k-regular if its orbit under the action of the operators of k-decimation remains in a finite dimensional vector space. Indeed, $\mathbb{Z}$ is included in fields such as $\mathbb{Q}$, $\mathbb{R}$, or $\mathbb{C}$. Thus the sequences can be seen as elements of $\mathbb{Q}^{\mathbb{N}}$ which is a $\mathbb{Q}$-vector space.

Remark 1.6.15. The original definition in [16] was formulated differently. Let R be a ring containing a commutative Noetherian ring R'. A sequence $s = (s(n))_{n\geq 0}$ in $R^{\mathbb{N}}$ is *(R',k)-regular* if there exists a finite number of sequences in $R^{\mathbb{N}}$

$$t_1 = (t_1(n))_{n\geq 0}, \ldots, t_\ell = (t_\ell(n))_{n\geq 0}$$

such that every sequence in $\mathrm{Ker}_k(s)$ is an R'-linear combination of $t_1, \ldots, t_\ell$. Thus the definition means that $\langle \mathrm{Ker}_k(s)\rangle \subseteq \langle t_1, \ldots, t_\ell\rangle$. Otherwise stated, $\langle \mathrm{Ker}_k(s)\rangle$ is a submodule of a finitely generated R'-module (in general, this does not imply that the submodule itself is finitely generated). Since R' is assumed to be Noetherian, one can show that every submodule of a finitely generated R'-module is finitely generated[2], and thus $\langle \mathrm{Ker}_k(s)\rangle$ is finitely generated. This was the point of view adopted in Definition 1.6.14. In particular, if the setting does not assume that R' is Noetherian (in particular, if R or R' is a semiring), then Definition 1.6.14 would be stronger than simply requiring $\langle \mathrm{Ker}_k(s)\rangle \subseteq \langle t_1, \ldots, t_\ell\rangle$.

Example 1.6.16. The base-2 sum-of-digits function s_2 gives the sequence

$$(s_2(n))_{n\geq 0} = 0, 1, 1, 2, 1, 2, 2, 3, 1, 2, 2, 3, 2, 3, 3, 4, 1, 2, 2, 3, 2, 3, 3, 4, 2, 3, 3, 4, \ldots .$$

(Notice that we can interchange the words function and sequence and also speak of k-regular functions when defined over $\mathbb{N}$.) Clearly this sequence is unbounded: $s_2(2^n - 1) = n$ for all n. Nevertheless, in view of (1.5), the $\mathbb{Z}$-module generated by its 2-kernel is generated by the sequence $(s_2(n))_{n\geq 0}$ itself and the constant sequence $(1)_{n\geq 0}$.

Obviously, every k-automatic sequence is k-regular.

Proposition 1.6.17. *Let s be a sequence taking finitely many different values, i.e., there exists a finite alphabet Σ such that $s \in \Sigma^\omega$. Let $k \geq 2$. The sequence is k-automatic if and only if it is k-regular.*

There is an intermediate class of sequences between k-automatic and k-regular sequences [130].

---

[2] An R'-module M is *Noetherian* if every submodule of M is finitely generated. Let R' be a Noetherian ring. An R'-module M is Noetherian if and only if it is finitely generated.

Definition 1.6.18. Let $k \geq 2$ be an integer. The map rep_k is extended to $\mathbb{N} \times \mathbb{N}$ as follows. For all $m, n \in \mathbb{N}$,

$$\text{rep}_k(m,n) = \left(0^{M-|\text{rep}_k(m)|}\text{rep}_k(m), 0^{M-|\text{rep}_k(n)|}\text{rep}_k(n)\right)$$

where $M = \max\{|\text{rep}_k(m)|, |\text{rep}_k(n)|\}$. The idea is that the shortest word is padded with leading zeroes to get two words of the same length.

A sequence $(s(n))_{n\geq 0}$ of integers is said to be *k-synchronized* if the language $\{\text{rep}_k(n, s(n)) \mid n \in \mathbb{N}\}$ is accepted by some finite automaton reading pairs of digits.

As an example, the complexity function $(p_{\mathbf{x}}(n))_{n\geq 0}$ of a k-automatic sequence $\mathbf{x}$ is k-synchronized [522]; we refer to Proposition 3.4.16. More results of this form are provided in Section 3.4. For results on the growth of regular sequences, see Section 2.3.

Proposition 1.6.19. *Let s be a sequence taking finitely many different values, i.e., there exists a finite alphabet Σ such that $s \in \Sigma^{\omega}$. Let $k \geq 2$. The sequence is k-automatic if and only if it is k-synchronized.*

Similarly to recognizable formal series, with every k-regular sequence $(s(n))_{n\geq 0} \in \mathbb{Z}^{\mathbb{N}}$ is associated *linear representation* (λ, μ, ν). There exist a positive integer r, a row vector $\lambda \in \mathbb{Z}^{1\times r}$ and a column vector $\nu \in \mathbb{Z}^{r\times 1}$, a matrix-valued morphism $\mu : \{0, \dots, k-1\} \to \mathbb{Z}^{r\times r}$ such that

$$s(n) = \lambda\mu(c_0 \cdots c_\ell)\nu$$

for all $c_\ell, \dots, c_0 \in \{0, \dots, k-1\}^*$ such that $\text{val}_k(c_\ell \cdots c_0) = \sum_{i=0}^{\ell} c_i k^i = n$. The converse also holds, if there exists a linear representation associated with the canonical k-ary expansion of integers (one has to take into account the technicality of representations with leading zeros), then the sequence is k-regular. See, for instance, [14, Theorem 16.2.3]. As a corollary, the nth term of a k-regular sequence can be computed with $\lfloor \log_k(n) \rfloor$ matrix multiplications.

Proof. Let $s = (s(n))_{n\geq 0} \in \mathbb{Z}^{\mathbb{N}}$ be a k-regular sequence. By definition, there exists a finite number of sequences $t_1, \dots, t_\ell$ such that $\langle \text{Ker}_k(s) \rangle = \langle t_1, \dots, t_\ell \rangle$. In particular, each t_j is a $\mathbb{Z}$-linear combination of elements in the k-kernel of s. We have finitely many t_js, so $t_1, \dots, t_\ell$ are linear combinations of finitely many elements in $\text{Ker}_k(s)$. Thus we can assume that $\langle \text{Ker}_k(s) \rangle$ is generated by finitely many elements from $\text{Ker}_k(s)$ itself. Without loss of generality, we will now assume that $t_1, \dots, t_\ell$ belong to $\text{Ker}_k(s)$.

From (1.4), for all $r \in \{0, \dots, k-1\}$ and all $i \in \{1, \dots, \ell\}$, $\rho_{k,r}(t_i)$ is a sequence in $\text{Ker}_k(s)$, and thus, there exist coefficients $(A_r)_{1,i}, \dots, (A_r)_{\ell,i}$ such that

$$\rho_{k,r}(t_i) = \sum_{j=1}^{\ell} (A_r)_{j,i}\, t_j\,.$$

Notice that A_r is an $\ell \times \ell$ matrix. Roughly, if we were in a vector space setting, this means that the matrices A_r represent the linear operators $\rho_{k,r}$ in the basis $t_1, \ldots, t_\ell$. Let $p \geq 0$ be an integer. Notice that if $\mathrm{rep}_k(p) = r_m \cdots r_0$, then $s(p)$ is the first term, i.e., corresponding to the index 0, of the sequence

$$(s(b^{m+1}n + p))_{n\geq 0} = \rho_{k,r_0} \circ \cdots \circ \rho_{k,r_m}((s(n))_{n\geq 0}) .$$

We will use the fact that $\rho_{k,r}$ is linear, i.e., if α, β are coefficients and v, w are two sequences, then $\rho_{k,r}(\alpha v + \beta w) = \alpha \rho_{k,r}(v) + \beta \rho_{k,r}(w)$. It is easy to see that

$$\rho_{k,r_0} \circ \cdots \circ \rho_{k,r_m}(t_i) = \sum_{j=1}^{\ell} (A_{r_0} \cdot \cdots \cdot A_{r_m})_{j,i}\, t_j .$$

If we have the following decomposition of s (in a vector space setting, we would have a unique decomposition of s in the basis $t_1, \ldots, t_\ell$)

$$s = \sum_{i=1}^{\ell} \sigma_i\, t_i$$

then, by linearity,

$$(s(b^{m+1}n + p))_{n\geq 0} = \sum_{i=1}^{\ell} \sigma_i \sum_{j=1}^{\ell} (A_{r_0} \cdot \cdots \cdot A_{r_m})_{j,i}\, (t_j(n))_{n\geq 0} = \sum_{j=1}^{\ell} \tau_j\, (t_j(n))_{n\geq 0}$$

where

$$\begin{pmatrix} \tau_1 \\ \vdots \\ \tau_\ell \end{pmatrix} = A_{r_0} \cdot \cdots \cdot A_{r_m} \begin{pmatrix} \sigma_1 \\ \vdots \\ \sigma_\ell \end{pmatrix} .$$

Consequently, $s(p)$ is obtained as

$$s(p) = \sum_{i=1}^{\ell} \tau_i\, t_i(0) = \big(t_1(0) \cdots t_\ell(0)\big)\, A_{r_0} \cdot \cdots \cdot A_{r_m} \begin{pmatrix} \sigma_1 \\ \vdots \\ \sigma_\ell \end{pmatrix} .$$

□

For a reader familiar with rational series, the previous result can be reformulated as follows. A sequence $s(n)$ is k-regular if and only if the formal series

$$\sum_{w\in\{0,\dots,k-1\}^*} s(\mathrm{val}_k(w))\, w$$

is recognizable (with the terminology of [77]; see Definition 3.4.1).

Example 1.6.20. For the sum-of-digits function given in Example 1.6.16, the sequence $s_2 = (s_2(n))_{n\geq 0}$ has a (base-2) linear representation given by

$$\lambda = \begin{pmatrix} 0 & 1 \end{pmatrix}, \ \mu(i) = \begin{pmatrix} 1 & 0 \\ i & 1 \end{pmatrix}, \ \nu = \begin{pmatrix} 1 \\ 0 \end{pmatrix}.$$

We let $\mathbf{1}$ denote the constant sequence. It does not belong to the 2-kernel of s_2, but it belongs to the $\mathbb{Z}$-module generated by it because it is equal to $\rho_{2,1}(s_2) - s_2$. Nevertheless, it is enough to see that $\rho_{2,0}(\mathbf{1}) = \rho_{2,1}(\mathbf{1}) = \mathbf{1}$ and take s_2 and $\mathbf{1}$ as generators to proceed as in the proof above. From the following relations we derive the two columns of matrix $\mu(0)$

$$\rho_{2,0}(s_2) = 1 \cdot s_2 + 0 \cdot \mathbf{1}, \quad \rho_{2,0}(\mathbf{1}) = 0 \cdot s_2 + 1 \cdot \mathbf{1}$$

and for $\mu(1)$

$$\rho_{2,1}(s_2) = 1 \cdot s_2 + 1 \cdot \mathbf{1}, \quad \rho_{2,1}(\mathbf{1}) = 0 \cdot s_2 + 1 \cdot \mathbf{1}.$$

The vector λ is given by $s_2(0) = 0$ ans $\mathbf{1}(0) = 1$. The vector ν is obtained from $s_2 = 1 \cdot s_2 + 0 \cdot \mathbf{1}$. To compute $s_2(19)$, observe that $\mathrm{rep}_2(19) = 10011$. Thus we compute

$$\begin{pmatrix} 0 & 1 \end{pmatrix} \mu(1)\mu(1)\mu(0)\mu(0)\mu(1) \begin{pmatrix} 1 \\ 0 \end{pmatrix} = 3.$$

Example 1.6.21. A less trivial example is considered in [201] by counting the number of odd numbers in the first n rows of the Pascal triangle. This sequence has a (base-2) linear representation given by

$$\lambda = \begin{pmatrix} 0 & 1 \end{pmatrix}, \ \mu(0) = \begin{pmatrix} 3 & 6 \\ 0 & 1 \end{pmatrix}, \ \mu(1) = \begin{pmatrix} 0 & -6 \\ 1 & 5 \end{pmatrix}, \ \nu = \begin{pmatrix} 1 \\ 0 \end{pmatrix}.$$

Remark 1.6.22. In [15, Section 6], a practical procedure to guess relations a possibly k-regular sequence will satisfy is described. Consider a sequence $(s(n))_{n\geq 0}$. The idea is to construct a matrix in which the rows represent truncated versions of elements of the k-kernel of $(s(n))_{n\geq 0}$, together with row reduction. Start with a matrix having a single row, say, corresponding to the first m elements of the sequence. Then repeatedly add subsequences of the form $(s(k^\ell n+r))_{n\geq 0}$ not linearly dependent of the previous stored sequences. From this, you have candidate relations that remain to be proven.

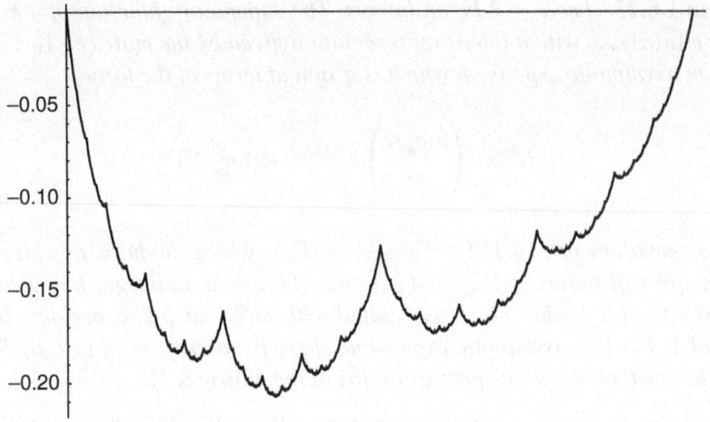

Fig. 1.8 The periodic function $\mathcal{G}$ on [0, 1].

Considering again the sum-of-digit function, Delange [191] showed that the summatory function of s_2 exhibits a particular behavior (also see [14, Thm. 3.5.4]).

$$\frac{1}{N}\sum_{j=0}^{N-1} s_2(j) = \frac{1}{2}\log_2 N + \mathcal{G}(\log_2 N) \tag{1.6}$$

where $\mathcal{G}$ is a continuous nowhere differentiable periodic function of period 1 (Figure 1.8).

General results do exist for summatory function of k-regular sequences. The result below can be found in [14, Thm. 16.4.1].

Theorem 1.6.23. *Let* $\mathsf{a} = (a(n))_{n\geq 0}$ *and* $\mathsf{b} = (b(n))_{n\geq 0}$ *be k-regular sequences. Then* $\mathsf{c} = \mathsf{a} \star \mathsf{b}$*, where, for all* $n \geq 0$, $c(n) = \sum_{i=0}^{n} a_i\, b_{n-i}$*, is k-regular.*

Corollary 1.6.24. *Let* $\mathsf{a} = (a(n))_{n\geq 0}$ *be a k-regular sequence. The sequence of partial sums*

$$\left(\sum_{i=0}^{n} a_i\right)_{n\geq 0}$$

is k-regular.

Proof. One simply takes for b the constant sequence $(1)_{n\geq 0}$ in Theorem 1.6.23. □

A linear representation of the summatory sequence can easily be deduced from the linear representation of the sequence itself, see [201, Lemma 1] or Proposition 2.2.11 in Chapter 2. Let us state the following result obtained by Dumas [201, 202] (see also Theorem 2.3.13).

Theorem 1.6.25. *Let $k \geq 2$ be an integer. The summatory function of a k-regular sequence $(u(n))_{n\geq 0}$ with a linear representation given by the matrices $\Gamma_0, \dots, \Gamma_{k-1}$ admits an asymptotic expansion which is a sum of terms of the form*

$$N^{\log_k \rho} \binom{\log_k N}{m} e^{i\theta \log_k N} \varphi(\log_k N)$$

for the eigenvalues $\rho e^{i\theta}$ of $\Gamma := \Gamma_0 + \cdots + \Gamma_{k-1}$ whose modulus ρ is larger than the joint spectral radius of $\Gamma_0, \dots, \Gamma_{k-1}$ and where m is an integer bounded by the maximal size of a Jordan block associated with $\rho e^{i\theta}$ and φ is a periodic function of period 1. For this asymptotic expansion, there is an error term in $\mathcal{O}(N^{\log_k r})$ for every r larger than the joint spectral radius of the matrices $\Gamma_0, \dots, \Gamma_{k-1}$.

Definition about the joint spectral radius will be given in Chapter 2; see also Chapter 11.8.1. Similar results are also discussed by Drmota and Grabner in [78, Theorem 9.2.15]. Let us also mention another result (see [14, Theorem 3.5.1]) with stronger assumptions but avoiding error terms. In this result, if v belongs to $\mathbb{C}^d$, then the notation $||v||$ stands for the Euclidean norm of v defined by $\left(\sum_{i=1}^{d} |v_i|^2\right)^{\frac{1}{2}}$. Moreover, if M is a square matrix of dimension d with entries in $\mathbb{C}$, then by $||M||$ we mean the L^2 norm, which is the matrix norm associated with the usual Euclidean norm on $\mathbb{C}^d$ by the formula $||M|| = \sup_{||x||=1} ||Mx||$.

Theorem 1.6.26. *Let $k \geq 2$ be an integer. Suppose there exist an integer $d \geq 1$, a sequence of vectors $(V_n)_{n\geq 0}$, $V_n \in \mathbb{C}^d$, defined by*

$$V_n = \begin{pmatrix} V_n^{(1)} \\ V_n^{(2)} \\ \vdots \\ V_n^{(d)} \end{pmatrix},$$

and k square matrices $\Gamma_0, \Gamma_0, \dots, \Gamma_{k-1}$ of dimension d such that

1. *$V_{kn+r} = \Gamma_r V_n$ for all $n \geq 0$ and all r, $0 \leq r < k$.*
2. *$||V_n|| = O(\log n)$.*
3. *There exist a $d \times d$ matrix Λ and a constant $c > 0$ such that either $||\Lambda|| < c$ or Λ is nilpotent, such that $\Gamma := \Gamma_0 + \Gamma_1 + \cdots + \Gamma_{k-1} = cI + \Lambda$.*

The matrix Γ being clearly invertible, if $||\Gamma^{-1}|| < 1$, then there exists a continuous function $G : \mathbb{R} \to \mathbb{C}^d$ of period 1 such that

$$\sum_{0\leq n<N} V_n = N^{\log_k c}(I + c^{-1}\Lambda)^{\log_k N} G(\log_k N).$$

1.7 Dynamical Systems

There are two main types of dynamical systems, namely, topological ones and measure-theoretic ones. Dynamical systems will be considered in particular in Chapters 8, 9, and 11.

1.7.1 Topological Dynamical Systems

Definition 1.7.1. A *topological dynamical system* (X,T) is defined as a compact metric space X together with a continuous map T defined onto the set X.

We are interested in iterating the map T, and we look at the *orbits* $\mathcal{O}(x)$ of $x \in X$ defined as

$$\mathcal{O}(x) = \{T^n(x) : n \in \mathbb{N}\}.$$

under the action T. The *trajectory* of $x \in X$ is the sequence $(T^n(x))_{n \geq 0}$.

A topological dynamical system (X,T) is *minimal* if, for all $\mathbf{x}$ in X, the orbit of $\mathbf{x}$, i.e., the set $\{T^n\mathbf{x} \mid n \in \mathbb{N}\}$, is dense in X. Let us note that if (X,S) is a subshift, and if X is furthermore assumed to be minimal, then X is periodic if and only if X is finite.

Two dynamical systems (X_1, T_1) and (X_2, T_2) are said to be *topologically conjugate* (or *topologically isomorphic*) if there exists an homeomorphism f from X_1 onto X_2 which conjugates T_1 and T_2, that is:

$$f \circ T_1 = T_2 \circ f.$$

If f is only onto, then (X_1, T_1) is said to factor onto (X_2, T_2), (X_2, T_2) is a factor of (X_1, T_1), and f is called a *factor map*.

1.7.2 Measure-Theoretic Dynamical Systems

We have considered here the notion of dynamical system, that is, a map acting on a given set, in a topological context. This notion can be extended to measurable spaces; we thus get measure-theoretic dynamical systems. For more details, one can refer, for instance, to [579]. See also Section 11.11.3.

Definition 1.7.2. A *measure-theoretic dynamical system* is defined as a system $(X, \mathcal{B}, \mu, T)$, where $\mathcal{B}$ is a σ-algebra, μ a probability measure defined on $\mathcal{B}$, and $T : X \to X$ is a measurable map which preserves the measure μ, i.e., for all $B \in \mathcal{B}$,

$\mu(T^{-1}(B)) = \mu(B)$. Such a measure is said to be *T-invariant* and the map T is said to preserve the measure μ.

The transformation T (or the system $(X, \mathcal{B}, \mu, T)$) is *ergodic* if for every $B \in \mathcal{B}$ such that $T^{-1}(B) = B$, then B has either zero measure or full measure.

Let (X, T) be a topological dynamical system. A topological system (X, T) always has an invariant probability measure. The case where there exists only one T-invariant measure is of particular interest. A topological dynamical system (X, T) is said to be *uniquely ergodic* if there exists one and only one T-invariant Borel probability measure over X. In particular, a uniquely ergodic topological dynamical system yields an ergodic measure-theoretic dynamical system.

A measure-theoretic ergodic dynamical system satisfies the *Birkhoff ergodic theorem*, also called *individual ergodic theorem*. Let us recall that the abbreviation a.e. stands for "almost everywhere": a property holds almost everywhere if the set of elements for which the property does not hold is contained in a set of zero measure.

Theorem 1.7.3. *Let $(X, \mathcal{B}, \mu, T)$ be a measure-theoretic dynamical system. Let $f \in L^1(X, \mathbb{R})$. Then the sequence $(\frac{1}{n}\sum_{k=0}^{n-1} f \circ T^k)_{n\geq 0}$ converges a.e. to a function $f^* \in L^1(X, \mathbb{R})$. One has $f^* \circ T = f^*$ a.e. and $\int_X f^*\, d\mu = \int_X f\, d\mu$. Furthermore, if T is ergodic, since f^* is a.e. constant, one has:*

$$\forall f \in L^1(X, \mathbb{R}), \quad \frac{1}{n}\sum_{k=0}^{n-1} f \circ T^k \xrightarrow[n\to\infty]{\mu-a.e.} \int_X f\, d\mu\,.$$

Note that the notions of conjugacy and factor between two topological dynamical systems extend in a natural way to the measure-theoretic context.

1.7.3 Symbolic Dynamics

Let us introduce some basic notions in symbolic dynamics. For expository books on the subject, see [167, 348, 381, 475] and [488]. For references on ergodic theory, also see, e.g., [579]. These notions will be central in particular in Chapters 8 and 9.

Let S denote the following map defined on Σ^ω, called the *one-sided shift*:

$$S((x_n)_{n\geq 0}) = (x_{n+1})_{n\geq 0}\,.$$

In particular, if $x = x_0x_1x_2\cdots$ is an infinite word over Σ, then for all $n \geq 0$, its suffix $x_nx_{n+1}\cdots$ is simply $S^n(x)$. The map S is uniformly continuous, onto but not one to one on Σ^ω. This notion extends in a natural way to $\Sigma^{\mathbb{Z}}$. In this latter case, the shift S is one to one. We thus get *symbolic dynamical systems*. Here symbolic refers to the fact that they are defined on words.

The definitions given below correspond to the *one-sided shift*, but they extend in a natural way to the *two-sided shift*.

Definition 1.7.4. Let x be an infinite word over the alphabet Σ. The *symbolic dynamical system* associated with $\mathbf{x}$ is then defined as the shift orbit closure $(\overline{\mathcal{O}(\mathbf{x})}, S)$, where $\overline{\mathcal{O}(\mathbf{x})} \subseteq \Sigma^\omega$ is the closure of the orbit $\mathcal{O}(\mathbf{x}) = \{S^n\mathbf{x} \mid n \in \mathbb{N}\}$ of x.

In the case of bi-infinite words, we similarly define $\mathcal{O}(\mathbf{x}) = \{S^n\mathbf{x} \mid n \in \mathbb{Z}\}$ where the (two-sided) shift map is defined on $\Sigma^{\mathbb{Z}}$. The set $X_\mathbf{x} := \overline{\mathcal{O}(\mathbf{x})}$ is a closed subset of the compact set Σ^ω; hence it is a compact space and S is a continuous map acting on it. One checks that, for every infinite word $\mathbf{y} \in \Sigma^\omega$, the word $\mathbf{y}$ belongs to $X_\mathbf{x}$ if and only if $L(\mathbf{y}) \subseteq L(\mathbf{x})$. For a proof, see [488] or Chapter 1 of [487]. Note that $\overline{\mathcal{O}(\mathbf{x})}$ is finite if and only if $\mathbf{x}$ is eventually periodic. Moreover, if $\mathbf{x}$ is an infinite word, $(X_\mathbf{x}, S)$ is minimal if and only if $\mathbf{x}$ is uniformly recurrent. Indeed, w is a factor of $\mathbf{x}$, we write

$$\overline{\mathcal{O}(\mathbf{x})} = \bigcup_{n \in \mathbb{N}} S^{-n}[w],$$

and we conclude by a compactness argument.

Generic examples of symbolic dynamical systems are provided by subshifts (also called shifts for short). Let Y be a closed subset of Σ^ω that is stable under the action of the shift S. The system (Y, S) is called a *subshift*. The *full shift* is defined as (Σ^ω, S). If Y is a subshift, there exists a set $\mathcal{F} \subset \Sigma^*$ of finite words such that an infinite word $\mathbf{x}$ belongs to X if and only if none of its factors belongs to $\mathcal{F}$. A subshift X is called a *subshift of finite type* if one can choose the set $\mathcal{F}$ to be finite. A subshift is said to be *sofic* if the set $\mathcal{F}$ is a regular language. A subshift (X, S) is said to be *periodic* if there exist $\mathbf{x} \in X$ and an integer k such that $X = \{\mathbf{x}, S\mathbf{x}, \ldots, S^k\mathbf{x} = \mathbf{x}\}$. Otherwise it is said to be *aperiodic*.

For the more general case of a group G acting on configurations in Σ^G, see Chapter 9. Elements of Σ^G can be considered as colorings of a group G by a finite alphabet Σ. The set of configurations Σ^G, endowed with the product topology, is a compact space on which we define the shift transformations: for every $g \in G$, the shift S^g translates a configuration $x \in \Sigma^G$ through $S^g(x)_h = x_{g^{-1}h}$ for every $h \in G$. In this framework, subshifts are exactly subsets of A^G that are both shift-invariant and closed for the product topology.

Example 1.7.5. The set of infinite words over $\{0, 1\}$ of Example 1.5.6 which do not contain the factor 11 is a subshift of finite type, whereas the set of infinite words over $\{0, 1\}$ having an even number of 1 between two occurrences of the letter 0 is a sofic subshift which is not of finite type.

Definition 1.7.6. Let Y be a subshift. For a word $w = w_0 \cdots w_r$, the *cylinder set* $[w]$ is the set $\{\mathbf{y} \in Y \mid y_0 = w_0, \ldots, y_r = w_r\}$.

The cylinder sets are *clopen* (open and closed) sets and form a basis of open sets for the topology of Y. Furthermore, one checks that a clopen set is a finite union of cylinders. In the bi-infinite case, the cylinders are the sets

$$[u.v]_Y = \{y \in Y \mid y_i = u_i, y_j = v_j, \; -|u| \leq i \leq -1, \; 0 \leq j \leq |v| - 1\}$$

and the same remark holds.

Then the *topological entropy* $h(X)$ of the symbolic dynamical system (X, S) measures the richness of its language L, defined as the set of factors of elements in X. It is defined as

$$h(X) = \lim_{n \to \infty} \frac{1}{n} \ln |L \cap \Sigma^n| .$$

It is closely related to the *growth rate* of the language L defined as $\limsup_{n \to \infty} |L \cap \Sigma^n|^{\frac{1}{n}}$ and considered in Chapter 5.

Chapter 2
Number Theoretic Aspects of Regular Sequences

Michael Coons and Lukas Spiegelhofer

Abstract We present a survey of results concerning regular sequences and related objects. Regular sequences were defined in the early 1990s by Allouche and Shallit as a combinatorially, algebraically, and analytically interesting generalization of automatic sequences. In this chapter, after an historical introduction, we follow the development from automatic sequences to regular sequences, and their associated generating functions, to Mahler functions. We then examine size and growth properties of regular sequences. The last half of the chapter focuses on the algebraic, analytic, and Diophantine properties of Mahler functions. In particular, we survey the rational-transcendental dichotomies of Mahler functions, due to Bézivin, and of regular numbers, due to Bell, Bugeaud, and Coons.

2.1 Introduction

The concept of "number" is central to mathematics and paramount to number theory. From the mathematical standpoint, one of the most important ways to view and treat numbers is algebraically, that is, to consider the integers as the ring $\mathbb{Z}$ under the operations addition and multiplication and the rationals $\mathbb{Q}$ as the field of fractions

The research of M. Coons was supported in part by Australian Research Council grant DE140100223. Lukas Spiegelhofer was supported by the Austrian Science Fund (FWF), projects F5502-N26 and F5505-N26, which are part of the Special Research Program "Quasi Monte Carlo Methods: Theory and Applications", and also by the ANR–FWF joint project MuDeRa (Multiplicativity, Determinism and Randomness). The authors thank the Erwin Schrödinger Institute for Mathematics and Physics where part of this chapter was written during the workshop on "Normal Numbers: Arithmetic, Computational and Probabilistic Aspects."

M. Coons (✉)
School of Mathematical and Physical Sciences, University of Newcastle, Callaghan, NSW, Australia
e-mail: michael.coons@newcastle.edu.au

L. Spiegelhofer
Institut für Diskrete Mathematik und Geometrie, Technische Universität Wien, Wien, Austria
e-mail: lukas.spiegelhofer@tuwien.ac.at

V. Berthé, M. Rigo (eds.), *Sequences, Groups, and Number Theory*,
Trends in Mathematics, https://doi.org/10.1007/978-3-319-69152-7_2

of $\mathbb{Z}$. Of course, from there interest is extended to the algebraic numbers, the field $\overline{\mathbb{Q}}$ of numbers, which are zeroes of polynomials with integer coefficients. The study of algebraic numbers and their properties is a continual fount of results and questions that for centuries has provided the foundational structures of mathematics and will—beyond doubt—form a significant part of these foundations for centuries to come.

The numbers of the preceding paragraph are abstract and in that sense do not really need to be represented. Yet, when one wishes to give an example of an integer, say 2 or 10 or 1729, one must write something down; if you wish to use only tick marks, treating the example 1729 will require large amounts of both time and space. Thus we have adopted the base system, with base 10—the number of fingers the average human has—as the most popular base for humans. The concept of "base expansion" is inseparable from modern computation and is fundamental to computer science. The use and importance of base expansions (predominantly binary) has become even more important with the advent of digital computers.

For those of us with interests at the interface of mathematics and theoretical computer science, the characterization of relationships between the algebraic viewpoint and the base-expansion viewpoint is an extremely important and interesting area of research. Two specific questions stand out here and form the backdrop of our chapter.

2.1.1 Two Important Questions

The first is an old question of Borel [99] concerning the probabilistic properties (probabilités dénombrables) of base expansions of real algebraic numbers.

Question 2.1.1 (Borel, 1909). Is the base expansion of an irrational algebraic real number normal?

Recall that a real number x is called *simply normal to the base k* (or *k-simply normal*) if each of $0, 1, \ldots, k-1$ occurs in the base-k expansion of x with equal frequency $1/k$. This number x is then called *normal to the base k* (or *k-normal*) provided it is k^m-simply normal for all positive integers m, and the number x is just called *normal* if this is true for all integers $k \geqslant 2$.

While Borel's question is asked from the standpoint of probability, Hartmanis and Stearns [285] were interested in the—at least morally related—question of computability. To state their question, we remind the reader that a real number x is *computable in real time* provided there is a multitape Turing machine that can compute the first n bits of x in time $\mathcal{O}(n)$.

Question 2.1.2 (Hartmanis and Stearns, 1965). Do there exist irrational algebraic real numbers which are computable in real time?

Presumably, the answers to these questions are "yes" and "no," respectively, though we stress here that our presumption is extremely presumptive. These presumptive answers reflect the well-observed notion that algebraic manipulations tend to do strange things to base expansions. In fact, compared to what is expected, very little is known about the digital properties of real algebraic numbers. For those interested, Bugeaud's recent work [119] provides a comprehensive exposition.

While Questions 2.1.1 and 2.1.2 are posed to study the digital properties of real algebraic numbers, in this chapter, we concern ourselves with a flipped version of these questions: *what are the number theoretic properties of real numbers whose expansions are highly structured?*

Real numbers with eventually periodic base expansions are the simplest numbers and sequences one can consider in our context. These numbers are not normal, are computable, and of course are algebraic—they are the zeroes of linear polynomials. This perceived exception to Questions 2.1.1 and 2.1.2 is why the word "irrational" appears in these questions. Indeed, the rational numbers are in many ways fundamentally different from the irrational algebraic numbers. For examples, see Dirichlet's approximation theorem and Roth's theorem [514] on the irrationality exponent of algebraic numbers. The digital properties of rational numbers have been almost completely classified (up to some deep questions about the orbits of primitive roots).

From a computational point of view, the next step is to consider real numbers whose base-k expansion is k-automatic[1] for some integer $k \geqslant 2$. This is where things become extremely interesting. In fact, here the base starts to matter. Recall that if a number is rational, then its base expansion is *eventually periodic in every base*. This is not true for numbers that are k-automatic for some integer $k \geqslant 2$. Cobham [155] showed that if a real number is both k-automatic and l-automatic for two integers k and l that are multiplicatively independent[2], then that real number is rational.

This difference from rationals continues with the complexity of base expansions. For a rational written in base k, the number of strings of digits of length n that occur in the expansion is bounded by a constant, while for a k-automatic real number, the number of strings can increase with n. But not too fast, this number is $\mathcal{O}(n)$ (see Theorem 1.6.9), and so an automatic number is not normal since a normal number must have all k^n possible strings occur.

For Borel's question, it may seem hopeful to then wonder if the set of automatic numbers contains an irrational algebraic number, but the negative answer to this question, which became known somewhat as the Cobham–Loxton–van der Poorten conjecture, was settled[3] by Adamczewski and Bugeaud in 2007 [3].

Theorem 2.1.3 (Adamczewski and Bugeaud). *The base expansion of an irrational real algebraic number cannot be output by a finite automaton.*

---

[1]For a detailed account of automatic sequences, see the monograph of Allouche and Shallit [14]. See also Section 1.6.1.

[2]Two integers k and l are multiplicatively independent provided $\log k/\log l$ is irrational.

[3]This result is inherent in the work of Cobham. In the 1980s, Loxton and van der Poorten [389] claimed to have proved that an automatic number is either rational or transcendental, but a few unresolvable flaws were found in their argument. This is why their name is associated with the conjecture.

2.1.2 Three (or Four) Hierarchies in One

According to Loxton [388], "the result about the decimal expansion of algebraic irrationals and finite automata suggests an alternative theoretical approach to randomness. We can try to assign a measure of computational complexity to a sequence by means of the following hierarchy:

(L0) [eventually] periodic sequences,
(L1) [. . .] sequences generated by finite automata,
(L2) sequences generated by automata with one push-down store,
(L3) sequences generated by non-deterministic automata with one push-down store, and
(L4) sequences generated by Turing machines.

Essentially, the n-th term of an [automatic] sequence is computed from the input n without any memory of earlier terms. A push-down store allows an arbitrary number of terms of the sequence to be stored and recalled later, the first one in being the last one out. Two push-down stores are equivalent to the doubly infinite tape of a Turing machine, which explains why the classification stops as it does. A random sequence is now one which cannot be generated by any machine less powerful than a Turing machine."

The well-informed reader will recognize Loxton's hierarchy as a subset of the Chomsky–Schützenberger hierarchy of formal languages (see also Section 1.5.1). This type of language-theoretical hierarchy, while classical and certainly of interest, lacks the mathematical structure to delve into such arithmetic questions that we will address here—especially at the higher levels of the hierarchy.

We present here a more natural hierarchy for such questions based on the work of Mahler and the generalization of automatic sequences presented by Allouche and Shallit. This hierarchy will be one of sequences, numbers, and functions simultaneously. From the standpoint of integer sequences, the *Mahler hierarchy* is as follows:

(M0) eventually periodic sequences,
(M1) automatic sequences,
(M2) regular sequences,
(M3) coefficient sequences of Mahler functions, and
(M4) integer sequences[4].

Levels (M0) and (M1) are taken from Loxton's hierarchy. Regular sequences were introduced in 1992 by Allouche and Shallit [17]. See also Section 1.6.2. Following

[4]We make no comment on the randomness properties of integer sequences, but will be content with their generality as is.

their treatment[5], let $\mathbb{C}$ denote the field of complex numbers and define the *k-kernel* of $f : \mathbb{Z}_{\geqslant 0} \to \mathbb{C}$ as the set

$$\mathrm{Ker}_k(f) := \left\{\{f(k^\ell n + r)\}_{n\geqslant 0} : \ell \geqslant 0, 0 \leqslant r < k^\ell\right\}. \tag{2.1}$$

Definition 2.1.4 (Allouche and Shallit). Let $k \geqslant 1$ be an integer. A sequence f taking values in $\mathbb{C}$ is called *k-regular* provided the $\mathbb{C}$-vector space $\langle \mathrm{Ker}_k(f)\rangle_{\mathbb{C}}$ spanned by $\mathrm{Ker}_k(f)$ is finite dimensional over $\mathbb{C}$.

Allouche and Shallit introduced regular sequences as a direct generalization of automatic sequences based on the k-kernel. Their generalization rests on a result of Eilenberg [211], who showed the following.

Theorem 2.1.5 (Eilenberg). *A sequence f is k-automatic if and only if* $\mathrm{Ker}_k(f)$ *is finite.*

While the notion of k-regularity is certainly worth studying in its own right, it becomes much more important when viewed as a bridge between the areas of theoretical computer science and number theory. As Allouche and Shallit showed, this notion is a direct extension of that of automatic sequences. Moreover, it is an extension that is algebraically, analytically, and arithmetically interesting and important.

The algebraic properties start with a correspondence between regular sequences and finite sets of matrices. Indeed, Allouche and Shallit [17, Lemma 4.1] (see also Section 1.6.2) showed that for a Noetherian ring R, *an R-valued sequence f is k-regular if and only if there exist a positive integer d, a finite set of matrices* $\mathscr{A}_f = \{\mathbf{A}_0, \ldots, \mathbf{A}_{k-1}\} \subseteq R^{d\times d}$, *and vectors* $\mathbf{v}, \mathbf{w} \in R^d$ *such that*

$$f(n) = \mathbf{w}^T \mathbf{A}_{i_0} \cdots \mathbf{A}_{i_s} \mathbf{v}, \tag{2.2}$$

where $(n)_k = i_s \cdots i_0$ *is the base-k expansion of n.*

The analytic importance comes via a result of Becker [61] relating regular sequences to Mahler functions. Recall the following definition; see the works of Mahler [403–405, 407].

Definition 2.1.6. A power series $F(z) \in \mathbb{C}[[z]]$ is *k-Mahler* for an integer $k \geqslant 2$ provided there is an integer $d \geqslant 1$ and polynomials $a_0(z), \ldots, a_d(z) \in \mathbb{C}[z]$ with $a_0(z)a_d(z) \neq 0$ such that

$$a_0(z)F(z) + a_1(z)F(z^k) + \cdots + a_d(z)F(z^{k^d}) = 0. \tag{2.3}$$

[5] Allouche and Shallit gave a more general treatment for sequences taking values in Noetherian rings. In our applications, the most important settings are those of the integers and complex numbers, depending on the type of result presented. For our purposes, for results on sequences and numbers, the integers will be the standard setting, and for results on power series those with complex coefficients will be the most important.

The minimal such d is called the *degree* of the Mahler function.

The above mentioned result of Becker states that *if* $\{f(n)\}_{n\geqslant 0}$ *is a* k*-regular sequence, then the generating function* $F(z) = \sum_{n\geqslant 0} f(n)z^n$ *is a* k*-Mahler function.* This established that those sequences in level (M3) contain those in (M2).

The arithmetic interest and importance of k-regular sequences are precisely the content of this chapter. We will present properties and results to this effect in the context of the Mahler hierarchy. It is important to note that while the Mahler hierarchy is stated in terms of sequences, it can be stated in term of numbers and functions as well.

Definition 2.1.7. If a sequence $\{f(n)\}_{n\geqslant 0}$ is k-automatic (resp. k-regular), then we call the generating function $F(z) = \sum_{n\geqslant 0} f(n)z^n$ *k-automatic* (resp. *k-regular*) as well and refer to $F(z)$ as a *k-automatic function* (resp. a *k-regular function*).

In this way, the levels (M1)–(M4) of the Mahler hierarchy can be translated to a hierarchy of functions as:

(M1) automatic functions,
(M2) regular functions,
(M3) Mahler functions, and
(M4) general power series.

The "number" version of the hierarchy is stated *mutatis mutandis* using the following definition.

Definition 2.1.8. Let $k \geqslant 2$ and $b \geqslant 2$ be integers. If $F(z)$ is a k-automatic function (resp. k-regular or k-Mahler), then we call the special value $F(1/b)$ a k-automatic number (resp. k-regular or k-Mahler).

Note that our notion of k-automatic number is more general than the traditional definitions; we call something an automatic number if it is the special value of an automatic function. In most of the literature, a real number is called k-automatic if its base-k expansion can be produced by an automaton. This is not the case for all of the numbers in our class. For example, the number $\sum_{n\geqslant 0} 3^{-2^n}$ is 2-automatic under our definition, though its base-2 expansion is not 2-automatic. Being able to treat such numbers is just one example of the strength and generality of using the framework of the Mahler hierarchy.

2.2 From Automatic to Regular to Mahler

In this section, we describe automatic and regular sequences based on their k-kernel and develop their properties as coefficient sequences of Mahler functions. We first recall the definitions from the context of the k-kernel with a little more generality than the previous section, then we give many simple properties and provide some examples.

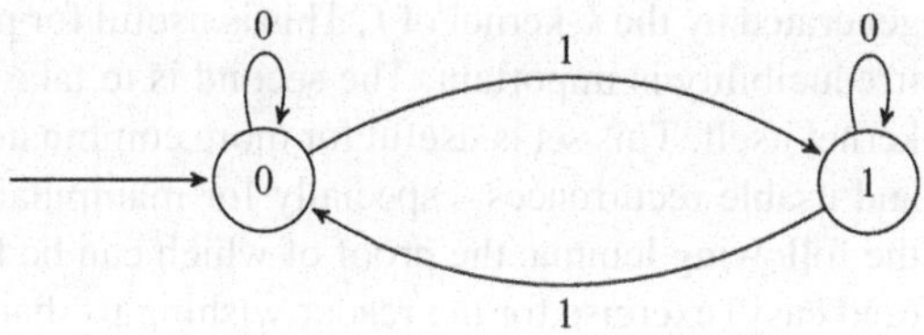

Fig. 2.1 The 2-automaton that produces the Thue–Morse sequence.

2.2.1 Definitions

We take Eilenberg's result (Theorem 2.1.5) as our definition of automaticity.

Notation 2.2.1. *Unless otherwise specified, a sequence f will be one that takes values in a commutative ring R, which when necessary to avoid complication will be taken as a subring of the complex numbers.*

Definition 2.2.2. A sequence f is k-automatic if and only if $\mathrm{Ker}_k(f)$ is finite.

Example 2.2.3. The canonical example of an automatic sequence is the *Thue–Morse sequence*. The Thue–Morse sequence $\{t(n)\}_{n\geqslant 0}$ over the alphabet $\{-1, 1\}$ is given by $t(n) := (-1)^{s(n)}$ where $s(n)$ is the number of 1s in the binary expansion of the number n. Using this definition, it is immediate that the sequence $\{t(n)\}_{n\geqslant 0}$ is 2-automatic. That is, there is a deterministic finite automaton that takes the binary expansion of n as input and outputs the value $t(n)$; see Figure 2.1.

To show that t is 2-automatic using the Eilenberg-inspired definition based on the k-kernel, it is enough to note that $t(2n) = t(n)$ and $t(2n + 1) = -t(n)$, so that $\mathrm{Ker}_2(t)$ has only two elements, namely, the sequences $t(n)$ and $-t(n)$.

As stated by Allouche in Shallit in their foundational paper [17], "unfortunately, the range of automatic sequences is necessarily finite, and this restricts their descriptive power."

Definition 2.2.4. The sequence f taking values in a ring R is k-regular provided the k-kernel of f is contained in a finitely generated R-module.

Example 2.2.5. Let $\{s(n)\}_{n\geqslant 0}$ be *Stern's diatomic sequence*, which is determined by the relations $s(0) = 0$, $s(1) = 1$, and for $n \geqslant 0$, by

$$s(2n) = s(n), \quad \text{and} \quad s(2n + 1) = s(n) + s(n + 1).$$

These recursions immediately imply that the 2-kernel of s is contained in the $\mathbb{Z}$-module generated by $\{s(n)\}_{n\geqslant 0}$ and $\{s(n + 1)\}_{n\geqslant 0}$, so that s is 2-regular. Note that s takes infinitely many values as well—$s(2^n+1) = n+1$—so that s is not 2-automatic.

The definition of k-regularity implies that there are a finite number of sequences $f_1, \ldots, f_d$ such that each element of the k-kernel of f is an R-linear combination of $f_1, \ldots, f_d$. This finite number of sequences can be taken in many ways, though two of these ways stand out. The first is to use an R-module basis for the R-module generated by the k-kernel of f. This is useful for proving results where minimality or irreducibility is important. The second is to take a spanning set directly from the k-kernel itself. This set is useful for more combinatorial results since it provides useful and usable recurrences, especially for manipulating sums. We record this result in the following lemma, the proof of which can be found in [17], though it is a worthy (and easy) exercise for the reader wishing to sharpen their teeth a bit on these ideas.

Lemma 2.2.6 (Allouche and Shallit). *The following are equivalent:*

(a) f is k-regular,
(b) the R-module generated by $\mathrm{Ker}_k(f)$ is generated by a finite number of elements of $\mathrm{Ker}_k(f)$,
(c) there exists an integer E such that for all $e_j > E$, each subsequence $f(k^{e_j}n + a_j)$ with $0 \leqslant a_j < k^{e_j}$ can be expressed as an R-linear combination

$$f(k^{e_j}n + a_j) = \sum_i c_{ij} f(k^{h_{ij}}n + b_{ij}),$$

where $h_{ij} \leqslant E$ and $0 \leqslant b_{ij} < k^{h_{ij}}$,
(d) there exist an integer d and d sequences $f = f_1, \ldots, f_d$ such that for $1 \leqslant i \leqslant d$ the k sequences $f_i(kn + a)$, $0 \leqslant a < k$, are R-linear combinations of the f_i,
(e) there exist an integer d, d sequences $f = f_1, \ldots, f_d$ and k matrices $\mathbf{A}_0, \ldots, \mathbf{A}_{k-1} \in R^{d\times d}$ such that if $\mathbf{v}(n) = [f_1, \ldots, f_d]^T$, then $\mathbf{v}(kn + a) = \mathbf{A}_a\mathbf{v}(n)$ for $0 \leqslant a < k$.

One of the most fundamental and important characterizations of k-regular sequence is their matrix formulation [17, Lemma 4.1] (see also Section 1.6.2).

Lemma 2.2.7 (Allouche and Shallit). *A sequence f is k-regular if and only if there exist a positive integer d, a finite set of matrices $\mathscr{A}_f = \{\mathbf{A}_0, \ldots, \mathbf{A}_{k-1}\} \subseteq R^{d\times d}$, and vectors $\mathbf{v}, \mathbf{w} \in R^d$ such that*

$$f(n) = \mathbf{w}^T \mathbf{A}_{i_0} \cdots \mathbf{A}_{i_s} \mathbf{v}, \tag{2.4}$$

where $(n)_k = i_s \cdots i_0$ is the base-k expansion of n.

Proof. We prove only the right-hand implication; the other is left as an exercise for the reader.

Suppose that f is k-regular and $(n)_k = i_s \cdots i_0$ is the base-k expansion of n. By Lemma 2.2.6(e), there exist an integer d, d sequences $f = f_1, \ldots, f_d$ and k matrices $\mathbf{A}_0, \ldots, \mathbf{A}_{k-1} \in R^{d\times d}$ such that if $\mathbf{v}(n) = [f_1, \ldots, f_d]^T$, then $\mathbf{v}(kn + a) = \mathbf{A}_a\mathbf{v}(n)$ for $0 \leqslant a < k$. Since $f = f_1$, setting $\mathbf{v} := \mathbf{v}(0)$ and $\mathbf{e}_1 := [1\ 0\ \cdots\ 0]^T$, we have that for each $n \geqslant 0$ that

$$f(n) = \mathbf{e}_1^T \mathbf{A}_{i_0} \cdots \mathbf{A}_{i_s} \mathbf{v}.$$

Setting $\mathbf{w} := \mathbf{e}_1$ gives the desired result. □

Definition 2.2.8. Let f be a k-regular sequence taking values in the ring R. If $\mathscr{A}_f = \{\mathbf{A}_0, \ldots, \mathbf{A}_{k-1}\} \subseteq R^{d\times d}$ is a finite set of matrices and $\mathbf{v}, \mathbf{w} \in R^d$ vectors such that

$$f(n) = \mathbf{w}^T \mathbf{A}_{i_0} \cdots \mathbf{A}_{i_s} \mathbf{v},$$

where $(n)_k = i_s \cdots i_0$ is the base-k expansion of n, then we call the tuple $(\mathbf{w}, \mathscr{A}_f, \mathbf{v})$ the *linear representation of* f.

Example 2.2.9. As we saw in a previous example, the Stern sequence is 2-regular. Using Lemma 2.2.6(e) and following the notation of Lemma 2.2.7, one can show that the Stern sequence has linear representation

$$\left([1\ 0], \{\mathbf{A}_0, \mathbf{A}_1\} = \left\{\begin{bmatrix} 1 & 1 \\ 0 & 1 \end{bmatrix}, \begin{bmatrix} 1 & 0 \\ 1 & 1 \end{bmatrix}\right\}, [1\ 0]\right).$$

We define the convolution of two sequences f and g by

$$f \star g(n) := \sum_{i+j=n} f(i)g(j).$$

The following result, which provides for the algebraic structure of the set of k-regular sequences, is due to Allouche and Shallit [17, Theorem 3.1 and Corollary 3.2], though we offer here a slightly different proof.

Theorem 2.2.10 (Allouche and Shallit). *The set of k-regular sequences forms a ring under standard addition and convolution.*

Proof. It is clear that the set of k-regular sequences forms a group under addition.

To see that the set is closed under convolution, let f and g be two k-regular sequences, whose k-kernels are contained in the R-modules generated by $f_1, f_2, \ldots, f_d$ and $g_1, g_2, \ldots, g_e$, respectively. To prove the theorem, it is enough to show that the k-kernel of $f \star g$ is contained in the R-module

$$C := \langle\{\{(f_i \star g_j)(n)\}_{n\geqslant 0} : 1 \leqslant i \leqslant d, 1 \leqslant j \leqslant e\}\rangle_R .$$

To see this, suppose that $c \in \mathrm{Ker}_k(f \star g)$ and that $\ell \geqslant 0$ and r $(0 \leqslant r < k^\ell)$ are such that $c(n) = (f \star g)(k^\ell n + r)$ for all $n \geqslant 0$. Then there are $\alpha_1, \ldots, \alpha_d, \beta_1, \ldots, \beta_e \in R$ such that

$$f(k^\ell n + r) = \sum_{i=0}^{d} \alpha_i f_i(n) \quad \text{and} \quad g(k^\ell n + r) = \sum_{j=0}^{e} \beta_j g_j(n).$$

Now

$$\begin{aligned}
c(n) = (f \star g)(k^\ell n + r) &= \sum_{a=0}^{n} f(k^\ell a + r) g(k^\ell (n-a) + r) \\
&= \sum_{a=0}^{n} \sum_{i=0}^{d} \alpha_i f_i(a) \sum_{j=0}^{e} \beta_j g_j(n-a) \\
&= \sum_{i=0}^{d} \sum_{j=0}^{e} \alpha_i \beta_j \sum_{a=0}^{n} f_i(a) g_j(n-a) \\
&= \sum_{i=0}^{d} \sum_{j=0}^{e} \alpha_i \beta_j (f_i \star g_j)(n) \qquad (2.5)
\end{aligned}$$

is an element of C, which proves the theorem. □

Equality (2.5) essentially gives a description of the matrix representation of the k-regular convolution $f \star g$, but working this out can in practice be extremely complicated—the bookkeeping involved is nothing short of a nightmare. From the number-theoretic perspective, the most useful special case of convolution is $1 \star f$, which is the sequence of partial sums of f. Fortunately, in this case the details are not so unfriendly. The following result is due to Dumas [201, Lemma 1], which we reproduce here with a few fixed typos.

Proposition 2.2.11 (Dumas). *Let f be a k-regular sequence, with matrix presentation as in* (2.4). *Then the sequence $g(m) = (1 \star f)(m) = \sum_{1 \leqslant n \leqslant m} f(n)$ is k-regular and*

$$g(m) = \mathbf{x}^T \mathbf{G}_{i_s} \cdots \mathbf{G}_{i_0} \mathbf{y},$$

where $(m)_k = i_s \ldots i_0$, $\mathbf{x}^T := [0_{1\times d}\ \mathbf{w}^T]$, $\mathbf{y}^T := [\mathbf{v}^T\ 0_{1\times d}]$ and for $b \in \{0, \ldots, k-1\}$,

$$\mathbf{G}_b := \begin{bmatrix} \mathbf{B}_0 & 0 \\ \mathbf{B}_0 - \mathbf{B}_{b+1} - \mathbf{A}_0 & \mathbf{I}_{d\times d} \end{bmatrix},$$

where $\mathbf{B}_b := \sum_{\ell=b}^{k-1} \mathbf{A}_\ell$ for $b = 0, \ldots, k-1$ and $\mathbf{B}_k := 0$.

Proof. Let $m \geqslant 1$ be an integer with $(m)_k = b_r b_{r-1} \cdots b_0$, and write $g(m) := \sum_{1 \leqslant n \leqslant m} f(n)$. It is quite clear that

$$g(m) = \mathbf{x}^T \left(\sum_{1 \leqslant n \leqslant m} \mathbf{A}_{(n)_k} \right) \mathbf{v} = \mathbf{x}^T \left(\sum_{0 \leqslant i \leqslant r} \left(\sum_{1 \leqslant j \leqslant b_i} \mathbf{A}_j \right) \sum_{\substack{|w| \leqslant i \\ w \in \{0,\ldots,k-1\}^*}} \mathbf{A}_w \right) \mathbf{v}, \qquad (2.6)$$

where we use that convention that if $b_i = 0$, then $\sum_{1\leqslant j\leqslant b_i} \mathbf{A}_j = 0$, and when $i = 0$, then $\sum_{|w|\leqslant i, w\in\{0,\dots,k-1\}^*} \mathbf{A}_w = \mathbf{I}_{d\times d}$.

Now, in the notation presented in the statement of the proposition, it is quite clear that

$$\mathbf{B}_0^i = \sum_{\substack{|w|\leqslant i\\ w\in\{0,\dots,k-1\}^*}} \mathbf{A}_w,$$

where our above convention is preserved since we understand $\mathbf{B}_0^0 = \mathbf{I}_{d\times d}$. Also, we note that

$$\mathbf{B}_0 - \mathbf{B}_{b_i+1} - \mathbf{A}_0 = \sum_{1\leqslant j\leqslant b_i} \mathbf{A}_j,$$

where again our above convention is preserved since for $b_i = 0$, we have $\mathbf{B}_0 - \mathbf{B}_1 - \mathbf{A}_0 = 0$.

With this information of the preceding paragraph, we interpret the equality (2.6) as

$$g(m) = \mathbf{x}^T \left(\sum_{0\leqslant i\leqslant r} (\mathbf{B}_0 - \mathbf{B}_{b_i+1} - \mathbf{A}_0)\, \mathbf{B}_0^i \right) \mathbf{v}. \tag{2.7}$$

But this is exactly the output of the matrix representation for $g(m)$ as described in the statement of the proposition. □

The importance of the ring structure under addition and convolution begins with the following immediate corollary of Theorem 2.2.10.

Corollary 2.2.12 (Allouche and Shallit). *The set of k-regular functions forms a ring under standard addition and multiplication.*

This importance continues with the relationship to Mahler functions as provided by Becker [61]. Following Becker, we require the following definition and lemma regarding the Cartier operators.

Definition 2.2.13. Given a positive integer $k \geqslant 2$, we define the *Cartier operators* $\Lambda_0, \dots, \Lambda_{k-1} : \mathbb{C}[[z]] \to \mathbb{C}[[z]]$ by

$$\Lambda_i\left(\sum_{n\geqslant 0} c(n)z^n\right) = \sum_{n\geqslant 0} c(kn+i)z^n,$$

for $i = 0, \dots, k-1$.

Lemma 2.2.14. *Let $F(z), G(z) \in \mathbb{C}[[z]]$. For $i = 0, \dots, k-1$ we have*

(a) $\Lambda_i(F(z^k)G(z)) = F(z)\Lambda_i(G(z))$, *and*
(b) $F(z) = \sum_{i=0}^{k-1} z^i \Lambda_i(F)(z^k)$,

where $\Lambda_i(F)(z^k)$ is understood as $\Lambda_i(F(z))$ evaluated at z^k, so that if $F(z) = \sum_{n\geqslant 0} f(n)z^n$, then $\Lambda_i(F)(z^k) = \sum_{n\geqslant 0} f(kn+i)z^{kn}$.

Proof. This is left as an exercise. □

Theorem 2.2.15 (Becker). *A k-regular function is a k-Mahler function.*

Proof. For convenience, we will assume that the k-regular function takes values in the complex numbers. This proof can be easily modified to give a result for any Noetherian ring R provided you work with the field of fractions of R.

Let $f = f_1, \ldots, f_d$ be a basis for the $\mathbb{C}$-vector space spanned by the k-kernel of f and set $F_i(z) := \sum_{n\geqslant 0} f_i(n)z^n$. Further, define the $\mathbb{C}(z)$-vector space V by

$$V := \langle \{F_i(z) : i = 1, \ldots, d\} \rangle_{\mathbb{C}(z)},$$

so that the set $\{F_i(z) : i = 1, \ldots, d\}$ is a basis for V, and define the operator

$$\Phi : V \to \mathbb{C}((z))$$

by $\Phi(G(z)) = G(z^k)$. We claim that $V = \Phi(V)$.

To show that $V \subset \Phi(V)$ we note that for each $i = 1, \ldots, d$

$$F_i(x) = \sum_{j=0}^{k-1} \sum_{n\geqslant 0} f_i(kn+j)(x^k)^n x^j,$$

and since each $\{f_i(kn+j)\}_{n\geqslant 0}$ is in the k-kernel of f, it is a $\mathbb{C}$-linear combination of the basis sequences $f_1, \ldots, f_d$. Thus we may write

$$F_i(x) = \sum_{j=1}^{d} p_{i,j}(x) F_j(x^k) \tag{2.8}$$

where for each i, j we have $p_{i,j}(x) \in \mathbb{C}[x]$ and $\deg p_{i,j}(x) \leqslant k-1$. But since $\{F_i(z) : i = 1, \ldots, d\}$ is a basis for V, we thus have that $\{F_i(z^k) : i = 1, \ldots, d\}$ spans $\Phi(V)$, and so the relationship in (2.8) shows that $V \subset \Phi(V)$.

For the other inclusion, we set $\mathbf{F}(x) := [F_1(x), \ldots, F_d(x)]^T$ and note that (2.8) gives

$$\mathbf{F}(x) = \mathbf{A}(x)\mathbf{F}(x^k), \tag{2.9}$$

where $\mathbf{A}(x) = (p_{i,j}(x))_{1\leqslant i,j\leqslant d} \in \mathbb{C}[x]^{d\times d}$. Also, since $\{F_i(z) : i = 1, \ldots, d\}$ is a basis for V, the matrix $\mathbf{A}(z)$ is nonsingular; if this were not the case, there would be a vector $\mathbf{v}(z) \in \mathbb{C}(z)^d$ such that $\mathbf{v}(z)\mathbf{A}(z) = 0$ so that by (2.9) we would have $\mathbf{v}(z)\mathbf{F}(z) = 0$, contradicting that the coordinates of $\mathbf{F}(z)$ are $\mathbb{C}(z)$-linear independent—they form a basis of V. Thus also

$$\mathbf{A}(z)^{-1}\mathbf{F}(z) = \mathbf{F}(z^k),$$

whence $\Phi(V) \subset V$, showing that $V = \Phi(V)$.

We note that the arguments of the previous two paragraphs also show that since V has dimension d, $F(z) \in V$, and $\Phi(V) \subset V$, the $d+1$ functions $F(z), F(z^k), \ldots, F(z^{k^d}) \in V$ are $\mathbb{C}(z)$-linearly dependent, meaning there are polynomials $a_0(z), \ldots, a_d(z) \in \mathbb{C}[z]$ such that

$$\sum_{i=0}^{d} a_i(z)F(z^{k^i}) = 0. \tag{2.10}$$

Of course, to prove the theorem, we must show that one has such a relationship with $a_0(z) \neq 0$.

Indeed, as Becker points out [61, p. 273], if one has a functional equation (2.10) with $a_j(z) \neq 0$ with $j > 0$ minimal, then we can just "shift" it down to one smaller j by applying one of the Cartier operators, since from Lemma 2.2.14(a) we have for $a = 0, \ldots, k-1$ that

$$0 = \Lambda_a \left(\sum_{i=j}^{d} a_i(z)F(z^{k^i}) \right) = \sum_{i=j}^{d} \Lambda_a\left(a_i(z)\right) F(z^{k^{i-1}}),$$

where we are guaranteed from Lemma 2.2.14(b) that for at least one $a = 0, \ldots, k-1$, the polynomial $\Lambda_a(a_j(z))$ is nonzero. □

This argument can be adjusted to prove the following stronger form of Becker's theorem, and so we state it here as a corollary.

Corollary 2.2.16. *If R is a Noetherian ring and $F(z) \in R[[z]]$ is k-regular for an integer $k \geqslant 2$, then there is an integer $d \geqslant 1$ and polynomials $a_0(z), \ldots, a_d(z) \in R[z]$ with $a_0(z)a_d(z) \neq 0$ such that*

$$a_0(z)F(z) + a_1(z)F(z^k) + \cdots + a_d(z)F(z^{k^d}) = 0.$$

That is, $F(z)$ is k-Mahler satisfying a Mahler functional equation with coefficients in the ring $R[z]$.

The most important case in the above corollary is the case of $R = \mathbb{Z}$.

Example 2.2.17. Let s again denote the Stern sequence and set $S(z) := \sum_{n\geqslant 0} s(n)z^n$. Using the definition of s, we have

$$\begin{aligned} zS(z) &= z\sum_{n\geqslant 0} s(2n)z^{2n} + z\sum_{n\geqslant 0} s(2n+1)z^{2n+1} \\ &= z\sum_{n\geqslant 0} s(n)z^{2n} + \sum_{n\geqslant 0} s(n)z^{2n+2} + \sum_{n\geqslant 0} s(n+1)z^{2n+2} \end{aligned}$$

$$= zS(z^2) + z^2S(z^2) + \sum_{n\geqslant 0} s(n)z^{2n}$$

$$= S(z^2)\left(1 + z + z^2\right),$$

which gives that the generating function $S(z)$ satisfies the 2-Mahler equation

$$zS(z) - (z^2 + z + 1)S(z^2) = 0.$$

2.2.2 *Some Comparisons Between Regular and Mahler Functions*

Becker's result, Theorem 2.2.15 above, shows that every regular function is a Mahler function. The converse of Becker's result is not true, which we can show as a consequence of the following result.

Proposition 2.2.18. *The sequence $\{a^n\}_{n\geqslant 0}$ is k-regular if and only if $a = 0$ or a is a root of unity.*

Proof. One direction is simple, since if $a = 0$ or a root of unity, the sequence of powers is periodic and hence k-regular.

For the other direction, assume $\{a^n\}_{n\geqslant 0}$ is k-regular. Then there exist an integer r and integers $\lambda_0, \dots, \lambda_{r-1}$, not all zero, such that

$$\sum_{j=0}^{r} \lambda_j a^{k^j n} = 0.$$

Now we use the Vandermonde determinant identity, which states that

$$\det \begin{bmatrix} 1 & b_0 & b_0^2 & \cdots & b_0^m \\ 1 & b_1 & b_1^2 & \cdots & b_1^m \\ \vdots & \vdots & \vdots & \ddots & \vdots \\ 1 & b_m & b_m^2 & \cdots & b_m^m \end{bmatrix} = \prod_{0\leqslant i<j\leqslant m} (b_j - b_i).$$

It follows that the sequences $\{b_j^n\}_{n\geqslant 0}$ are linearly independent if and only if the numbers $b_0, b_1, \dots, b_m$ are distinct. Hence the numbers $1, a^k, a^{k^2}, \dots, a^{k^r}$ are not all distinct, and we must have $a^{k^j} = a^{k^l}$ for some $j \neq l$. Thus either $a = 0$ or a is a root of unity. □

Example 2.2.19. The function $1/(1-2z)$ is k-Mahler for every k but is not k-regular for any k. To see this, note that inside the disk of radius $1/2$ centered at zero, we have that

$$F(z) := \frac{1}{1-2z} = \sum_{n\geqslant 0} 2^n z^n.$$

By Proposition 2.2.18, the sequence $\{2^n\}_{n\geqslant 0}$ is not k-regular for any k. But it is quite easy to check that $F(z) = 1/(1-2z)$ satisfies the Mahler equation

$$(1-2z)F(z) - (1-2z^k)F(z^k) = 0,$$

for any k, so that $F(z)$ is k-Mahler for each k.

In fact, Example 2.2.19 suggests the following result concerning the degree of rational Mahler functions.

Proposition 2.2.20. *If $R(z)$ is a nonzero rational function, then it is a k-Mahler function of degree 1 for every positive integer $k \geqslant 2$.*

Proof. Now write $R(z) = p(z)/q(z)$ for nonzero polynomials $p(z)$ and $q(z)$. Then $R(z)$ satisfies the k-Mahler equation

$$p(z^k)q(z)R(z) - p(z)q(z^k)R(z^k) = 0,$$

which is of degree 1. □

While not all Mahler functions are regular functions, there are some describable families. For example, Becker showed that if $F(z)$ is k-Mahler and the coefficient $a_0(z)$ of $F(z)$ in the functional equation is a nonzero constant, then $F(z)$ is k-regular.

Theorem 2.2.21 (Becker [61]). *Let $F(z) \in \mathbb{C}[[z]]$ be a k-Mahler function satisfying*

$$\sum_{i=0}^{d} a_i(z)F(z^{k^i}) = 0,$$

where $0 \neq a_0(z) \in \mathbb{C}$ and $a_1(z), \ldots, a_d(z) \in \mathbb{C}[z]$. Then $F(z)$ is k-regular.

Proof. Without loss of generality, we may assume that $a_0(z) = -1$, since we may just divide by the appropriate complex number if needed. Thus,

$$F(z) = \sum_{i=1}^{d} a_i(z)F(z^{k^i}). \tag{2.11}$$

Set $H := \max\{\deg a_i(z) : i = 1, \ldots, d\}$, and let V be the $\mathbb{C}$-vector space generated by the functions

$$G_{ij}(z) := z^i F(z^{k^j}) \qquad (i = 0, \ldots, H; j = 0, \ldots, d).$$

For $a = 0, \dots, k-1$ we have $\Lambda_a(G_{ij}(z)) \in V$. To see this, note that if $j = 1, \dots, d$, then by Lemma 2.2.14(a) we have

$$\Lambda_a(z^i F(z^{k^j})) = F(z^{k^{j-1}})\Lambda_a(z^i) \in V,$$

since $\Lambda_a(z^i)$ is a monomial (possibly a constant) of degree at most H. If $j = 0$, then we use the functional equation (2.11) and Lemma 2.2.14(a) to obtain

$$\Lambda_a(z^i F(z)) = \sum_{\ell=1}^{d} \Lambda_a\left(z^i a_\ell(z) F(z^{k^\ell})\right) = \sum_{\ell=1}^{d} \Lambda_a\left(z^i a_\ell(z)\right) F(z^{k^{\ell-1}}).$$

Since for each combination of i and ℓ, $\deg z^i a_\ell(z) \leqslant 2H$, we have $\deg \Lambda_a(z^i a_\ell(z)) \leqslant 2H/k \leqslant H$, so that $\Lambda_a(z^i F(z)) \in V$.

Since $\Lambda_a(V) \subset V$ for each $a = 0, \dots, k-1$, we have that V is mapped into itself for any element in the semigroup $\Lambda := \langle\{\Lambda_0, \dots, \Lambda_{k-1}\}\rangle$. Since V is finite dimensional and $F(z) \in V$, we have that the set $\Lambda(F(z))$ (the semigroup Λ evaluated at $F(z)$ for each element) generates a finite-dimensional $\mathbb{C}$-vector space. But, using the definitions of regularity and the Cartier operators, this is possible if and only if $F(z)$ is k-regular. □

Theorem 2.2.21 is a simplified version of the following result of Dumas, which we will use in the proof of Theorem 2.2.24. Its proof can be attained by an argument almost identical to the proof of Theorem 2.2.21; for details see Dumas's thesis [200, Theorem 24].

Theorem 2.2.22 (Dumas). *Let $F(z) \in \mathbb{C}[[z]]$ be a power series satisfying*

$$\sum_{i=0}^{d} a_i(z) F(z^{k^i}) = E(z),$$

where $0 \neq a_0(z) \in \mathbb{C}$, $a_1(z), \dots, a_d(z) \in \mathbb{C}[z]$, and $E(z)$ is k-regular. Then $F(z)$ is k-regular.

Sometimes functions satisfying a Mahler functional equation with $a_0(z) = 1$ are called k-Becker; for example, see Adamczewski and Bell [2]. Becker conjectured that a result very similar to Theorem 2.2.21 holds for all regular functions.

Conjecture 2.2.23 (Becker). If $F(z)$ is a k-regular function, then there exists a k-regular rational function $r(z)$ such that the function $F(z)/r(z)$ satisfies a Mahler functional equation with $a_0(z) = 1$.

Theorem 2.2.24 (Structure Theorem, Dumas [200]). *A k-Mahler function is the quotient of a series and an infinite product which are k-regular. That is, if $F(z)$ is the solution of the Mahler functional equation*

$$a_0(z)F(z) + a_1(z)F(z^k) + \cdots + a_d(z)F(z^{k^d}) = 0,$$

where $a_0(z)a_d(z) \neq 0$, the $a_i(z)$ are polynomials, then there exists a k-regular series $H(z)$ such that

$$F(z) = \frac{H(z)}{\prod_{j\geqslant 0} \Gamma(z^{k^j})},$$

where $a_0(z) = \rho z^{\delta}\Gamma(z)$, with $\rho \neq 0$ and $\Gamma(0) = 1$.

Proof. Suppose that $F(z) = \sum_{n\geqslant 0} f(n)z^n$ satisfies

$$a_0(z)F(z) + a_1(z)F(z^k) + \cdots + a_d(z)F(z^{k^d}) = 0,$$

where $a_0(z)a_d(z) \neq 0$, the $a_i(z)$ are polynomials, and for each $i = 0, \ldots, d$ let δ_i be the order of $a_i(z)$ at $z = 0$, where we let $\delta_i = 0$ if $a_i(z) = 0$, and define the polynomials $b_i(z)$ by $a_i(z) = z^{\delta_i} b_i(z)$. Further, let

$$D := \max\left\{\delta_0, \left\lfloor \frac{k\delta_0 - \delta_1}{k-1} \right\rfloor, \left\lfloor \frac{k^2\delta_0 - \delta_2}{k^2-1} \right\rfloor, \ldots, \left\lfloor \frac{k^d\delta_0 - \delta_d}{k^d-1} \right\rfloor\right\},$$

and define the polynomial

$$p(z) := \sum_{n=0}^{D-\delta_0} f(n)z^n,$$

so that there is a power series $F_D(z)$ such that

$$F(z) = p(z) + z^{D-\delta_0+1}F_D(z). \tag{2.12}$$

Combining this with the Mahler functional equation and separating the $i = 0$ term, we have

$$z^{D+1}b_0(z)F_D(z) = -\sum_{i=0}^{d} a_i(z)p(z^{k^i}) - \sum_{i=0}^{d} z^{\lambda_i} b_i(z)F_D(z^{k^i}), \tag{2.13}$$

where

$$\lambda_i = \delta_i + k^i(D - \delta_0 + 1).$$

We claim that $\lambda_i \geqslant D + 1$ for each $i = 1, \ldots, d$. To see this, note that for each $i = 0, \ldots, d$ we have

$$D \geqslant \left\lfloor \frac{k^i\delta_0 - \delta_i}{k^i - 1} \right\rfloor \geqslant \frac{k^i\delta_0 - \delta_i}{k^i - 1} + \frac{1}{k^i - 1} - 1,$$

which gives the desired lower bound on λ_i after some rearrangement.

Since each $\lambda_i \geqslant D+1$ and the left-hand side of (2.13) is divisible by z^{D+1}, we have that the polynomial $\sum_{i=0}^{d} a_i(z)p(z^{k^i})$ is also divisible by z^{D+1}, so we may write

$$\sum_{i=0}^{d} a_i(z)p(z^{k^i}) = z^{D+1}E(z)$$

for some polynomial $E(z)$. Thus we have that

$$b_0(z)F_D(z) = -E(z) - \sum_{i=0}^{d} z^{\lambda_i-(D+1)}b_i(z)F_D(z^{k^i}). \tag{2.14}$$

Now let ρ be the nonzero number such that

$$a_0(z) = z^{\delta_0}b_0(z) = \rho z^{\delta_0}\Gamma(z),$$

with $\Gamma(0) = 1$, and set

$$G(z) := F_D(z)\prod_{j\geqslant 0}\Gamma(z^{k^j}).$$

Thus we may write (2.14) as

$$G(z) = -\rho^{-1}E(z)\prod_{j\geqslant 1}\Gamma(z^{k^j}) - \rho^{-1}\sum_{i=0}^{d} z^{\lambda_i-(D+1)}\left(b_i(z)\prod_{j=0}^{i}\Gamma(z^{k^j})\right)G(z^{k^i}). \tag{2.15}$$

The infinite product $P(z) := \prod_{j\geqslant 0}\Gamma(z^{k^j})$ is k-regular by Theorem 2.2.21 since it satisfies the Mahler functional equation

$$P(z) - \Gamma(z)P(z^k) = 0.$$

Combining this with Theorem 2.2.22, (2.15) gives that $G(z)$ is k-regular.

Using the definition of $G(z)$ and (2.12), we have

$$F(z) = p(z) + z^{D-\delta_0+1}\frac{G(z)}{\prod_{j\geqslant 0}\Gamma(z^{k^j})}.$$

Setting $H(z) := p(z)\prod_{j\geqslant 0}\Gamma(z^{k^j}) + z^{D-\delta_0+1}G(z)$, we have both that $H(z)$ is k-regular, since the set of k-regular functions form a ring, and also that

$$F(z) = \frac{H(z)}{\prod_{j\geqslant 0}\Gamma(z^{k^j})},$$

which is the desired result. □

2.3 Size and Growth

The range of automatic sequences is finite, so questions of size and growth concerning automatic sequences are typically uninteresting. Regular sequences can take an infinite number of values. Three immediate questions that arise are as follows: (1) How slow can an unbounded regular sequence grow? (2) Are there good upper bounds for such sequences? (3) What is the maximum possible growth?

2.3.1 Lower Bounds

When considering the question of the growth of a regular sequence, from the lower bound perspective, it is worth noting that any such result will be an "infinitely often" result at best. For example, there are regular sequences that are unbounded, yet take the value 1 infinitely. The Stern sequence s is a great witness to this property. As we have stated previously, $s(2^n+1)=n+1$, so that the Stern sequence is unbounded, yet also, $s(2^n) = 1$ for all n. Similar results hold for the valuation function $\nu_k(n)$, which is the largest integer m such that k^m divides n; ν_k is clearly unbounded, and it takes each nonnegative integer value an infinite number of times.

In 2014, an "infinitely often" lower bound-type result was given by Bell, Coons, and Hare [65]. We present their result with proof here.

Theorem 2.3.1 (Bell, Coons, and Hare). *Let $k \geqslant 2$. If $f:\mathbb{N}\to\mathbb{Z}$ is an unbounded k-regular sequence, then there exists $c>0$ such that $|f(n)|>c\log n$ infinitely often.*

Lemma 2.3.2. *Let $k\geqslant 2$ be an integer, let $\mathbf{A}_0,\ldots,\mathbf{A}_{k-1}$ be $d\times d$ integer matrices, and let $\mathscr{B}$ be the semigroup generated by $\mathbf{A}_0,\ldots,\mathbf{A}_{k-1}$. Then either $\mathscr{B}$ is finite or there is some $\mathbf{S}\in\mathscr{B}$ and fixed vectors $\mathbf{v}$ and $\mathbf{w}\in\mathbb{C}^d$ such that $|\mathbf{w}^T\mathbf{S}^n\mathbf{v}|\geqslant n$ for all sufficiently large n.*

Proof. Suppose that $\mathscr{B}$ is infinite. Then since $\mathscr{B}$ is finitely generated, a result of McNaughton and Zalcstein [414] gives that there is some $\mathbf{S}$ in $\mathscr{B}$ such that the matrices $\mathbf{S},\mathbf{S}^2,\mathbf{S}^3,\ldots$ are all distinct. Let $p(x)$ be the characteristic polynomial of $\mathbf{S}$. Then $p(x)$ is a monic integer polynomial. If $p(x)$ has a root λ that is strictly greater than 1 in modulus, then $\mathbf{S}$ has an eigenvector $\mathbf{v}$ such that $\mathbf{S}\mathbf{v}=\lambda\mathbf{v}$. Pick a nonzero vector $\mathbf{w}$ such that $\mathbf{w}^T\mathbf{v}=C\neq 0$. Then $|\mathbf{w}^T\mathbf{S}^n\mathbf{v}|=|C|\cdot|\lambda|^n\geqslant n$ for n sufficiently large.

If, on the other hand, all the roots of $p(x)$ are at most 1 in modulus, then all nonzero eigenvalues of $\mathbf{S}$ are algebraic integers with all conjugates having modulus 1; hence, they are roots of unity. Let $\mathbf{Y}$ be a matrix in $\mathrm{GL}_d(\mathbb{C})$ such that $\mathbf{T}:=\mathbf{Y}^{-1}\mathbf{S}\mathbf{Y}$ is in Jordan form, where we take Jordan blocks to be upper triangular. Then each Jordan block in $\mathbf{T}$ is of the form $\mathbf{J}_i(\lambda)$ with λ either zero or a root of unity and $i\geqslant 1$. Since $\mathbf{S}$ does not generate a finite subsemigroup of $\mathscr{B}$, there is some root of unity ω and some $m>1$ such that $\mathbf{T}$ has a block of the form $\mathbf{J}_m(\omega)$. We may assume, without loss of generality, that $\mathbf{J}_m(\omega)$ is the first block occurring in $\mathbf{T}$. Then the $(1,2)$-entry of $\mathbf{T}^n$ is $n\omega^{n-1}$ and so $|e_1^T\mathbf{T}^n e_2| = n$ for every n. In particular, we have

$$|e_1^T\mathbf{Y}^{-1}\mathbf{S}^n\mathbf{Y}e_2| \geqslant n$$

for every n. Taking $\mathbf{w}^T = e_1^T\mathbf{Y}^{-1}$ and $\mathbf{v} = \mathbf{Y}e_2$ gives the result. □

Proof (of Theorem 2.3.1). Let $k \geqslant 2$ be an integer, and suppose that $f : \mathbb{N} \to \mathbb{Z}$ is an unbounded k-regular sequence. Given a word $w = i_s \cdots i_0 \in \{0, \ldots, k-1\}^*$, as stated previously, we let $[w]_k$ denote the natural number $n = i_s k^s + \cdots + i_1 k + i_0$. The $\mathbb{Z}$-submodule of all $\mathbb{Z}$-valued sequences spanned by $\mathrm{Ker}_k(f)$ is a finitely generated torsion-free module and hence free of finite rank. Let $\{\{g_1(n)\}_{n\geqslant 0}, \ldots, \{g_d(n)\}_{n\geqslant 0}\}$ be a $\mathbb{Z}$-module basis for the $\mathbb{Z}$-module spanned by $\mathrm{Ker}_k(f)$. Then for each $i \in \{0, 1, \ldots, k-1\}$, the functions $g_1(kn+i), \ldots, g_d(kn+i)$ can be expressed as $\mathbb{Z}$-linear combinations of $g_1(n), \ldots, g_d(n)$, and hence there are $d \times d$ integer matrices $\mathbf{A}_0, \ldots, \mathbf{A}_{k-1}$ such that

$$[g_1(n), \ldots, g_d(n)]\mathbf{A}_i = [g_1(kn+i), \ldots, g_d(kn+i)]$$

for $i = 0, \ldots, k-1$ and all $n \geqslant 0$. In particular, if $i_s \cdots i_0$ is the base-k expansion of n, then $[g_1(0), \ldots, g_d(0)]\mathbf{A}_{i_s} \cdots \mathbf{A}_{i_0} = [g_1(n), \ldots, g_d(n)]$. (We note that this holds even if we pad the base-k expansion of n with zeros at the beginning.) We claim that the $\mathbb{Q}$-span of the vectors $[g_1(i), \ldots, g_d(i)]^T$, as i ranges over all natural numbers, must be all of $\mathbb{Q}^d$. Indeed, if this were not the case, then their span would be a proper subspace of $\mathbb{Q}^d$, and hence the span would have a nontrivial orthogonal complement. In particular, there would exist integers $c_1, \ldots, c_d$, not all zero, such that

$$c_1 g_1(n) + \cdots + c_d g_d(n) = 0$$

for every n, contradicting the fact that $g_1(n), \ldots, g_d(n)$ are linearly independent sequences.

Let $\mathscr{A}$ denote the semigroup generated by $\mathbf{A}_0, \ldots, \mathbf{A}_{k-1}$. Then we have just shown that there exist words $\mathbf{X}_1, \ldots, \mathbf{X}_d$ in $\mathscr{A}$ such that

$$[g_1(0), \ldots, g_d(0)]\mathbf{X}_1, \ldots, [g_1(0), \ldots, g_d(0)]\mathbf{X}_d$$

span $\mathbb{Q}^d$. Now, if $\mathscr{A}$ is finite, then $\{g_1(n)\}_{n\geqslant 0}, \ldots, \{g_d(n)\}_{n\geqslant 0}$ take only finitely many distinct values. Since $\{f(n)\}_{n\geqslant 0}$ is a $\mathbb{Z}$-linear combination of $\{g_1(n)\}_{n\geqslant 0}, \ldots, \{g_d(n)\}_{n\geqslant 0}$, we see that it too takes only finitely many distinct values, which contradicts our assumption that it is unbounded. Thus $\mathscr{A}$ must be infinite. By Lemma 2.3.2, there exist $\mathbf{Y} \in \mathscr{A}$ and vectors $\mathbf{x}, \mathbf{y} \in \mathbb{C}^d$ such that $|\mathbf{x}^T\mathbf{Y}^n\mathbf{y}| \geqslant n$ for all n sufficiently large.

By construction, we may write $\mathbf{x}^T = \sum_j \alpha_j [g_1(0), \ldots, g_d(0)]\mathbf{X}_j$ for some complex numbers α_j. Then

$$\mathbf{x}^T\mathbf{Y}^n = \sum_j \alpha_j[g_1(0),\ldots,g_d(0)]\mathbf{X}_j\mathbf{Y}^n.$$

Let u_j be the word in $\{0,1,\ldots,k-1\}^*$ corresponding to $\mathbf{X}_j$, and let y be the word in $\{0,\ldots,k-1\}^*$ corresponding to $\mathbf{Y}$; that is, $u_j = i_s\cdots i_0$ where $\mathbf{X}_j = \mathbf{A}_{i_s}\cdots\mathbf{A}_{i_0}$ and similarly for y. Then we have

$$[g_1(0),\ldots,g_d(0)]\mathbf{X}_j\mathbf{Y}^n = [g_1([u_jy^n]_k),\ldots,g_d([u_jy^n]_k)]^T.$$

Write $\mathbf{y}^T = [\beta_1,\ldots,\beta_d]$. Then

$$\mathbf{x}^T\mathbf{Y}^n\mathbf{y} = \sum_{i,j}\alpha_i\beta_jg_j([u_iy^n]_k).$$

By assumption, each of $\{g_1(n)\}_{n\geqslant 0},\ldots,\{g_d(n)\}_{n\geqslant 0}$ is in the $\mathbb{Z}$-module generated by $\mathrm{Ker}_k(f)$, and hence there exist natural numbers $p_1,\ldots,p_t$ and $q_1,\ldots,q_t$ with $0\leqslant q_m < k^{p_m}$ for $m=1,\ldots,t$ such that for each $s=1,\ldots,d$ we have $g_s(n) = \sum_{i=1}^t \gamma_{i,s}f(k^{p_i}n+q_i)$ for some integers $\gamma_{i,s}$. Then

$$\mathbf{x}^T\mathbf{Y}^n\mathbf{y} = \sum_{i,j,\ell}\alpha_i\beta_j\gamma_{\ell,j}\, f([u_iy^nv_\ell]_k),$$

where v_ℓ is the unique word in $\{0,1,\ldots,k-1\}^*$ of length p_ℓ such that $[v_\ell]_k = q_\ell$. Let $K = \sum_{i,j,\ell}|\alpha_i|\cdot|\beta_j|\cdot|\gamma_{\ell,j}|$. Then since $|\mathbf{x}^T\mathbf{Y}^n\mathbf{y}|\geqslant n$ for all n sufficiently large, there is some $N_0>0$ such that for $n>N_0$ some element from

$$\{\{|f([u_iy^nv_j]_k)|\}_{n\geqslant 0}: i=1,\ldots,d, j=1,\ldots,t\}\}$$

is at least n/K.

We let M denote the maximum of the lengths of $u_1,\ldots,u_d,y,v_1,\ldots,v_t$. Then each of $[u_iy^nv_j]_k < k^{2Mn}$ for $n\geqslant 2$. Hence we have constructed an infinite set of natural numbers $N=N_n := [u_iy^nv_j]_k$ such that $|f(N)| > \log_k(N)/2K$ and so taking $c = (2MK\log k)^{-1}$, we see that $|f(N)| > c\log N$ for infinitely many N. □

The above proof actually shows something a bit more specific. It shows for an unbounded k-regular sequence that there exist words $u_1,\ldots,u_m,y,v_1,\ldots,v_m\in\{0,1,\ldots,k-1\}^*$ and a constant $c_0>0$ such that for all sufficiently large n there exist an i and j such that $|f([u_iy^nv_j]_k)|\geqslant c_0n$. Here for a word $w = i_s\cdots i_0\in\{0,1,\ldots,k-1\}^*$, we have written $[w]_k = i_sk^s+\cdots+i_0$. This can be thought of as a type of "pumping lemma" for attaining unbounded growth. This argument will prove quite useful when we consider good upper bounds in the next section.

2.3.2 Upper Bounds

The question of upper bounds was first addressed by Allouche and Shallit [17, Theorem 2.10] in their original paper introducing regular sequences.

Theorem 2.3.3 (Allouche and Shallit). *Let f be a k-regular sequence with values in $\mathbb{C}$. Then there is a constant c such that $f(n) = \mathcal{O}(n^c)$.*

Proof. We use the matrix version of regular sequences as given by Lemma 2.2.7. In particular, let d be a positive integer, $\mathscr{A}_f = \{\mathbf{A}_0, \ldots, \mathbf{A}_{k-1}\} \subseteq \mathbb{C}^{d\times d}$, and $\mathbf{v}, \mathbf{w} \in \mathbb{C}^d$ be vectors such that

$$f(n) = \mathbf{w}^T \mathbf{A}_{i_0} \cdots \mathbf{A}_{i_s} \mathbf{v},$$

where $(n)_k = i_s \cdots i_0$ is the base-k expansion of n.

Let $\|\cdot\|$ be a (submultiplicative) matrix norm, $i_0 \cdots i_s$ be the base k expansion of n, and

$$c := \max\{\|\mathbf{v}\|, \|\mathbf{w}\|, \|\mathbf{A}_0\|, \ldots, \|\mathbf{A}_{k-1}\|\}.$$

Then

$$|f(n)| \leqslant \|\mathbf{v}\| \cdot \|\mathbf{w}\| \cdot \prod_{j=0}^{s} \|\mathbf{A}_{i_j}\| \leqslant c^{s+3}.$$

Using the bound $s \leqslant \log_k n$ with some rearrangement gives the result. □

In recent work, Coons [163] determined the optimal constant c for which Theorem 2.3.3 holds. Its description requires a few definitions, the first of which formalizes what is meant by "optimal" in this context.

Definition 2.3.4. Let $k \geqslant 1$ be an integer and $f : \mathbb{Z}_{\geqslant 0} \to \mathbb{C}$ be a (not eventually zero) k-regular sequence. We define the *growth exponent of f*, denoted GrExp(f), by

$$\mathrm{GrExp}(f) := \limsup_{\substack{n\to\infty \\ f(n)\neq 0}} \frac{\log|f(n)|}{\log n}.$$

Definition 2.3.5. The *spectral radius* of a square matrix is the maximal absolute value of eigenvalues of the matrix. The *joint spectral radius* of a finite set of matrices $\mathscr{A} = \{\mathbf{A}_0, \mathbf{A}_1, \ldots, \mathbf{A}_{k-1}\}$, denoted $\rho(\mathscr{A})$, is defined as the real number

$$\rho(\mathscr{A}) = \limsup_{n\to\infty} \max_{0\leqslant i_0, i_1, \ldots, i_{n-1}\leqslant k-1} \|\mathbf{A}_{i_0}\mathbf{A}_{i_1}\cdots\mathbf{A}_{i_{n-1}}\|^{1/n},$$

where $\|\cdot\|$ is any (submultiplicative) matrix norm.

The joint spectral radius was introduced by Rota and Strang [513] and has a wide range of applications. See Rota and Strang [513] also for details about the independence of the matrix norm in the definition. For an extensive treatment, see Jungers's monograph [315].

Theorem 2.3.6 (Coons). *Let $k \geqslant 1$ and $d \geqslant 1$ be integers and $f : \mathbb{Z}_{\geqslant 0} \to \mathbb{C}$ be a (not eventually zero) k-regular sequence. If $\mathscr{A}_f$ is any collection of k integer matrices associated to a basis of the $\mathbb{C}$-vector space $\langle \mathrm{Ker}_k(f) \rangle_{\mathbb{C}}$, then*

$$\log_k \rho(\mathscr{A}_f) = \mathrm{GrExp}(f),$$

where $\log_k$ denotes the base-k logarithm.

Before moving on with the needed preliminary results for the proof of this theorem, we describe what it means for a collection of k integer matrices to be associated to a basis of the $\mathbb{C}$-vector space $\langle \mathrm{Ker}_k(f) \rangle_{\mathbb{C}}$. This is all taken in the context of Lemma 2.2.7 that provides for a set of matrices $\mathscr{A}_f$ coming from Lemma 2.2.6(e). In particular, given a word $w = i_s \cdots i_0 \in \{0, \dots, k-1\}^*$, we let $[w]_k$ denote the natural number n such that $(n)_k = w$. Let $\{\{f(n)\}_{n\geqslant 0} = \{g_1(n)\}_{n\geqslant 0}, \dots, \{g_d(n)\}_{n\geqslant 0}\}$ be a basis for the $\mathbb{C}$-vector space $\langle \mathrm{Ker}_k(f) \rangle_{\mathbb{C}}$. Then for each $i \in \{0, 1, \dots, k-1\}$, the sequences $\{g_1(kn+i)\}_{n\geqslant 0}, \dots, \{g_d(kn+i)\}_{n\geqslant 0}$ can be expressed as $\mathbb{C}$-linear combinations of $\{g_1(n)\}_{n\geqslant 0}, \dots, \{g_d(n)\}_{n\geqslant 0}$, and hence there is a set of $d \times d$ matrices $\mathscr{A}_f = \{\mathbf{A}_0, \dots, \mathbf{A}_{k-1}\}$ with entries in $\mathbb{C}$ such that

$$\mathbf{A}_i [g_1(n), \dots, g_d(n)]^T = [g_1(kn+i), \dots, g_d(kn+i)]^T$$

for $i = 0, \dots, k-1$ and all $n \geqslant 0$. In particular, if $i_s \cdots i_0$ is the base-k expansion of n, then $\mathbf{A}_{i_0} \cdots \mathbf{A}_{i_s} [g_1(0), \dots, g_d(0)]^T = [g_1(n), \dots, g_d(n)]^T$. (We note that this holds even if we pad the base-k expansion of n with zeros at the beginning.)

Definition 2.3.7. We call a set of matrices $\mathscr{A}_f$, as constructed in the previous paragraph, *a set of matrices associated to a basis of* $\langle \mathrm{Ker}_k(f) \rangle_{\mathbb{C}}$. In general, if $\mathscr{B}_f$ is any set of matrices for which there are vectors $\mathbf{w}$ and $\mathbf{v}$ such that f has linear representation $(\mathbf{w}, \mathscr{B}_f, \mathbf{v})$, then we call the set $\mathscr{B}_f$ *a set of matrices associated to f*.

The first step in the proof of Theorem 2.3.6 is to modify the proof of Theorem 2.3.3 to include the notion of the joint spectral radius. This is done by appealing to a result, which we record here as Lemma 2.3.8; it can be found as Proposition 4 of Blondel *et al.* [90], though it was first given in the original paper of Rota and Strang [513].

Lemma 2.3.8. *Let $k \geqslant 1$ be an integer and $\mathscr{A} = \{\mathbf{A}_0, \mathbf{A}_1, \dots, \mathbf{A}_{k-1}\}$ be a finite set of matrices. Given $\varepsilon > 0$ then there is a submultiplicative matrix norm $\|\cdot\|$ such that $\|\mathbf{A}_i\| < \rho(\mathscr{A}) + \varepsilon$ for each $i \in \{0, 1, \dots, k-1\}$.*

With this lemma in hand, it is quite easy to give a tight upper bound for the optimal constant for Theorem 2.3.3.

Proposition 2.3.9. *Let $k \geqslant 2$ be an integer and $f : \mathbb{Z}_{\geqslant 0} \to \mathbb{C}$ be a k-regular function. For any $\varepsilon > 0$, there is a constant $c = c(\varepsilon) > 0$ such that for all $n \geqslant 1$,*

$$\frac{|f(n)|}{n^{\log_k(\rho(\mathscr{A}_f)+\varepsilon)}} \leqslant c,$$

where $\mathscr{A}_f$ is any set of matrices associated to f.

Proof. Let $\varepsilon > 0$ be given and let $\|\cdot\|$ be a matrix norm such that the conclusion of Lemma 2.3.8 holds. Then

$$|f(n)| \leqslant \|\mathbf{v}\| \cdot \|\mathbf{w}\| \cdot \prod_{j=0}^{s} \|\mathbf{A}_{i_j}\| \leqslant \|\mathbf{v}\| \cdot \|\mathbf{w}\| \cdot (\rho(\mathscr{A}) + \varepsilon)^{s+1},$$

where the base-k expansion of n is $i_s \cdots i_0$. Using the bound $s \leqslant \log_k n$ with some rearrangement gives the result. □

As it turns out, if $\mathscr{B}_f$ is any set of matrices associated to f and $\mathscr{A}_f$ is any set of matrices associated to a basis of $\langle \mathrm{Ker}_k(f)\rangle_{\mathbb{C}}$, then $\rho(\mathscr{A}_f) \leqslant \rho(\mathscr{B}_f)$, though the proof of this statement is only apparent after validating Theorem 2.3.6.

Lemma 2.3.10. *Let $k \geqslant 1$ be an integer and $\mathscr{A} = \{\mathbf{A}_0, \mathbf{A}_1, \ldots, \mathbf{A}_{k-1}\}$ be a finite set of matrices. If $\varepsilon > 0$ is a real number, then there is a positive integer m and a matrix $\mathbf{A}_{i_0} \cdots \mathbf{A}_{i_{m-1}}$, such that*

$$(\rho(\mathscr{A}) - \varepsilon)^m < \rho(\mathbf{A}_{i_0} \cdots \mathbf{A}_{i_{m-1}}) < (\rho(\mathscr{A}) + \varepsilon)^m.$$

Proof. By using the properties of limits, this is a direct consequence of the definition of the joint spectral radius. Details are left as an exercise. □

Restricting to a set of matrices associated to a basis of $\langle \mathrm{Ker}_k(f)\rangle_{\mathbb{C}}$ allows us to provide the lower bound analogue of Proposition 2.3.9.

Proposition 2.3.11. *Let $k \geqslant 2$ be an integer and $f : \mathbb{Z}_{\geqslant 0} \to \mathbb{C}$ be a k-regular function. For any $\varepsilon > 0$, there is a constant $c = c(\varepsilon) > 0$ such that for infinitely many $n \geqslant 1$,*

$$\frac{|f(n)|}{n^{\log_k(\rho(\mathscr{A}_f)-\varepsilon)}} \geqslant c,$$

where $\mathscr{A}_f$ is any set of matrices associated to a basis of $\langle \mathrm{Ker}_k(f)\rangle_{\mathbb{C}}$.

Proof. As in the proof of Theorem 2.3.1, we follow the argument of Bell, Coons, and Hare (see p. 198 of [65]).

Let $k \geqslant 2$ be an integer, suppose that $f : \mathbb{Z}_{\geqslant 0} \to \mathbb{C}$ is an unbounded k-regular sequence, and $\mathscr{A}_f = \{\mathbf{A}_0, \ldots, \mathbf{A}_{k-1}\}$ be a set of matrices associated to a basis $\{\{f(n)\}_{n\geqslant 0} = \{g_1(n)\}_{n\geqslant 0}, \ldots, \{g_d(n)\}_{n\geqslant 0}\}$ of the $\mathbb{C}$-vector space $\langle \mathrm{Ker}_k(f)\rangle_{\mathbb{C}}$.

Let $\varepsilon > 0$ be given. Then by Lemma 2.3.10 there is a positive integer m and a matrix $\mathbf{A} = \mathbf{A}_{i_0} \cdots \mathbf{A}_{i_{m-1}}$ such that $\rho(\mathbf{A}) > (\rho(\mathscr{A}_f) - \varepsilon)^m$. Let λ be an eigenvalue of $\mathbf{A}$ with $|\lambda| = \rho(\mathbf{A})$. Then there is an eigenvector $\mathbf{y}$ such that $\mathbf{Ay} = \lambda\mathbf{y}$. Pick a vector $\mathbf{x}$ such that $\mathbf{x}^T\mathbf{y} = c_1 \neq 0$. Then

$$\left|\mathbf{x}^T\mathbf{A}^n\mathbf{y}\right| = |c_1| \cdot |\lambda|^n = |c_1| \cdot \rho(\mathbf{A})^n > |c_1| \cdot \left(\rho(\mathscr{A}_f) - \varepsilon\right)^{nm}. \tag{2.16}$$

We claim that the $\mathbb{C}$-span of the vectors $[g_1(i), \ldots, g_d(i)]$, as i ranges over all natural numbers, must span all of $\mathbb{C}^d$. If this were not the case, then their span would be a proper subspace of $\mathbb{C}^d$, and hence the span would have a nontrivial orthogonal complement. In particular, there would exist $c_1, \ldots, c_d \in \mathbb{C}$, not all zero, such that

$$c_1 g_1(n) + \cdots + c_d g_d(n) = 0$$

for every n, contradicting the fact that $g_1(n), \ldots, g_d(n)$ are $\mathbb{C}$-linearly independent sequences.

Let $\langle\mathscr{A}_f\rangle$ denote the semigroup generated by the elements of $\mathscr{A}_f$. We have just shown that there exist words $\mathbf{X}_1, \ldots, \mathbf{X}_d$ in $\langle\mathscr{A}_f\rangle$ such that

$$[g_1(0), \ldots, g_d(0)]\mathbf{X}_1, \ldots, [g_1(0), \ldots, g_d(0)]\mathbf{X}_d$$

span $\mathbb{C}^d$.

Now consider $\mathbf{x}^T\mathbf{A}^n\mathbf{y}$ as described in the paragraph ending with (2.16). The following lines are as in the proof of Theorem 2.3.1. By construction, we may write $\mathbf{x}^T = \sum_j \alpha_j [g_1(0), \ldots, g_d(0)]\mathbf{X}_j$ for some complex numbers α_j. Then

$$\mathbf{x}^T\mathbf{A}^n = \sum_j \alpha_j [g_1(0), \ldots, g_d(0)]\mathbf{X}_j\mathbf{A}^n.$$

Let u_j be the word in $\{0, 1, \ldots, k-1\}^*$ corresponding to $\mathbf{X}_j$ and let $y = i_{m-1} \cdots i_0$ be the word in $\{0, \ldots, k-1\}^*$ corresponding to $\mathbf{A}$; that is, $y = i_{m-1} \cdots i_0$ where $\mathbf{A} = \mathbf{A}_{i_0} \cdots \mathbf{A}_{i_{m-1}}$ and similarly for u_j. Then we have

$$[g_1(0), \ldots, g_d(0)]\mathbf{X}_j\mathbf{A}^n = [g_1([u_j y^n]_k), \ldots, g_d([u_j y^n]_k)]^T.$$

Write $\mathbf{y}^T = [\beta_1, \ldots, \beta_d]$. Then

$$\mathbf{x}^T\mathbf{A}^n\mathbf{y} = \sum_{i,j} \alpha_i \beta_j g_j([u_i y^n]_k).$$

By assumption, each of $\{g_1(n)\}_{n\geqslant 0}, \ldots, \{g_d(n)\}_{n\geqslant 0}$ is in the $\mathbb{C}$-vector space generated by $\mathrm{Ker}_k(f)$, and hence there exist natural numbers $p_1, \ldots, p_t$ and $q_1, \ldots, q_t$ with $0 \leqslant q_\ell < k^{p_\ell}$ for $\ell = 1, \ldots, t$ such that for each $j = 1, \ldots, d$, we have $g_j(n) = \sum_{\ell=1}^{t} \gamma_{\ell,j} f(k^{p_\ell} n + q_\ell)$ for some constants $\gamma_{\ell,j} \in \mathbb{C}$. Then

$$\mathbf{x}^T\mathbf{A}^n\mathbf{y} = \sum_{i,j,\ell} \alpha_i\beta_j\gamma_{\ell,j} f([u_i y^n v_\ell]_k),$$

where v_ℓ is the unique word in $\{0, 1, \ldots, k-1\}^*$ of length p_ℓ such that $[v_\ell]_k = q_\ell$. Let $K = \sum_{i,j,\ell} |\alpha_i| \cdot |\beta_j| \cdot |\gamma_{\ell,j}|$. Then since $|\mathbf{x}^T\mathbf{A}^n\mathbf{y}| \geqslant |c_1| \cdot \left(\rho(\mathscr{A}_f) - \varepsilon\right)^{nm}$ for all n, some element from

$$\left\{|f([u_i y^n v_\ell]_k)| : i = 1, \ldots, d, \ell = 1, \ldots, t\}\right\}$$

is at least $(|c_1|/K) \cdot \left(\rho(\mathscr{A}_f) - \varepsilon\right)^{nm}$ for each n. Set $c_2 := |c_1|/K$.

If $M = \max\{|u_i|, |v_\ell| : i = 1, \ldots, d,\ \ell = 1, \ldots, t\}$, then

$$N = [u_i(i_{m-1} \cdots i_0)^n v_\ell]_k < k^{2M+nm},$$

so that $\log_k(N) - 2M < nm$. Thus, by the finding of the previous paragraph, there are infinitely many N such that

$$\frac{|f(N)|}{N^{\log_k(\rho(\mathscr{A}_f)-\varepsilon)}} = \frac{|f(N)|}{(\rho(\mathscr{A}_f) - \varepsilon)^{\log_k N}} > \frac{c_2}{(\rho(\mathscr{A}_f) - \varepsilon)^{2M}},$$

which is the desired result. □

Proof (of Theorem 2.3.6). For a given $\varepsilon > 0$, Proposition 2.3.9 implies that

$$\lim_{n\to\infty} \frac{|f(n)|}{n^{\log_k(\rho(\mathscr{A}_f)+2\varepsilon)}} = 0,$$

and Proposition 2.3.11 implies that

$$\limsup_{n\to\infty} \frac{|f(n)|}{n^{\log_k(\rho(\mathscr{A}_f)-2\varepsilon)}} = \infty.$$

Taken together these give

$$\log_k(\rho(\mathscr{A}_f) - 2\varepsilon) \leqslant \mathrm{GrExp}(f) \leqslant \log_k(\rho(\mathscr{A}_f) + 2\varepsilon).$$

Since ε can be taken arbitrarily small, this proves the theorem. □

Example 2.3.12. For the Stern sequence s, one has

$$\mathrm{GrExp}(s) = \log_2 \varphi,$$

where $\varphi = (1 + \sqrt{5})/2$ is the golden ratio. This follows from work of Reznick [501, Theorem 5.13]. See also, Calkin and Wilf [124] and Coons and Tyler [165].

Before moving on, we note the works of Dumas [201, 202] concerning the asymptotic expansion of the summatory functions of regular sequences. Among many results and useful algorithms, his results have the flavor of the following theorem [201, Theorem 1].

Theorem 2.3.13 (Dumas). *Let f be a k-regular function with linear representation* $(\mathbf{w}, \mathscr{A}_f, \mathbf{v})$. *Then*

$$s(n) := \sum_{j\leqslant n} f(n) \sim \sum_{\substack{\alpha>\alpha_* \\ \ell\geqslant 0,\vartheta}} n^{\alpha} \binom{\log_k n}{\ell} \exp(i\vartheta \log_k n)\Psi_{\alpha,\ell,\vartheta}(\log_k n) + \mathscr{O}(n^{\alpha_*}),$$

where exponents α and angular variables ϑ are real numbers, the numbers ℓ are nonnegative integers, and the functions Ψ are 1-periodic functions. Specific details can be found in Dumas' work [201].

2.3.3 Maximum Values and the Finiteness Property

Determining the maximum values of regular sequences remains a mysterious area, though it is related to a very interesting and important open question regarding the joint spectral radius. As examples and results surrounding this area are sparse, in this section, we will present a motivating extended example—Stern's sequence—as a way to frame some questions.

Recall from Example 2.2.5 that Stern's diatomic sequence is 2-regular and defined by the relations $s(0) = 0$, $s(1) = 1$, and for $n \geqslant 0$, by

$$s(2n) = s(n), \quad \text{and} \quad s(2n+1) = s(n) + s(n+1).$$

The first few values of the sequence are

$$0, 1, 1, 2, 1, 3, 2, 3, 1, 4, 3, 5, 2, 5, 3, 4, 1, 5, 4, 7, 3, 8, 5, 7, 2, 7, 5, 8, 3, 7, 4, 5, 1, \ldots.$$

The Stern sequence, like essentially all observed regular sequences, has a limiting distribution between consecutive powers of 2 (powers of k^r for k-regular sequences for some appropriate r). In fact, if one looks at the plot of the points $(n, s(n))$ for n between consecutive powers of 2, the picture seems to have asymptotically stabilized; see Figures 2.2 and 2.3.

In particular, note the stabilizing two maximums in the each of the plots in Figures 2.2 and 2.3. It is easy to show that the Stern sequence is palindromic between consecutive powers of 2, so we may focus on just the first maximum. (The maximum is in fact attained at at most two points, which we state here without proof.) We will use the defining recursions to classify and get a bound on this maximum value.

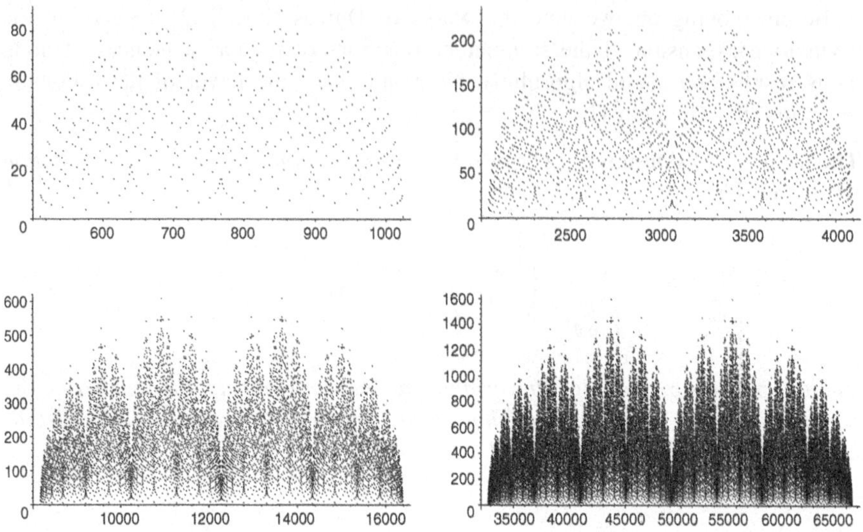

Fig. 2.2 Stern's diatomic sequence in the intervals $[2^n, 2^{n+1}]$ for $n = 9, 11, 13, 15$.

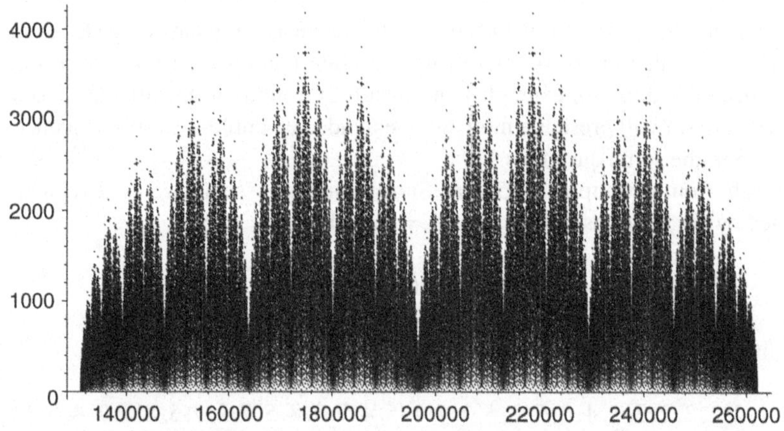

Fig. 2.3 Stern's diatomic sequence in the interval $[2^{17}, 2^{18}]$.

To this end, for $m \geqslant 0$ define

$$M_m := \max\{s(n) : n \in [2^m, 2^{m+1})\}.$$

Then, by observation, we have that $M_0 = 1$, $M_1 = 2$, and $M_2 = 3$.

For $m \geqslant 3$ we note that $s(2n+1) \geqslant s(n) = s(2n)$, so that the maximum value always occurs at an odd index $2n+1 \in [2^m, 2^{m+1})$. Of course, like for all numbers, for this value $2n+1$, one of n or $n+1$ is even, so that the recursion for odd indices gives

$$M_m \leqslant M_{m-1} + M_{m-2}.$$

But, combining this inequality with the fact that $M_0 = 1$ and $M_1 = 2$, gives that

$$M_m \leqslant F_{m+2},$$

where F_k is the k-th Fibonacci number. This inequality is actually an equality, which we will now show.

Proposition 2.3.14. *The maximal value of the Stern sequence in the interval* $[2^m, 2^{m+1})$ *is the Fibonacci number* F_{m+2}, *and this value occurs at* $n = (2^{m+2} - (-1)^{m+2})/3$.

Proof. We have already shown above that $M_m \leqslant F_{m+2}$, so it remains only to show that there is an integer $n \in [2^m, 2^{m+1})$ such that $s(n) = F_{m+1}$.

To this end, set $\alpha_m := (2^{m+2} - (-1)^{m+2})/3$. It is clear that $\alpha_m \in [2^m, 2^{m+1})$ and that

$$\alpha_{m+1} = \begin{cases} 2\alpha_m + 1 & \text{if } m \text{ is even;} \\ 2\alpha_m - 1 & \text{if } m \text{ is odd,} \end{cases}$$

therefore by the recurrence for s we have

$$\begin{aligned} s(\alpha_{m+1}) &= \begin{cases} s(\alpha_m) + s(\alpha_m + 1) = s(\alpha_m) + s(2\alpha_{m-1}) & \text{if } m \text{ is even;} \\ s(\alpha_m - 1) + s(\alpha_m) = s(2\alpha_{m-1}) + s(\alpha_m) & \text{if } m \text{ is odd} \end{cases} \\ &= s(\alpha_{m-1}) + s(\alpha_m). \end{aligned}$$

By induction, it follows that $s(\alpha_m) = F_{m+2}$, which is exactly what we set out to show. □

The binary forms

$$\alpha_m := \begin{cases} [(\texttt{10})^{m/2}\texttt{1}]_2 & \text{if } m \text{ is even;} \\ [(\texttt{10})^{(m-1)/2}\texttt{11}]_2 & \text{if } m \text{ is odd} \end{cases}$$

of the integers α_m here are a point of interest. They are of the form $w^k u$ for some words u and w and some integer k. This implies something even more interesting for the normalized graph of the Stern sequence between consecutive powers of two. To be clear, we state the generalizations of these ideas as a series of formal questions.

Question 2.3.15. Let f be a k-regular sequence. Is there an integer $M \geqslant 1$, such that f (suitably normalized to the box $[0, 1]^2$) has a limit when taken between powers of k^M? That is, the normalized picture of the points $(n, f(n))$, where $n \in [k^{Mj}, k^{M(j+1)}]$, converges.

Question 2.3.16. Let f be an integer-valued k-regular sequence. If f is not an automatic sequence, is there a positive integer M such that

$$\max_{k^{Mm}\leqslant n\leqslant k^{M(m+1)}-1} |f(n)| < \max_{k^{M(m+1)}\leqslant n\leqslant k^{M(m+2)}-1} |f(n)| \ ?$$

Question 2.3.17. Suppose that Question 2.3.16 has a positive answer and that f is an integer-valued k-regular sequence. Is it true that there are words $u, w \in \{0, \dots, k-1\}^*$ such that one of the maximum values $\alpha_{f,m}$ of $|f(n)|$ in $[k^{Mm}, k^{M(m+1)}]$ satisfies $\alpha_{f,m} = w^{n_m}u$ for some increasing sequence of integers n_m and infinitely many m?

The careful reader will notice that Questions 2.3.16 and 2.3.17 have the added assumption that f is integer-valued. This assumption cannot be removed completely as the questions have negative answers when one looks at general real-valued sequences. In fact, this line of questioning is related to an open question regarding the joint spectral radius (see Definition 2.3.5) of a finite set of matrices.

Definition 2.3.18. A finite set of matrices $\mathscr{A}$ is said to have the *finiteness property* provided there is a specific finite product $\mathbf{A}_{i_0}\cdots\mathbf{A}_{i_{m-1}}$ of matrices from $\mathscr{A}$ such that $\rho(\mathbf{A}_{i_0}\cdots\mathbf{A}_{i_{m-1}})^{1/m} = \rho(\mathscr{A})$.

Arising from the work of Daubechies and Lagarias [179], Lagarias and Wang [367] conjectured that the finiteness property holds for all finite sets of real matrices, though this was shown to be false—hence the negative answer to the generalization of Question 2.3.17 for real-valued regular sequences. The existence of counterexamples was first shown by Bousch and Mairesse [102] (see also [91, 360]), and a constructive counterexample was recently given by Hare, Morris, Sidorov, and Theys [284]. Their counterexample is reminiscent of the Stern sequence, and so we give it here to add a little connective flavor to the questions.

Example 2.3.19 (Hare, Morris, Sidorov, and Theys). Let τ denote the sequence of integers defined by $\tau_0 = 1$, $\tau_1, \tau_2 = 2$, and $\tau_{n+1} = \tau_n\tau_{n-1} - \tau_{n-2}$ for all $n \geqslant 2$, and let F_n be the nth Fibonacci number for $n \geqslant 0$. Define the real number $\alpha_* \in (0, 1]$ by

$$\alpha_* := \prod_{n\geqslant 1}\left(1 - \frac{\tau_{n-1}}{\tau_n\tau_{n+1}}\right)^{(-1)^n F_{n+1}}.$$

Then this infinite product converges unconditionally, and the set

$$\left\{\begin{bmatrix}1 & 1\\ 0 & 1\end{bmatrix}, \alpha_*\begin{bmatrix}1 & 0\\ 1 & 1\end{bmatrix}\right\}$$

does not have the finiteness property.

Note that the number

$$\alpha_* = 0.749326546330367557943961948091344672091327370236064317 3\dots,$$

and it is unknown if α_* is irrational, though it is suspected.

It is an open and interesting question to determine if all finite sets of rational matrices satisfy the finiteness property. The current best result toward this conjecture is that of Jungers and Blondel [316], who showed that the finiteness property holds for all finite sets of rational matrices provided it holds for all pairs of matrices with entries in $\{-1, 0, 1\}$. Restricting to the case of nonnegative rational matrices, Jungers and Blondel [316] could reduce $\{-1, 0, 1\}$ to the set $\{0, 1\}$.

As a fact related to Question 2.3.15, we want to show that there are only few large values of $s(n)$ in the interval $[2^m, 2^{m+1})$, compared to the maximal value $M_m = F_{m+2}$. First, we note that the mean value of $s(n)$ in such an interval equals $(3/2)^m$, which can be proved by induction. What we want to show is that there are in fact exponentially few integers n in $[2^m, 2^{m+1})$ such that $s(n) \geqslant \varepsilon M_m$, for any $\varepsilon > 0$. By definition of the mean value and the nonnegativity of $s(n)$, the number N of such integers satisfies $N\varepsilon M_m/2^m \leqslant (3/2)^m$; therefore, $N \leqslant 3^m/(M_m\varepsilon) \ll (3/\varphi)^m/\varepsilon$, where φ is the golden ratio. Since φ is strictly larger than $3/2$, there are exponentially few integers n such that $s(n)$ is large. This leads us to the following proposition for the graph of $s(n)$ in dyadic intervals $[2^m, 2^{m+1})$, normalized to $[0, 1]^2$. We define functions f_m from $[0, 1]$ to $[0, 1]$ by

$$f_m(x) = \frac{1}{F_{m+2}} s\big(2^m + \lfloor 2^m x \rfloor\big).$$

Proposition 2.3.20. *The sequence $\{f_m\}_{m \geqslant 0}$ of functions converges to zero almost everywhere.*

Proof. By the above considerations, there is an $K < 1$ such that

$$\lambda(\{x \in [0, 1] : f_m(x) \geqslant \varepsilon\}) \leqslant K^m/\varepsilon,$$

where λ is the Lebesgue measure. It follows that

$$\begin{aligned}
&\lambda\big(\{x \in [0, 1] : \exists m \geqslant M \text{ such that } f_m(x) \geqslant \varepsilon\}\big) \\
&\quad = \lambda\left(\bigcup_{m \geqslant M} \{x \in [0, 1] : f_m(x) \geqslant \varepsilon\}\right) \leqslant \sum_{m \geqslant M} \lambda(\{x \in [0, 1] : f_m(x) \geqslant \varepsilon\}) \\
&\quad \leqslant \frac{1}{\varepsilon} \sum_{m \geqslant M} K^m = \frac{1}{\varepsilon} \frac{K^M}{1 - K}.
\end{aligned}$$

Setting $A_M(\varepsilon) = \{x \in [0, 1] : f_m(x) < \varepsilon \text{ for all } m \geqslant M\}$, we obtain $\lambda(A_M) \geqslant 1 - K^M/(\varepsilon(1 - K))$. It follows that

$$1 = \lambda\left(\bigcup_{M \geqslant 1} A_M\right) = \lambda(B_\varepsilon),$$

where

$$B_\varepsilon = \{x \in [0,1] : \exists M \geqslant 1 \text{ such that } f_m(x) < \varepsilon \text{ for all } m \geqslant M\}.$$

Therefore

$$\lambda(\{x \in [0,1] : f_m(x) \to 0 \text{ as } m \to \infty\}) = \lambda\left(\bigcap_{\varepsilon>0} B_\varepsilon\right) = \lambda\left(\bigcap_{n\geqslant 1} B_{1/n}\right) = 1. \quad \square$$

In fact, we conjecture the following more precise statement.

Conjecture 2.3.21. The sequence $\{f_m\}_{m\geqslant 0}$ of functions converges pointwise, and the limit is nonzero if and only if $x \in [0,1]$ is of the form $x = a/(3\cdot 2^s)$ for some integers $a \geqslant 1$ and $s \geqslant 0$.

Another interesting question concerns values of $s(n)$ near the mean value $(3/2)^m$. Lansing [369] studies the quantity

$$H(\lambda, m) = \frac{1}{2^m}\left|\{2^m \leqslant n < 2^{m+1} : s(n) \geqslant \lambda(3/2)^m\}\right|$$

and notes that the data "suggests that $H(\lambda, m)$ converges to a smooth function, but it is not clear if it actually does." This statement is based on the behavior for some small values of m. We used randomly chosen integers in the interval $\left[2^m, 2^{m+1}\right)$ for some larger m in order to guess the asymptotic behavior. Our experiments suggest that $H(\lambda, m)$ converges to zero for all $\lambda > 0$.

We finish this section with a remark concerning the distribution of the values of $s(n)$. Heuristically, the method of obtaining $s(n)$ by a matrix product is (formally) similar to studying the product of independent identically distributed random variables. The question therefore suggests itself: is the distribution of the values $s(n)$ in dyadic intervals $\left[2^m, 2^{m+1}\right)$ close to a log-normal distribution? We leave this as another open question.

2.4 Analytic and Algebraic Properties of Mahler Functions

In this section, we consider the properties of regular functions and Mahler functions viewed as functions of a complex variable. In particular, we will address questions of convergence, analytic behavior, and rationality. In particular, the results will lead to a proof of Bézivin's theorem [83] that an irrational Mahler function is transcendental. The arguments in this section follow closely those of Bell, Coons, and Rowland [66], who gave an alternative proof of Bézivin's result.

2.4.1 Analytic Properties of Mahler Functions

Allouche and Shallit's upper bound on regular sequences, Theorem 2.3.3, yields the following as an immediate corollary.

Proposition 2.4.1. *A regular function $F(z)$ converges inside the unit circle.*

This proposition can be used to give an alternative proof that there are Mahler functions that are not regular.

Example 2.4.2 (Example 2.2.19 Revisited). Recall from Example 2.2.19, the function $1/(1-2z)$ is k-Mahler for each k. But $z = 1/2$ is a singularity of the function, so it does not converge everywhere inside the unit circle. Hence it is not k-regular for any k by Proposition 2.4.1.

Dumas' structure theorem, Theorem 2.2.24, yields the following immediate corollary, which we note here as a proposition.

Proposition 2.4.3. *Let $k \geqslant 2$ be an integer and let $F(z) \in \mathbb{C}[[z]]$ be a k-Mahler function. Then $F(z)$ has a positive radius of convergence.*

Proof. Denote by $B(0, r)$ the open ball of radius $r > 0$ centered at the origin. Let $k \geqslant 2$ be an integer and $F(z) \in \mathbb{C}[[z]]$ be a k-Mahler function satisfying, say,

$$\sum_{j=0}^{d} a_j(z)F(z^{k^j}) = 0,$$

for $a_j(z) \in \mathbb{C}[z]$, $a_0(z)a_d(z) \neq 0$. Proposition 2.4.1 states that a k-regular series is analytic in the unit disk, so Theorem 2.2.24 gives that $F(z)$ converges in $B(0, r)$, where $r \in (0, 1)$ is the minimal distance from 0 to a nonzero root of $a_0(z)(z-1)$. □

It is quite easy to see that all polynomials are regular functions, and so they are all Mahler functions as well. As it turns out, polynomials are precisely the set of entire Mahler functions—and so also the set of entire regular functions.

Theorem 2.4.4. *Let $k \geqslant 2$ be an integer and $F(z) \in \mathbb{C}[[z]]$ be a k-Mahler function. If $F(z)$ is entire, then $F(z)$ is a polynomial.*

Proof. Let $k \geqslant 2$ be an integer and $F(z) \in \mathbb{C}[[z]]$ be an entire k-Mahler function satisfying

$$\sum_{j=0}^{d} a_j(z)F(z^{k^j}) = 0,$$

for $a_j(z) \in \mathbb{C}[z]$ with $a_0(z)a_d(z) \neq 0$. Write

$$F(z^{k^d}) = -\sum_{j=0}^{d-1} \frac{a_j(z)}{a_d(z)} F(z^{k^j}). \tag{2.17}$$

Pick $L > 1$ such that all of the zeros of $a_d(z)$ are in the open disk, $B(0, L)$, of radius L centered at the origin. Notice that since the $a_i(z)$ are polynomials, there is an $N > 1$ and a constant $C > 1$ such that for $|z| \geqslant L$, we have

$$\max_{0 \leqslant i \leqslant d-1} \left\{ \left| \frac{a_i(z)}{a_d(z)} \right| \right\} < C|z|^N; \tag{2.18}$$

in particular, the value $N = \max_{0 \leqslant i \leqslant d-1}\{\deg a_i(z), 2\}$ is sufficient.

For $\ell \geqslant 0$ denote

$$M_\ell := \max\left\{|F(z)| : |z| = L^{k^\ell}\right\},$$

where L is as chosen above. Using (2.17), (2.18), and the maximum modulus theorem, we have for $j \geqslant d$ that

$$M_j \leqslant (d+1)C\left(L^{k^{j-d}}\right)^N M_{j-1} \leqslant C(d+1)L^{Nk^j} M_{j-1}.$$

Thus recursively, we have for each $n \geqslant d$ that

$$M_n \leqslant M_{d-1}(C(d+1))^n L^{Nk^{n+1}}.$$

But since $L > 1$, this implies that there is some constant $b > 0$ such that for $n \geqslant d$ we have

$$M_n \leqslant L^{bk^n}.$$

Now let $m \geqslant b + 2$ be a natural number, fix an $\alpha \in \mathbb{C}$ and consider

$$F^{(m-1)}(\alpha) = \frac{1}{2\pi i} \int_{\gamma_n} \frac{F(z)}{(z-\alpha)^m} dz,$$

where γ_n is the circle of radius L^{k^n} with n large enough so that α is inside the circle of radius $L^{k^n}/2$ centered at the origin. Then for all z on γ_n, we have that

$$\frac{|z|}{2} \leqslant |z - \alpha|.$$

Thus for n large enough, we have

$$|F^{(m-1)}(\alpha)| \leqslant \frac{1}{2\pi} \cdot 2\pi L^{k^n} \cdot \frac{2^m M_n}{(L^{k^n})^m} = \frac{2^m M_n}{(L^{k^n})^{m-1}} \leqslant 2^m L^{k^n(b-m+1)}.$$

Recall that $m \geqslant b + 2$ so that the above gives that

$$|F^{(m-1)}(\alpha)| \leqslant \frac{2^m}{L^{k^n}}.$$

Since n can be taken arbitrarily large, we have that $F^{(m-1)}(\alpha) = 0$. But $\alpha \in \mathbb{C}$ was arbitrary, and so $F^{(m-1)}(z)$ is identically zero; hence $F(z)$ is a polynomial. □

2.4.2 Rational-Transcendental Dichotomy of Mahler Functions

Using Theorem 2.4.4 one can prove a rational-transcendental dichotomy of Mahler functions; see Bézivin [83].

Theorem 2.4.5 (Bézivin). *Let $k \geqslant 2$ be an integer and $F(z) \in \mathbb{C}[[z]]$ be a k-Mahler function. If $F(z)$ is algebraic, then $F(z)$ is a rational function.*

In fact, since algebraic functions have only a finite number of singularities (see Flajolet and Sedgewick [224, Section VII.7.1]), Theorem 2.4.5 is a consequence of the upcoming Theorem 2.4.7. First we record a lemma, the proof of which is left as an exercise, though it can be found in the paper of Bell, Coons, and Rowland [66].

Lemma 2.4.6. *Let $k \geqslant 2$ be an integer and let $F(z) \in \mathbb{C}[[z]]$ be a k-Mahler function. The function $F(z)$ is meromorphic if and only if it has finitely many singularities.*

Theorem 2.4.7. *Let $k \geqslant 2$ be an integer and $F(z) \in \mathbb{C}[[z]]$ be a k-Mahler function. If $F(z)$ has only finitely many singularities, then $F(z)$ is a rational function.*

Proof. Let $k \geqslant 2$ be an integer and $F(z) \in \mathbb{C}[[z]]$ be a k-Mahler function satisfying

$$\sum_{j=0}^{d} a_j(z)F(z^{k^j}) = 0, \tag{2.19}$$

for $a_j(z) \in \mathbb{C}[z]$ with $a_0(z)a_d(z) \neq 0$. If $F(z)$ has only finitely many singularities, then since by Lemma 2.4.6 it is meromorphic, there is a nonzero polynomial $q(z) \in \mathbb{C}[z]$ such that $q(z)F(z)$ is entire. For $j \in \{0, \ldots, d-1\}$ set

$$q_j(z) := \frac{1}{q(z^{k^j})} \prod_{i=0}^{d} q(z^{k^i}) \in \mathbb{C}[z].$$

Multiplying (2.19) by $\prod_{i=0}^{d} q(z^{k^i}) \in \mathbb{C}[z]$, we then have that

$$\sum_{j=0}^{d} a_j(z)q_j(z)q(z^{k^j})F(z^{k^j}) = 0,$$

where since $q(z)$ is not identically zero we have that $a_0(z)q_0(z)a_d(z)q_d(z) \neq 0$. Hence $q(z)F(z)$ is an entire k-Mahler function and thus, by the preceding lemma, a polynomial. This proves that $F(z)$ is a rational function. □

One can actually do a lot better as Randé showed in his thesis [492].

Theorem 2.4.8 (Randé). *Let $k \geqslant 2$ be an integer and $F(z) \in \mathbb{C}[[z]]$ be a k-Mahler function. Then $F(z)$ is a rational function, or it has the unit circle as a natural boundary.*

Recall that a function is *differentiably finite* (or *D-finite*) provided it satisfies a linear homogeneous differential equation with polynomial coefficient. Since D-finite functions can have only a finite number of singularities (see Flajolet and Sedgewick [224, Section VII.9.1]), Randé's result implies the following corollary.

Corollary 2.4.9. *Let $k \geqslant 2$ be an integer and $F(z) \in \mathbb{C}[[z]]$ be a k-Mahler function. If $F(z)$ is D-finite, then $F(z)$ is a rational function.*

It is an open and very interesting question to determine where Mahler functions fall in the diffeo-algebraic hierarchy. Of particular interest is whether an irrational Mahler function can satisfy an algebraic differential equation. A function that does not satisfy an algebraic differential equation is called *hypertranscendental*.

Question 2.4.10. Is it true that an irrational Mahler function is hypertranscendental?

For Mahler functions of degree one, this question has been mostly answered by Bundschuh [120], though any sort of general result for other degrees remains open.

2.5 Rational-Transcendental Dichotomy of Regular Numbers

While the rational-transcendental dichotomy of regular (and Mahler) functions is more or less straightforward as shown in the previous section, the dichotomy at the level of their special values was much more elusive.

Adamczewski and Bugeaud [3] showed that a real automatic irrational number is transcendental, and Bell, Bugeaud, and Coons [63] generalized their result to show that if $F(z)$ is a regular function, then the value $F(1/b)$, for any integer $b \geqslant 2$, is either rational or transcendental. In this section, we provide a simplified version of the result of Bell, Bugeaud, and Coons.

Theorem 2.5.1 (Bell, Bugeaud, and Coons). *Let $F(z) \in \mathbb{Z}[[z]]$ be a k-regular power series and $b \geqslant 2$ be a positive integer. Then either $F(1/b)$ is rational or it is transcendental.*

We take as our starting point Equation (2.9). To this end, let $F(z)$ be a k-regular function and let $\mathbf{F}(z) := [F(z) = F_1(z), \ldots, F_d(z)]^T$ be the vector of functions that form a basis for the $\mathbb{Q}(z)$-vector space V in the proof of Theorem 2.2.15, and recall that (2.9) gives

$$\mathbf{F}(z) = \mathbf{A}(z)\mathbf{F}(z^k), \tag{2.20}$$

where $\mathbf{A}(z) = (a_{i,j}(z)/B)_{1\leqslant i,j\leqslant d} \in \mathbb{Q}[z]^{d\times d}$ is a nonsingular matrix of polynomials $a_{i,j}(z) \in \mathbb{Z}(z)$ of degree at most $k-1$ and B is a nonzero positive integer.

We will require some additional notation. In particular, in this section we take all complex matrix norms $\|\cdot\|$ to be the operator norm, i.e., $\|A\| = \sup_{\|\mathbf{v}\|=1} \|A\mathbf{v}\|$, where the norm of a vector $\mathbf{v}$ is the ordinary Euclidean norm. Also, we let $\nu : \mathbb{Q}((x)) \to \mathbb{Z} \cup \{\infty\}$ be the valuation defined by $\nu(0) = \infty$ and

$$\nu\left(\sum_{n\geqslant -m} c_n x^n\right) := \inf\{i : c_i \neq 0\}$$

when $\sum_{n\geqslant -m} c_n x^n \in \mathbb{Q}((x))$ is a nonzero Laurent power series (this valuation will also be used in further sections).

Lemma 2.5.2. *Let* $\mathbf{F}(z)$ *satisfy* (2.20) *and* $H := \max_{1\leqslant i,j\leqslant d}\{\deg a_{i,j}(z)\}$. *Then there are* $\varepsilon > 0$, *polynomials* $P_1(z), \dots, P_d(z), Q(z) \in \mathbb{Z}[z]$ *of degree at most* $(d-1)(d+2)H$ *with* $Q(0) = 1$, *and a positive constant* $C = C(\varepsilon)$ *such that for* $i \in \{1, \dots, d\}$ *we have*

$$|F_i(t) - P_i(t)/Q(t)| \leqslant Ct^{d(d+2)H}$$

for $t \in (0, \varepsilon)$.

Proof. For $i \in \{1, 2, \dots, d\}$, the theory of simultaneous Padé approximation (see the monograph *Rational Approximations and Orthogonal Polynomials* by Nikishin and Sorokin [445, Chapter 4] for details) provides polynomials $P_i(z)$ and $Q(z)$ of degree each bounded by $(d-1)(d+2)H$, and $Q(0) = 1$, such that

$$\nu\left(Q(z)F_i(z) - P_i(z)\right) \geqslant d(d+2)H.$$

For $i \in \{1, \dots, d\}$, we thus have

$$\nu\left(F_i(z) - \frac{P_i(z)}{Q(z)}\right) \geqslant d(d+2)H.$$

Since $Q(0) = 1$ and by Proposition 2.4.1 each of $F_1(z), \dots, F_d(z)$ converges inside the unit disk, $F_i(z) - P_i(z)/Q(z)$ is analytic inside $B(0, r)$ for $i \in \{1, \dots, d\}$ for some $r > 0$ since $Q(0) = 1$. Hence there exist power series $G_1(z), \dots, G_d(z)$ that are analytic inside $B(0, r)$ such that

$$F_i(z) - \frac{P_i(z)}{Q(z)} = z^{d(d+2)H} G_i(z)$$

for $i \in \{1, \dots, d\}$. Let $\varepsilon \in (0, r)$. Then there is a positive constant C such that

$$|G_1(z)|, \dots, |G_d(z)| \leqslant C$$

for $|z| \leqslant \varepsilon$. Thus for $i \in \{1, \dots, d\}$,

$$\left| F_i(t) - \frac{P_i(t)}{Q(t)} \right| \leqslant C t^{d(d+2)H}$$

whenever $t \in (0, \varepsilon)$. □

Having established the first rational approximations to our vector of regular functions, we now establish a family of good rational approximations, which will be used in the proof of Theorem 2.5.1.

Lemma 2.5.3. *Let* $\mathbf{F}(z)$ *satisfy* (2.20) *and* $H := \max_{1 \leqslant i,j \leqslant d}\{\deg a_{i,j}(z)\}$ *and let* $t \in (0, 1)$. *Then for each* $n \geqslant 0$, *there are polynomials* $P_{1,n}(z), \dots, P_{d,n}(z), Q_n(z) \in \mathbb{Z}[z]$ *satisfying:*

(i) $\max_{1 \leqslant i \leqslant d}\{\deg P_{i,n}(z), \deg Q_n(z)\} \leqslant ((d + 2)(d - 1) + 1)Hk^n$;
(ii) $Q_n(z) = B^n Q_0(z^{k^n})$;
(iii) there exists an $\varepsilon > 0$ *and positive constants* $C_0 = C_0(\varepsilon)$ *and* $C_1 = C_1(\varepsilon)$, *not depending on t, such that for* $i \in \{1, \dots, d\}$ *and for all n sufficiently large we have* $Q_n(t) \neq 0$ *and*

$$|F_i(t) - P_{i,n}(t)/Q_n(t)| \leqslant C_1 C_0^n t^{d(d+2)Hk^n},$$

whenever $t \in (0, \varepsilon)$ *and in particular the order of vanishing of* $F_i(t) - P_{i,n}(t)/Q(t)$ *at* $t = 0$ *is at least* $d(d + 2)Hk^n$.

Proof. By Lemma 2.5.2, there are $\varepsilon > 0$, polynomials $P_{1,0}(z), \dots, P_{d,0}(z), Q_0(z) \in \mathbb{Z}[z]$ of degree at most $(d + 2)(d - 1)H$ with $Q_0(0) = 1$, and a positive constant C such that for $i \in \{1, \dots, d\}$ we have

$$|F_i(t) - P_i(t)/Q_0(t)| \leqslant C t^{d(d+2)H}$$

whenever $t \in (0, \varepsilon)$.

We define

$$\mathbf{R}_0(z) := [P_{1,0}(z)/Q_0(z), \dots, P_{d,0}(z)/Q_0(z)]^T \tag{2.21}$$

and for $n \geqslant 1$, we take

$$\mathbf{R}_n(z) = \mathbf{A}(z)\mathbf{R}_{n-1}(z^k). \tag{2.22}$$

We note that there exist integer polynomials $P_{i,n}(z)$ for $i \in \{1, \dots, d\}$ and $Q_n(z)$ such that

(a) $\mathbf{R}_n(z) = [P_{1,n}(z)/Q_n(z), \dots, P_{d,n}(z)/Q_n(z)]^T$;
(b) $Q_n(z) = B \cdot Q_{n-1}(z^k)$ for $n \geqslant 1$.

From (b), we immediately get $Q_n(z) = B^n Q_0(z^{k^n})$. Since the entries of $\mathbf{A}(z)$ are all polynomials of degree at most H, we see that if we define

$$d_n := \max_{1\leqslant i\leqslant d} \{\deg P_{i,n}(z), \deg Q_n(z)\},$$

then (2.22) gives $d_n \leqslant kd_{n-1} + H$. By induction we see, using the fact that $d_0 \leqslant (d-1)(d+2)H$, that

$$\begin{aligned} d_n &\leqslant d(d+2)H \cdot k^n + H(k^{n-1} + \cdots + k + 1) \\ &= k^n d_0 + H \cdot \frac{k^n - 1}{k-1} \leqslant ((d-1)(d+2)+1)Hk^n. \end{aligned} \tag{2.23}$$

By assumption,

$$\mathbf{F}(x) = \mathbf{A}(x)\mathbf{F}(x^k),$$

and hence for $n \geqslant 1$ we have

$$\mathbf{F}(x) - \mathbf{R}_n(x) = \mathbf{A}(z)\mathbf{A}(z^k)\cdots\mathbf{A}(z^{k^{n-1}})\left(\mathbf{F}(z^{k^n}) - \mathbf{R}_0(z^{k^n})\right).$$

Then for n sufficiently large, we have $t^{k^n} < \varepsilon$. Hence if e_i denotes the $d \times 1$ column vector whose i-th coordinate is 1 and all other coordinates are zero, then

$$\begin{aligned} |F_i(t) - P_{i,n}(t)/Q_n(t)| &= \left\| e_i^T (\mathbf{F}(t) - \mathbf{R}_n(t)) \right\| \\ &= \left\| e_i^T \mathbf{A}(t)\mathbf{A}(t^k)\cdots\mathbf{A}(t^{k^{n-1}})(\mathbf{F}(t^{k^n}) - \mathbf{R}_0(t^{k^n})) \right\| \\ &\leqslant \left\| (\mathbf{F}(t^{k^n}) - \mathbf{R}_0(t^{k^n})) \right\| \cdot \prod_{\ell=0}^{n-1} \left\| \mathbf{A}(t^{k^\ell}) \right\| \\ &\leqslant C\sqrt{d} t^{d(d+2)Hk^n} \cdot \prod_{\ell=0}^{n-1} \left\| \mathbf{A}(t^{k^\ell}) \right\|. \end{aligned}$$

Since each of the entries in $\mathbf{A}(z)$ is a polynomial with rational coefficients, there is a positive constant C_0 (independent of t) such that

$$\prod_{\ell=0}^{n-1} \left\| \mathbf{A}(t^{k^\ell}) \right\| < C_0^n$$

for all $n \geqslant 1$ and any $t \in (0, 1)$. Hence we have

$$|F_i(t) - P_{i,n}(t)/Q_n(t)| < C\sqrt{d} C_0^n t^{d(d+2)Hk^n}$$

for all $i \in \{1, \ldots, d\}$ and all n sufficiently large. To see that this gives the statement about the order of vanishing at $t = 0$, note that if $F_i(t) - P_{i,n}(t)/Q_n(t)$ has a zero of order ℓ at $t = 0$, then we can write $F_i(t) - P_{i,n}(t)/Q_n(t)$ as $t^\ell G(t)$ where $G(0) \neq 0$. It follows that there is a neighborhood of zero such that $|t^\ell G(t)| > |G(0)||t|^\ell/2$ for t in this neighborhood. Letting t approach 0 from the right and using the fact that

$$|F_i(t) - P_{i,n}(t)/Q_n(t)| < C\sqrt{d}C_0^n t^{d(d+2)Hk^n}$$

gives $\ell \geqslant d(d+2)Hk^n$ and so $F_i(t) - P_{i,n}(t)/Q_n(t)$ has a zero at $t = 0$ of order at least $d(d+2)Hk^n$. □

With these preliminaries in hand, we are ready to proceed with the proof of Theorem 2.5.1. We will use the following version of the p-adic Schmidt subspace theorem due to Schlickewei [528].

Theorem 2.5.4 (p-Adic Schmidt Subspace Theorem). *Let $n \geqslant 2$, $\varepsilon > 0$, and let $p_1, \ldots, p_s$ be distinct prime numbers. Further, let $L_{1,\infty}, \ldots, L_{n,\infty}$ be linearly independent linear forms in $X_1, \ldots, X_n$ with algebraic coefficients in $\mathbb{C}$, and for $j = 1, \ldots, s$, let $L_{1,p_j}, \ldots, L_{n,p_j}$ be linearly independent forms in $X_1, \ldots, X_n$ with algebraic coefficients in $\overline{\mathbb{Q}}_{p_j}$. Consider the inequality*

$$|L_{1,\infty}(\mathbf{x}), \ldots, L_{n,\infty}(\mathbf{x})| \cdot \prod_{j=1}^{s} |L_{1,p_j}(\mathbf{x}), \ldots, L_{n,p_j}(\mathbf{x})|_{p_j} \leqslant \|\mathbf{x}\|^{-\varepsilon}, \tag{2.24}$$

with $\mathbf{x} \in \mathbb{Z}^n$. There are a finite number of proper linear subspaces $T_1, \ldots, T_t$ of $\mathbb{Q}^n$ such that all solutions of (2.24) *lie in $T_1 \cup \cdots \cup T_t$.*

Proof (of Theorem 2.5.1). Let $\mathbf{F}(z)$ satisfy (2.20). By Lemma 2.5.3, there exist polynomials $P_{1,n}(x), \ldots, P_{d,n}(x), Q_n(x) \in \mathbb{Z}[x]$ such that

$$Q_n(x) = B^n Q_0(x^{k^n}), \tag{2.25}$$

and constants $C_1, C_2 > 0$ such that for $i \in \{1, \ldots, d\}$ and for sufficiently large n, we have $Q_n(1/b) \neq 0$ and

$$\left| F(1/b) - \frac{P_{d,n}(1/b)}{Q_n(1/b)} \right| \leqslant \frac{C_1 C_0^n}{b^{d(d+2)Hk^n}}.$$

Let D be the smallest positive integer such that

$$p_n := b^{Dk^n} P_{d,n}(1/b) \quad \text{and} \quad q_n := b^{Dk^n} Q_n(1/b)$$

are both integers, and so we have

$$|q_n \cdot F(1/b) - p_n| \leqslant \frac{C_1 C_0^n |q_n|}{b^{d(d+2)Hk^n}}. \tag{2.26}$$

Recall, by Lemma 2.5.3, we have

$$\deg P_{d,n}(x) \leqslant \deg Q_n(x) \leqslant d(d+1)H$$

so that also $D < d(d+1)H$. Also by (2.25), we have that $\deg Q_0(x^{k^n}) = D$. Write

$$Q_0(x^{k^n}) := \sum_{i=0}^{D} a_i x^{ik^n},$$

and assume, without loss of generality, $a_i \neq 0$ for each i (the general case follows *mutatis mutandis*). Note that by (2.25) we have that

$$|q_n| = B^n \left| \sum_{i=0}^{D} a_i b^{(D-i)k^n} \right| \leqslant B^n \sum_{i=0}^{D} |a_i| b^{(D-i)k^n} \leqslant C_2 B^n b^{Dk^n},$$

where $C_2 \geqslant \sum_{i=0}^{D} |a_i| > 0$ is a positive constant. Thus for n large enough, since $d \geqslant 2$ we have

$$|q_n F(1/b) - p_n| \leqslant \frac{C_1 C_2 (C_1 B)^n b^{d(d+1)Hk^n}}{b^{d(d+2)Hk^n}} = \frac{C_1 C_2 (C_1 B)^n}{b^{dHk^n}} < \frac{1}{b^{Hk^n}}. \tag{2.27}$$

We now setup to apply the p-adic Schmidt subspace theorem, suppose that $\xi := F(1/b)$ is algebraic and for $\mathbf{x} = (x_1, \ldots, x_{D+2}) \in \mathbb{Z}^{D+2}$ set

$$L_{i,\infty}(\mathbf{x}) := x_i \qquad (i \in \{1, \ldots, D+1\}),$$

and

$$L_{D+2,\infty}(\mathbf{x}) := \xi \sum_{i=1}^{D+1} x_i + x_{D+2}.$$

Also for each prime p dividing b set

$$L_{i,p}(\mathbf{x}) := x_i \qquad (i \in \{1, \ldots, D+2\}).$$

For $n \in \mathbb{N}$ denote

$$\mathbf{s}_n := (B^n a_0 b^{Dk^n}, B^n a_1 b^{(D-1)k^n}, \ldots, B^n a_D, -p_n) \in \mathbb{Z}^{D+2}.$$

Then (2.27) gives for large enough n that

$$|L_{D+2,\infty}(\mathbf{s}_n)| < \frac{1}{b^{Hk^n}}.$$

Also, we have that

$$|L_{1,\infty}(\mathbf{s}_n)\cdots L_{D+1,\infty}(\mathbf{s}_n)| = \prod_{i=0}^{D} B^n|a_i|b^{ik^n} \leqslant C_3 B^{Dn} b^{\frac{D(D+1)}{2}k^n},$$

where $C_3 := \prod_{i=1}^{D+1}|a_i| + 1 \geqslant 0$ is a positive constant.

For primes p dividing b, we have

$$\begin{aligned}\prod_{i=1}^{D+2}\prod_{p|b}|L_{i,p}(\mathbf{s}_n)|_p &\leqslant \prod_{i=0}^{D}\prod_{p|b}|B^n a_i b^{ik^n}|_p \\ &\leqslant \prod_{i=0}^{D}\prod_{p|b}|b^{ik^n}|_p = \prod_{i=0}^{D}\prod_{p|b}p^{-\nu_p(b)\cdot ik^n} = b^{-\frac{D(D+1)}{2}k^n},\end{aligned}$$

where for $\prod_{p|b}|L_{D+2,p}(\mathbf{s}_n)|_p$ we used the trivial bound of 1.

To bound $\|\mathbf{s}_n\|$, we note first that since $|L_{D+2,\infty}(\mathbf{s}_n)| < b^{-Hk^n}$, we have

$$|p_n| \leqslant |\xi|B^n\left|\sum_{i=0}^{D} a_i b^{(D-i)k^n}\right| + b^{-Hk^n}.$$

Thus

$$\begin{aligned}\|\mathbf{s}_n\|^{D+2} &= \sum_{i=0}^{D}\left|B^n a_i b^{(D-i)k^n}\right|^{D+2} + |p_n|^{D+2} \\ &< \sum_{i=0}^{D}\left|B^n a_i b^{(D-i)k^n}\right|^{D+2} + \left(|\xi|B^n\left|\sum_{i=0}^{D} a_i b^{(D-i)k^n}\right| + b^{-Hk^n}\right)^{D+2} \\ &\leqslant \left(\sum_{i=0}^{D}\left|B^n a_i b^{(D-i)k^n}\right| + |\xi|B^n\left|\sum_{i=0}^{D} a_i b^{(D-i)k^n}\right| + b^{-Hk^n}\right)^{D+2},\end{aligned}$$

and so there is constant $C_4 > 0$ such that $\|\mathbf{s}_n\| \leqslant C_4 B^n b^{Dk^n}$. Thus for a given $\varepsilon > 0$, we have that

$$\frac{1}{C_4^{\varepsilon} B^{\varepsilon n} b^{\varepsilon D k^n}} \leqslant \|\mathbf{s}_n\|^{-\varepsilon}.$$

Now set $\varepsilon = \frac{1}{2D}$. Then putting these bounds together gives for n large enough that

$$|L_{1,\infty}(\mathbf{s}_n),\ldots,L_{D+2,\infty}(\mathbf{s}_n)|\cdot\prod_{p|b}|L_{1,p}(\mathbf{s}_n),\ldots,L_{D+2,p}(\mathbf{s}_n)|_p$$

$$<\frac{C_3B^{Dn}}{b^{Hk^n}}=\frac{C_3B^{Dn}}{b^{Hk^n}}\cdot\frac{C_4^{\varepsilon}B^{\varepsilon n}b^{\varepsilon Dk^n}}{C_4^{\varepsilon}B^{\varepsilon n}b^{\varepsilon Dk^n}}\leqslant\frac{C_3B^{Dn}C_4^{\varepsilon}B^{\varepsilon n}}{b^{(H-\varepsilon D)k^n}}\cdot\|\mathbf{s}_n\|^{-\varepsilon}\leqslant\|\mathbf{s}_n\|^{-\varepsilon},$$

for n large enough, since $H\geqslant 1$ as long as $F(x)$ is not identically 1 (in which case $F(1/b)$ is rational and the theorem holds anyway).

Thus for n large enough, the $(D+2)$-tuples $\mathbf{s}_n$ are solutions to the system,

$$|L_{1,\infty}(\mathbf{s}_n),\ldots,L_{D+2,\infty}(\mathbf{s}_n)|\cdot\prod_{p|b}|L_{1,p}(\mathbf{s}_n),\ldots,L_{D+2,p}(\mathbf{s}_n)|_p\leqslant\|\mathbf{s}_n\|^{-\frac{1}{2D}},$$

which, by the p-adic Schmidt subspace theorem, lie in finitely many proper linear subspaces of $\mathbb{Q}^{D+2}$. Hence there exists a nonzero $(D+2)$-tuple $(\alpha_0,\ldots,\alpha_{D+1})\in\mathbb{Q}^{D+2}$, such that for n large enough

$$\alpha_0B^na_0b^{Dk^n}+\sum_{i=1}^{D}\alpha_iB^na_ib^{(D-i)k^n}-\alpha_{D+1}p_n=0.$$

Dividing by q_n and taking the limit as $n\to\infty$, we have

$$\alpha_0-\alpha_{D+1}\xi=0,$$

so that $\xi=F(1/b)\in\mathbb{Q}$, which completes the proof of the theorem. □

Remark 2.5.5. In very recent work, Adamczewski and Faverjon [4] have extended the results of Adamczewski and Bugeaud [3] and Bell, Bugeaud, and Coons [63] to the best possible. They have shown that a Mahler function evaluated at an algebraic number is either rational or transcendental. Moreover, their proof avoided the use of Schmidt's subspace theorem!

2.6 Diophantine Properties of Mahler Functions

In our final section, we look at the Diophantine properties of Mahler functions. We first look at how well a Mahler function can be approximated by rational functions. We then use that information to present the universal transcendence test for Mahler functions due to Bell and Coons [64]. Finally, we focus on the approximation of Mahler functions with algebraic functions.

2.6.1 Rational Approximation of Mahler Functions

Suppose we have a rational solution to (2.3). Our first result of this section gives bounds on the degrees of the numerator and the denominator of a rational Mahler function. This result can be found in Bell and Coons [64, Proposition 2].

Proposition 2.6.1. *Let $F(z) = P(z)/Q(z)$ be a rational k-Mahler function satisfying* (2.3) *with* $\gcd(P(z), Q(z)) = 1$ *and set* $H := \max\{\deg a_i(z) : i = 0, \ldots, d\}$. *Then*

$$\deg Q(z) \leqslant \lfloor H(k-1)/(k^{d+1} - 2k^d + 1) \rfloor,$$

and

$$\deg P(z) \leqslant \deg Q(z) + \lfloor H/k^{d-1}(k-1) \rfloor.$$

Proof. Write $F(z) = P(z)/Q(z)$ with $\gcd(P(z), Q(z)) = 1$. Since $F(z)$ is a power series, $Q(0) \neq 0$. Then we have

$$\sum_{i=0}^{d} a_i(z) P(z^{k^i})/Q(z^{k^i}) = 0.$$

In particular, if we multiply both sides by

$$R(z) := \prod_{j=0}^{d-1} Q(z^{k^j}),$$

we see that $Q(z^{k^d})$ must divide $a_d(z)P(z^{k^d})R(z)$. Since $\gcd(P(z), Q(z)) = 1$, we then have that $Q(z^{k^d})$ divides $a_d(z)R(z)$. Let D denote the degree of $Q(z)$. Then considering degrees, we have

$$k^d D \leqslant \deg a_d(z) + \deg R(z) \leqslant H + D + kD + \cdots + k^{d-1} D.$$

In other words, $(k^d - k^{d-1} - \cdots - 1)D \leqslant H$. Since

$$k^d - k^{d-1} - \cdots - 1 = k^d - (k^d - 1)/(k-1) \geqslant k^d(k-2)/(k-1),$$

if $k > 2$, we have

$$D \leqslant H(k-1)/k^d(k-2).$$

If $k = 2$, then all we get is $D \leqslant H$. In any case, setting

$$A(H,k,d) := \lfloor H(k-1)/(k^{d+1} - 2k^d + 1) \rfloor,$$

we have $D = \deg Q(z) \leqslant A(H,k,d)$.

Similarly, we can bound the degree of $P(z)$, but this is slightly more subtle. Suppose that $F(z) = P(z)/Q(z)$ has a pole at $z = \infty$ of order $M > 0$ with $Mk^{d-1} + H < Mk^d$. Since $F(z)$ satisfies (2.3), we have

$$F(z^{k^d})a_d(z) = -\sum_{i=0}^{d-1} a_i(z)F(z^{k^i}). \tag{2.28}$$

Now, the right-hand side of (2.28) has a pole at $z = \infty$ of order at most $k^{d-1}M + H$, and the left-hand side of (2.28) has a pole at $z = \infty$ of order at least $k^d M$. Since the equality (2.28) must hold, we conclude that $Mk^{d-1} + H \geqslant Mk^d$ and so $M \leqslant H/(k^d - k^{d-1})$. In other words,

$$\deg P(z) \leqslant \deg Q(z) + H/k^{d-1}(k-1),$$

which finishes the proof. □

2.6.2 A Transcendence Test for Mahler Functions

While we can bound the degrees of the numerator and the denominator of a rational Mahler function, unfortunately, deciding whether a general power series is a rational function is still not effectively determinable. After all, one can imagine that the function is very close to some rational function, and one must go very far out when looking at its coefficients to see that it is irrational. Fortunately, as Bell and Coons showed [64, Lemma 1], deciding whether a Mahler function is a rational function is effective.

Lemma 2.6.2. *Let $F(z)$ be a Mahler function satisfying* (2.3) *and as before set $H := \max\{\deg a_i(z) : i = 0, \ldots, d\}$. If $P(z)/Q(z)$ is a rational function with $Q(0) \neq 0$ and the degrees of $P(z)$ and $Q(z)$ are strictly less than some positive integer κ, then $F(z) - P(z)/Q(z)$ is either identically zero or it has a nonzero coefficient of z^i for some $i \leqslant H + \kappa \cdot k^{d+1}/(k-1)$.*

Proof. Suppose not. Then $F(z) - P(z)/Q(z) = z^M T(z)$ for some nonzero power series $T(z)$ with $T(0)$ nonzero and some $M > H + \kappa \cdot k^{d+1}/(k-1)$. Then we have

$$\sum_{i=0}^{d} a_i(z)P(z^{k^i})/Q(z^{k^i}) = \sum_{i=0}^{d} a_i(z) z^{Mk^i} T(z^{k^i}). \tag{2.29}$$

Notice the right-hand side of (2.29) has a zero of at least order M at $z = 0$. On the other hand, we can write the left-hand side of (2.29) as a rational function with denominator $Q(z)Q(z^k)\cdots Q(z^{k^d})$ and numerator

$$\sum_{i=0}^{d} a_i(z)P(z^{k^i})R_i(z),$$

where $R_i(z) := \prod_{j\neq i} Q(z^{k^j})$. Thus the numerator of the left-hand side of (2.29) when written in lowest terms has degree at most $H + \kappa(k^d + \cdots + k + 1)$. But this can occur only if the left-hand side of (2.29) is identically zero since $M > H + \kappa(k^{d+1} - 1)/(k-1)$, a contradiction. □

Proof (of Universal Test for Transcendence of Mahler Functions in Figure 2.4). Let **M** be the matrix formed in Step 2 of the universal transcendence test described in Figure 2.4.

Suppose that **M** does not have full rank. Then there is a nonzero row vector $\mathbf{q} := [q_0, q_1, \ldots, q_\kappa]$ such that $\mathbf{q} \cdot \mathbf{M} = 0$. In other words,

$$(q_\kappa + q_{\kappa-1}z + \cdots + q_0 z^\kappa)F(z)$$

has the property that 0 is the coefficient of z^i for $i = \kappa, \ldots, \kappa + H + \kappa(k^{d+1} - 1)/(k-1)$; that is, there is a polynomial $P(z)$ of degree less than κ such that

---

Universal test for transcendence of Mahler functions.

Let $k \geqslant 2$ and $d \geqslant 1$ be integers and $F(z)$ be a k-Mahler function satisfying

$$a_0(z)F(z) + a_1(z)F(z^k) + \cdots + a_d(z)F(z^{k^d}) = 0,$$

for polynomials $a_0(z), \ldots, a_d(z) \in \mathbb{C}[z]$. Set $H := \max\{\deg a_i(z) : i = 0, \ldots, d\}$ and

$$\kappa := \lfloor H(k-1)/(k^{d+1} - 2k^d + 1)\rfloor + \lfloor H/k^{d-1}(k-1)\rfloor + 1.$$

Step 1. Compute the coefficient, $f(i)$, of z^i of $F(z)$ for

$$i = 0, 1, \ldots, \kappa + H + \kappa(k^{d+1} - 1)/(k-1).$$

Step 2. Form the

$$(1+\kappa) \times (1 + H + \kappa(k^{d+1} - 1)/(k-1))$$

matrix **M** whose (i, j)-entry is $f(i + j - 2)$.

Step 3. Put this matrix in echelon form and verify whether it has full rank (*i.e.*, rank equal to $1+\kappa$).

Step 4. If it does, then $F(z)$ is transcendental; otherwise it is rational.

Fig. 2.4 Universal test for transcendence of Mahler functions of Bell and Coons.

$$(q_\kappa + q_{\kappa-1}z + \cdots + q_0 z^\kappa)F(z) - P(z)$$

has a zero of order at least $\kappa + H + \kappa(k^{d+1} - 1)/(k-1)$ at $z = 0$. Then $P(z)$ must have an order of zero at $z = 0$ that is at least as great as the order of zero of $Q(z) := q_\kappa + q_{\kappa-1}z + \cdots + q_0 z^\kappa$ at $z = 0$. This means that $P(z)/Q(z)$ can be reduced to be written as a ratio of polynomials of degree less than κ with the denominator being nonzero at $z = 0$ and such that $F(z) - P(z)/Q(z)$ has a zero at $z = 0$ of order at least $H + \kappa(k^{d+1} - 1)/(k-1)$. Lemma 2.6.2 gives then that $F(z) - P(z)/Q(z)$ is identically zero and hence $F(z)$ is rational.

Conversely, if $F(z)$ is rational, then we write $F(z) = P(z)/Q(z)$ with the degree of $P(z)$ and $Q(z)$ less than κ and use $Q(z)$ to provide a nonzero row vector $\mathbf{q}$ as above with $\mathbf{q} \cdot \mathbf{M} = 0$. □

2.6.3 Algebraic Approximation of Mahler Functions

The main result presented in this subsection is the recent result of Coons [164] concerning a zero order estimate for the difference of a Mahler function with an algebraic function.

As before, let $\nu : \mathbb{C}((z)) \to \mathbb{Z} \cup \{\infty\}$ be the valuation defined by $\nu(0) := \infty$ and

$$\nu\left(\sum c_n z^n\right) := \min\{i : c_i \neq 0\}$$

when $\sum_n c_n z^n$ is nonzero. Also, for $G(z)$ an algebraic function with minimal polynomial $P(z, y) \in \mathbb{C}[z, y]$, we call the value $\deg_y P(z, y)$ the *degree* of $G(z)$, and we call the value $\exp\left(\deg_z P(z, y)\right)$ the *height* of $G(z)$.

Theorem 2.6.3 (Coons). *If $F(z)$ is an irrational k-Mahler function of degree d_F and height A_F, and $G(z)$ is an algebraic function of degree at most n and height at most H_G, then*

$$\nu\big(F(z) - G(z)\big) \leqslant (d_F + 1) \cdot A_F \cdot n^{d_F+1} + \frac{k^{d_F+1} - 1}{k - 1} \cdot \log H_G \cdot n^{d_F}.$$

The order of Coons's bound is very similar to that of previous results on zero estimates of Mahler functions, though those focused on upper bounds for $\nu(Q(z, F(z)))$ for polynomials $Q(z, y) \in \mathbb{C}[z, y]$ and used quite deep methods, relying on the elimination-theoretic method of Nesterenko [443, 444]; see Becker [60], Nishioka [447], and Töpfer [566]. The approach taken by Coons is quite elementary and easily lends itself to exposition.

The case of rational functions was given by Bell's and Coons's result of the previous section (see Proposition 2.6.1). It is translated to the language of Theorem 2.6.3 as the following.

Lemma 2.6.4 (Bell and Coons). *Let $F(z)$ be an irrational k-Mahler function of degree d_F and height A_F, and let $P(z)/Q(z)$ be any rational function with $Q(0) \neq 0$. Then*

$$\nu\left(F(z) - \frac{P(z)}{Q(z)}\right) \leqslant A_F + \frac{k^{d_F+1} - 1}{k-1} \cdot \max\{\deg P(z), \deg Q(z)\}.$$

Theorem 2.6.3 is the generalization of this result to approximation by algebraic functions of arbitrary degree. To prove this generalization, we use a resultant argument.

Lemma 2.6.5. *Let $f(z)$ and $g(z)$ be two algebraic functions of degrees at least 2 satisfying polynomials of degrees Δ_f and Δ_g with coefficients of degree at most δ_f and δ_g, respectively. Then the algebraic function $f(z) + g(z)$ satisfies a polynomial of degree*

$$\Delta_{f+g} \leqslant \Delta_f \Delta_g$$

with coefficients of degree

$$\delta_{f+g} \leqslant \delta_f \Delta_g + \delta_g \Delta_f.$$

Proof. This result follows by using the Sylvester matrix to calculate a certain resultant. For R a ring and $P, Q \in R[y]$ with

$$P(y) = \sum_{i=0}^{\deg_y P} p_i y^i \quad \text{and} \quad Q(y) = \sum_{i=0}^{\deg_y Q} q_i y^i,$$

the resultant of P and Q with respect to the variable y is denoted by $\mathrm{res}_y(P, Q)$ and may be calculated as the determinant of the $(\deg_y Q + \deg_y P) \times (\deg_y Q + \deg_y P)$ Sylvester matrix; that is

$$\mathrm{res}_y(P, Q) := \det \begin{pmatrix} p_0 & p_1 & p_2 & \cdots & p_{\deg_y P} & & & \\ & p_0 & p_1 & p_2 & \cdots & p_{\deg_y P} & & \\ & & \ddots & \ddots & \ddots & & \ddots & \\ & & & p_0 & p_1 & p_2 & \cdots & p_{\deg_y P} \\ q_0 & q_1 & q_2 & \cdots & q_{\deg_y Q} & & & \\ & q_0 & q_1 & q_2 & \cdots & q_{\deg_y Q} & & \\ & & \ddots & \ddots & \ddots & & \ddots & \\ & & & q_0 & q_1 & q_2 & \cdots & q_{\deg_y Q} \end{pmatrix},$$

where there are $\deg_y Q$ rows of the coefficients of P and $\deg_y P$ rows of the coefficients of Q. Now suppose $R = \mathbb{C}[z, x]$, so that the entries of the above Sylvester matrix are polynomials in the variables z and x, and set $D(x, z) := \mathrm{res}_y(P, Q)$. Since polynomial degrees are additive, using the Leibniz formula for the determinant, we have immediately that

$$\deg_z D(x, z) \leqslant \deg_y Q \deg_z P + \deg_y P \deg_z Q \tag{2.30}$$

and

$$\deg_x D(x, z) \leqslant \deg_y Q \deg_x P + \deg_y P \deg_x Q. \tag{2.31}$$

The lemma now follows immediately by combining (2.30) and (2.31) with the fact that given algebraic functions $f(z), g(z) \in \mathbb{C}[[z]]$ and polynomials $P_f(z, y)$, $P_g(z, y) \in \mathbb{C}[z, y]$ with $P_f(z, f) = P_g(z, g) = 0$, the algebraic function $f(z) + g(z)$ is a root of the polynomial $\mathrm{res}_y(P_f(z, y), P_g(z, x - y))$ viewed as a polynomial in x. □

Using Lemma 2.6.4 as the result for algebraic functions of degree 1, we now focus on algebraic functions of degree at least 2.

Lemma 2.6.6. *Let $a_0(z), \ldots, a_d(z)$ be polynomials of degree at most A. If $G(z) \in \mathbb{C}[[z]]$ is an algebraic function of degree $\Delta_G \geqslant 2$ satisfying a minimal polynomial with coefficients of degree at most δ_g, then the function*

$$M_G(z) := \sum_{i=0}^{d} a_i(z) G(z^{k^i})$$

is an algebraic function satisfying a polynomial of degree $\Delta_{M_G} \leqslant \Delta_G^{d+1}$ whose coefficients have degree

$$\delta_{M_G} \leqslant (d + 1)A \cdot \Delta_G^{d+1} + \frac{k^{d+1} - 1}{k - 1} \cdot \delta_G \cdot \Delta_G^d.$$

Proof. Since $G(z)$ is an algebraic function, so is $\sum_{i=0}^{d} a_i(z) G(z^{k^i})$. One can easily gain information about the sum using the theory of resultants.

To get an upper bound on $\nu(M_G(z))$, we apply the idea of the previous paragraph by including the terms $G_i(z) := a_i(z) G(z^{k^i})$ one at a time. To do this, let

$$P_G(z, y) := g_{\Delta_G} y^{\Delta_G} + \cdots + g_1 y + g_0$$

be the minimal polynomial of $G(z)$. Here we have denoted the degree of $G(z)$ by Δ_G. Set $\delta_G := \deg_z P_G(z, y)$. Then

$$P_{G_i}(z, y) = a_i(z)^{\Delta_G} P_G(z^{k^i}, y/a_i(z))$$

is a polynomial with $P_{G_i}(z, G_i(z)) = 0$, where, of course, we only form this polynomial when $a_i(z) \neq 0$. Here, we have that $P_{G_i}(z, y)$ is still minimal with respect to the degree of y, but there is no guarantee that it is minimal with respect to the degree of z for this degree of y. However, we do have that the minimal polynomial of $G_i(z)$ divides $P_{G_i}(z, y)$ and the quotient is just a polynomial in z. In any case, the above gives that

$$\Delta_{G_i} := \deg_y P_{G_i}(z, y) = \deg_y P_G(z, y) = \Delta_G \tag{2.32}$$

and

$$\delta_{G_i} := \deg_z P_{G_i}(z, y) \leqslant A\Delta_G + k^i\delta_G. \tag{2.33}$$

The lemma now follows by combining (2.32) and (2.33) with Lemma 2.6.5. □

Lemma 2.6.7. *Let $G(z) \in \mathbb{C}[[z]]$ be an algebraic function of degree at least 2 satisfying the polynomial $P_G(z, y) = a_n(z)y^n + a_{n-1}(z)y^{n-1} + \cdots + a_1(z)y + a_0(z)$, with $a_0(z) \neq 0$. Then $\nu(G(z)) \leqslant \nu(a_0(z))$. In particular, $\nu(G(z)) \leqslant \deg_z P_G(z, y)$.*

Proof. Since $P_G(z, y)$ is a minimal polynomial, we have $a_0(z) \neq 0$. We thus have, identically,

$$\left(a_n(z)G(z)^{n-1} + a_{n-1}(z)G(z)^{n-2} + \cdots + a_1(z)\right) G(z) = -a_0(z).$$

The fact $G(z), a_n(z), \ldots, a_0(z) \in \mathbb{C}[[z]]$ then gives

$$\nu\left(a_n(z)G(z)^{n-1} + a_{n-1}(z)G(z)^{n-2} + \cdots + a_1(z)\right) + \nu(G(z)) = \nu(a_0(z)),$$

which proves the lemma, since each of the terms is a nonnegative integer. □

Proof (of Theorem 2.6.3). Let $F(z)$ be a k-Mahler function satisfying (2.3) of degree d_F and height A_F and let $G(z)$ be an algebraic function of degree at most n and height at most H_G. Since by Lemma 2.6.4, the theorem holds for $n = 1$, we may assume without loss of generality that $n \geqslant 2$.

Set $M := \nu(F(z) - G(z))$, and write

$$F(z) - G(z) = z^M T(z),$$

where $T(z) \in \mathbb{C}[[z]]$ with $T(0) \neq 0$. Then also

$$\sum_{i=0}^{d} a_i(z)F(z^{k^i}) - \sum_{i=0}^{d} a_i(z)G(z^{k^i}) = \sum_{i=0}^{d} a_i(z)z^{k^iM}T(z^{k^i}),$$

which since $F(z)$ satisfies (2.3) reduces to

$$M_G(z) := \sum_{i=0}^{d} a_i(z)G(z^{k^i}) = -\sum_{i=0}^{d} a_i(z)z^{k^iM}T(z^{k^i}).$$

This immediately implies that

$$\nu(F(z) - G(z)) = M \leqslant \nu\,(M_G(z)) \leqslant \delta_{M_G},$$

where the last inequality follows from Lemma 2.6.7. By definition, $\delta_G = \log H_G$, hence applying Lemma 2.6.6 proves the theorem. □

The most important term in the inequality of Theorem 2.6.3 is the rightmost term. One of the most important questions in the algebraic approximation of Mahler functions concerns the degree of n in this term. The current best known upper bound is d_F, but a lower value may be true. In particular, one may expect a "Roth-type" upper bound.

Question 2.6.8. If $F(z)$ is an irrational Mahler function and $G(z)$ is an algebraic function of degree at most n and height at most H_G where $\log H_G \geqslant n \geqslant 1$, then is there a constant $c > 0$ such that $\nu(F(z) - G(z)) \leqslant c \cdot \log H_G \cdot n$?

Chapter 3
First-Order Logic and Numeration Systems

Émilie Charlier

Abstract The Büchi-Bruyère theorem states that a subset of $\mathbb{N}^d$ is b-recognizable if and only if it is b-definable. This result is a powerful tool for showing that many properties of b-automatic sequences are decidable. Going a step further, first-order logic can be used to show that many enumeration problems of b-automatic sequences can be described by b-regular sequences. The latter sequences can be viewed as a generalization of b-automatic sequences to integer-valued sequences. These techniques were extended to two wider frameworks: U-recognizable subsets of $\mathbb{N}^d$ and β-recognizable subsets of $\mathbb{R}^d$. In the second case, real numbers are represented by infinite words, and hence, the notion of β-recognizability is defined by means of Büchi automata. Again, logic-based characterization of U-recognizable (resp. β-recognizable) sets allows us to obtain various decidability results. The aim of this chapter is to present a survey of this very active research domain.

3.1 Introduction

In computer science and in mathematics in general, we are concerned with the following questions: How do we have sets of numbers at our disposal? How can we manipulate them? Which sets of numbers should be considered simple? In which sense? In order to approach such questions, we first need to represent numbers. The basic consideration is as follows: properties of numbers are translated into syntactical (or combinatorial) properties of their representations. This is where numeration systems come into play. For example, the famous theorem of Cobham (and Semenov for its multidimensional version) tells us that nontrivial properties of numbers are dependent on the base we choose.

In this chapter, we will consider multidimensional subsets of numbers whose sets of representations are accepted by finite automata. Representations of numbers will always be taken from one of the following families of numeration systems: the unary

É. Charlier (✉)
Department of Mathematics, University of Liège, Allée de la découverte 12 (B37), B-4000 Liège, Belgium
e-mail: echarlier@ulg.ac.be

V. Berthé, M. Rigo (eds.), *Sequences, Groups, and Number Theory*,
Trends in Mathematics, https://doi.org/10.1007/978-3-319-69152-7_3

systems, the integer bases $b \geq 2$, and, more generally, the positional numeration systems based on increasing sequences $U = (U_n)_{n\geq 0}$, the abstract numeration systems S based on regular languages, and finally the real bases $\beta > 1$. Depending on the cases, we shall refer to such sets as 1-recognizable sets, b-recognizable sets, U-recognizable sets, S-recognizable sets, and β-recognizable sets.

Many descriptions of recognizable sets were given in various works [78, 96, 113, 114, 211, 503]. Here, we focus on characterizations of recognizable subsets (first of $\mathbb{N}^d$ and then of $\mathbb{R}^d$) in terms of first-order logic. We start by presenting the Büchi-Bruyère theorem, which states that a subset of $\mathbb{N}^d$ is b-recognizable if and only if it is b-definable, that is, definable by a first-order formula of the structure $\langle \mathbb{N}, +, V_b \rangle$ where V_b is a base-dependent predicate (see below for formal definitions). We explain how this result turns out to be a powerful tool for showing that many properties of b-automatic sequences are decidable. We illustrate our purpose with many examples of decidable problems on b-automatic sequences. Going a step further, we show that first-order logic can also be used to prove that many enumeration problems of b-automatic sequences can be described by b-regular sequences. The latter sequences are at the core of Chapters 2 and 4. First-order logic is also mentioned in Chapters 9 and 10 in the context of the domino problem and of Wang tiles.

In the last (and longest) part of this chapter, we give an extensive presentation of (multidimensional) β-recognizable sets of real numbers. Those sets are defined by means of Büchi automata. Again, we give a logic-based characterization of these sets and show how we can use it to obtain various decidability results. We end by showing the links between the so-called β-self-similar sets, the attractors of some (base-dependent) graph-directed iterated function systems, and certain sets recognizable by Büchi automata. Let us mention here that the numeration systems in real bases $\beta > 1$ are referred to as the main motivation of Chapter 7.

Besides these logic-based characterizations and their applications, we mention (usually without proofs) various results concerning recognizable sets. Among them, in the vein of Eilenberg's result [211], we explicitly list the possible growth functions of (unidimensional) S-recognizable sets. Let us emphasize that this is done in the very general framework of abstract numeration systems and, thus, encompasses the previous known results about b-recognizable sets only. In particular, this result permits us to exclude right away a huge amount of (unidimensional) subsets from the class of S-recognizable sets, and further, it also permits us to exhibit many subsets which are never S-recognizable, that is, no matter which abstract numeration system we choose. Finally, let us mention that along the lines, we present four open problems.

3.2 Recognizable Sets of Nonnegative Integers

Finite automata may be seen as the simplest devices. They have only finite memory, and they are only able to read words and accept or reject them in the end. Regular languages, *i.e.*, languages accepted by finite automata, form the bottom level of

Chomsky–Schützenberger hierarchy. For this reason, it makes sense to consider the following definition of "simple sets" of numbers. A subset X of $\mathbb{N}$ is said to be recognizable with respect to a given numeration system rep: $\mathbb{N} \to A^*$ if the language

$$\{\text{rep}(n) \mid n \in X\} \subseteq A^*$$

is accepted by a finite automaton.

In order to be able to recognize multidimensional sets of numbers by means of finite automata, we need to represent tuples of numbers by finite words. The classical way to manage this is to introduce a padding symbol, which allows each component to be represented by words of the same length. A subset X of $\mathbb{N}^d$ is recognizable with respect to a numeration system rep: $\mathbb{N} \to A^*$ if the language

$$\{(\text{rep}(n_1), \ldots, \text{rep}(n_d))^\# \mid (n_1, \ldots, n_d) \in X\} \subseteq ((A \cup \{\#\})^d)^*,$$

where # is some padding symbol, is accepted by a finite automaton.

Formally, for alphabets $A_1, \ldots, A_d$ and for a letter #, the *padding map* $(\cdot)^\#: A_1^* \times \cdots \times A_d^* \to ((A_1 \cup \{\#\}) \times \cdots \times (A_d \cup \{\#\}))^*$ is defined by

$$(w_1, \ldots, w_d)^\# = (\#^{m-|w_1|} w_1, \ldots, \#^{m-|w_d|} w_d),$$

where $m = \max\{|w_1|, \ldots, |w_d|\}$. In this way, from a subset R of the monoid $A_1^* \times \cdots \times A_d^*$, we create a language

$$R^\# = \{(w_1, \ldots, w_d)^\# \mid (w_1, \ldots, w_d) \in R\} \subseteq ((A_1 \cup \{\#\}) \times \cdots \times (A_d \cup \{\#\}))^*.$$

In particular, if $\# \notin \cup_{i=1}^d A_i$, then no word in $R^\#$ contains the letter $(\#, \ldots, \#)$.

Here and throughout the chapter, d designates a dimension, *i.e.*, an integer greater than or equal to 1. We will also use the notation

$$\boldsymbol{\#} = \underbrace{(\#, \ldots, \#)}_{d \text{ times}}, \quad \mathbf{0} = \underbrace{(0, \ldots, 0)}_{d \text{ times}}, \quad \boldsymbol{\star} = \underbrace{(\star, \ldots, \star)}_{d \text{ times}},$$

where # and $\star$ are fixed symbols.

3.2.1 Unary Representations

Perhaps the simplest way of representing a natural number n is to repeat a symbol n times. This approach presents an obvious drawback: it requires way too much memory space in practice to store a number and, even worse, to do computations with them. Even though they are highly unpractical, unary representations are of some theoretical interest, for example in computability theory. Let a be some fixed symbol. The *unary numeration system* $\mathrm{rep}_1\colon \mathbb{N} \to a^*$ is defined by $\mathrm{rep}_1(n) = a^n$ for all $n \in \mathbb{N}$. The set of all possible representations is $L_1 = \mathrm{rep}_1(\mathbb{N}) = a^*$.

Definition 3.2.1. A subset X of $\mathbb{N}^d$ is 1-*recognizable* if the language $\mathrm{rep}_1(X)$ is regular.

In dimension 1, the 1-recognizable sets are exactly the finite union of arithmetic progressions, as they correspond to regular languages over a unary alphabet. In the multidimensional case, it is already more complicated to capture the essence of 1-recognizable sets; see Section 3.2.5.

3.2.2 Integer Bases

Throughout this chapter, b designates an integer greater than or equal to 2.

The *integer base b numeration system* $\mathrm{rep}_b\colon \mathbb{N} \to A_b^*$ is defined as follows: positive integers n are represented by finite words $\mathrm{rep}_b(n) = c_\ell \cdots c_1 c_0$ over the alphabet $A_b = \{0, 1, \ldots, b-1\}$ obtained from the greedy algorithm:

$$n = \sum_{i=0}^{\ell} c_i\, b^i.$$

By convention $\mathrm{rep}_b(0) = \varepsilon$. The greedy algorithm only imposes having a nonzero leading digit c_ℓ, and the set of all *greedy* (or *canonical*) *b-representations* is

$$L_b = \mathrm{rep}_b(\mathbb{N}) = A_b^* \setminus 0A_b^*.$$

We may also consider non-greedy b-representations. The *evaluation map* $\mathrm{val}_b\colon \mathbb{N}^* \to \mathbb{N}$ is defined by $\mathrm{val}_b(c_\ell \cdots c_1 c_0) = \sum_{i=0}^{\ell} c_i\, b^i$. Any word $c_\ell \cdots c_1 c_0 \in \mathbb{N}^*$ such that $\mathrm{val}_b(c_\ell \cdots c_1 c_0) = n$ is called a *b-representation* of n.

We extend the definitions of the functions rep_b and val_b to the multidimensional setting as follows (and we keep the same notation):

$$\mathrm{rep}_b\colon \mathbb{N}^d \to (A_b^{\,d})^*,\ (n_1, \ldots, n_d) \mapsto (\mathrm{rep}_b(n_1), \ldots, \mathrm{rep}_b(n_d))^0$$

$$\mathrm{val}_b\colon (\mathbb{N}^d)^* \to \mathbb{N}^d,\ (w_1, \ldots, w_d) \mapsto (\mathrm{val}_b(w_1), \ldots, \mathrm{val}_b(w_d)).$$

Let us emphasize that the components of $\mathrm{rep}_b(\mathbf{n})$ are padded with zeros. Also note that $(w_1, \ldots, w_d) \in (\mathbb{N}^d)^*$ implies that $|w_1| = \cdots = |w_d|$.

The following proposition is a generalization of Proposition V.3.1 in [211].

Proposition 3.2.2. *Let # be a symbol not belonging to A_b. For any subset X of $\mathbb{N}^d$, the following are equivalent:*

1. *The language $\mathrm{rep}_b(X)$ is regular.*
2. *The language $\mathbf{0}^*\mathrm{rep}_b(X)$ is regular.*

3. *There exists a regular language* $L \subseteq (A_b^d)^*$ *such that* $\mathbf{0}^*(\mathbf{0}^*)^{-1}L = \mathbf{0}^*\mathrm{rep}_b(X)$.
4. *There exists a regular language* $L \subseteq (A_b^d)^*$ *such that* $\mathrm{val}_b(L) = X$.
5. *The language* $\{(\mathrm{rep}_b(n_1), \dots, \mathrm{rep}_b(n_d))^\# \mid (n_1, \dots, n_d) \in X\}$ *is regular.*
6. *The language* $\#^*\{(\mathrm{rep}_b(n_1), \dots, \mathrm{rep}_b(n_d))^\# \mid (n_1, \dots, n_d) \in X\}$ *is regular.*
7. *There exists a regular language* $L \subseteq ((A_b \cup \{\#\})^d)^*$ *such that*

$$\#^*(\#^*)^{-1}L = \#^*\{(\mathrm{rep}_b(n_1), \dots, \mathrm{rep}_b(n_d))^\# \mid (n_1, \dots, n_d) \in X\}.$$

Proof. If no word of a language $L \subseteq A^*$ starts with a specific letter $a \in A$, then L is regular if and only if a^*L is as well. This shows 1 $\iff$ 2 and 5 $\iff$ 6. For 1 $\implies$ 4, take $L = \mathrm{rep}_b(X)$. For 4 $\implies$ 3, observe that if $X = \mathrm{val}_b(L)$ for some regular language $L \subseteq (A_b^d)^*$, then $\mathbf{0}^*(\mathbf{0}^*)^{-1}L = \mathbf{0}^*\mathrm{rep}_b(X)$. 3 $\implies$ 2 is clear. For 5 $\implies$ 7, take $L = \{(\mathrm{rep}_b(n_1), \dots, \mathrm{rep}_b(n_d))^\# \mid (n_1, \dots, n_d) \in X\}$. 7 $\implies$ 6 is clear. Finally we show 1 $\iff$ 5. Given a DFA accepting $\{(\mathrm{rep}_b(n_1), \dots, \mathrm{rep}_b(n_d))^\# \mid (n_1, \dots, n_d) \in X\}$, we modify it by replacing every # with 0 in every transition. The resulting automaton is an NFA accepting $\mathrm{rep}_b(X)$. Now suppose that $\mathscr{A}$ is a DFA accepting $\mathrm{rep}_b(X)$. We modify $\mathscr{A}$ by replacing every transition labeled $(a_1, \dots, a_d) \in A_b^d$ with k components equal to 0 with 2^k transitions obtained by placing either 0 or # in every component where there was a 0. Let $\mathscr{B}$ denote the resulting DFA. Now we can build a DFA $\mathscr{C}$ accepting the words in $((A_b \cup \{\#\})^d)^*$ such that, in every component, each occurrence of # is preceded by # or by nothing, and the last occurrence of # is not followed by 0. The language $\{(\mathrm{rep}_b(n_1), \dots, \mathrm{rep}_b(n_d))^\# \mid (n_1, \dots, n_d) \in X\}$ is the intersection of the languages accepted by $\mathscr{B}$ and $\mathscr{C}$; hence, it is regular. □

Definition 3.2.3. A subset X of $\mathbb{N}^d$ is *b-recognizable* if any of the assertions of Proposition 3.2.2 is satisfied.

Remark 3.2.4. The integer base b numeration systems have the remarkable property that $\mathbb{N}^d$ is b-recognizable since $\mathbf{0}^*\mathrm{rep}_b(\mathbb{N}^d) = (A_b^d)^*$. It is also true that $\mathrm{val}_b^{-1}(X) = \mathbf{0}^*\mathrm{rep}_b(X)$ for any subset X of $\mathbb{N}^d$. The latter fact was actually used in the proof of Proposition 3.2.2 (it is needed in the implication 4 $\implies$ 3).

It is equivalent to say that the characteristic sequence $\chi_X \colon \mathbb{N}^d \to \{0, 1\}$ is b-automatic:

Definition 3.2.5. A sequence $x \colon \mathbb{N}^d \to \mathbb{N}$ is *b-automatic* if there exists a finite deterministic automaton with output (DFAO for short) $\mathscr{M} = (Q, q_0, A_b^d, \delta, A, \tau)$ such that, for all $\mathbf{n} \in \mathbb{N}^d$, $x(\mathbf{n}) = \tau(\delta(q_0, \mathrm{rep}_b(\mathbf{n})))$.

Note that a DFAO being finite by definition, the image of a b-automatic sequence is necessarily finite. Therefore, b-automatic sequences may be viewed as multidimensional infinite words over a finite alphabet A.

Example 3.2.6. The DFAO of Figure 3.1 generates the sequence $011010111\cdots$ when reading the greedy 2-representations of the nonnegative integers.

Proposition 3.2.7. *Let X be a subset of $\mathbb{N}^d$. Then X is b-recognizable if and only if χ_X is b-automatic.*

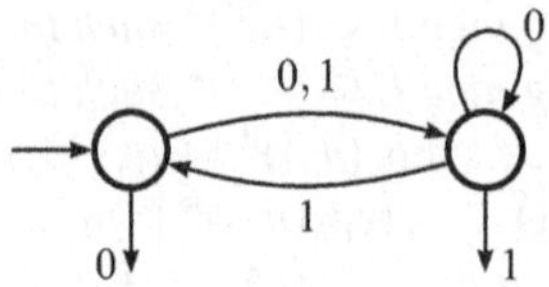

Fig. 3.1 A DFAO generating some 2-recognizable set

Proof. In order to build a DFAO generating χ_X starting from a DFA accepting $\text{rep}_b(X)$, it suffices to output 1 when ending in a terminal state and to output 0 when ending in a nonterminal state. In particular, the obtained DFAO outputs 0 if we enter a non-greedy b-representation. The other direction works well because $\mathbb{N}^d$ is b-recognizable. By declaring terminal those states outputting 1 and nonterminal those states outputting 0, we obtain a DFA that might accept non-greedy b-representations as well. But if L is the accepted language of this DFA, then $\text{val}_b(L) = X$ (which is the fourth item in Proposition 3.2.2). □

Similarly, we have the following result.

Proposition 3.2.8. *Let A be a finite alphabet and let $x: \mathbb{N}^d \to A$. Then x is b-automatic if and only if every subset $x^{-1}(a)$ of $\mathbb{N}^d$ (for $a \in A$) is b-recognizable.*

Proof. In order to build DFAs accepting a language L such that $\text{val}(L) = x^{-1}(a)$ starting from a DFAO generating x, it suffices to declare a state to be final if and only if the corresponding output is a. For the other direction, let $A = \{a_1, \ldots, a_k\}$, and for each i, let $\mathcal{M}_i = (Q_i, q_{0,i}, F_i, A_b^d, \delta_i)$ be a DFA accepting $\mathbf{0}^*\text{rep}_b(x^{-1}(a_i))$. Let $\mathcal{M} = \mathcal{M}_1 \times \cdots \times \mathcal{M}_k$. For all $\mathbf{n} \in \mathbb{N}^d$, the state reached from the initial state $(q_{0,1}, \ldots, q_{0,k})$ after reading $\text{rep}_b(\mathbf{n})$ contains exactly one final component (in some $\mathcal{M}_i$). We define $\tau(q_1, \ldots, q_k) = a_i$ if there is exactly one i such that $q_i \in F_i$ (τ is undefined on other states). Then the DFAO obtained from $\mathcal{M}$ and τ generates x. □

One way to describe the b-recognizable sets is to study their growth functions.

Definition 3.2.9. For a subset X of $\mathbb{N}$, we let $t_X(n)$ denote the $(n+1)$st term of X. The map $t_X: \mathbb{N} \to \mathbb{N}$ is called the *growth function* of X.

Theorem 3.2.10. *[211] Any b-recognizable subset X of $\mathbb{N}$ satisfies either*

$$\limsup_{n\to+\infty} (t_X(n+1) - t_X(n)) < +\infty, \text{ or}$$

$$\limsup_{n\to+\infty} \frac{t_X(n+1)}{t_X(n)} > 1.$$

Thanks to this result, examples of sets that are not b-recognizable for any b have been exhibited. The set $\{n^2 : n \in \mathbb{N}\}$ of squares is such an example.

There are several equivalent definitions of b-recognizable sets using logic, morphisms, finiteness of the b-kernel, or formal series. We refer the reader to the survey [114] for an extensive presentation. The equivalence with b-definable sets will be discussed in Section 3.3.

3.2.3 Positional Numeration Systems

A *positional numeration system* $\text{rep}_U\colon \mathbb{N} \to A_U^*$ is based on an increasing sequence $U\colon \mathbb{N} \to \mathbb{N}$ such that $U(0) = 1$ and $C_U = \sup_{i\geq 0} \lceil \frac{U(i+1)}{U(i)} \rceil < +\infty$. Positive integers n are represented by finite words $\text{rep}_U(n) = c_\ell \cdots c_1 c_0$ over the alphabet $A_U = \{0, 1, \ldots, C_U - 1\}$ obtained from the greedy algorithm:

$$n = \sum_{i=0}^{\ell} c_i\, U(i).$$

By convention $\text{rep}_U(0) = \varepsilon$. The greedy algorithm imposes having a nonzero leading digit c_ℓ and that, for every $0 \leq j \leq \ell$, $\sum_{i=0}^{j} c_i U(i) < U(j+1)$. A description of the set of all *greedy* (or *canonical*) *U-representations* $L_U = \text{rep}_U(\mathbb{N})$ highly depends on the base sequence U. The evaluation map is $\text{val}_U\colon \mathbb{N}^* \to \mathbb{N},\ c_\ell \cdots c_1 c_0 \mapsto \sum_{i=0}^{\ell} c_i\, U(i)$. Any word $c_\ell \cdots c_1 c_0 \in \mathbb{N}^*$ such that $\text{val}_U(c_\ell \cdots c_1 c_0) = n$ is called a *U-representation* of n.

Example 3.2.11. If $U\colon i \mapsto b^i$, then we recover the integer base b numeration systems presented in the previous section.

Example 3.2.12. The positional numeration system rep_F based on the Fibonacci sequence $F\colon \mathbb{N} \to \mathbb{N}$ defined by $F(0) = 1,\ F(1) = 2$ and $F(i+2) = F(i+1) + F(i)$ for $i \in \mathbb{N}$, is called *the Zeckendorf numeration system* [592]. Zeckendorf proved that the set of all greedy F-representations is the language of the finite words over $\{0, 1\}$ that do not begin in 0 and that do not contain the word 11 as a factor: $L_F = 1\{0, 01\}^* \cup \{\varepsilon\}$. This language is accepted by the DFA of Figure 3.2.

Again, we extend the definitions of rep_U and val_U to the multidimensional setting:

$$\text{rep}_U\colon \mathbb{N}^d \to (A_U^d)^*,\ (n_1, \ldots, n_d) \mapsto (\text{rep}_U(n_1), \ldots, \text{rep}_U(n_d))^0$$
$$\text{val}_U\colon (\mathbb{N}^d)^* \to \mathbb{N}^d,\ (w_1, \ldots, w_d) \mapsto (\text{val}_U(w_1), \ldots, \text{val}_U(w_d)).$$

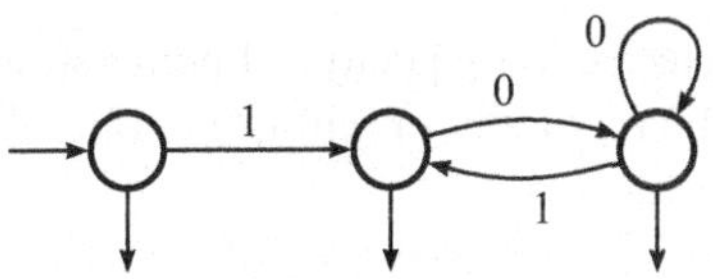

Fig. 3.2 A DFA accepting the Zeckendorf representations of nonnegative integers

We give the statement of the following proposition without proof since it is similar to that of Proposition 3.2.2.

Proposition 3.2.13. *Let # be a symbol not belonging to A_U. For any subset X of $\mathbb{N}^d$, the following are equivalent:*

1. *The language* $\mathrm{rep}_U(X)$ *is regular.*
2. *The language* $\mathbf{0}^*\mathrm{rep}_U(X)$ *is regular.*
3. *There exists a regular language* $L \subseteq (A_U^d)^*$ *such that*

$$\mathbf{0}^*(\mathbf{0}^*)^{-1}L = \mathbf{0}^*\mathrm{rep}_U(X). \tag{3.1}$$

4. *The language* $\{(\mathrm{rep}_U(n_1), \ldots, \mathrm{rep}_U(n_d))^{\#} \mid (n_1, \ldots, n_d) \in X\}$ *is regular.*
5. *The language* $\#^*\{(\mathrm{rep}_U(n_1), \ldots, \mathrm{rep}_U(n_d))^{\#} \mid (n_1, \ldots, n_d) \in X\}$ *is regular.*
6. *There exists a regular language* $L \subseteq ((A_U \cup \{\#\})^d)^*$ *such that*

$$\#^*(\#^*)^{-1}L = \#^*\{(\mathrm{rep}_U(n_1), \ldots, \mathrm{rep}_U(n_d))^{\#} \mid (n_1, \ldots, n_d) \in X\}.$$

Observe that we lost the fourth characterization of Proposition 3.2.2. For integer bases, the non-greedy representations are only those with leading zeros. For positional numeration systems, there are other kinds of non-greedy representations. For example, 100 and 11 are both F-representations of 3. In general, if $X \subseteq \mathbb{N}$ and $L \subseteq (A_U^d)^*$ are such that $X = \mathrm{val}_U(L)$, then we do not know that (3.1) holds for the same L.

Definition 3.2.14. A subset X of $\mathbb{N}^d$ is *U-recognizable* if any of the assertions of Proposition 3.2.13 is satisfied.

Let us mention two open problems concerning positional numeration systems. The first one was already reported in [78, Chapter 2]. As far as we know, the best results achieved in this area are those of [297].

Problem 3.2.15. Characterize those positional numeration systems rep_U such that $\mathbb{N}$ is U-recognizable.

Here we propose another related problem. However, an answer to any of these two problems does not seem to provide a straightforward answer to the other. We first give a remark.

Remark 3.2.16. For any subset X of $\mathbb{N}^d$, we have $\mathrm{rep}_U(X) = \mathrm{val}_U^{-1}(X) \cap \mathrm{rep}_U(\mathbb{N}^d)$. Therefore, whenever $\mathbb{N}$ is U-recognizable (and hence $\mathbb{N}^d$ is as well), then for any subset X of $\mathbb{N}^d$, the regularity of $\mathrm{val}_U^{-1}(X)$ implies that of $\mathrm{rep}_U(X)$. However, there is no evidence that the converse should be true.

Problem 3.2.17. Characterize those positional numeration systems rep_U such that, for any subset X of $\mathbb{N}^d$, the regularity of $\mathrm{rep}_U(X)$ implies that of $\mathrm{val}_U^{-1}(X)$.

3.2.4 Abstract Numeration Systems

In this very general framework, the question is reversed. We first choose a language L, the basic assumption being that L is regular, and then we declare L to form the set of all valid representations of nonnegative integers, with the rule:

$$\forall n, m \in L,\ n < m \iff rep_S(n) \prec rep_S(m).$$

Formally, an *abstract numeration system* S is given by a regular language L over a totally ordered alphabet $(A, <)$. A nonnegative integer n is represented by the $(n+1)$st word in L in radix (or genealogical) order $\prec$. The question is now to – efficiently – describe the map $n \mapsto \mathrm{rep}_S(n)$, which of course depends on the choice of S.

Definition 3.2.18. A subset X of $\mathbb{N}^d$ is *S-recognizable* if the language

$$\{(\mathrm{rep}_S(n_1), \ldots, \mathrm{rep}_S(n_d))^\# \mid (n_1, \ldots, n_d) \in X\} \subseteq ((A \cup \{\#\})^d)^*$$

is regular, where # is some padding symbol not contained in the numeration alphabet A.

Note that, for a fixed S, the choice of padding the representations to the right or to the left is arbitrary and gives two different notions of S-recognizability. At first glance, one could think that we just have to consider the reversed representations, but the numeration language L might not be closed under reversal, and even if it were, then the order of the representations could change. Recall that if $w = a_1 \cdots a_{|w|}$, then $\widetilde{w} = a_{|w|} \cdots a_1$.

Example 3.2.19. Consider $S = (a^*b^* \cup a^*c^*,\ a < b < c)$. Then the pair $(6, 9)$ is represented by $\binom{\#ac}{aaa} = \binom{\#}{a}\binom{a}{a}\binom{c}{a}$. If we had chosen a right padding instead, $(6, 9)$ would have been represented by $\binom{ac\#}{aaa}$, which is not equal to $\binom{\#ac}{aaa}^{\sim} = \binom{ca\#}{aaa}$. In fact, the latter word is not even the S-representation of any pair of nonnegative integers since ca does not belong to the numeration language.

Abstract numeration systems encompass positional numeration systems having a regular numeration language; see Problem 3.2.15. The next example illustrates that the converse is not true.

Example 3.2.20. We saw that the set $X = \{n^2 : n \in \mathbb{N}\}$ is not b-recognizable for any b. However, this set is S-recognizable for the abstract numeration system S of Example 3.2.19 since $\mathrm{rep}_S(X) = a^*$.

More generally, we have the following result.

Theorem 3.2.21 ([502, 554]). *For any polynomial $P \in \mathbb{Q}[x]$ such that $P(\mathbb{N}) \subseteq \mathbb{N}$, there exists an abstract numeration system S such that P is S-recognizable.*

Describing the S-recognizable subsets of $\mathbb{N}^d$ is not easy in general. In the vein of Theorem 3.2.10, the following result, which we give without proof, lists the possible growth orders of such sets. These growth orders depend on the growth of the numeration language, which is either polynomial or exponential as shown by the following lemma.

For any language L over an alphabet A and any nonnegative integer n, we let $\mathbf{v}_L(n)$ denote the number of words of length less than or equal to n in L.

Lemma 3.2.22. *For all regular languages L, there exist $p, c \in \mathbb{N}$ and $\alpha, a_0, \ldots, a_{p-1} \in \mathbb{R}_{\geq 0}$ with $p, \alpha \geq 1$ such that*

$$\forall i \in \{0, \ldots, p-1\},\ \mathbf{v}_L(np+i) \sim a_i n^c \alpha^n \quad (n \to +\infty). \tag{3.2}$$

Proof. The formal series $\sum_{n\geq 0} \mathbf{v}_L(n) x^n$ are $\mathbb{N}$-recognizable for all regular languages L; see, for instance, [77]; also see Section 3.4.1. Since $(\mathbf{v}_L(n))_{n\geq 0}$ are nondecreasing sequences, the lemma follows from [520, Theorem II.10.2]. □

Theorem 3.2.23 ([147]). *Let $S = (L, A, <)$ be an abstract numeration system and let X be an infinite S-recognizable subset of $\mathbb{N}$. Suppose that* (3.2) *holds and that*

$$\forall j \in \{0, \ldots, q-1\},\ \mathbf{v}_{\mathrm{rep}_S(X)}(nq+j) \sim b_j n^d \beta^n \quad (n \to +\infty), \tag{3.3}$$

for some $q, d \in \mathbb{N}$ and some $\beta, b_0, \ldots, b_{q-1} \in \mathbb{R}_{\geq 0}$ with $q, \beta \geq 1$. Whenever $\beta > 1$, we have

$$t_X(n) = \Theta\big((\log(n))^{c-df} n^f\big), \ \textit{with}\ f = \frac{\log(\sqrt[p]{\alpha})}{\log(\sqrt[q]{\beta})}.$$

If $\beta = 1$, then

$$t_X(n) = \Theta\big(n^{\frac{c}{d}} (\sqrt[p]{\alpha})^{\Theta(n^{1/d})}\big).$$

If moreover $q = 1$, then

$$t_X(n) = \Theta\big(n^{\frac{c}{d}} (\sqrt[p]{\alpha})^{(1+o(1))(\frac{n}{b_0})^{1/d}}\big).$$

Definition 3.2.24. Two real numbers α and β different from 1 are said to be *multiplicatively dependent* if $\alpha = \beta^r$ for some $r \in \mathbb{Q}$, or, equivalently, if $\frac{\log \alpha}{\log \beta} \in \mathbb{Q}$. Otherwise, α and β are said to be *multiplicatively independent*.

The following corollary of Theorem 3.2.23 considers the case of a polynomial numeration language.

Corollary 3.2.25. *Let $S = (L, A, <)$ be an abstract numeration system built on a polynomial regular language, and let X be an infinite S-recognizable subset of $\mathbb{N}$. Then $t_X(n) = \Theta(n^r)$ for some rational $r \geq 1$.*

Proof. By Lemma 3.2.22, the growth functions $\mathbf{v}_L(n)$ and $\mathbf{v}_{\mathrm{rep}_S(X)}(n)$ satisfy (3.2) and (3.3), respectively. The fact that L is polynomial means that $\alpha = 1$. As $1 \le \beta \le \alpha$, we have $\beta = 1$ as well. Then from Theorem 3.2.23, we obtain $t_X(n) = \Theta(n^{\frac{c}{d}})$. □

By Theorem 3.2.21, we know that any set of the form $\{n^k \mid n \in \mathbb{N}\}$, with $k \in \mathbb{N}$, is S-recognizable for some S. In the constructions of [502, 554], the numeration languages are of polynomial growth. Consider the base 4 numeration system, whose numeration language is of exponential growth. By Theorem 3.2.23, if $X = \mathrm{val}_4(\{1,3\}^*)$, then $t_X(n) = \Theta(n^2)$. Indeed, with the notation of Theorem 3.2.23, we have $\alpha = 4$, $\beta = 2$, $p = q = 1$ (hence $f = 2$), and $c = d = 0$.

Proposition 3.2.26. *For every rational number $r \ge 1$, there exists an abstract numeration system S built on a polynomial regular language and an infinite S-recognizable subset X of $\mathbb{N}$ such that $t_X(n) = \Theta(n^r)$.*

Proof. Fix a rational number $r \ge 1$. Write $r = \frac{c}{d}$ where c and d are positive integers. Define $\mathcal{B}_\ell$ to be the bounded language $a_1^* a_2^* \cdots a_\ell^*$. We have $\mathbf{v}_{\mathcal{B}_\ell}(n) = \binom{n+\ell}{\ell}$ for all $\ell \ge 1$ and $n \in \mathbb{N}$ (e.g., see [149, Lemma 1]). Let S be the abstract numeration system built on $\mathcal{B}_c$ with the order $a_1 < a_2 < \cdots < a_c$, and let $X = \mathrm{val}_S(\mathcal{B}_d)$ (since $c \ge d$, we have $\mathcal{B}_d \subseteq \mathcal{B}_c$). Hence we have $\mathbf{v}_{\mathcal{B}_c}(n) = \binom{n+c}{c}$ and $\mathbf{v}_{\mathrm{rep}_S(X)}(n) = \binom{n+d}{d}$ for all $n \in \mathbb{N}$. Then from Theorem 3.2.23, we obtain $t_X(n) = \Theta(n^{\frac{c}{d}}) = \Theta(n^r)$. □

Theorem 3.2.23 also allows us to exhibit subsets of $\mathbb{N}$ which are not S-recognizable for any abstract numeration system S. For example, let $C = \{C_n \mid n \in \mathbb{N}\}$ denote the set of Catalan numbers $C_n = \frac{1}{n+1}\binom{2n}{n}$. As is well known, we have $C_n \sim \frac{4^n}{n^{3/2}\sqrt{\pi}}$ $(n \to +\infty)$, which does not correspond to any of the forms described by Theorem 3.2.23. Hence, for all S, the set C is not S-recognizable.

3.2.5 *The Cobham–Semenov Theorem*

So far we have introduced several numeration systems and have considered the question of describing recognizable sets of nonnegative integers within a fixed numeration system. The celebrated theorem of Cobham concerns, on the contrary, sets of numbers that are simultaneously recognizable in different integer bases. Cobham's theorem and its numerous generalizations are the subject of several surveys [114, 206]. Nevertheless, due to the importance of this result and its relevance to the subject of the present chapter, we briefly discuss it in this short section.

Definition 3.2.27. *Semi-linear subsets* of $\mathbb{N}^d$ are the finite unions of sets of the form $\mathbf{p}_0 + \mathbf{p}_1\mathbb{N} + \cdots + \mathbf{p}_\ell\mathbb{N}$, where $\mathbf{p}_0, \mathbf{p}_1, \ldots, \mathbf{p}_\ell \in \mathbb{N}^d$.

Recall that b and b' are multiplicatively independent if $\frac{\log(b)}{\log(b')} \notin \mathbb{Q}$; see Definition 3.2.24.

Theorem 3.2.28 (Cobham–Semenov [155, 536]). *Let b and b' be multiplicatively independent integer bases. If a subset of $\mathbb{N}^d$ is simultaneously b-recognizable and b'-recognizable, then it is semi-linear.*

As semi-linear sets are b-recognizable for all integer bases b, we obtain that a subset of $\mathbb{N}^d$ is b-recognizable for all $b \geq 2$ if and only if it is semi-linear. Note that we cannot replace $b \geq 2$ by $b \geq 1$ as, for example, the linear set $X = \{(n, 2n) \mid n \in \mathbb{N}\} = (1, 2)\mathbb{N}$ is not 1-recognizable.

We have just argued that the family of 1-recognizable sets is distinct from that of semi-linear sets. It is worth noticing that 1-recognizable sets also do not correspond to the so-called *recognizable subsets* of $\mathbb{N}^d$, which are the subsets X of $\mathbb{N}^d$ for which the equivalence relation $\sim_X$ over $\mathbb{N}^d$ defined by

$$x \sim_X y \iff (\forall z \in \mathbb{N}^d,\ x + z \in X \iff y + z \in X)$$

has finite index. For example, the diagonal $D = \{(n, n) \mid n \in \mathbb{N}\}$ is 1-recognizable but not recognizable as $(m, 0) \sim_D (n, 0)$ if and only if $m = n$. On the other hand, it is true that the recognizable subsets of $\mathbb{N}^d$ are all 1-recognizable. More precisely, we have the following result.

Theorem 3.2.29 ([144]). *A subset X of $\mathbb{N}^d$ is S-recognizable for all abstract numeration systems S if and only if it is 1-recognizable.*

3.3 First-Order Logic and b-Automatic Sequences

In this section, we present an equivalent definition of b-automatic sequences in terms of logic. It is given by the Büchi-Bruyère theorem. This criterion is of high interest since it represents a powerful tool in order to show that many properties of b-automatic sequences are decidable.

3.3.1 *b-Definable Sets of Integers*

A *(logical) structure* $\mathcal{S} = \langle S, (R_i) \rangle$ consists of a set S, called the domain of the structure, and countably many relations $R_i \subseteq S^{d_i}$, where the d_i's are positive integers, called the arities of the R_i's.

A *first-order formula* is defined recursively from

- variables $x_1, x_2, x_3, \ldots$ describing elements of the domain S,
- the equality $=$,
- the relations given in the structure $\mathcal{S}$,
- the connectives $\vee, \wedge, \implies, \iff, \neg$,
- the quantifiers $\forall, \exists$ on variables.

Example 3.3.1. The Presburger arithmetic is described by the first-order formulæ of the structure $\langle \mathbb{N}, + \rangle$. See Section 3.7.

Let $\mathscr{S}$ be a logical structure whose domain is S. For a first-order formula $\varphi(x_1, \ldots, x_d)$ of $\mathscr{S}$, we let

$$X_\varphi = \{(s_1, \ldots, s_d) \in S^d \mid \mathscr{S} \models \varphi(s_1, \ldots, s_d)\}.$$

A subset X of S^d is *definable in* $\mathscr{S}$ if there exists a first-order formula $\varphi(x_1, \ldots, x_d)$ of $\mathscr{S}$ such that $X = X_\varphi$, *i.e.*, such that, for all $(s_1, \ldots, s_d) \in S^d$, $\varphi(s_1, \ldots, s_d)$ is true if and only if $(s_1, \ldots, s_d) \in X$.

We shall use particular notation for constant relations and for functional relations. A constant relation is a relation of the form $\{c\}$. It will be simply denoted c. A functional relation is a binary relation R such that for any $s \in S$, there is at most one $t \in S$ with $(s, t) \in R$. Such a relation R will be denoted $f: S \to S$ where it is understood that $f(s) = t$ if there exists $t \in S$ such that $(s, t) \in R$ and $f(s)$ is undefined otherwise.

Definition 3.3.2. A subset X of $\mathbb{N}^d$ is *b-definable* if it is definable in the logical structure $\langle \mathbb{N}, +, V_b \rangle$, where $+$ is the ternary relation defined by $x + y = z$ and V_b is the function defined by $V_b(0) = 1$, and for x a positive integer, $V_b(x)$ is the largest power of b dividing x.

Example 3.3.3. One has $V_2(9) = 1$ and $V_2(24) = 8$.

3.3.2 *The Büchi-Bruyère Theorem*

Theorem 3.3.4 ([112, 115]). *A subset X of $\mathbb{N}^d$ is b-recognizable if and only if it is b-definable. Moreover, both directions are effective.*

For a detailed proof of this result, we refer the reader to [114]. We only sketch the idea of their proof here. They work with automata accepting reversed b-representations of numbers. From a DFA recognizing X least significant digit first, that is, such that it accepts a language $L \subseteq (A_b^d)^*$ satisfying $X = \{\mathrm{val}_b(\widetilde{w}) \mid w \in L\}$, they construct a first-order formula φ of the structure $\langle \mathbb{N}, +, V_b \rangle$ defining X. Conversely, given a first-order formula φ of the structure $\langle \mathbb{N}, +, V_b \rangle$ defining X, they build a DFA accepting all the reversed b-representations of the elements in X, that is, accepting the language $(\mathrm{rep}_b(X))^{\sim}\mathbf{0}^*$.

3.3.3 *The First-Order Theory of $\langle \mathbb{N}, +, V_b \rangle$ Is Decidable*

As a corollary of the Büchi-Bruyère theorem, the first-order theory of $\langle \mathbb{N}, +, V_b \rangle$ is decidable: given any closed first-order formula of $\langle \mathbb{N}, +, V_b \rangle$, we can decide whether it is true or false in $\mathbb{N}$. As this corollary has a nice short proof, we give it here.

Since there is no constant in the structure, a closed formula of $\langle \mathbb{N}, +, V_b \rangle$ is necessarily of the form $\exists x \varphi(x)$ or $\forall x \varphi(x)$. The set X_φ is b-recognizable by the Büchi-Bruyère theorem. This means that we can effectively construct a DFA accepting $\text{rep}_b(X_\varphi)$. The closed formula $\exists x \varphi(x)$ is true if $\text{rep}_b(X_\varphi)$ is nonempty and false otherwise. As the emptiness of a regular language is decidable [301], we can decide if $\exists x \varphi(x)$ is true.

The case $\forall x \varphi(x)$ reduces to the previous one since $\forall x \varphi(x)$ is logically equivalent to $\neg \exists x \neg \varphi(x)$. We can again construct a DFA accepting the b-representations of $X_{\neg\varphi}$. The language it accepts is empty if and only if the closed formula $\forall x \varphi(x)$ is true.

3.3.4 Applications to Decidability Questions for b-Automatic Sequences

Proposition 3.3.5. *If we can express a property $P(n)$ of an integer n using quantifiers, logical operations, the operations of addition and subtraction, and comparison of integers or elements of a b-automatic sequence x, then $\exists n P(n)$, $\exists^\infty n P(n)$, and $\forall n P(n)$ are decidable.*

We just have to convince ourselves that those properties P can all be expressed by a first-order formula of $\langle \mathbb{N}, +, V_b \rangle$. If $x \colon \mathbb{N}^d \to \mathbb{N}$ is b-automatic, then, for all letters a occurring in x, the subsets $x^{-1}(a)$ of $\mathbb{N}^d$ are b-recognizable by Proposition 3.2.8. Hence they are definable by some first-order formulæ ψ_a of $\langle \mathbb{N}, +, V_b \rangle$ by the Büchi-Bruyère theorem: $\psi_a(n_1, \dots, n_d)$ is true if and only if $x(n_1, \dots, n_d) = a$. Therefore, we can express that $x(m_1, \dots, m_d) = x(n_1, \dots, n_d)$ by the first-order formula $\varphi(m_1, \dots, m_d, n_1, \dots, n_d)$ of $\langle \mathbb{N}, +, V_b \rangle$:

$$\varphi(m_1, \dots, m_d, n_1, \dots, n_d) \equiv \bigvee_a (\psi_a(m_1, \dots, m_d) \wedge \psi_a(n_1, \dots, n_d)).$$

In practice, given a DFAO $\mathcal{A}$ computing $x \colon \mathbb{N}^d \to \mathbb{N}$, we can directly compute a DFA recognizing the tuples $(m_1, \dots, m_d, n_1, \dots, n_d) \in \mathbb{N}^{2d}$ such that $x(m_1, \dots, m_d) = x(n_1, \dots, n_d)$. We compute the product of automata $\mathcal{A} \times \mathcal{A}$, thus reading tuples of size $2d$, and simulate $(m_1, \dots, m_d)$ on the first component and $(n_1, \dots, n_d)$ on the second component, and we accept if the outputs of $\mathcal{A}$ after reading $\text{rep}_b(m_1, \dots, m_d)^\#$ and $\text{rep}_b(n_1, \dots, n_d)^\#$ are the same and reject otherwise.

In fact, Theorem 3.3.4 allows us to prove a stronger result than the decidability of such properties of b-automatic sequences. What we obtain is that the characteristic sequences of those properties are themselves b-automatic. The following proposition is far from being exhaustive. It only aims to give a flavor of the properties that can be handled by using this technique. For similar results, we refer to [12, 148]. A finite word is *unbordered* if no proper prefix equals a suffix. A *palindrome* is a finite word equal to its reversal: $w = \widetilde{w}$.

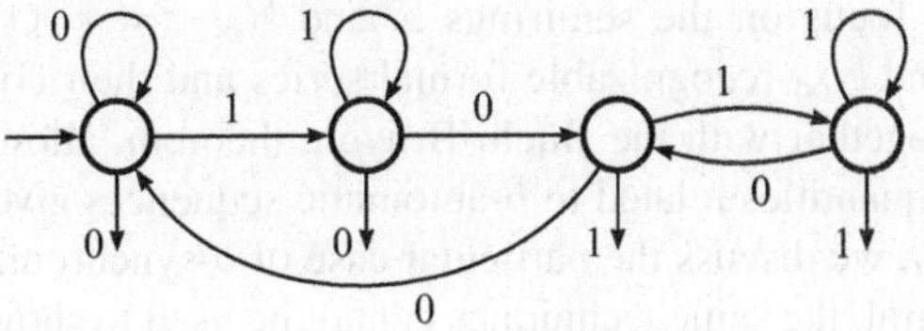

Fig. 3.3 DFAO generating the regular paperfolding sequence

Proposition 3.3.6. *Let $x\colon \mathbb{N} \to \mathbb{N}$ be a b-automatic sequence. Then the following sequences $y\colon \mathbb{N} \to \{0, 1\}$ are also b-automatic:*

- $y(i) = 1$ *if and only if x has an overlap at position i*
- $y(i) = 1$ *if and only if x has an unbordered factor of length i*
- $y(i) = 1$ *if and only if x has a square at position i*
- $y(i) = 1$ *if and only if x has a palindrome at position i.*

Some properties of interest of automatic sequences are not expressible by a first-order formula of $\langle \mathbb{N}, +, V_b \rangle$ as the following proposition shows. The regular paperfolding sequence

$$0010011000110110001001110011011000100111000110111001001 1\cdots$$

is the 2-automatic sequence computed by the DFAO of Figure 3.3.

Proposition 3.3.7 ([523]). *If x is the paperfolding sequence, then the predicate "x has an abelian square at position i of length 2n" is not expressible in $\langle \mathbb{N}, +, V_2 \rangle$.*

This method for deciding first-order expressible properties of b-automatic sequences is very bad in terms of complexity. In the worst case, we have a tower of exponentials in the number of states of the given DFAO whose height is the number of alternating quantifiers of the first-order predicate. Nevertheless, this procedure was implemented by Mousavi and works efficiently in many cases. His open source software package is called `Walnut` [426]. It can be used in practice in order to prove (and reprove) many results about some particular b-automatic sequences, in a purely mechanical way [248–250].

3.4 Enumeration

The object of this section is to study enumeration problems about b-automatic sequences. It turns out that the sequences $(a(m))_{m\in\mathbb{N}}$ that count the number of $n \in \mathbb{N}$ such that $P(m, n)$ is true, for any first-order predicate P of the logical structure $\langle \mathbb{N}, +, V_b \rangle$, are indeed *b-regular sequences*; this is Theorem 3.4.15. We first introduce b-regular sequences over an arbitrary semiring K (also see Chapters 2 and 4). Then we focus on the semirings $\mathbb{N}$ and $\mathbb{N}_\infty := \mathbb{N} \cup \{\infty\}$. We discuss $\mathbb{N}$-recognizable and $\mathbb{N}_\infty$-recognizable formal series and their connections to finite automata. This, together with the Büchi-Bruyère theorem, allows us to prove that counting various quantities related to b-automatic sequences gives rise to b-regular sequences. Finally, we discuss the particular case of b-synchronized sequences and show that, in general, the same techniques cannot be used to show that the obtained sequences are b-synchronized: some of them are, whereas some others are not.

3.4.1 *b*-Regular Sequences

A formal series S is a map from A^* to K, where A is a finite alphabet and K is a semiring. The image of a word w is denoted (S, w), as is customary. We also use the notation $S = \sum_{w \in A^*} (S, w)\, w$.

Definition 3.4.1. Let A be a finite alphabet and K be a semiring. A formal series $S: A^* \to K$ is *K-recognizable* if there exist an integer $m \geq 1$, vectors $\lambda \in K^{1 \times m}$, $\gamma \in K^{m \times 1}$, and a morphism of monoids $\mu: A^* \to K^{m \times m}$ such that, for all $w \in A^*$, $(S, w) = \lambda \mu(w) \gamma$. The triple (λ, μ, γ) is called a *linear representation* of S, and we say it is of *size*, or of *dimension*, m.

The family of K-recognizable series has many stability properties. We list here (without proofs) only those we will explicitly use for our purpose. For more on K-recognizable series, we refer the reader to [77].

The characteristic series of a language $L \subseteq A^*$ is $\chi_L := \sum_{w \in L} w$. It can be viewed as a map from A^* to K for any semiring K (as any semiring contains 0 and 1).

Proposition 3.4.2. *For any language L, the following assertions are equivalent.*

1. *L is regular.*
2. *χ_L is $\mathbb{N}$-recognizable.*
3. *For all semirings K, χ_L is K-recognizable.*

The *Hadamard product* of two formal series S and T is their term-wise product: $S \odot T = \sum_{w \in A^*} (S, w)(T, w)\, w$. In particular, $S \odot \chi_L = \sum_{w \in L} (S, w)\, w$.

Proposition 3.4.3. *If $S: A^* \to K$ is a K-recognizable series and $L \subseteq A^*$ is a regular language, then $S \odot \chi_L$ is K-recognizable.*

Proposition 3.4.4. *Every formal series $S: A^* \to K$ with only finitely many terms $(S, w) \neq 0$ is K-recognizable.*

It follows from the previous two propositions that two formal series that differ only in a finite number of words are either both K-recognizable or both not K-recognizable.

We will need the following lemma.

Lemma 3.4.5. *Let $S: A^* \to K$ be a K-recognizable series, $B \subseteq A$ be a nonempty sub-alphabet, and $\pi: A \to B$ be a letter-to-letter morphism. Then the series $T: B^* \to K$ defined by*

$$T = \sum_{u \in A^*} (S, u)\, \pi(u) = \sum_{w \in B^*} \Big(\sum_{\substack{u \in A^* \\ \pi(u)=w}} (S, u) \Big) w$$

is K-recognizable.

Proof. Let (λ, μ, γ) be a linear representation of S, say of size m. Define a morphism $\mu': B^* \to K^{m \times m}$ by $\mu'(b) = \sum_{a \in A, \pi(a)=b} \mu(a)$ for each $b \in B$. By induction on $|w|$, we easily get that $\mu'(w) = \sum_{u \in A^*, \pi(u)=w} \mu(u)$ for all $w \in B^*$. Therefore, (λ, μ', γ) is a linear representation of T: for all $w \in B^*$,

$$\lambda \mu'(w) \gamma = \sum_{\substack{u \in A^* \\ \pi(u)=w}} \lambda \mu(u) \gamma = \sum_{\substack{u \in A^* \\ \pi(u)=w}} (S, u) = (T, w).$$

□

By an abuse of notation, we sometimes write $\sum_{\mathbf{n} \in \mathbb{N}^d} x(\mathbf{n})\, \mathrm{rep}_b(\mathbf{n})$ instead of $\sum_{w \in \mathrm{rep}_b(\mathbb{N}^d)} x(\mathrm{val}_b(w))\, w$. Similarly, $\sum_{n_1, \ldots, n_d \in \mathbb{N}} x(n_1, \ldots, n_d)\, (\mathrm{rep}_b(n_1), \ldots, \mathrm{rep}_b(n_d))^\#$ is the series $S: ((A_b \cup \{\#\})^d)^* \to K$ defined by $(S, w) = x(n_1, \ldots, n_d)$ if $w = (\mathrm{rep}_b(n_1), \ldots, \mathrm{rep}_b(n_d))^\#$ for some $n_1, \ldots, n_d \in \mathbb{N}$ and $(S, w) = 0$ otherwise.

Proposition 3.4.6. *Let # be a symbol not belonging to A_b. For any sequence $x: \mathbb{N}^d \to K$, the following assertions are equivalent.*

1. *$\sum_{w \in (A_b{}^d)^*} x(\mathrm{val}_b(w))\, w$ is K-recognizable.*
2. *$\sum_{\mathbf{n} \in \mathbb{N}^d} x(\mathbf{n})\, \mathrm{rep}_b(\mathbf{n})$ is K-recognizable.*
3. *There exists a K-recognizable series $S: (A_b{}^d)^* \to K$ such that, for all $\mathbf{n} \in \mathbb{N}^d$, $(S, \mathrm{rep}_b(\mathbf{n})) = x(\mathbf{n})$.*
4. *There exists a K-recognizable series $T: ((A_b \cup \{\#\})^d)^* \to K$ such that, for all $n_1, \ldots, n_d \in \mathbb{N}$, $(T, (\mathrm{rep}_b(n_1), \ldots, \mathrm{rep}_b(n_d))^\#) = x(n_1, \ldots, n_d)$.*
5. *$\sum_{n_1, \ldots, n_d \in \mathbb{N}} x(n_1, \ldots, n_d)\, (\mathrm{rep}_b(n_1), \ldots, \mathrm{rep}_b(n_d))^\#$ is K-recognizable.*

Proof. 1 $\Longrightarrow$ 2: We have

$$G_0 := \sum_{\mathbf{n} \in \mathbb{N}^d} x(\mathbf{n})\, \mathrm{rep}_b(\mathbf{n}) = \sum_{w \in (A_b{}^d)^*} x(\mathrm{val}_b(w))\, w \odot \chi_{\mathrm{rep}_b(\mathbb{N}^d)}.$$

As $\mathbb{N}^d$ is b-recognizable, we obtain 1 $\Longrightarrow$ 2 from Proposition 3.4.3.

The implication 2 $\Longrightarrow$ 3 is clear.

3 $\Longrightarrow$ 1 $\wedge$ 4: Assume that 3 holds and let $S: (A_b{}^d)^* \to K$ be a K-recognizable series such that, for all $\mathbf{n} \in \mathbb{N}^d$, $(S, \mathrm{rep}_b(\mathbf{n})) = x(\mathbf{n})$. Let (λ, μ, γ) be a linear representation of S, say of size m.

First, let $\lambda' = [1\ 0\ \cdots\ 0] \in \mathbb{N}^{1\times(m+1)}$, $\mu'\colon (A_b{}^d)^* \to \mathbb{N}^{(m+1)\times(m+1)}$ be the morphism defined by

$$\mu'(\mathbf{0}) = \begin{bmatrix} 1 & 0 \cdots 0 \\ 0 & \\ \vdots & \begin{bmatrix} \mu(0) \end{bmatrix} \\ 0 & \end{bmatrix}, \quad \mu'(a) = \begin{bmatrix} 0 & [\lambda\mu(a)] \\ 0 & \\ \vdots & \begin{bmatrix} \mu(a) \end{bmatrix} \\ 0 & \end{bmatrix}, \text{ for } a \neq \mathbf{0},$$

and $\gamma' = [\lambda\gamma \quad \gamma]^T \in \mathbb{N}^{(m+1)\times 1}$. Then, for all $w \in (A_b{}^d)^*$, $\lambda'\mu'(w)\gamma' = x(\mathrm{val}_b(w))$; hence, $\sum_{w\in(A_b{}^d)^*} x(\mathrm{val}_b(w))\, w$ is K-recognizable, which is 1.

Second, we define a morphism $\mu''\colon ((A_b \cup \{\#\})^d)^* \to K^{m\times m}$ by $\mu''(a) = \mu(\pi(a))$ for all $a \in (A_b \cup \{\#\})^d$, where $\pi\colon (A_b \cup \{\#\})^d \to A_b{}^d$ is the letter-to-letter morphism defined by

$$(\pi(a_1, \ldots, a_d))_i = \begin{cases} a_i, & \text{if } a_i \neq \#, \\ 0, & \text{if } a_i = \#. \end{cases}$$

Then, for all $w \in ((A_b \cup \{\#\})^d)^*$, we have $\lambda\mu''(w)\gamma = \lambda\mu(\pi(w))\gamma = (S, \pi(w))$. Thus 4 holds, as the formal series

$$T = \sum_{w\in((A_b\cup\{\#\})^d)^*} (S, \pi(w))\, w$$

is K-recognizable and such that, for all $\mathbf{n} = (n_1, \ldots, n_d) \in \mathbb{N}^d$,

$$(T, (\mathrm{rep}_b(n_1), \ldots, \mathrm{rep}_b(n_d))^\#) = (S, \mathrm{rep}_b(\mathbf{n})) = x(\mathbf{n}).$$

4 $\Longrightarrow$ 5: Let $T\colon ((A_b \cup \{\#\})^d)^* \to K$ be such that, for all $\mathbf{n} = (n_1, \ldots, n_d) \in \mathbb{N}^d$, $(T, (\mathrm{rep}_b(n_1), \ldots, \mathrm{rep}_b(n_d))^\#) = x(\mathbf{n})$. Since $\mathbb{N}^d$ is b-recognizable, the language

$$L_\# := \{(\mathrm{rep}_b(n_1), \ldots, \mathrm{rep}_b(n_d))^\# \mid n_1, \ldots, n_d \in \mathbb{N}\}$$

is regular by Proposition 3.2.2. As the formal series

$$G_\# := \sum_{n_1,\ldots,n_d\in\mathbb{N}} x(n_1, \ldots, n_d)\, (\mathrm{rep}_b(n_1), \ldots, \mathrm{rep}_b(n_d))^\#$$

satisfies $G_\# = T \odot \chi_{L_\#}$, it is K-recognizable if T is as well by Proposition 3.4.3.

5 $\Longrightarrow$ 2: Suppose that $G_\#$ is K-recognizable. By Lemma 3.4.5, the series

$$R = \sum_{u\in((A_b\cup\{\#\})^d)^*} (G_\#, u)\, \pi(u) = \sum_{w\in(A_b{}^d)^*} \Big(\sum_{\substack{u\in((A_b\cup\{\#\})^d)^* \\ \pi(u)=w}} (G_\#, u) \Big)\, w$$

is K-recognizable. As, for all $\mathbf{n} = (n_1, \ldots, n_d) \in \mathbb{N}^d$,

$$(R, \mathrm{rep}_b(\mathbf{n})) = \sum_{\substack{u \in ((A_b \cup \{\#\})^d)^* \\ \pi(u) = \mathrm{rep}_b(\mathbf{n})}} (G_\#, u) = \sum_{\substack{u \in L_\# \\ \pi(u) = \mathrm{rep}_b(\mathbf{n})}} (G_\#, u)$$
$$= (G_\#, (\mathrm{rep}_b(n_1), \ldots, \mathrm{rep}_b(n_d))^\#) = x(\mathbf{n}),$$

we obtain $G_0 = R \odot \chi_{\mathrm{rep}_b(\mathbb{N}^d)}$; hence, G_0 is K-recognizable by Proposition 3.4.3. □

Definition 3.4.7. A sequence $x \colon \mathbb{N}^d \to K$ is (K, b)-*regular* if any of the assertions of Proposition 3.4.6 is satisfied.

Thanks to the following elementary lemma, we can equivalently consider reversals of representations, *i.e.*, starting with the least significant digit. Here $\widetilde{\alpha}$ denotes the transpose of the matrix α, and $\widetilde{\mu}$ is the morphism defined by $\widetilde{\mu}(a) = \widetilde{\mu(a)}$ for each letter a.

Lemma 3.4.8. *If a formal series* $S \colon A^* \to K$ *admits the linear representation* (λ, μ, γ)*, then the reversal series* $\widetilde{S} := \sum_{w \in A^*} (S, \widetilde{w})\, w$ *admits the linear representation* $(\widetilde{\gamma}, \widetilde{\mu}, \widetilde{\lambda})$.

Proof. For all $w = a_1 \cdots a_{|w|} \in A_b^*$, $\widetilde{\mu(w)} = (\mu(a_{|w|} \cdots a_1))^\sim = (\mu(a_{|w|}) \cdots \mu(a_1))^\sim = \widetilde{\mu(a_1)} \cdots \widetilde{\mu(a_{|w|})} = \widetilde{\mu}(a_1) \cdots \widetilde{\mu}(a_{|w|}) = \widetilde{\mu}(a_1 \cdots a_{|w|}) = \widetilde{\mu}(w)$, hence $(\widetilde{S}, w) = (S, \widetilde{w}) = \lambda \mu(\widetilde{w}) \gamma = (\lambda \mu(\widetilde{w}) \gamma)^\sim = \widetilde{\gamma}\, \widetilde{\mu(\widetilde{w})}\, \widetilde{\lambda} = \widetilde{\gamma}\, \widetilde{\mu}(w)\, \widetilde{\lambda}$. □

In what follows, the semiring K will be either $\mathbb{N}$ or $\mathbb{N}_\infty = \mathbb{N} \cup \{\infty\}$ with $0 \cdot \infty = 0$. Let us mention the following result, a proof of which can be found in Chapter 2.

Proposition 3.4.9 ([17]). *If* $x \colon \mathbb{N} \to \mathbb{N}$ *is an* $(\mathbb{N}, b)$*-regular sequence, then there exists some* $c \in \mathbb{N}$ *such that* $x(n) \in O(n^c)$.

3.4.2 $\mathbb{N}$-Recognizable and $\mathbb{N}_\infty$-Recognizable Formal Series

We have the following useful characterizations of $\mathbb{N}$-recognizable and $\mathbb{N}_\infty$-recognizable formal series. Here π_i denotes the projection onto the ith component.

Theorem 3.4.10. *Let* $S \colon A^* \to \mathbb{N}$. *The following assertions are equivalent.*

1. *S is $\mathbb{N}$-recognizable.*
2. *There exists a regular language $L \subseteq (A \times \Delta)^*$ (where Δ is a finite alphabet) such that, for all $w \in A^+$, (S, w) equals the number of $z \in L$ with $\pi_1(z) = w$.*

Proof. 1 $\Longrightarrow$ 2. Suppose that S is $\mathbb{N}$-recognizable. We consider

$$S' \colon A^* \to \mathbb{N},\ w \mapsto \begin{cases} (S, w), & \text{if } w \neq \varepsilon, \\ 0, & \text{if } w = \varepsilon. \end{cases}$$

Then S' is $\mathbb{N}$-recognizable as it is a finite modification of S. Let (λ, μ, γ) be a linear representation of S' of size n. We may suppose, without loss of generality, that $\lambda = [1\ 0 \cdots 0]$ and $\gamma = [0 \cdots 0\ 1]^T$. Indeed, let $\lambda' = [1\ 0 \cdots 0] \in \mathbb{N}^{1\times(n+2)}$, $\gamma' = [0 \cdots 0\ 1]^T \in \mathbb{N}^{(n+2)\times 1}$, and $\mu' \colon A^* \to \mathbb{N}^{(n+2)\times(n+2)}$ be the morphism defined by

$$\mu'(a) = \begin{bmatrix} 0 & [\lambda\mu(a)] & [\lambda\mu(a)\gamma] \\ 0 & & \\ \vdots & \begin{bmatrix} \\ \mu(a) \\ \\ \end{bmatrix} & \begin{bmatrix} \\ \mu(a)\gamma \\ \\ \end{bmatrix} \\ 0 & & \\ 0 & [0 \cdots 0] & 0 \end{bmatrix}, \text{ for } a \in A.$$

Then $(\lambda', \mu', \gamma')$ is a linear representation of S' of size $n + 2$.

Let $\mathcal{M} = (Q, q_0, F, A \times Q, \delta)$ be the DFA defined as follows. Let

$$m = \max_{\substack{a\in A \\ 1\le i,j\le n}} \mu(a)_{ij}$$

and

$$\begin{aligned} Q &= \{(i, r) \mid 1 \le i \le n,\ 1 \le r \le m\} \\ q_0 &= (1, 1) \\ F &= \{(n, r) \mid 1 \le r \le m\} \\ \delta((i, r), (a, (j, s))) &= (j, s) \text{ if } 1 \le s \le \mu(a)_{ij}. \end{aligned}$$

So $\delta((i, r), (a, (j, s)))$ is not defined if $s > \mu(a)_{ij}$, and not every state in Q is necessarily accessible. We show by induction on $|w|$ that $\mu(w)_{ij}$ equals the number of paths of label z with $\pi_1(z) = w$ from (i, r) to $\{(j, s) \mid 1 \le s \le m\}$ (for any $1 \le r \le m$). Let $P_{i,j}(w)$ denote the number of such paths. The base case $w = \varepsilon$ is clear as $\mu(\varepsilon)$ is the identity matrix of size n, and $P_{i,j}(\varepsilon)$ is equal to 1 if $i = j$ and to 0 else. For $a \in A$ and $x \in A^*$,

$$P_{i,k}(a) = \mathrm{Card}\{s \mid \delta((i, r), (a, (k, s))) = (k, s)\} = \mu(a)_{ik}.$$

and

$$\mu(ax)_{ij} = \sum_{k=1}^{n} \mu(a)_{ik}\mu(x)_{kj} = \sum_{k=1}^{n} P_{i,k}(a)P_{k,j}(x) = P_{i,j}(ax),$$

where we have applied the induction hypothesis to x. Now if L is the language accepted by $\mathcal{M}$, then, for all $w \in A^+$, $(S, w) = (S', w) = \lambda\mu(w)\gamma = \mu(w)_{1n} = P_{1,n}(w) = \mathrm{Card}\{z \in L \mid \pi_1(z) = w\}$.

$2 \implies 1$. Suppose that $\mathcal{M} = (Q, q_1, F, A \times Q, \delta)$ is a DFA accepting a language $L \subseteq (A \times \Delta)^*$ such that, for all $w \in A^+$, (S, w) equals the number of $z \in L$ with $\pi_1(z) = w$. Let $Q = \{q_1, \ldots, q_n\}$. Define $\lambda = [1\ 0 \cdots 0] \in \mathbb{N}^{1\times n}$, $\gamma \in \mathbb{N}^{n\times 1}$ be such that $\gamma_{i1} = 1$ if $q_i \in F$ and $\gamma_{i1} = 0$ if $q_i \notin F$. Let $\mu(a)_{ij}$ be the number of paths of label z with $\pi_1(z) = a$ from q_i to q_j, and let $\mu: A^* \to \mathbb{N}^{n\times n}$ be the induced morphism. It is easy to see that, for all $w \in A^*$, $\mu(w)_{ij}$ is the number of paths of label z with $\pi_1(z) = w$ from q_i to q_j. Then, for all $w \in A^+$,

$$\lambda\mu(w)\gamma = \sum_{\substack{1\leq j\leq n \\ q_j \in F}} \mu(w)_{1j} = \mathrm{Card}\{z \in L \mid \pi_1(z) = w\} = (S, w).$$

This proves that S is $\mathbb{N}$-recognizable (whatever the value (S, ε) is). □

We sometimes want to count quantities that might be unbounded in certain entries, as, for example, the length of the longest square (or k-power, overlap, palindrome, unbordered factor, etc.) beginning at position i.

Proposition 3.4.11. *If $S: A^* \to \mathbb{N}$ is $\mathbb{N}_\infty$-recognizable, then it is $\mathbb{N}$-recognizable.*

Proof. Let $n \geq 1$, $\lambda \in \mathbb{N}_\infty^{1\times n}$, $\gamma \in \mathbb{N}_\infty^{n\times 1}$, and a morphism of monoids $\mu: A^* \to \mathbb{N}_\infty^{n\times n}$ such that, for all $w \in A^*$, $(S, w) = \lambda\mu(w)\gamma$. As, for all $w \in A^*$, $(S, w) \in \mathbb{N}$, any occurrence of ∞ in the computation of $\lambda\mu(w)\gamma$ must belong to a multiplication with 0. Hence we can modify λ, γ, μ to λ', γ', μ' by replacing any occurrence of ∞ by 0. In this way, $\lambda' \in \mathbb{N}^{1\times n}$, $\gamma' \in \mathbb{N}^{n\times 1}$, $\mu': A^* \to \mathbb{N}^{n\times n}$, and, for all $w \in A^*$, $(S, w) = \lambda'\mu'(w)\gamma'$. This shows that S is $\mathbb{N}$-recognizable. □

Lemma 3.4.12. *If $S: A^* \to \mathbb{N}_\infty$ is $\mathbb{N}_\infty$-recognizable, then the language $\{w \in A^* \mid (S, w) = \infty\}$ is regular.*

Proof. Let (λ, μ, γ) be a linear representation of S. Consider the set $\{0, p, \infty\}$ (where p is any symbol, intended to represent positive integers). We endow this set with a structure of commutative semiring as follows: $0+0 = 1$, $p+0 = p+p = p$, $p + \infty = \infty + \infty = \infty$, $0 \cdot 0 = p \cdot 0 = \infty \cdot 0 = 0$, $p \cdot p = p$, and $p \cdot \infty = \infty \cdot \infty = \infty$. Define a morphism of semirings $\tau: \mathbb{N}_\infty \to \{0, p, \infty\}$ by $\tau(0) = 0$, $\tau(n) = p$ for $n \in \mathbb{N} \setminus \{0\}$, and $\tau(\infty) = \infty$. Now we define a DFA $\mathcal{M} = (Q, q_0, F, A, \delta)$ as follows: $Q = \{0, p, \infty\}^{1\times n}$, $q_0 = [\tau(\lambda_{11}) \cdots \tau(\lambda_{1n})]$, $F = \{q \in Q \mid q\,[\tau(\gamma_{11}) \cdots \tau(\gamma_{n1})]^T\} = \infty\}$, and $\delta(q, a) = q\,(\tau(\mu(a)_{ij}))_{1\leq i,j\leq n}$. We have $\delta(q_0, w) \in F \iff \tau(\lambda\mu(w)\gamma) = \infty \iff (S, w) = \lambda\mu(w)\gamma = \infty$. This proves that $\mathcal{M}$ accepts $\{w \in A^* \mid (S, w) = \infty\}$. □

Theorem 3.4.13. *Let $S: A^* \to \mathbb{N}_\infty$. The following assertions are equivalent.*

1. *S is $\mathbb{N}_\infty$-recognizable.*
2. *There exists a regular language $L \subseteq ((A \cup \{\#\}) \times \Delta)^*$ (where $\# \notin A$ and Δ is a finite alphabet) such that, for all $w \in A^+$, (S, w) equals the number of $z \in L$ with $\tau_\#(\pi_1(z)) = w$, where $\tau_\#$ is the morphism defined by $a \mapsto a$ for $a \in A$ and $\# \mapsto \varepsilon$.*

Proof. 1 $\Longrightarrow$ 2. Suppose that S is $\mathbb{N}_\infty$-recognizable. By Lemma 3.4.12, the language $L_1 = \{w \in A^* \mid (S,w) = \infty\}$ is regular. Now, the series $S' = S \odot \chi_{\{w\in A^* \mid (S,w)\neq\infty\}}$ is $\mathbb{N}$-recognizable by Propositions 3.4.3 and 3.4.11. From Theorem 3.4.10, we get a regular language $L_2 \subseteq (A \times \Delta)^*$ (for some alphabet Δ) such that, for all $w \in A^+$, $(S', w) = \mathrm{Card}\{z \in L_2 \mid \pi_1(z) = w\}$. Let $a \in \Delta$ and let $\# \notin A$. Then define $L_3 = \{z \in (A \cup \{\#\}) \times \Delta)^* \mid \pi_1(z) \in L_1\#^*, \pi_2(z) \in a^*\}$. Clearly L_3 is regular, and, for all $w \in A^+$, $(S,w) = \mathrm{Card}\{z \in L_2 \cup L_3 \mid \tau_\#(\pi_1(z)) = w\}$.

2 $\Longrightarrow$ 1. Suppose that $L \subseteq ((A \cup \{\#\}) \times \Delta)^*$ is such that, for all $w \in A^+$, $(S,w) = \mathrm{Card}\{z \in L \mid \tau_\#(\pi_1(z)) = w\}$ and that $\mathcal{M} = (Q, q_1, F, A, \delta)$ is a DFA accepting L. Let $Q = \{q_1, \dots, q_n)$. For each $a \in A \cup \{\#\}$, we define a matrix $D_a \in \mathbb{N}^{n\times n}$ as follows: $(D_a)_{ij}$ equals the number of letters $b \in \Delta$ such that $\delta(q_i, \binom{a}{b}) = q_j$. Then any finite path in $\mathcal{M}$ labeled z with $\tau_\#(\pi_1(z)) = a_1 \cdots a_\ell$, with the a_i's in A, is of the form

$$\binom{\#}{\star}, \dots, \binom{\#}{\bullet}, \binom{a_1}{\bullet}, \binom{\#}{\bullet}, \dots, \binom{\#}{\bullet}, \binom{a_2}{\bullet}, \dots, \binom{\#}{\bullet}, \dots, \binom{\#}{\bullet}, \binom{a_\ell}{\bullet}, \binom{\#}{\bullet}, \dots, \binom{\#}{\bullet},$$

where $\bullet$ could be anything. Now, let $D = \sum_{i\geq 0} D_\#^i \in \mathbb{N}_\infty^{n\times n}$. Then the computation $(DD_{a_1}DD_{a_2}\cdots DD_{a_\ell}D)_{ij}$ returns the number of such paths from q_i to q_j. Let $\lambda = [1\ 0 \cdots 0] \in \mathbb{N}^{1\times n}$ and $\gamma \in \mathbb{N}^{n\times 1}$ be defined by $\gamma_{i1} = 1$ if $q_i \in F$ and $\gamma_{i1} = 0$ if $q_i \notin F$. Let $\mu\colon A^* \to \mathbb{N}_\infty^{n\times n}$ be the morphism defined by $\mu(a) = DD_a$ for $a \in A$, and let $\gamma' = D\gamma$. Then, for all $w = a_1 \cdots a_\ell \in A^+$,

$$\begin{aligned}(S,w) &= \mathrm{Card}\{z \in L \mid \tau_\#(\pi_1(z)) = w\}\\ &= \sum_{1\leq j\leq n,\, q_j\in F} (DD_{a_1}DD_{a_2}\cdots DD_{a_\ell}D)_{1j}\\ &= \lambda\mu(a_1\cdots a_\ell)D\gamma\\ &= \lambda\mu(w)\gamma'.\end{aligned}$$

This shows that S is $\mathbb{N}_\infty$-recognizable (whatever the value (S,ε) is). □

3.4.3 Counting b-Definable Properties of b-Automatic Sequences Is b-Regular

We are now able to prove the main result of this section, namely, Theorem 3.4.15.

Proposition 3.4.14. *If $x\colon \mathbb{N}^d \to \mathbb{N}$ is $(\mathbb{N}_\infty, b)$-regular, then it is $(\mathbb{N}, b)$-regular.*

Proof. Suppose that $x\colon \mathbb{N}^d \to \mathbb{N}$ is $(\mathbb{N}_\infty, b)$-regular. Then $\sum_{w\in(A_b^d)^*} x(\mathrm{val}_b(w))\, w$ is $\mathbb{N}_\infty$-recognizable. By Proposition 3.4.11, the latter formal series is indeed $\mathbb{N}$-recognizable since $(S,w) = x(\mathrm{val}_b(w)) \in \mathbb{N}$ for all $w \in (A_b^d)^*$. Hence x is $(\mathbb{N}, b)$-regular. □

Theorem 3.4.15. *If X is a b-definable subset of $\mathbb{N}^{d+1}$, then the sequence $a\colon \mathbb{N}^d \to \mathbb{N}_\infty$ defined by*

$$\forall (n_1, \ldots, n_d) \in \mathbb{N}^d,\ a(n_1, \ldots, n_d) = \mathrm{Card}\{m \in \mathbb{N} \mid (n_1, \ldots, n_d, m) \in X\}, \qquad (3.4)$$

is $(\mathbb{N}_\infty, b)$-regular. If moreover $a(\mathbb{N}^d) \subseteq \mathbb{N}$, then a is $(\mathbb{N}, b)$-regular.

Proof. By Theorem 3.3.4, X is b-recognizable. So, the language

$$L = \{(\mathrm{rep}_b(n_1), \ldots, \mathrm{rep}_b(n_{d+1}))^{\#} \mid (n_1, \ldots, n_{d+1}) \in X\}$$

is regular by Proposition 3.2.2. Then, for all $n_1, \ldots, n_d \in \mathbb{N}$,

$$a(n_1, \ldots, n_d) = \mathrm{Card}\{z \in L \mid \pi_{1,\ldots,d}(z) \in \#^*(\mathrm{rep}_b(n_1), \ldots, \mathrm{rep}_b(n_d))^{\#}\}$$

where $\pi_{1,\ldots,d}$ denotes the projection onto the first d components. Let $A = (A_b \cup \{\#\})^d \setminus \{\#\}$ and

$$S\colon A^* \to \mathbb{N}_\infty,\ w \mapsto \mathrm{Card}\{z \in L \mid \pi_{1,\ldots,d}(z) \in \#^* w\}.$$

Observe that, for all $w \in A^*$ and $z \in L$, we have $\pi_{1,\ldots,d}(z) \in \#^* w \iff \tau_\#(\pi_{1,\ldots,d}(z)) = w$ since w does not contain the letter $\#$. Here $\tau_\#$ is the morphism defined by $a \mapsto a$ for $a \in A$ and $\# \mapsto \varepsilon$. Then S is $\mathbb{N}_\infty$-recognizable by Theorem 3.4.13. As $(S, (\mathrm{rep}_b(n_1), \ldots, \mathrm{rep}_b(n_d))^{\#}) = a(n_1, \ldots, n_d)$ for all $n_1, \ldots, n_d \in \mathbb{N}$, we obtain that a is $(\mathbb{N}_\infty, b)$-regular by Proposition 3.4.6.

The fact that a is $(\mathbb{N}, b)$-regular if $a(\mathbb{N}) \subseteq \mathbb{N}$ follows from Proposition 3.4.14. □

As an application, the factor complexity of a b-automatic sequence is $(\mathbb{N}, b)$-regular.

Proposition 3.4.16. *The factor complexity $n \mapsto \mathrm{Card}(L_n(x))$ of a b-automatic sequence $x\colon \mathbb{N} \to \mathbb{N}$ is $(\mathbb{N}, b)$-regular.*

Proof. Let $x\colon \mathbb{N} \to \mathbb{N}$ be a b-automatic sequence. For all $n \in \mathbb{N}$, $\mathrm{Card}(L_n(x)) = \mathrm{Card}\{i \in \mathbb{N} \mid \forall j < i,\ x(j) \cdots x(j + n - 1) \neq x(i) \cdots x(i + n - 1)\}$. Now let $X = \{(i, n) \mid \forall j < i,\ \exists t < n,\ x(j + t) \neq x(i + t)\}$. Then $X \subseteq \mathbb{N}^2$ is b-definable by Theorem 3.3.4, and, for all $n \in \mathbb{N}$, we have $\mathrm{Card}(L_n(x)) = \mathrm{Card}\{i \in \mathbb{N} \mid (i, n) \in X\}$; hence, the factor complexity of x is $(\mathbb{N}, b)$-regular by Theorem 3.4.15. □

In a similar manner, we can show the following. In order not to overburden the text, we do not define these counting functions here and refer the interested reader to [148].

Proposition 3.4.17. *Let $x\colon \mathbb{N} \to \mathbb{N}$ be a b-automatic sequence.*

- *The function that maps n to the number of squares (or palindromes, unbordered factors, k-powers) of x beginning at position n is $(\mathbb{N}_\infty, b)$-regular.*

- *The recurrence function of x is* $(\mathbb{N}_\infty, b)$*-regular.*
- *The appearance function of x is* $(\mathbb{N}, b)$*-regular.*
- *The separator length function of x is* $(\mathbb{N}, b)$*-regular.*
- *The permutation complexity of x is* $(\mathbb{N}, b)$*-regular.*
- *The periodicity function of x is* $(\mathbb{N}_\infty, b)$*-regular.*
- *The function that maps n to the number of unbordered factors of length n of x is* $(\mathbb{N}, b)$*-regular.*

Using the same technique, it can be shown that all these quantities are either $O(n)$ or infinite for at least one n.

Proposition 3.4.18. *Let X be a b-definable subset of $\mathbb{N}^2$, and let $a\colon \mathbb{N} \to \mathbb{N}_\infty$ be the sequence defined by $a(n) = \mathrm{Card}\{m \in \mathbb{N} \mid (n, m) \in X\}$ for all $n \in \mathbb{N}$. Then either $a(n) = \infty$ for some $n \in \mathbb{N}$ or $a(n) = O(n)$.*

Proof. If $L \in \mathbb{N}$ is such that for all $(m, n) \in X$, $|\mathrm{rep}_b(m)| \leq |\mathrm{rep}_b(n)| + L$, then for all $n \in \mathbb{N}$, $a(n) \leq b^L n$. If a is not $O(n)$, then for all $L \in \mathbb{N}$, there exists $(m, n) \in X$ such that $|\mathrm{rep}_b(m)| > |\mathrm{rep}_b(n)| + L$. Therefore $(\mathrm{rep}_b(m), \#^K \mathrm{rep}_b(n)) \in \mathrm{rep}_b(X)^\#$ for some $K > L$. As X is b-definable, there is a DFA accepting $\mathrm{rep}_b(X)^\#$. By choosing L equal to the number of states of this DFA and applying the pumping lemma, we obtain infinitely many elements (m', n) in X. This means that $a(n) = \infty$. □

It seems more difficult to obtain similar enumeration results in the multidimensional setting. For example, what about the following question?

Problem 3.4.19. Must the function $f\colon \mathbb{N}^2 \to \mathbb{N}$ that counts the number of rectangular factors of size $m \times n$ in a bidimensional b-automatic sequence be $(\mathbb{N}, b)$-regular?

3.4.4 b-Synchronized Sequences

The family of b-synchronized sequences lies in between the families of b-automatic sequences and b-regular sequences; see Proposition 3.4.21 and Theorem 3.4.23 below. Therefore, a natural question in the context developed in the present chapter is whether (various) enumeration problems about b-automatic sequences can or cannot be described by b-synchronized sequences.

Definition 3.4.20. A sequence $x\colon \mathbb{N}^d \to \mathbb{N}$ is *b-synchronized* if its graph, *i.e.*, the subset

$$G_x := \{(n_1, \ldots, n_d, x(n_1, \ldots, n_d)) \mid n_1, \ldots, n_d \in \mathbb{N}\}$$

of $\mathbb{N}^{d+1}$, is b-recognizable.

Proposition 3.4.21. *Let A be a finite subset of $\mathbb{N}$ and let $x\colon \mathbb{N}^d \to A$. Then x is b-synchronized if and only if it is b-automatic.*

Proof. For each $a \in A$, $x^{-1}(a) = \{(n_1, \ldots, n_d) \in \mathbb{N}^d \mid (n_1, \ldots, n_d, a) \in G_x\}$ and $G_x = \bigcup_{a \in A} (x^{-1}(a) \times \{a\})$. Therefore, the result follows from Proposition 3.2.8 and Theorem 3.3.4. □

Note that the use of Theorem 3.3.4 in the previous proof is somewhat superfluous since we could easily build finite automata recognizing the fibers $x^{-1}(a)$ and the graph G_x.

We have the following useful lemma.

Lemma 3.4.22. *If $x: \mathbb{N}^d \to \mathbb{N}$ is a b-synchronized sequence, then there is a b-definable subset X of $\mathbb{N}^{d+1}$ such that, for all $n_1, \ldots, n_d \in \mathbb{N}$, $x(n_1, \ldots, n_d) = \mathrm{Card}\{m \in \mathbb{N} \mid (n_1, \ldots, n_d, m) \in X\}$.*

Proof. Let x be a b-synchronized sequence. Then G_x is b-definable by Theorem 3.3.4. Therefore, the subset $X = \{(n_1, \ldots, n_d, m) \in \mathbb{N}^{d+1} \mid m < x(n_1, \ldots, n_d)\} = \{(n_1, \ldots, n_d, m) \in \mathbb{N}^{d+1} \mid \exists \ell\ (n_1, \ldots, n_d, \ell) \in G_x \text{ and } m < \ell\}$ is b-definable as well, and of course $x(n_1, \ldots, n_d) = \mathrm{Card}\{m \in \mathbb{N} \mid (n_1, \ldots, n_d, m) \in X\}$ for all $n_1, \ldots, n_d \in \mathbb{N}$. □

Theorem 3.4.23. *Any b-synchronized sequence is $(\mathbb{N}, b)$-regular.*

Proof. This is a consequence of Lemma 3.4.22 and Theorem 3.4.15. □

Proposition 3.4.24. *If $x: \mathbb{N} \to \mathbb{N}$ is b-synchronized, then $x(n)$ is $O(n)$.*

Proof. The result is a consequence of Lemma 3.4.22 and Proposition 3.4.18. □

We saw in Proposition 3.4.16 that the factor complexity of a b-automatic sequence is $(\mathbb{N}, b)$-regular. In fact, we have the more precise following result, which we give without proof.

Proposition 3.4.25 ([522]). *Let $x: \mathbb{N} \to \mathbb{N}$ be a b-automatic sequence. Then the factor complexity of x is b-synchronized.*

In view of Propositions 3.4.18 and 3.4.24, one might think that all the quantities of Proposition 3.4.17 are in fact b-synchronized. However, it is not the case.

Proposition 3.4.26. *Let $X = \{2^i \mid i \in \mathbb{N}\}$. Then χ_X is 2-automatic, but the function that counts the number of unbordered factors of length n of χ_X is not 2-synchronized.*

Proof. As $\mathrm{rep}_b(X) = 10^*$, we get that χ_X is 2-automatic. Let $y: \mathbb{N} \to \mathbb{N}$ be the function that maps n to the number of unbordered factors of length n of χ_X. Suppose that y is 2-synchronized, *i.e.*, that its graph $G_y = \{(n, y(n)) \mid n \in \mathbb{N}\}$ is 2-recognizable. Then $\mathrm{rep}_2(G)$ is accepted by some DFA $\mathcal{M}$. For all integers $n \geq 2$, we have $y(2^n + 1) = n + 2$; hence, $(10^{n-1}1, 0^{n-\lfloor \log_2(n+2) \rfloor}\mathrm{rep}_2(n+2))$ is accepted by $\mathcal{M}$. By choosing $n - \lfloor \log_2(n+2) \rfloor$ to be larger than the size of $\mathcal{M}$, the result follows from an application of the pumping lemma. □

3.5 First-Order Logic and U-Automatic Sequences

In order to be able to provide a logical framework for positional numeration systems, we encounter two major problems:

- In general, $\mathbb{N}$ is not U-recognizable.
- In general, the addition is not recognized by finite automaton.

Theorem 3.5.3 below shows that a nice setting is given by the so-called Pisot numeration systems.

Definition 3.5.1. A *Pisot number* is an algebraic integer greater than 1 such that all of its Galois conjugates have moduli less than 1.

Definition 3.5.2. A positional numeration system rep_U is *Pisot* if the base sequence U satisfies a linear recurrence whose characteristic polynomial is the minimal polynomial of a Pisot number.

Theorem 3.5.3 ([113, 232]). *If* rep_U *is a Pisot numeration system, then the sets* $\mathbb{N}$ *and* $\{(x,y,z) \in \mathbb{N}^3 \mid x + y = z\}$ *are U-recognizable.*

Definition 3.5.4. A subset of $\mathbb{N}^d$ is *U-definable* if it is definable in the logical structure $\langle \mathbb{N}, +, V_U \rangle$, where $V_U(0) = 1$, and for x a positive integer, $V_U(x)$ denotes the smallest U_i occurring in the greedy U-representation of x with a nonzero coefficient.

Example 3.5.5. We have $V_F(11) = 3$ and $V_F(26) = 5$.

Theorem 3.5.6 ([113]). *If* rep_U *is a Pisot numeration system, then a subset of* $\mathbb{N}^d$ *is U-recognizable if and only if it is U-definable. Consequently, the first-order theory of* $\langle \mathbb{N}, +, V_U \rangle$ *is decidable.*

As an application, one can prove (and reprove, or verify) many results about the Fibonacci word

$$f = 0100101001001010010100100101 0010 \cdots$$

(which is the fixed point of $0 \mapsto 01,\ 1 \mapsto 0$). Indeed, the Fibonacci word f is an F-automatic sequence as it is generated by the DFAO of Figure 3.4 whenever the inputs are the Zeckendorf representations of nonnegative integers.

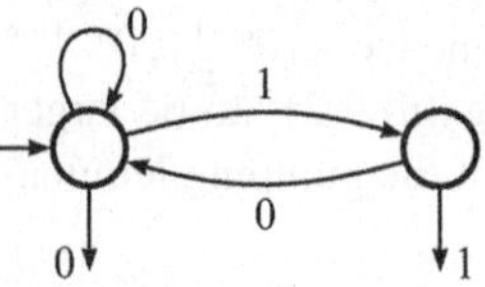

Fig. 3.4 A DFAO generating the Fibonacci word

Here are some concrete applications (among many others), all of which have been shown in a purely mechanical way [427]. Again, in order not to overburden the text, we give no definition but one. An infinite word is *linearly recurrent* if there exists a constant C such that the distance between any two occurrences of any factor x is at most $C|x|$. For the missing definitions (neither included here nor in Chapter 1), we refer the interested reader to [427].

- f is not ultimately periodic.
- f contains no fourth powers.
- f is reversal invariant.
- f is linearly recurrent.
- Characterizations of the squares (or cubes, antisquares, palindromes, antipalindromes) occurring in f.
- Characterizations of the least periods of factors (or unbordered factors, Lyndon factors, special factors) of f.
- Computation of the critical exponent and initial critical exponent of f.
- The lexicographically least element in the shift orbit closure $\mathscr{S}(f)$ is $0f$.

In a similar fashion, one can also obtain results concerning the Tribonacci word

$$t = 0102010010200102010010102010010201010 2\cdots$$

(which is the fixed point of $0 \mapsto 01,\ 1 \mapsto 02,\ 2 \mapsto 0$) [428]. In this case, we work within the positional numeration system based on the sequence $U \colon \mathbb{N} \to \mathbb{N}$ defined by $U(0) = 1,\ U(1) = 2,\ U(2) = 3$ and $U(n+3) = U(n+2) + U(n+1) + U(n)$ for $n \in \mathbb{N}$.

We end this section by a problem.

Problem 3.5.7. Do the results on enumeration of b-automatic sequences described in this section extend to Pisot numeration systems?

3.6 First-Order Logic and Real Numbers

In general real numbers are represented by infinite words. In this context, we consider Büchi automata, which allows us to define a notion of (base-related) recognizability of multidimensional sets of reals. In the continuity of the ideas developed so far, we will show that the so-called β-recognizable sets can again be characterized in terms of first-order logic, which will provide us with decision procedures for various problems concerning those sets.

3.6.1 Büchi Automata

Büchi automata are defined as NFAs, but the acceptance criterion has to be adapted: an infinite word is accepted if it labels a path going infinitely many times through an accepting state. In the present chapter, we always assume that a Büchi automaton is finite. Without loss of generality, we also always assume that there is only one initial state.

Example 3.6.1. The Büchi automaton of Figure 3.5 accepts the infinite words over $\{a,b\}$ containing finitely many a's.

Subsets of $A^{\mathbb{N}}$ are called *ω-languages*, and *ω-regular languages* are defined as ω-languages which are accepted by (finite) Büchi automata. Regular languages and ω-regular languages share some important properties: their families are closed under Boolean operations, morphic image and inverse image under a morphism. Nevertheless, they differ in some other aspects. One of them is determinism. As with DFAs, we can define deterministic Büchi automata. But one has to be careful as the family of ω-languages that are accepted by deterministic Büchi automata is strictly included in that of ω-regular languages.

Example 3.6.2. No deterministic Büchi automaton accepts the ω-language accepted by the Büchi automaton of Figure 3.5.

For more on automata reading infinite words, see [476]. Let us stress that, contrary to the present chapter, Büchi automata are not considered finite by default in [476].

3.6.2 Real Bases β

Throughout the text, β designates a real number greater than 1. For a real number x, any infinite word $u = u_\ell \cdots u_1 u_0 \star u_{-1} u_{-2} \cdots$ with $\ell \geq 0$, $u_i \in C$ for all $i \leq \ell$ where C is a finite subset of $\mathbb{Z}$ and such that

$$\mathrm{val}_\beta(u) := \sum_{-\infty < i \leq \ell} u_i \beta^i = x$$

is a *β-representation* of x. In general, this is not unique.

Note that β-numeration systems are also presented in Chapter 8.1.

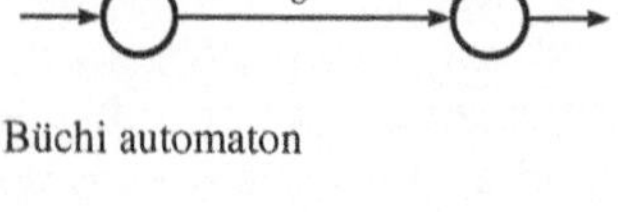

Fig. 3.5 A (nondeterministic) Büchi automaton

Example 3.6.3. Consider $x = \phi^{-1}$, where ϕ is the Golden Ratio. The words $u = 0 \star 001111\cdots$, $v = 0 \star 0101010\cdots$, and $w = 0 \star 10^{\omega}$ are all β-representations of x.

For $x \geq 0$, among all β-representations of x, we distinguish the *β-expansion*

$$d_{\beta}(x) = x_{\ell} \cdots x_1 x_0 \star x_{-1} x_{-2} \cdots$$

which is obtained by the greedy algorithm: we fix the minimal $\ell \in \mathbb{N}$ such that

$$x = \sum_{-\infty < i \leq \ell} x_i \beta^i \text{ and, for all } i \leq \ell,\ x_i \geq 0,$$

and, for all $k \leq \ell$,

$$\sum_{-\infty < i \leq k} x_i \beta^i < \beta^{k+1}.$$

The digits x_i then belong to the alphabet $A_{\beta} = \{0, \ldots, \lceil \beta \rceil - 1\}$. One has $x_{\ell} \neq 0$ if and only if $x \geq 1$ and real numbers in $[0, 1)$ have a β-expansion of the form $0 \star u$ with $u \in A_{\beta}^{\mathbb{N}}$. In particular, $d_{\beta}(0) = 0 \star 0^{\omega}$.

In order to deal with negative numbers, $\bar{a}$ denotes the integer $-a$ for all $a \in \mathbb{Z}$. Moreover we write $\overline{u\,v} = \bar{u}\,\bar{v}$, $\overline{u \star v} = \bar{u} \star \bar{v}$, and $\bar{\bar{u}} = u$. For $x < 0$, the β-expansion of x is defined as

$$d_{\beta}(x) = \overline{d_{\beta}(-x)}.$$

We let $\bar{A}_{\beta} = \{\bar{0}, \bar{1}, \ldots, \overline{\lceil \beta \rceil - 1}\}$ and $\tilde{A}_{\beta} = A_{\beta} \cup \bar{A}_{\beta}$ (with $\bar{0} = 0$).

Now let us define the β-expansion of a vector $\mathbf{x}$ of $\mathbb{R}^d$.

Definition 3.6.4. Let $\mathbf{x} = (x_1, \ldots, x_d)$ be a vector in $\mathbb{R}^d$. We define the β-expansion of $\mathbf{x}$ as being the word $d_{\beta}(\mathbf{x})$ over the alphabet $\tilde{A}_{\beta}^{d} \cup \{\star\}$ that belongs to $0^* d_{\beta}(x_1) \times 0^* d_{\beta}(x_2) \times \cdots \times 0^* d_{\beta}(x_d)$ and that does not start with $\mathbf{0}$ except if $|x_i| < 1$ for all i, in which case we consider the word starting with $\mathbf{0}\star$.

Otherwise stated, the β-expansions of each component are synchronized by possibly using some leading zeros in such a way that all the $\star$ symbols occur at the same position in every β-expansion.

Example 3.6.5. Consider $\mathbf{x} = (x_1, x_2) = (\frac{1+\sqrt{5}}{4}, 2 + \sqrt{5})$. We have

$$d_{\phi}(\mathbf{x}) = \begin{matrix} 0\,0\,0\,0 \star 1\,0\,0\,1\,0\,0 \cdots \\ 1\,0\,0\,0 \star 0\,0\,0\,0\,0\,0 \cdots \end{matrix}$$

where the first ϕ-expansion is padded with some leading zeros. Now we consider an example where all the components have moduli less than one. With $\mathbf{y} = (x_1, x_2) =$

$(\frac{1+\sqrt{5}}{4}, -\frac{1}{2})$, we get

$$d_\phi(\mathbf{y}) = \begin{matrix} 0 \star 1\,0\,0\,1\,0\,0 \cdots \\ 0 \star 0\,\bar{1}\,0\,0\,\bar{1}\,0 \cdots \end{matrix}$$

where the two ϕ-expansions start with one symbol 0 followed by $\star$.

We let $S_\beta(\mathbb{R}^d)$ be the topological closure of $\mathbf{0}^* d_\beta(\mathbb{R}^d)$. For $u \star v \in (\mathbb{Z}^d)^+ \star (\mathbb{Z}^d)^{\mathbb{N}}$ with finitely many possible digits, we define $\mathrm{val}_\beta(u \star v)$ to be the vector in $\mathbb{R}^d$ obtained by evaluating each component of $u \star v$.

Definition 3.6.6. For $X \subseteq \mathbb{R}^d$, we define $S_\beta(X)$ as $S_\beta(X) = S_\beta(\mathbb{R}^d) \cap \mathrm{val}_\beta^{-1}(X)$. For $\mathbf{x} \in \mathbb{R}^d$, the elements in $S_\beta(\mathbf{x})$ are called the *quasi-greedy β-representations of* $\mathbf{x}$.

Here and throughout the text, we write $S_\beta(\mathbf{x})$ instead of $S_\beta(\{\mathbf{x}\})$. Note that β-expansions are particular quasi-greedy β-representations.

Remark 3.6.7. If β is an integer, then $S_\beta(X)$ is the set of *all* β-representations of elements in X. Otherwise stated, when β is an integer, any β-representation is a quasi-greedy β-representation.

Proposition 3.6.8. *Let $X \subseteq \mathbb{R}^d$. Then X is closed if and only if $S_\beta(X)$ is closed.*

Proof. Suppose first that X is closed. Then $\mathrm{val}_\beta^{-1}(X)$ is closed since the function $\mathrm{val}_\beta \colon (\tilde{A}_\beta^d)^+ \star (\tilde{A}_\beta^d)^\omega \to \mathbb{R}^d$ is continuous. As $S_\beta(\mathbb{R}^d)$ is closed by definition, we obtain that $S_\beta(X) = S_\beta(\mathbb{R}^d) \cap \mathrm{val}_\beta^{-1}(X)$ is closed as well.

Conversely, suppose that $S_\beta(X)$ is closed, and let $\mathbf{x}^{(n)}$ be a sequence of X converging to some $\mathbf{x}$. By the pigeonhole principle, there exists a subsequence $\mathbf{x}^{(k(n))}$ of $\mathbf{x}^{(n)}$ such that, for all n, $\mathbf{x}^{(k(n))} - \mathbf{x}$ has a constant sign (potentially 0) on each component. Then the sequence $d_\beta(\mathbf{x}^{(k(n))})$ converges to some $u \star v \in S_\beta(X)$. The function val_β being continuous, we have $\mathrm{val}_\beta(u \star v) = \mathbf{x}$, and hence $\mathbf{x} \in X$. This proves that X is closed. □

As usual, we let $d_\beta^*(1)$ denote the lexicographically greatest $w \in \mathbb{N}^{\mathbb{N}}$ not ending in 0^ω and such that $\mathrm{val}_\beta(0 \star w) = 1$. The infinite word $d_\beta^*(1)$ has the property of being the supremum of all its shifted sequences; see, for instance, [386]. For all bases $\beta > 1$, one has $d_\beta(1) = 1 \star 0^\omega$, whereas the definition of $d_\beta^*(1)$ indeed depends on β. The following theorem is known as Parry's theorem or Parry's criterion. A proof of this result can be found in [386].

Theorem 3.6.9 (Parry [469]). *Let $u = u_\ell \cdots u_1 u_0 \star u_{-1} u_{-2} \cdots$ with $\ell \geq 0$ and $u_i \in \mathbb{N}$ for all $i \leq \ell$. Then*

$$u \in 0^* d_\beta(\mathbb{R}^{\geq 0}) \iff \forall k \leq \ell,\ u_k u_{k-1} \cdots < d_\beta^*(1), \text{ and}$$

$$u \in S_\beta(\mathbb{R}^{\geq 0}) \iff \forall k \leq \ell,\ u_k u_{k-1} \cdots \leq d_\beta^*(1).$$

Example 3.6.10. We continue Example 3.6.3. We have $d_\phi^*(1) = (10)^\omega$. Thanks to Parry's theorem, the ϕ-expansions of real numbers in $[0, 1)$ are of the form $0 \star u$, where $u \in \{0, 1\}^{\mathbb{N}}$ does not contain 11 as a factor and does not end in $(10)^\omega$. So the ϕ-expansion of x is w, but both v and w belong to $S_\beta(x)$.

The following proposition characterizes which real numbers admit quasi-greedy β-representations other than those of the form $0^\ell d_\beta(x)$: they are exactly the real numbers in the set $\{\frac{x}{\beta^i} \mid x \in \mathbb{Z}_\beta,\ i \in \mathbb{N}\}$, where $\mathbb{Z}_\beta$ is the set of the so-called β-integers. The notion of β-integers will be central in Section 3.6.5 and thus deserves a proper definition.

Definition 3.6.11. A real number x is a *β-integer* if $d_\beta(x)$ is of the kind $u \star 0^\omega$. The set of β-integers is denoted by $\mathbb{Z}_\beta$.

Proposition 3.6.12. *Let $x \in [0, 1)$. If $d_\beta(x) = 0 \star x_1 \cdots x_k 0^\omega$ with $k \geq 1$ and $x_k \neq 0$, then $S_\beta(x) = 0^*\{d_\beta(x), 0 \star x_1 \cdots x_{k-1}(x_k - 1)d_\beta^*(1)\}$, and $S_\beta(x) = 0^* d_\beta(x)$ otherwise.*

Proof. Let $u \star v \in S_\beta(x)$. As $x \in [0, 1)$, we have $u \in 0^+$. If v does not end in $d_\beta^*(1)$, then $u \star v \in 0^* d_\beta(x)$ by Theorem 3.6.9. Suppose now that v ends in $d_\beta^*(1) = d_1 d_2 \cdots$. Let $m \geq 0$ be minimal such that $v = v_1 \ldots v_m d_\beta^*(1)$. Then $m \geq 1$ and $v_m < d_1$. We claim that $d_\beta(x) = 0 \star v_1 \cdots v_{m-1}(v_m + 1)0^\omega$. By minimality of m, for all $1 \leq j \leq m$, we have

$$v_j \ldots v_m d_\beta^*(1) < d_\beta^*(1) \leq d_1 \cdots d_{m-j+1} d_\beta^*(1),$$

hence $v_j \ldots v_m < d_1 \cdots d_{m-j+1}$. If $v_j \ldots v_{m-1} < d_1 \cdots d_{m-j}$, then $v_j \ldots v_{m-1}(v_m + 1)0^\omega < d_\beta^*(1)$. If $v_j \ldots v_{m-1} = d_1 \cdots d_{m-j}$, then $v_m < d_{m-j+1}$ and $v_j \ldots v_{m-1}(v_m + 1)0^\omega \leq d_1 \cdots d_{m-j+1} 0^\omega < d_\beta^*(1)$. As $\mathrm{val}_\beta(0 \star v_1 \cdots v_{m-1}(v_m + 1)0^\omega) = x$, we obtain the claim by Theorem 3.6.9.

Now we suppose that $d_\beta(x) = 0 \star x_1 \cdots x_k 0^\omega$ with $k \geq 1$ and $x_k \neq 0$ (in particular, $x > 0$). From the previous paragraph, we obtain that $S_\beta(x) \subseteq 0^*\{d_\beta(x), 0 \star x_1 \cdots x_{k-1}(x_k - 1)d_\beta^*(1)\}$. The other inclusion holds by Theorem 3.6.9.

If $d_\beta(x) = 0 \star 0^\omega$, then $x = 0$ and $S_\beta(0) = 0^+ \star 0^\omega$. Finally we suppose that $d_\beta(x)$ does not end in 0^ω. From the first paragraph, we obtain that if $u \star v \in S_\beta(x)$, then $u \in 0^+$ and v does not end in $d_\beta^*(1)$. This proves $S_\beta(x) = 0^* d_\beta(x)$. □

Corollary 3.6.13. *Let $x \in \mathbb{R}^{\geq 0}$ and let $d_\beta^*(1) = d_1 d_2 \cdots$.*

- *If $d_\beta(x) = x_\ell \cdots x_0 \star x_{-1} \cdots x_{-k} 0^\omega \in A_\beta^+ \star A_\beta^\omega$ with $x_{-k} \neq 0$, then*

$$S_\beta(x) = 0^*\{d_\beta(x), x_\ell \cdots x_0 \star x_{-1} \cdots x_{-k+1}(x_{-k} - 1)d_1 d_2 \cdots\}$$

- *If $d_\beta(x) = x_\ell \cdots x_k 0^k \star 0^\omega \in A_\beta^+ \star A_\beta^\omega$ with $x_k \neq 0$, then*

$$S_\beta(x) = 0^*\{d_\beta(x), x_\ell \cdots x_{k+1}(x_k - 1)d_1 \cdots d_k \star d_{k+1} d_{k+2} \cdots\}.$$

- $S_\beta(x) = 0^* d_\beta(x)$ *in all other cases.*

Moreover, we have $S_\beta(-x) = \overline{S_\beta(x)}$.

3.6.3 β-Recognizable Sets of Real Numbers

Definition 3.6.14. A subset X of $\mathbb{R}^d$ is *β-recognizable* if $S_\beta(X)$ is ω-regular.

The following result shows that leading zeros do not affect the β-recognizability of a subset. We omit the proof as it is similar to that of Proposition 3.2.2.

Proposition 3.6.15. *Let* $X \subseteq \mathbb{R}^d$. *The following are equivalent:*

- *X is β-recognizable.*
- $S_\beta(X) \cap \big((\tilde{A}_\beta^d \setminus \{\mathbf{0}\})(\tilde{A}_\beta^d)^* \star (\tilde{A}_\beta^d)^\omega \cup \mathbf{0}\star(\tilde{A}_\beta^d)^\omega\big)$ *is ω-regular.*
- *There exists an ω-regular language* $L \subseteq (\tilde{A}_\beta^d)^+ \star (\tilde{A}_\beta^d)^\omega$ *such that* $\mathbf{0}^*(\mathbf{0}^*)^{-1}L = S_\beta(X)$.

We also have the following nice criterion.

Proposition 3.6.16. *Two β-recognizable subsets of* $\mathbb{R}^d$ *coincide if and only if they have the same ultimately periodic quasi-greedy β-representations.*

Proof. The result follows from the well-known fact that two ω-regular languages are equal if and only if they have the same ultimately periodic elements [476]. □

In the case of closed subsets of $\mathbb{R}^d$, we can require additional conditions on the Büchi automata recognizing them.

Proposition 3.6.17. *A β-recognizable subset X of* $\mathbb{R}^d$ *is closed if and only if* $S_\beta(X)$ *is accepted by a deterministic Büchi automaton all of whose states are final.*

Proof. It is easily seen that an ω-regular language L is closed if and only if it is accepted by a deterministic Büchi automaton in which each state is final (see, e.g., [476, Proposition 3.9]). Then the result follows from Proposition 3.6.8. □

We note that, in our context of Büchi automata recognizing sets of real numbers, the final/non-final status of the states occurring before an edge labeled ★ has no impact on the accepted language.

Definition 3.6.18. A *Parry number* is a real number β greater than 1 for which $d_\beta^*(1)$ is ultimately periodic.

Proposition 3.6.19. *If β is Parry, then a subset X of* $\mathbb{R}^d$ *is β-recognizable if and only if* $d_\beta(X)$ *is ω-regular.*

Proof. For the sake of clarity, we do the proof for $d = 1$. Let $X \subseteq \mathbb{R}$. First note that $d_\beta(X)$ is ω-regular if and only if $0^* d_\beta(X)$ is as well. By Corollary 3.6.13, we have

$$0^* d_\beta(X) = S_\beta(X) \setminus \{u \star v \in \tilde{A}_\beta^+ \star \tilde{A}_\beta^\omega \mid uv \text{ ends in } d_\beta^*(1) \text{ or in } \overline{d_\beta^*(1)}\}.$$

As β is a Parry number, $\{u \star v \in \tilde{A}_\beta^+ \star \tilde{A}_\beta^\omega \mid uv \text{ ends in } d_\beta^*(1) \text{ or in } \overline{d_\beta^*(1)}\}$ is an ω-regular language. This shows that $d_\beta(X)$ is ω-regular if $S_\beta(X)$ is as well.

Conversely, as $d_\beta^*(1) = d_1 d_2 \cdots$ is ultimately periodic, the two ω-languages

$$L_1 = \{\left(\begin{smallmatrix} u \\ u \end{smallmatrix}\right)\left(\begin{smallmatrix} \star \\ \star \end{smallmatrix}\right)\left(\begin{smallmatrix} v \\ v \end{smallmatrix}\right)\left(\begin{smallmatrix} a \\ a-1 \end{smallmatrix}\right)\left(\begin{smallmatrix} 0 \\ d_1 \end{smallmatrix}\right)\left(\begin{smallmatrix} 0 \\ d_2 \end{smallmatrix}\right)\cdots \mid u \in A_\beta^+,\ v \in A_\beta^*,\ a \neq 0\}$$
$$L_2 = \{\left(\begin{smallmatrix} u \\ u \end{smallmatrix}\right)\left(\begin{smallmatrix} a \\ a-1 \end{smallmatrix}\right)\left(\begin{smallmatrix} 0 \\ d_1 \end{smallmatrix}\right)\cdots\left(\begin{smallmatrix} 0 \\ d_k \end{smallmatrix}\right)\left(\begin{smallmatrix} \star \\ \star \end{smallmatrix}\right)\left(\begin{smallmatrix} 0 \\ d_{k+1} \end{smallmatrix}\right)\left(\begin{smallmatrix} 0 \\ d_{k+2} \end{smallmatrix}\right)\cdots \mid u \in A_\beta^*,\ a \neq 0,\ k \in \mathbb{N}\}.$$

are ω-regular. By Corollary 3.6.13, we have

$$S_\beta(X) = 0^* d_\beta(X) \ \cup\ \pi_2\big((L_1 \cup L_2 \cup \overline{L_1} \cup \overline{L_2}) \cap (0^* d_\beta(X) \times (\tilde{A}_\beta \cup \{\star\})^\omega)\big).$$

This proves that $S_\beta(X)$ is ω-regular if $d_\beta(X)$ is as well. □

As a consequence of Propositions 3.6.16 and 3.6.19, we obtain the following result.

Proposition 3.6.20. *If β is Parry, then two β-recognizable subsets of $\mathbb{R}^d$ coincide if and only if they have the same ultimately periodic β-expansions.*

3.6.4 Weakly β-Recognizable Sets of Real Numbers

We now consider particular β-recognizable sets of real numbers, namely, the weakly β-recognizable subsets. We note that we have chosen to respect the original terminology of [95, 384], even though the property of being weakly β-recognizable is in fact stronger than being β-recognizable. This terminology comes from the fact that weak Büchi automata are less expressive than Büchi automata: not all ω-regular languages are accepted by weak Büchi automata.

Definition 3.6.21. A Büchi automaton is said to be *weak* if each of its strongly connected components contains either only final states or only nonfinal states.

Definition 3.6.22. A subset X of $\mathbb{R}^d$ is *weakly β-recognizable* if $S_\beta(X)$ is accepted by a weak deterministic Büchi automaton.

The advantage of weak deterministic Büchi automata is that they admit a canonical form [384, 551]. Therefore, they can be viewed as the analogues of DFAs for infinite words. Moreover, the family of ω-languages accepted by weak deterministic Büchi automata is closed under the Boolean operations of union, intersection, and complementation [408, 551]. However, let us stress that weak Büchi automata *cannot* be determinized. For example, the Büchi automaton of Figure 3.5 is clearly weak, but as already pointed out, there is no deterministic Büchi automaton accepting the same ω-language. This has important consequences

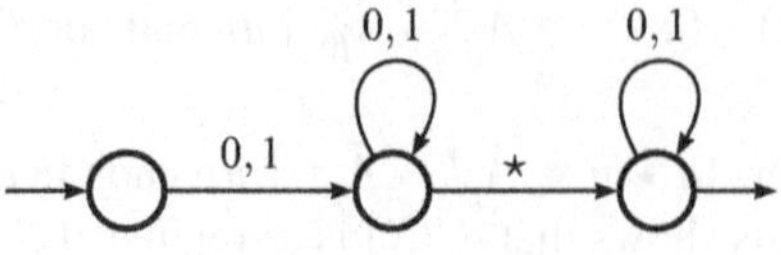

Fig. 3.6 A weak deterministic Büchi automaton accepting $S_2(\mathbb{R})$

in our work, namely, for the choice of Definition 3.6.22, which is highlighted by the following remark.

Remark 3.6.23. It is not true that a subset X of $\mathbb{R}^d$ is weakly β-recognizable if and only if $d_\beta(X)$ is accepted by a weak deterministic Büchi automaton, even when β is an integer base. Indeed, the set $\mathbb{R}$ is weakly 2-recognizable as $S_2(\mathbb{R})$ is accepted by the weak deterministic Büchi automaton of Figure 3.6. Yet we have $S_2(\mathbb{R}) \setminus 0^*d_2(\mathbb{R}) = \{0,1\}^+ \star \{0,1\}^* 1^\omega$. Since the family of ω-languages accepted by weak deterministic Büchi automata is closed under intersection and complementation, if $d_2(\mathbb{R})$ were accepted by a weak deterministic Büchi automaton, then $\{0,1\}^* 1^\omega$ would be as well, which is known to be not true as already mentioned in Example 3.6.2. This remark has to be compared with Proposition 3.6.19.

It is interesting to note that, for closed subsets of $\mathbb{R}^d$, the concepts of β-recognizability and weak β-recognizability actually coincide.

Proposition 3.6.24. *A closed subset of $\mathbb{R}^d$ is β-recognizable if and only if it is weakly β-recognizable.*

Proof. This is a straightforward consequence of Proposition 3.6.17. □

The following result is a consequence of Theorem 3.6.9. We first fix some notation that will be useful here and in the proof of Theorem 3.6.29 below. For $r \in \mathbb{R}$, we define $\text{sign}(r)$ to be $+$ if $r \geq 0$ and $-$ else. If $\mathbf{x} = (x_1, \ldots, x_d) \in \mathbb{R}^d$, then $\text{sign}(\mathbf{x}) = (\text{sign}(x_1), \ldots, \text{sign}(x_d))$. For $X \subseteq \mathbb{R}^d$ and $\mathbf{s} \in \{+,-\}^d$, we define $X_\mathbf{s} = \{\mathbf{x} \in X \mid \text{sign}(\mathbf{x}) = \mathbf{s}\})$.

Proposition 3.6.25. *If β is a Parry number, then $\mathbb{R}^d$ is weakly β-recognizable.*

Proof. As a consequence of Theorem 3.6.9, a DFA $\mathcal{A}_\beta$ is canonically associated with any Parry number β. For details on the construction of $\mathcal{A}_\beta$, we refer the reader to [386]. This DFA accepts the language of factors of those infinite words u such that $0 \star u = d_\beta(x)$ for some $x \in [0,1)$. All states of $\mathcal{A}_\beta$ are final (as any prefix of a factor is again a factor). Moreover, $\mathcal{A}_\beta$ has a loop labeled 0 on its initial state.

Given $\mathbf{s} \in \{+,-\}^d$, we build a weak deterministic Büchi automaton $\mathcal{A}_{\beta,\mathbf{s}}$ accepting $S_\beta\big((\mathbb{R}^d)_\mathbf{s}\big)$. Then the union of those 2^d ω-languages will be $S_\beta\big(\mathbb{R}^d\big)$, which will still be accepted by a weak deterministic Büchi automaton since the class of ω-languages accepted by such automata is closed under union.

We construct the automaton $\mathcal{A}_{\beta,\mathbf{s}}$ by considering two copies of $\mathcal{A}_\beta \times \cdots \times \mathcal{A}_\beta$ (d times), one for the β-integer part and one for the β-fractional part of the β-representations. For each state q of $\mathcal{A}_\beta \times \cdots \times \mathcal{A}_\beta$, we let (q, int) (resp. (q, frac))

denote the state of $\mathcal{A}_{\beta,\mathbf{s}}$ that corresponds to q in the β-integer (resp. β-fractional) part copy. In all labels of transitions of both copies of $\mathcal{A}_\beta \times \cdots \times \mathcal{A}_\beta$, we replace the ith component by its opposite value if $s_i = -$, and we leave it unchanged otherwise.

The initial state of $\mathcal{A}_{\beta,\mathbf{s}}$ is a new additional state i and, for each transition labeled $a \in \tilde{A}_\beta^d$ from the initial state to any state (q, int) of the β-integer part copy of $\mathcal{A}_\beta \times \cdots \times \mathcal{A}_\beta$, there is a new transition labeled a from i to (q, int). The terminal states are all states (q, frac). We complete $\mathcal{A}_{\beta,\mathbf{s}}$ by adding, for each state q of $\mathcal{A}_\beta \times \cdots \times \mathcal{A}_\beta$, a transition from (q, int) to (q, frac) labeled $\star$. □

Example 3.6.26. The canonical DFA $\mathcal{A}_\phi$ is depicted in Figure 3.7. The deterministic Büchi automaton depicted in Figure 3.8 accepts the ω-language $S_\phi(\mathbb{R}^{\geq 0})$. Note that the two ϕ-representations v and w of ϕ^{-1} of Example 3.6.3 are accepted as they are both quasi-greedy, whereas u is not.

Theorem 3.6.27 provides a decomposition of weakly β-recognizable subsets into their β-integer and β-fractional parts. In the case where the base β is an integer, this decomposition is in fact independent of the chosen integer base; this is Theorem 3.6.29.

To express this decomposition, we introduce the following notation. For $\mathbf{x} \in \mathbb{Z}_\beta^d$, we let $\text{rep}_\beta(\mathbf{x})$ be defined by $d_\beta(\mathbf{x}) = \text{rep}_\beta(\mathbf{x})\star\mathbf{0}^\omega$. Note that by Corollary 3.6.13, we have that, for all $\mathbf{x} \in \mathbb{Z}_\beta^d$, $S_\beta(\mathbf{x}) \cap ((\tilde{A}_\beta^d)^*\star\mathbf{0}^\omega) = \mathbf{0}^*\text{rep}_\beta(\mathbf{x})\star\mathbf{0}^\omega$. Symmetrically, for $u \in A_\beta^+$, we let $\text{val}_\beta(u) = \text{val}_\beta(u\star\mathbf{0}^\omega)$.

Recall that a Büchi automaton is said to be *trim* if it is accessible and coaccessible, *i.e.*, each state can be reached from the initial state and from each state

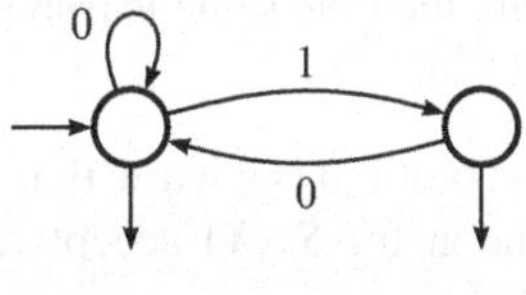

Fig. 3.7 The canonical DFA $\mathcal{A}_\phi$

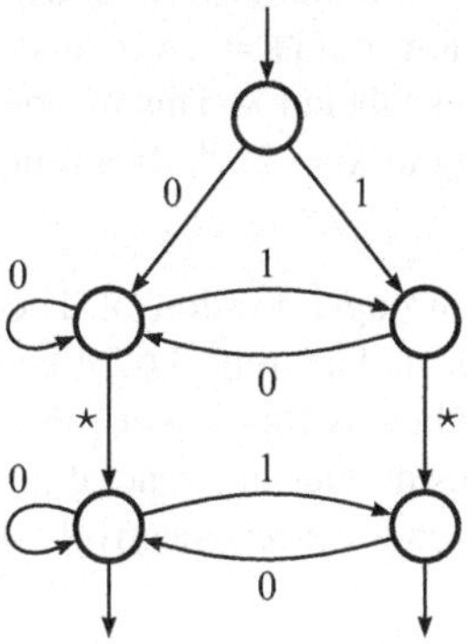

Fig. 3.8 A deterministic Büchi automaton accepting $S_\phi(\mathbb{R}^{\geq 0})$

starts an infinite accepting path. From any given Büchi automaton, we can easily build another Büchi automaton which is trim and accepts the same ω-language. Moreover, if the original Büchi automaton is weak (resp. deterministic), the obtained trim Büchi automaton is as well.

Theorem 3.6.27. *Any weakly β-recognizable subset X of $\mathbb{R}^d$ is a finite union of sets of the form $X^I + X^F$ where $X^I \subseteq \mathbb{Z}_\beta^d$ is such that $\mathrm{rep}_\beta(X^I) \subseteq (A_\beta^d)^*$ is regular and $X^F \subseteq [0,1]^d$ is weakly β-recognizable.*

Proof. Let $X \subseteq \mathbb{R}^d$ and let $\mathscr{A} = (Q, q_0, \tilde{A}_\beta^d \cup \{\star\}, F, \delta)$ be a trim deterministic Büchi automaton accepting $S_\beta(X)$. No infinite path (starting from any state) of $\mathscr{A}$ contains more than one occurrence of the letter $\star$. Hence, the set of states Q can be divided into two parts: Q_1 containing the states occurring before transitions labeled $\star$ and Q_2 containing the states occurring after those transitions. Note that $F \subseteq Q_2$. Let $q_1, \ldots, q_m$ be the states of Q_2 that can be reached (in one step) by reading the letter $\star$. Without loss of generality, we assume that the ω-languages accepted from $q_1, \ldots, q_m$ are pairwise distinct. This implies that, for all $u \in \mathbf{0}^*\mathrm{rep}_\beta(X \cap \mathbb{Z}_\beta^d)$ and all $\ell \in \mathbb{N}$, $q_0 \cdot \mathbf{0}^\ell u\star = q_0 \cdot u\star$. For each i, $1 \le i \le m$, we define $X_i^I = \{\mathrm{val}_\beta(u) \mid q_0 \cdot u\star = q_i\}$, and $X_i^F = \{\mathrm{val}_\beta(\mathbf{0}\star v) \mid v \text{ is accepted from } q_i\}$. We have $X = \bigcup_{i=1}^m X_i^I + X_i^F$. Now, for each i, $1 \le i \le m$, we consider the DFA $\mathscr{D}_i = (Q_1, q_0, \tilde{A}_\beta^d, F_i, \delta_1)$ and the Büchi automaton $\mathscr{B}_i = (Q_2, q_i, \tilde{A}_\beta^d, F, \delta_2)$, where $F_i = \{q \in Q_1 \mid q \cdot \star = q_i\}$ and δ_1 (resp. δ_2) is equal to the original transition function δ restricted to the domain $Q_1 \times \tilde{A}_\beta^d$ (resp. $Q_2 \times \tilde{A}_\beta^d$). Then the language accepted by $\mathscr{D}_i$ is $\mathbf{0}^*\mathrm{rep}_\beta(X_i^I)$ and the ω-language accepted by $\mathscr{B}_i$ is $S_\beta(X_i^F) \cap (\mathbf{0}\star\tilde{A}_\beta^\omega)$. It is now easy to modify $\mathscr{B}_i$ to obtain a deterministic Büchi automaton accepting $S_\beta(X_i^F)$. Finally, if in addition $\mathscr{A}$ has the property of being weak, then the same is true for the obtained deterministic Büchi automata accepting $S_\beta(X_j^F)$. □

Remark that, in the previous proof, it is not true that the union $X = \bigcup_{i=1}^m (X_i^I + X_i^F)$ is disjoint as a Büchi automaton for $S_\beta(X)$ accepts all quasi-greedy β-representations of elements in X.

Example 3.6.28. In the Büchi automaton of Figure 3.8, the infinite paths corresponding to the ϕ-representations $d_\phi(1) = 1 \star 0^\omega$ and $0 \star d_\phi^*(1) = 0 \star (01)^\omega$ of 1 go through the two different edges labeled $\star$. This means that, in the decomposition of Theorem 3.6.27 corresponding to $X = \mathbb{R}^{\ge 0}$, the number 1 belongs to all of the sets $X^I + X^F$.

The following result is a stronger version of Theorem 3.6.27 in the restricted case of integer bases. Indeed, in Theorem 3.6.29 below, the sets in the union are independent of the base b, whereas this is not the case in the previous theorem. Unfortunately, this stronger result does not generalize to real bases as in general $\mathbb{Z}_\beta$ differs from $\mathbb{Z}_{\beta'}$ if $\beta \neq \beta'$, even for multiplicatively dependent β, β'. For example, $2 \in \mathbb{Z}_{\varphi^2} \setminus \mathbb{Z}_\varphi$.

Theorem 3.6.29. *Any subset X of $\mathbb{R}^d$ is a finite union of sets of the form $X^I + X^F$ with $X^I \subseteq \mathbb{Z}^d$ and $X^F \subseteq [0,1]^d$ and such that $\operatorname{rep}_b(X^I)$ is regular and X^F is weakly b-recognizable for all b for which X is weakly b-recognizable.*

Proof. Let $X \subseteq \mathbb{R}^d$. With the notation introduced before Proposition 3.6.25, we have

$$X = \cup_{\mathbf{s}\in\{+,-\}^d} X_{\mathbf{s}},$$

and if $\mathcal{A}$ is a deterministic Büchi automaton accepting $S_b(X)$, then the ω-languages $S_b(X_{\mathbf{s}})$ are accepted by the deterministic Büchi automata obtained from $\mathcal{A}$ by only keeping those edges whose labels have sign $\mathbf{s}$. For the sake of simplicity, we suppose that $X \subseteq (\mathbb{R}^{\geq 0})^d$. (If we had $X \subseteq (\mathbb{R}^d)_{\mathbf{s}}$ for some $\mathbf{s} \neq (+,\ldots,+)$ (d times), then we would have to discuss the sign of each component separately, which is just a tedious adaptation of what follows.)

For $\mathbf{i} \in \mathbb{N}^d$, we define $F(X,\mathbf{i}) = \{\mathbf{x} \in [0,1]^d \mid \mathbf{i}+\mathbf{x} \in X\}$ and $I(X,\mathbf{i}) = \{\mathbf{i}' \in \mathbb{N}^d \mid F(X,\mathbf{i}) = F(X,\mathbf{i}')\}$. Then let $C(X) = \{I(X,\mathbf{i}) \mid \mathbf{i} \in \mathbb{N}^d \text{ and } F(X,\mathbf{i}) \neq \emptyset\}$. We have

$$X = \bigcup_{I(X,\mathbf{i})\in C(X)} I(X,\mathbf{i}) + F(X,\mathbf{i}).$$

Now suppose that $\mathcal{A}$ is a weak trim deterministic Büchi automaton accepting $S_b(X)$. Let q_0 be the initial state of $\mathcal{A}$ and let $q_1,\ldots,q_m$ be the states of $\mathcal{A}$ that can be reached (in one step) by reading the letter $\star$. Without loss of generality, we may suppose that the ω-languages accepted from the states $q_1,\ldots,q_m$ are pairwise distinct. We consider the same decomposition of X as in the proof of Theorem 3.6.27:

$$X = \bigcup_{j=1}^{m} (X_j^I + X_j^F),$$

with $X_j^F = \{\operatorname{val}_b(\mathbf{0}\star v) \mid v \text{ is accepted from } q_j\}$ and $X_j^I = \{\operatorname{val}_b(u) \mid q_0 \cdot u\star = q_j\}$. From the proof of Theorem 3.6.27, we know that, for each j, $\operatorname{rep}_b(X_j^I)$ is regular and that X_j^F is weakly b-recognizable.

Let us show that the two exhibited decompositions of X are actually the same. In particular, the obtained decomposition will be independent of the base b, which will prove the result. To obtain the correspondence between the two decompositions, it is enough to show that, for all $u \in A_b^*$ and all $j \in \{1,\ldots,m\}$, the following assertions are equivalent:

1. $q_0 \cdot u\star = q_j$.
2. $F(X, \operatorname{val}_b(u)) = X_j^F$.
3. $I(X, \operatorname{val}_b(u)) = X_j^I$.

As $\mathcal{A}$ accepts all the b-representations of the elements of X and the ω-languages accepted from the states $q_1,\ldots,q_m$ are pairwise distinct, the subsets $X_1^F,\ldots,X_m^F$

and $X_1^I, \ldots, X_m^I$ are pairwise distinct. Therefore, we only have to show $1 \implies 2 \wedge 3$. Suppose that $q_0 \cdot u\star = q_j$. If $x = \mathrm{val}_b(\mathbf{0}\star v)$ and v is accepted from q_j, then $u\star v \in L(\mathscr{A})$; hence, $x + \mathrm{val}_b(u) \in X$. Conversely, let $x \in [0,1]^d$ such that $x + \mathrm{val}_b(u) \in X$. Then there exists $v \in A_b^\omega$ such that $x = \mathrm{val}_b(\mathbf{0}\star v)$. As $\mathscr{A}$ accepts all the b-representations of the elements of X, we have $u\star v \in L(\mathscr{A})$. Because $\mathscr{A}$ is deterministic, v is necessarily accepted from q_j; hence, $x \in X_j^F$. This proves 2; hence, we have obtained $1 \iff 2$. Now, let $\mathbf{i} \in I(X, \mathrm{val}_b(u))$. Then $F(X, \mathbf{i}) = F(X, \mathrm{val}_b(u)) = X_j^F$. From $2 \implies 1$, we obtain $q_0 \cdot \mathrm{rep}_b(\mathbf{i})\star = q_j$; hence, $\mathbf{i} \in X_j^I$. Finally, let $\mathbf{i} \in X_j^I$. Then $\mathbf{i} = \mathrm{val}_b(u')$ with $q_0 \cdot u'\star = q_j$. From $1 \implies 2$, we obtain $F(X, \mathbf{i}) = F(X, \mathrm{val}_b(u))$; hence, $\mathbf{i} \in I(X, \mathrm{val}_b(u))$. Hence we have 3, which ends the proof. □

3.6.5 First-Order Theory for Mixed Real and Integer Variables in Base β and Büchi Automata

In order to obtain an analogue of the Büchi-Bruyère theorem for real numbers represented in base β, we need a suitable logical structure for defining the so-called β-definable subsets of $\mathbb{R}^d$. In this section we present the chosen logical structure.

Definition 3.6.30. For $a \in \tilde{A}_\beta$, we define a binary relation $X_{\beta,a}$ as follows. Suppose that $x, y \in \mathbb{R}$ with $d_\beta(x) = x_\ell \cdots x_0 \star x_{-1}x_{-2}\cdots$, then $X_{\beta,a}(x,y)$ if and only if $y = \beta^i$ for some $i \in \mathbb{Z}$, and either $i > \ell$ and $a = 0$ or $i \le \ell$ and $x_i = a$.

In other words, $X_{\beta,a}(x,y)$ is true whenever y is an integer power of the base β and the digit in ${}^\omega 0 d_\beta(x)$ corresponding to this power is a. The notation ${}^\omega 0$ means that we add infinitely many zeros to the right of the greedy representation $d_\beta(x)$. Note that here we use the notation $X_{\beta,a}(x,y)$ for $(x,y) \in X_{\beta,a}$.

Recall that $\mathbb{Z}_\beta$ is the set of β-integers; see Definition 3.6.11.

Definition 3.6.31. A subset of $\mathbb{R}^d$ is *β-definable* if it is definable by a first-order formula of

$$\langle \mathbb{R}, +, \le, \mathbb{Z}_\beta, X_\beta \rangle,$$

where X_β is the finite collection of binary predicates $\{X_{\beta,a} \mid a \in \tilde{A}_\beta\}$.

Remark 3.6.32. $x = 0$ is defined by $x + x = x$.

Remark 3.6.33. The property of being an integer power of β is definable in $\langle \mathbb{R}, +, \le\ , X_\beta \rangle$: x is a power of $\beta \iff X_{\beta,1}(x,x)$. Note that the letter 1 always belong to A_β since $\beta > 1$. If x is a power of β, then one can define the next (or the previous) power of β as follows:

$$\begin{aligned} x' = \beta x \iff & (x' \text{ is a power of } \beta) \\ & \wedge (x' > x) \\ & \wedge\ (\forall y)((y \text{ is a power of } \beta \wedge y > x) \implies y \geq x'). \end{aligned}$$

By adding the constant 1 to the structure, we can also define the properties of being a positive or negative power of β by adding $x > 1$ or $x < 1$, respectively. Consequently, any constant power of β is definable in $\langle \mathbb{R}, +, \leq, 1, X_\beta \rangle$.

Lemma 3.6.34. *The structures* $\langle \mathbb{R}, +, \leq, 1, X_\beta \rangle$ *and* $\langle \mathbb{R}, +, \leq, \mathbb{Z}_\beta, X_\beta \rangle$ *are equivalent.*

Proof. On the one hand, $z = 1$ can be defined in $\langle \mathbb{R}, +, \leq, \mathbb{Z}_\beta, X_\beta \rangle$ by the formula

$$z \in \mathbb{Z}_\beta \wedge \big[(\forall x)\big((x \in \mathbb{Z}_\beta \wedge x > 0) \implies x \geq z\big)\big].$$

On the other hand, the set $\mathbb{Z}_\beta$ can be defined in $\langle \mathbb{R}, +, \leq, 1, X_\beta \rangle$:

$$z \in \mathbb{Z}_\beta \iff (\forall y)\big[(y \text{ is a negative power of } \beta) \implies X_{\beta,0}(z, y)\big].$$

□

Remark 3.6.35. Multiplication (or division) by β is β-definable:

$$y = \beta x \Leftrightarrow (\forall b)\Big[\bigwedge_{a \in \tilde{A}_\beta} (X_{\beta,a}(x, b) \implies X_{\beta,a}(y, \beta b))\Big].$$

Note that $X_{\beta,a}(x, b)$ implies that b is an integer power of β. Consequently, multiplication (or division) by a constant power of β is β-definable.

Remark 3.6.36. The structures $\langle \mathbb{R}, +, \leq, 1 \rangle$ and $\langle \mathbb{R}, +, \leq, \mathbb{Z} \rangle$ are *not* logically equivalent : $z = 1$ is definable in $\langle \mathbb{R}, +, \leq, \mathbb{Z} \rangle$, whereas $z \in \mathbb{Z}$ is not definable in $\langle \mathbb{R}, +, \leq, 1 \rangle$; see Proposition 3.6.38.

Let us characterize the subsets of $\mathbb{R}^d$ that are definable in $\langle \mathbb{R}, +, \leq, 1 \rangle$ and in $\langle \mathbb{R}, +, \leq, \mathbb{Z} \rangle$, respectively. We will make use of the following important result.

Theorem 3.6.37 ([221]). *The structure* $\langle \mathbb{R}, +, \leq, 1 \rangle$ *admits the elimination of quantifiers.*

A *rational polyhedron* of $\mathbb{R}^d$ is the intersection of finitely many half-spaces whose borders are hyperplanes whose equations have integer coefficients. These sets are sometimes referred to as *convex polytopes*. Note that a rational polyhedron is not necessarily bounded.

Proposition 3.6.38. *The subsets of* $\mathbb{R}^d$ *which are definable in* $\langle \mathbb{R}, +, \leq, 1 \rangle$ *are the finite unions of rational polyhedra. In particular, the subsets of* $\mathbb{R}$ *which are definable in* $\langle \mathbb{R}, +, \leq, 1 \rangle$ *are the finite unions of intervals with rational endpoints.*

Proof. From Theorem 3.6.37, a subset X of $\mathbb{R}^d$ is definable in $\langle \mathbb{R}, +, \leq, 1 \rangle$ if and only if it can be expressed by a finite Boolean combination of linear constraints with rational coefficients. Now consider an equivalent formula in disjunctive normal form. This gives us the desired result. □

We end this section by a characterization of those subsets X of $\mathbb{R}^d$ which are definable in $\langle \mathbb{R}, +, \leq, \mathbb{Z} \rangle$. Note that the proof of this characterization depends on a subsequent result (namely, Theorem 3.6.44).

Theorem 3.6.39. *A subset X of $\mathbb{R}^d$ is definable in $\langle \mathbb{R}, +, \leq, \mathbb{Z} \rangle$ if and only if it is a finite union of sets of the form $X^I + X^F$, with $X^I \subseteq \mathbb{Z}^d$ definable in $\langle \mathbb{Z}, +, \leq \rangle$ and $X^F \subseteq [0, 1]^d$ definable in $\langle \mathbb{R}, +, \leq, 1 \rangle$.*

Proof. Suppose that $X = X^I + X^F$ where $X^I \subseteq \mathbb{Z}^d$ is definable in $\langle \mathbb{Z}, +, \leq \rangle$ and $X^F \subseteq [0, 1]^d$ is definable in $\langle \mathbb{R}, +, \leq, 1 \rangle$. By Remark 3.6.36, X^F is definable in $\langle \mathbb{R}, +, \leq, \mathbb{Z} \rangle$. If $\phi(y_1, \ldots, y_d)$ is a first-order formula of $\langle \mathbb{Z}, +, \leq \rangle$ defining X^I, then $\phi(y_1, \ldots, y_d) \wedge y_1 \in \mathbb{Z} \wedge \cdots \wedge y_d \in \mathbb{Z}$ is a first-order formula of $\langle \mathbb{R}, +, \leq, \mathbb{Z} \rangle$ defining X^I. Thus the predicate $(x_1, \ldots, x_d) \in X$ is definable in $\langle \mathbb{R}, +, \leq, \mathbb{Z} \rangle$ by $(\exists y_1) \cdots (\exists y_d)(\exists z_1) \cdots (\exists z_d)(x_1 = y_1 + z_1 \wedge \cdots \wedge x_d = y_d + z_d \wedge (y_1, \ldots, y_d) \in X^I \wedge (z_1, \ldots, z_d) \in X^F)$. Finite unions of definable sets are always definable, in any structure.

For the other direction, suppose that $X \subseteq \mathbb{R}^d$ is definable in $\langle \mathbb{R}, +, \leq, \mathbb{Z} \rangle$. By Theorem 3.6.44, X is weakly b-recognizable for all b. By Theorem 3.6.29, X is a finite union of sets of the form $X^I + X^F$, where $X^I \subseteq \mathbb{Z}^d$ is such that $\mathrm{rep}_b(X^I)$ is regular and $X^F \subseteq [0, 1]^d$ is b-recognizable for all b. Then, by Theorem 3.2.28 (which can be adapted to $\mathbb{Z}^d$ in a straightforward way), each X^I is semi-linear, hence definable in $\langle \mathbb{Z}, +, \leq \rangle$, and by Theorem 3.6.45, each X^F is definable in $\langle \mathbb{R}, +, \leq, 1 \rangle$. □

Note that we have used Theorem 3.6.29, which is a stronger version of Theorem 3.6.27. Indeed, we need the sets in the decomposition of X to be independent of the base b.

Finally, in the particular case of bounded subsets of $\mathbb{R}^d$, we have the following characterizations.

Corollary 3.6.40. *For any bounded subset X of $\mathbb{R}^d$, the following assertions are equivalent.*

1. *X is definable in $\langle \mathbb{R}, +, \leq, \mathbb{Z} \rangle$.*
2. *X is definable in $\langle \mathbb{R}, +, \leq, 1 \rangle$.*
3. *X is a finite union of rational polyhedra.*

Proof. This follows from Proposition 3.6.38 and Theorem 3.6.39. □

3.6.6 Characterizing β-Recognizable Sets Using Logic

The following theorem can be viewed as an analogue of Theorem 3.3.4 for real numbers represented in real bases β. Let us emphasize that the base β needs be a Pisot number in order to recognize the addition. We do not present here the details of the normalization in real Pisot bases, but the interested reader is referred to [145, 231].

Theorem 3.6.41 ([145]).

- *If β is a Parry number, then every β-recognizable subset of $\mathbb{R}^d$ is β-definable.*
- *If β is a Pisot number, then every β-definable subset of $\mathbb{R}^d$ is β-recognizable.*

In the context of the present chapter, the relevant direction is given by the second assertion. Indeed, our aim is to build suitable DFAs starting from formulæ expressing various properties of β-recognizable sets of numbers, in order to decide whether a given set satisfies a given property. For this reason, we only give a proof of the second assertion of Theorem 3.6.41. The interested reader will find a proof of the other direction in [145].

Proof (of the second assertion). The proof goes by induction on the length of the formula defining X. It is sufficient to discuss the logical operations $\neg\varphi, \varphi \vee \psi, \exists x\phi$ as all others can be obtained from these three. At each step of the induction, we need to obtain Büchi automata for $S_\beta(X_1), \ldots, S_\beta(X_n)$, where $X_1, \ldots, X_n$ are the current subsets of $\mathbb{R}^d$ in the recursive definition of X. Let φ, ψ be such that $X_\varphi, X_\psi \subseteq \mathbb{R}^d$. We have $S_\beta(X_{\neg\varphi}) = S_\beta(\mathbb{R}^d) \setminus S_\beta(X_\varphi)$ and $S_\beta(X_{\varphi\vee\psi}) = S_\beta(X_\varphi) \cup S_\beta(X_\psi)$. If $\mathscr{B}$ is a Büchi automata accepting $S_\beta(X_\phi)$ where ϕ contains a free variable called x, then the ω-language L accepted by the Büchi automata obtained from $\mathscr{B}$ by deleting the component corresponding to x in every label is such that $\mathbf{0}^*(\mathbf{0}^*)^{-1}L = S_\beta(X_{\exists x\phi})$. The induction step then follows from Propositions 3.6.15 and 3.6.25 and from the stability of ω-regular languages under Boolean operations and projection on components.

Let us verify that the atomic formulæ of $\langle \mathbb{R}, + \leq, \mathbb{Z}_\beta, X_\beta \rangle$ are all β-recognizable. We need β to be a Pisot number only for the addition to be β-recognizable [229]. Now we suppose that β is a Parry number. By Proposition 3.6.25, $\mathbb{R}^d$ is β-recognizable for any dimension d. Let $\mathscr{G}$ be a Büchi automaton accepting $d_\beta(\mathbb{R}^2)$ (such an automaton exists by Proposition 3.6.19). The ω-languages $d_\beta(\{(x, y) \in \mathbb{R}^2 \mid x = y\})$ and $d_\beta(\{(x, y) \in \mathbb{R}^2 \mid x < y\})$ are accepted by the intersections of $\mathscr{G}$ with the Büchi automata of Figures 3.9 and 3.10, respectively. We have $d_\beta(\mathbb{Z}_\beta) = d_\beta(\mathbb{R}) \cap (\tilde{A}_\beta)^+ \star 0^\omega$. For each $a \in \tilde{A}_\beta$, $d_\beta(X_{\beta,a})$ is accepted by the intersection of $\mathscr{G}$ with the Büchi automaton represented in Figure 3.11. Finally, in order to start the induction process, we have to build Büchi automata accepting $S_\beta(\{(x, y, z) \in \mathbb{R}^3 \mid x + y = z\})$, $S_\beta(\{(x, y) \in \mathbb{R}^2 \mid x < y\})$, $S_\beta(\mathbb{Z}_\beta)$, and $S_\beta(X_{\beta,a})$, which can be done thanks to Proposition 3.6.19. □

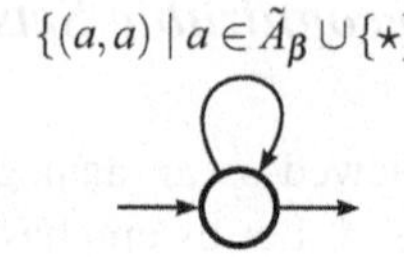

Fig. 3.9 A Büchi automaton for the equality

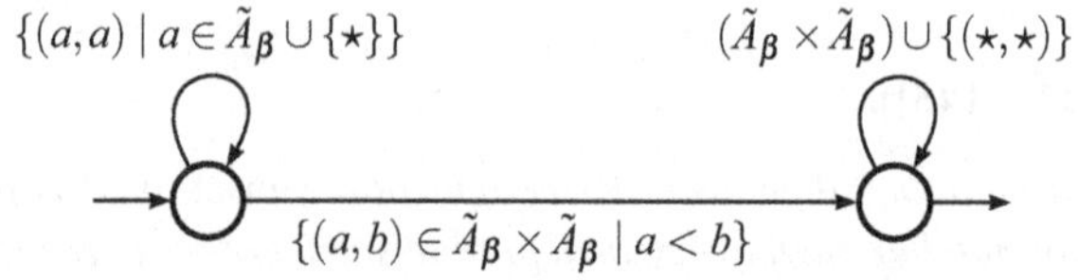

Fig. 3.10 A Büchi automaton for the order

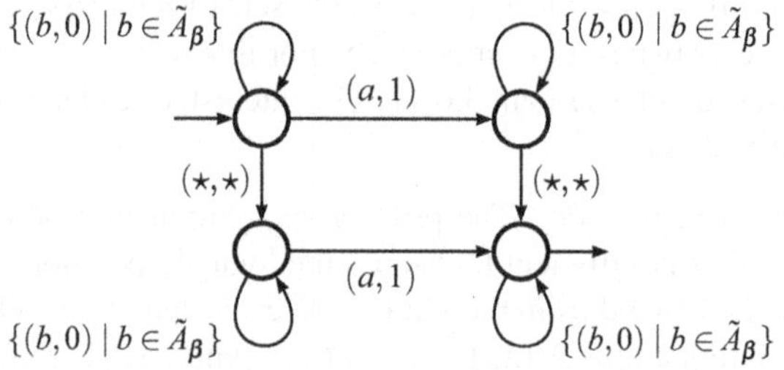

Fig. 3.11 A Büchi automaton for $X_{\beta,a}$

Corollary 3.6.42. *If β is a Pisot number, then the first-order theory of $\langle \mathbb{R}, +, \leq, \mathbb{Z}_\beta, X_\beta \rangle$ is decidable.*

Proof. A closed first-order formula of $\langle \mathbb{R}, +, \leq, \mathbb{Z}_\beta, X_\beta \rangle$ is of the form $\exists x \varphi(x)$ or $\forall x \varphi(x)$. By Theorem 3.6.41, the sets $X_\varphi = \{x \in \mathbb{R} \mid \varphi(x) \text{ is true}\}$ and $X_{\neg\varphi} = \{x \in \mathbb{R} \mid \varphi(x) \text{ is false}\}$ are β-recognizable. As the emptiness of an ω-regular language is decidable [476], we can decide whether X_φ is nonempty (resp. $X_{\neg\varphi}$ is empty) and, thus, whether $\exists x \varphi(x)$ (resp. $\forall x \varphi(x)$) is true. □

Like Theorem 3.3.4, this result has many applications: any property of β-recognizable sets that can be expressed by a first-order predicate in the structure $\langle \mathbb{R}, +, \leq, \mathbb{Z}_\beta, X_\beta \rangle$ is decidable. For example, it is decidable whether a β-recognizable subset of $\mathbb{R}^d$ is a subgroup of $\mathbb{R}^d$ with respect to the addition. As another example, we are also able to decide topological properties of β-recognizable sets. Note that, in this context, interesting examples of compact β-recognizable sets are given by a class of fractal sets, called β-self-similar sets [1]. Indeed, it follows from Theorem 3.6.56 below that β-self-similar sets are β-recognizable when β is Pisot. This fact is highlighted in Remark 3.6.59.

Proposition 3.6.43. *If β is Pisot, then the following properties of β-recognizable subsets X of $\mathbb{R}^d$ are decidable: X has a nonempty interior, X is open, X is closed, X is bounded, X is compact, X is dense.*

Proof. Suppose that β is Pisot and let X be a β-recognizable subset X of $\mathbb{R}^d$ and let φ be a first-order formula of $\langle \mathbb{R}, +, \leq, \mathbb{Z}_\beta, X_\beta \rangle$ defining X. For $(x_1, \ldots, x_d) \in \mathbb{R}^d$ and $\varepsilon > 0$, we let $B(x_1, \ldots, x_d, \varepsilon)$ denote the set $\{(y_1, \ldots, y_d) \in \mathbb{R}^d \mid -\varepsilon < x_1 - y_1 < \varepsilon \wedge \cdots \wedge -\varepsilon < x_d - y_d < \varepsilon\}$. Clearly, the predicate $(y_1, \ldots, y_d) \in B(x_1, \ldots, x_d, \varepsilon)$ is expressible by a first-order formula of $\langle \mathbb{R}, +, \leq, \mathbb{Z}_\beta, X_\beta \rangle$. Then, we can express that X has a nonempty interior by the formula

$$(\exists x_1) \cdots (\exists x_d) \Big(\varphi(x_1, \ldots, x_d) \wedge \Big[(\exists \varepsilon > 0)\, (\forall y_1) \cdots (\forall y_d) \\ \big((y_1, \ldots, y_d) \in B(x_1, \ldots, x_d, \varepsilon) \implies \varphi(y_1, \ldots, y_d) \big) \Big] \Big).$$

It is open if and only if

$$(\forall x_1) \cdots (\forall x_d) \Big(\varphi(x_1, \ldots, x_d) \implies \Big[(\exists \varepsilon > 0)\, (\forall y_1) \cdots (\forall y_d) \\ \big((y_1, \ldots, y_d) \in B(x_1, \ldots, x_d, \varepsilon) \implies \varphi(y_1, \ldots, y_d) \big) \Big] \Big).$$

It is closed if and only if it is not open. It is bounded if and only if

$$(\exists R > 0)\, (\forall x_1) \cdots (\forall x_d) \Big(\varphi(x_1, \ldots, x_d) \implies (x_1, \ldots, x_d) \in B(0, \ldots, 0, R) \Big).$$

It is compact if and only if it is closed and bounded. Finally, it is dense if and only if

$$(\forall x_1) \cdots (\forall x_d) (\forall \varepsilon > 0) (\exists y_1) \cdots (\exists y_d) (\varphi(y_1, \ldots, y_d) \wedge (y_1, \ldots, y_d) \in B(x_1, \ldots, x_d, \varepsilon)).$$

As those properties of X are all expressible by a closed first-order formula of $\langle \mathbb{R}, +, \leq, \mathbb{Z}_\beta, X_\beta \rangle$, they are decidable by Corollary 3.6.42. □

We note that, thanks to Proposition 3.6.17, the property of being closed can be directly verified from a Büchi automaton recognizing the set under consideration. Indeed, given a Büchi automaton accepting $S_\beta(X)$, we can effectively compute a DFA accepting $\mathrm{Pref}(S_\beta(X))$. Then, by Proposition 3.6.8, this DFA seen as a Büchi automaton accepts $S_\beta(X)$ if and only if X is closed. As it is decidable if two Büchi automata accept the same ω-language [476], we can decide whether a β-recognizable set X is closed.

3.6.7 Analogues of the Cobham–Semenov Theorem for Real Numbers

Several analogues of Cobham's theorem were obtained in the context of integer base b representations of real numbers. In this section, we list some of them without proof. We will show the connections between these results, as well as with Theorem 3.6.61. This connection is achieved by using graph-directed iterated function systems (GDIFS) and allows us to provide extensions of the abovementioned results: Theorem 3.6.50 extends to $\mathbb{R}^d$, Theorem 3.6.61 extends to a large class of GDIFS, and the logical characterization of b-recognizable sets of reals used for proving Theorem 3.6.46 extends to the so-called Pisot real bases.

Theorem 3.6.44 ([95]). *Let b and b' be integer bases with different sets of prime divisors. A subset of $\mathbb{R}^d$ is simultaneously b-recognizable and b'-recognizable if and only if it is definable in $\langle \mathbb{R}, +, \leq, \mathbb{Z} \rangle$.*

The hypothesis of sharing no prime divisors is stronger than that of being multiplicatively independent. In order to obtain an analogue of the Cobham theorem for multiplicatively independent integer bases, we need an extra hypothesis, which is the *weak* b-recognizability.

Theorem 3.6.45 ([94]). *Let b and b' be multiplicatively independent integer bases. A subset of $[0, 1]^d$ is simultaneously weakly b-recognizable and weakly b'-recognizable if and only if it is definable in $\langle \mathbb{R}, +, \leq, 1 \rangle$.*

Note that, together with Theorems 3.6.29 and 3.2.28, Theorem 3.6.45 implies the following result.

Theorem 3.6.46 ([94]). *Let b and b' be multiplicatively independent integer bases. A subset of $\mathbb{R}^d$ is simultaneously weakly b-recognizable and weakly b'-recognizable if and only if it is definable in $\langle \mathbb{R}, +, \leq, \mathbb{Z} \rangle$.*

In the particular case where $d = 1$ and we consider only compact subsets of $[0, 1]$, Theorem 3.6.45 is indeed another formulation of Theorem 3.6.50 below. To state this result, we need a definition first.

Definition 3.6.47. A subset X of $[0, 1]^d$ is *b-self-similar* if it is closed, and there are finitely many sets of the form

$$(b^a X - \mathbf{t}) \cap [0, 1]^d$$

for $a \in \mathbb{N}$ and $\mathbf{t} \in ([0, b^a) \cap \mathbb{Z})^d$.

Example 3.6.48. The Pascal triangle modulo 2 (see Figure 3.12) is 2-self-similar. It is the closure of the set $\{\frac{1}{2^\ell}(m, n) \mid \binom{n}{m} \equiv 1 \bmod 2,\ \ell \geq |\mathrm{rep}_2(m, n)|\}$.

Example 3.6.49. The Menger sponge (see Figure 3.13) is 3-self-similar. It is the closure of the set of points $\mathbf{x} \in [0, 1]^3$ such that $\mathrm{rep}_3(\mathbf{x})$ does not contain any of the digits $(0, 1, 1), (1, 0, 1), (1, 1, 0), (1, 1, 1)$.

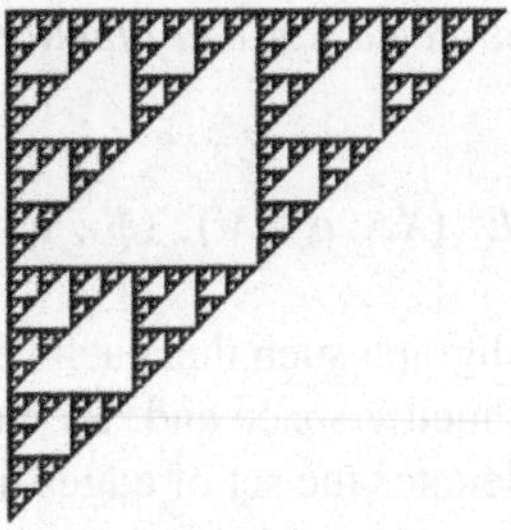

Fig. 3.12 The Pascal triangle modulo 2

Fig. 3.13 The Menger sponge

Theorem 3.6.50 ([1]). *Let $b, b' \geq 2$ be multiplicatively independent integers. A compact subset of* $[0, 1]$ *is simultaneously b-self-similar and b′-self-similar if and only if it is a finite union of closed intervals with rational endpoints.*

The object of the next section is to study the connection between Theorems 3.6.45 and 3.6.50.

3.6.8 Linking Büchi Automata, β-Self-Similarity and GDIFS

We generalize Definition 3.6.47 to real bases β. The set of polynomials in β with integer coefficients is denoted by $\mathbb{Z}[\beta]$. Note that it is not equal to the set $\mathbb{Z}_\beta$ of β-integers as, for example, $d_\varphi(\varphi - 1) = 0 \star 10^\omega$, hence $\varphi - 1 \in \mathbb{Z}[\varphi] \setminus \mathbb{Z}_\varphi$.

Definition 3.6.51. A subset X of $\left[0, \frac{\lceil\beta\rceil-1}{\beta-1}\right]^d$ is *β-self-similar* if it is closed, and there are only finitely many sets of the form

$$(\beta^a X - \mathbf{t}) \cap \left[0, \frac{\lceil\beta\rceil - 1}{\beta - 1}\right]^d,$$

for $a \in \mathbb{N}$ and $\mathbf{t} \in \left(\left[0, \frac{(\lceil\beta\rceil-1)}{\beta-1}\beta^a\right) \cap \mathbb{Z}[\beta]\right)^d$.

Definition 3.6.52. A *graph-directed iterated function system* (*GDIFS* for short) is given by a 4-tuple

$$(V,\ E,\ (X_v,\ v \in V),\ (\phi_e,\ e \in E))$$

where (V, E) is a connected digraph such that each vertex has at least one outgoing edge, for each $v \in V$, X_v is a metric space and, for each $e \in E_{uv}$, $\phi_e : X_v \to X_u$ is a contraction map, where E_{uv} denotes the set of edges in E from u to v.

Theorem 3.6.53 ([208, 306]). *For each GDIFS* $(V,\ E,\ (X_v,\ v \in V),\ (\phi_e,\ e \in E))$ *on complete metric spaces* X_v*, there is a unique list of nonempty compact subsets* $(K_u,\ u \in V)$ *such that, for all* $u \in V$*,* $K_u \subseteq X_u$ *and*

$$K_u = \bigcup_{v \in V} \bigcup_{e \in E_{uv}} \phi_e(K_v).$$

Definition 3.6.54. The *attractor* of a GDIFS on complete metric spaces is the list of nonempty compact subsets from Theorem 3.6.53.

We will use the following result.

Theorem 3.6.55 ([70, 231]). *The sets* $\left[\frac{-c}{\beta-1}, \frac{c}{\beta-1}\right] \cap \mathbb{Z}[\beta]$ *are finite for all* $c \in \mathbb{N}$ *if and only if* β *is a Pisot number.*

Theorem 3.6.56 ([145]). *Let* β *be a Pisot number. For any compact subset* X *of* $\left[0, \frac{\lceil\beta\rceil-1}{\beta-1}\right]^d$*, the following are equivalent:*

1. *There is a Büchi automaton* $\mathcal{A}$ *over the alphabet* A_β^d *such that* $\mathrm{val}_\beta(\mathbf{0}{\star}L(\mathcal{A})) = X$.
2. *X belongs to the attractor of a GDIFS on* $\mathbb{R}^d$ *whose contraction maps are of the form* $\mathbf{x} \mapsto \frac{\mathbf{x}+\mathbf{t}}{\beta}$ *with* $\mathbf{t} \in A_\beta^d$.
3. *X is β-self-similar.*

Proof. 1 $\Longrightarrow$ 2. Let $\mathcal{A} = (Q, q_0, F, A_\beta^d, \delta)$ be a trim Büchi automaton such that $\mathrm{val}_\beta(\mathbf{0}{\star}L(\mathcal{A})) = X$. Because X is closed and val_β is continuous, we may suppose that $Q = F$, *i.e.*, that all states are final. The GDIFS on $\mathbb{R}^d$ we build is obtained from $\mathcal{A}$ by replacing each label $\mathbf{t} \in A_\beta^d$ by the contraction map $\mathbf{x} \mapsto \frac{\mathbf{x}+\mathbf{t}}{\beta}$. For all $q \in Q$, let L_q denote the set of infinite words accepted from q in $\mathcal{A}$, and let $X_q = \{\mathrm{val}_\beta(\mathbf{0}{\star}w) \mid w \in L_q\}$. We claim that $(X_q,\ q \in Q)$ is the attractor of this GDIFS. This is sufficient as $X = X_{q_0}$. The fact that $\mathcal{A}$ is trim and contains only final states implies that the subsets X_q are closed and nonempty. Moreover, they satisfy $X_q \subseteq \left[0, \frac{\lceil\beta\rceil-1}{\beta-1}\right]^d$. Then, by Theorem 3.6.53, it suffices to show that the list $(X_q,\ q \in Q)$ satisfies

$$\forall q \in Q,\quad X_q = \bigcup_{p \in Q} \bigcup_{q \xrightarrow{\mathbf{t}} p} \frac{1}{\beta}(X_p + \mathbf{t}).$$

This follows from the following two observations:

$$\forall q \in Q,\ L_q = \bigcup_{p\in Q} \bigcup_{q \xrightarrow{\mathbf{t}} p} \mathbf{t}L_p$$

$$\forall \mathbf{w} \in (A_\beta^d)^\omega, \forall \mathbf{t} \in A_\beta^d,\ \mathrm{val}_\beta(\mathbf{0}\star\mathbf{tw}) = \frac{\mathrm{val}_\beta(\mathbf{0}\star\mathbf{w}) + \mathbf{t}}{\beta}.$$

$2 \implies 1$. Let $(K_v,\ v \in V)$ be the attractor of a GDIFS on $\mathbb{R}^d$ whose contraction maps are of the form $\mathbf{x} \mapsto \frac{\mathbf{x}+\mathbf{t}}{\beta}$ with $\mathbf{t} \in A_\beta^d$ and suppose that $X = K_{v_0}$ for some $v_0 \in V$. Let $\mathcal{A}$ be the Büchi automaton $(V, v_0, V, A_\beta^d, \delta)$ where the transitions correspond to the edges of the GDIFS in which we have replaced the labels $\frac{\mathbf{x}+\mathbf{t}}{\beta}$ by $\mathbf{t}$. As the underlying digraph of a GDIFS is connected and such that there is at least one outgoing edge starting from each vertex, the Büchi automaton $\mathcal{A}$ is trim. Then, from the proof of $1 \implies 2$, we obtain that $K_v = \{\mathrm{val}_\beta(\mathbf{0}\star w) \mid w \in L_v\}$ for all $v \in V$ (where L_v is defined as before); hence, $X = \{\mathrm{val}_\beta(\mathbf{0}\star w) \mid w \in L(\mathcal{A})\}$.

$2 \implies 3$. Let $(K_v,\ v \in V)$ be the attractor of a GDIFS on $\mathbb{R}^d$ whose contraction maps are of the form $S_\mathbf{t}\colon \mathbf{x} \mapsto \frac{\mathbf{x}+\mathbf{t}}{\beta}$ with $\mathbf{t} \in A_\beta^d$, and suppose that $X = K_{v_0}$ for some $v_0 \in V$. For all vertices u and v, we let E_{uv}^ℓ denote the set of words of length ℓ over A_β^d that label a path from u to v (where the labels $\frac{\mathbf{x}+\mathbf{t}}{\beta}$ are replaced by $\mathbf{t}$). For all $u \in V$ and $\ell \in \mathbb{N}$, we have

$$K_u = \bigcup_{v\in V} \bigcup_{\mathbf{t}_1\cdots\mathbf{t}_\ell \in E_{uv}^\ell} S_{\mathbf{t}_1} \circ \cdots \circ S_{\mathbf{t}_\ell}(K_v).$$

For the sake of conciseness, we let $r_\beta = \frac{\lceil\beta\rceil - 1}{\beta - 1}$. By setting $u = v_0$ and $\ell = a$ in the previous equality, we obtain that

$$(\beta^a X - \mathbf{t}) \cap [0, r_\beta]^d = \bigcup_{v\in V} \bigcup_{\mathbf{t}_1\cdots\mathbf{t}_a \in E_{v_0 v}^a} \left(\beta^a (S_{\mathbf{t}_1} \circ \cdots \circ S_{\mathbf{t}_a}(K_v) - \mathbf{t}) \cap [0, r_\beta]^d\right)$$

for all $a \in \mathbb{N}$ and $\mathbf{t} \in \mathbb{R}^d$. Observe that

$$\beta^a (S_{\mathbf{t}_1} \circ \cdots \circ S_{\mathbf{t}_a}(K_v)) - \mathbf{t} = K_v + (\mathbf{t}_a + \beta\mathbf{t}_{a-1} + \cdots + \beta^{a-1}\mathbf{t}_1) - \mathbf{t}$$

and, if $\mathbf{t} \in (\mathbb{Z}[\beta])^d$ and, for each $1 \le i \le a$, $\mathbf{t}_i \in A_\beta^d$, then

$$\mathbf{t}_a + \beta\mathbf{t}_{a-1} + \cdots + \beta^{a-1}\mathbf{t}_1 - \mathbf{t} \in (\mathbb{Z}[\beta])^d.$$

For all $v \in V$, the set K_v is included in $[0, r_\beta]^d$; hence, the sets $(K_v + \mathbf{x}) \cap [0, r_\beta]^d$ are empty for all $\mathbf{x} \notin [-r_\beta, r_\beta]^d$. Since β is Pisot, Theorem 3.6.55 implies that

$[-r_\beta, r_\beta] \cap \mathbb{Z}[\beta]$ is finite. Consequently, there are finitely many sets of the form $(K_v + \mathbf{x}) \cap [0, r_\beta]^d$, with $\mathbf{x} \in (\mathbb{Z}[\beta])^d$ and $v \in V$. As any set $(\beta^a X - \mathbf{t}) \cap [0, r_\beta]^d$ with $a \in \mathbb{N}$ and $\mathbf{t} \in ([0, r_\beta \beta^a) \cap \mathbb{Z}[\beta])^d$ is a finite union of such sets, this proves that X is β-self-similar.

3 $\implies$ 2. Suppose that X is β-self-similar. Again, we let $r_\beta = \frac{\lceil \beta \rceil - 1}{\beta - 1}$. Define a GDIFS on $\mathbb{R}^d$ as follows: the vertices of the underlying digraph are the nonempty compact sets among

$$N_{a,\mathbf{t}}(X) := (\beta^a X - \mathbf{t}) \cap [0, r_\beta]^d,$$

with $a \in \mathbb{N}$ and $\mathbf{t} \in ([0, r_\beta \beta^a) \cap \mathbb{Z}[\beta])^d$ and, for each $\mathbf{s} \in A_\beta^d$, there is an edge labeled $\frac{\mathbf{x}+\mathbf{s}}{\beta}$ from $N_{a,\mathbf{t}}(X)$ to $N_{a+1,\beta\mathbf{t}+\mathbf{s}}(X)$ if both are nonempty. For all $a \in \mathbb{N}$ and $\mathbf{t} \in ([0, r_\beta \beta^a) \cap \mathbb{Z}[\beta])^d$, we have

$$\begin{aligned}
\bigcup_{\mathbf{s} \in A_\beta^d} \frac{1}{\beta}(N_{a+1,\beta\mathbf{t}+\mathbf{s}}(X) + \mathbf{s}) &= \bigcup_{\mathbf{s} \in A_\beta^d} \frac{1}{\beta}\Big(\Big[(\beta^{a+1} X - \beta\mathbf{t} - \mathbf{s}) \cap [0, r_\beta]^d\Big] + \mathbf{s}\Big) \\
&= \bigcup_{\mathbf{s} \in A_\beta^d} \Big((\beta^a X - \mathbf{t}) \cap \frac{1}{\beta}([0, r_\beta]^d + \mathbf{s})\Big) \\
&= (\beta^a X - \mathbf{t}) \cap \Big(\bigcup_{\mathbf{s} \in A_\beta^d} \frac{1}{\beta}([0, r_\beta]^d + \mathbf{s})\Big) \\
&= (\beta^a X - \mathbf{t}) \cap [0, r_\beta]^d \\
&= N_{a,\mathbf{t}}(X),
\end{aligned}$$

hence the sets $N_{a,\mathbf{t}}(X)$ form the attractor of this GDIFS. To conclude with the proof, observe that $X = N_{0,\mathbf{0}}(X)$. □

Note that, in the previous theorem, the Pisot condition is needed only for the implications 2 $\implies$ 1 and 2 $\implies$ 3. Also note that all the equivalences are effective, meaning that from any of the hypotheses 1, 2, or 3, we can effectively construct a Büchi automaton for 1, a GDIFS for 2, and the β-kernel for 3.

We are now able to show the connection between Theorems 3.6.45 and 3.6.50.

Proposition 3.6.57. *Any b-self-similar subset of* $[0, 1]^d$ *is weakly b-recognizable.*

Proof. Let X be a b-self-similar subset of $[0, 1]^d$. By Theorem 3.6.56, there is a Büchi automaton $\mathcal{A}$ over the alphabet A_b^d such that $\mathrm{val}_b(\mathbf{0} \star L(\mathcal{A})) = X$. We show that this implies that X is weakly b-recognizable. For the sake of clarity, we discuss the case $d = 1$ (the general case is just a tedious adaptation of the same arguments as we have to consider each component separately). We have

$$S_b(X) = 0^+ \star L(\mathscr{A}) \cup 0^* \pi_1\big(M \cap [(A_b \cup \{\star\})^\omega \times (0 \star L(\mathscr{A}))]\big)$$

where

$$M = \{\tbinom{0}{0}\tbinom{\star}{\star}\tbinom{u}{u}\tbinom{a+1}{a}\tbinom{0}{b-1}^\omega \mid u \in A_b^*,\ 0 \le a \le b-2\} \cup \{\tbinom{1}{0}\tbinom{\star}{\star}\tbinom{0}{b-1}^\omega\}$$
$$\cup \{\tbinom{0}{0}\tbinom{\star}{\star}\tbinom{u}{u}\tbinom{a}{a+1}\tbinom{b-1}{0}^\omega \mid u \in A_b^*,\ 0 \le a \le b-2\}$$

This shows that $S_b(X)$ is an ω-regular language; hence, X is b-recognizable. Since X is closed, it is weakly b-recognizable by Proposition 3.6.24. □

The following result generalizes Theorem 3.6.50 to the multidimensional setting.

Theorem 3.6.58 ([142, 145]). *Let $b, b' \geq 2$ be two multiplicatively independent integers. A compact subset of $[0, 1]^d$ is simultaneously b-self-similar and b'-self-similar if and only if it is a finite union of rational polyhedra.*

Proof. The result is a consequence of Proposition 3.6.38, Theorem 3.6.45, and Proposition 3.6.57. □

Remark 3.6.59. We have seen in Proposition 3.6.57 that any b-self-similar subset of $[0, 1]^d$ is weakly b-recognizable. By using Theorem 3.6.56 and the fact that the normalization is realizable by a letter-to-letter transducer [229, 231], we obtain that this fact also holds for Pisot bases β: any β-self-similar subset of $[0, 1]^d$ is weakly β-recognizable. However, the converse is not true as, for every base $\beta > 1$, there exist weak β-recognizable subsets of $[0, 1]^d$ which are not closed. For example, any interval of the form $[r, s[$ is weakly β-recognizable for all bases $\beta > 1$. Hence the hypothesis of b-self-similarity is strictly stronger than that of b-recognizability.

We also obtain the following analogue of the Cobham–Semenov theorem for GDIFS.

Theorem 3.6.60. *Let $b, b' \geq 2$ be multiplicatively independent integers. A compact subset of $\mathbb{R}^d$ is the attractor of two GDIFS, one with contraction maps of the form $\mathbf{x} \mapsto \frac{\mathbf{x}+\mathbf{t}}{b}$ with $\mathbf{t} \in A_b^d$ and the other with contraction maps of the form $\mathbf{x} \mapsto \frac{\mathbf{x}+\mathbf{t}}{b'}$ with $\mathbf{t} \in A_{b'}^d$, if and only if it is a finite union of rational polyhedra.*

Proof. The result is a consequence of Theorem 3.6.56, Proposition 3.6.57, Theorem 3.6.46, and Corollary 3.6.40. □

The previous result has to be compared with the following theorem. Here $\dim_H$ denote the Hausdorff dimension, and an iterated function system (IFS for short) is a GDIFS whose graph contains only one vertex. An IFS $\Phi = (\phi_1, \dots, \phi_k)$ is said to satisfy *the open set condition* if there exists a nonempty open set V such that $\phi_1(V), \dots, \phi_k(V)$ are pairwise disjoint subsets of V.

Theorem 3.6.61 ([220]). *Suppose that a compact subset K of $\mathbb{R}$ is the attractor of two IFS Φ and Ψ all of whose contraction maps are affinities sharing the same* *contraction ratios, denoted r_Φ and r_Ψ, respectively, and suppose that Φ satisfies the open set condition.*

- *If* $\dim_H(K) < 1$ *then* $\frac{\log|r_\Psi|}{\log|r_\Phi|} \in \mathbb{Q}$.
- *If* $\dim_H(K) = 1$ *and K is not a finite union of intervals, then* $\frac{\log|r_\Psi|}{\log|r_\Phi|} \in \mathbb{Q}$.

We note that Theorem 3.6.60 is more general than Theorem 3.6.61 in two ways as it concerns the more general setting of GDIFS and it is formulated for the d-dimensional Euclidean space. It is also weaker as the contraction ratios must be of the form $\frac{1}{b}$ and $\frac{1}{b'}$.

3.7 Exercises

The following exercises are related to Section 3.2.

Exercise 3.7.1. Consider the abstract numeration system S built on the language L accepted by the automaton of Figure 3.14. The set $X = \mathrm{val}_S(L')$, where $L' = \{\varepsilon\} \cup 2\{0,2\}^*$. By definition, X is S-recognizable. Show that $t_X(n) = \Theta(n\log(n))$. Proceed by using two different methods: first, in a direct way by characterizing the elements in X and, second, by showing $\mathbf{v}_L(n) = (n+1)2^n$ and $\mathbf{v}_{L'}(n) = 2^n$ for all $n \in \mathbb{N}$ and then by using Theorem 3.2.23.

Exercise 3.7.2. Consider the base 4 numeration system. Let $X = \mathrm{val}_4(L)$ where L is the language accepted by the automaton of Figure 3.14. It is 4-recognizable by construction. Show that $t_X(n) = \Theta((\frac{n}{\log(n)})^2)$.

Exercise 3.7.3. Consider the 4-recognizable set $X = \mathrm{val}_4(\{1,2,3\}^*)$ and show that $t_X(n) = \Theta\left(n^{\frac{\log(4)}{\log(3)}}\right)$.

Exercise 3.7.4. Define $L_F = \{\varepsilon\} \cup 1(0+01)^*$ to be the language of the Zeckendorf numeration system, and let $X = \mathrm{val}_4(L_F)$. Show that

$$\forall n \in \mathbb{N},\ \mathbf{v}_{L_F}(n) = \frac{5+3\sqrt{5}}{10}\phi^n + \frac{5-3\sqrt{5}}{10}(1-\phi)^n$$

and that $t_X(n) = \Theta\left(n^{\frac{\log(4)}{\log(\varphi)}}\right)$.

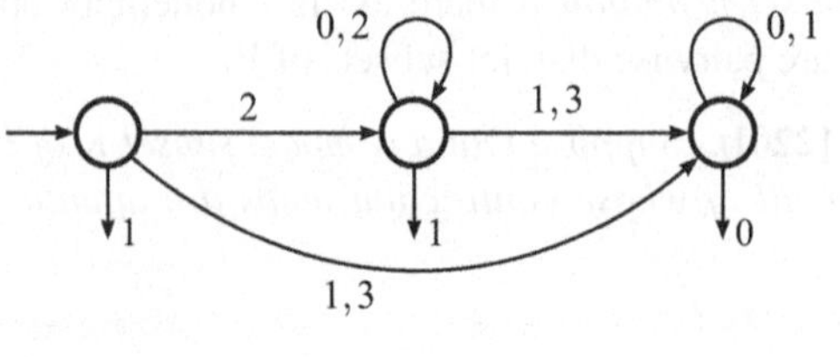

Fig. 3.14 The trim minimal automaton of L

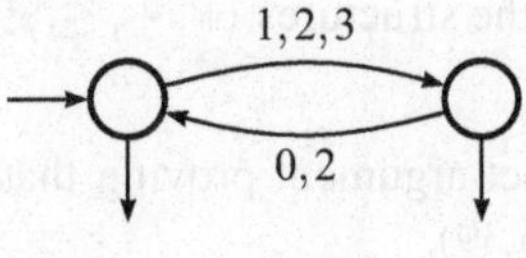

Fig. 3.15 The trim minimal automaton of L

Exercise 3.7.5. Define L to be the language accepted by the DFA depicted in Figure 3.15. Let $X = \mathrm{val}_4(L)$ and show that $\mathbf{v}_L(2n) \sim \frac{9}{5}6^n$ and $\mathbf{v}_L(2n+1) \sim \frac{24}{5}6^n$ $(n \to +\infty)$ and that $t_X(n) = \Theta\left(n^{\frac{\log(4)}{\log(\sqrt{6})}}\right)$.

Exercise 3.7.6. Let $X = \mathrm{val}_2(1^*0^*)$. Show that $\mathbf{v}_{1^*0^*}(n) = \binom{n+2}{2}$ for all $n \in \mathbb{N}$ and that $t_X(n) = 2^{(1+o(1))\sqrt{2n}}$.

The following exercises are related to Section 3.2.5.

Exercise 3.7.7. Show that semi-linear sets are b-recognizable for all b.

An *ultimately periodic subset of* $\mathbb{N}$ is a subset X of $\mathbb{N}$ for which there exist integers $i \geq 0$ (the *preperiod*) and $p \geq 1$ (the *period*) such that, for all $x \in \mathbb{N}$, $x \in X$ if and only if $x + p \in X$.

Exercise 3.7.8. Let X be a subset of $\mathbb{N}$. Show that the following assertions are equivalent:

- X is a finite union of arithmetic progressions.
- X is ultimately periodic.
- X is semi-linear.
- X is a recognizable subset of $\mathbb{N}$.
- X is 1-recognizable.
- X is b-recognizable for all integers bases b.
- X is S-recognizable for all abstract numeration systems S.

The following exercise is related to Section 3.3.

Exercise 3.7.9. Show the following assertions

- $x \leq y$ is definable in $\langle \mathbb{N}, + \rangle$ but not in $\langle \mathbb{Z}, + \rangle$.
- $x = y$ is definable in $\langle \mathbb{N}, + \rangle$ but not in $\langle \mathbb{Z}, + \rangle$.
- $x = 0$ is definable in $\langle \mathbb{N}, + \rangle$ and in $\langle \mathbb{Z}, + \rangle$.
- $x = 1$ is definable in $\langle \mathbb{N}, + \rangle$ but not in $\langle \mathbb{Z}, + \rangle$.
- For every $c \in \mathbb{N}$, $x = c$ is definable in $\langle \mathbb{N}, + \rangle$.
- The arithmetic progressions are definable in $\langle \mathbb{N}, + \rangle$.
- A subset X of $\mathbb{N}$ is definable in $\langle \mathbb{N}, + \rangle$ if and only if it is a finite union of arithmetic progressions.
- A subset X of $\mathbb{N}^d$ is definable in $\langle \mathbb{N}, + \rangle$ if and only if it is semi-linear.

The following exercises are related to Section 3.6.

Exercise 3.7.10. Show that the structures $\langle \mathbb{R}, +, \leq, \mathbb{Z} \rangle$ and $\langle \mathbb{R}, +, \leq, \mathbb{N} \rangle$ are equivalent.

Exercise 3.7.11. Find a direct argument proving that $\mathbb{Z}$ is not definable in $\langle \mathbb{R}, +, \leq, 1 \rangle$ (not using Theorem 3.6.39).

Exercise 3.7.12. Show that the finite unions of rational polyhedra are b-self-similar for all b.

3.8 Bibliographic Notes

Abstract numeration systems were introduced in [372]. A characterization of S-recognizable subsets of $\mathbb{N}^d$ was given in [143]. More precise asymptotics than those of Theorem 3.2.23 are given in [147].

Recognizable sets of $\mathbb{N}^d$ are a particular case of recognizable sets of a general monoid; see [211]. Kleene's theorem [301] holds only for free monoids, so it holds for $\mathbb{N}$ but not for $\mathbb{N}^d$, $d \geq 2$. In the case of the free monoid A^*, recognizable sets are regular languages over A.

Other applications of the decidability of the first-order theory of $\langle \mathbb{N}, +, V_b \rangle$ than those presented in Section 3.3.4 were obtained in [524].

For more on automatic sequences, see the book [14].

The definition of (K, b)-regular sequences given in this chapter is that of Berstel and Reutenauer [77]. It differs from the original one of Allouche and Shallit [17] since K is an arbitrary semiring, hence not necessarily a Nœtherian ring. As here we are specifically interested in $K = \mathbb{N}$ and $K = \mathbb{N}_\infty$, we need this more general framework.

The notion of b-synchronized sequences was introduced in [130]. Related works are [129, 522]. In particular, see [522] for a proof of Proposition 3.4.25.

Most of the results presented in Section 3.4 come from [148]. The enumeration of properties of automatic sequences were also discussed in the surveys [540, 541].

The definition of β-recognizability in the present chapter differs from that of [145]. I believe that Definition 3.6.14 gives the right notion of β-recognizability as it allows us to prove that $\mathbb{R}^d$ is β-recognizable when β is a Parry number and to correct some mistakes in [145], in particular Fact 1 and Lemma 23. This choice is also justified by Remark 3.6.23. Finally, this definition is coherent with the original definition of b-recognizable sets of reals [96].

The proof of Theorem 3.6.29 is from [93, 111].

Theorem 3.6.55 is stated in [231] in a more general form. Let us also mention the related recent paper [230].

The complexity of all the algorithms provided by the methods presented in this chapter is given by a tower of exponentials whose height is given by the number of alternating quantifiers. However, some specific problems concerning b-recognizable sets of integers or real numbers have been shown to be decidable in an efficient way. I refer the interested reader to the following related works: [374, 413, 416].

In [220], a more general version of Theorem 3.6.61 is given. For a generalization of this result to $\mathbb{R}^d$, but under a more restrictive separation condition, see [213].

Acknowledgements I thank Julien Leroy and Narad Rampersad for a careful reading of this chapter and for many helpful comments.

Chapter 4
Some Applications of Algebra to Automatic Sequences

Jason Bell

Abstract We give an overview of the theory of rings satisfying a polynomial identity and use this to give a proof of a characterization due to Berstel and Reutenauer of automatic and regular sequences in terms of two properties, which we call the shuffle property and the power property. These properties show that if one views an automatic sequence f as a map on a free monoid on k-letters to a finite subset of a ring, then the values of f are closely related to values of f on related words obtained by permuting letters of the word. We use this characterization to give answers to three questions from Allouche and Shallit, two of which have not appeared in the literature. The final part of the chapter deals more closely with the shuffle property, and we view this as giving a generalization of regular sequences. We show that sequences with the shuffle property are closed under the process of taking sums and taking products; in addition we show that there is closure under a noncommutative product, which turns the collection of shuffled sequences into a noncommutative algebra. We show that this algebra is very large, in the sense that it contains a copy of a free associative algebra on countably many generators. We conclude by giving some open questions, which we hope will begin a more careful study of shuffled sequences.

4.1 Introduction

We recall that, given a finite set Δ, a sequence $f : \mathbb{N} \to \Delta$ is said to be *k-automatic* if the nth term of this sequence is generated by a finite-state machine taking the base-k expansion of n as input, starting with the least significant digit. Automatic sequences appear in many different contexts (see, e.g., [14, 189]), and there are many examples of their application to algebra, such as their involvement in the characterization of algebraic power series over finite fields [14, Chapter 12]. The purpose of this chapter, however, is to look at applications of algebra, in particular the theory of

J. Bell (✉)
Department of Pure Mathematics, University of Waterloo, 200 University Ave. W., Waterloo, ON, Canada N2L 3G1
e-mail: jpbell@uwaterloo.ca

V. Berthé, M. Rigo (eds.), *Sequences, Groups, and Number Theory*, Trends in Mathematics, https://doi.org/10.1007/978-3-319-69152-7_4

polynomial identities, in giving a characterization of automatic sequences due to Berstel and Reutenauer [77, Chapter 3] and to use this characterization to answer some open questions about automatic and regular sequences.

Another way of defining the k-automatic property comes from looking at the k-*kernel* of a sequence. The k-kernel of a sequence $f(n)$ is defined to be the collection of sequences of the form $f(k^i n + j)$ where $i \geq 0$ and $0 \leq j < k^i$. A sequence is k-automatic if and only if its k-kernel is finite. Using this definition of k-automatic sequences, Allouche and Shallit [14, 16] generalized the notion of k-automatic sequences, defining *k-regular sequences.*

Let K be a field. Given a sequence $f(n)$ taking values in K, we create a vector subspace of $K^{\mathbb{N}}$, $V(f(n); k)$, which is defined to be the subspace spanned by all sequences $f(k^i n + j)$, where $i \geq 0$ and $0 \leq j < k^i$.

Definition 4.1.1. A sequence is *k-regular* if $V(f(n); k)$ is a finite-dimensional K-vector space.

Since the k-kernel of a sequence $f(n)$ spans $V(f(n); k)$ as a K-vector space, we see that a k-automatic sequence is necessarily k-regular. We remark that one can adopt a slightly more general viewpoint by replacing the field K by a commutative ring and asking that submodules generated by elements of the kernel be finitely generated. Everything we do in this chapter holds in this more general setting, but for ease of exposition we restrict to the case where our sequences are K-valued with K a field.

In the first half of this chapter, we take a ring theoretic look at automatic sequences and regular sequences and prove a result due to Berstel and Reutenauer [77, Chapter 3], which is not as well known as it probably should be. Part of the reason for this is that their result is written in the context of noncommutative rational series, so we give a straightforward translation of these results into our setting. In fact, we shall give a quantitative version of their result, which we hope will be of future use. We then give some applications of this result.

Let $\mathscr{A}$ be a finite alphabet. We let $\mathscr{A}^*$ denote the *free monoid* on $\mathscr{A}$; that is, $\mathscr{A}$ is the collection of all words on $\mathscr{A}$. We let ε denote the empty word on $\mathscr{A}$.

We are interested in maps $f : \mathscr{A}^* \to K$, where $\mathscr{A}$ is some finite alphabet and K is a field. In the case that $\mathscr{A} = \{1, 2, \ldots, k\}$ for some positive integer k, we can regard a word in $\mathscr{A}$ as an integer as follows. We associate 0 to the empty word ε, and for nontrivial words $a_m \cdots a_0 \in \mathscr{A}^*$ with $1 \leq a_i \leq k$, we associate a positive integer using the correspondence

$$a_m \cdots a_0 \mapsto a_m k^m + a_{m-1} k^{m-1} + a_{m-2} k^{m-2} + \cdots + a_0. \tag{4.1}$$

Observe that this gives a bijection between $\mathscr{A}^*$ and the natural numbers and hence, in this case, we may think of f as being a sequence indexed by the nonnegative integers taking values in K. We note that it is more common to use the alphabet $\{0, 1, \ldots, k-1\}$ and instead restrict to the regular sublanguage of the monoid $\{0, \ldots, k-1\}$ consisting of words whose first letter is not zero when dealing with

k-automatic and k-regular sequences. The point of view we adopt is completely equivalent, but we find the framework used in this chapter easier to deal with.

Using this correspondence and the k-kernel definition of a k-automatic sequence, we see that we can think of a k-automatic sequence taking values in a field K as being a map $f : \{1, 2, \ldots, k\}^* \to K$ with the property that the collection of maps $f^u(w) := f(wu)$ obtained by taking a word $u \in \{1, 2, \ldots, k\}^*$ is finite. Similarly, if the vector space of set maps from $\{1, \ldots, k\}^*$ to K generated by the maps of the form f^u is finite-dimensional, then f is k-regular. We shall give an overview of the work of Berstel and Reutenauer [77, Chapter 3], describing k-automatic and k-regular sequences, in terms of two additional properties, which for the purposes of this chapter we shall call the *shuffle property* and the *power property*. These properties are defined in Section 4.2 and Section 4.3, respectively. In Section 4.4 we present some results from the theory of rings satisfying a polynomial identity which will be necessary in giving the aforementioned characterization of k-regular sequences. In Section 4.5 we show that for a sequence, possessing these two properties is equivalent to being regular. If in addition to having these two properties, the sequence only takes on a finite number of values, the sequence is automatic. In Section 4.6 and Section 4.7, we use our characterization to answer some questions of Allouche and Shallit about whether certain sequences are k-regular and k-automatic. In Section 4.8 we develop the basic properties of sequences with the shuffle property and show that they have nice closure properties. In Section 4.9, we conclude with some open problems and some useful remarks.

4.2 The Shuffle Property

In this section we define the *shuffle property* and show that regular sequences necessarily possess this property. Let $\mathscr{A}$ be a finite alphabet and let K be a field. Given a map $f : \mathscr{A}^* \to K$, we say that f has the *d-shuffle property* if for any words $w, w_1, \ldots, w_d, w' \in \mathscr{A}^*$ we have

$$\sum_{\sigma \in S_d} \operatorname{sgn}(\sigma) f(w w_{\sigma(1)} w_{\sigma(2)} \cdots w_{\sigma(d)} w') = 0. \tag{4.2}$$

We define the dth*-shuffle function*

$$\operatorname{Shuf}_d(f; w, w_1, \ldots, w_d, w') := \sum_{\sigma \in S_d} \operatorname{sgn}(\sigma) f(w w_{\sigma(1)} w_{\sigma(2)} \cdots w_{\sigma(d)} w'). \tag{4.3}$$

Example 4.2.1. Let $\mathscr{A} = \{0, 1\}$, let K be the field of rational numbers, and define $f(w)$ to be the number of ones in the word $w \in \mathscr{A}^*$. Then f has the 2-shuffle property.

Example 4.2.2. Let $\mathscr{A} = \{0, 1\}$ and define $f(w)$ to be the nonnegative integer whose binary expansion is equal to w. Then f has the 4-shuffle property, but does not have the 3-shuffle property.

We postpone the proof of the fact that f has the 4-shuffle property till after the proof of Proposition 4.2.8. To see that f does not have the 3-shuffle property, take $w_1 = 0$, $w_2 = 1$, $w_3 = 10$. Then $f(w_1w_2w_3) = [0110]_2 = 6$. Similarly, $f(w_1w_3w_2) = 5$, $f(w_2w_3w_1) = 12$, $f(w_2w_1w_3) = 10$, $f(w_3w_1w_2) = 9$, $f(w_3w_2w_1) = 10$. Thus

$$\sum_{\sigma \in S_3} \mathrm{sgn}(\sigma) f(w_{\sigma(1)}w_{\sigma(2)}w_{\sigma(3)}) = 6 - 5 + 12 - 10 + 9 - 10 = 2 \neq 0.$$

The motivation for this shuffle property definition comes from the following theorem of Amitsur and Levitzki [20].

Theorem 4.2.3 (Amitsur-Levitzki [20]). *Let $\mathbf{A}_1, \ldots, \mathbf{A}_{2m}$ be $m \times m$ matrices with entries in a commutative ring C. Then*

$$\mathsf{S}_d(\mathbf{A}_1, \ldots, \mathbf{A}_{2m}) := \sum_{\sigma \in S_{2m}} \mathrm{sgn}(\sigma) \mathbf{A}_{\sigma(1)} \cdots \mathbf{A}_{\sigma(2m)} = 0. \tag{4.4}$$

Before we prove this result, we need a basic result about matrices.

Lemma 4.2.4. *Let C be a commutative $\mathbb{Q}$-algebra and let $\mathbf{Y} \in M_n(C)$. Suppose that $\mathbf{Y}$ has the property that the trace of $\mathbf{Y}^i$ is zero for $i = 1, 2, \ldots, n$. Then $\mathbf{Y}^n = 0$.*

Proof. Let $y_{i,j}$ denote the (i,j)-entry of $\mathbf{Y}$. Then we may assume without loss of generality that C is generated by the $y_{i,j}$. Now let $R = \mathbb{Q}[x_{i,j}\colon 1 \leq i, j \leq n]$ and let $\mathbf{Z} \in M_n(R)$ be the matrix whose (i,j)-entry is $x_{i,j}$. We note that if we impose the constraints on the $x_{i,j}$ that give that the trace of $\mathbf{Z}^i$ is equal to zero for $i = 1, 2, \ldots, n$, then we obtain a homomorphic image R' of R and by hypothesis the map $x_{i,j} \mapsto y_{i,j}$ gives a surjective map $\mathbb{Q}$-algebra homomorphism from R' to C. Thus it is sufficient to show that $\mathbf{Z}^n = 0$ in $M_n(R')$. We note that there is a $\mathbb{Q}$-algebra S that contains R and is a finite R-module and that contains all the eigenvalues of the matrix $\mathbf{Z}$. Then the relations imposed by setting the trace of $\mathbf{Z}^i$ to zero for $i = 1, 2, \ldots, n$ give a homomorphic image S' of S that is a finite R'-module. Now $\mathbf{Z}$ is triangularizable in $M_n(S')$ and so we may assume that $\mathbf{Z}$ is upper triangular. Then if $\lambda_1, \ldots, \lambda_n \in S'$ are the diagonal entries of $\mathbf{Z}$, then we have $\sum_{i=1}^n \lambda_i^j = 0$ for $j = 1, 2, \ldots, n$. We now claim that $\lambda_1, \lambda_2, \cdots, \lambda_n$ are nilpotent elements of S'. Once we establish this claim, we see that some power of $\mathbf{Z}$ is strictly upper triangular and so $\mathbf{Z}$ is nilpotent. Then the Cayley-Hamilton theorem gives that $\mathbf{Z}^n = 0$ in $M_n(S')$ and hence in $M_n(R')$, and so we get $\mathbf{Y}^n = 0$.

To establish the claim, we note that it is sufficient to show that if C is a finitely generated commutative $\mathbb{Q}$-algebra and $\lambda_1, \ldots, \lambda_n \in C$ satisfy $\sum_{i=1}^n \lambda_i^j = 0$ for $j = 1, 2, \ldots, n$ then $\lambda_1, \ldots, \lambda_n$ are nilpotent. Let N denote the nil radical of C (the set of all nilpotent elements of C). Then we may replace C by C/N, and we see that it is sufficient to show that if C is a finitely generated reduced (i.e., it has no nonzero nilpotent elements) commutative $\mathbb{Q}$-algebra and $\lambda_1, \ldots, \lambda_n \in C$ satisfy $\sum_{i=1}^{n} \lambda_i^j = 0$ for $j = 1, 2, \ldots, n$, then $\lambda_1, \ldots, \lambda_n$ are all zero. Since a reduced commutative ring embeds in a direct product of integral domains, we then see we can use projection maps to reduce to the case of an integral domain. So we prove the more general fact that if C is a commutative $\mathbb{Q}$-algebra that is an integral domain and $\lambda_1, \ldots, \lambda_n \in C$ are distinct and nonzero and $\beta_1, \ldots, \beta_n \in C$ are nonzero then $\sum \beta_i \lambda_i^j \neq 0$ for some $j = 1, \ldots, n$. We note that the original claim follows easily from this more general statement (but this requires characteristic zero). To see this general claim, we note that since we are in an integral domain, we get the result for $n = 1$. Now suppose that the result holds for $n < d$ and consider the case where $n = d$. Then if $\sum_{i=1}^{d} \beta_i \lambda_i^j = 0$ for $j = 1, 2, \ldots, d$, we see that

$$\sum_{i=1}^{d} \beta_i \lambda_i^j \lambda_d - \sum_{i=1}^{d} \beta_i \lambda_i^{j+1} = 0$$

for $j = 1, 2, \ldots, d-1$. Simplifying, we see that

$$\sum_{i=1}^{d-1} \beta_i (\lambda_i - \lambda_d) \lambda_i^j = 0,$$

for $j = 1, \ldots, d-1$. But this contradicts the induction hypothesis, and we get the claim. This finishes the proof. □

Proof (Proof of Theorem 4.2.3). We only prove the case when C is an integral domain, which is sufficient for the considerations of this chapter. Here we give an argument due to Rosset [512] (we note that Rosset does the more general case of a commutative ring). We note that one can immediately reduce to the case when C is a field of characteristic zero, since an integral domain C of positive characteristic is a homomorphic image of an integral domain C' of characteristic zero, and one can show that if the result holds for the field of fractions of C' then it holds for C. Now let $\mathbf{A}_1, \ldots, \mathbf{A}_{2n}$ be $2n$ elements in $M_n(C)$ and let V be a $2n$-dimensional C-vector space with basis $\mathbf{e}_1, \ldots, \mathbf{e}_{2n}$. For each $i \geq 0$, we recall that there is a d-th exterior product, $\wedge^d V$ is formed by taking the vector space $V_d := \otimes_{i=1}^{d} V$ and then forming the quotient V_d / W_d where W_d is the subspace of V_d spanned by elements of the form $\mathbf{e}_{i_1} \otimes \cdots \otimes \mathbf{e}_{i_d} - \operatorname{sgn}(\sigma) \mathbf{e}_{\sigma(i_1)} \otimes \cdots \otimes \mathbf{e}_{\sigma(i_d)}$, where $\sigma \in S_d$ and $1 \leq i_1, \ldots, i_d \leq 2n$. Then we let $\mathbf{e}_{i_1} \wedge \cdots \wedge \mathbf{e}_{i_d}$ denote the image of $\mathbf{e}_{i_1} \otimes \cdots \otimes \mathbf{e}_{i_d}$ in this quotient. Then one can check that $\wedge^d V$ is a $\binom{2n}{d}$-dimensional space with basis consisting of elements of the form $\mathbf{e}_{i_1} \wedge \cdots \wedge \mathbf{e}_{i_d}$ with $i_1 < i_2 < \cdots < i_d$.

Now let $E = \bigoplus_{d=0}^{2n} \wedge^i(V)$; this is called the exterior algebra on V. Notice that E is a ring with multiplication formed by taking the natural bilinear "wedge" map $\wedge^i(V) \times \wedge^j(V) \to \wedge^{i+j}(V)$ given by

$$\left((\mathbf{e}_{p_1} \wedge \cdots \wedge \mathbf{e}_{p_i}), (\mathbf{e}_{q_1} \wedge \cdots \wedge \mathbf{e}_{q_j}) \right) \mapsto \mathbf{e}_{p_1} \wedge \cdots \wedge \mathbf{e}_{p_i} \wedge \mathbf{e}_{q_1} \wedge \cdots \wedge \mathbf{e}_{q_j}.$$

Then E is generated as a C-algebra by $\mathbf{e}_1, \ldots, \mathbf{e}_{2n}$. Let E_e denote the subalgebra of E consisting of the direct sum of $\wedge^i V$ with i even. Then E_e is generated by elements of the form $\mathbf{e}_i \wedge \mathbf{e}_j$ as an algebra, and it is straightforward to check, using the relations given above, that these elements commute with one another and so E_e is a commutative ring of characteristic zero. Now consider $B := M_n(C) \otimes_C E \cong M_n(E)$. Let $\mathbf{X} = \mathbf{A}_1 \otimes \mathbf{e}_1 + \cdots + \mathbf{A}_{2n} \otimes \mathbf{e}_{2n} \in B$. Then

$$\mathbf{Y} := \mathbf{X}^2 = \sum_{i<j} (\mathbf{A}_i\mathbf{A}_j - \mathbf{A}_j\mathbf{A}_i)\mathbf{e}_i \wedge \mathbf{e}_j \in M_n(E_e).$$

Notice that $\mathbf{Y}$ has trace zero, as it is an E_e-linear combination of commutators. More generally,

$$\mathbf{Y}^i = \sum_{1 \leq j_1, \ldots, j_{2i} \leq 2n} \mathbf{A}_{j_1} \cdots \mathbf{A}_{j_{2i}} \mathbf{e}_{j_1} \wedge \cdots \wedge \mathbf{e}_{j_{2i}},$$

which is equal to

$$\sum_{1 \leq j_1 < j_2 < \cdots < j_{2i} \leq 2n} \mathsf{S}_{2i}(\mathbf{A}_{j_1}, \ldots, \mathbf{A}_{j_{2i}}) \mathbf{e}_{j_1} \wedge \cdots \wedge \mathbf{e}_{j_{2i}}.$$

It is straightforward to check that $\mathsf{S}_{2i}(\mathbf{A}_{j_1}, \ldots, \mathbf{A}_{j_{2i}})$ always has trace zero for $i \geq 1$ and so we see that the trace of $\mathbf{Y}^i$ is zero for $i = 1, 2, \ldots, n$. Thus $\mathbf{Y}$ is an $n \times n$ matrix over the commutative $\mathbb{Q}$-algebra E_e, and it has the property that the trace of all of its powers is equal to zero. By Lemma 4.2.4, we have that $\mathbf{Y}^n = 0$. Thus $\mathbf{Y}^n = \mathbf{X}^{2n} = 0$. As before, we have $\mathbf{X}^{2n} = \mathsf{S}_{2n}(\mathbf{A}_1, \ldots, \mathbf{A}_{2n})\mathbf{e}_1 \wedge \cdots \wedge \mathbf{e}_{2n}$ and so we get the desired result. □

We now show that the d-shuffle property implies all larger shuffle properties hold.

Proposition 4.2.5. *If f has the d-shuffle property, then f has the e-shuffle property for all $e \geq d$.*

Proof. By induction, it is sufficient to prove this when $e = d + 1$. Notice that

$$\begin{aligned} &\mathrm{Shuf}_{d+1}(f; w, w_1, \ldots, w_{d+1}, w') \\ &= \sum_{i=1}^{d+1} (-1)^{i-1} \mathrm{Shuf}_d(f, ww_i, w_1, \ldots, \widehat{w_i}, \ldots, w_{d+1}, w'), \end{aligned}$$

where $\widehat{w_i}$ means that w_i is omitted from the list. Hence if f has the d-shuffle property, it must also have the $(d + 1)$-shuffle property. □

We now introduce some notation. Given a finite alphabet $\mathscr{A}$, a field K, a word $w \in \mathscr{A}^*$ and a function $f : \mathscr{A} \to K$, we define two functions $f_w, f^w : \mathscr{A} \to \mathscr{A}$ by

$$f_w(u) := f(wu) \quad \text{and} \quad f^w(u) = f(uw). \tag{4.5}$$

We now give some definitions.

Definition 4.2.6. Given a finite alphabet $\mathscr{A}$, a field K, and a function $f : \mathscr{A}^* \to K$, we say that f is left (resp. right) $\mathscr{A}$-regular if the K-vector space spanned by the functions $\{f_w \mid w \in \mathscr{A}^*\}$ (resp. $\{f^w \mid w \in \mathscr{A}^*\}$) is finite-dimensional. If in addition to being left (resp. right) $\mathscr{A}$-regular, the range of f is a finite subset of K, we say that f is left (resp. right) $\mathscr{A}$-automatic.

Later, we will give Kleene's theorem, which states that being left $\mathscr{A}$-regular is equivalent to being right $\mathscr{A}$-regular. Thus we will omit the words left and right and just use the term $\mathscr{A}$-regular. In the case that $\mathscr{A} = \{1, 2, \ldots, k\}$, we shall say that a right $\mathscr{A}$-regular function is k-regular and shall say that a right $\mathscr{A}$-automatic function is k-automatic.

Notice that this definition of k-regular coincides with the definition of k-regular given by Allouche and Shallit [14] and the definition of k-automatic coincides with the conventional definitions of the k-automatic property.

For convenience, we use the following notation. Given a finite alphabet $\mathscr{A}$, a field K, and a function $f : \mathscr{A}^* \to K$, we let $L(f)$ denote the K-vector space spanned by the functions $\{f_w \mid w \in \mathscr{A}^*\}$, and we let $R(f)$ denote the vector space spanned by $\{f^w \mid w \in \mathscr{A}^*\}$.

Proposition 4.2.7. *Let $\mathscr{A}$ be a finite alphabet and let f be a left (resp. right) $\mathscr{A}$-regular function taking values in a field K. Let m denote the dimension of $L(f)$ (resp. the dimension of $R(f)$). Then there exist some $m \geq 1$, $m \times m$ matrices $\{\mathbf{A}_a \mid a \in \mathscr{A}\}$ with entries in K, and $\mathbf{v}, \mathbf{w} \in K^{d\times 1}$ such that*

$$f(x_1 \cdots x_i) = \mathbf{w}^{\mathrm{T}} \mathbf{A}_{x_1} \cdots \mathbf{A}_{x_i} \mathbf{v}$$

for all words $x_1 \cdots x_i \in \mathscr{A}^*$.

Proof. Choose $\varepsilon = w_1, \ldots, w_m \in \mathscr{A}^*$ such that $f_{w_1}, \ldots, f_{w_m}$ span the vector space $L(f)$. Given $x \in \mathscr{A}$, for each i pick $c_{i,j} \in K, j \leq m$, such that

$$f_{xw_i} = \sum_{j=1}^{m} c_{i,j} f_{w_j}.$$

Define the $m \times m$ matrix $\mathbf{A}_x$ whose (i,j)-entry is given by $c_{i,j}$. Take $\mathbf{v}$ to be the $m \times 1$ column vector whose i^{th} coordinate is $f(w_i)$. Notice that if $x_1, \ldots, x_i \in \mathscr{A}$, then

$$\mathbf{e}_1^{\mathrm{T}} \mathbf{A}_{x_1} \cdots \mathbf{A}_{x_i} \mathbf{v} = f(x_1 \cdots x_i).$$

In the case that f is right $\mathscr{A}$-regular, an analogous construction gives the same result. □

Proposition 4.2.8. *Let $\mathscr{A}$ be a finite alphabet, let K be a field, and let $f : \mathscr{A}^* \to K$ be a left (resp. right) $\mathscr{A}$-regular sequence. Then f has the d-shuffle property for $d = 2\dim(L(f))$ (resp. $d = 2\dim(R(f))$).*

Proof. Let m denote the dimension of $L(f)$. Since f is left $\mathscr{A}$-regular, there exist $m \times m$ matrices $\{\mathbf{A}_x \mid x \in \mathscr{A}\}$ with entries in K and some vector $\mathbf{v} \in K^{d\times 1}$ such that

$$f(x_1 \cdots x_m) = \mathbf{e}_1^{\mathrm{T}} \mathbf{A}_{x_1} \cdots \mathbf{A}_{x_m} \mathbf{v}$$

for all words $x_1 \cdots x_m \in \mathscr{A}^*$. Let $d = 2m$. We claim that f has the d-shuffle property. To see this, let $w_1, \ldots, w_d, w, w'$ be words in $\mathscr{A}^*$. Let $\mathscr{S}$ denote the monoid on $\{\mathbf{A}_x \mid x \in \mathscr{A}^*\}$. Then there exist matrices $\mathbf{U}_1, \ldots, \mathbf{U}_d, \mathbf{U}, \mathbf{U}'$ in $\mathscr{S}$ which correspond to $w_1, \ldots, w_d, w, w'$, respectively, given by the correspondence $\varepsilon \mapsto \mathbf{I}_d$ and

$$x_1 \cdots x_m \in \mathscr{A}^* \mapsto \mathbf{A}_{x_1} \cdots \mathbf{A}_{x_m}.$$

Then

$$f(ww_{\sigma(1)} \cdots w_{\sigma(d)}w') = \mathbf{e}_1^{\mathrm{T}} \mathbf{U} \mathbf{U}_{\sigma(1)} \cdots \mathbf{U}_{\sigma(d)} \mathbf{U}' \mathbf{v}.$$

Hence

$$\begin{aligned}\sum_{\sigma \in S_d} \operatorname{sgn}(\sigma) f(ww_{\sigma(1)} \cdots w_{\sigma(d)}w') &= \sum_{\sigma \in S_d} \mathbf{e}_1^{\mathrm{T}} \mathbf{U} \left(\sum_{\sigma \in S_d} \mathbf{U}_{\sigma(1)} \cdots \mathbf{U}_{\sigma(d)} \right) \mathbf{U}' \mathbf{v} \\ &= 0,\end{aligned}$$

where the last step follows from the Amitsur-Levitzki theorem. □

We now prove that the function given in Example 4.2.2 has the 4-shuffle property. We note that if $f : \{0, 1\}^* \to \mathbb{Q}$ is the function which maps a word w to $[w]_2$, where $[w]_2$ is the natural number whose binary expansion is equal to w, then the $\mathbb{Q}$-vector space $R(f)$ is two-dimensional, spanned by f and the constant function g which sends every word to 1. To see this, notice that $f^w(u) = [uw]_2 = 2^{\mathrm{length}(w)}[u]_2 + [w]_2$ and so

$$f^w = 2^{\mathrm{length}(w)} f + [w]_2 g.$$

Clearly f and g are both in $R(f)$ and are linearly independent. Thus $\dim(R(f)) = 2$ and so f has the 4-shuffle property.

4.3 The Power Property

In this section we define the *power property*, which is the second ingredient of the characterization of automatic and regular sequences of Berstel and Reutenauer. Given a finite alphabet $\mathscr{A}$ and a field K, we say that $f : \mathscr{A} \to K$ has the *d-power property* if for any word w_0, there exists a polynomial $\Phi(t) \in K[t]$ of degree at most

d with constant coefficient 1 such that for any words w, w' we have

$$\Phi(t)\left(\sum_{i=0}^{\infty} f(ww_0^i w')\right)t^i,$$

is a polynomial in $K[t]$ of degree at most d. We shall say that f has the *power property* if f has the d-power property for some $d \geq 1$.

Lemma 4.3.1. *Let K be a field and let $\mathbf{X}$ be a $d \times d$ matrix with entries in K. Then*

$$\sum_{i=0}^{\infty} \mathbf{X}^i t^i = \mathbf{Y}(t)\det(1 - t\mathbf{X})^{-1},$$

for some matrix $\mathbf{Y}$ with entries in $K[t]$ of degree at most $d-1$.

Proof. Let $\mathbf{Y}(t)$ denote the classical adjoint of $1 - t\mathbf{X}$. Then $\mathbf{Y}(t)$ is a matrix with entries given by polynomials in t of degree at most $d-1$ and

$$\mathbf{Y}(t)(1 - t\mathbf{X}) = \det(1 - t\mathbf{X})\mathbf{I}_d.$$

Notice that

$$\mathbf{Y}(t)(1 - t\mathbf{X})\sum_{i=0}^{\infty} \mathbf{X}^i t^i = \mathbf{Y}(t)$$

and so

$$\sum_{i=0}^{\infty} \mathbf{X}^i t^i = \mathbf{Y}(t)\det(1 - t\mathbf{X})^{-1}.$$

□

Proposition 4.3.2. *Let f be a $\mathscr{A}$-regular function taking values in a field K. Then f has the m-power property, where m is the dimension of $L(f)$.*

Proof. It suffices to show that for any word w_0 and w, w' that

$$\sum_{i=1}^{\infty} f(ww_0^i w')t^i$$

is a rational function in t whose numerator and denominator have degrees that are at most m (with denominator independent of w, w' and depending only upon w_0 and having constant coefficient 1). By Proposition 4.2.7, there exists some m and $m \times m$ matrices $\mathbf{U}, \mathbf{U}_0, \mathbf{U}'$ such that $f(ww_0^i w') = \mathbf{e}_1^{\mathrm{T}}\mathbf{U}\mathbf{U}_0^i\mathbf{U}'\mathbf{v}$. We then have

$$\sum_{i=0}^{\infty} f(ww_0^i w')t^i$$
$$= \sum_{i=0}^{\infty} \mathbf{e}_1^{\mathrm{T}}\mathbf{U}\left(\mathbf{U}_0^i\right)\mathbf{U}'\mathbf{v}$$
$$= \mathbf{e}_1^{\mathrm{T}}\mathbf{U}\left(\sum_{i=0}^{\infty} \mathbf{U}_0^i t^i\right)\mathbf{U}'\mathbf{v}$$
$$= \mathbf{e}_1^{\mathrm{T}}\mathbf{U}\mathbf{Y}(t)\mathbf{U}'\mathbf{v} \cdot \det(1 - \mathbf{U}_0 t)^{-1},$$

where $\mathbf{Y}(t)$ is the classical adjoint of $1 - \mathbf{U}_0 t$. This expression is easily seen to be a rational function of the form

$$P(t)\Phi(t)^{-1},$$

and the numerator and denominator have degree at most m and $\Phi(0) = 1$ and Φ depends only upon $\mathbf{U}_0$ and hence only on w_0. □

4.4 Shirshov's Height Theorem

We now introduce an important result in combinatorial ring theory: Shirshov's theorem. We have seen that $\mathscr{A}$-regular functions have both the shuffle and power properties. In fact, it is the case that the shuffle and power properties characterize regular sequences. To deduce this we need a famous combinatorial result due to Shirshov. We first take a detour and survey the beautiful field of polynomial identity algebras.

Definition 4.4.1. Let K be a field and let B be a K-algebra. We say that B is a *polynomial identity* ring if there exists a nonzero, noncommutative polynomial $p(x_1, \ldots, x_d)$ with coefficients in K, such that $p(b_1, \ldots, b_d) = 0$ for all $b_1, \ldots, b_d \in B$. In this case the polynomial p is called a *polynomial identity* for B. The total degree of the polynomial identity p of B of least positive degree is called the *PI degree* of B.

Example 4.4.2. Any commutative algebra B is a polynomial identity ring since it satisfies the identity $x_1x_2 - x_2x_1 = 0$.

Example 4.4.3 (Wagner). The ring of 2×2 matrices over a field K satisfies the identity $[x_1, [x_2, x_3]^2] = 0$, where $[a, b] = ab - ba$.

Proof. Notice that if $\mathbf{X}$ is a 2×2 matrix then by the Cayley-Hamilton theorem

$$\mathbf{X}^2 - \mathrm{Tr}(\mathbf{X})\mathbf{X} + \det(\mathbf{X})\mathrm{I}_2 = 0.$$

Hence if **X** has trace 0, then its square is a scalar matrix. In particular the square of a commutator must commute with every 2×2 matrix. □

An important fact is that if a ring B satisfies a nontrivial polynomial identity, then it in fact satisfies a homogeneous multilinear identity (i.e., each monomial occurring in the identity has degree precisely one in each variable).

Proposition 4.4.4. *Let K be a field and let B be a K-algebra satisfying a polynomial identity of degree at most d. Then B satisfies a multilinear homogeneous identity of degree at most d.*

Proof. Let $m(f)$ be the maximum degree of a variable appearing in f. Among all nonzero polynomial identities, we pick one with the property that $m(f)$ is minimal. Let us call this minimum m. Among all such identities with $m(f) = m$, we pick one with the property that the number of variables of degree m is minimal. Let $f(x_1, \ldots, x_d)$ be such a minimal polynomial identity for the ring B. By permuting the variables, if necessary, we may assume that m is the degree of x_1 in f. Consider the identity $g(x_1, y_1, x_2, \ldots, x_d) := f(x_1 + y_1, \ldots, x_d) - f(x_1, \ldots, x_d) - f(y_1, \ldots, x_d) \in K\{x_1, y_1, x_2, \ldots, x_d\}$. We note that this is an identity for B. Then it is straightforward to see that this transforms any monomial of degree m in x_1 to a monomial of total degree m in x_1 and y_1 and no terms of degree m in just x_1 or just y_1. That means that either $m(g) < m$ or $m(g) = m$ but the number of variables of degree m in g is strictly less than that of f. By minimality of f we have that $g = 0$. But this occurs only if $m = m(f) = 1$. So having $m = 1$ says that every monomial appears with degree at most 1. Now pick a monomial occurring in f with nonzero coefficient of smallest degree, say $r \leq d$, By relabeling indices, we may assume that the monomial is $x_1 \cdots x_r$. Then consider $f(x_1, \ldots, x_r, 0, \ldots, 0)$. This is nonzero and must be homogeneous by minimality of r.

Notice this process yields an algorithm to convert a polynomial identity into a homogeneous multilinear identity. One begins with an identity, and if it is of degree strictly greater than one in some variable, say x_1, then we add a new variable y_1, and we look at $f(x_1 + y_1, \ldots) - f(x_1, \ldots) - f(y_1, \ldots)$. This creates an additional variable, but the total degree does not increase. If we keep repeating this process, the argument above shows that it must terminate and so we obtain an identity in some set of variables in which each variable occurs with degree at most 1. Then we pick a monomial of minimal length and set all variables not occurring in this monomial equal to zero to get a homogeneous multilinear identity. Notice that the total degree never increases at any step, so we see the total degree of the homogeneous multilinear identity we ultimately produce is at most that of the original identity. □

Remark 4.4.5. We note that in the case that the polynomial $p(x_1, \ldots, x_d)$ is multilinear and homogeneous, to check that a K-algebra B satisfies this identity, it is sufficient to check that $p(b_1, \ldots, b_d) = 0$ for $(b_1, \ldots, b_d) \in Y^d$ where Y is a K-spanning set for B.

Proof. To see this, we remark that if $a_1, \ldots, a_d \in B$ then we can write each a_i as a K-linear combination of elements of Y. Then using multilinearity, we can write

$p(a_1, \ldots, a_d)$ as a K-linear combination of elements of the form $p(b_1, \ldots, b_d)$ with $(b_1, \ldots, b_d) \in Y^d$. Thus if the latter all vanish, then p is necessarily an identity for B. □

We will make use of this fact for algebras B of the form $B := K\{x_1, \ldots, x_k\}/I$, where I is a two-sided ideal of the free algebra on noncommuting variables $x_1, \ldots, x_k$. In this case, we can take the images of the words in $x_1, \ldots, x_k$ to be our spanning set, and so to check that a homogeneous multilinear identity holds for an algebra B, it suffices to check that it vanishes when evaluated at elements from the semigroup generated by a finite set of generators.

Arguably, the most important types of identities studied in the theory of polynomial identities are the *standard identities*, which we already encountered, albeit in a different form, when addressing the shuffle property. The nth standard identity is defined as follows:

$$\mathsf{S}_n(x_1, \ldots, x_n) = \sum_{\sigma \in S_n} \operatorname{sgn}(\sigma) x_{\sigma(1)} \cdots x_{\sigma(n)}. \tag{4.6}$$

The Amitsur-Levitzki theorem shows that the ring of $n \times n$ matrices over a commutative ring satisfies the $2n^{\text{th}}$ standard identity. In fact, every finitely generated algebra satisfying a polynomial identity must satisfy one of these standard identities.

Theorem 4.4.6 (Braun). *Let K be a field and let B be a finitely generated K-algebra satisfying an identity. Then B satisfies the identity S_n for some $n \geq 1$.*

Proof. See Braun [105] or Amitsur and Small [21, Corollary 1.2.8A]. □

We recall that if we are given a free monoid $\mathscr{A}^*$ with $\mathscr{A} = \{x_1, \ldots, x_k\}$, we can put a *pure lexicographic order* $\preceq$ on $\mathscr{A}^*$ by declaring that

$$x_1 \prec x_2 \prec \cdots \prec x_k.$$

In this order we declare that if w is a proper initial factor of w' then $w \prec w'$. So, for example, $x_1 \preceq x_1x_3$ and $x_2x_3x_4 \prec x_3x_1$. We note that this lexicographic order extends to right infinite words on $\mathscr{A}$. If we alter the order somewhat and use the pure lexicographic order to order words of the same length and for words of different length declare that the longer word is bigger, then this new order is called the *degree lexicographic order*.

Using intricate combinatorial techniques, Shirshov proved the following beautiful result.

Theorem 4.4.7 (Shirshov). *Let $\mathscr{A} = \{x_1, \ldots, x_k\}$ be a finite alphabet, and let m be a positive integer. If w is a right-infinite word over the alphabet $\mathscr{A}$, then either there is some nontrivial word $w_0 \in \mathscr{A}^*$ such that w_0^d is a factor of w for every $d \geq 1$ or w contains a finite factor of the form $w_1w_2 \cdots w_m$ where $w_1 \succ w_2 \succ \cdots \succ w_m$ are nontrivial words in $\mathscr{A}^*$ with no w_i equal to a prefix of w_j for $i \neq j$ and $\succ$ is the pure lexicographic order induced by $x_1 \succ x_2 \succ \cdots \succ x_k$.*

We postpone the proof of Shirshov's theorem until we have developed a few basic combinatorial tools.

To prove Theorem 4.4.7, we require two basic combinatorial tools. The first is König's infinity lemma; the second is a theorem of Furstenberg.

Theorem 4.4.8 (König's infinity lemma). *Let $\mathscr{A}$ be a finite alphabet, and let $\mathscr{S}$ be an infinite subset of $\mathscr{A}^*$ that is closed under the process of taking factors. Then there exists $w \in \mathscr{A}^{\mathbb{N}}$ such that every finite factor of w is in $\mathscr{S}$.*

Proof. We define $w = a_1a_2\cdots$ as follows. Since $\mathscr{S}$ is infinite, there is some $a_1 \in \mathscr{A}$ such that a_1 is the first letter of infinitely many elements of $\mathscr{S}$; now suppose that for every $i \geq 1$ we have defined a word $a_1a_2\cdots a_i$ with the property that it is a prefix of infinitely many elements of $\mathscr{S}$. Then since $\mathscr{A}$ is finite, there is some $a_{i+1} \in \mathscr{A}$ such that $a_1a_2\cdots a_ia_{i+1}$ is a prefix of infinitely many elements of $\mathscr{S}$. Continuing in this way, we obtain an infinite word w with the desired property. □

The next result is Furstenberg's theorem, which is part of a more general result in Ergodic theory. We give an algebraic proof in the case in which we are interested. We recall that a right-infinite word is *uniformly recurrent* if each factor u has the property that there is some natural number $N = N(u)$ such that whenever u occurs as a factor, its next occurrence is at most N positions later in our right-infinite word. As far as we know, this rather simple proof (which requires the axiom of choice) does not appear in the literature.

Theorem 4.4.9 (Furstenberg's theorem). *Let $\mathscr{A}$ be a finite alphabet and let $w \in \mathscr{A}^{\mathbb{N}}$. Then there is a uniformly recurrent word $u \in \mathscr{A}^{\mathbb{N}}$ such that every finite factor of u is a factor of w.*

Proof. Let K be a field. Write $\mathscr{A} = \{x_1, \ldots, x_k\}$, and consider the algebra $B = K\{x_1, \ldots, x_k\}/I$, where I is the ideal generated by all monomials that do not occur as a factor of w. Then B is infinite-dimensional as a K-vector space since w is infinite and the images of the factors of w are linearly independent in B. Now let $\mathscr{S}$ denote the collection of ideals J of $K\{x_1, \ldots, x_k\}$ that are generated by monomials, contain I, and have the property that B/J is infinite-dimensional. Then $\mathscr{S}$ is nonempty since I is in $\mathscr{S}$. We note that $\mathscr{S}$ is closed under unions of chains, for if L is the union of a chain in $\mathscr{S}$, then L is certainly generated by monomials and contains I; if $K\{x_1, \ldots, x_k\}/L$ is finite-dimensional, then there is some N such that L contains all monomials of length $\geq N$. In particular, L is finitely generated as it generated by all monomials of length exactly N along with the finite set of monomials of length $< N$ that are in L. Since L is the union of a chain in $\mathscr{S}$, and L is finitely generated, there is some element in our chain that must be equal to L. Thus we can pick a maximal element L of $\mathscr{S}$ by Zorn's lemma. Now since L is generated by monomials, $K\{x_1, \ldots, x_k\}/L$ is spanned by the images of all monomials over $x_1, \ldots, x_k$ that are not in the ideal L. Since $K\{x_1, \ldots, x_k\}/L$ is infinite-dimensional and the collection of words that are not in L is closed under taking factors, we see by König's infinity lemma that there is a right-infinite word u with the property that all factors of u are not in L. In particular, they are not in I, and so all factors of u are factors of w.

We claim that u is uniformly recurrent. If not, there is some finite factor u_0 such that there are arbitrarily long factors of u that avoid u_0. Then let L' be the ideal generated by u_0 and L. Then since there are arbitrarily long words in u that avoid u_0, we see that these words have nonzero image in the ring $K\{x_1,\dots,x_k\}/L'$, and so $K\{x_1,\dots,x_k\}/L'$ is infinite-dimensional. But this contradicts maximality of L in $\mathscr{S}$. The result follows. □

Proof (Proof of Theorem 4.4.7). We follow the proof of Pirillo [484]. By Furstenberg's theorem, there is some uniformly recurrent right-infinite word v such that every factor of v is a factor of w. Now if v is eventually periodic, then $v = v_0 w_0^{\omega}$ for some v_0, w_0 with w_0 nontrivial, and we get the claim. If v is not eventually periodic, then v must have at least m distinct factors of length $m-1$ (see [14, Theorem 10.2.6]). Let $w_1,\dots,w_m$ be these distinct factors and suppose that $w_1 \succ w_2 \succ \cdots \succ w_m$. Since v is uniformly recurrent, there is a factor v_0 of v that can be written in the form $v_0 = v_1 \cdots v_m$ where w_i is a prefix of v_i. Then we see that $v_{\sigma(1)} \cdots v_{\sigma(m)} \prec v_1 \cdots v_m$, where the inequality is strict whenever σ is not the identity. □

Shirshov's theorem has as an immediate application the following result about algebras satisfying a polynomial identity.

Corollary 4.4.10 (Shirshov). *Let K be a field and let B be a finitely generated K-algebra with generators $x_1,\dots,x_k$ that satisfies a polynomial identity. Then either B is finite-dimensional as a K-vector space or there is some word $w \in \{x_1,\dots,x_d\}^*$ such that the images of $1, w, w^2,\dots$ in the algebra B are linearly independent over K.*

Proof. We put the degree lexicographic order, $\prec$, on the monomials in $\{x_1,\dots,x_k\}^*$ induced by the $x_1 \succ x_2 \succ \cdots \succ x_k$. We let $\prec_p$ be the corresponding pure lexicographic order. Let I be the two-sided ideal of $K\{x_1,\dots,x_k\}$ generated by the collection of monomials $w \in \{x_1,\dots,x_k\}^*$ with the property that the image of w in B can be expressed as a K-linear combination of the image of monomials that are strictly smaller than w in the degree lexicographic order. Since B satisfies a polynomial identity, we know that it satisfies a homogeneous multilinear identity by Proposition 4.4.4. We may write this identity as

$$f(x_1,\dots,x_m) = x_1 \cdots x_m + \sum_{\sigma \in S_m \setminus \mathrm{id}} c_\sigma x_{\sigma(1)} \cdots x_{\sigma(m)}.$$

It then follows that if $w_1 \succ_p w_2 \succ_p w_3 \succ_p \cdots \succ_p w_m$ with no w_i equal to a prefix of w_j for $i \neq j$ then the identity f shows that $w_1 w_2 \cdots w_m$ can be written as a K-linear combination of strictly smaller words in the degree lexicographic order.

Now if $K\{x_1,\dots,x_k\}/I$ is finite-dimensional, then B is finite-dimensional since by construction the images in B of words over $\{x_1,\dots,x_k\}^*$ that are not in I span B as a K-vector space. So if B is infinite-dimensional over K, then $K\{x_1,\dots,x_k\}/I$ must be infinite-dimensional. Then there are arbitrarily long words on $\{x_1,\dots,x_k\}^*$ that are not in the ideal I, and so by König's infinite lemma, we see that there is a right-infinite word w whose factors all lie outside of the ideal I. But by the remarks above, we see that w cannot contain any factor of the form $w_1 w_2 \cdots w_m$ with $w_1 \succ_p w_2 \succ_p w_3 \succ_p \cdots \succ_p w_m$ with no w_i equal to a prefix of w_j for $i \neq j$. Thus, by Shirshov's theorem, there is a nontrivial word w_0 such that for every $d \geq 1$, w_0^d is a factor of w. This means that $w_0, w_0^2, \ldots$ are not in I. By definition of I this means that their images in B are linearly independent in B. □

In fact, there is an even stronger version of Shirshov's theorem for rings satisfying a polynomial identity, which we give now.

Theorem 4.4.11 (Strong version of Shirshov's theorem). *Let C be a commutative ring, and let B be a finitely generated C-algebra with generators $x_1, \ldots, x_k$ and which satisfies a polynomial identity of degree d. Then every element in B is a C-linear combination of elements of the form*

$$w_1^{i_1} \cdots w_m^{i_m},$$

where $m \leq d^5 k^d$ and $w_1, \ldots, w_m$ are words of length less than d.

Proof. See Theorem 1.2.2 of Amitsur and Small [21]. □

We have the following unexpected application of Shirshov's theorem to functions with a shuffle property.

Corollary 4.4.12. *Let K be a field and let $f : \mathscr{A} \to K$ have the d-shuffle property. Then the vector space $L(f)$ (resp. $R(f)$) is spanned by elements of the form $f_{w_1^{j_1} \cdots w_m^{j_m}}$ (resp. $f^{w_1^{j_1} \cdots w_m^{j_m}}$) with $m \leq d^5 |\mathscr{A}|^d$ and $w_1, \ldots, w_m$ words of length at most d.*

Proof. Write $\mathscr{A} = \{x_1, x_2, \ldots, x_k\}$. Let $B = K\{x_1, \ldots, x_k\}$ be the free K-algebra on k-variables, *i.e.*, the ring of "noncommutative polynomials" in k variables with coefficients in K.

We define an ideal $I \subseteq K\{x_1, \ldots, x_k\}$ as follows. Given words $w_1, \ldots, w_m$ in $x_1, \ldots, x_k$, we declare that

$$c_1 w_1 + \cdots + c_m w_m \ \in I$$

if

$$c_1 f_{uw_1} + \cdots + c_m f_{uw_m} \qquad \text{is identically 0 for all } u \in \mathscr{A}^*.$$

We note that all such relations form an ideal. Notice that since f has the d-shuffle property, for any d words $w_1, \ldots, w_d$, we have

$$\mathsf{S}_d(w_1, \ldots, w_d) \ = \ \sum_{\sigma \in S_d} \operatorname{sgn}(\sigma) w_{\sigma(1)} \cdots w_{\sigma(d)} \ \in I.$$

Since the function

$$\mathsf{S}_d(x_1, \ldots, x_d)$$

is multilinear, we see that $S_d(b_1,\ldots,b_d) \in I$ for all $b_1,\ldots,b_d$ in B. Consequently, B/I satisfies a polynomial identity. It follows from Theorem 4.4.11 that every element of B/I is a K-linear combination of the images of elements of the form $w_1^{j_1}\cdots w_m^{j_m}$ with $m \le d^5|\mathscr{A}|^d, j_1,\ldots,j_m \ge 0$ and $w_1,\ldots,w_m$ words of length at most d. It follows that any word $w \in \{x_1,\ldots,x_k\}^*$ there is some $b \in I$ such that $w-b$ can be written as a linear combination of words of the form $w_1^{j_1}\cdots w_m^{j_m}$ with $m \le d^5|\mathscr{A}|^d$ and $w_1,\ldots,w_m$ of length at most d. By definition of I we then have

$$f_w \in \mathrm{Span}_K\{f_{w_1^{j_1}\cdots w_m^{j_m}} \mid m \le d^5|\mathscr{A}|^d, \mathrm{length}(w_i) \le d \text{ for } i \le m\}.$$

We thus obtain the desired result. A similar argument works for the vector space $R(f)$. □

4.5 Characterization of $\mathscr{A}$-Regular Sequences

In this section, we are finally able to give the general version of Kleene's theorem along with the structure result of Berstel and Reutenauer.

Lemma 4.5.1. *Let K be a field, let d and e be positive integers, and let $f : \mathscr{A}^* \to K$ be a function with the e-power property and let m be at most $d^5|\mathscr{A}|^d$ and let $w_1,\ldots w_m$ be words of length at most d. Then there exists $e \ge 0$ such that the K-vector space spanned by $\{f_{w_1^{i_1}\cdots w_m^{i_m}} \mid i_1,\ldots,i_m \ge 0\}$ (resp. $f^{w_1^{i_1}\cdots w_m^{i_m}} \mid i_1,\ldots,i_m \ge 0\}$) is spanned by*

$$\{f_{w_1^{i_1}\cdots w_m^{i_m}} \mid 0 \le i_1,\ldots,i_m \le e\}$$

(resp. $\{f_{w_1^{i_1}\cdots w_m^{i_m}} \mid 0 \le i_1,\ldots,i_m \le e\}$).

Proof. By the e-power property, there exist polynomials $\Phi_1,\ldots,\Phi_m$ of degree at most e with constant coefficient 1 such that for every word u we have

$$\Phi_i(t)\sum_{s=0}^{\infty} f(w_1^{i_1}\cdots w_j^{s}\cdots w_m^{i_m}u)t^{i_1}\cdots t^{i_m}$$

is a polynomial in $K[t]$ of degree at most e. Let e be the maximum of the degrees of $P, \Phi_1,\ldots,\Phi_m$. Now suppose that it is not the case that $\{f_{w_1^{i_1}\cdots w_m^{i_m}} \mid 0 \le i_1,\ldots,i_m \le e\}$ spans $\{f_{w_1^{i_1}\cdots w_m^{i_m}} \mid i_1,\ldots,i_m \ge 0\}$. Then there exist $j_1,\ldots,j_m$ with $j_i > e$ for some i such that $f_{w_1^{j_1}\cdots w_m^{j_m}}$ is not in the span of $\mathscr{S}$. It is no loss of generality to assume that $(j_1,\ldots,j_m)$ has the property that if $(j_1',\ldots,j_m')$ satisfies:

- $j_k' \le j_k$ for $1 \le k \le m$; and
- $(j_1',\ldots,j_m') \ne (j_1,\ldots,j_m)$,

then

$$f_{w_1^{j'_1}\cdots w_m^{j'_m}} \in \{f_{w_1^{i_1}\cdots w_m^{i_m}} \mid 0 \le i_1,\ldots,i_m \le e\}.$$

By assumption $j_i > e$ and by the e-power property, we have

$$\sum_{k=0}^{e} c_k f_{w_1^{j_1}\cdots w_i^{j_i-k}\cdots w_m^{j_m}} = 0,$$

where c_k is the coefficient of x^k in Φ_i. Hence

$$f_{w_1^{j_1}\cdots w_m^{j_m}} \in \mathrm{Span}_K\{f_{w_1^{j_1}\cdots w_i^{j_i-k}\cdots w_m^{i_m}} \mid 1 \le k \le j_i\}\},$$

since Φ_i has constant coefficient 1. By minimality, we deduce that $f_{w_1^{j_1}\cdots w_m^{j_m}}$ is in the span of $\{f_{w_1^{i_1}\cdots w_m^{i_m}} \mid 0 \le i_1,\ldots,i_m \le e\}$, a contradiction. The result follows. □

Theorem 4.5.2 (Berstel-Reutenauer). *Let K be a field and let $f : \mathscr{A}^* \to K$. Then the following are equivalent:*

1. *f is right k-regular;*
2. *f is left k-regular;*
3. *f has the d-shuffle and the d-power property for some d;*
4. *f has the d-shuffle and the d-power property for all sufficiently large d.*

Proof. From Propositions 4.2.5, 4.2.8, and 4.3.2, we have that if f is either right or left k-regular, then f has the d-shuffle and the d-power property for all sufficiently large d. Hence (1) and (2) both imply (4). Clearly (4) implies (3). Thus, it is sufficient to show that if f has the d-shuffle and the d-power property for some d, then f is both left and right regular. Suppose that f has the d-shuffle and the d-power property. We claim that the K-vector space $L(f)$ is finite-dimensional. To see this, notice that by Corollary 4.4.12, the vector space $L(f)$ is spanned by elements of the form $f_{w_1^{i_1}\cdots w_m^{i_m}}$ with $m \le d^5|\mathscr{A}|^d$ and $w_1,\ldots,w_m$ having length at most d. Since f has the d-power property, by Lemma 4.5.1 for any words $w_1,\ldots,w_m$ of length at most d with $m \le d^5|\mathscr{A}|^d$, the K-vector space spanned by $\{f_{w_1^{i_1}\cdots w_m^{i_m}} \mid i_1,\ldots,i_m \ge 0\}$ is spanned by

$$\{f_{w_1^{i_1}\cdots w_m^{i_m}} \mid 0 \le i_1,\ldots,i_m \le d\}.$$

It now follows that $L(f)$ is in the K-span of the set

$$\bigcup_{m\le d^5|\mathscr{A}|^d} \bigcup_{\substack{\mathrm{length}(w_j)\le d,\\ 1\le j\le m}} \{f_{w_1^{i_1}\cdots w_m^{i_m}} \mid i_1,\ldots,i_m \le d\}.$$

Thus $L(f)$ is finite-dimensional and so f is left $\mathscr{A}$-regular. A similar argument shows that f is right $\mathscr{A}$-regular and hence (3) implies both (1) and (2). □

As a result of the equivalence between left and right regularity, we drop the words left and right and talk only of *$\mathscr{A}$-regular functions* from now on.

Corollary 4.5.3. *Let $f : \mathscr{A}^* \to A$. Then the following are equivalent:*

1. *f is k-automatic;*
2. *f has the d-shuffle and the d-power property for some d and has a finite range;*
3. *f has the d-shuffle and the d-power property for all sufficiently large d and has a finite range.*

We make the following remark that follows from the proof of Theorem 4.5.2. This gives a bound on the dimension of the vector space in terms of the quantities d and e for which one has the d-shuffle and e-power property, which is undoubtedly far from optimal, but has the advantage of being explicit in terms of d and e.

Remark 4.5.4. If $|\mathscr{A}| \geq 2$ and $f : \mathscr{A}^* \to K$ has the d-shuffle property and the e-power property, then $L(f)$ and $R(f)$ are both at most N-dimensional where

$$N = (e+1)^{d^5|\mathscr{A}|^d} \mathscr{A}|^{2d^6|\mathscr{A}|^d}.$$

Proof. The proof of Theorem 4.5.2 shows that $L(f)$ is in the K-span of the set

$$\bigcup_{m \leq d^5|\mathscr{A}|^d} \bigcup_{\substack{\text{length}(w_j) \leq d, \\ 1 \leq j \leq m}} \{f_{w_1^{i_1} \cdots w_m^{i_m}} \mid i_1, \ldots, i_m \leq e\}.$$

Since there are at most $|\mathscr{A}|^{d+1}$ words of length at most d, we see that the number of m-tuples of words of length at most d with $m \leq d^5|\mathscr{A}|^d$ is at most

$$\sum_{m=1}^{d^5|\mathscr{A}|^d|} |\mathscr{A}|^{(d+1)m}| \leq |\mathscr{A}|^{2d^6|\mathscr{A}|^d}.$$

Now for each such m-tuple, we pick up a space of dimension at most $(e+1)^m$ in our spanning set, and so the dimension of $L(f)$ is at most $(e+1)^{d^5|\mathscr{A}|^d} |\mathscr{A}|^{2d^6|\mathscr{A}|^d}$. A similar argument works for $R(f)$. □

4.6 Sandwich Functions

In this section we introduce a special type of function that is produced from a K-valued function on a free monoid, where K is a field; these functions will be called, for reasons that will soon become apparent, *sandwich functions*. We will then characterize all automatic sandwich functions.

Let $\mathscr{A} = \{x_1, \ldots, x_k\}$ be a finite alphabet. We put a pure lexicographic order $\preceq$ on $\mathscr{A}^*$ by declaring that

$$x_1 \prec x_2 \prec \cdots \prec x_k.$$

We note that this lexicographic order extends to right infinite words over the alphabet $\mathscr{A}$.

Let w_1 and w_2 be two (possibly right-infinite) words on $\mathscr{A}$ with $w_1 \preceq w_2$. In dealing with right-infinite words, it will be convenient to use w^ω to denote the right-infinite word $wwww\cdots$. We define $f : \mathscr{A} \to \{0, 1\}$ by $f(w) = 1$ if $w_1 \preceq w \preceq w_2$ and $f(w) = 0$ otherwise. We call f a *sandwich function* since the words which get mapped to 1 are sandwiched between w_1 and w_2.

Example 4.6.1. Let $\mathscr{A} = \{x_1, \ldots, x_k\}$. Then the constant function 1 and the constant function 0 are both sandwich functions.

Proof. To get the constant function 1, take $w_1 = \varepsilon$ and take $w_2 = x_k^\omega$. To get the constant function 0, take $w_1 = w_2 = x_k^\omega$. □

Example 4.6.2. Let $\mathscr{A} = \{x_1, \ldots, x_k\}$. Then the function f which sends words beginning with x_1 to 1 and all other words to 0 is a sandwich function.

Proof. Take $w_1 = x_1$ and take $w_2 = x_1 x_k^\omega$. Then the sandwich function given by these words is f. □

We now characterize $\mathscr{A}$-automatic sandwich functions. To do this we first prove some basic results. As notation for the following lemma, we introduce the function χ which inputs a statement and outputs 1 if the statement is true and 0 if the statement is false.

Lemma 4.6.3. *Let w be a (possibly right-infinite) word on a finite alphabet $\mathscr{A}$ which is either finite or eventually periodic. Then the functions $f(u) := \chi(u \preceq w)$ and $g(u) = \chi(u \succeq w)$ are both $\mathscr{A}$-automatic.*

Proof. If w is finite, notice that if v is a word whose length is greater than the length of w, then

$$f_v(u) = f(vu) = \chi(vu \prec w) = \chi(v \preceq w).$$

Hence the vector space $L(f)$ is contained in the space spanned by the constant function 1 and the functions $\{f_v \mid \text{length}(v) \leq \text{length}(w)\}$. Thus $L(f)$ is finite-dimensional. A similar argument shows that $L(g)$ is finite-dimensional. If w is eventually periodic, then we can write $w = w_1 w_2^\omega$. Notice that if v is not an initial factor of w, then

$$f_v(u) = f(vu) = \chi(vu \preceq w) = \chi(v \preceq w),$$

which is a constant function. If v is an initial factor of w of length at least $\text{length}(w_1) + 2\text{length}(w_2)$, then $f_v = f_{v'}$, where v' is an initial factor of v obtained by removing the last $\text{length}(w_2)$ letters from v. Hence $L(f)$ is again contained in a finite-dimensional vector space and hence must be finite-dimensional. A similar argument works for $L(g)$. □

We need a lemma about monotonic subsequences.

Lemma 4.6.4. *Let w be a right-infinite word on a finite alphabet $\mathscr{A}$ which is not eventually periodic. Then for any $d \geq 1$, there exist finite words $u_1, \ldots, u_d$ such that:*

- $\text{length}(u_1) < \text{length}(u_2) < \cdots < \text{length}(u_d)$;
- *the sequence $u_1, \ldots, u_d$ is monotonic with respect to $\prec$;*
- *for $2 \leq i \leq d$, there exists a word $u_i' \neq u_i$ of the same length as u_i which does not have u_{i-1} as an initial factor and has the property that u_{i-1}, u_i', u_i is monotonic;*
- *$u_1 u_2 \cdots u_d$ is a factor of w.*

Proof. For $i \geq 0$ define w_i to be the right-infinite word obtained by removing the first i letters from w. Since w is not eventually periodic, the words $w_0, w_1, \ldots$ are all distinct. It follows that they are totally ordered by $\prec$. It follows that there is a monotonic (with respect to $\prec$) subsequence $w_{i_1}, w_{i_2}, \ldots$ of $w_0, w_1, \ldots$. Without loss of generality $w_{i_1} \prec w_{i_2} \prec \cdots$. Let $v_j = w_{i_j}$ for $j \geq 1$. Pick $k_1 = 1$. Since $v_1 \prec v_2$ and v_2 is not eventually periodic, there exists some number m_1 such that the first m_1 letters of v_1 differ from the first m_1 letters of v_2 and the first m_1 letters of v_2 differ with the first m_1 letters of v_3. Choose $j_1 \geq m_1$ such that if we remove the first j_1 letters of v_1 be obtain a word v_{k_2} with $k_2 \geq 3$. Define u_1 to be the first j_1 letters of v_1. Now $v_{k_2} \prec v_{k_2+1}$ and hence there is some m_2 such that the first m_2 letters of v_{k_2} differ from the first m_2 letters of v_{k_2+1} and the first m_2 letters of v_{k_2+1} differ from the first m_2 letters of v_{k_2+2}. As before, we choose $j_2 > \max(j_1, m_2)$ such that if we remove the first j_2 letters of v_{k_2} we obtain some word v_{k_3} with $k_3 \geq k_2 + 2$. We define u_2 to be the first j_2 letters of v_{k_2} and u_2' to be the first j_2 letters of v_2. Notice that $u_1 \prec u_2' \prec u_2$ and $u_2' \neq u_2$, and it cannot have u_1 as an initial factor of u_1. Continuing in this manner, we see that we can write

$$w = uu_1u_2u_3\cdots$$

with u some initial factor and words $u_1, u_2, \ldots, u_d$, $u_1', \ldots, u_d'$ satisfying the conditions in the statement of the lemma. □

Theorem 4.6.5. *Let w_1 and w_2 be (possibly right-infinite) words on a finite alphabet. Then the sandwich function f corresponding to w_1 and w_2 is $\mathscr{A}$-automatic if and only if w_1 and w_2 are both either finite or ultimately periodic.*

Proof. Suppose that w_1 is neither finite nor ultimately periodic. Write $w_1 = w_1' w_1''$ where w_1' is a finite word which is not an initial factor of w_2 and w_1'' is a right-infinite word. Then w_1'' is not eventually periodic and hence by Lemma 4.6.4, for any d we can find words $u, u_1, u_2, \ldots, u_d$ such that $w_1 = w_1' u u_1 \cdots u_d u'$ for some right-infinite word u'; $u_1, \ldots, u_d$ is monotonic with respect to $\prec$;

$$\text{length}(u_1) < \cdots < \text{length}(u_d);$$

and for $i \geq 2$ there exist words u_i' with u_{i-1}, u_i', u_i monotonic with $u_i' \neq u_i$, $\text{length}(u_i') = \text{length}(u_i)$ and u_{i-1} not an initial factor of u_i'. We have two cases:

Case I $u_1 \prec u_2 \prec \cdots \prec u_d$.

In this case take $v_d = u_d'$ and for $1 \leq i \leq d-1$ take $v_i = u_i$. Consider

$$\text{Shuf}_d(f; w_1' u, v_1, \ldots, v_{d-1}, v_d, \epsilon).$$

Notice that since $w_1' u$ is lexicographically less than the initial factor of w_2 of the same length and since $v_1 \prec v_2 \prec \cdots \prec v_d$, we have that

$$w_1 \prec w_1' u v_{\sigma(1)} \cdots v_{\sigma(d)} \prec w_2$$

unless σ is the identity. When σ is the identity, we have $w_1 \not\prec w_1' u v_1 \cdots v_d$ and hence

$$\sum_{\sigma \in S_d} \text{sgn}(\sigma) f(w_1' u v_{\sigma(1)} \cdots v_{\sigma(d)}) \equiv 1 (\text{mod } 2).$$

Thus f cannot have the d-shuffle property. Since d is arbitrary we conclude that f is not $\mathscr{A}$-automatic.

Case II: $u_1 \succ u_2 \succ \cdots \succ u_d$.

In this case take $v_d = u_d'$ and for $1 \leq i \leq d-1$ take $v_i = u_i$. In this case we have

$$f(w_1' u v_{\sigma(1)} \cdots v_{\sigma(d)}) = 1 \qquad \text{if and only if} \sigma \neq \text{id}.$$

Thus the result again holds in this case.

In either case, the d-shuffle property fails to hold for any d, and so our function cannot be $\mathscr{A}$-regular. A similar argument shows that f is not $\mathscr{A}$-regular if w_2 is right-infinite and not eventually periodic.
Next consider what happens if w_1 and w_2 are finite or eventually periodic. Then by the lemma

$$f(w) = \chi(w \succ w_1) \chi(w \prec w_2)$$

and hence f is the coordinate-wise product of two $\mathscr{A}$-automatic sequences. It follows that f is automatic. □

4.7 Applications

4.7.1 The Logarithm and Automaticity

We now give an application of our description of automatic sandwich functions to answer a question of Allouche and Shallit. We note that this question had been answered via a different method by Yossi [425] in 2008.

Proposition 4.7.1. *Let $\alpha \in \mathbb{R}$. Then the sequence given by $f(0) = 1$ and*

$$f(n) = \lfloor \log_k n + \alpha \rfloor \qquad \text{for } n \geq 1$$

is k-regular if and only if k^α is rational.

Proof. We may assume that $\alpha \in [0, 1)$. It is easy to verify that the function $g(0) = 0$ and $g(n) = \lfloor \log_k n \rfloor$ for $n \geq 1$ is k-regular. Let us consider the function $f(n) - g(n)$. This function takes values in $\{0, 1\}$ and is 1 at an integer $n \geq 1$ if and only if there is some integer m such that

$$\log_k n + \alpha \geq m > \log_k n.$$

Equivalently, we must have

$$k^\alpha n \geq k^m \geq n.$$

Write

$$k^{-\alpha} = \sum_{i=1}^{\infty} a_i / k^i$$

with $a_i \in \{1, 2, 3, \ldots, k\}$. Let $w_1 = \varepsilon$ and let w_2 be the right-infinite word $a_1 a_2 a_3 \cdots$. Let w be the word in $\{1, 2, \ldots, k\}$ which corresponds to n using the correspondence described in equation (4.1). Then if we regard $h := f - g$ as a function on $\{1, 2, \ldots, k\}^*$, then $h(w) = 1$ if and only if $w \preceq w_2$, where $\prec$ is the lexicographic order induced by taking $1 \prec 2 \prec \cdots \prec k$; equivalently, $h(w) = 1$ if $w_1 \preceq w \preceq w_2$ and is 0 otherwise. Hence h is a sandwich function. Notice that w_2 is a right-infinite word which is eventually periodic if and only if k^α is rational. Thus h is k-regular if and only if k^α is rational. Since $f = g + h$ and g is k-regular, we see that f is k-regular if and only if h is k-regular. The result now follows. □

Allouche and Shallit [14] ask whether the sequence $\lfloor \log_2 n + \frac{1}{2} \rfloor$ is 2-automatic; since $\sqrt{2}$ is irrational, we deduce that it is not.

4.7.2 The 2-Adic Behavior of the Logarithm

Allouche and Shallit [14, Section 16.7, Q. 4] ask whether the sequence

$$f(n) \;=\; \min_{i \geq n+1} i - v_2(i)$$

is 2-regular, where $v_2(i)$ is the 2-adic valuation; that is, $v_2(i)$ satisfies $2^{v_2(i)} | i$ but $2^{v_2(i)+1} \not| i$. We show that it is not. Here we use the fact that right regularity and left regularity are equivalent properties to answer a question which is difficult to handle using the traditional definition of regularity in terms of right regularity, but which is easily handled if one uses the notion of left regularity.

Proposition 4.7.2. *Let*

$$f(n) \;=\; \min_{i \geq n+1} i - v_2(i).$$

Then $f(n)$ is not 2-regular.

Proof. Let $\mathscr{A} = \{0, 1\}$. Let $w_i = 1^i 0 \in \mathscr{A}^*$. Consider the subspace of $L(f)$ generated by $\{f_{w_i} \mid i \geq 0\}$. Suppose that f is $\mathscr{A}$-regular. Then since $L(f)$ is finite-dimensional, there exists some $m \geq 3$ such that this subspace is spanned by $\{f_{w_i} \mid i \leq m\}$. Let $d = 2^{2^m}$. By assumption, there exist integers $c_0, \ldots, c_m$ such that

$$f_{w_d} \;=\; \sum_{i=0}^{m} c_i f_{w_i}.$$

In particular, we have

$$\begin{aligned} f(w_d w_j) &= f_{w_d}(w_j) \\ &= \sum_{i=0}^{m} c_i f_{w_i}(w_j) \\ &= \sum_{i=0}^{m} c_i f(w_i w_j) \end{aligned} \tag{4.7}$$

for all $j \geq 0$. Observe that for $[w_i w_j]_2 = 2^{i+j+2} - 2 - 2^j$. Hence

$$f(w_i w_j) \;=\; \begin{cases} 2^{i+j+2} - 2^j - j & \text{if } 2^j \geq i+2; \\ 2^{i+j+2} - i - j - 2 & \text{if } 2^j \leq i+2. \end{cases}$$

In particular, if $m \le j \le 2^m$, we have $f(w_d w_j) = 2^{d+j+2} - d - j - 2$ and $f(w_i w_j) = 2^{i+j+2} - 2^j - j$ for $1 \le i \le m$. Using equation (4.7), we see that for $m \le j \le 2^m$ we have

$$\begin{aligned} 2^{d+j+2} - d - j - 2 &= f(w_d w_j) \\ &= \sum_{i=0}^{m} c_i f(w_i w_j) \\ &= \sum_{i=0}^{m} c_i (2^{i+j+2} - 2^j - j). \end{aligned}$$

Simplifying, we deduce

$$2^j \left(2^{d+2} - \sum_{i=0}^{m} c_i (2^{i+2} - 1) \right) + j \left(-1 + \sum_{i=0}^{m} c_i \right) = d + 2,$$

for $m \le j \le 2^m$. Let

$$A = 2^{d+2} - \sum_{i=0}^{m} c_i (2^{i+2} - 1)$$

and let

$$B = -1 + \sum_{i=0}^{m} c_i.$$

Then we have

$$2^j A + jB = d + 2,$$

for $m \le j \le 2^m$. Consider the function $G(x) = 2^x A + xB - (d+2)$. We have $G(m) = G(m+1) = \cdots = G(2^m) = 0$ and hence by Rolle's theorem $G'(x) = 2^x A \log 2 + B$ must have at least $2^m - m \ge 2$ real zeros. This implies that $A = B = 0$ since the function $G'(x)$ is monotonic. Then the fact that $G(m) = 0$ along with $A = B = 0$ now implies that $d + 2 = 0$, which is a contradiction. It follows that $f(n)$ is not a 2-regular function. □

4.7.3 Nim Sums and Nim Products

In this subsection we consider the 2-regularity of sequences constructed using nim sums and nim products. Given nonnegative integers n and m, we define the *nim sum*, $n \oplus m$, of n and m as follows. We write $n = [a_d \cdots a_0]_2$ with $a_i \in \{0, 1\}$ and $m = [b_d \cdots b_0]_2$ with $b_i \in \{0, 1\}$, where we may pad the binary expansion of either a or b with zeros at the beginning to ensure that they have the same length. We then define

$$n \oplus m := [(a_d + b_d \bmod 2) \cdots (a_0 + b_0 \bmod 2)]_2.$$

For example, if $m = 12$ and $n = 21$, then

$$m \oplus n = [01100]_2 \oplus [10101]_2 = [11001]_2 = 25.$$

The *nim product*, $\otimes$, is defined as follows:

$$2^{2^a} \otimes 2^{2^b} = \begin{cases} 2^{2^a} \cdot 2^{2^b} & \text{if } a \neq b; \\ 3 \cdot 2^{2^a - 1} & \text{if } a = b. \end{cases} \tag{4.8}$$

The product is then defined for all pairs of natural numbers using associativity and distributivity. For example,

$$\begin{aligned} 8 \otimes 3 &= (2^{2^1} \otimes 2^{2^0}) \otimes (2^{2^0} \oplus 1) \\ &= 2^{2^1} \otimes (2^{2^0} \otimes 2^{2^0}) \oplus (2^{2^1} \otimes 2^{2^0}) \\ &= 2^{2^1} \otimes 3 \oplus 8 \\ &= 2^{2^1} \otimes (2^{2^0} \oplus 1) \oplus 8 \\ &= 8 \oplus 4 \oplus 8 \\ &= 4. \end{aligned}$$

The nim sum and nim products have the following properties, which can be found in Conway [162, Chap. 6]:

- $2^{2^a} \otimes x = 2^{2^a} x$ for $0 \leq x < 2^{2^a}$;
- the set of nonnegative numbers less than 2^{2^a} is a field under $\oplus$ and $\otimes$.

Allouche and Shallit [14, Section 16.7, Q. 5,6] ask the following questions:

- Is the nim sum of two 2-regular sequences a 2-regular sequence?
- Is the sequence $\{n \otimes n\}$ a 2-regular sequence?

We show the answer to these questions is "no." In fact, Allouche and Shallit ask questions involving 2-dimensional arrays of numbers, but negative answers to the above questions imply negative answers to their questions about arrays. This time, we use the power property and show that it fails to hold.

Lemma 4.7.3. *Let i be a nonnegative integer. Then $2^i \otimes 2^i = 3 \cdot 2^{i-1}$ if and only if i is a power of 2.*

Proof. If i is a power of 2, then $2^i \otimes 2^i = 3 \cdot 2^{i-1}$ by the definition of the nim product. Next suppose that i is not a power of 2. Then there is some a such that $i = 2^a + j$ with $0 < j < 2^a$. Then

$$\begin{aligned} 2^i \otimes 2^i &= (2^{2^a} \otimes 2^j) \otimes (2^{2^a} \otimes 2^j) \\ &= (2^{2^a} \otimes 2^{2^a}) \otimes (2^j \otimes 2^j) \\ &= (2^{2^a} \oplus 2^{2^a-1}) \otimes (2^j \otimes 2^j) \\ &= \left(2^{2^a} \otimes (2^j \otimes 2^j)\right) \oplus \left(2^{2^a-1} \otimes (2^j \otimes 2^j)\right) \\ &= \left(2^{2^a} \cdot (2^j \otimes 2^j)\right) \oplus \left(2^{2^a-1} \otimes (2^j \otimes 2^j)\right), \end{aligned}$$

where we are using the two facts from Conway mentioned above to obtain these equalities.

Observe that if $2^i \otimes 2^i = 3 \cdot 2^{i-1}$, then the binary expansion of $2^i \otimes 2^i$ cannot have any 1's appearing in the $2^a + j - 1$ least significant digits. In particular it must have 0's appearing in the 2^a least significant digits. Since $2^{2^a} \otimes (2^j \otimes 2^j)$ has this property, $2^{2^a-1} \otimes (2^j \otimes 2^j)$ must also have this property if $2^i \otimes 2^i = 3 \cdot 2^{i-1}$, as the nim sum of these two numbers is $2^i \otimes 2^i$. But since $\{0, 1, \ldots, 2^{2^a} - 1\}$ is a field under $\otimes$ and $\oplus$, we see that $2^{2^a-1} \otimes (2^j \otimes 2^j)$ is a nonzero number less than 2^{2^a}. We conclude that $2^i \otimes 2^i \neq 3 \cdot 2^{i-1}$. □

Proposition 4.7.4. *The sequence $\{m \otimes m\}$ is not 2-regular.*

Proof. To do this, we show that the sequence does not have the power property. It is sufficient to show that the power series

$$F(x) = \sum_{i=0}^{\infty} (2^i \otimes 2^i) x^i$$

is not rational. Suppose that $F(x)$ is rational. Then

$$G(x) = F(x) - \frac{3}{2}(1-2x)^{-1} = \sum_{i=0}^{\infty} (2^i \otimes 2^i - 3 \cdot 2^{i-1}) x^i$$

must also be rational. Notice that by Lemma 4.7.3 the coefficient of x^n is zero in $G(x)$ if and only if n is a power of 2. If follows from the Skolem-Mahler-Lech theorem (see [282]) that $G(x)$ is not rational. We conclude that F is not rational and so $\{n \otimes n\}$ is not 2-regular. □

Proposition 4.7.5. *There exist 2-regular sequences $f(n)$ and $g(n)$ such that $f(n) \oplus g(n)$ is not 2-regular.*

Proof. Let $f(n)$ be defined by

$$f(n) = \begin{cases} \frac{4^m-1}{3} & \text{if } n = 2^m \\ 0 & \text{otherwise.} \end{cases}$$

Then it is easy to check that $f(n)$ is 2-shuffled and has the power property. Hence f is 2-regular. Let

$$g(n) = \begin{cases} m & \text{if } n = 2^m \\ 0 & \text{otherwise.} \end{cases}$$

Then $g(n)$ is 2-regular. Observe that $f(2^m) \oplus g(2^m) = f(2^m) - g(2^m)$ if and only if the binary expansion of m has 0's in all the even positions (beginning the count from the least significant digit). Suppose that $f(n) \oplus g(n)$ is 2-regular. Then the power property gives that the sequence

$$f(2^m) \oplus g(2^m)$$

must satisfy a linear recurrence. If this is the case, then by the Skolem-Mahler-Lech theorem (see [282]), the set of m such that $f(2^m) \oplus g(2^m) = \frac{4^m-1}{3} - m$ must contain an infinite arithmetic progression. But

$$\#\{m \le 4^N \mid f(2^m) \oplus g(2^m) = f(2^m) - g(2^m)\} \le 2^N.$$

In other words, the density of the set of m such that $f(2^m) \oplus g(2^m) = \frac{4^m-1}{3} - m$ is 0. But this is a contradiction, since by the Skolem-Mahler-Lech theorem, the density must be positive. It follows that $f \oplus g$ is not 2-regular. □

We note that if $f(n)$ and $g(n)$ are k-automatic, then both $f(n) \otimes g(n)$ and $f(n) \oplus g(n)$ are k-automatic; this follows easily from the fact that k-automatic sequences assume only finitely many different values.

4.8 Shuffled Sequences

In this section we develop the basic properties of sequences possessing the shuffle property. We will find it more useful to work with abelian groups rather than fields in obtaining some of our closure properties, so we shall adopt this point of view in this section.

Definition 4.8.1. Let $\mathscr{A}$ be a finite alphabet and let A be an abelian group. We say that $f : \mathscr{A}^* \to A$ is an *$\mathscr{A}$-shuffled sequence* if f has the d-shuffle property for some positive integer d. In the case that $\mathscr{A} = \{1, 2, \ldots, k\}$, we say that f is *k-shuffled*. We point out that being k-shuffled is not the same as having the k-shuffle property; when we use the word *shuffled*, the k is making reference to the underlying alphabet and when we use the word *shuffle*, the k is making reference to the shuffle identity satisfied by a function.

In the case that no confusion will arise, we will drop the alphabet and refer to a sequence as being a *shuffled sequence*. We note that the shuffle property, although originally defined in the case where our abelian group is a field (with group law given by addition), makes sense in this more general setting.

We have the following containments:

$$\begin{aligned}\mathscr{A}\text{-automatic sequences} &\subseteq \mathscr{A}\text{-regular sequences}\\ &\subseteq \mathscr{A}\text{-shuffled sequences}.\end{aligned}$$

It is well known [14, Theorem 16.3.1] that a k-regular sequence $f : \mathbb{N} \to \mathbb{Z}$ has polynomially bounded growth in the sense that there is some $d > 0$ such that $|f(n)| \leq n^d$ for all sufficiently large n. The following example shows that no such growth restriction on shuffled sequences holds.

Example 4.8.2. A k-shuffled sequence can have arbitrarily rapid growth.

Proof. Let $a_1, a_2, \ldots$ be a sequence of integers and let $f : \{1, 2\}^* \to \mathbb{C}$ be defined by

$$f(w) = \begin{cases} 0 & \text{if 2 appears in } w \\ a_d & \text{if } w = 1^d. \end{cases}$$

Then $f(w)$ has the 2-shuffle property and hence is a shuffled sequence. Since the a_d are arbitrary, we see that shuffled sequences can have arbitrarily rapid growth. □

We now give some closure properties for shuffled sequences.

Proposition 4.8.3. *Let A and B be two abelian groups, and let $\mathscr{A}$ be a finite alphabet. If $f : \mathscr{A}^* \to A$ and $g : \mathscr{A}^* \to B$ are shuffled sequences, then so are $(f \oplus g) : \mathscr{A}^* \to A \oplus B$ and $f \otimes g : \mathscr{A}^* \to A \otimes_{\mathbb{Z}} B$. (Here $f \oplus g)(w) = f(w) \oplus g(w)$ and $f \otimes g(w) = f(w) \otimes g(w)$.)*

Proof. Create the ideals I_1 and I_2 in $\mathbb{Z}\{x_1, \ldots, x_k\}$ as follows. Let I_1 be the set of elements of the form $c_1 w_1 + \cdots c_d w_d$ such that

$$c_1 f(ww_1w') + \cdots + c_d f(ww_dw') = 0$$

for all $w, w' \in \mathscr{A}^*$. Similarly, define I_2 to be the set of elements of the form $c_1 w_1 + \cdots c_d w_d$ with

$$c_1 g(ww_1w') + \cdots + c_d g(ww_dw') = 0$$

for all $w, w' \in \mathscr{A}^*$. Then I_1 and I_2 are ideals and $R_1 := \mathbb{Z}\{x_1, \ldots, x_k\}/I_1$ and $R_2 := \mathbb{Z}\{x_1, \ldots, x_k\}/I_2$ both satisfy the identity S_d and hence satisfy polynomial identities. It follows from Regev's theorem [498], that $R_1 \otimes_{\mathbb{Z}} R_2$ also satisfies a polynomial identity. Since it is finitely generated as a $\mathbb{Z}$-algebra, it satisfies the standard identity S_m for some $m \geq 0$. Now suppose that the image of $\sum c_{i,j} w_i \otimes w_j$ is zero in $R_1 \otimes_{\mathbb{Z}} R_2$. Then by construction, we have

$$\sum c_{i,j} f(ww_i w') \otimes g(ww_j w') = 0$$

for all words w, w'. Let $R \subseteq R_1 \otimes_{\mathbb{Z}} R_2$ be the subalgebra generated by the images of $x_1 \otimes x_1, \ldots, x_k \otimes x_k$. Then R must also satisfy the identity S_m, since it is a subring of $R_1 \otimes R_2$. Consequently, the sequence $f \otimes g$ must have the m-shuffle property. □

Remark 4.8.4. Let $\mathscr{A}$ be a finite alphabet and let A and B be abelian groups. If $f : \mathscr{A}^* \to A$ is a shuffled sequence and $\phi : A \to B$ is a homomorphism of abelian groups, then $\phi \circ f : \mathscr{A}^* \to B$ is a shuffled sequence.

Corollary 4.8.5. *Let $\mathscr{A}$ be a finite group and let A be an abelian group. If $f, g : \mathscr{A} \to A$ are shuffled, then $(f + g)$ is shuffled. If, in addition, A is a ring then $f \cdot g$ is shuffled.*

Proof. For the first part, use Proposition 4.8.3 and Remark 4.8.4, taking the homomorphism $\phi : A \oplus A \to A$ given by $\phi(a, a') = a + a'$. For the product, again use Proposition 4.8.3 and Remark 4.8.4, this time taking the homomorphism $\phi : A \otimes A \to A$ given by $\phi(a \otimes a') = aa'$. □

Sometimes it is easier to verify that a sequence is shuffled by showing that an identity holds other than the standard identity of the shuffle property holds. We make this more precise in the next theorem.

Theorem 4.8.6. *Let n be a nonnegative integer. Suppose that there exist integers $\{c_\sigma \mid \sigma \in S_n\}$, at least one of which is equal to 1, such that*

$$\sum_{\sigma \in S_n} c_\sigma f(ww_{\sigma(1)} \cdots w_{\sigma(n)} w') = 0.$$

for all words $w, w_1, \ldots, w_n, w' \in \mathscr{A}^*$. Then f is shuffled.

Proof. Write $\mathscr{A} = \{x_1, \ldots, x_k\}$. We define an ideal I in the algebra $\mathbb{Z}\{x_1, \ldots, x_k\}$ as follows. We declare that $\sum a_w w \in I$ if

$$\sum a_w f(w' w w'') = 0$$

for all $w', w'' \in \mathscr{A}^*$. Notice that $R = \mathbb{Z}\{x_1, \ldots, x_k\}/I$ satisfies a polynomial identity, as it satisfies the multilinear, homogeneous polynomial identity $\sum_{\sigma \in S_n} c_\sigma t_{\sigma(1)} \cdots t_{\sigma(n)}$. Since R is finitely generated, it satisfies a standard identity S_m for some m (cf. Theorem 4.4.6 and see, also, Braun [105]). Consequently,

$$\sum_{\sigma \in S_m} \operatorname{sgn}(\sigma) w_{\sigma(1)} \cdots w_{\sigma(m)} \in I$$

for all $(w_1, \ldots, w_m) \in (\mathscr{A}^*)^m$. Hence

$$\sum_{\sigma \in S_m} \operatorname{sgn}(\sigma) f(w w_{\sigma(1)} \cdots w_{\sigma(m)} w') = 0$$

for all $w, w_1, \ldots, w_m, w' \in \mathscr{A}^*$. □

Let $\mathscr{A}$ be a finite alphabet and let R be a ring. The set of maps $f : \mathscr{A}^* \to R$ has a multiplication defined as follows: Given $x_1, \ldots, x_d \in \mathscr{A}$, we define

$$f \star g(x_1 \cdots x_d) := \sum_{i=0}^{d} f(x_1 \cdots x_i) g(x_{i+1} \cdots x_d), \tag{4.9}$$

where $f(x_1 \cdots x_i)$ is taken to mean $f(\varepsilon)$ when $i = 0$ and we take $g(x_{i+1} \cdots x_d)$ to be $g(\varepsilon)$ when $i = d$. This product, along with ordinary sum, turns the set of maps $f : \mathscr{A}^* \to R$ into an associative ring [77]. We show that the shuffled sequences form a subalgebra.

Proposition 4.8.7. *If $f : \mathscr{A}^* \to A$ and $f : \mathscr{A}^* \to A$ are shuffled, then $f \star g$ is shuffled.*

Proof. We may pick d such that f and g both have the d-shuffled property. We claim that if $w_1, \ldots, w_d, u_1, \ldots, u_d, w, w' \in \mathscr{A}^*$, then

$$\sum_{\sigma \in S_d} \sum_{\tau \in S_d} \operatorname{sgn}(\sigma) \operatorname{sgn}(\tau) (f \star g)(w w_{\sigma(1)} \cdots w_{\sigma(d)} u_{\tau(1)} \cdots u_{\tau(d)} w') = 0.$$

To see this, let

$$v(\sigma, \tau) = w w_{\sigma(1)} \cdots w_{\sigma(d)} u_{\tau(1)} \cdots u_{\tau(d)} w'.$$

Then

$$\sum_{\sigma,\tau \in S_d} \operatorname{sgn}(\sigma) \operatorname{sgn}(\tau) (f \star g)(v(\sigma, \tau)) = \sum_{\sigma,\tau \in S_d} \sum_{\substack{v_1 v_2 = \\ v(\sigma,\tau)}} \operatorname{sgn}(\sigma) \operatorname{sgn}(\tau) f(v_1) g(v_2).$$

If $v_1 v_2 = v(\sigma, \tau)$ then either $v_1 = v_1(\sigma, \tau)$ contains $w w_{\sigma(1)} \cdots w_{\sigma(d)}$ as an initial factor, or $v_2 = v_2(\sigma, \tau)$ contains $u_{\tau(1)} \cdots u_{\tau(d)} w'$ as a terminal factor (or both). Notice that if $v_1 = w w_{\sigma(1)} \cdots w_{\sigma(d)} v_1'$, then v_1' and v_2 depend only on τ. Hence

$$\sum_{\sigma \in S_d} \sum_{\tau \in S_d} \operatorname{sgn}(\sigma) \operatorname{sgn}(\tau) f(v_1) g(v_2)$$

$$= \sum_{\tau \in S_d} \sum_{\sigma \in S_d} \mathrm{sgn}(\sigma)\mathrm{sgn}(\tau) f(ww_{\sigma(1)} \cdots w_{\sigma(d)} v_1') g(v_2)$$

$$= \sum_{\tau \in S_d} \mathrm{sgn}(\tau) g(v_2) \left(\sum_{\sigma \in S_d} \mathrm{sgn}(\sigma) f(ww_{\sigma(1)} \cdots w_{\sigma(d)} v_1') \right)$$

$$= 0,$$

since f has the d-shuffled property. Similarly, if v_2 contains the terminal factor $u_{\tau(1)} \cdots u_{\tau(d)} w'$, then the fact that g has the d-shuffled property guarantees that

$$\sum_{\sigma \in S_d} \sum_{\tau \in S_d} f(v_1) g(v_2) = 0.$$

Observe that $S_d \times S_d$ embeds in S_{2d} by taking an ordered pair (σ, τ) and letting σ act on $\{1, 2, \ldots, d\}$ and letting τ act on $\{d+1, \ldots, 2d\}$. Given $\mu \in S_{2d}$, we define $c_\mu = \mathrm{sgn}(\sigma)\mathrm{sgn}(\tau)$ if μ corresponds to (σ, τ) under this inclusion and we take $c_\mu = 0$ otherwise. Then from Theorem 4.8.6, we see that $f \star g$ is shuffled. □

We let $\mathscr{S}_k$ denote the collection of k-shuffled sequences taking values in $\mathbb{C}$. Then we have just shown that $(\mathscr{S}_k, +, \star)$ is a $\mathbb{C}$-algebra. The following proposition shows that in some sense this algebra is very large.

Proposition 4.8.8. *The algebra* $(\mathscr{S}_k, +, \star)$ *contains a copy of the free* $\mathbb{C}$*-algebra on infinitely many generators.*

Proof. Since the free algebra on two generators contains a copy of the free algebra on infinitely many generators, it is sufficient to do show we contain a copy of the free algebra on two generators. Let $f_0 : \{1, 2, \ldots, k\}^* \to \mathbb{C}$ be defined to be 1 on words of the form 12^i and be defined to be 0 on all other words. Let $f_1 : \{1, 2, \ldots, k\}^* \to \mathbb{C}$ be defined to be i on the word 12^i and to be 0 on words not of the form 12^i. Then it is easy to check that f_0 and f_1 are k-shuffled sequences. We claim that f_0 and f_1 generate a free algebra. Suppose that

$$G := \sum_{k=1}^{d} \sum_{(i_1, \ldots, i_k) \in \{0,1\}^k} \alpha_{i_1, \ldots, i_k} f_{i_1} \star f_{i_2} \star \cdots \star f_{i_k} = 0$$

for some d with $\alpha_{i_1, \ldots, i_d} \neq 0$ for some $(i_1, \ldots, i_d) \in \{0, 1\}^d$. Notice that if w is a word of the form $12^{a_1} 12^{a_2} \cdots 12^{a_d}$ then $f_{i_1} \star \cdots \star f_{i_k}(w) = 0$ for $k < d$. Hence

$$G(12^{a_1} 12^{a_2} \cdots 12^{a_d})$$

$$= \sum_{(i_1, \ldots, i_d) \in \{0,1\}^d} \alpha_{i_1, \ldots, i_d} f_{i_1} \star f_{i_2} \star \cdots \star f_{i_d} (12^{a_1} 12^{a_2} \cdots 12^{a_d})$$

$$= \sum_{(i_1,\dots,i_d)\in\{0,1\}^d} \alpha_{i_1,\dots,i_d} a_1^{i_1} \cdots a_d^{i_d}.$$

by assumption G is identically 0 and hence the polynomial

$$H(x_1,\dots,x_d) := \sum_{(i_1,\dots,i_d)\in\{0,1\}^d} \alpha_{i_1,\dots,i_d} x_1^{i_1} \cdots x_d^{i_d}$$

vanishes on all points $(x_1,\dots,x_d) \in \mathbb{N}^d$. But this implies that H is the zero polynomial, which contradicts the fact that $\alpha_{i_1,\dots,i_d} \neq 0$ for some $(i_1,\dots,i_d) \in \{0,1\}^d$. We conclude that the algebra generated by f_0 and f_1 is free. This completes the proof. □

4.9 Open Problems and Concluding Remarks

One of the fundamental theorems from the theory of automatic sequences is Cobham's theorem. Two integers p and q are multiplicatively independent if $p^a \neq q^b$ for $(a,b) \neq (0,0)$. Cobham's theorem [14, Chapter 11] states that if a sequence is p-automatic and q-automatic and p and q are multiplicatively independent, then the sequence is eventually periodic. Given a sequence $f(n)$ taking values in an abelian group, by the correspondence described in item 4.1, it makes sense to talk about the sequence being a k-shuffled sequence.

Question 4.9.1. Suppose a $\mathbb{Z}$-valued sequence $f(n)$ is both p-shuffled and q-shuffled for two multiplicatively independent integers p and q. What can be said about $f(n)$? For instance, does $f(n)$ satisfy a linear recurrence?

Another question comes from looking at closure properties. In Section 4.8 we showed that shuffled sequences are closed under ordinary products and under the $\star$ product. We now ask if shuffled sequences are closed under Cauchy products.

Question 4.9.2. Given integer-valued sequences $f(n)$ and $g(n)$ that are k-shuffled, is the Cauchy product of $f(n)$ and $g(n)$ also k-shuffled?

In Section 4.8 we showed that the set $\mathscr{S}_k$ of k-shuffled sequences taking values in $\mathbb{C}$ forms a $\mathbb{C}$-algebra under the $\star$ product.

Question 4.9.3. Can one find nice generating sets for the $\mathbb{C}$-algebra $(\mathscr{S}_k, +, \star)$?

Question 4.9.4. Can one extend the notion of shuffled sequences to nonconstant length substitutions? See, for example, Shallit [542] and Allouche, Scheicher, and Tichy [13].

Two final questions we pose come from the nim sum and nim product. In Section 4.7 we studied sequences defined using the nim sum and the nim product. In each example, we showed that the power property failed to hold. We did not show, however, that the shuffle property fails to hold.

Question 4.9.5. Let $\oplus$ and $\otimes$ denote, respectively, the nim sum and the nim product. Is the sequence $\{m \otimes m\}$ a 2-shuffled sequence? If $f(n)$ and $g(n)$ are 2-shuffled sequences, is $f(n) \oplus g(n)$ also a 2-shuffled sequence?

Question 4.9.6. Does $\{2^n \oplus 3^n\}$ satisfy a linear recurrence? We note that this is related to Mahler's study of Z-numbers [406].

We note that throughout we have been looking at sequences taking values in a field or occasionally in an abelian group. In fact, we can work more generally by using a commutative ring C and look at sequences taking values in a C-module. Allouche and Shallit [16] define the more general notion of a (C, k)-regular sequence. We note that since Shirshov's height theorem is over a more general base ring C, all the results in this paper which relate to k-regular sequences have analogues in this more general context of (C, k)-regular sequences.

Acknowledgements I thank Jean-Paul Allouche and Jeffrey Shallit for many helpful comments. I also thank Jean-Paul Allouche for raising Question 4.9.4.

Chapter 5
Avoiding or Limiting Regularities in Words

Pascal Ochem, Michaël Rao, and Matthieu Rosenfeld

Abstract It is commonly admitted that the origin of combinatorics on words goes back to the work of Axel Thue in the beginning of the twentieth century, with his results on repetition-free words. Thue showed that one can avoid cubes on infinite binary words and squares on ternary words. Up to now, a large part of the work on the theoretic part of combinatorics on words can be viewed as extensions or variations of Thue's work, that is, showing the existence (or nonexistence) of infinite words avoiding, or limiting, a repetition-like pattern. The goal of this chapter is to present the state of the art in the domain and also to present general techniques used to prove a positive or a negative result. Given a repetition pattern P and an alphabet, we want to know if an infinite word without P exists. If it exists, we are also interested in the size of the language of words avoiding P, that is, the growth rate of the language. Otherwise, we are interested in the minimum number of factors P that a word must contain. We talk about limitation of usual, fractional, abelian, and k-abelian repetitions and other generalizations such as patterns and formulas. The last sections are dedicated to the presentation of general techniques to prove the existence or the nonexistence of an infinite word with a given property.

5.1 Introduction

It is commonly admitted that the origin of combinatorics on words goes back to the work of Axel Thue in the beginning of the twentieth century [562, 563], with his results on repetition-free words. A *word* is a (possibly infinite) sequence of letters, taken in a finite alphabet. A *factor* of a word is a subsequence of consecutive letters in the word. A *square* (resp. *cube*) is a nonempty factor of the form uu (resp. uuu).

P. Ochem (✉)
CNRS & LIRMM, Montpellier, F-34095 Montpellier Cedex 5, France
e-mail: ochem@lirmm.fr

M. Rao · M. Rosenfeld
LIP, ENS de Lyon, 46 allée d'Italie, F-69007 Lyon, France
e-mail: michael.rao@ens-lyon.fr; matthieu.rosenfeld@ens-lyon.fr

V. Berthé, M. Rigo (eds.), *Sequences, Groups, and Number Theory*,
Trends in Mathematics, https://doi.org/10.1007/978-3-319-69152-7_5

Thue showed that one can avoid cubes on infinite binary words and squares on ternary words.

Up to now, a large part of the work on the theoretic part of combinatorics on words can be viewed as extensions or variations of Thue's work, that is, showing the existence (or nonexistence) of infinite words avoiding, or limiting, a repetition-like pattern.

The goal of this chapter is to present the state of the art in the domain and also to present general techniques used to prove a positive or a negative result. Given a repetition pattern P and an alphabet, we want to know if an infinite word without P exists, and if it exists, we are also interested in the size of the language of words avoiding P, that is, the growth rate of the language.

In Section 5.2 we talk about the avoidability and the limitation of usual repetitions, that is, word equality, and its generalizations like the fractional repetition threshold. We also talk about avoidability of patterns and formulas. In Section 5.3, we talk about avoidability of abelian repetitions and its generalizations. Finally, in Section 5.4 and Section 5.5, we present general techniques which can be used to prove the existence or the nonexistence of an infinite word with a given property.

5.2 Usual Repetitions

In this section, we focus on the avoidability and limitation of repetitions when the equivalence relation is the equality, that is, the "usual" case. Some results have already been discussed in [79, Chapter 4]. In this case we only give main results and redirect the reader to that chapter.

5.2.1 Thue's Results and Ternary Square-Free Words

Thue showed that one can avoid squares on ternary words and overlaps on binary words [562, 563] (for a translation, see [74]).

Let $\sigma : \{0, 1\}^* \to \{0, 1\}^*$ be the morphism such that $\sigma(0) = 01$ and $\sigma(1) = 10$. The fixed point of σ with first letter 0 is known as the Thue–Morse (or Prouhet–Thue–Morse) word:

$$w_{TM} = 0110100110010110\ldots.$$

Thue showed that this word avoids *overlaps*, that is, factors $xuxux$ where u is a (maybe empty) word and x a letter. In consequence, the Thue–Morse word avoids cubes.

Let τ be the morphism $0 \to 012, 1 \to 02, 2 \to 1$. The fixed point of τ is known as the *ternary Thue–Morse word* or the *Hall word*:

$$w_{TTM} = 012021012102012021020121\ldots.$$

Since $\tau' \circ \tau = \sigma \circ \tau'$, where τ' is the morphism $0 \to 011, 1 \to 01, 2 \to 0$, w_{TTM} is also the preimage of w_{TM} by τ'. One can easily show that if w has a square uu, then $\tau'(w)$ has an overlap $\tau(u)\tau(u)0$. Thus, since w_{TM} is overlap-free, then w_{TTM} is square-free.

The *growth rate* of a language L on an alphabet $\mathscr{A}$ is $\limsup_{n\to\infty} |L \cap \mathscr{A}^n|^{\frac{1}{n}}$. Some authors prefer the terminology of *entropy*, which is, in the case of languages on words, the logarithm of the growth rate. The growth rate of overlap-free binary words is 1, since there are only polynomially many such words (see [79, Section 4.2.3]), and the growth rate of cube-free binary words is between 1.45697 and 1.4576 [209, 356]. The growth rate of ternary square-free words is between 1.30173 and 1.30179 [356, 452]. We will see in Sections 5.4 and 5.5 some techniques to compute lower and upper bounds of the growth rate of a language.

5.2.2 Erdős's Question: Avoiding Long Squares

In 1957 and 1961, Erdős asked two questions on the avoidability of squares in words [216, 217]. Firstly, he asked if one can avoid arbitrarily long squares in binary words. Secondly, he asked if one can avoid factors uv over a finite alphabet, where v is a permutation of the letters of u. (This notion is the *abelian equivalence* and will be discussed in Section 5.3.)

Erdős thought that the answer to the first question was negative. Entringer, Jackson, and Schatz showed the opposite: it is possible to construct an infinite binary word without squares of size 6 and more [214]. This result has been improved by Fraenkel and Simpson: it is possible to construct an infinite binary word with only three squares: 00, 11, and 1010 (and this is the best we can do) [228]. Perhaps the simplest construction of such a word is given by Badkobeh and Crochemore:

Theorem 5.2.1 ([28]). *Let* $\eta : 0 \to 01001110001101, 1 \to 0011, 2 \to 000111$. *Then* $\eta(w_{TTM})$ *contains only* 3 *squares:* 00, 11, *and* 1010.

5.2.3 Fractional Repetitions and Dejean's Conjecture

A *repetition* in a word w is a pair of words (p, q) such that pq is a factor of w, p is nonempty, and q is a prefix of pq. The *exponent* of a repetition (p, q) is $\frac{|pq|}{|p|}$, and its *period* is $|p|$. Squares are thus repetitions of exponent 2.

A word is said *x-free* (resp. x^+-free) if it does not contain a repetition of exponent y with $y \geq x$ (resp. $y > x$). For an integer $k \geq 2$, the *repetition threshold* for k letters, denoted by $\mathrm{R}(k)$, is the infimum over the set of x such that there exists an infinite

x-free word over a k-letter alphabet or equivalently the smallest x such that there exists an infinite x^+-free word over a k-letter alphabet.

Since squares cannot be avoided on binary words, but overlaps can, one has $\mathrm{R}(2) = 2$. Dejean [187] conjectured that for every $k \geq 2$, we have:

$$\mathrm{R}(k) = \begin{cases} \frac{7}{4} & \text{if } k = 3 \\ \frac{7}{5} & \text{if } k = 4 \\ \frac{k}{k-1} & \text{otherwise.} \end{cases}$$

This conjecture is now completely solved and already discussed in [79, Section 4.3]. Notice that the proof is constructive, but none of the constructions are morphic words, except for $k \in \{2, 3\}$, and use Pansiot's encoding. We call a *Dejean word* an $\mathrm{R}(k)^+$-free k-ary word.

We know that there are exponentially many Dejean words on k letters, for $k \in \{3, 4\}$ [449], $k \in \{5, \ldots, 10\}$ [331], and every odd $k \in \{7, \ldots, 101\}$ [568]. Moreover it is conjectured that $\lim_{k \to \infty} g_k = 1.242\ldots$, where g_k is the growth rate of Dejean words on k letters [545].

For $k = 2$, $\mathrm{R}(k) = 2$, and we are in the case of overlap-free binary words. As we already know, the growth rate is 1. More generally, there are polynomially many $\frac{7}{3}$-free binary words (their growth rate is then 1), and there are exponentially many $\frac{7}{3}^+$-free binary words [331]. The growth rate of $\frac{7}{3}^+$-free binary words is estimated at $1.2206448\ldots$ [544].

Looking at a stronger version of Dejean's question, one has the following. It is known that for every $k \geq 3$, there is an infinite $\mathrm{R}(k)^+$-free word on k letters with only finitely many $\mathrm{R}(k)$-repetitions [27, 29, 568]. Ochem proposed the following stronger version of Dejean's conjecture.

Conjecture 5.2.2 ([450]).

(1) For every $k \geq 5$, there exists an infinite $\frac{k}{k-1}^+$-free word over k letters with letter frequency $\frac{1}{k+1}$.
(2) For every $k \geq 6$, there exists an infinite $\frac{k}{k-1}^+$-free word over k letters with letter frequency $\frac{1}{k-1}$.

This conjecture has already been proved for several cases when $k < 9$ [140, 450, 493].

5.2.4 Generalized Repetition Threshold

A word is (α^+, ℓ)-free if it contains no repetition of exponent greater than α and period at least ℓ. The *generalized repetition threshold* $\mathrm{R}(k, \ell)$ is the smallest real α such that there exists an infinite (α^+, ℓ)-free word on k letters [308]. The case $\ell = 1$ corresponds to Dejean's repetition threshold. The general behavior of $\mathrm{R}(k, \ell)$

is known when ℓ tends to infinity [223], but the exact values, especially for small ℓ, are still unknown or conjectured.

Conjecture 5.2.3 ([308]). For $\ell \geq 2$, $\mathrm{R}(3,\ell) = 1 + \frac{1}{\ell}$ and $\mathrm{R}(4,\ell) = 1 + \frac{1}{\ell+2}$.

The following theorem gives a partial answer to the previous conjecture (joint work of Kolpakov and Rao).

Theorem 5.2.4. *For all $\ell \geq 11$, $\mathrm{R}(3,\ell) \geq 1 + \frac{1}{\ell}$.*

Proof. Suppose that $\mathrm{R}(3,\ell) < 1 + \frac{1}{\ell}$ for a $\ell > 0$. Then for every $n > 0$ and $0 < m \leq \ell$, one can find n ternary words $u_1, u_2, \ldots, u_n$ of size m with the following property: for every $e > 0$ and $1 \leq i < j \leq n$ and $1 \leq x, y \leq m + 1 - e$ such that $u_i[x : x + e - 1] = u_j[y : y + e - 1]$, one has:

- $e < j - i + 1$ if $x < y$
- $e < j - i$ if either $x = y$ or $x > y$ and $i + 1 < j$.

To show this, it suffices to take $u_1 = w[1 : m], u_2 = [1 + \ell : m + \ell], \ldots, u_n = w[1 + (n-1)\ell : m + (n-1)\ell]$, where w is an infinite $(1 + 1/\ell, \ell)$-free ternary word.

A backtracking algorithm shows that such $u_1, u_2, \ldots, u_n$ do not exist for $n = 10$ and $m = 11$. Thus $\mathrm{R}(3,\ell) \geq 1 + \frac{1}{\ell}$. □

5.2.5 Limiting Occurrences and Letters

Erdős's question is about the limitation of squares as factor in binary words. If one wants to limit squares as occurrences in binary words, one has the following.

Theorem 5.2.5 ([363, 465]). *The minimal density of square occurrences in an infinite binary word is $\frac{103}{187} = 0.55080213\ldots$.*

The upper bound is given by a morphic word [363] and the lower bound is proven using a general technique presented in Section 5.4.

A related question is about the possible frequencies of a letter in a power-free language. For ternary square-free words, one has the following.

Theorem 5.2.6 ([346, 450]).

- *The minimal density of a letter in an infinite ternary square-free word is $\frac{883}{3215} = 0.27465007\ldots$.*
- *The maximal density of a letter in an infinite ternary square-free word is $\frac{255}{653} = 0.39050535\ldots$.*

Let $\rho(x)$ (resp. $\rho^+(x)$) be the minimal frequency of a letter in an x-free (resp. x^+-free) word. The function ρ is defined in [357] and also studied in [450, 465]. For example, one has $\rho(2^+) = \rho(7/3) = 1/2$, and $\rho(7/3^+) = 327/703 = 0.4651493598\ldots$. The minimal frequency of a letter in a cube-free binary word is approximately $0.40636\ldots$, but its exact value is still unknown.

The behavior between $\frac{7}{3}^+$ and $\frac{5+\sqrt{5}}{2}$ is complicated. It seems that $\rho(x)$ is a piecewise constant function and that the set of x such that $\rho(x) \neq \rho(x^+)$ has $\frac{5+\sqrt{5}}{2}$ as an accumulation point. It seems also that, for $x \geq \frac{5+\sqrt{5}}{2}$, the optimal word is always a Sturmian word. More precisely, one can conjecture the following.

Conjecture 5.2.7. For every integer $n \geq 4$:

- $\rho([n-1, \overline{1, n-3}]) = \rho(n) = [0, n-1, \overline{1, n-3}]$,
- for all $k \in \mathbb{N}$, $\rho(U_{n,k}^+) = \rho(U_{n,k+1}) = [0, n(, 1, n-2)^k, \overline{1, n-3}]$,

where $[a, b, c, \ldots]$ denotes the continued fraction $a + 1/(b + 1/(c + 1/(\ldots)))$, and $U_{n,k} = n + 1 - \frac{D_{n,k-1}+2}{D_{n,k}}$, $D_{n,-1} = -1$, $D_{n,0} = 1$, and $D_{n,k+1} = nD_{n,k} - D_{n,k-1}$.

5.2.6 Patterns and Formulas

Let $\mathscr{A}_k$ denote the alphabet $\{0, \ldots k-1\}$. A *pattern* p is a nonempty finite word over an alphabet $\Delta = \{A, B, C, \ldots\}$ of capital letters called *variables*. An *occurrence* of p in a (finite or infinite) word w is a non-erasing morphism $h : \Delta^* \to \mathscr{A}^*$ such that $h(p)$ is a factor of w. We say that w avoids p if it contains no occurrence of p. The *avoidability index* $\lambda(p)$ of a pattern p is the size of the smallest alphabet $\mathscr{A}$ such that there exists an infinite word avoiding p over $\mathscr{A}$. Bean, Ehrenfeucht, and McNulty [48] and Zimin [595] characterized unavoidable patterns, i.e., such that $\lambda(p) = \infty$. We say that a pattern p is *t-avoidable* if $\lambda(p) \leq t$. They also give an algorithm to test whether a pattern is unavoidable.

Cassaigne [132] began and Ochem [449] finished the determination of the avoidability index of every pattern with at most three variables. Two words $u \in \mathscr{A}^*$ and $v \in \mathscr{B}^*$ are *isomorphic* if there exists a bijection $b : \mathscr{A} \to \mathscr{B}$ such that $u = b(v)$.

Theorem 5.2.8. *Let p be a pattern over the variables* $\{A, B, C\}$.

- *If p contains an occurrence of a pattern in Table 5.1, or the mirror of a pattern in Table 5.1, then* $\lambda(p) = 2$.
- *If p is isomorphic to a pattern in* $\{A, AB, ABA, ABAC, ABACA, ABACAB, ABACABA, ABACB, ABACBA, ABC\}$, *then* $\lambda = \infty$.
- *Otherwise,* $\lambda(p) = 3$.

A variable that appears only once in a pattern is said to be *isolated*. Following Cassaigne [132], we associate to a pattern p the *formula* f obtained by replacing every isolated variable in p by a dot. The factors between the dots are called *fragments*.

An *occurrence* of a formula f in a word w is a non-erasing morphism $h : \Delta^* \to \mathscr{A}^*$ such that the h-image of every fragment of f is a factor of w. Every other notion related to patterns is similarly defined for formulas. Clearly, if a formula f is associated to a pattern p, every word avoiding f also avoids p, so $\lambda(p) \leq \lambda(f)$. Notice

Table 5.1 Minimally two avoidable patterns over three variables.

AAA, ABCABC, AABBA, ABAAB, ABABA, AABABB, ABCBABC, AABAACBAAB, AABACCB, AABBCABBA, AABBCAC, AABBCBC, AABBCC, AABCBC, AABCCAB, AABCCBA, ABAACBC, ABAACCB, ABACACB, ABACBC, ABACCAB, ABACCBA, ABBACCA, ABBACCB, ABBCACB, ABBCBAC, ABBCBCA, ABBCCAB, ABCAACB, ABCACAB, ABCACB, ABCBBAC, AABAACBAB, AABAACBB, AABABCBB, AABACABBA, AABACBABA, AABACBABB, AABACBBAB, AABBCABA, AABBCBAB, AABBCBBAA, AABCCACA, ABAACABAB, ABAACABBA, ABAACBBAB, ABABCABBA, ABABCBAAB, ABABCBC, ABACBBC, ABBACBAAB, ABBCABC

also that every recurrent word avoiding p also avoids f. Moreover, it is not to hard to see that there exists a recurrent word avoiding p over $\mathcal{A}_{\lambda(p)}$, so that $\lambda(p) = \lambda(f)$.

Without loss of generality, a formula is such that no variable is isolated and no fragment is a factor of another fragment.

A *doubled* pattern contains every variable at least twice. Thus, a doubled pattern is a formula with exactly one fragment. Every doubled pattern is 3-avoidable [451]. A formula is said to be *binary* if it has at most two variables. The avoidability index of every binary formula has been recently determined [453]. Moreover, for every 2-avoidable binary formula f, it is known whether f is avoided by exponentially many binary words [453]. We say that a formula f is *divisible* by a formula f' if f does not avoid f', that is, there is a non-erasing morphism h such that the image of every fragment of f' by h is a factor of a fragment of f. If f is divisible by f', then every word avoiding f' also avoids f and thus $\lambda(f) \le \lambda(f')$. Moreover, the reverse f^R of a formula f satisfies $\lambda(f^R) = \lambda(f)$. For example, the fact that *ABA.AABB* is 2-avoidable implies that *ABAABB* and *BAB.AABB* are 2-avoidable.

Finally, Clark [152, 153] has obtained several examples of formulas with avoidability index 5. The simplest one is *AB.BA.AC.BC.CDA.DCD*.

Clark [152] has introduced the notion of *n-avoidance basis* for formulas, which is the smallest set of formulas with the following property: for every $i \le n$, every avoidable formula with i variables is divisible by at least one formula with at most i variables in the n-avoidance basis.

From the definition, it is not hard to obtain that the 1-avoidance basis is $\{AA\}$ and the 2-avoidance basis is $\{AA,\ ABA.BAB\}$. Clark obtained the 3-avoidance basis. The formulas in the 3-avoidance basis are given below with their avoidability index:

- *AA* ($\lambda = 3$ [562])
- *ABA.BAB* ($\lambda = 3$ [132])
- *ABCA.BCAB.CABC* ($\lambda = 3$ [237])
- *ABCBA.CBABC* ($\lambda = 2$ [237])
- *ABCA.CABC.BCB* ($\lambda = 3$ [237])
- *ABCA.BCAB.CBC* ($\lambda = 3$ [237])
- *AB.AC.BA.CA.CB* ($\lambda = 4$ [32])

The following properties of the avoidance basis are derived:

- The n-avoidance basis is a subset of the $(n + 1)$-avoidance basis.

- The n-avoidance basis is closed under reverse. (In particular, $ABCA.BCAB.CBC$ is the reverse of $ABCA.CABC.BCB$.)
- Two formulas in the n-avoidance basis with the same number of variables are incomparable by divisibility. (However, $AB.AC.BA.CA.CB$ divides AA.)
- The n-avoidance basis is computable.

Let us comment on the last property. Every pattern with n variables and length at least 2^n has a factor that is a doubled pattern, which is 3-avoidable. Thus, every formula in the n-avoidance basis has either one fragment of length at most 2^n or all of its fragments have length at most $2^n - 1$. So, to compute the n-avoidance, we have to consider only finitely many formulas and select the ones that are not divisible by another formula with n variables.

The *circular formula* C_t is the formula over $t \geq 1$ variables $A_0, \ldots, A_{t-1}$ containing the t fragments of the form $A_i A_{i+1} \ldots A_{i+t}$ such that the indices are taken modulo t. Thus, the first three formulas in the 3-avoidance basis, namely, $C_1 = AA$, $C_2 = ABA.BAB$, and $C_3 = ABCA.BCAB.CABC$, are also the first three circular formulas. More generally, for every $t \leq n$, the n-avoidance basis contains C_t. It is known that $\lambda(C_i) = 2$ for every $i \geq 4$ [237].

We conclude this part with some open problems. Minor open problems include:

1. Prove that long enough doubled patterns are 2-avoidable.
2. Prove that every formula in the n-avoidance basis has at least two fragments except AA.
3. Find two 2-avoidable formulas f_1 and f_2 such that no infinite ternary word avoids f_1 and f_2 simultaneously.

Major open problems include:

1. Cassaigne's conjecture: for every formula f, there exists a morphic word in $\mathscr{A}^*_{\lambda(f)}$ that avoids f.
2. Is there an algorithm that decides whether $\lambda(f) \leq k$, given a formula f and an integer k?
3. Is there an avoidable formula f such that $\lambda(f) \geq 6$?
4. Is there an infinite family of avoidable formulas $f_1, f_2, f_3, \ldots$ such that $\lambda(f_1) < \lambda(f_2) < \lambda(f_3) < \ldots$?

Notice that the major open problems are related to each other. A negative answer to (3) would restrain problem (2) to the cases $2 \leq k \leq 4$. Also, a positive answer to (1) restrains problem (2) to a question about morphic words.

5.3 Abelian and Sum Equivalence

Abelian powers are a commutative version of usual powers. The avoidability of abelian repetitions has been studied since a question from Erdős in 1957 [216, 217].

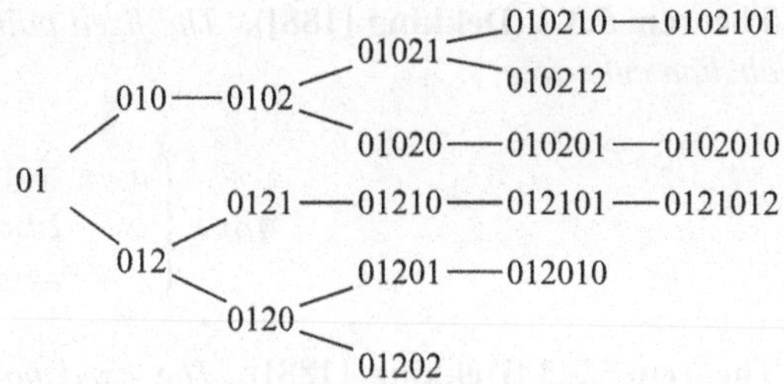

Fig. 5.1 Exhaustive search of every ternary word avoiding abelian squares. (See Section 5.4.1)

Let $\mathscr{A}$ be an alphabet. The *Parikh vector* $\Psi(w)$ of a word $w \in \mathscr{A}^*$ is the vector indexed over $\mathscr{A}$ such that for all $a \in \mathscr{A}$, $\Psi(w)_a = |w|_a$. For instance, over $A = \{a, b, c\}$, we have $\Psi(abac) = \begin{pmatrix} 2 \\ 1 \\ 1 \end{pmatrix}$ and $\Psi(bbc) = \begin{pmatrix} 0 \\ 2 \\ 1 \end{pmatrix}$. Two words u and w are *abelian equivalent*, denoted by $u \approx_a v$, if and only if $\Psi(u) = \Psi(v)$. Equivalently, two words are abelian equivalent if and only if they are permutations of each other. Given an integer $k \geq 2$, the word $w \in \mathscr{A}^+$ is an *abelian k-th power* if there are $w_1, w_2, \ldots, w_k \in \mathscr{A}^+$ such that $w = w_1 w_2 \ldots w_k$ and for all $i \in \{2, \ldots, k\}$, $w_i \approx_a w_1$. An *abelian square* (resp. *abelian cube*) is an abelian 2nd power (resp. 3rd power).

One can easily check that every ternary word of length at least 8 contains an abelian square (see Figure 5.1) and every binary word of length 10 contains an abelian cube.

Erdős asked whether there is an infinite abelian square-free word over an alphabet of size 4 [216, 217]. After some intermediary results (alphabet of size 25 by Evdokimov [219] and size 5 by Pleasant [486]), Keränen gave a positive answer:

Theorem 5.3.1 (Keränen [342]). *The fixed points of the following 85-uniform morphism are abelian square-free:*

$$\sigma_K: \begin{cases} a \to \texttt{abcacdcbcdcadcdbdabacabadbabcbdbcbacbcdcacbabdabacadcbcdcacdbcb} \\ \qquad \texttt{acbcdcacdcbdcdadbdcbca} \\ b \to \texttt{bcdbdadcdadbadacabcbdbcbacbcdcacdcbdcdadbdcbcabcbdbadcdadbdacdc} \\ \qquad \texttt{bdcdadbdadcadabacadcdb} \\ c \to \texttt{cdacabadabacbabdbcdcacdcbdcdadbdadcadabacadcdbcdcacbadabacabdad} \\ \qquad \texttt{cadabacabadbabcbdbadac} \\ d \to \texttt{dabdbcbabcbdcbcacdadbdadcadabacabadbabcbdbadacdadbdcbabcbdbcaba} \\ \qquad \texttt{dbabcbdbcbacbcdcacbabd}. \end{cases}$$

Moreover, Carpi showed that the number of abelian-square-free words over four letters is exponential [128]. The best known lower bound on the growth rate, due to Keränen, is 1.02306 [344].

Besides that, Dekking answered the question of the avoidability of abelian n-th powers, for $n \geq 3$.

Theorem 5.3.2 (Dekking [188]). *The fixed points of the following morphism are abelian cube-free:*

$$\sigma_{D3}: \begin{cases} a \to \mathtt{aabc} \\ b \to \mathtt{bbc} \\ c \to \mathtt{acc}. \end{cases}$$

Theorem 5.3.3 (Dekking [188]). *The fixed point of the following morphism is abelian 4th power-free:*

$$\sigma_{D2}: \begin{cases} a \to \mathtt{abb} \\ b \to \mathtt{aaab}. \end{cases}$$

Moreover, the growth rate of abelian-cube-free words over 3 letters is at least $3^{1/19} = 1.059526\ldots$ [494], and the growth rate of abelian-4th-power-free binary words is at least $2^{1/16} = 1.044273\ldots$ [172].

In fact, we can show that the morphisms of Theorems 5.3.1, 5.3.2, and 5.3.3 are *abelian k-th power-free* for the corresponding k, that is, the image of an abelian k-th power-free word by the given morphism is also abelian k-th power-free. Dekking gave sufficient conditions for a morphism to be abelian k-th power-free and used them to show Theorems 5.3.2 and 5.3.3.

Carpi gave stronger sufficient conditions for a morphism to be abelian k-th power-free [127]. These conditions can be used to show Theorem 5.3.1. In order to check the conditions, the help of a computer is needed. This set of conditions is conjectured to be a characterization of abelian k-th power-free morphisms.

Nonetheless, there are non-abelian k-th power-free morphisms whose fixed point is abelian k-th power-free. This is the case for the morphism σ_4 used to prove that $\mathbb{Z}$ is not uniformly 3-repetitive in Theorem 5.3.30 or for σ_6 used in Theorem 5.3.16 and Theorem 5.3.22.

In [79, Section 4.6.3], the authors present the algorithm of [173] which decides if the fixed point of a given morphism is abelian-k-power-free. This algorithm can be applied to the morphisms of Keränen and Dekking, but not to σ_4 and σ_6.

We explain here the algorithm of [497], which generalizes the algorithm of [173] and which can decide for a wider class of morphisms, including σ_4 and σ_6. We first introduce some definitions.

Definition 5.3.4 (Matrix Associated to a Morphism). To a morphism h on $\mathscr{A}^*$, we associate a matrix M_h on $\mathscr{A} \times \mathscr{A}$ such that $(M_h)_{a,b} = |h(b)|_a$. The *eigenvalues of* h are the eigenvalues of M_h.

Using this matrix we have the following property:

$$\forall w, \Psi(h(w)) = M_h \Psi(w).$$

The notion of template was first introduced in [173] and is useful both for proving and formulating the results of this section.

Definition 5.3.5 (*k*-Template and Realization). Fix k an integer and $\mathscr{A}$ an alphabet of size n. A *k-template* over $\mathscr{A}$ is a $(2k)$-tuple $t = [a_1, \ldots, a_{k+1}, \mathbf{d}_1, \ldots, \mathbf{d}_{k-1}]$ where for all i, $a_i \in \mathscr{A} \cup \{\varepsilon\}$ and $\mathbf{d}_i \in \mathbb{Z}^n$.

A word $w = a_1w_1a_2w_2 \ldots w_ka_{k+1}$, where $w_i \in \mathscr{A}^*$, is a *realization* of (or *realizes*) the template t if for all $i \in \{1, \ldots, k-1\}$, $\Psi(w_{i+1}) - \Psi(w_i) = \mathbf{d}_i$.

For instance *aabaac* is a realization of the 2-template $[\varepsilon, b, c, (0, 0, 0)]$. A word is an abelian k-th power if and only if it realizes the k-template $[\varepsilon, \ldots, \varepsilon, \vec{0}, \ldots, \vec{0}]$.

Definition 5.3.6. Let h be a morphism and let $t' = [a'_1, \ldots, a'_{k+1}, \mathbf{d}'_1, \ldots, \mathbf{d}'_{k-1}]$ and $t = [a_1, \ldots, a_{k+1}, \mathbf{d}_1, \ldots, \mathbf{d}_{k-1}]$ be two k-templates. We say that t' *is a parent by h of t* if there are $p_1, s_1, \ldots, p_{k+1}, s_{k+1} \in \mathscr{A}^*$ such that:

- $\forall i \in \{1, \ldots, k+1\}$, $h(a'_i) = p_ia_is_i$,
- $\forall i \in \{1, \ldots, k-1\}$, $\mathbf{d}_i = M_h\mathbf{d}'_i + \Psi(s_{i+1}p_{i+2}) - \Psi(s_ip_{i+1})$.

We denote by $\mathrm{Par}_h(t)$ the set of parents of t by h.

Definition 5.3.7. The set of *ancestors* of a k-template t_0, denoted by $\mathrm{Anc}_h(t_0)$, is the smallest set S such that:

- $t_0 \in S$,
- for all $t \in S$, $\mathrm{Par}_h(t) \subseteq S$.

The "is an ancestor" relation is the transitive, reflexive closure of the "is a parent" relation.

In [173], the authors use the notion of template to show that if the matrix M_h^{-1} is defined and has induced euclidean norm smaller than one, then on can decide if the fixed point of h is abelian k-th power-free. We present the following generalization. We say that a morphism h is *primitive* if there exists $k \in \mathbb{N}$ such that for all $a \in \mathscr{A}$, $h^k(a)$ contains all the letters of $\mathscr{A}$ (i.e., M_{h^k} is positive). Moreover for any primitive morphism h and any letter $a \in \mathscr{A}$ such that $h(a) = au$ with $u \in \mathscr{A}^*$, we denote by $h^\omega(a)$ the fixed point of h which is the limit of the sequence $(h^i(a))_{i\in\mathbb{N}}$.

Theorem 5.3.8. *For any primitive morphism h with no eigenvalue of absolute value 1 and for any template t_0, it is possible to decide whether $h^\omega(a)$ avoids t_0.*

As a corollary, one can decide for every integer k, alphabet $\mathscr{A}$ and primitive morphism $h : \mathscr{A}^* \mapsto \mathscr{A}^*$ with no eigenvalue of absolute value 1, if a fixed point of h is abelian k-th power-free.

We will first give some useful lemmas about parents:

Lemma 5.3.9. *Let t_p be a parent of a template t_0, and $w \in \mathscr{A}^*$. If w realizes t_p, then $h(w)$ realizes t_0.*

The proof of this lemma can be found in [79, Section 4.6.3]. Templates and parents are defined so that they have this property.

For any k-template $t = [a_1, \ldots, a_{k+1}, \mathbf{d}_1, \ldots, \mathbf{d}_{k-1}]$, let $\Delta(t) = \max_{i=1}^{k-1} ||\mathbf{d}_i||_1$ and $\delta = \max_{a\in\mathscr{A}} |h(a)|$.

Lemma 5.3.10. *Let t be a k-template and $w \in \mathrm{Fact}(h^\omega(a))$ a word which realizes t. If $|w| > k\left(\frac{(k-1)\Delta(t)}{2} + \delta + 1\right) + 1$, then for every w' such that $w \in \mathrm{Fact}(h(w'))$, there is a parent t' of t such that a factor of w' realizes t'.*

Proof. Let $t = [a_1, \ldots, a_{k+1}, \mathbf{d}_1, \ldots, \mathbf{d}_{k-1}]$ be a k-template and $w \in \mathrm{Fact}(h^\omega(a))$ a word which realizes t such that $|w| > k\left(\frac{(k-1)\Delta(t)}{2} + \delta + 1\right) + 1$. Then there are $w_1, \ldots, w_n \in \mathscr{A}^*$ such that $w = a_1 w_1 a_2 w_2 \ldots w_k a_{k+1}$ and $\forall i \in \{1, \ldots, k-1\}$, $\Psi(w_{i+1}) - \Psi(w_i) = \mathbf{d}_i$. Thus for any $i, j \in \{1, \ldots, k\}$ such that $j < i$, $\Psi(w_i) = \Psi(w_j) + \sum_{m=j}^{i-1} \mathbf{d}_m$ and, by triangular inequality, we have:

$$|w_i| - |w_j| = ||\Psi(w_i)||_1 - ||\Psi(w_j)||_1 \leq \sum_{m=j}^{i-1} ||\mathbf{d}_m||_1 \leq (i-j)\Delta(t).$$

Therefore for any $i, j \in \{1, \ldots, k\}$, $|w_j| \leq |i-j|\Delta(t) + |w_i|$, and for any i, $|w| \leq \sum_{m=1}^{k}(|i-m|\Delta(t) + |w_i|) + k + 1$. Thus $|w| \leq \frac{k(k-1)}{2}\Delta(t) + k|w_i| + k + 1$. Then $k\left(\frac{(k-1)\Delta(t)}{2} + |w_i| + 1\right) + 1 \geq |w| > k\left(\frac{(k-1)\Delta(t)}{2} + \delta + 1\right) + 1$, and consequently $\forall i$, $|w_i| > \delta = \max_{a \in \mathscr{A}} |h(a)|$. We also know that $\forall i$, $w_i \in \mathrm{Fact}(h^\omega(a))$ so there are $w'_1, \ldots, w'_k \in \mathscr{A}^*$, $a'_1, \ldots, a'_{k+1} \in \mathscr{A}$, $p_1, \ldots, p_{k+1} \in \mathrm{Pref}(h)$ and $s_1, \ldots, s_{k+1} \in \mathrm{Suff}(h)$ such that:

- $\forall i$, $w_i = s_i h(w'_i) p_{i+1}$,
- $\forall i$, $h(a'_i) = p_i a_i s_i$,
- $w' = a'_1 w'_1 a'_2 \ldots a'_k w'_k a'_{k+1}$.

w' realizes the template $t' = [a'_1, \ldots, a'_{k+1}, \Psi(w'_2) - \Psi(w'_1), \ldots, \Psi(w'_k) - \Psi(w'_{k-1})]$. Moreover for all i:

$$\mathbf{d}_i = \Psi(w_{i+1}) - \Psi(w_i)$$
$$\mathbf{d}_i = \Psi(s_{i+1} h(w'_{i+1}) p_{i+2}) - \Psi(s_i h(w'_i) p_{i+1})$$
$$\mathbf{d}_i = M_h \Psi(w'_{i+1}) - M_h \Psi(w'_i) + \Psi(s_{i+1} p_{i+2}) - \Psi(s_i p_{i+1})$$
$$\mathbf{d}_i = M_h(\Psi(w'_{i+1}) - \Psi(w'_i)) + \Psi(s_{i+1} p_{i+2}) - \Psi(s_i p_{i+1})$$

Thus t' is a parent of t and t' is realized by w'. □

Definition 5.3.11 (Small Realization). A *small realization* of a k-template t is a realization w of t such that $|w| > k\left(\frac{(k-1)\Delta(t)}{2} + \delta + 1\right) + 1$.

Using Lemmas 5.3.9 and 5.3.10, one can easily show the following proposition.

Proposition 5.3.12. *Let h be a primitive morphism and t_0 a k-template. Then $h^\omega(a)$ avoids t_0 if and only if $h^\omega(a)$ avoids every small realizations of every elements of $\mathrm{Anc}_h(t_0)$.*

Any template has finitely many small realizations, and we only need to compute small factors of $h^\omega(a)$ to compute them. If $\mathrm{Anc}_h(t_0)$ is finite and computable, then the set of small realizations of the templates of $\mathrm{Anc}_h(t_0)$ is also computable, and we can decide if $h^\omega(a)$ avoids t_0.

In particular it is proven in [173] that if M_h^{-1} is defined and has induced euclidean norm smaller than 1, then $\mathrm{Anc}_h(t_0)$ is finite. And in order to compute $\mathrm{Anc}_h(t_0)$, we can recursively add the missing parents. The parents of a template t can be obtained by considering all the possible choices of letters, suffixes, and prefixes and using M_h^{-1} to compute the corresponding vectors.

In the setting of Theorem 5.3.8, M_h is not necessarily invertible which implies that t_0 has infinitely many parents. Thus we need to find a way to discard many elements of $\mathrm{Anc}_h(t_0)$. In fact, using the Jordan normal form of M_h, we can find conditions on the vectors of the templates of $\mathrm{Anc}_h(t_0)$.

Let h be a primitive morphism and $PJP^{-1} = M_h$ be a Jordan decomposition of M_h. We can write $\Psi(h(w)) = M_h\Psi(w) = PJP^{-1}\Psi(w)$, which implies $P^{-1}\Psi(h(w)) = JP^{-1}\Psi(w)$. Let $B = (b_1, \ldots, b_n)$ be the base where b_j is the j-th line of P^{-1}. We get the two following lemmas.

Lemma 5.3.13. *For all $i \in [1, n]$ such that $|J_{i,i}| < 1$, there exists $c_i \in \mathbb{R}^+$ such that: for all $w \in \mathrm{Fact}(h^\omega(a))$, $|b_i \cdot \Psi(w)| < c_i$.*

The complete proof can be found in Proposition 3.3 of [497]. We give a sketch of the proof in the special case where J is diagonal.

Let $i \in [1, n]$ such that $|J_{i,i}| < 1$. For any factor w of $h^\omega(a)$ which is not the factor of the image of a letter, there is a prefix p of the image of a letter by h, a suffix s of the image of a letter by h, and a factor w' of $h^\omega(a)$ such that $w = sh(w')p$. Since J is diagonal, $b_i \cdot \Psi(h(w')) = J_{i,i}(b_i \cdot \Psi(w'))$. This implies $b_i \cdot \Psi(w) = J_{i,i}b_i \cdot \Psi(w') + b_i \cdot \Psi(sp)$ or w is the factor of the image of a letter. Now let $max_{sp} = \max_{a,b,s \in \mathrm{Suff}(h(a)), p \in \mathrm{Pref}(h(b))} |b_i \cdot \Psi(sp)|$ and $max_f = \max_{a, f \in \mathrm{Fact}(h(a))} |b_i \Psi(f)|$, we get $|b_i \cdot \Psi(w)| \leq |J_{i,i}b_i \cdot \Psi(w')| + max_{sp}$ or $|b_i \cdot \Psi(w)| \leq max_f$. This implies $|b_i \cdot \Psi(w)| \leq \max(|J_{i,i}b_i \cdot \Psi(w')| + max_{sp}, max_f)$ and thus by induction for all $w \in \mathrm{Fact}(h^\omega(a))$:

$$|b_i \cdot \Psi(w)| \leq \max\left(\frac{max_{sp}}{1 - |J_{i,i}|}, max_f\right).$$

If J is not diagonal, we can work Jordan block by Jordan block instead of coordinate by coordinate, and we also get an explicit formula.

Lemma 5.3.14. *For all k-template t_0, and for all $i \in [1, n]$ such that $|J_{i,i}| > 1$, there exists $c_i \in \mathbb{R}^+$ such that for all $v \in \mathbb{Z}^n$ which appears in a least one template of $\mathrm{Anc}_h(t_0)$, $|b_i \cdot v| < c_i$.*

The proof of this lemma is similar to the proof of the last lemma. One uses the fact that t is the parent of an ancestor of t_0 (or is t_0) and shows by induction that there is a bound.

It is easy to deduce from Lemma 5.3.13 that for any vector v of a template t which is realized by a factor of $h^\omega(a)$, for all i such that $|J_{i,i}| < 1$, $|b_i v| > 2c_i$.

Thus the two lemmas together imply that, if there is no i such that $|J_{i,i}| = 1$, then the norms of the vectors of a template $t \in \mathrm{Anc}(T_0)$ which is realized by a factor of $h^{\omega}(a)$ are bounded (since it is bounded by c_i or $2c_i$ on each coordinate).

Using this fact and Proposition 5.3.12, we get the following.

Proposition 5.3.15. *Let h be a primitive morphism with no eigenvalues of norm 1 and t_0 a k-template. Then $h^{\omega}(a)$ avoids t_0 if and only if $h^{\omega}(a)$ avoids every small realization of every element of* $S_{bounded}(t_0) = \mathrm{Anc}_h(t_0) \cap \{t \mid$ *for every vector* $v \in t, \forall i, |b_i \cdot v| \leq 2c_i\}$.

Proof. If $t \in \mathrm{Anc}_h(t_0) \setminus S_{bounded}(t_0)$, then there exists i such that $|b_i \cdot v| > 2c_i$ and from Lemma 5.3.13 there is no factor of $h^{\omega}(a)$ realizing t. This implies that there is a small realization of a template from $S_{bounded}(Tt_0)$ if and only if there is a small realization of a template from $\mathrm{Anc}_h(t_0)$. So the result is now equivalent to Proposition 5.3.12. □

There are only finitely many possible values for any vector of any $t \in S_{bounded}(t_0)$ since they have integer coordinates and are bounded. Thus $S_{bounded}(t_0)$ is a finite set. If M_h is invertible, one can easily compute the parents of a given template and thus compute $S_{bounded}(t_0)$ by iteratively adding the missing parents.

In the case where M_h is not invertible, we can first generate exhaustively the set $S_1 = \{t \mid t \text{ is a k-template and for every vector } v \in t, \forall i, |b_i \cdot v| \leq c_i\}$. This set is finite and easy to generate since there are finitely many letters and finitely many vectors. Now for each template t, compute all the templates $T \in S_1$ such that t is a parent of T (for this direction we do not need M_h to be invertible). Finally construct $S_{Bounded}(t_0)$ by inductively adding the missing parents obtained from the previous computation.

This can be done more efficiently by computing inductively the parents of t_0 that respect the bounds without computing S_1. In order to compute the parents of a given template t, one can first compute the Smith decomposition to find an integer basis B_k of the kernel of M_h. There are finitely many choices for the letters, suffixes, and prefixes when computing a parent, so we try all of them exhaustively. For a given choice, we need to be able to compute all the pre-images of a vector $\mathbf{v}$ that respect the bounds. Using the Smith decomposition, one can find a particular solution $\mathbf{u}$ to $M_h\mathbf{u} = \mathbf{v}$. Then we know that all the other solutions are of the form $\mathbf{k} = \mathbf{u} + B_k\mathbf{x}$ where $\mathbf{x}$ is an integer vector, and since we know that $\mathbf{k}$ is bounded, we can deduce bounds on B_k. And so we can generate exhaustively all such vectors and compute all the parents of a given template. Note that we do not compute exactly $S_{bounded}(t_0)$ as defined in the theorem, since we might lose some parents that were only accessible by elements that do no respect the bound, but the theorem is still true for this set.

This concludes the proof of Theorem 5.3.8. The algorithm first computes the Jordan normal form of M_h and the bounds from Lemma 5.3.13. Then it recursively computes the parents of the template t_0 included in $S_{bounded}(t_0)$. Finally, for each template in $S_{bounded}(t_0)$, we check if there are small realizations of this template.

This algorithm can be used for the proof of Theorems 5.3.1, 5.3.2 and 5.3.3. We can also use it to show that the fixed points of the morphism given in [133] are abelian-cube-free or to get the following result.

Theorem 5.3.16. *Let σ_6 be the following morphism:*

$$\sigma_6 : \begin{cases} a \to ace,\ b \to adf \\ c \to bdf,\ d \to bdc \\ e \to afe,\ f \to bce. \end{cases}$$

Then $\sigma_6^\omega(a)$ is abelian-square-free.

The eigenvalues of the matrix of σ_6 are 0 (with algebraic multiplicity 3, and geometric multiplicity 2), 3, $\sqrt{3}$, and $-\sqrt{3}$. Thus the matrix of σ_6 is neither invertible nor diagonalizable. Moreover, by Lemma 5.3.13, the Parikh vectors of the factors of $\sigma_6^\omega(a)$ are at bounded distance from a subspace of $\mathbb{R}^6$ of dimension 3. This property will be important to prove Theorem 5.3.22.

5.3.1 Mäkelä's Questions

Erdős asked if it is possible to avoid arbitrarily long ordinary squares on binary words. This question was answered positively by Entringer, Jackson, and Schatz [214]. In the same spirit, Mäkelä asked the following two questions about the avoidability of long abelian cubes (resp. squares) on a binary (resp. ternary) alphabet:

Problem 5.3.17 (Mäkelä (See [343])). Can you avoid abelian cubes of the form uvw where $|u| \geq 2$ over two letters? You can do this at least for words of length 250.

Problem 5.3.18 (Mäkelä (See [343])). Can you avoid abelian squares of the form uv where $|u| \geq 2$ over three letters? Computer experiments show that you can avoid these patterns at least in words of length 450.

One can show that the answer to the first question is negative:

Theorem 5.3.19 ([496]). *There is no infinite word over a binary alphabet avoiding abelian cubes of period at least* 2.

The proof of the this theorem is explained in Section 5.4. Then one can then reformulate the questions from Mäkela and ask:

Problem 5.3.20. Is there a $p \in \mathbb{N}$ such that one can avoid abelian squares of period at least p over three letters? If yes what is the smallest such p?

Problem 5.3.21. Is there a $p \in \mathbb{N}$ such that one can avoid abelian cubes of period at least p over two letters? If yes what is the smallest such p?

Let σ_3 be the following morphism:

$$\sigma_3 : \begin{cases} a \to \texttt{bbbaabaaac} \\ b \to \texttt{bccacccbcc} \\ c \to \texttt{ccccbbbcbc} \\ d \to \texttt{cccccccaa} \\ e \to \texttt{bbbbbcabaa} \\ f \to \texttt{aaaaaaabaa}. \end{cases}$$

Theorem 5.3.22 ([497]). *$\sigma_3(\sigma_6^\omega(a))$ does not contain any square of period at least 6.*

Thus we know that $2 \leq p \leq 6$ in Problem 5.3.20 and $p \geq 3$ in Problem 5.3.21. In order to prove Theorem 5.3.22, we use the same technique as for Theorem 5.3.8. If there is an abelian square in $\sigma_3(\sigma_6^\omega(a))$, then there is a factor of $\sigma_3(\sigma_6^\omega(a))$ that realizes the 2-template $T_0 = [\varepsilon, \varepsilon, \varepsilon, \mathbf{0}]$. By Lemma 5.3.10 if the length of this abelian square is at least 25, then there is a parent t of T_0 by σ_3 which is realized by a factor of $\sigma_6^\omega(a)$. So if we can show that there is no parent of T_0 by σ_3 which is realized by a factor of $\sigma_6^\omega(a)$, then we know that $\sigma_3(\sigma_6^\omega(a))$ only contains small abelian squares, and we can compute the exact value by looking at the small factors of $\sigma_3(\sigma_6^\omega(a))$. Given a parent t of T_0 by σ_3, we can use Theorem 5.3.8 to decide if it is avoided by $\sigma_6^\omega(a)$. But since M_{σ_3} is not invertible, T_0 has infinitely many parents by σ_3, so we need to find a way to eliminate most of them.

For any $a_1, a_2, a_3 \in \mathscr{A} \cup \{\varepsilon\}$, if $t = [a_0, a_1, a_2, v]$ is a parent of T_0 by σ_3, then there are $p_1, s_1, p_2, s_2, p_3, s_3$ such that:

- $\forall i \in \{1, 2, 3\}, h(a_i') = p_i s_i$,
- $\mathbf{0} = M_{\sigma_3}\mathbf{v} + \Psi(s_1 p_2) - \Psi(s_2 p_3)$.

There are only finitely many values for the a_i, s_i, and p_i, so we only need to be able to find the set of solutions for fixed a_i, s_i, and p_i. We need to find vectors v such that $M_{\sigma_3} v = \Psi(s_2 p_3) - \Psi(s_1 p_2)$. One can use the Smith normal form to find an integer base $K = (k_1, k_2, k_3)$ of the kernel of M_{σ_3} and a particular integer solution v_0 such that $M_{\sigma_3} v_0 = \Psi(s_2 p_3) - \Psi(s_1 p_2)$. Thus all the solutions are of the form $v = v_0 + Kx$ where x is an integer vector.

Lemma 5.3.13 tells us that there are three vectors $b_1, b_2, b_3 \in \mathbb{C}^6$ and three constants $c_1, c_2, c_3 \in \mathbb{R}^+$ such that if $t = [a_0, a_1, a_2, v]$ is realized by a factor of $\sigma_6^\omega(a)$ then for all $i \in \{1, 2, 3\}$, $|b_i \cdot v| < c_i$. Using that, one can check that in fact x is bounded by a constant C. Thus there are finitely many possible v such that t is realizable by a factor of $\sigma_6^\omega(a)$ and they can be computed.

5.3.2 *Abelian Patterns*

A nice way to define an abelian analog of the notion of pattern is to use an alternative definition of the occurrence of a pattern.

Theorem 5.3.23. *For any word $w \in \mathscr{A}^*$ and pattern P in Δ, the three following are equivalent:*

1. *w is an occurrence of the pattern P;*
2. *there exists a morphism $h : \Delta^* \mapsto \mathscr{A}^*$ such that $h(P) = w$;*
3. *there exists $u_1, u_2, \ldots, u_{|P|} \in \mathscr{A}^*$ such that $w = u_1 u_2 \ldots u_{|P|}$ and for all $i, j \in [1, |P|]$, $P_i = P_j \implies u_i = u_j$.*

With this alternative definition in mind, we say that a word $w \in \mathscr{A}^*$ is an *abelian occurrence* of a pattern $P \in \Delta$ if there are $u_1, u_2, \ldots, u_{|P|} \in \mathscr{A}^*$ such that $w = u_1 u_2 \ldots u_{|P|}$ and for all $i, j \in [1, |P|]$, $P_i = P_j \implies u_i \approx_a u_j$. Note that if a word w is an occurrence of a pattern P, then w is also an abelian occurrence of P. A word w *contains an abelian occurrence* of P if one of its factors is an abelian occurrence of P. If w does not contain an abelian occurrence of P, we say that w *avoids P* in the abelian sense. We say that a pattern is *abelian-k-avoidable* if there is an infinite word from an alphabet of size k that avoids this pattern. For any pattern $P \in \Delta^*$, the *abelian-avoidability index* of P (denoted by $\lambda_a(P)$) is the smallest integer k such that P is abelian-k-avoidable or ∞ if there is no such k.

It is left to the reader to verify that the relation "contains an abelian occurrence of" over $\mathscr{A}^* \cup \Delta^*$ is transitive. It implies that if a pattern P contains an abelian occurrence of a pattern P' then $\lambda_a(P') \le \lambda_a(P)$. This is really similar to the notion of divisibility between patterns or formulas but in the abelian framework, and this is useful to show the following first result.

Theorem 5.3.24. *Let $P \in \{A, B\}^+$. If $P \in \{A, B, AB, BA, ABA, BAB\}$, then $\lambda_a(P) = \infty$, otherwise $\lambda_a(P) \le 4$.*

Proof. Let $F = \{A, B, AB, BA, ABA, BAB\}$.

Recall that abelian squares are avoidable over four letters which implies that $\lambda_a(AA) = 4$. One can check that F is the set of nonempty binary words avoiding abelian squares. Then for any $P \in \{A, B\}^+ \setminus F$, P contains an abelian occurrence of AA which implies $\lambda_a(P) \le \lambda_a(AA) = 4$.

Let w be an infinite word over a finite alphabet. There is at least one letter, say a, which occurs infinitely many times. Thus there is a nonempty word u such that aua is a factor of w and then w contains an abelian occurrence of ABA. This implies that $\lambda_a(ABA) = \infty$. Moreover, for all $P \in F$, ABA contains an abelian occurrence of P which implies $\lambda_a(ABA) \le \lambda_a(P)$ and $\lambda_a(P) = \infty$. □

Using the same idea and the fact that $\lambda_a(AAA) = 3$, we can show the following.

Theorem 5.3.25. *Binary patterns of length at least 9 are abelian-3-avoidable.*

Proof. This is easy to check exhaustively that every pattern of length at least 9 contains an abelian cube. Using $\lambda_a(AAA) = 3$, we get the result. □

There are only finitely many binary patterns with abelian-avoidability index at least 4.

Finding one pattern P such that $\lambda_a(P) = 2$ is not enough to get the same kind of result on the binary alphabet since there are infinitely many binary patterns avoiding P. The solution is to find a set S of patterns, such that $\forall P \in S, \lambda_a(P) = 2$ and such that there are only finitely many patterns avoiding all the elements of S in the abelian sense.

Let w_n be the fixed point of the morphism τ_n such that $\tau_n(0) = 0^{n+1}1, \tau_n(1) = 01^n$. Dekking showed that w_2 avoids $AAAA$ in the abelian sense (see Theorem 5.3.3). It is possible to generalize the conditions from Dekking [188] to obtain a set of binary patterns avoided by at least one of the w_i in the abelian sense and to use this set to show that binary patterns of length at least 119 are avoidable in the abelian sense over the binary alphabet [174].

It is in fact possible to decide under some restriction whether the fixed point of a given morphism avoids a pattern in the abelian sense.

Theorem 5.3.26. *For any alphabets Δ and $\mathscr{A}$, pattern $P \in \Delta^*$, morphism $h : \mathscr{A}^* \mapsto \mathscr{A}^*$, and any letter $a \in \mathscr{A}$ such that*

- $h(a) = as$ *for some s,*
- M_h*, the matrix associated to h, is invertible,*
- $\|M_h^{-1}\|_2 < 1$,
- $\forall \sigma \in \mathscr{A}, |h(\sigma)| > 1$,

it is possible to decide whether $h^\omega(a)$ contains an abelian occurrence of P.

The algorithm is based on a generalization of the notion of templates to patterns. The details can be found in [511]. Note that one could generalize this result to matrix with no eigenvalues of module 1 by using the same technique than for Theorem 5.3.8. An implementation of this decision algorithm can check the following lemma.

Lemma 5.3.27 ([511]). *The patterns in Table 5.2 are avoidable in the abelian sense over a binary alphabet.*

Table 5.2 Patterns avoidable in the abelian sense over a binary alphabet

- Avoided by $a \mapsto aabaa, b \mapsto bbabb$: *AABBBAAAB*, *ABAAABBBA*, *AAABABABBB*, *AAABABBABB*, *AAABABBBAB*, *AABBBABAAB*, *AABBBABABA*, *ABAABABBBA*, *ABAABBBABA*, *ABABAABBBA*, *ABBBABAAAB*, *AABAABBBAB*, *AABBBAABAB*, *AABBBAABAAB*, *AAABABBAAAB*, *AABBBABBBAA*, *ABABABBBABA*, *ABABBABBABA*, *AAABAAABBAB*, *AAABBABAAAB*, *AAABAABAABAB*, *AAABABAAABAB*, *AABAAABABAAB*, *AAABAAABABBA*, *AAABAABABAAB*, *AAABABAABAAB*, *ABBABAAABAAB*, *ABABBBABBBABA*.
- Avoided by $a \mapsto aaaab, b \mapsto abbab$: *ABAABBBAAB*, *AAABBABABB*, *AAABBABBAB*, *AABAABBABB*, *AABABABBBA*, *AABABBABBA*, *AABABBBAAB*, *AABABBBABA*, *AABBAABBBA*, *AABBABABBA*, *AABBABBAAB*, *AABBABBABA*, *AABBBAABBA*, *ABAABBABBA*, *AABBABABBBA*, *AABABBBABBBA*.
- Avoided by $a \mapsto abb, b \mapsto aaab$: *AAAA*, *AAABAABBB*, *AAABBBABB*, *AABBABBBA*, *AABBBABBA*, *AAABBAAABB*, *AABABAAABB*, *ABBBAABBBA*, *AAABAABBAB*, *AAABAABAABB*, *AAABBAABAAB*, *AABAABAABBA*, *AABAABBAAAB*, *AABABABAAAB*, *AAABBAAABAB*, *AABAAABABAB*, *AABAAABBAAB*, *AAABAABAAABAB*.
- Avoided by $a \mapsto aaab, b \mapsto bbba$: *AAABABBBAA*, *AAABBAABBB*, *AAABBABBAA*, *ABABAAABBB*, *ABABBBAABBA*, *AABABBAAABA*, *AABBBABAAABA*.
- Avoided by $a \mapsto abaa, b \mapsto babb$: *AABBABBABBA*, *AABABBABBBA*, *AABBBABBBABA*, *ABABBABBABBA*, *ABABBABBBABA*, *ABABBBABABBA*, *ABBABABBBABBA*.
- Avoided by $a \mapsto aaaba, b \mapsto babbb$: *ABAABBBAAA*, *AABABBBAAA*.
- Avoided by $a \mapsto aababbaaaba, b \mapsto babbbaababb$: *AABAAABAAABAB*, *ABBBABBBABBBA*, *AAABAAABAAABAAA*.

Moreover it is possible to check by exhaustive search that any pattern of length at least 15 contains an occurrence of one of the patterns from Table 5.2. Using that and the fact that the relation "contains an abelian occurrence of" is transitive, we get the following lemma.

Theorem 5.3.28. *Binary patterns of length at least 15 are abelian-2-avoidable.*

It seems hard to show that a binary pattern is not abelian-2-avoidable and only few of the patterns that avoid a pattern from Lemma 5.3.27 are proven to be abelian-2-unavoidable.

5.3.3 Powers Modulo Φ, Additive Powers, and k-Repetitive Groups

Let $(G, +)$ be a semigroup and $\Phi : (\mathscr{A}^*, .) \to (G, +)$ be a morphism. For any $k \geq 2$, a *k-th power modulo* Φ is a word $w = w_1 w_2 \dots w_k$ with for all $i \in \{2, \dots, k\}$, $\Phi(w_i) = \Phi(w_1)$. If moreover $|w_1| = |w_2| = \dots = |w_k|$ then it is a *uniform k-th power modulo* Φ. We say that $(G, +)$ is *k-repetitive* (resp., *uniformly k-repetitive*) if for any alphabet $\mathscr{A}$ and any morphism $\Phi : (\mathscr{A}^*, .) \to (G, +)$ every infinite word over $\mathscr{A}$ contains a k-power modulo Φ (resp., a uniform k-power modulo Φ). Note that if $(G, +)$ is abelian, then every abelian k-power is a uniform k-power modulo Φ. Since usual squares are avoidable over 3 letters, we know that the free group on three generators is not 2-repetitive, and we deduce that the free group on two generators is not 2-repetitive.

One of the most important question related to this notion is to know which groups are (uniformly) k-repetitive for a given k. An important result is the following theorem from Pirillo and Varricchio.

Theorem 5.3.29 ([485]). *Let k be an integer greater than 1. The following statements are equivalent:*

1. *$\mathbb{Z}$ is not uniformly k-repetitive,*
2. *any finitely generated and uniformly k-repetitive semigroup is finite.*

It is a long-standing question whether $\mathbb{Z}$ is uniformly 2-repetitive or not. Recently Cassaigne *et al.* showed that $\mathbb{Z}$ is not uniformly 3-repetitive. We deduce that any finitely generated infinite group is not uniformly 3-repetitive.

In the rest, we only consider groups $(G, +) = (\mathbb{Z}^d, +)$ for some $d > 0$. Uniform k-th powers modulo Φ are usually called *additive k-th powers*, without mention of the morphism Φ, if the value of $\Phi(a)$ is clear in the context (i.e., $\mathscr{A} \subseteq G$). For instance, if we take $\mathscr{A} = \{0, 1, 2, 3\} \subseteq \mathbb{Z}$, then 012030 is an additive square, and in fact one can check by exhaustive search that additive squares are not avoidable over this particular subset of $\mathbb{Z}$. An application of Szemerédi's theorem shows that for $d = 1$, for any finite alphabet $\mathscr{A}$ and $k \in \mathbb{N}$, it is not possible to avoid k-th power modulo Φ over $\mathscr{A}$, that is, $(\mathbb{Z}, +)$ is k-repetitive for any k.

It was recently showed that:

- $\mathbb{Z}$ is not uniformly 3-repetitive;
- $\mathbb{Z}^2$ is not uniformly 2-repetitive.

Those two facts are corollaries of the following theorems.

Theorem 5.3.30 ([133]). *The fixed point of* σ_4 *is additive-cube-free, with*

$$\sigma_4 : \begin{cases} 0 \to 03 \\ 1 \to 43 \\ 3 \to 1 \\ 4 \to 01. \end{cases}$$

Let Φ be the morphism such that

$$\Phi : \begin{cases} a \to (0,0),\ b \to (1,1) \\ c \to (2,1),\ d \to (0,1) \\ e \to (2,0),\ f \to (1,0). \end{cases}$$

Theorem 5.3.31 ([497]). $\Phi(\sigma_6^{\omega}(a))$ *does not contain two consecutive blocks of the same size and the same sum.*

The proof of Theorem 5.3.30 given in [133] is really specific to the morphism σ_4 and relies on linear algebra and a lot of computation done by computer. In [497], the authors gave a procedure deciding if the fixed points of a given morphism are additive-square-free. The algorithm uses some of the idea of [133] and the ideas of the decision procedure for abelian power freeness already used for Theorem 5.3.8. The algorithm is also based on templates.

In order to compute which templates correspond to an additive power, we first need to associate a matrix M_Φ to Φ such that for any word w, $\Phi(w) = M_\Phi \Psi(w)$. It is easy to see that two factors u and v have the same sum if $M_\Phi(\Psi(u) - \Psi(v)) = 0$. Thus a factor w is an additive square if and only if it realizes a template $t = [\varepsilon, \varepsilon, \varepsilon, v]$ with $M_\Phi v = 0$. It is possible to compute an integer base K of the kernel of M_Φ using Smith normal form and then $v \in \{Kx \mid x \in \mathbb{Z}^3\}$. One can then use Lemma 5.3.13 and some linear algebra to show that if t is realizable, then x and thus v are bounded [497]. It is then easy to compute a finite superset of all the realizable templates corresponding to an additive cube. And we can check using Theorem 5.3.8 that all these templates are avoided by $\sigma_6^{\omega}(a)$. It implies that $\Phi(\sigma_6^{\omega}(a))$ does not contain additive square. This proof technique can also be applied to Theorem 5.3.30 or to other similar results.

$$\text{Let } \tau_4 : \begin{cases} 0 \to 001 \\ 1 \to 041 \\ 2 \to 41 \\ 4 \to 442 \end{cases},\ \tau_4' : \begin{cases} 0 \to 03 \\ 2 \to 53 \\ 3 \to 2 \\ 5 \to 02 \end{cases} \text{ and } \tau_4'' : \begin{cases} 0 \to 03 \\ 2 \to 63 \\ 3 \to 2 \\ 5 \to 02. \end{cases}$$

Theorem 5.3.32. *$\tau_4^{\omega}(0)$, $\tau_4'^{\omega}(0)$, and $\tau_4''^{\omega}(0)$ avoid additive cubes.*

Rao conjectured that for any integers $0 < i < j$ such that i and j are coprime and $j \geq 6$, additive cubes are avoidable over $\{0, i, j\}$, since it seems easy to find a morphism τ from $\{0, 1, 3, 4\}^*$ to $\{0, i, j\}^*$ such that $\tau(w)$ avoids additive cubes if and only if w avoids additive cubes, using the decision algorithm presented in [494]. Moreover, he showed that this conjecture is true for $6 \leq j \leq 9$ and that additive cubes are also avoidable over $\{0, 1, 5\}$. Thus using these observations and Theorem 5.3.32, $\{0, 1, 2, 3\}$, $\{0, 1, 4\}$, and $\{0, 2, 5\}$ might be the only alphabets over which additive cubes are not avoidable.

Problem 5.3.33. Are additive cubes avoidable over $\{0, 1, 2, 3\}$? $\{0, 1, 4\}$? $\{0, 2, 5\}$?

5.3.4 k-Abelian Equivalence

One recent generalization of the abelian equivalence is the k-abelian equivalence introduced by Karhumäki *et al.* [329, 330]. Let $k \geq 1$. For any words u and w, we denote by $|u|_w$ the number of occurrences of w as a factor of u. Let $k \in \mathbb{N} \cup \{+\infty\}$ and u, v be two words over the alphabet $\mathcal{A}$. We let $\mathcal{A}^{\leq k}$ denote the set of words of length at most k over $\mathcal{A}$. Two words $u \in \mathcal{A}^*$ and $v \in \mathcal{A}^*$ are *k-abelian equivalent*, if for every $w \in \mathcal{A}^{\leq k}$, $|u|_w = |v|_w$. A word u is a *k-abelian n-th power*, $n \geq 2$, if $u = u_1 u_2 \ldots u_n$ such that $u_i \approx_{a,k} u_{i+1}$ for every $i \in \{1, \ldots, n-1\}$. A *k-abelian square* (resp. *k-abelian cube*) is a k-abelian 2nd power (resp. k-abelian 3rd power). This notion is between the abelian equivalence (which is the 1-abelian equivalence) and the usual equality between words (which can be viewed as the ∞-abelian equivalence). Since cubes are avoidable over the binary alphabet (e.g., in the Prouhet–Thue–Morse word), but are not avoidable in the abelian sense, it is natural to ask for the smallest k for which k-abelian cubes are avoidable over the binary alphabet.

Theorem 5.3.34 ([304]). *2-abelian squares are not avoidable on ternary words.*

We will show in Section 5.4 how to prove this result. On the other hand, we have the following.

Theorem 5.3.35 ([494]).

- *One can avoid* 2*-abelian cubes on binary words.*
- *One can avoid* 3*-abelian squares on ternary words.*

The proof follows from the fact that for any abelian square-free word (resp. abelian-cube-free word) w, the morphic word $\tau_3(w)$ (resp. $\tau_2(w)$) is 3-abelian square-free (resp. 2-abelian cube-free), where τ_2 and τ_3 are given in Table 5.3. It also implies that there are exponentially many such words.

Following Erdős' and Mäkelä's questions, we can ask to avoid only long k-abelian repetitions. One has the following.

Table 5.3 Morphisms for k-abelian n-th power-free words

2-abelian cube-free morphism:

$$\tau_2 : \begin{cases} 0 \to 001001010010110010010100100110010011001011011011 \\ 1 \to 001001100100110011011001101100100110011011011011 \\ 2 \to 001101101011010010110101101001010010011011011011 \end{cases}$$

3-abelian square-free morphism:

$$\tau_3 : \begin{cases} 0 \to 0102012021012010201210212 \\ 1 \to 0102101201021201210120212 \\ 2 \to 0102101210212021020120212 \\ 3 \to 0121020120210201210120212 \end{cases}$$

Theorem 5.3.36 ([496, 497]). *Let $g(k)$ be the least integer such that there exists an infinite binary word with only $g(k)$ distinct k-abelian squares. Then $g(1) = \infty$, $5 \leq g(2) \leq 734$, $g(4) = g(3) = 4$, and $g(k) = 3$ for every $k \geq 5$.*

The determination of the exact value $g(2)$ is still open and has the same order of difficulty as Problem 5.3.20 and Problem 5.3.21.

5.3.5 *k*-Binomial Equivalence

Another generalization of the abelian equivalence has been introduced: the k-binomial equivalence [505]. Let u and v be two words. Let $\binom{u}{v}$ denotes the number of times v occurs as a subsequence of u (meaning as a "scattered" subword). We say that u and v are *k-binomially equivalent* if

$$\forall w \in \mathscr{A}^{\leq k}, \binom{u}{w} = \binom{v}{w}.$$

We write $u \approx_{b,k} v$ if u and v are k-binomially equivalent. Note that the 1-binomial equivalence is the abelian equivalence. Moreover, this notion is not comparable with the k-abelian equivalence, except for $k = 1$.

A 2-*binomial square* (resp. 2-*binomial cube*) is a nonempty word of the form xy where $x \approx_{b,2} y$ (resp. $x \approx_{b,2} y \approx_{b,2} z$).

Theorem 5.3.37 ([495]). *The ternary Thue–Morse word avoids 2-binomial squares.*

Consider the morphism $h : 0 \mapsto 001$ and $h : 1 \mapsto 011$.

Theorem 5.3.38 ([495]). *For every 2-binomial-cube-free word $w \in \{0, 1\}^*$, $h(w)$ is 2-binomial-cube-free.*

Corollary 5.3.39. *The infinite word $h^{\omega}(0) = 001001011\cdots$ fixed point of h avoids 2-binomial cubes.*

5.4 Techniques for Negative Results

5.4.1 *Exhaustive Search and Backtracking*

If one wants to prove that a factorial language is finite, a simple and often effective method is to do an exhaustive search to find every word within the language.

Algorithm 1 Exhaustive search

1: **procedure** RECURSE(w)
2:     **for** every $x \in \mathscr{A}$ **do**
3:         **if** $wx \in \mathscr{L}$ **then** RECURSE(wx)

Algorithm 1 terminates if and only if the language is finite. Note that the number of call of RECURSE is exactly the size of the language.

If $\mathscr{L}$ is defined by a set of forbidden factors, we only need to check (line 3) the suffixes of w, since we know that every proper prefix of w is in $\mathscr{L}$. Moreover, we may use a good data structure to check efficiently if a suffix of w is forbidden. For example, one can remember the Parikh vector of every prefix of w to check in $O(n)$ if a suffix of w is an abelian square.

Example 5.4.1. There are no infinite 2-abelian square-free words over 3 letters. The Algorithm 1 terminates and finds 81 217 678 words. The larger word in the language has length 537 [304].

This exhaustive search can be shortened using the lexicographic order. We fix now an arbitrary order on $\mathscr{A}$.

Algorithm 2 Search of lexicographic least element

1: **procedure** RECURSE(w)
2:     **for** i from 1 to $|w| - 1$ **do**
3:         **if** $w[1:i] > w[|w| - i + 1 : |w|]$ **then return**
4:     **for** every $x \in \mathscr{A}$ **do**
5:         **if** $wx \in \mathscr{L}$ **then** RECURSE(wx)

Let $\overline{\mathscr{L}}$ be the set of infinite words w such that every finite factor in w is in $\mathscr{L}$. If $\mathscr{L}$ is infinite, then the set $\overline{\mathscr{L}}$ is nonempty, by König's lemma. Then the condition of line 3 in Algorithm 2 fails for every prefix of the lexicographic least element in $\overline{\mathscr{L}}$. Thus, Algorithm 2 terminates if and only if $\mathscr{L}$ is finite.

This can be improved again if $\mathscr{L}$ is stable by permutations of letters, that is, if $\mathscr{L} = h(\mathscr{L})$ for every bijection $h : \mathscr{A} \to \mathscr{A}$ (see [153]).

Again, the condition of line 4 in Algorithm 3 fails for every prefix of the lexicographic least element in $\overline{\mathscr{L}}$, and the procedure terminates if and only if $\mathscr{L}$ is finite.

Algorithm 3 Search of lexicographic least element up to permutations

procedure RECURSE(w)
    for i from 1 to $|w| - 1$ **do**
        for every bijection h on $\mathscr{A}$ **do**
            if $w[1:i] > h(w[|w| - i + 1 : |w|])$ **then return**
    for every $x \in \mathscr{A}$ **do**
        if $wx \in \mathscr{L}$ **then** RECURSE(wx)

Example 5.4.2. On the case of 2-abelian-square-free words, the Algorithm 1 terminates after 81 217 678 calls, the Algorithm 2 terminates after 354 802 calls, and the Algorithm 3 terminates after 139 962 calls.

Example 5.4.3. There are no infinite binary words avoiding abelian cubes of period at least 2. The Algorithm 3 terminates after 2 873 166 727 calls, and the longest word in the exploration has size 289.

5.4.2 Bounds on Densities by Exhaustive Searches

Exhaustive searches can also be used to get bounds on densities.

Definition 5.4.4. A *generalized occurrence function* on a factorial language $\mathscr{L}$ is a function $f : \mathscr{L} \to \mathbb{R}^+ \cup \{\infty\}$ such that there is a $g : \mathscr{L} \to \mathbb{R}^+ \cup \{\infty\}$, and $f(w) = \sum_{1 \le i \le j \le |w|} g(w[i:j])$.

This notion generalizes the notion of occurrences of repetitions or letters that we discussed in Section 5.2. One can easily verify the following.

Proposition 5.4.5. *The function f is a generalized occurrence function if and only if the two following conditions are fulfilled:*

(c1) $f(\epsilon) = 0$, *and*
(c2) for every $u, v, w \in \mathscr{A}^*$, $f(uvw) + f(v) \ge f(uv) + f(vw)$.

Let $\mathscr{L}$ be a factorial language and f be a generalized occurrence function. We define

$$\ell_f^{\mathscr{L}} = \lim_{n \to +\infty} \min_{u \in \mathscr{L} \cap \mathscr{A}^n} \frac{f(u)}{n},$$

which is well defined, since for every integer n and k, $m_{kn+l} \ge \frac{k}{k+1} m_n$, where $m_n = \min_{u \in \mathscr{L} \cap \mathscr{A}^n} \frac{f(u)}{n}$. One can easily show the following.

Lemma 5.4.6. *There is an infinite word w in $\overline{\mathscr{L}}$ such that*

$$\lim_{n \to \infty} \frac{f(w[1:n])}{n} = \ell_f^{\mathscr{L}}.$$

Proof. The proof follows the ideas presented in [559]. One has directly, for every infinite word w in $\overline{\mathscr{L}}$, $\liminf_{n\to\infty} \frac{f(w[1:n])}{n} \geq \ell_f^{\mathscr{L}}$. Suppose that $\ell_f^{\mathscr{L}} < \infty$ and let $A = \left\{w \in \mathscr{L} : \frac{f(w[1..n])}{n} \leq \ell_f^{\mathscr{L}} \text{ for all } n \in \{1,\ldots,|w|\}\right\}$. Suppose that A is finite. Then for every word $w \in \mathscr{L}$, there is a sequence of words $w_1,\ldots,w_k$ such that $w = w_1w_2\ldots w_k$, for every $i \in \{1,\ldots,k\}$, $|w_i| \leq m+1$ (where $m = \max_{w\in A}|w|$), and for every $i \in \{1,\ldots,k-1\}$, $f(w_i) \geq (\ell_f^{\mathscr{L}}+\delta)|w_i|$, for a $\delta > 0$. This implies that there is a $\delta' > 0$ such that the set $\left\{w \in \mathscr{L} \mid \frac{f(w)}{|w|} \leq \ell_f^{\mathscr{L}} + \delta'\right\}$ is finite, and we have a contradiction.

Thus A is infinite, and by König's lemma, there is an infinite word w such that $\lim_{n\to\infty} \frac{f(w[1:n])}{n} \leq \ell_f^{\mathscr{L}}$. □

A word w is k-biprolongable in $\mathscr{L}$ if there exists a word $lwr \in \mathscr{L}$ with $|l| = |r| = k$. A set $S \subseteq \mathscr{L}$ is a *suffix cover* of $\mathscr{L}$ if there exists an integer k such that for every k-biprolongable word $w \in \mathscr{L}$, there is a word in S which is a suffix of w.

Let f be a generalized occurrence function. For every $u \in \mathscr{L}$, let

$$A_u(q) = \left\{w \in \mathscr{L} : uw \in \mathscr{L} \text{ and for every prefix } w' \text{ of } w, \frac{f(uw') - f(u)}{|w'|} < q\right\}.$$

Proposition 5.4.7. *Let $q \in \mathbb{R}$ and S be a suffix cover of $\mathscr{L}$. If for every $u \in S$, $A_u(q)$ is finite, then for every infinite word w in $\overline{\mathscr{L}}$, $\liminf_{n\to\infty} \frac{f(w[1:n])}{n} \geq q$.*

Proof. Let k be such that every k-biprolongable word in $\mathscr{L}$ has a suffix in S. Let w be an infinite word in $\overline{\mathscr{L}}$, and let $w = w_0w_1\ldots$ be such that w_0 is the smallest prefix of w which has a k-biprolongable suffix in S, and for every $i \geq 0$, w_{i+1} the smallest prefix of $w[1+|w_0\ldots w_i| : \infty]$ not in $A_{u_i}(q)$, where u_i is a suffix of $w_0\ldots w_i$ in S. Let w' be a prefix of w and j be the least integer such that w' is a prefix of $w_0w_1\ldots w_j$. Then $f(w') \geq \sum_{i=1}^{j-1} f(u_iw_i) - f(u_i) \geq q|w_1\ldots w_{j-1}|$. Since each w_i has bounded size, one has $\liminf_{n\to\infty} \frac{f(w[1:n])}{n} \geq q$. □

Example 5.4.8. If one wants to prove that the maximum frequency of a letter in a ternary square-free word is at most $\frac{255}{653}$, one can show that the sets $A_u\left(\frac{398}{653}\right)$ are finite for every $u \in \{0, 01, 021, 0121\}$ and the occurrence function $f(w) = |w|_1 + |w|_2$. The largest set is $A_{0121}\left(\frac{398}{653}\right)$, with 188 614 words of maximal size 10 090. Since 1 and 2 play a symmetric role, and $\{0, 01, 021, 0121, 012, 02, 0212\}$ is a suffix cover of the square-free words, we have the result.

5.4.3 *Mean Cycles and Rauzy Graphs*

A *weighted graph* is a triplet (V, A, p), where V is the finite set of *vertices*, $A \subseteq V\times V$ is the set of *arcs*, and $p : A \to \mathbb{R}^+ \cup \{\infty\}$ is the *weight* function. We only work with weighted graphs, so we generally call them *graphs*. If X is a set of arcs, its weight $p(X)$ is the sum of the weights in X.

Let $G = (V, A, p)$ be a graph. A *path* in a graph is a (possibly infinite) sequence of arcs $e_1, \ldots, e_k$ such that for every $i \in \{1, \ldots, k\}$, $e_i = (x_i, x_{i+1})$ and $e_{i+1} = (x_{i+1}, x_{i+2})$ (where k is the cardinality of the sequence). A *cycle* in a graph is a sequence of arcs $e_1, \ldots, e_k$ such that for every $i \in \{1, \ldots, k\}$, $e_i = (x_i, x_{i+1})$ and $e_{i+1} = (x_{i+1}, x_{i+2})$ (indices are taken modulo k).

The *mean cycle* of a graph G is $\Gamma(G) = \min \frac{w(C)}{|C|}$ over all cycles C in G. One can easily see that one takes the minimum over cycles which only pass at most one time through each vertex. The mean cycle is a well-known parameter in optimization area which is closely related to the maximum eigenvalue in max-plus algebra, and there are efficient algorithms to compute it (see, e.g., [337]).

The *Rauzy graph* of order n of a language $\mathscr{L}$ on the alphabet $\mathscr{A}$, where n is a positive integer, is the graph of vertex set $\mathscr{L} \cap \mathscr{A}^n$ and of arc set $\{(u, v) : u[2:n] = v[1:n-1]$and $u[1]v \in \mathscr{L}\}$ (see also [78, Section 4.5.5]).

Let f be a generalized occurrence function on $\mathscr{L}$. Let $n > 1$ and (V, A) be the Rauzy graph of order n of $\mathscr{L}$.

Let p be the weight function such that $p(u, v) = f(uv[n]) - f(u)$. Let $G = (V, A, p)$.

Lemma 5.4.9. *There is a constant c_G such that for every $n' \geq 0$, $\min_{w \in \mathscr{A}^{n'}} f(w) + c_G \geq \Gamma(G) \times n'$. In particular, we have $\ell_f^{\mathscr{L}} \geq \Gamma(G)$.*

Proof. Suppose w.l.o.g. that $n' \geq n$. Let $k = n' - n$, and let $P = ((x_1, x_2), \ldots (x_k, x_{k+1}))$ be the path in G such that for every $i \in \{1, \ldots k+1\}$, x_i is $w[i : i+n]$. Since f is generalized occurrence function, and by the definition of p, $f(w) \geq \sum_{i \in \{0,\ldots,k-1\}} p((x_i, x_{i+1}))$. Now P can be decomposed into $P_0 C_1 P_1 \ldots C_l P_l$, where for every $i \in \{1, \ldots l\}$, C_i is a cycle in G, and $P_0 \ldots P_l$ is a path with $\sum_{i \in \{0,\ldots,l\}} |P_i| \leq |V|$. Thus $f(w) \geq \Gamma(G) \times (n' - n - V)$. □

5.4.4 Upper Bound on the Growth Rate

The mean cycle is an analogue of the maximum eigenvalue in the (max, +) algebra. Lemma 5.4.9 recalls the following well-known fact.

Lemma 5.4.10. *Let $n > 1$ and let $G = (V, A)$ be the Rauzy graph of order n of $\mathscr{L}$. Then the spectral radius of the incidence matrix of G is an upper bound for the growth rate of $\mathscr{L}$.*

More generally, Lemma 5.4.10 is also true for every automaton which recognizes a superset of $\mathscr{L}$. The next section presents an iterative construction of such an automaton.

5.4.5 Nonuniform Rauzy Graphs

The mean cycle method to compute bounds is often much more effective if we use a nonuniform version of Rauzy graphs, called *suffix graphs* in this section.

Let $\mathcal{A}$ be a finite alphabet, and let f be a generalized occurrence function on $\mathcal{A}^*$. One can always suppose that $\mathcal{L} = \mathcal{A}^*$, by taking $f(w) = \infty$ for every $w \notin \mathcal{L}$. Thus we may omit the superscript $\mathcal{L}$ in the following.

Definition 5.4.11. A *good suffix cover on* $\mathcal{A}$ is a finite set V of words on the alphabet $\mathcal{A}$ such that:

(a) V is a complete suffix code, i.e., for every $u, v \in V$ with $u \neq v$, u is not a suffix of v, and every left-infinite word has a suffix in V.
(b) For every $ux \in V$ with $u \in \mathcal{A}^*$ and $x \in \alpha$, there is a $v \in V$ such that u is a suffix of v.

Definition 5.4.12. A *suffix graph of* $(\mathcal{A}, f)$ is a graph (V, A, p) such that:

- V is a good suffix cover,
- $(u, v) \in A$ if v is a suffix of ux for an $x \in \mathcal{A}$, and $p((u, v)) = f(ux) - f(u)$.

Note that a suffix graph is uniquely determined by the good suffix cover and by f. One can easily show that Lemma 5.4.9 is also true for suffix graphs. Figure 5.2 gives an example of a suffix graph on $\{0, 1\}^*$ and the occurrence function f which counts the number of squares. Thus, the suffix graph of Figure 5.2 shows that $\ell_f \geq \frac{1}{3}$, that is, the minimal density of squares in a binary word is at least $\frac{1}{3}$.

5.4.5.1 Automatic Method to Construct a Suffix Graph

We present now an automatic method which constructs a suffix graph. Throughout this section, we fix a finite alphabet $\mathcal{A}$ and a generalized occurrence function f.

Proposition 5.4.13. *Let* $G = (V, A, p)$ *be a suffix graph, and let* $(u, v) \in A$ *such that* $|u| < |v|$. *Then* $|u| = |v| - 1$, *and there is no* $w \in V \setminus \{u\}$ *such that* $(w, v) \in A$.

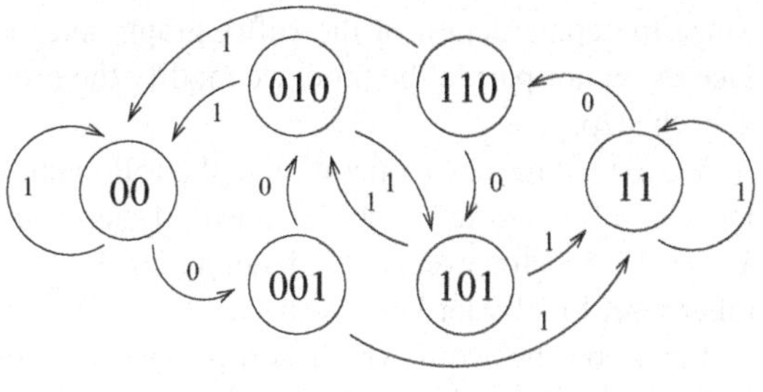

Fig. 5.2 A suffix graph of $(0, 1^*, f)$, where f counts the number or squares. One has $\Gamma(G) = \frac{1}{3}$

We say that a vertex $v \in V$ is *critical* in $G = (V, A, p)$ if there exists $u \in V$ such that u is the suffix of v of length $|v| - 1$. For example, the critical vertices of the graph in Figure 5.2 are 001 and 110.

Lemma 5.4.14. *Let* $G = (V, A, w)$ *be a suffix graph, and let* $v \in V$ *be a noncritical vertex in* V. *Then* $V * v = (V \setminus \{v\}) \cup \{xv \mid x \in \mathscr{A}\}$ *is a good suffix cover.*

Proof. Clearly, $V' = V * v$ is a maximal suffix code. Suppose that (b) is not fulfilled and let $uy \in V'$ such that u is not a suffix of any word in V'. Then $uy \in V' \setminus V = \{xv : x \in \mathscr{A}\}$, and $uy = xv = xwy$ for a $x \in \mathscr{A}$. Let $w' \in V$ be such that either w' is a suffix of u or u is a suffix of w'. Such a w' always exists, since V is a complete suffix code. We have $w' \neq v$, otherwise $u = v$ will be a suffix of $zv \in V'$. Thus $w' \in V'$, u is not a suffix of w', and w' is a suffix of u. Then v is critical. Contradiction. □

We denote by $G * v$ the suffix graph with vertex set $V * v$.

We describe now the algorithm used to obtain a lower bound. We start with the suffix graph G with suffix cover $V = \mathscr{A}$.

In an infinite loop, we take a circuit C in G of ratio $\frac{p(C)}{|C|} = \Gamma(G)$. We take a vertex v in C of minimum length, and we replace G by $G * v$. Note that a vertex of minimum length on the cycle cannot be critical.

The mean cycle cannot decrease and occasionally increases. At each cycle in the infinite loop, we get a lower bound for ℓ_f. We are only limited by the amount of memory needed to store the suffix graph.

Example 5.4.15. Let f denote the number of square occurrences in a word w. If we apply the previous algorithm for the function f on the binary words, it gives the lower bound $\frac{103}{187}$ after approximately 120000 iterations and 8 hours of single core computation on a 3 GHz CPU. The longest word in V has size 275. In [363], we show that if we count only squares of size at most 274, then the minimum ratio is at most $\frac{38}{69}$. Thus every suffix graph which proves the $\frac{103}{187}$ bound has a vertex of size at least 275. If we want to prove the exact bound with a uniform Rauzy graph, the graph would have at least 2^{275} vertices. This shows why the nonuniform generalization is crucial here.

5.4.5.2 Improvement of the Extension Phase

Since the construction of the suffix graph and the computation of the mean cycle take the major part of the time, we modify the procedure to make several extensions at each step.

We extend the $V * v$ operation in the following way. Let V be a good suffix cover, let $v \in \mathscr{A}^+$, and let v_i be the suffix of size i in v for every $i \in \{1, \ldots, |v|\}$. Let $V_0 = V$, and for every $i \geq 1$, let $V_i = V_{i-1} * v_i$ if $v_i \in V_{i-1}$, and $V_i = V_{i-1}$ otherwise. Finally let $V * v = V_{|v|}$.

Let w be the word which corresponds to a cycle of minimal ratio in G, that is, if the cycle is $((x_0, x_1), \ldots, (x_{k-1}, x_k))$, $w[i]$ is the last letter of x_i for every $i \in \{1, \ldots, k\}$. $\Gamma(G) \times k$ is a lower bound for $f(w^l w) - f(w^l)$ for a certain l (e.g., when

$(l-1)\times k \geq n+1$, where n is the size of the larger word in the good suffix cover). The circular word w may contain an occurrence of a factor u such that $g(u) > 0$ (using notation of Definition 5.4.4) and such that u is not counted in this lower bound (i.e., the occurrence of u is not an occurrence in x_i, for every $i \in \{1, \ldots, k\}$).

At each iteration of the algorithm, we choose the smallest such u in a cycle C of minimal ratio, and we replace G by $G * u$. Informally speaking, we kill the cycle C by adding an occurrence of a factor which increases the weight of the cycle. In all the weight functions f we study here, we have $\lim_{l\to\infty} f(u^l) = +\infty$; thus such a u always exists.

5.4.5.3 Computation of the Mean Cycle

The mean cycle is computed by a Howard-like algorithm (see [239]).

Before computing the mean cycle, we simplify the graph. We first remove every arc a with $p(a) = \infty$, since it cannot be part of a cycle with minimal ratio when $\ell_f < \infty$. We also remove vertices with out-degree zero or in-degree zero. Then we work on *doubly valuated graphs*, that is, graphs with two weight functions p and l, where $l(a)$ is the *length* of the arc a, in which we allow multiple arcs (several copies of the same arc (u, v) with different weights). On doubly valuated graphs, the mean cycle is $\Gamma(G) = \min \frac{p(C)}{l(C)}$ over every cycles. Clearly, the mean cycle of a graph is the mean cycle of the doubly valuated graph with $l((u, v)) = 1$ for every arc (u, v).

Our simplified procedure is the following: while there is a $v \in V$ of in-degree or out-degree at most 1 and such that $(v, v) \notin A$, we contract v. That is, we remove v and every arc incident to v in the graph, and we add (u, w) in A for every $(u, v), (v, w) \in A^2$, with $p((u, w)) = p((u, v)) + p((v, w))$ and $l((u, w)) = l((u, v)) + l((v, w))$. It is not hard to see that the mean cycle of the contracted graph is the mean cycle of the original graph. Moreover, Howard algorithm also works for doubly valuated graphs. Note that this procedure can introduce multiple arcs. For example, in Figure 5.2, one can contract the vertex 001: one removes the vertex 001, and we add the arc (00, 010) of weight 0 and length 2, and the arc (00, 11) of weight 1 and length 2.

Example 5.4.16. Let f be the number of square occurrences and $\mathcal{A} = \{0, 1\}$. With the previous improvements, the lower bound $\frac{103}{187}$ for ℓ_s takes only 39 iterations and less than one second to compute.

5.5 Techniques for Positive Results

We present general techniques which can be used to prove the existence of an infinite word with a given property P (in our case, avoiding a repetition-like patterns). There are two main classes of techniques to show that an infinite word exists: either we give an explicit construction of a word with the property or we show it by some nonconstructive way that such a word must exist, since the function which counts the number of words of size n with the property is growing quickly.

5.5.1 Finding a Candidate Morphism

Most of explicit constructions are pure morphic words, morphic words, or the image by a morphism of an already known word. Finding such constructions is usually done in two steps. The first one is to find a candidate morphism σ, that is, a morphism σ such that a long prefix of $\sigma(u)$ has the desired property P, where u is a well-chosen infinite word. This word u can be, for example, a fixed point of σ (and in this case, we search for a pure morphic word with the property), a word with the property P on a larger alphabet, or a word with another property P'. For example, if one wants to limit squares on binary words, a natural choice for P' is to be square-free, that is, one can search for a morphism $\sigma : \{0, 1, 2\} \to \{0, 1\}$ such that $\sigma(w_{TTM})$ has the property.

Once a candidate is found, the second step is to show that the infinite word $\sigma(u)$ has the desired property. This is usually done by a decision algorithm, which depends on P. Several of these algorithms have already been presented (e.g., in Section 5.3 for abelian powers or in [79, Section 4.2.5] for usual powers), and this second step will not be discussed here.

These two steps may have variable difficulties. For example, one can easily decide if a morphic word avoids abelian powers or long abelian powers, but finding such a morphic word is usually a hard task. One can cite that the simplest known construction of an abelian-square-free word on a 4-letter alphabet is the fixed point of the 85-uniform morphism from Keränen.

We focus here on the first step, finding a candidate morphism. If one wants to find a morphism σ such that $\sigma^\omega(a)$ or $\sigma(u)$ has a hereditary property P, one can use one of the following techniques.

First, one can try to construct a long word w with the property P, with a backtracking algorithm (as Algorithm 1). To avoid border effect, one can remove from w long enough prefix and suffix.

Sometimes, w already looks like a morphic word. If one can find a small set of words $\{w_1, \ldots, w_k\}$ such that a long factor w' of w is the image of a word v by the morphism $\sigma : \{1, \ldots, k\}^* \to \mathcal{A}^*$ such that $\sigma(i) = w_i$, then one can guess that there is an infinite word u such that $\sigma(u)$ has the property P. It suffices then to find the property P' that u must have, and then use an already known construction, or apply recursively the technique on P' and the alphabet $\mathcal{A}_k$.

If the previous approach fails (that can happen in particular if too many words have the property P), a second approach is to explicitly try to find a σ among a large class of morphisms, such that either $\sigma^\omega(a)$ has the property or $\sigma(u)$ has the property, for a previously specified u with a good property P'. One can start by choosing a set of candidate images $U \subseteq \mathcal{A}^*$ and try among $|U|^{|\mathcal{A}|}$ morphisms (if one is looking for a fixed point) or $|U|^k$ morphisms if u is on a k letter alphabet (if one is looking for word $\sigma(u)$ where $u \in \mathcal{A}_k^*$).

This set U can be the set of all words in $\mathcal{A}^*$ of size at most n with the property P, and one can increase n until a candidate morphism is found. Of course, one is quickly restricted by a combinatorial explosion. Thus one can restrict the set U to some promising words. For example, one can take only words with a common prefix or words often appearing in a long word with the property P.

5.5.2 Avoiding Patterns and Formulas

We address the problem of finding the avoidability index of a given formula f. First, we ensure that f is avoidable using Zimin's algorithm. We use the techniques given in Section 5.4 to obtain a lower bound k on $\lambda(f)$, which is also the conjectured value of $\lambda(f)$.

Now, we try to guess whether f is avoided by exponentially many words. This may be done by looking at the number of avoiding words of length, say, 5, 10, 15, 20, 25, 30. If f is doubled, then we can try the nonconstructive method described in Section 5.5.4.

The *avoidability exponent* $AE(p)$ of a formula f is the largest real α such that every α-free word avoids f. If f is avoided by exponentially many words, $AE(f) > 1$, and either f has at least two fragments or the nonconstructive method does not give the optimal upper bound, then we can try to find a suitable uniform morphism with the method in Section 5.5.3. Finally, f may not be avoided by exponentially many words. We say that two infinite words are *equivalent* if they have the same set of recurrent factors. Then there exists a finite set S of morphic words that *essentially avoids* f. This means that every infinite word over $\mathcal{A}_{\lambda(f)}$ avoiding f is equivalent to a word in S. Examples of such formulas include:

- $\{g_x(b_3), g_t(b_3)\}$ essentially avoids *ABA.AABB* [453].
- $\{b_3, b'_3, b''_3\}$ essentially avoids *ABCAB.BAC.ACA* [453].
- $\{b_4, b'_4, b''_4\}$ essentially avoids *AB.BA.AC.CA.BC* [32].

where

- b_3 is the fixed point of $\texttt{0} \mapsto \texttt{012}, \texttt{1} \mapsto \texttt{02}, \texttt{2} \mapsto \texttt{1}$ (i.e., the ternary Thue–Morse word).
- b_3, b'_3, and b''_3 are the three non-equivalent words obtained from b_3 by permutations of $\mathcal{A}_3$.
- b_4 is the fixed point of $\texttt{0} \mapsto \texttt{01}, \texttt{1} \mapsto \texttt{03}, \texttt{2} \mapsto \texttt{21}, \texttt{3} \mapsto \texttt{23}$.
- b_4, b'_4, and b''_4 are the three non-equivalent words obtained from b_3 by permutations of $\mathcal{A}_4$.
- $g_x(\texttt{0}) = \texttt{01110}$, $\quad g_t(\texttt{0}) = \texttt{01011011010}$,
  $g_x(\texttt{1}) = \texttt{0110}$, $\quad g_t(\texttt{1}) = \texttt{01011010}$,
  $g_x(\texttt{2}) = \texttt{0}$. $\quad g_t(\texttt{2}) = \texttt{010}$.

5.5.3 The Dejean Method

In this section, we consider another useful tool in pattern avoidance that has been defined in [451] and already used implicitly in [449]. We extend this definition to formulas. It is not hard to see that $AE(C_i) = 1 + \frac{1}{i}$.

Let us show that $AE(ABCBA.CBABC) = \frac{4}{3}$. Suppose for contradiction that a $\frac{4}{3}$-free word contains an occurrence h of $ABCBA.CBABC$. We write $a = |h(A)|$, $b = |h(B)|$, and $c = |h(C)|$. The factor $h(ABCBA)$ is a repetition with period $|h(ABCB)|$. So we have $\frac{a+b+c+b+a}{a+b+c+b} < \frac{4}{3}$. This simplifies to $2a < 2b + c$. Similarly, $CBABC$ gives $2c < a + 2b$, BAB gives $2b < a$, and BCB gives $2b < c$. Summing up these four inequalities gives $2a + 4b + 2c < 2a + 4b + 2c$, which is a contradiction. On the other hand, the word 01234201567865876834201234 is $\left(\frac{4}{3}^+\right)$-free and contains an occurrence of $ABCBA.CBABC$ with $A = 01$, $B = 2$, and $C = 34$.

As an exercise, prove that $AE(ABCA.CABC.BCB) = \frac{5-\sqrt{5}}{2}$.

The avoidability exponent depends on the repetitions induced by f. We have $AE(f) = 1$ for formulas such as $f = AB.BA.AC.CA.BC$ or $f = AB.BA.AC.BC.CDA.DCD$ that do not have enough repetitions. That is, for every $\epsilon > 0$, there exists a $(1 + \epsilon)$-free word that contains an occurrence of f.

Recall that the repetition threshold $RT(n)$ is the smallest real number α such that there exists an infinite a^+-free word over $\mathcal{A}_n$. The proof of Dejean's conjecture established that $RT(2) = 2$, $RT(3) = \frac{7}{5}$, $RT(4) = \frac{7}{4}$, and $RT(n) = \frac{n}{n-1}$ for every $n \geq 5$.

If $AE(f) > 1$, we consider the smallest integer n such that $RT(n) < AE(f)$. Thus, every $RT(n)^+$-free word over $\mathcal{A}_n$ avoids f (which already gives $\lambda(f) \leq n$). Then, for increasing values of q, we look for a q-uniform morphism $m : \mathcal{A}_n^* \to \mathcal{A}_k^*$ such that $m(w)$ avoids f for every $RT(n)^+$-free word $w \in \mathcal{A}_n^\ell$, where ℓ is a trade-off between impatience and paranoia.

Given such a candidate morphism m, we use Lemma 2.1 in [449] to show that for every $RT(n)^+$-free word $w \in \mathcal{A}_n^*$, the image $m(w)$ is (β^+, t)-free. The pair (β, t) is chosen such that $RT(n) < \beta < AE(f)$ and t is the smallest possible for the corresponding β. If $\beta < AE(f)$, then every occurrence h of f in a (β^+, t)-free word is such that the length of the h-image of every variable of f is bounded above by a function of t and f only. Thus, the h-image of every fragment of f has bounded length, and we can check that f is avoided by inspecting a finite set of factors of words of the form $m(w)$.

5.5.4 A Power Series Method

Several nonconstructive methods have been already used to prove the existence of words with a given property. One can cite the Lovász local lemma, the entropy compression, or some power series methods. A selection of such methods was already presented in [8, Chapter 4].

The next method generalizes some previously known methods.

Let $L \subset \mathcal{A}_k^*$ be a factorial language defined by a set F of forbidden factors of length at least 2. We denote the factor complexity of L by $n_i = |L \cap \mathcal{A}_k^i|$. We define L' as the set of words w such that w is not in L and the prefix of length $|w|-1$ of w is in L. For every forbidden factor $f \in F$, we choose a number $1 \le s_f \le |f|$. Then, for every $i \ge 1$, we define an integer a_i such that

$$a_i \ge \max_{u \in L} \left|\left\{v \in \mathcal{A}_k^i \mid uv \in L',\ uv = br,\ f \in F,\ s_f = i\right\}\right|. \tag{5.1}$$

We consider the formal power series $P(x) = 1 - kx + \sum_{i\ge1} a_i x^i$.

Lemma 5.5.1. *If $P(x)$ has a positive real root x_0, then $n_i \ge x_0^{-i}$ for every $i \ge 0$.*

Proof. Let us rewrite that $P(x_0) = 1 - kx_0 + \sum_{i\ge1} a_i x_0^i = 0$ as

$$k - \sum_{i\ge1} a_i x_0^{i-1} = x_0^{-1} \tag{5.2}$$

Since $n_0 = 1$, we prove by induction that $\frac{n_i}{n_{i-1}} \ge x_0^{-1}$ in order to obtain that $n_i \ge x_0^{-i}$ for every $i \ge 0$. By using (5.2), we obtain the base case: $\frac{n_1}{n_0} = n_1 = k \ge x_0^{-1}$. Now, for every length $i \ge 1$, there are:

- k^i words in $\mathcal{A}_k^i$,
- n_i words in L,
- at most $\sum_{1\le j\le i} n_{i-j} a_j$ words in L',
- $k(k^{i-1} - n_{i-1})$ words in $\mathcal{A}_k^i \setminus \{L \cup L'\}$.

This gives $n_i + \sum_{1\le j\le i} n_j a_{i-j} + k(k^{i-1} - n_{i-1}) \ge k^i$, that is, $n_i \ge kn_{i-1} - \sum_{1\le j\le i} n_{i-j} a_j$.

$$\begin{aligned}
\frac{n_i}{n_{i-1}} &\ge k - \textstyle\sum_{1\le j\le i} a_j \frac{n_{i-j}}{n_{i-1}} \\
&\ge k - \textstyle\sum_{1\le j\le i} a_j x_0^{j-1} && \text{By induction.} \\
&\ge k - \textstyle\sum_{j\ge1} a_j x_0^{j-1} \\
&= x_0^{-1} && \text{By (5.2).}
\end{aligned}$$

□

For example, one can use this method to show that shuffle squares are avoidable over seven letters. A *shuffle square* is a word whose letters can be partitioned into two identical subwords. For example, every square is a shuffle square, **a***a***b***b***c***c* and **ab***a***c***bc* are shuffle squares with the subword *abc*, and *c***c***bc***baca** is a shuffle square with the subword *cbca*.

Theorem 5.5.2 ([277]). *There exist at least 5.59^n words of length n over the 7-letter alphabet containing no shuffle square as a factor.*

Proof. The *q-prefix* (resp. *q-suffix*) of a word is its prefix (resp. suffix) of length q. A shuffle square is *minimal* if it does not contain a smaller shuffle square as a factor. A shuffle square is *small* if its length is two and is *large* otherwise. The set F of forbidden factors contains every minimal shuffle square. We set $s_f = 1$ if $f \in F$ is small and $s_f = |f| - 2$ otherwise.

We set $a_1 = 1$ because $s_f = 1$ only for small shuffle squares and there is only one way to extend a prefix by one letter to obtain a suffix xx with $x \in \mathcal{A}_m$. To obtain reasonable upper bounds a_t for $t \geq 2$, we need to bound the number of large minimal shuffle squares. To every shuffle square f of a word w of length i, we associate the *height function* h: $[0, \dots, 2i] \to \mathbb{Z}$ defined as follows:

- $h(0) = 0$.
- For $0 < j \leq 2i$, $h(j) = h(j-1) + 1$ if the j-th letter of f belongs to the subword w containing the first letter of f, and $h(j) = h(j-1) - 1$ otherwise.

Since f is a shuffle square, we have $h(2i) = 0$. Moreover, if $h(j) = 0$ for some $0 < j < 2i$, then the prefix of length j of f is a shuffle square. So, if h is the height function of a minimal shuffle square, then $h(j) > 0$ for every $0 < j < 2i$. Thus, every height function of a minimal shuffle square is associated to a unique Dyck word of length $2i - 2$. The number of height functions is thus at most $\frac{(2i-2)!}{i!\,(i-1)!}$. According to (5.1), we need to bound the number of solutions to $uv = bf$ such that u is fixed and $|v| = s_f = |f| - 2 = 2i - 2$. The 2-prefix of f is fixed since it corresponds to the 2-suffix of u. Notice that the 2-prefix of a large minimal shuffle square of a word w is equal to the 2-prefix of w, so the 2-prefix of w is also fixed. Thus, there are at most m^{i-2} possibilities for w. Since f is determined by w and its height function, there are at most $m^{i-2}\frac{(2i-2)!}{i!\,(i-1)!}$ possibilities for f. So we set $a_{2i-2} = m^{i-2}\frac{(2i-2)!}{i!\,(i-1)!}$ and consider the polynomial

$$
\begin{aligned}
P(x) &= 1 - mx + x + \textstyle\sum_{i\geq 2} m^{i-2}\frac{(2i-2)!}{i!\,(i-1)!}x^{2i-2} \\
&= 1 - (m-1)x + \left(\frac{2x}{1+\sqrt{1-4mx^2}}\right)^2 .
\end{aligned}
$$

For $m = 6$, $P(x)$ has no positive root. For $m = 7$, we have $P(x_0) = 0$ with $x_0 = 0.1788487593\dots$. So there exist at least α^n words of length n over $\mathcal{A}_7$ that avoid shuffle squares, where $\alpha = x_0^{-1} = 5.5913163944\dots$ □

5.5.5 *Kolpakov's Method*

Kolpakov presented another nonconstructive method to give lower bounds on the growth rate of some repetition-free languages [356]. This method strongly uses the fact that one wants to avoid usual powers and can give a bound very close to the actual growth rate. We present here the idea of the method on square-free ternary words. This method can be easily adapted to other (fractional) powers. In particular, this is used for Dejean words in [358].

We fix a positive integer m. Let G be the Rauzy graph of order m of the square-free words over the alphabet $\mathscr{A} = \{0, 1, 2\}$. We remove in G every vertex which is not in a cycle and call this new graph G'.

Let $\mathscr{F}$ be the language of square-free words such that every factor of size m is a vertex in G', and let $\mathscr{F}(n) = \mathscr{F} \cap \mathscr{A}^n$. Let $s = |\mathscr{F}(m)|$ (i.e., the number of vertices in G') and $\mathscr{F}(m) = \{w_1, \ldots w_s\}$.

We denote by $\pi(w)$ the set of ancestors of w in G'. Let Δ be the adjacency matrix of G', r its spectral radius (which is a positive real number, by Perron–Frobenius theorem), and let $\tilde{x} = (x_1, \ldots, x_s)$ be a nonnegative eigenvector which corresponds to r. Note that r is an upper bound for the growth rate of $\mathscr{F}$.

For a word $w \in \mathscr{F}(m)$ and $n \geq m$, we denote by $\mathscr{F}^w(n)$ the number of words in $\mathscr{F}(n)$ with suffix w. Let $S(n) = \sum_{i=1}^{s} x_i \cdot |\mathscr{F}^{w_i}(n)|$. The goal is to find an $\alpha > 1$ such that, assuming $S(k+1) \geq \alpha S(k)$ for every $k < n$, one has $S(n+1) \geq \alpha S(n)$, and thus α will be a lower bound for the growth rate of $\mathscr{F}$. From now on, we suppose that such $\alpha > 1$ exists. For small values of n, the fact is checked by computing explicitly $\mathscr{F}^{w_i}(n)$ and $S(n)$.

We estimate $\mathscr{F}^w(n+1)$. Let $\mathscr{H}(n)$ be the set of words w such that $w[1 : n-1]$ and $w[n-m : n]$ are square-free and that w contains a square as a suffix. Let $\mathscr{H}^w(n)$ be the set of words of $\mathscr{H}(n)$ with suffix w. One has

$$|\mathscr{F}^w(n+1)| = \sum_{w' \in \pi(w)} |\mathscr{F}^{w'}(n)| - |\mathscr{H}^w(n+1)|.$$

And thus:

$$S(n+1) = \sum_{i=1}^{s} x_i \cdot |\mathscr{F}^{w_i}(n+1)|$$

$$= \sum_{i=1}^{s} \sum_{w' \in \pi(w_i)} x_j \cdot |\mathscr{F}^{w'}(n)| - \sum_{i=1}^{s} x_i \cdot |\mathscr{H}^{w_i}(n+1)|$$

$$S(n+1) = r \cdot S(n) - \sum_{i=1}^{s} x_j \cdot |\mathscr{H}^{w_i}(n+1)|. \tag{5.3}$$

For $w \in \mathscr{H}^w(n)$, let $\lambda(w)$ be the smallest period of a square in w. Let $\mathscr{H}_j^w(n) = \{w \in \mathscr{H}^w(n) \mid \lambda(w) = j\}$. One has

$$|\mathscr{H}^w(n)| = \sum_{j=\lceil \frac{m}{2} \rceil + 1}^{\lfloor n/2 \rfloor} |\mathscr{H}_j^w(n)|.$$

The computation of $|\mathcal{H}^w(n)|$ is cut into two part. We fix an integer $p \geq m$. Let $A^w(n) = \sum_{j=\lceil \frac{m}{2} \rceil+1}^{p} |\mathcal{H}_j^w(n)|$ and $B^w(n) = \sum_{j=p+1}^{\lfloor n/2 \rfloor} |\mathcal{H}_j^w(n)|$. For large values of j, and thus for B^w, one uses the easy bound $|\mathcal{H}_j^w(n)| \leq |\mathcal{F}^w(n-j)|$. Thus:

$$\sum_{i=1}^{s} x_i \cdot B^{w_i}(n+1) \leq \sum_{j=p+1}^{\lfloor (n+1)/2 \rfloor} \sum_{i=1}^{s} x_i \cdot |\mathcal{F}^w(n-j+1)| = \sum_{j=p+1}^{\lfloor (n+1)/2 \rfloor} S(n-j+1)$$

$$\leq \sum_{j=p+1}^{\infty} \frac{S(n)}{\alpha^j} = \frac{S(n)}{\alpha^{p-1}(\alpha-1)}. \qquad (5.4)$$

For A^w, one needs to be more precise. For a $w_i, w_{i'} \in \mathcal{F}(m)$, a $j \geq \lceil \frac{m}{2} \rceil + 1$, and a $l \geq j$, one computes $\eta_{i,i'}$, the number of words w of size $l+m$, with prefix $w_{i'}$, suffix w_i, such that $w[1 : |w|-1]$ is square-free, and w has a square of period j as a suffix, and has no square of period less than j. One has $|\mathcal{H}^{w_i}(n+1)| \leq \sum_{i'=1}^{s} \eta_{i,i'} |\mathcal{F}^{w_{i'}}(n-l)|$. Of course, the bound is better when l is larger. But the size of the computations is exponential in l. Kolpakov used $l = 2j$ in [356].

For convenience, one can rewrite $\sum_{i=1}^{s} x_i \cdot A^{w_i}$ as:

$$\sum_{i=1}^{s} x_i \cdot A^{w_i} = \sum_{d=d_0}^{q} \sum_{i=1}^{s} \eta_i'(d) \cdot |\mathcal{F}^{w_i}(n-d)|.$$

Let $\eta''(d_0) = \eta'(d_0)$, and for every $d \in \{d_0, \ldots q-1\}$, let $\rho_d = \min_i(\eta_i'(d)/x_i)$ and $\eta''(d+1) = \eta'(d) + \delta\mu$, where $\mu_i = \eta_i'(d) - \rho_d \cdot x_i$. Let $\rho_q = \max(\eta_i'(d)/x_i)$. By induction of d', one has

$$\sum_{i=1}^{s} x_i \cdot A^{w_i} \leq \sum_{d=d_0}^{d'} \rho_d S(n-d)| + \sum_{i=1}^{s} \eta_i''(d'+1)|\mathcal{F}^{w_i}(n-d'+1)|$$

$$+ \sum_{d=d'+2}^{q} \sum_{i=1}^{s} \eta_i'(d+1)|\mathcal{F}^{w_i}(n-d)|.$$

Thus $\sum_{i=1}^{s} x_i \cdot A^{w_i} \leq P(1/\alpha)S(n)$, where $P(y) = \sum_{d=d_0}^{q} \rho_d y^d$. Putting together with equations (5.3) and (5.4), one has

$$S(n+1) \geq S(n)\left(r + P\left(\frac{1}{\alpha}\right) + \frac{1}{\alpha^{p-1}(\alpha-1)}\right).$$

Then if $\alpha \leq r + P(1/\alpha) + 1/(\alpha^{p-1}(\alpha-1))$, and $S(k+1) \geq \alpha S(k)$ for every $k < n$, one has $S(n+1) \geq \alpha S(n)$.

The computation can be greatly reduced using symmetries: one has $|\mathcal{F}^w(n)| = |\mathcal{F}^{w'}(n)|$ if w and w' are isomorphic. In [356], with $m = 45$, $p = 52$, and $q = 60$, Kolpakov got the lower bound $\alpha = 1.30173$ for the growth rate of ternary square-free words.

Chapter 6
Coloring Problems for Infinite Words

Caïus Wojcik and Luca Q. Zamboni

Abstract Given a finite coloring (or finite partition) of the free semigroup $\mathscr{A}^+$ over a set $\mathscr{A}$, we consider various types of monochromatic factorizations of right-sided infinite words $x \in \mathscr{A}^\omega$. In 2006 T. Brown asked the following question in the spirit of Ramsey theory: Given a nonperiodic infinite word $x = x_1x_2x_3\cdots$ with values in a set $\mathscr{A}$, does there exist a finite Coloring $\varphi : \mathscr{A}^+ \to C$ relative to which x does not admit a φ-monochromatic factorization, i.e., a factorization of the form $x = u_1u_2u_3\cdots$ with $\varphi(u_i) = \varphi(u_j)$ for all $i, j \geq 1$? We give an optimal affirmative answer to this question by showing that if $x = x_1x_2x_3\cdots$ is an infinite nonperiodic word with values in a set $\mathscr{A}$, then there exists a 2-coloring $\varphi : \mathscr{A}^+ \to \{0, 1\}$ such that for any factorization $x = u_1u_2u_3\cdots$, we have $\varphi(u_i) \neq \varphi(u_j)$ for some $i \neq j$. Some stronger versions of the usual notion of monochromatic factorization are also introduced and studied. We establish links, and in some cases equivalences, between the existence of these factorizations and fundamental results in Ramsey theory including the infinite Ramsey theorem, Hindman's finite sums theorem, partition regularity of IP-sets, and the Milliken–Taylor theorem.

6.1 Introduction

The following question was independently posed by T. Brown in [110] and by the second author in [591][1]:

Question 6.1.1. Let $x \in \mathscr{A}^{\mathbb{N}}$ be nonperiodic. Does there exist a finite coloring of $\mathscr{A}^+$ (the free semigroup set generated by $\mathscr{A}$) relative to which x does not admit a monochromatic factorization.

[1]The original formulation of the question was stated in terms of finite colorings of the set of all factors of x.

C. Wojcik (✉) · L. Q. Zamboni
Institut Camille Jordan, Université Lyon 1, CNRS UMR 5208, 43 boulevard du 11 novembre 1918, F-69622 Villeurbanne Cedex, France
e-mail: caius.wojcik@gmail.com; zamboni@math.univ-lyon1.fr

V. Berthé, M. Rigo (eds.), *Sequences, Groups, and Number Theory*, Trends in Mathematics, https://doi.org/10.1007/978-3-319-69152-7_6

In other words, given a nonperiodic word $x \in \mathscr{A}^{\mathbb{N}}$, does there exist a finite nonempty set C and a mapping $\varphi : \mathscr{A}^+ \to C$ such that for each factorization $x = u_1u_2u_3\cdots$, there exist $i, j \geq 1$ such that $\varphi(u_i) \neq \varphi(u_j)$? If such a finite coloring $\varphi : \mathscr{A}^+ \to C$ exists, we say that φ is a *separating coloring* (or *separating* $|C|$-*coloring*) for x.

Question 6.1.1 belongs to the class of Ramsey-type problems in which one tries to show that some abstract form of Ramsey's theorem does not hold in certain settings. For instance, the infinite version of Ramsey's theorem [491] (for coloring of pairs) states that whenever the set $\Sigma_2(\mathbb{N})$ of all two-element subsets of $\mathbb{N}$ is finitely colored, there exists an infinite set $X \subseteq \mathbb{N}$ with $\Sigma_2(X)$ monochromatic. Hence the same applies when $\mathbb{N}$ is replaced by $\mathbb{R}$. On the other hand, W. Sierpiński [548] showed that there exists a finite coloring of $\Sigma_2(\mathbb{R})$ such that there does not exist an uncountable set X with $\Sigma_2(X)$ monochromatic. In other words, Ramsey's theorem does not extend to the uncountable setting in $\mathbb{R}$. Similarly, by a straightforward application of Ramsey's theorem, one deduces that given any finite coloring of $\mathbb{N}$, there exists an infinite $X \subseteq \mathbb{N}$ all of whose pairwise sums $\{n+m : n, m \in X, n \neq m\}$ is monochromatic. Again it follows the same is true with $\mathbb{N}$ replaced by $\mathbb{R}$. On the other hand, N. Hindman, I. Leader, and D. Strauss [292] recently exhibited (using the Continuum Hypothesis) the existence of a finite coloring of $\mathbb{R}$ such that there does not exist an uncountable set with all its pairwise sums monochromatic. In other words, this additive formulation of Ramsey's Theorem also fails in the uncountable setting in $\mathbb{R}$. A related question in these sorts of problems concerns the least number of colors necessary to avoid the presence of monochromatic subsets of a certain kind. For instance, N. Hindman [291] showed that there exists a 3-coloring of $\mathbb{N}$ such that there does not exist an infinite subset X with $X+X$ monochromatic, and it is an open question of Owings [464] whether the same result may be obtained with only two colors. Again, using the Continuum Hypothesis and a few more colors (288 to be precise), it is possible to extend Hindman's result to the reals (see Theorem 2.8 in [292]). But again it is not known whether the same result can be obtained with only two colors. Thus a stronger version of Question 6.1.1 would read: Does every nonperiodic word admit a separating 2-coloring?

Various partial results in support of an affirmative answer to Question 6.1.1 were obtained in [73, 390, 392, 393, 519]. In this chapter we give a complete answer to this question and study various variations. We begin with some preliminaries in which we recall some basic notions in combinatorics on words which will be relevant. Then in Section 6.3 we give a solution to Question 6.1.1 by showing that every aperiodic word admits a separating 2-coloring. We also give a reformulation of this fact in terms of ultrafilters. More precisely, let $\beta\mathscr{A}^+$ denote the Stone-Čech compactification of the discrete semigroup $\mathscr{A}^+$ which we regard as the set of all ultrafilters on $\mathscr{A}^+$, identifying the points of $\mathscr{A}^+$ with the principal ultrafilters. As is well known, the operation of concatenation on $\mathscr{A}^+$ extends uniquely to $\beta\mathscr{A}^+$ making $\beta\mathscr{A}^+$ a compact right topological semigroup with $\mathscr{A}^+$ contained in its topological center (see, for instance, [293]). In particular by the Ellis–Numakura lemma, $\beta\mathscr{A}^+$ contains an idempotent element, i.e., an element p verifying $p \cdot p = p$.

We show that an infinite word $x = x_1x_2x_3\cdots \in \mathscr{A}^{\mathbb{N}}$ is periodic if and only if there exists $p \in \beta\mathscr{A}^+$ such that for each $A \in p$, there exists a factorization $x = u_1u_2u_3\cdots$ with each $u_i \in A$. Moreover p as above may be taken to be an idempotent element of $\beta\mathscr{A}^+$. In Section 6.4 we consider different variations of Question 6.1.1 relative to stronger monochromatic properties. For example, given a finite coloring of $\mathscr{A}^+$ and a word $x \in \mathscr{A}^{\mathbb{N}}$, we say a factorization $x = u_1u_2u_3\cdots$ is φ-sequentially monochromatic (resp., φ-super monochromatic) if $\{u_i\cdots u_j : 1 \le i \le j\}$ (resp., if $\{u_{n_1}u_{n_2}\cdots u_{n_k} : n_1 < n_2 < \cdots < n_k, k \ge 1\}$ is φ-monochromatic. We establish links, and in some cases equivalences, between the existence of these factorizations and fundamental results in Ramsey theory including the infinite Ramsey theorem, Hindman's theorem, partition regularity of IP-sets and the Milliken–Taylor theorem. We show that given a finite coloring φ of $\mathscr{A}^+$ and a word $x \in \mathscr{A}^{\mathbb{N}}$, there exists a suffix x' of x admitting a φ-sequentially monochromatic factorization and a point z in the shift orbit closure of x admitting a φ-super monochromatic factorization. We also prove that for each finite coloring $\varphi : \mathscr{A}^+ \to C$, for almost all words $x \in \mathscr{A}^{\omega}$, there exists z in the subshift generated by x admitting a factorization $z = u_1u_2u_3\cdots$ with the property that the set consisting of all unrepeated concatenations of the u_i is φ-monochromatic.

6.2 Preliminaries

Let us briefly recall some basic notions on words. See also Section 1.3. Throughout this chapter, $\mathscr{A}$ denotes a nonempty finite set (although most of the results extend to the case that $\mathscr{A}$ is infinite). We let $\mathscr{A}^*$ denote the set of all finite words $u = u_1u_2\cdots u_n$ with $u_i \in \mathscr{A}$. We call n the *length* of u and denote it $|u|$. The empty word is denoted ε and by convention $|\varepsilon| = 0$. For $u \in \mathscr{A}^*$ and $a \in \mathscr{A}$, we let $|u|_a$ denote the number of occurrences of a in u. We also let $\mathscr{A}^+ = \mathscr{A}^* \setminus \{\varepsilon\}$ denote the free semigroup generated by $\mathscr{A}$ consisting of all finite (non-empty) words $u_1u_2\cdots u_n$ with $u_i \in \mathscr{A}$. Given words $u, v \in \mathscr{A}^+$ we say v is a *border* of u if v is both a proper prefix and a proper suffix of u. In case u admits a border, we say u is *bordered*. Otherwise u is called *unbordered*.

Let $\mathscr{A}^{\mathbb{N}}$ denote the set of all infinite words $x = x_1x_2x_3\cdots$ with $x_i \in \mathscr{A}$. We endow $\mathscr{A}^{\mathbb{N}}$ with the topology generated by the metric

$$d(s,t) = \frac{1}{2^n} \text{ where } n = \inf\{k \mid s_k \neq t_k\}$$

whenever $s = (s_n)_{n\ge1}$ and $t = (t_n)_{n\ge1}$ are two elements of $\mathscr{A}^{\mathbb{N}}$. The resulting topology is generated by the collection of *cylinders* $[a_1,\ldots,a_n]$, where for each $n > 0$ and $a_i \in \mathscr{A}$, $1 \le i \le n$,

$$[a_1,\ldots,a_n] = \{s \in \mathscr{A}^{\omega} \mid s_i = a_i \text{ for } 1 \le i \le n\}.$$

It can also be described as being the product topology on $\mathscr{A}^{\mathbb{N}}$ with the discrete topology on $\mathscr{A}$. In particular this topology is compact. Each probability vector $\mathbf{p} = (p_i)_{i \in \mathscr{A}} \in [0;1]^{\mathscr{A}}$ determines a measure $\mu_{\mathbf{p}}$ (*Bernoulli measure* corresponding to $\mathbf{p}$) defined as the unique measure on the σ-algebra of $\mathscr{A}^{\omega}$ such that $\mu_{\mathbf{p}}([a_1, \ldots, a_n]) = p_{a_1} \cdots p_{a_n}$. By the *standard Bernoulli measure* on $\mathscr{A}^{\mathbb{N}}$, we mean the Bernoulli measure corresponding to the probability vector $\mathbf{p} = (\frac{1}{d}, \ldots, \frac{1}{d})$, with $d = \mathrm{Card}(\mathscr{A})$. For $x \in A^{\mathbb{N}}$, we let $\Omega(x)$ denote the shift orbit closure of x, i.e., the closure in $\mathscr{A}^{\mathbb{N}}$ of $\{x_n x_{n+1} x_{n+2} \cdots : n \geq 1\}$.

Given $x = x_1 x_2 x_3 \cdots \in \mathscr{A}^{\mathbb{N}}$ and $u \in \mathscr{A}^+$, let $x\big|_u$ denote the set of all occurrences of u in x, i.e.,

$$x\big|_u = \{n \mid x_n x_{n+1} \cdots x_{n+|u|-1} = u\}.$$

A word $u \in \mathscr{A}^+$ is called a *factor* of x if $x\big|_u \neq \emptyset$. A factor u of x is called *recurrent* if $x\big|_u$ is infinite and *uniformly recurrent* if $x\big|_u$ is *syndetic*, of bounded gaps. An infinite word x is called *recurrent* (resp., *uniformly recurrent*) if each of its factors is recurrent (resp., uniformly recurrent).

We say $x \in \mathscr{A}^{\mathbb{N}}$ is *periodic* if $x = u^{\omega} = uuu \cdots$ for some $u \in \mathscr{A}^+$, and *eventually periodic* if $x = uv^{\omega}$ for some $u, v \in \mathscr{A}^+$. We say x is *aperiodic* if x is not eventually periodic. An infinite word $x \in \mathscr{A}^{\mathbb{N}}$ is called *Lyndon* if there exists a linear order on $\mathscr{A}$ with respect to which x is lexicographically smaller than all its proper suffixes. In particular, Lyndon words are not periodic, although they may be ultimately periodic, e.g., ab^{ω}.

A word $x \in \{0,1\}^{\mathbb{N}}$ is called *Sturmian* (*cf.*, Chapter 2 in [385]) if it is aperiodic and *balanced*, i.e., for all factors u and v of x such that $|u| = |v|$ one has

$$||u|_a - |v|_a| \leq 1, \; a \in \{0,1\}.$$

It follows that each Sturmian word contains exactly one of the two factors 00 and 11. Alternatively, a binary infinite word x is Sturmian if x has a unique left (or equivalently right) special factor of length n for each integer $n \geq 0$. This is equivalent to saying that for each $n \geq 0$ the number of distinct factors of x of length n is exactly equal to $n + 1$. As a consequence one derives that a Sturmian word x is closed under reversal, i.e., if u is a factor of x, then so is its reversal $u^{\sim}$ (see, for instance, Proposition 2.1.19 in [385]). The most famous Sturmian word is the Fibonacci word $f = 0100101001001010010 \cdots$ which is the fixed point of the morphism F defined by $F : 0 \mapsto 01, 1 \mapsto 0$.

6.3 A Coloring Problem

We begin this section with some simple examples illustrating the content of Question 6.1.1. Let $x \in \mathscr{A}^{\mathbb{N}}$ be a Lyndon word and define $\varphi : \mathscr{A}^+ \to \{0,1\}$ by

$$\varphi(u) = \begin{cases} 0 & \text{if } u \text{ is a prefix of } x; \\ 1 & \text{otherwise} \end{cases} \tag{6.1}$$

We claim that no factorization of x is φ-monochromatic. In fact, suppose to the contrary that $x = u_1u_2u_3\cdots$ is a φ-monochromatic factorization of x. Since u_1 is a prefix of x, it follows that $\varphi(u_1) = 0$ and hence $\varphi(u_i) = 0$ for all $i \geq 1$. In other words, each u_i is a prefix of x and hence is lexicographically smaller than any other factor of x of equal length. It follows that $u_2u_3u_4\cdots$ is lexicographically less or equal to x, a contradiction. Thus φ defines a separating 2-coloring of x.

The coloring rule φ in (6.1) defines a separating coloring for any infinite word which does not admit a prefixal factorization. This includes, for instance, all words $x \in \mathscr{A}^{\mathbb{N}}$ in which the first symbol of x does not occur in x in bounded gaps:

Lemma 6.3.1. *Let $x \in \mathscr{A}^{\omega}$ be an infinite word having a prefixal factorization. Then the first letter of x is uniformly recurrent.*

Proof. Let $x = u_1u_2u_3\cdots$ be a prefixal factorization and let a denote the first letter of x. Then every factor u of x of length $> |u_1|$ contains an occurrence of a. In fact, the first occurrence of u cannot be contained in any u_i, $i \geq 1$, and hence this first occurrence of u must either contain some u_i or must overlap two adjacent u_j's. In either case, a occurs in u. Thus the first letter of x is uniformly recurrent.

The following gives a characterization of words admitting a prefixal factorization:

Lemma 6.3.2. *A word $x \in \mathscr{A}^{\mathbb{N}}$ admits a prefixal factorization if and only if x begins with only finitely many unbordered prefixes.*

Proof. Suppose that x admits a prefixal factorization $x = u_1u_2u_3\cdots$. Any prefix u of x of length greater than $|u_1|$ is bordered. Indeed, we can write $u = u_1\cdots u_ku'$ for a suitable $k \geq 1$ and u' a proper prefix of u_{k+1} and then of x. If $u' \neq \varepsilon$, then u', as $|u'| < |u|$, is a proper prefix and suffix of u, i.e., a border of u. If $u' = \varepsilon$, then u_k is a border of u. Hence, u is bordered. For the converse, let u be the longest unbordered prefix of x, and put $m = |u|$. Let S be the set of all nonempty prefixes of x of length at most m. Then every prefix v of x can be written as a product $v = v_1v_2\cdots v_k$ where each v_i, $1 \leq i \leq k$, is in S. This is clear if $|v| \leq m$. If $|v| > m$, then v is bordered so we can write $v = v'v''$ where v' and v'' are both nonempty prefixes of x. By induction on the length of v, each of v' and v'' is a product of elements of S, whence v is a product of elements of S. By the pigeonhole principle, infinitely many prefixes of x begin with the same u_1 in S. Of those infinitely many begin with the same u_1u_2 with u_2 in S. Of those infinitely many begin with the same $u_1u_2u_3$ with u_3 in S. Continuing we get $x = u_1u_2u_3\cdots$ where each u_i, $i \geq 1$, is in S (note that this proof is just an application of the usual König infinity lemma).

In contrast, the 2-coloring rule φ in (6.1) does not define a separating coloring of the Fibonacci word

$$x = 0100101001001010010100\cdots$$

In fact the factorization

$$x = 01 \cdot 0 \cdot 01 \cdot 01 \cdot 0 \cdot 01 \cdot 0 \cdot 01 \cdot 01 \cdot 0 \cdot 01 \cdots$$

according to first returns to 0 is φ-monochromatic. On the other hand, let us consider the 3-coloring $\varphi : \{0, 1\}^+ \to \{0, 1, 2\}$ defined by:

$$\varphi(u) = \begin{cases} 0 & \text{if } u \text{ is a prefix of } x \text{ ending with } 0; \\ 1 & \text{if } u \text{ is a prefix of } x \text{ ending with } 1; \\ 2 & \text{if } u \text{ is not a prefix of } x. \end{cases} \tag{6.2}$$

We claim that no factorization of x is φ-monochromatic. In fact, suppose that $x = u_1u_2u_3 \cdots$ is a φ-monochromatic factorization of x. Then each u_i is a prefix of x terminating in the same letter $a \in \{0, 1\}$. Whence the factorization $au_1a^{-1} \cdot au_2a^{-1} \cdot au_3a^{-1} \cdots$ defines a prefixal factorization of the Lyndon word ax, a contradiction.

Next consider the *Thue–Morse* infinite word

$$x = 0110100110010110100101\cdots$$

where the nth term of x (starting from $n = 0$) is defined as the sum modulo 2 of the digits in the binary expansion of n. The origins of this word go back to the beginning of the last century with the works of A. Thue [562, 563] in which he proves among other things that x is *overlap-free*, i.e., x contains no factor of the form uuu' where u' is a nonempty prefix of u. We claim that the coloring rule $\varphi : \{0, 1\}^+ \to \{0, 1, 2\}$ in (6.2) also defines a separating 3-coloring for the Thue–Morse word. In fact, suppose to the contrary that $x = u_1u_2u_3 \cdots$ is a φ-monochromatic factorization of x. Since u_1 is a prefix of x, it follows that $\varphi(u_1) \in \{0, 1\}$, i.e., there exists $a \in \{0, 1\}$ such that each u_i is a prefix of x terminating with a. Pick $i \geq 2$ such that $|u_i| \leq |u_{i+1}|$. Then as each u_i is a prefix of x, it follows that u_i is a prefix of u_{i+1} and hence au_iu_i is a factor of x. Writing $u_i = va$, (with v empty or in $\{0, 1\}^+$), we have $au_iu_i = avava$ which is an overlap, contradicting that x is overlap-free. This proves that there exists a separating 3-coloring for the Thue–Morse word. S. Avgustinovich and O. Parshina proved that it is possible to color $\{0, 1\}^+$ using only two colors in such a way that no factorization of the Thue–Morse word is monochromatic.

Let us remark that since the Thue–Morse word x is not periodic, each proper suffix of x begins in some factor which is not a prefix of x. This means that each proper suffix x' of x may be written as an infinite concatenation $x' = u_1u_2u_3 \cdots$ with $\varphi(u_i) = 2$ for all $i \geq 1$, where φ is the 3-coloring of $\{0, 1\}^+$ defined above. Moreover, this monochromatic factorization of x' has an even stronger monochromatic property: The set $\{u_n : n \geq 1\}^+$ is also φ-monochromatic (each element has φ color equal to 2). This is because each element of $\{u_n : n \geq 1\}$ is a non-prefix of x and hence the same is true of any concatenation formed by elements from this set. As we shall see later, a weaker version of this phenomenon is true in greater generality: Given any $x \in A^{\mathbb{N}}$ and any finite coloring $\varphi : \mathscr{A}^+ \to C$, one can always find a suffix x' of x which admits a factorization $x' = u_1u_2u_3\cdots$ where $\varphi(u_i\cdots u_j) = \varphi(u_1)$ for all $1 \leq i \leq j$. This fact may be obtained via a straightforward application of the infinite Ramsey theorem [491] (see [110, 393] or see [532] for a proof by M. P. Schützenberger which does not use Ramsey's theorem).

Various partial results in support of an affirmative answer to Question 6.1.1 were obtained in [73, 390, 393, 393, 519]. For instance, in [390], it is shown that Question 6.1.1 admits an affirmative answer for all nonuniformly recurrent words and various classes of uniformly recurrent words including Sturmian words. In [519], V. Salo and I. Törmä proved that for every aperiodic linearly recurrent word $x \in \mathscr{A}^{\mathbb{N}}$, there exists a finite coloring of $\mathscr{A}^+$ relative to which x does not admit a monochromatic factorization into factors of increasing lengths. And recently A. Bernardino, R. Pacheco and M. Silva [73] proved that Question 6.1.1 admits an affirmative answer for fixed points of strongly recognizable primitive substitutions. In addition to the fact that these partial results apply only to a restricted class of nonperiodic words (e.g., Sturmian words or fixed points of certain primitive substitutions), in most cases the number of colors required to color $\mathscr{A}^+$ to avoid a monochromatic factorization of x is determined to be quite large. For instance, in [519], the authors proved that if $x \in \mathscr{A}^{\mathbb{N}}$ is an aperiodic linearly recurrent word, then there exists a constant $K \geq 2$ and a coloring $\varphi : \mathscr{A}^+ \to C$ with $\mathrm{Card}(C) = 2 + \sum_{i=0}^{K^5-1} 2K^i(K+1)^{2i}$, such that no factorization of $x = u_1u_2u_3\cdots$, verifying the additional constraint that $|u_i| \leq |u_{i+1}|$ for each $i \geq 1$, is φ-monochromatic. The constant K above is chosen such that for every factor u of x, every first return w to u satisfies $|w| \leq K|u|$ (see, for instance, [205]). A similar large bound depending on the recognizability index of a substitution is obtained in [73] in the context of fixed points of strongly recognizable substitutions. In contrast, it is shown in [390] that every Sturmian word admits a separating 3-coloring.

The following theorem proved in [589] gives a complete and optimal affirmative answer to Question 6.1.1 by showing that for every nonperiodic word $x = x_1x_2x_3\cdots \in \mathscr{A}^{\mathbb{N}}$, there exists a 2-coloring $\varphi : \mathscr{A}^+ \to \{0, 1\}$ relative to which no factorization of x is φ-monochromatic. Moreover, this is a characterization of periodicity of infinite words:

Theorem 6.3.3. *Let $x = x_1x_2x_3\cdots \in \mathscr{A}^{\mathbb{N}}$ be an infinite word. Then x is periodic if and only if for every 2-coloring $\varphi : \mathscr{A}^+ \to \{0, 1\}$ there exists a φ-monochromatic factorization of x, i.e., a factorization $x = u_1u_2u_3\cdots$ such that $\varphi(u_i) = \varphi(u_j)$ for all $i, j \geq 1$.*

Proof. First assume $x = x_0x_1x_2\cdots \in \mathscr{A}^{\mathbb{N}}$ is periodic, i.e., $x \in \{u\}^{\mathbb{N}}$ for some $u \in \mathscr{A}^+$. Then the factorization $x = u_1u_2u_3\cdots$ with each $u_i = u$ is φ-monochromatic for any choice of $\varphi : \mathscr{A}^+ \to \{0, 1\}$. Next assume x is not periodic. We define a 2-coloring $\varphi : \mathscr{A}^+ \to \{0, 1\}$ with the property that no factorization of x

is φ-monochromatic. Pick any total order on the set $\mathscr{A}$ and let $\prec$ denote the induced lexicographic order on $\mathscr{A}^+$ and $\mathscr{A}^{\mathbb{N}}$. For $u, v \in \mathscr{A}^+$ with $|u| = |v|$, we write $u \preccurlyeq v$ if either $u \prec v$ or $u = v$. For each $n \geq 1$, let $P_x(n)$ denote the prefix of x of length n, and for each $y \in \mathscr{A}^{\mathbb{N}}$, let $x \wedge y$ denote the longest common prefix of x and y. Define $\varphi : \mathscr{A}^+ \to \{0, 1\}$ by the following rule:

If u is not a prefix of x, then set

$$\varphi(u) = \begin{cases} 0 & \text{if } u \prec P_x(|u|) \\ 1 & \text{if } P_x(|u|) \prec u. \end{cases} \tag{6.3}$$

If u is a prefix of x, say $x = uy$ with $y \in \mathscr{A}^{\mathbb{N}}$, then set

$$\varphi(u) = \begin{cases} 0 & \text{if } x \prec y \\ 1 & \text{if } y \prec x. \end{cases} \tag{6.4}$$

Since x is not periodic, for every proper suffix y of x, we have either $x \prec y$ or $y \prec x$, whence φ is well defined. We observe that φ has the following key property: If u is any nonempty prefix of x, say $x = uy$ with $y \in \mathscr{A}^{\mathbb{N}}$, then there exists $N \geq 0$ such that $\varphi(u) \neq \varphi(v)$ for every prefix v of y with $|v| > N$. In fact, let $N = |x \wedge y|$. Then if v is a prefix of y with $|v| > N$, then $v \neq P_x(|v|)$. So if $\varphi(u) = 0$, then $x \prec y$ and hence $P_x(|v|) \prec v$, whence $\varphi(v) = 1$. Similarly if $\varphi(u) = 1$, then $y \prec x$ and hence $v \prec P_x(|v|)$, whence $\varphi(v) = 0$. The other key property of φ is given by the following lemma:

Lemma 6.3.4. *For all $u \in \mathscr{A}^+$, if $u = u_1u_2\cdots u_k$ with $u_i \in \mathscr{A}^+$, then there exists $1 \leq i \leq k$ such that $\varphi(u_i) = \varphi(u)$.*

Proof. Fix $a \in \{0, 1\}$. We will prove by induction on k that if $u = u_1u_2\cdots u_k$ and $\varphi(u_i) = a$ for all $1 \leq i \leq k$, then $\varphi(u) = a$. Without loss of generality, we may assume $a = 0$ for otherwise we could replace $\prec$ by the reverse order. For $k = 1$ the result is immediate. Next we consider the case $k = 2$, i.e., we assume $u = u_1u_2$ and $\varphi(u_1) = \varphi(u_2) = 0$, and we will show that $\varphi(u) = 0$. If u_1 is not a prefix of x, then $u_1 \prec P_x(|u_1|)$ and hence $u \prec P_x(|u|)$, whence $\varphi(u) = 0$. Thus we may suppose that u_1 is a prefix of x. Let u_2' be such that u_1u_2' is a prefix of x and $|u_2| = |u_2'|$. If $u_2 \prec u_2'$, then $u = u_1u_2 \prec u_1u_2' = P_x(|u|)$, whence $\varphi(u) = 0$. Thus we may suppose that $u_2' \preceq u_2$. Also since $\varphi(u_1) = 0$, it follows that $P_x(|u_2'|) \preceq u_2'$, whence $P_x(|u_2'|) \preceq u_2' \preceq u_2$. If any one of these two inequalities were strict, then $P_x(|u_2|) \prec u_2$ which would imply that $\varphi(u_2) = 1$, a contradiction. Thus we must have $u_2 = u_2' = P_x(|u_2|)$, i.e., both u and u_2 are prefixes of x. Let $x' \in \mathscr{A}^{\mathbb{N}}$ be such that $x = u_1x'$ and let $w \in \mathscr{A}^+$ denote the longest common prefix between x and x'. Since $\varphi(u_1) = 0$, there exist symbols $a, b \in \mathscr{A}$ with $a \prec b$ such that u_1wb and wa are each a prefix of x. Also since u_2 is a prefix of both x and x', we have $|u_2| \leq |w|$ and hence we can write $w = u_2v$ for some $v \in \mathscr{A}^*$. So each of $u_1u_2vb = uvb$ and u_2va is a prefix of x. Finally since $\varphi(u_2) = 0$, we have that $P_x(|va|) \preceq va \prec vb$, whence $\varphi(u) = 0$. This completes the case $k = 2$. Now, let $n > 2$ and suppose the result of the lemma holds for all $k < n$, and suppose $u = u_1u_2\cdots u_n$ with $\varphi(u_i) = a$ for all $1 \le i \le n$. Then by considering the factorization $u = (u_1u_2)u_3\cdots u_n$ of length $n-1$ and the fact that $\varphi(u_1u_2) = a$ (which comes from the case $k = 2$), we deduce by induction hypothesis that $\varphi(u) = a$ as required.

To complete the proof of Theorem 6.3.3, we will show that no factorization of x is φ-monochromatic. Let $x = u_1u_2u_3\cdots$ with $u_i \in \mathscr{A}^+$. Pick $N \ge 0$ such that $\varphi(v) \ne \varphi(u_1)$ for all prefixes v of $u_2u_3\cdots$ with $|v| > N$. Pick $k \ge 2$ such that $|u_2\cdots u_k| > N$. By the previous lemma, there exists $2 \le i \le k$ such that $\varphi(u_i) = \varphi(u_2\cdots u_k)$. Hence $\varphi(u_i) = \varphi(u_2\cdots u_k) \ne \varphi(u_1)$.

Theorem 6.3.3 has several immediate consequences. Let $x \in \mathscr{A}^{\mathbb{N}}$ and $a \in \mathscr{A}$. A factor u of x is said to be *rich* in a if $|u|_a \ge |v|_a$ for all factors v of x with $|v| = |u|$. Theorem 6.7 in [390] states that a Sturmian word $x \in \{0,1\}^{\mathbb{N}}$ does not admit a factorization of the form $x = u_1u_2u_3\cdots$ where each u_i is a prefix of x rich in the same letter $a \in \{0,1\}$. The following consequence of Theorem 6.3.3 generalizes this result to all binary nonperiodic words:

Corollary 6.3.5. *Let $x \in \{0,1\}^{\mathbb{N}}$ and $a \in \{0,1\}$. Suppose x admits a prefixal factorization $x = u_1u_2u_3\cdots$ with each u_i rich in a, i.e., $|u_i|_a \ge |v|_a$ whenever v is a factor of x of length $|v| = |u_i|$. Then x is periodic.*

Proof. Suppose to the contrary that x is not periodic. Let $\varphi : \{0,1\}^+ \to \{0,1\}$ be the separating 2-coloring for x defined in (6.3) and (6.4) relative to the order on $\{0,1\}$ where a is taken to be the least element. For each $i \ge 1$, writing $x = u_iy_i$ with $y_i \in \{0,1\}^{\mathbb{N}}$, we claim that $x \prec y_i$. Otherwise if $y_i \prec x$, then we can write $x = zbx' = u_izay'$ for some $z \in \{0,1\}^*$, $x', y' \in \{0,1\}^{\mathbb{N}}$ and where $\{a,b\} = \{0,1\}$. But then the factor u_i' of length $|u_i|$ immediately preceding the suffix y' of x would contain one more occurrence of the symbol a than u_i, contradicting that u_i was rich in a. Having established that $x \prec y_i$, it follows that $\varphi(u_i) = 0$ for all $i \ge 1$ contradicting that φ is a separating 2-coloring for x.

The following consequence of Corollary 6.3.5 is proved in [589]:

Corollary 6.3.6. *Let $\mathscr{A}$ be an arbitrary nonempty set, $x \in \mathscr{A}^{\mathbb{N}}$ and $a \in \mathscr{A}$. Suppose x admits a prefixal factorization $x = u_1u_2u_3\cdots$ with each u_i rich in a. Then $\{n \in \mathbb{N} : x_n = a\}$ is a finite union of (infinite) arithmetic progressions.*

The following two corollaries are also immediate consequences of Theorem 6.3.3 (see [589]):

Corollary 6.3.7. *Let $x = x_1x_2x_3\cdots \in \mathscr{A}^{\mathbb{N}}$ and k be a positive integer. Let $B \subseteq \mathscr{A}^+$ with* $\mathrm{Card}(B) \ge 2k-1$. *Suppose that x factors over every k-element subset A of B. Then x is periodic.*

Proof. Let us assume to the contrary that x is not periodic. By Theorem 6.3.3, there exists a 2-coloring $\varphi : \mathscr{A}^+ \to \{0,1\}$ relative to which no factorization of x is φ-monochromatic. Let $\Sigma_k(B)$ denote the set of all k-element subsets of B. By assumption x factors over each $A \in \Sigma_k(B)$. On the other hand, since

$\mathrm{Card}(B) \geq 2k-1$, it follows that there exists a φ-monochromatic subset $A \in \Sigma_k(B)$. This gives rise to a φ-monochromatic factorization of x, a contradiction.

Corollary 6.3.8. *Let $x = x_1x_2x_3 \cdots \in \mathscr{A}^{\mathbb{N}}$ and k be a positive integer. Let $u_1, u_2, u_3, \ldots, u_{2k+1}$ be words in A^+ and suppose $x \in \{u_1, u_2\}^{\mathbb{N}} \cap \{u_2, u_3\}^{\mathbb{N}} \cap \cdots \cap \{u_{2k}, u_{2k+1}\}^{\mathbb{N}} \cap \{u_{2k+1}, u_1\}^{\mathbb{N}}$. Then x is periodic.*

Proof. Suppose to the contrary that x is not periodic. Pick any separating 2-coloring $\varphi : \mathscr{A}^+ \to \{0, 1\}$ for x. Then $\varphi(1) = \varphi(2i+1)$ for each $1 \leq i \leq k$. Thus $x \notin \{u_{2k+1}, u_1\}^{\mathbb{N}}$, a contradiction.

Theorem 6.3.3 may be reformulated in the language of ultrafilters. Given a nonempty set S, let $\mathscr{P}(S)$ denote the set of all subsets of S. Recall that a subset $p \subseteq \mathscr{P}(S)$ is called a *filter* on S if

- $S \in p$ and $\emptyset \notin p$
- If $A \in p$ and $B \in p$ then $A \cap B \in p$
- If $A \subseteq B$ and $A \in p$ then $B \in p$.

A filter p on S is called an *ultrafilter* if for all $A \in \mathscr{P}(S)$ either $A \in p$ or $A^c \in p$ where A^c denotes the complement of A, i.e., $A^c = S \setminus A$. Equivalently, a filter p is an ultrafilter if for each $A \in p$ whenever $A = A_1 \cup \cdots \cup A_n$, we have that at least one $A_i \in p$. Each $x \in S$ determines an ultrafilter $e(x)$ on S defined by $e(x) = \{A \subseteq S : x \in A\}$. An ultrafilter p on S is called *principal* if $p = e(x)$ for some $x \in S$. Otherwise p is said to be *free*. Let βS denote the collection of all ultrafilters p on S. By identifying each $x \in S$ with the principal ultrafilter $e(x)$, we regard $S \subseteq \beta S$. If S is infinite, then a straightforward application of Zorn's lemma guarantees the existence of free ultrafilters on S.

Given $A \subseteq S$, we put $\overline{A} = \{p \in \beta S : A \in p\}$. Then $\{\overline{A} : A \subseteq S\}$ defines a basis for a topology on βS relative to which βS is both compact and Hausdorff and the mapping $x \mapsto e(x)$ defines an injection $S \hookrightarrow \beta S$ whose image is dense in βS. In fact, if S is given the discrete topology, then βS is identified with the Stone-Čech compactification of S : Any continuous mapping from $f : S \to K$, where K is a compact Hausdorff space, lifts uniquely to a continuous mapping $\beta f : \beta S \to K$. Of special interest is the case in which S is a discrete semigroup. In this case the operation on S extends uniquely to βS making βS a right topological semigroup with S contained in its topological center. This means that $\rho_p : \beta S \to \beta S$, defined by $\rho_p(q) = q \cdot p$, is continuous for each $p \in \beta S$ and $\lambda_s : \beta S \to \beta S$, defined by $\lambda_s(q) = s \cdot q$, is continuous for each $s \in S$. The operation $\cdot$ on βS is defined as follows: For $p, q \in \beta S$

$$p \cdot q = \{A \subseteq S : \{s \in S : s^{-1}A \in q\} \in p\}, \tag{6.5}$$

where $s^{-1}A = \{t \in S : st \in A\}$. As a consequence of the Ellis–Numakura lemma, βS contains an idempotent element, i.e., an element p verifying $p \cdot p = p$ (see, for instance, [293]). Subsets $A \subseteq S$ belonging to idempotents in βS have rich combinatorial structures: Let $\mathrm{Fin}(\mathbb{N})$ denote the set of all finite subsets of $\mathbb{N}$. Given an infinite sequence $\langle s_n \rangle_{n \in \mathbb{N}}$ in S, let

$$FP(\langle s_n\rangle_{n\in\mathbb{N}}) = \{\prod_{n\in F} s_n : F \in \mathrm{Fin}(\mathbb{N})\}$$

where for each $F \in \mathrm{Fin}(\mathbb{N})$, the product $\prod_{n\in F} s_n$ is taken in increasing order of indices. A subset A of S is called an *IP-set* if A contains $FP(\langle s_n\rangle_{n\in\mathbb{N}})$ for some infinite sequence $\langle s_n\rangle_{n\in\mathbb{N}}$ in S. IP-sets are characterized as belonging to idempotent elements: $A \subseteq S$ is an IP-set if and only if A belongs to some idempotent element of βS (see, for instance, Theorem 5.12 in [293]).

For $x = x_1x_2x_3\cdots \in \mathscr{A}^{\mathbb{N}}$, we set

$$\mathscr{F}(x) = \{A \subseteq \mathscr{A}^+ : x \in A^{\mathbb{N}}\}$$

and

$$\mathscr{U}(x) = \{p \in \beta\mathscr{A}^+ : p \subseteq \mathscr{F}(x)\}.$$

Thus $p \in \mathscr{U}(x)$ if and only if x factors over every $A \subseteq \mathscr{A}^+$ belonging to p. The following reformulation of Theorem 6.3.3 states that x is periodic if and only if $\mathscr{U}(x)$ is non-empty:

Theorem 6.3.9. *Let $x = x_1x_2x_3\cdots \in \mathscr{A}^{\mathbb{N}}$ be an infinite word. Then x is periodic if and only if there exists $p \in \beta\mathscr{A}^+$ such that for each $A \in p$, there exists a factorization $x = u_1u_2u_3\cdots$ with each $u_i \in A$.*

Proof. Suppose x is periodic, i.e., $x \in \{u\}^{\mathbb{N}}$ for some $u \in \mathscr{A}^+$. Then the principal ultrafilter $e(u) = \{A \subseteq \mathscr{A}^+ : u \in A\}$ clearly belongs to $\mathscr{U}(x)$. We note that $\mathscr{U}(x)$ also contains free elements of $\beta\mathscr{A}^+$. In fact, for each $i \geq 1$, let $A_i = \{u^j : j \geq i\}$. Then $\{A_i : i \geq 1\}$ satisfies the finite intersection property whence there exists $p \in \beta\mathscr{A}^+$ containing $\{A_i : i \geq 1\}$, and any such p is a free ultrafilter belonging to $\mathscr{U}(x)$. Conversely, suppose there exists $p \in \beta\mathscr{A}^+$ such that for each $A \in p$, there exists a factorization $x = u_1u_2u_3\cdots$ with each $u_i \in A$. Let $\varphi : \mathscr{A}^+ \to \{0, 1\}$ be any 2-coloring of $\mathscr{A}^+$. We will show that x admits a φ-monochromatic factorization. The result then follows from Theorem 6.3.3. Consider the partition $\mathscr{A}^+ = \varphi^{-1}(0) \cup \varphi^{-1}(1)$. Since $\mathscr{A}^+ \in p$, it follows that $\varphi^{-1}(a) \in p$ for some $a \in \{0, 1\}$. Thus there exists a factorization $x = u_1u_2u_3\cdots$ with each $u_i \in \varphi^{-1}(a)$. In other words, x admits a φ-monochromatic factorization.

The ultrafilter p in Theorem 6.3.9 may be taken to be an idempotent element of $\beta\mathscr{A}^+$.

Theorem 6.3.10. *Let $x = x_0x_1x_2\cdots \in \mathscr{A}^{\mathbb{N}}$. Then the following are equivalent:*

i) x is periodic.
ii) $\mathscr{U}(x)$ is a closed sub-semigroup of $\beta\mathscr{A}^+$.
iii) $\mathscr{U}(x)$ contains an idempotent element.
iv) The set Pref(x) *consisting of all prefixes of x is an IP-set.*

Proof. We first note that for each infinite word x, the set $\mathscr{U}(x)$ is a closed subset of $\beta\mathscr{A}^+$. In fact, suppose $p \in \beta\mathscr{A}^+ \setminus \mathscr{U}(x)$, then there exists $A \in p$ with $A \notin \mathscr{F}(x)$. Then $\overline{A}$ is an open neighborhood of p and any $q \in \overline{A}$ contains the set A and hence is not in $\mathscr{U}(x)$. To see that *i)* $\Longrightarrow$ *ii)*, suppose that x is periodic. By Theorem 6.3.9, we have $\mathscr{U}(x) \neq \emptyset$. It remains to show that $p \cdot q \in \mathscr{U}(x)$ whenever $p, q \in \mathscr{U}(x)$. Pick the shortest $u \in \mathscr{A}^+$ such that $x \in \{u\}^{\mathbb{N}}$ and set $\mathfrak{A} = \{u^j : j \in \mathbb{N}\}$. Then $p \in \mathscr{U}(x)$ if and only if $\mathfrak{A} \in p$. Let $A \in p \cdot q$ with $p, q \in \mathscr{U}(x)$. Then by (6.5) we have $\{s \in \mathscr{A}^+ : s^{-1}A \in q\} \in p$. Since $p \in \mathscr{U}(x)$ it follows that $\{s \in \mathscr{A}^+ : s^{-1}A \in q\} \cap \mathfrak{A} \neq \emptyset$. Pick $n \in \mathbb{N}$ such that $u^n \in \{s \in \mathscr{A}^+ : s^{-1}A \in q\}$. Then $\{t \in \mathscr{A}^+ : u^n t \in A\} \in q$ and since $q \in \mathscr{U}(x)$ it follows that there exists $m \in \mathbb{N}$ with $u^m \in \{t \in \mathscr{A}^+ : u^n t \in A\}$. In other words $u^{n+m} \in A$ and hence x factors over A. Thus $p \cdot q \in \mathscr{U}(x)$. The implication *ii)* $\Longrightarrow$ *iii)* follows from the Ellis–Numakura lemma. To see *iii)* $\Longrightarrow$ *iv)* pick an idempotent element $p \in \mathscr{U}(x)$. We note that Pref(x) belongs to every $q \in \mathscr{U}(x)$. In fact, suppose that Pref$(x) \notin q$, for some $q \in \mathscr{U}(x)$. Then $\mathscr{A}^+ \setminus$ Pref$(x) \in q$ which implies that x factors over $\mathscr{A}^+ \setminus$ Pref(x). But this is a contradiction since in any factorization of x, the first term occurring in the factorization belongs to Pref(x). Thus in particular Pref$(x) \in p$. Since p is an idempotent, it follows that Pref(x) is an IP-set. Finally, to see that *iv)* $\Longrightarrow$ *i)* assume that Pref(x) is an IP-set. Then Pref(x) contains $FP(\langle s_n \rangle_{n\in\mathbb{N}})$ for some infinite sequence $\langle s_n \rangle_{n\in\mathbb{N}}$ of prefixes of x. This means that for each $n \geq 2$, both $s_1 s_2 \cdots s_n$ and $s_2 \cdots s_n$ are prefixes of x. Thus $x = s_1 x$ which implies that x is periodic.

6.4 Variations on the Coloring Problem

Let $x = 0110100110010110100 10\cdots \in \{0,1\}^{\mathbb{N}}$ denote the Thue–Morse infinite word. Recall that the coloring rule $\varphi : \{0,1\}^+ \to \{0,1,2\}$ in (6.2) satisfies the following very strong monochromatic property: For every proper suffix x' of x, there exists a factorization $x' = u_1 u_2 u_3 \cdots$ such that the free semigroup $S = \{u_i : i \geq 1\}^+$ is φ-monochromatic. In general, this is way too strong of a condition. In fact, given any nonempty set $\mathscr{A}$, let us consider the 2-coloring of $\nu : \mathscr{A}^+ \to \{0,1\}$ given by $\nu(u) = \nu_2(|u|) \bmod 2$ where $\nu_2(|u|)$ denotes the 2-adic valuation of $|u|$. Then for any $u \in \mathscr{A}^+$, we have that $\nu(u) \neq \nu(u^2)$ and hence no sub-semigroup of $\mathscr{A}^+$ is ν−monochromatic. However, we mentioned earlier that a weaker monochromatic property does hold in greater generality. Let $x \in \mathscr{A}^{\mathbb{N}}$ and $\varphi : \mathscr{A}^+ \to C$ be a finite coloring of $\mathscr{A}^+$. A factorization $x = u_1 u_2 u_3 \cdots$ with $u_i \in \mathscr{A}^+$ is said to be *sequentially monochromatic* if $\exists c \in C$ such that $\varphi(V_i V_{i+1} \cdots V_{i+j}) = c$ for all $i, j \geq 0$. The following result was first proved in [393]:

Theorem 6.4.1. *The following statements are equivalent:*

i) For any finite coloring $\varphi : \mathscr{A}^+ \to C$ and any word $x \in \mathscr{A}^{\mathbb{N}}$, there exists a suffix x' of x which admits a φ-sequentially monochromatic factorization.

ii) For any finite coloring $\varphi : \Sigma_2(\mathbb{N}) \to C$*, where* $\Sigma_2(\mathbb{N})$ *denotes the set of all two-element subsets of* $\mathbb{N}$*, there exist* $c \in C$ *and an infinite set* $\mathscr{N} \subseteq \mathbb{N}$ *such that* $\Sigma_2(\mathscr{N}) \subseteq \varphi^{-1}(c)$.

Proof. We note that item ii) is a special case of the Infinite Ramsey's Theorem (see [491]). We begin by showing that *ii)* $\Longrightarrow$ *i)*. Let $x \in \mathscr{A}^{\mathbb{N}}$ and $\varphi : \mathscr{A}^+ \to C$ be any finite coloring. Then φ induces a finite coloring $\varphi' : \Sigma_2(\mathbb{N}) \to C$ given by $\varphi'(\{m < n\}) = \varphi(x_m x_{m+1} \cdots x_{n-1})$. By (2) there exists $c \in C$ and an infinite subset $\mathscr{N} = \{n_0 < n_1 < n_2 < \cdots\}$ of $\mathbb{N}$ such that for all $m, n \in \mathscr{N}$ with $m < n$ we have $\varphi'(\{m < n\}) = c$. It follows that the factorization of the suffix $x' = x_{n_0} x_{n_0+1} x_{n_0+2} \cdots$ given by $x' = V_0 V_1 V_2 \cdots$ where $|V_i| = n_{i+1} - n_i$ is φ-sequentially monochromatic.

To see that *i)* $\Longrightarrow$ *ii)*, let $\varphi : \Sigma_2(\mathbb{N}) \to C$ be any finite coloring of $\Sigma_2(\mathbb{N})$. Let $x \in \{0, 1\}^\omega$ be any aperiodic word. Then φ induces a finite coloring $\varphi' : \mathscr{A}^+ \to C \cup \{*\}$, where $*$ denotes a symbol not in C, defined as follows: For each $u \in \mathscr{A}^+$, if $u \notin \mathrm{Fact}(x)$, then set $\varphi'(u) = *$. Otherwise, let $m(u)$ be the least natural number m such that $u = x_m x_{m+1} \cdots x_{m+|u|-1}$, that is, $m(u)$ is the first occurrence of u in x. Then we put

$$\varphi'(u) = \varphi(\{m(u), m(u) + |u|\}).$$

By (1) there exists $n \geq 0$ such that the suffix $x' = x_n x_{n+1} x_{n+2} \cdots$ of x admits a φ'-sequentially monochromatic factorization $x' = V_0 V_1 V_2 \cdots$. Put $c = \varphi'(V_0)$. Since $V_0 \in \mathrm{Fact}(x)$ we have $c \in C$. Also, as x is aperiodic, there exists $s \geq 0$ such that $n = m(V_0 V_1 \cdots V_s)$. Indeed, set $x = Ux'$ with $|U| = n$. The statement is clear if $n = 0$. Otherwise, if for each s, $m(V_0 V_1 \cdots V_s) < n$, then by the pigeonhole principle, there exists $0 \leq k < n$ such that $k = m(V_0 V_1 \cdots V_s)$ for infinitely many values of s. This implies that $x = T^k(x) = T^n(x)$, whence x is purely periodic and hence x is ultimately periodic, a contradiction.

Similarly, for each $r \geq 1$ there exists $s \geq r$ such that $m(V_r \cdots V_s) = n + \sum_{i=0}^{r-1} |V_i|$. Given any increasing sequence $0 = n_0 < n_1 < n_2 < \cdots$, put $W_k = V_{n_k} V_{n_k+1} \cdots V_{n_{k+1}-1}$. Then clearly the factorization $x' = W_0 W_1 W_2 \cdots$ is also φ'-sequentially monochromatic. Thus we can assume that x' admits a φ'-sequentially monochromatic factorization $x' = V_0 V_1 V_2 \cdots$, such that $m(V_0) = n$ and $m(V_r) = n + \sum_{i=0}^{r-1} |V_i|$ for each $r \geq 1$. Setting

$$\mathscr{N} = \{n < n + |V_0| < \cdots < n + \sum_{i=0}^{r} |V_i| < \cdots\},$$

we have $\Sigma_2(\mathscr{N}) \subseteq \varphi^{-1}(c)$ as required.

We consider the following strengthening of the notion of sequentially monochromatic factorizations:

Definition 6.4.2. Let $x \in \mathscr{A}^{\mathbb{N}}$ and $\varphi : \mathscr{A}^+ \to C$ be a finite coloring of $\mathscr{A}^+$. A factorization $x = u_1 u_2 u_3 \cdots$ with each $u_i \in \mathscr{A}^+$ is called

- φ*- super monochromatic* if $\exists c \in C$ such that $\varphi(u_{n_1}u_{n_2}\cdots u_{n_k}) = c$ for all $k \geq 1$ and $n_1 < n_2 < \cdots < n_k$.
- φ*-ultra monochromatic* if $\exists c \in C$ such that $\varphi(u_{n_{\sigma(1)}}u_{n_{\sigma(2)}}\cdots u_{n_{\sigma(k)}}) = c$ for all $k \geq 1$ and all $n_1 < n_2 < \cdots < n_k$ and all permutations σ of $\{1, 2, \cdots, k\}$.

Clearly any φ-super monochromatic factorization is φ-sequentially monochromatic, and any φ-ultra monochromatic factorization is φ-super monochromatic.

The following result, established jointly by Lionel Nguyen Van Thé and the second author, shows that given any infinite word $x \in \mathscr{A}^{\mathbb{N}}$ and any finite coloring $\varphi : \mathscr{A}^{\mathbb{N}} \to C$, there exists an element of the subshift generated by x admitting a φ-super monochromatic factorization:

Theorem 6.4.3. *Given an infinite word $x \in \mathscr{A}^{\mathbb{N}}$ and a finite coloring $\varphi : \mathscr{A}^+ \to C$, there exists an element z in the shift orbit closure of x admitting a φ-super monochromatic factorization.*

Proof. Fix an infinite word $x \in \mathscr{A}^{\mathbb{N}}$ and a finite coloring $\varphi : \mathscr{A}^+ \to C$. Pick a recurrent point $y \in \Omega(x)$ (shift orbit closure of x). As y is recurrent, there exists a factorization $y = u_1u_2u_3\cdots$ with $u_i \in \mathscr{A}^+$ such that for each $n \geq 1$, we have that $u_1\cdots u_n$ is a suffix of u_{n+1}. It follows by a simple induction argument that $u_{n_1}u_{n_2}\cdots u_{n_k}$ is a factor of y (and hence of x) for each $k \geq 1$ and $n_1 < n_2 < \cdots < n_k$. Let $\mathrm{Fin}(\mathbb{N})$ denote the set of all finite nonempty subsets of $\mathbb{N}$. Define a partial order on $\mathrm{Fin}(\mathbb{N})$ by $F < G$ if and only if $\max F < \min G$. For each $F = \{n_1, n_2, \ldots, n_k\} \in \mathrm{Fin}(\mathbb{N})$, with $n_1 < n_2 < \cdots < n_k$, set $u(F) = u_{n_1}u_{n_2}\cdots u_{n_k}$. Then for all $F, G \in \mathrm{Fin}(\mathbb{N})$, if $F < G$ then $u(F \cup G) = u(F)u(G)$. The finite coloring $\varphi : \mathscr{A}^+ \to C$ induces a finite coloring $\eta : \mathrm{Fin}(\mathbb{N}) \to C$ given by $\eta(F) = \varphi(u(F))$. We now make use of the so-called Finite Unions Theorem which is an immediate consequence of Hindman's theorem (see Corollary 5.17 in [293]) :

Theorem 6.4.4 (Infinite Finite Unions Theorem). *Let $\eta : \mathrm{Fin}(\mathbb{N}) \to C$. Then there exists a sequence $\langle F_t\rangle_{t=1}^{\infty}$ in $\mathrm{Fin}(\mathbb{N})$ with $F_1 < F_2 < F_3 < \cdots$ such that $\{\bigcup_{t\in H} F_t : H \in \mathrm{Fin}(\mathbb{N})\}$ is η-monochromatic.*

Applying Theorem 6.4.4 to the induced coloring $\eta : \mathrm{Fin}(\mathbb{N}) \to C$, we obtain an infinite sequence $\langle F_t\rangle_{t=1}^{\infty}$ in $\mathrm{Fin}(\mathbb{N})$ with $F_1 < F_2 < F_3 < \cdots$ such that $\{u(\bigcup_{t\in H} F_t) : H \in \mathrm{Fin}(\mathbb{N})\}$ is φ-monochromatic. In other words, $\{u(F_{n_1})u(F_{n_2})\cdots u(F_{n_k}) : n_1 < n_2 < \cdots < n_k, k \geq 1\}$ is φ-monochromatic. The result now follows by taking $z = u(F_1)u(F_2)u(F_3)\cdots$ which belongs to $\Omega(x)$.

The following proposition shows that the previous result does not extend to ultra monochromatic factorizations:

Proposition 6.4.5. *Let $r \in \mathbb{N}$ and $x \in \{0, 1\}^{\mathbb{N}}$ be a r-power-free Sturmian word. Define $\varphi : \{0, 1\}^+ \to \{0, 1\}$ by*

$$\varphi(u) = \begin{cases} 0 & \text{if } u \text{ is a factor of } x; \\ 1 & \text{otherwise} \end{cases}$$

Then no $y \in \Omega(x)$ admits a φ-ultra monochromatic factorization.

Proposition 6.4.5 follows immediately from the following lemma:

Lemma 6.4.6. *Let $r \in \mathbb{N}^+$ and $x \in \{0,1\}^\omega$ be a r-power-free Sturmian word. Then for each infinite sequence $\omega = V_0, V_1, V_2, \cdots$ with $V_i \in \{0,1\}^+$, there exist $k \geq 1$, $0 \leq n_1 < n_2 < \cdots < n_k$, and a permutation σ of $\{1,2,\ldots,k\}$ such that $V_{n_{\sigma(1)}} V_{n_{\sigma(2)}} \cdots V_{n_{\sigma(k)}} \notin Fact(x)$.*

Proof. For each $\omega = V_0, V_1, V_2, \cdots$ with $V_i \in \{0,1\}^+$, $i \geq 0$, set $N(\omega) = |V_0 V_1 \cdots V_r|$. We proceed by induction on $N(\omega)$ to show that for each $\omega = V_0, V_1, V_2, \cdots$ with $V_i \in \{0,1\}^+$, and each r-power-free Sturmian word x, there exist $k \geq 1$, $0 \leq n_1 < n_2 < \cdots < n_k$ and a permutation σ of $\{1,2,\ldots,k\}$ such that $V_{n_{\sigma(1)}} V_{n_{\sigma(2)}} \cdots V_{n_{\sigma(k)}} \notin \text{Fact}(x)$.

The base case of the induction is when $N(\omega) = r+1$, i.e., $|V_0| = |V_1| = \cdots = |V_r| = 1$. Let x be a r-power-free Sturmian word. For $a \in \{0,1\}$, put $\bar{a} = 1 - a$ so that $\{a, \bar{a}\} = \{0,1\}$. Fix $a \in \{0,1\}$ so that $\bar{a}\bar{a} \notin \text{Fact}(x)$. First suppose that $V_i = V_j = \bar{a}$ for some $0 \leq i < j$. In this case $V_i V_j \notin \text{Fact}(x)$. Thus we can assume that at most one $V_i = \bar{a}$. In this case, there exist $0 \leq n_1 < n_2 < \cdots < n_r \leq r$ such that $V_{n_i} = a$ for each $1 \leq i \leq r$. It follows that $V_{n_1} V_{n_2} \cdots V_{n_r} = a^r \notin \text{Fact}(x)$.

For the inductive step, let $N > r+1$, and suppose that for each $\omega = V_0, V_1, V_2, \cdots$ with $V_i \in \{0,1\}^+$ and $N(\omega) < N$ and for each r-power-free Sturmian word x, there exist $k \geq 1$, $0 \leq n_1 < n_2 < \cdots < n_k$, and a permutation σ of $\{1,2,\ldots,k\}$ such that $V_{n_{\sigma(1)}} V_{n_{\sigma(2)}} \cdots V_{n_{\sigma(k)}} \notin \text{Fact}(x)$. Now let $\omega = V_0, V_1, V_2, \cdots$ with $V_i \in \{0,1\}^+$, $i \geq 0$ and $N(\omega) = N$ and let x be a r-power-free Sturmian word. Without loss of generality, we may assume $11 \notin \text{Fact}(x)$ and that x begins with 0. Note that if $11 \notin \text{Fact}(x)$ and x begins with 1, we can replace x with $0x$ which is Sturmian and r-power-free. We claim that for some $k \geq 1$, and $0 \leq n_1 < n_2 < \cdots < n_k$ and permutation σ of $\{1,2,\ldots,k\}$, we have $V_{n_{\sigma(1)}} V_{n_{\sigma(2)}} \cdots V_{n_{\sigma(k)}} \notin \text{Fact}(x)$. Suppose to the contrary that for every $k \geq 1$, $0 \leq n_1 < n_2 < \cdots < n_k$ and permutation σ of $\{1,2,\ldots,k\}$, we have $V_{n_{\sigma(1)}} V_{n_{\sigma(2)}} \cdots V_{n_{\sigma(k)}} \in \text{Fact}(x)$. Since x is r-power-free, we have $\limsup_{n \to \infty} |V_n| = +\infty$.

Suppose first that for some $a \in \{0,1\}$, there exist $0 \leq i < j$ such that V_i begins with a and V_j begins with $\bar{a}$. Pick $j < m < n$ such that $|V_n| > r|V_m|$. Since $V_m V_n V_i$, $V_m V_n V_j$, $V_n V_m V_i$, and $V_n V_m V_j$ are each factors of x, it follows that each of $V_m V_n$ and $V_n V_m$ is a right special factor of x. But since $|V_m V_n| = |V_n V_m|$ and x has exactly one right special factor of each length, it follows that $V_m V_n = V_n V_m$, from which one easily derives that V_m^r is a prefix of V_n and hence in particular $V_m^r \in \text{Fact}(x)$, a contradiction. Thus we may suppose that all V_i begin with the same letter $a \in \{0,1\}$. A similar argument shows that all V_i terminate with the same letter $b \in \{0,1\}$. Moreover, as $11 \notin \text{Fact}(x)$, either a or b must equal 0. Since $\text{Fact}(x)$ is closed under reversal, short of replacing each V_i in ω by its reversal, we may suppose that $a = 0$, i.e., each V_i begins with 0. Thus $V_i 0 \in \text{Fact}(x)$ for each $i \geq 0$.

Now consider the morphism $L_0 : 0 \mapsto 0$, and $1 \mapsto 01$. For each $i \geq 0$, define $V_i' \in \{0,1\}^+$ by $L_0(V_i') = V_i$ and put $\omega' = V_0', V_1', V_2', \ldots$. Finally, as x begins with 0, define $x' \in \{0,1\}^\omega$ by $L_0(x') = x$. Then, as is well known, x' is a Sturmian word. Moreover, since x is r-power-free, so is x' and at least one V_i with $0 \leq i \leq r-1$ must contain an occurrence of 1. Thus $N(\omega') < N(\omega)$. For each $k \geq 1$, $0 \leq n_1 < n_2 < \cdots < n_k$ and permutation σ of $\{1, 2, \ldots, k\}$, we have $V_{n_{\sigma(1)}} V_{n_{\sigma(2)}} \cdots V_{n_{\sigma(k)}} 0 \in \mathrm{Fact}(x)$. Thus $V'_{n_{\sigma(1)}} V'_{n_{\sigma(2)}} \cdots V'_{n_{\sigma(k)}} \in \mathrm{Fact}(x')$, and this is a contradiction to our inductive hypothesis.

We mention that A. Frid has extended the validity of previous lemma to the case of any infinite word of linear factor complexity.

We next show that for every finite coloring $\varphi : \mathscr{A}^+ \to C$, from a measure theoretic point of view, almost all infinite words x, there exists a point $z \in \Omega(x)$ admitting a φ-ultra monochromatic factorization. First we consider the case in which x is periodic which turns out to be equivalent to Hindman's theorem in [290].

Theorem 6.4.7. *The following statements are equivalent:*

i) For every finite coloring $\varphi : \mathscr{A}^+ \to C$, each periodic word $x \in \mathscr{A}^\omega$ admits a φ-ultra monochromatic factorization.
ii) For each finite coloring $\varphi : \mathbb{N} \to C$ of the positive integers, there exist $c \in C$ and an infinite sequence $(n_k)_{k=1}^\infty$ such that $FS((n_k)_{k=1}^\infty) = \{\sum_{i \in F} n_i \,|\, F \in Fin(\mathbb{N})\} \subseteq \varphi^{-1}(c)$.

Proof. We note that item *ii)* is the statement of Hindman's theorem. We begin by showing that *i)* $\Longrightarrow$ *ii)*. Let $\varphi : \mathbb{N} \to C$ be a finite coloring of the positive integers, and let x be the periodic word $x = a^\omega$, with $a \in \mathscr{A}$. Then φ induces a finite coloring $\varphi' : \{a\}^+ \to C$ given by $\varphi'(a^n) = \varphi(n)$. By *i)* there exists a φ'-ultra monochromatic factorization $x = u_1 u_2 u_3 \cdots$. Put $c = \varphi'(u_1)$. For $k \geq 1$, set $n_k = |u_k|$ so that each $u_k = a^{n_k}$. Then for each finite subset F of $\mathbb{N}$, we have

$$\varphi\left(\sum_{i \in F} n_i\right) = \varphi\left(\sum_{i \in F} |u_i|\right) = \varphi\left(|\prod_{i \in F} u_i|\right) = \varphi'\left(\prod_{i \in F} u_i\right) = c$$

since $\prod_{i \in F} u_i = a^{\sum_{i \in F} n_i}$ and hence is a factor of x. Whence $\mathrm{FS}((n_k)_{k=0}^\infty) = \{\sum_{i \in F} n_i \,|\, F \in \mathrm{Fin}(\mathbb{N})\} \subseteq \varphi^{-1}(c)$.

To see that *ii)* $\Longrightarrow$ *i)*, let $\varphi : \mathscr{A}^+ \to C$, $u \in \mathscr{A}^+$, and $x = u^\omega$. Define $\varphi' : \mathbb{N}^+ \to C$ by $\varphi'(n) = \varphi(u^n)$. By (2) there exist $c \in C$ and an infinite sequence $(n_k)_{k=1}^\infty$ such that $\mathrm{FS}((n_k)_{k=1}^\infty) = \{\sum_{i \in F} n_i \,|\, F \in \mathrm{Fin}(\mathbb{N})\} \subseteq \varphi'^{-1}(c)$. For each $k \geq 1$ set $u_k = u^{n_k}$. Then clearly the factorization $x = u_1 u_2 u_3 \cdots$ is φ-ultra monochromatic.

As an immediate consequence, we obtain:

Corollary 6.4.8. *Let $\mathscr{A}$ be a finite set, and let μ be the Bernoulli measure on $\mathscr{A}^\omega$. Let $\varphi : \mathscr{A}^+ \to C$ be any finite coloring. Then for μ-almost all $x \in \mathscr{A}^\omega$, there exists $y \in \Omega(x)$ which admits a φ-ultra monochromatic factorization.*

Proof. As is well known almost all words $x \in \mathscr{A}^\omega$ with respect to the measure μ are of full complexity, meaning $\mathrm{Fact}(x) = \mathscr{A}^+$ (see, for instance, [14, Theorem 10.1.6]). (As an example, *normal words* [448, Chap. 8] are of full complexity.) Thus for almost all words $x \in \mathscr{A}^\omega$, relatively to measure μ, there exists $a \in \mathscr{A}$ such that $a^\omega \in \Omega(x)$. The result now follows from Theorem 6.4.7.

The above results suggest the following question:

Question 6.4.9. Let $x \in \mathscr{A}^\omega$ be a uniformly recurrent word. Suppose that for each $\varphi : \mathscr{A}^+ \to C$, there exists $y \in \Omega(x)$ admitting a φ-ultra monochromatic factorization. Then does it follow that x is periodic?

Let $\varphi : \mathscr{A}^+ \to C$ be a finite coloring of $\mathscr{A}^+$, $x \in \mathscr{A}^\omega$, and k be a positive integer. A φ-monochromatic (resp., φ-sequentially monochromatic, φ-super monochromatic, φ-ultra monochromatic) factorization $x = V_0V_1V_2\cdots$ is said to be *k-shift invariant* if for each $1 \leq j \leq k$ the induced factorization $T^j(x) = W_0W_1W_2\cdots$ with $|W_i| = |V_i|$, $i \geq 0$, is φ-monochromatic (resp., φ-sequentially monochromatic, φ-super monochromatic, φ-ultra monochromatic). A φ-monochromatic (resp., φ-sequentially monochromatic, φ-super monochromatic, φ-ultra monochromatic) factorization $x = V_0V_1V_2\cdots$ is called *shift invariant* if for each positive integer j, the induced factorization $T^j(x) = W_0W_1W_2\cdots$ with $|W_i| = |V_i|$ is φ-monochromatic (resp., φ-sequentially monochromatic, φ-super monochromatic, φ-ultra monochromatic).

The following simple variation of the infinite Ramsey theorem is a simple iterated application of the usual version of Ramsey's theorem.

Proposition 6.4.10. *Let $\varphi : \Sigma_2(\mathbb{N}) \to C$ be a finite coloring and k a nonnegative integer. There exists an infinite set $\mathscr{N} \subseteq \mathbb{N}$ and a sequence $(c_i)_{i=0}^k$ such that for each $0 \leq i \leq k$, we have $c_i \in C$ and*

$$\Sigma_2(\mathscr{N} + i) \subseteq \varphi^{-1}(c_i).$$

As an immediate consequence we deduce that

Corollary 6.4.11. *Let $\varphi : \mathscr{A}^+ \to C$, $x \in \mathscr{A}^\omega$, and $k \geq 1$. Then there exists a suffix x' of x which admits a k-shift invariant φ-sequentially monochromatic factorization.*

Proof. As in the proof of Theorem 6.4.1, we apply the above variation of Ramsey's theorem to the coloring $\varphi' : \Sigma_2(\mathbb{N}) \to C$ given by $\varphi'(\{m < n\}) = \varphi(x_m x_{m+1} \cdots x_{n-1})$.

Proposition 6.4.12. *A word $x \in \mathscr{A}^\omega$ is ultimately periodic if and only if for every finite coloring $\varphi : \mathscr{A}^+ \to C$, there exists a suffix of x which admits a shift invariant φ-monochromatic factorization.*

Proof. Clearly, if x is ultimately periodic, and hence of the form $x = uv^\omega$ for some $u, v \in \mathscr{A}^*$ with $v \neq \varepsilon$, then for any $\varphi : \mathscr{A}^+ \to C$, the factorization $v \cdot v \cdot v \cdot \cdots$ of the suffix v^ω is shift invariant φ-monochromatic. Conversely, suppose x is aperiodic. Choose a recurrent word $y \in \Omega(x)$. Thus each prefix of y occurs infinitely often in x. Let $\varphi : \mathscr{A}^+ \to \{0, 1\}$ be given by $\varphi(u) = 0$ if u is a prefix of y and $\varphi(u) = 1$ otherwise. Let x' be any suffix of x. We claim that x' does not admit a shift invariant φ-monochromatic factorization. In fact, suppose to the contrary that x' admits a shift invariant φ-monochromatic factorization $x' = V_0V_1V_2\cdots$. Since y is recurrent and each prefix of y occurs infinitely often in x, there exist $0 \leq i < j$ such that if we consider the shifted factorizations $T^i(x') = W_0W_1W_2\cdots$ and $T^j(x') = W'_0W'_1W'_2\cdots$, where $|W_i| = |W'_i| = |V_i|$ for each $i \geq 0$, both W_0 and W'_0 are prefixes of y. It follows that W_i and W_i' are prefixes of y for each $i \geq 0$. But since they are of equal length, we have $W_i = W_i'$ for each $i \geq 0$. Thus $T^i(x') = T^j(x')$ which implies that x is ultimately periodic, a contradiction.

We recall that a subset A of $\mathbb{N}^+$ is called an *IP-set* if A contains $\mathrm{FS}((n_i)_{i=1}^{\infty})$ for some infinite sequence $(n_i)_{i=1}^{\infty}$. In terms of IP-sets, Hindman's theorem states that any finite coloring of $\mathbb{N}^+$ contains a monochromatic IP-set. By using the so-called Finite Unions Theorem, which is equivalent to Hindman's Finite Sums Theorem (*cf.* [47, 417]), one can show that IP-sets in $\mathbb{N}^+$ are *partition regular*, i.e., if A is an IP-set and $A = \bigcup_{i=1}^{k} A_i$, then there exists $1 \leq i \leq k$ such that A_i is an IP-set. We recall also the following well-known theorem of Milliken–Taylor [417, 561]:

Theorem 6.4.13. *Let k be a positive integer and $\varphi : \Sigma_k(\mathbb{N}^+) \to C$ a finite coloring. Then there exist $c \in C$ and an infinite sequence $(n_i)_{i=1}^{\infty}$ such that*

$$\left\{\sum_{i \in F_1} n_i, \sum_{i \in F_2} n_i, \ldots, \sum_{i \in F_k} n_i\right\} \in \varphi^{-1}(c)$$

for each $F_1 < F_2 < \cdots < F_k$ with $F_i \in Fin(\mathbb{N}^+)$, $1 \leq i \leq k$.

The next theorem shows that for each finite coloring $\varphi : \mathscr{A}^+ \to C$, and each periodic word $x \in \mathscr{A}^{\omega}$, there exists a shift invariant φ-ultra monochromatic factorization of x. We present two proofs, one uses the fact that IP-sets are partition regular and the other uses the Milliken–Taylor Theorem.

Theorem 6.4.14. *For each finite coloring $\varphi : \mathscr{A}^+ \to C$, each periodic word $x \in \mathscr{A}^{\omega}$ admits a shift invariant φ-ultra monochromatic factorization.*

Proof. (*First proof*) Let $\varphi : \mathscr{A}^+ \to C$ be given. Let $u = u_1 u_2 \cdots u_k \in \mathscr{A}^+$, $u_i \in \mathscr{A}$, $i = 1, \ldots, k$ and $x = u^{\omega}$. Consider the coloring $\varphi_1 : \mathbb{N}^+ \to C$ defined by $\varphi_1(n) = \varphi(u^n)$. Then by Hindman's theorem, there exists an infinite sequence $(n_i^{(1)})_{i=1}^{\infty}$ and $c_1 \in C$ such that $\mathrm{FS}((n_i^{(1)})_{i=1}^{\infty}) \subseteq \varphi_1^{-1}(c_1)$. This implies that the factorization

$$x = u^{n_1^{(1)}} \cdot u^{n_2^{(1)}} \cdot u^{n_3^{(1)}} \cdots$$

is φ-ultra monochromatic. Next consider the coloring $\varphi_2 : \mathrm{FS}((n_i^{(1)})_{i=1}^{\infty}) \to C$ defined by $\varphi_2(n) = \varphi((u_2 \cdots u_k u_1)^n)$. By partition regularity of IP-sets, it follows that there exists an infinite sequence $(n_i^{(2)})_{i=1}^{\infty}$ and $c_2 \in C$ such that $\mathrm{FS}((n_i^{(2)})_{i=1}^{\infty}) \subseteq \varphi_2^{-1}(c_2)$. It follows that the factorizations

$$x = u^{n_1^{(2)}} \cdot u^{n_2^{(2)}} \cdot u^{n_3^{(2)}} \cdots$$

and

$$T(x) = (u_2 \cdots u_k u_1)^{n_1^{(2)}} (u_2 \cdots u_k u_1)^{n_2^{(2)}} (u_2 \cdots u_k u_1)^{n_3^{(2)}} \cdots$$

are both φ-ultra monochromatic. Continuing in this way up to stage k, we can find an infinite sequence $(n_i^{(k)})_{i=1}^{\infty}$ such that for each $0 \leq i \leq k-1$ the factorization

$$T^i(x) = (u_{i+1}\cdots u_k u_1 \cdots u_i)^{n_1^{(k)}} (u_{i+1}\cdots u_k u_1 \cdots u_i)^{n_2^{(k)}} (u_{i+1}\cdots u_k u_1 \cdots u_i)^{n_3^{(k)}} \cdots$$

is φ-ultra monochromatic. Since $T^k(x) = x$ the result now follows.

(*Second proof*) As before let $\varphi : \mathscr{A}^+ \to C$ be given, $u = u_1 u_2 \cdots u_k \in \mathscr{A}^+$, $u_i \in \mathscr{A}$, $i = 1, \ldots, k$, and $x = u^\omega$. Then φ induces a finite coloring

$$\Psi : \Sigma_k(\mathbb{N}^+) \to C^k$$

defined by

$$\begin{aligned}\Psi(\{n_1 < n_2 < \cdots < n_k\}) \\ = (\varphi((u_1 u_2 \cdots u_k)^{n_1}), \varphi((u_2 u_3 \cdots u_k u_1)^{n_2}), \ldots, \varphi((u_k u_1 \cdots u_{k-1})^{n_k})).\end{aligned}$$

By Theorem 6.4.13, there exist $c = (c_1, c_2, \ldots, c_k) \in C^k$ and $(n_i)_{i=1}^{\infty}$ such that

$$\Psi\left(\left\{\sum_{i\in F_1} n_i, \sum_{i\in F_2} n_i, \ldots, \sum_{i\in F_k} n_i\right\}\right) = c \qquad (*)$$

for each $F_1 < F_2 < \cdots < F_k$ with $F_i \in \mathrm{Fin}(\mathbb{N}^+)$, $1 \leq i \leq k$.

Fix $1 \leq j \leq k$ and $F \in \mathrm{Fin}(\{k, k+1, k+2, \ldots\})$. We claim that

$$\varphi((u_j \cdots u_k u_1 \cdots u_{j-1})^{\sum_{i\in F} n_i}) = c_j.$$

This is a consequence of $(*)$ by taking $F_i = \{i\}$ for $1 \leq i < j$, $F_j = F$, and $F_{j+i} = \{M+i\}$ for $1 \leq i \leq k-j$ where $M = \max(F)$. It follows that the factorization $x = u^{n_k} u^{n_{k+1}} u^{n_{k+2}} \cdots$ is shift invariant φ-ultra monochromatic.

Chapter 7
Normal Numbers and Computer Science

Verónica Becher and Olivier Carton

Abstract Émile Borel defined normality more than 100 years ago to formalize the most basic form of randomness for real numbers. A number is normal to a given integer base if its expansion in that base is such that all blocks of digits of the same length occur in it with the same limiting frequency. This chapter is an introduction to the theory of normal numbers. We present five different equivalent formulations of normality, and we prove their equivalence in full detail. Four of the definitions are combinatorial, and one is, in terms of finite automata, analogous to the characterization of Martin-Löf randomness in terms of Turing machines. All known examples of normal numbers have been obtained by constructions. We show three constructions of numbers that are normal to a given base and two constructions of numbers that are normal to all integer bases. We also prove Agafonov's theorem that establishes that a number is normal to a given base exactly when its expansion in that base is such that every subsequence selected by a finite automaton is also normal.

7.1 Introduction

Flip a coin a large number of times, and roughly half of the flips will come up heads and half will come up tails. *Normality* makes analogous assertions about the digits in the expansion of a real number. Precisely, let b be an integer greater than or equal to 2. A real number is *normal* to base b if each of the digits $0, \ldots, b-1$ occurs in its expansion with the same asymptotic frequency $1/b$, each of the blocks of two

V. Becher (✉)
Departamento de Computación, Facultad de Ciencias Exactas y Naturales, Universidad de Buenos Aires. CONICET, Pabellón, Ciudad Universitaria, C1428EGA Buenos Aires, Argentina
e-mail: vbecher@dc.uba.ar

O. Carton
IRIF, UMR 8243, CNRS & Université Paris Diderot, Case 7014, F-75205 Paris Cedex 13, France
e-mail: Olivier.Carton@irif.fr

V. Berthé, M. Rigo (eds.), *Sequences, Groups, and Number Theory*,
Trends in Mathematics, https://doi.org/10.1007/978-3-319-69152-7_7

digits occurs with frequency $1/b^2$, each of the blocks of three digits occurs with frequency $1/b^3$, and so on, for every block length. A number is *absolutely normal* if it is normal to every base. Émile Borel [99] defined normality more than 100 years ago to formalize the most basic form of randomness for real numbers. Many of his questions are still open, such as whether any of π, e, or $\sqrt{2}$ is normal in some base, as well as his conjecture that the irrational algebraic numbers are absolutely normal [100].

In this chapter, we give an introduction to the theory of normal numbers. We start by considering five different equivalent formulations of normality, and we prove their equivalence in full detail. These proofs have not appeared all together in the literature before. Four of the definitions are combinatorial, and one is, in terms of finite automata, analogous to the characterization of Martin-Löf randomness [198] in terms of Turing machines. This characterization of normality holds for various enrichments of finite automata [57, 131], but the relation with deterministic push-down automata remains unsolved. We also briefly mention another well-known equivalent definition of normality, in terms of uniform distribution modulo 1, that will be further considered in Chapter 8.

All known examples of normal numbers have been obtained by constructions. We first focus in three selected constructions of numbers that are normal to a given base. We then present two constructions of absolutely normal numbers, one is a slightly simplified version of the pioneer work done by Alan Turing and the other is a simplified version of the polynomial time algorithm in [53].

Finally we consider the problem of preserving normality by selection by finite automata of a subsequence of a give sequence. We give the proof of Agafonov's theorem [6] showing that a number is normal to a given base exactly when its expansion in that base is such that every subsequence selected by a finite automata is also normal.

Notation Let A be finite set of symbols that we refer as the alphabet. We write A^ω for the set of all infinite words in alphabet A, A^* for the set of all finite words, $A^{\leq k}$ for the set of all words of length up to k, and A^k for the set of words of length exactly k. The length of a finite word w is denoted by $|w|$. The positions of finite and infinite words are numbered starting at 1. To denote the symbol at position i of a word w, we write $w[i]$, and to denote the substring of w from position i to j, we write $w[i \dots j]$. The empty word is denoted by λ.

For two words w and u, the number $|w|_u$ of *occurrences* of u in w and the number $\|w\|_u$ of *aligned occurrences* of u in w are, respectively, given by

$$|w|_u = |\{i : w[i \dots i + |u| - 1] = u\}|,$$

$$\|w\|_u = |\{i : w[i \dots i + |u| - 1] = u \text{ and } i \equiv 1 \mod |u|\}|.$$

For example, $|aaaaa|_{aa} = 4$ and $\|aaaaa\|_{aa} = 2$. Notice that the definition of aligned occurrences has the condition $i \equiv 1 \mod |u|$ instead of $i \equiv 0 \mod |u|$, because the positions are numbered starting at 1. When a word u is just a symbol, $|w|_u$ and

$\|w\|_u$ coincide. Counting aligned occurrences of a word of length r over alphabet A is the same as counting occurrences of the corresponding symbol over alphabet A^r. Precisely, consider alphabet A, a length r, and an alphabet B with $|A|^r$ symbols. The set of words of length r over alphabet A and the set B are isomorphic, as witnessed by the isomorphism $\pi : A^r \to B$ induced by the lexicographic order in the respective sets. Thus, for any $w \in A^*$ such that $|w|$ is a multiple of r, $\pi(w)$ has length $|w|/r$ and $\pi(u)$ has length 1, as it is just a symbol in B. Then, for any $u \in A^r$, $\|w\|_u = |\pi(w)|_{\pi(u)}$.

7.2 Borel's Definition of Normality

A *base* is an integer greater than or equal to 2. For a real number x in the unit interval, the *expansion* of x in base b is a sequence $a_1 a_2 a_3 \ldots$ of integers from $\{0, 1, \ldots, b-1\}$ such that

$$x = \sum_{k \geq 1} a_k b^{-k} = 0.a_1 a_2 a_3 \ldots$$

To have a unique representation of all rational numbers, we require that expansions do not end with a tail of $b-1$. We will abuse notation, and whenever the base b is understood, we will denote the first n digits in the expansion of x with $x[1 \ldots n]$.

Definition 7.2.1 (Strong Aligned Normality, Borel [99]). A real number x is simply normal to base b if, in the expansion of x in base b, each digit d occurs with limiting frequency equal to $1/b$,

$$\lim_{n \to \infty} \frac{|x[1 \ldots n]|_d}{n} = \frac{1}{b}$$

A real number x is normal to base b if each of the reals $x, bx, b^2x, \ldots$ are simply normal to bases $b^1, b^2, b^3, \ldots$. A real x is absolutely normal if x is normal to every integer base greater than or equal to 2.

Theorem 7.2.2 (Borel [99]). *Almost all real numbers (with respect to Lebesgue measure) are absolutely normal.*

Are the usual mathematical constants, such as π, e, or $\sqrt{2}$, absolutely normal? Or at least simply normal to *some* base? The question remains open.

Conjecture 7.2.3 (Borel [100]). Irrational algebraic numbers are absolutely normal.

The most famous example of a normal number is due to Champernowne [141]. He proved that the number

$$0.1234567891011121314151617181920212223242526 27\ldots$$

is normal to base 10. The construction can be done in any base, obtaining a number normal to that base. It is unknown whether Champernowne numbers are normal to the bases that are multiplicatively independent to the base used in the construction. Champernowne's construction has been generalized in many interesting ways. There are also some other methods to obtain examples of numbers that are normal to a given base. In Section 7.7, we comment on the different methods, and we present three selected constructions.

All known examples of absolutely normal numbers are given by constructions. The oldest were not even computable. The first computable construction is due to A. Turing [52, 570]. In Section 7.8, we give references of known constructions, and we present two of them.

7.3 Equivalences Between Combinatorial Definitions of Normality

Borel's original definition of normality turned out to be redundant. Pillai in 1940, see [118, Theorem 4.2], proved the equivalence between Definition 7.2.1 and the following.

Definition 7.3.1 (Aligned Normality). A real number x is normal to base b if x is simply normal to bases $b^1, b^2, b^3, \ldots$.

Niven and Zuckerman in 1951, see [118, Theorem 4.5], proved yet another equivalent formulation of normality by counting occurrences of blocks but not aligned. This formulation was stated earlier by Borel himself, without proof.

Definition 7.3.2 (Non-aligned Normality). A real number x is normal to base b if for every block u,

$$\lim_{n\to\infty} \frac{|x[1\ldots n]|_u}{n} = \frac{1}{b^{|u|}}.$$

We will prove that Definitions 7.2.1, 7.3.1 and 7.3.2 are equivalent. The following lemma gives a central limit theorem that bounds the number of words in alphabet A of length k having too few or too many occurrences of some block w.

Definition 7.3.3. Let A be an alphabet of b symbols. We define the set of words of length k such that a given word w has a number of occurrences that differs from the expected value in plus or minus εk,

$$Bad(A, k, w, \varepsilon) = \left\{ v \in A^k : \left| \frac{|v|_w}{k} - b^{-|w|} \right| \geq \varepsilon \right\}.$$

Lemma 7.3.4 (Adapted from Hardy and Wright [283, proof of Theorem 148]). *Let b be an integer greater than or equal to 2, and let k be a positive integer. If $6/k \le \varepsilon \le 1/b$, then for every $d \in A$,*

$$|Bad(A, k, d, \varepsilon)| < 4b^k e^{-b\varepsilon^2 k/6}.$$

Proof. Observe that for any $d \in A$,

$$Bad(A, k, d, \varepsilon) = \sum_{n \le k/b - \varepsilon k} \binom{k}{n} (b-1)^{k-n} + \sum_{n \ge k/b + \varepsilon k} \binom{k}{n} (b-1)^{k-n}$$

Fix b and k and write $N(n)$ for

$$\binom{k}{n} (b-1)^{k-n}.$$

For all $n < k/b$, we have that $N(n) < N(n+1)$ and the quotients

$$\frac{N(n)}{N(n+1)} = \frac{(n+1)(b-1)}{k-n}$$

decrease as n increases. Similarly, for all $n > k/b$, we have that $N(n) < N(n-1)$ and the quotients

$$\frac{N(n)}{N(n-1)} = \frac{k-n+1}{n(b-1)}$$

increase as n decreases. The strategy will be to "shift" each of the sums m positions.

We bound the first sum as follows. For any n we can write

$$N(n) = \frac{N(n)}{N(n+1)} \cdot \frac{N(n+1)}{N(n+2)} \cdot \ldots \cdot \frac{N(n+m-1)}{N(n+m)} \cdot N(n+m)$$

Let

$$m = \lfloor \varepsilon k/2 \rfloor \text{ and } p = \lfloor k/b - \varepsilon k \rfloor$$

For each n such that $n \le p + m - 1$, we have that $n + m < k/b$, so

$$\begin{aligned} \frac{N(n)}{N(n+1)} &\le \frac{N(p+m-1)}{N(p+m)} \\ &= \frac{(p+m)(b-1)}{k-p-m+1} \\ &< \frac{(k/b - \varepsilon k/2)(b-1)}{k - k/b + \varepsilon k/2} \end{aligned}$$

$$
\begin{aligned}
&= 1 - \frac{\varepsilon b/2}{1 - 1/b + \varepsilon/2} \\
&< 1 - \varepsilon b/2 \qquad \text{(using the hypothesis } \varepsilon \le 1/b\text{).} \\
&< e^{-b\varepsilon/2}.
\end{aligned}
$$

Then,

$$
\begin{aligned}
N(n) &< \left(e^{-b\varepsilon/2}\right)^m N(n+m) \\
&\le e^{-b\varepsilon(\varepsilon k/2-1)/2} N(n+m) \\
&\le 2e^{-b\varepsilon^2 k/4} N(n+m), \qquad \text{(the hypothesis } \varepsilon \le 1/b \text{ implies } e^{b\varepsilon/2} < 2\text{)}
\end{aligned}
$$

We obtain,

$$
\sum_{n \le u} N(n) < 2e^{-b\varepsilon^2 k/2} \sum_{n \le u} N(n+m) \le 2\, b^k e^{-b\varepsilon^2 k/4}.
$$

We now bound the second sum, shifting it m positions. For any n we can write

$$
N(n) = \frac{N(n)}{N(n-1)} \cdot \frac{N(n-1)}{N(n-2)} \cdot \ldots \cdot \frac{N(n-m+1)}{N(n-m)} \cdot N(n-m)
$$

Let

$$
m = \lfloor \varepsilon k/2 \rfloor \text{ and } q = \lceil k/b + \varepsilon k \rceil.
$$

For each n such that $n \ge q - m + 1$, we have $n - m > k/b$, so

$$
\begin{aligned}
\frac{N(n)}{N(n-1)} &\le \frac{N(q-m+1)}{N(q-m)} \\
&= \frac{k-q+m}{(q-m+1)(b-1)} \\
&= \frac{k - \lceil k/b + \varepsilon k \rceil + \lfloor \varepsilon k/2 \rfloor}{(\lceil k/b + \varepsilon k \rceil - \lfloor \varepsilon k/2 \rfloor + 1)(b-1)} \\
&\le \frac{k - k/b - \varepsilon k/2}{(k/b + \varepsilon k/2 + 1)(b-1)} \\
&< \frac{1 - 1/b - \varepsilon/2}{(1/b + \varepsilon/2)(b-1)}
\end{aligned}
$$

Now

$$\begin{aligned}
&\tfrac{1-1/b-\varepsilon/2}{(1/b+\varepsilon/2)(b-1)} \le 1 - b\varepsilon/3 \\
\Leftrightarrow\quad & 1 - 1/b - \varepsilon/2 \le (1 - b\varepsilon/3)(1/b + \varepsilon/2)(b-1) \\
\Leftrightarrow\quad & (b-1)/b - \varepsilon/2 \le (1/b + \varepsilon/2)(b-1) - (b\varepsilon/3)(1/b + \varepsilon/2)(b-1) \\
\Leftrightarrow\quad & (b\varepsilon/3)(1/b + \varepsilon/2)(b-1) \le b\varepsilon/2 \\
\Leftrightarrow\quad & (1/b + \varepsilon/2)(b-1) \le 3/2.
\end{aligned}$$

Since $\varepsilon \le 1/b$, we obtain the required inequality,

$$(1/b + \varepsilon/2)(b-1) \le (1/b + 1/(2b))(b-1) = 3/(2b)(b-1) \le 3/2$$

We conclude,

$$\frac{N(n)}{N(n-1)} \le 1 - b\varepsilon/3 \le e^{-b\varepsilon/3}.$$

Then,

$$\begin{aligned}
N(n) &< \left(e^{-b\varepsilon/3}\right)^m N(n-m) \\
&\le e^{-b\varepsilon\lfloor \varepsilon k/2 \rfloor/3} N(n-m) \\
&\le e^{-b\varepsilon(\varepsilon k/2 - 1)/3} N(n-m) \\
&\le 2\, e^{-b\varepsilon^2 k/6} N(n-m), \qquad \text{(the hypothesis } \varepsilon \le 1/b \text{ implies } e^{b\varepsilon/3} < 2).
\end{aligned}$$

Thus,

$$\sum_{n \ge q} N(n) < 2\, b^k e^{-b\varepsilon^2 k/6}.$$

This completes the proof.

The next lemma bounds the number of words of k symbols in alphabet A that contain too many or too few occurrences of some block of length ℓ, with respect to a toleration specified by ε.

Lemma 7.3.5. *Let A be an alphabet of b symbols. Let k, ℓ be positive integers and ε a real such that $6/\lfloor k/\ell \rfloor \le \varepsilon \le 1/b^\ell$. Then,*

$$\left| \bigcup_{w \in A^\ell} Bad(A, k, w, \varepsilon) \right| < 2\ell\, b^{k+2\ell}\, e^{-b^\ell \varepsilon^2 k/(6\ell)}.$$

Proof. Split the set $\{1, 2, \ldots, k\}$ into the congruence classes modulo ℓ. Each of these classes contains either $\lfloor k/\ell \rfloor$ or $\lceil k/\ell \rceil$ elements. Let M_0 denote the class of all indices which leave remainder zero when being reduced modulo ℓ. Let $n_0 = |M_0|$.

For each x in A^k, consider the word in $(A^\ell)^{n_0}$

$$x[i_1 \ldots (i_1 + \ell - 1)]x[i_2 \ldots (i_2 + \ell - 1)] \ldots x[i_{n_0} \ldots (i_{n_0} + \ell - 1)]$$

for $i_1, \ldots i_{n_0} \in M_0$. By Lemma 7.3.4, we have

$$\left|Bad(A^\ell, n_0, w, \varepsilon)\right| < 4\,(b^\ell)^{n_0} e^{-b^\ell \varepsilon^2 n_0/6}.$$

Clearly, similar estimates hold for the indices in the other residue classes. Let $n_1, \ldots, n_{\ell-1}$ denote the cardinalities of these other residue classes. By assumption $n_0 + \cdots + n_{\ell-1} = k$. Then,

$$\begin{aligned}
|Bad(A, k, w, \varepsilon)| &\leq \sum_{j=0}^{\ell-1} \left|Bad(A^\ell, n_j, w, \varepsilon)\right| \\
&\leq \sum_{j=0}^{\ell-1} 4(b^\ell)^{n_j} e^{-b^\ell \varepsilon^2 n_j/6} \\
&\leq \sum_{j=0}^{\ell-1} 4(b^\ell)^{k/\ell+1} e^{-b^\ell \varepsilon^2 k/(6\ell)} \\
&= 4\,\ell\, b^{k+\ell}\, e^{-b^\ell \varepsilon^2 k/(6\ell)}.
\end{aligned}$$

The last inequality holds because

$$(b^\ell)^{\lceil k/\ell \rceil} e^{-b^\ell \varepsilon^2 \lceil k/\ell \rceil/6} < (b^\ell)^{k/\ell+1} e^{-b^\ell \varepsilon^2 k/(6\ell)}$$

and $\varepsilon \leq 1/b^\ell$ ensures

$$(b^\ell)^{\lfloor k/\ell \rfloor} e^{-b^\ell \varepsilon^2 \lfloor k/\ell \rfloor/6} \leq b^k e^{-b^\ell \varepsilon^2 k/(6\ell)} e^{1/(6b^\ell)} \leq b^k e^{-b^\ell \varepsilon^2 k/(6\ell)} b^\ell.$$

Now, summing up over all $w \in A^\ell$, we obtain

$$\left|\bigcup_{w \in A^\ell} Bad(A, k, w, \varepsilon)\right| < 2\ell\, b^{k+2\ell} e^{-b^\ell \varepsilon^2 k/(6\ell)}.$$

Instead of the factor 4, we can put the factor 2 because if a word $w \in A^\ell$ occurs fewer times than expected in a given word $x \in A^k$, then there is another word $v \in A^\ell$ that occurs in x more times than expected.

Lemma 7.3.6. *Let $(x_{1,n})_{n\geq 0}, (x_{2,n})_{n\geq 0}, \ldots, (x_{k,n})_{n\geq 0}$ be sequences of real numbers such that $\sum_{i=1}^{k} x_{i,n} = 1$, and let $c_1, c_2, \ldots, c_k$ be real numbers such that $\sum_{i=1}^{k} c_i = 1$. Then,*

1. *If for each i, $\liminf_{n\to\infty} x_{i,n} \geq c_i$ then for each i, $\lim_{n\to\infty} x_{i,n} = c_i$.*
2. *If for each i, $\limsup_{n\to\infty} x_{i,n} \leq c_i$ then for each i, $\lim_{n\to\infty} x_{i,n} = c_i$.*

Proof. For any i in $\{1, \ldots, k\}$,

$$\begin{aligned}\limsup_{n\to\infty} x_{i,n} &= \limsup_{n\to\infty}(1 - \sum_{j\neq i} x_{j,n}) \\ &= 1 + \limsup_{n\to\infty}(-\sum_{j\neq i} x_{j,n}) \\ &= 1 - \liminf_{n\to\infty}(\sum_{j\neq i} x_{j,n}) \\ &\leq 1 - \sum_{j\neq i} \liminf_{n\to\infty} x_{j,n} \\ &\leq 1 - \sum_{j\neq i} c_j \\ &= c_i.\end{aligned}$$

Since

$$\liminf \leq \limsup \text{ and } \limsup_{n\to\infty} x_{i,n} \leq c_i \leq \liminf_{n\to\infty} x_{i,n},$$

necessarily,

$$\liminf_{n\to\infty} x_{i,n} = \limsup_{n\to\infty} = c_i \text{ and } \lim_{n\to\infty} x_{i,n} = c_i.$$

Theorem 7.3.7. *Definitions 7.2.1, 7.3.1 and 7.3.2 are equivalent.*

Proof. Let x be a real number. We use the fact that for every block $w \in A^*$,

$$\lim_{n\to\infty} \frac{|x[1\ldots n]|_w}{n} = b^{-|w|}$$

if and only if there is a positive integer r such that

$$\lim_{n\to\infty} \frac{|x[1\ldots nr]|_w}{nr} = b^{-|w|}.$$

A similar fact is true for the limit of $\|x[1 \ldots n\ell]\|_w/n$.

1. We show that *strong aligned normality* implies *non-aligned normality*. Observe that for any $w \in A^\ell$,

$$|x[1 \ldots n]|_w = \sum_{i=0}^{\ell-1} \|(b^i x)[1 \ldots n-i]\|_w$$

Then,

$$\lim_{n\to\infty} \frac{|x[1 \ldots n]|_w}{n} = \sum_{i=0}^{\ell-1} \lim_{n\to\infty} \frac{\|(b^i x)[1 \ldots n-i]\|_w}{n} = \sum_{i=0}^{\ell-1} b^{-\ell}/\ell = b^{-\ell}.$$

2. We prove that *non-aligned normality* implies *aligned normality*. Define

$$\|v\|_{w,r} = |\{i : v[i..i+|w|-1] = w \text{ and } i = r \bmod |w|\}|.$$
$$\|v\|_{w,*} = \max_{1\le r\le |w|} \|v\|_{w,r}$$
$$V(w,k,\varepsilon) = \{v \in A^{k|w|-1} : \|v\|_{w,*} > (k-1)(b^{-|w|} + \varepsilon)\}$$

Given $w \in A^*$, let d be corresponding digit in $A^{|w|}$, and observe that for each $v \in V(w,k,\varepsilon)$, there is $\tilde{v} \in Bad(A^{|w|}, k-1, d, \varepsilon)$ and there are words $s, t \in A^*$ such that $|s| + |t| = |w| - 1$ and $v = s\tilde{v}t$. Thus,

$$|V(w,k,\varepsilon)| \le |w| b^{|w|-1} |Bad(A^{|w|}, k-1, d, \varepsilon)|.$$

So by Lemma 7.3.5, for every positive real δ, there is k_0 such that for every $k > 0$,

$$|V(w,k,\varepsilon)|\, b^{-(k|w|-1)} < \delta.$$

Fix ℓ and assume $w \in A^\ell$. Then, for any $k \ge \max(2, k_0)$,

$$\limsup_{n\to\infty} \frac{\|x[1 \ldots n\ell]\|_w}{n} \le \limsup_{n\to\infty} \frac{1}{n(k-1)\ell} \sum_{t=1}^{n\ell-\ell+1} \|x[t \ldots t+(k-1)\ell+\ell-2]\|_{w,2-t}$$

$$\le \limsup_{n\to\infty} \frac{1}{n(k-1)\ell} \sum_{t=1}^{n\ell-\ell+1} \|x[t \ldots t+(k-1)\ell+\ell-2]\|_{w,*}$$

$$= \limsup_{n\to\infty} \sum_{v\in A^{k\ell-1}} \frac{|x[1 \ldots (n+k-1)\ell-1]|_v}{n\ell} \frac{\|v\|_{w,*}}{k-1}$$

$$\leq \sum_{v\in A^{k\ell-1}} \left(\limsup_{n\to\infty} \frac{|x[1\ldots(n+k-1)\ell-1]|_v}{n\ell}\right) \frac{\|v\|_{w,*}}{k-1}$$

$$= \sum_{v\in A^{k\ell-1}} \left(\limsup_{n\to\infty} \frac{|x[1\ldots n\ell]|_v}{n\ell}\right) \frac{\|v\|_{w,*}}{k-1}$$

$$= \sum_{v\in A^{k\ell-1}} b^{-(k\ell-1)} \frac{\|v\|_{w,*}}{k-1}$$

$$= \sum_{v\in A^{k\ell-1}\setminus V(w,k,\varepsilon)} b^{-(k\ell-1)} \frac{\|v\|_{w,*}}{k-1} + \sum_{v\in V(w,k,\varepsilon)} b^{-(k\ell-1)} \frac{\|v\|_{w,*}}{k-1}$$

$$\leq (b^{-\ell}+\varepsilon) \sum_{v\in A^{k\ell-1}\setminus V(w,k,\varepsilon)} b^{-(k\ell-1)} + \sum_{v\in A^{k\ell-1}\setminus V(w,k,\varepsilon)} b^{-(k\ell-1)}$$

$$\leq b^{-\ell}+\varepsilon+\delta.$$

To obtain the inequality in the second line, observe that each aligned occurrence of w in a position $j\ell+1$, where $k-1\leq j<n$, is counted $(k-1)\ell$ times by $\|x[t\ldots t+k\ell-2]\|_{w,2-t}$ for $(j+1-k)\ell+1\leq t\leq j\ell+1$.

Since the last inequality is true for any $\delta,\varepsilon>0$, we conclude that

$$\limsup_{n\to\infty} \frac{\|x[1\ldots n\ell]\|_w}{n} \leq b^{-\ell}.$$

Applying Lemma 7.3.6, we conclude,

$$\lim_{n\to\infty} \frac{\|x[1\ldots n\ell]\|_w}{n} = b^{-|w|}.$$

3. We prove that *aligned normality* implies *strong aligned normality*. It is sufficient to prove that if x has aligned normality, then bx also has aligned normality. Define

$$U(k,w,i) = \{u\in A^k : u[i\ldots i+|w|-1]=w\}.$$

Fix a positive integer ℓ. For any $w\in A^\ell$ and for any positive integer r,

$$\liminf_{n\to\infty} \frac{\|(bx)[1\ldots nr\ell]\|_w}{nr} \geq \liminf_{n\to\infty} \frac{1}{r}\sum_{k=0}^{r-2} \sum_{u\in U(r\ell,w,2+\ell k)} \frac{\|x[1\ldots nr\ell]\|_u}{n}$$

$$= \frac{1}{r}\sum_{k=0}^{r-2} \sum_{u\in U(\ell r,w,2+\ell k)} b^{-r\ell}$$

$$= \frac{r-1}{r} b^{-\ell}.$$

For every r, the following equality holds:

$$\liminf_{n\to\infty} \frac{\|(bx)[1\ldots n\ell]\|_w}{n} = \liminf_{n\to\infty} \frac{\|(bx)[1\ldots nr\ell]\|_w}{nr}.$$

Then, using the inequality obtained above, we have

$$\liminf_{n\to\infty} \frac{\|(bx)[1\ldots n\ell]\|_w}{n} \geq \frac{r-1}{r} b^{-\ell}.$$

Since this last inequality holds for every r, we obtain,

$$\liminf_{n\to\infty} \frac{\|(bx)[1\ldots n\ell]\|_w}{n} \geq b^{-\ell}.$$

Finally, this last inequality is true for every $w \in A^\ell$; hence, by Lemma 7.3.6,

$$\lim_{n\to\infty} \frac{\|(bx)[1\ldots n\ell]\|_w}{n} = b^{-\ell}.$$

7.4 Normality as a Seemingly Weaker Condition

The following result is due to Piatetski-Shapiro in 1957 [481] and was rediscovered later by Borwein and Bailey [101] who called it the hot spot lemma. In Theorem 7.4.1, we present two versions of this result, one with non-aligned occurrences and one with aligned occurrences. The theorem has been extended relaxing the constant C to a sublinear function; see [118] for the references.

Theorem 7.4.1. *Let x be a real and let b be an integer greater than or equal to 2. Let $A = \{0, \ldots, b-1\}$. The following conditions are equivalent,*

1. *The real x is normal to base b.*
2. *There is a constant C such that for infinitely many lengths ℓ and for every w in A^ℓ*

$$\limsup_{n\to\infty} \frac{\|x[1\ldots n|w|]\|_w}{n} < C \cdot b^{-|w|}.$$

3. *There is a constant C such that for infinitely many lengths ℓ and for every w in A^ℓ*

$$\limsup_{n\to\infty} \frac{|x[1\ldots n]|_w}{n} < C \cdot b^{-|w|}.$$

Proof. The implications $1 \Rightarrow 2$ and $1 \Rightarrow 3$ follow from Theorem 7.3.7.

We now prove $2 \Rightarrow 1$. Define,

$$\widetilde{Bad}(A^{|w|}, k, w, \varepsilon) = \left\{ v \in A^{k|w|} : \left| \frac{\|v\|_w}{k} - b^{-|w|} \right| > \varepsilon \right\}$$

Lemma 7.3.5 implies that the size of $\widetilde{Bad}(A^{|w|}, k, w, \varepsilon)$ shrinks exponentially as k increases. Suppose there is C such that for infinitely many lengths ℓ and for every $w \in A^\ell$,

$$\limsup_{n\to\infty} \frac{\|x[1\ldots n\ell]\|_w}{n} < C \cdot b^{-\ell}.$$

Fix ℓ and $w \in A^\ell$. Fix $\varepsilon > 0$ and take k large enough.

$$\begin{aligned}
\liminf_{n\to\infty} \frac{\|x[1\ldots nk\ell\|_w}{nk} &= \liminf_{n\to\infty} \sum_{v\in A^{k\ell}} \frac{\|x[1\ldots nk\ell\|_v}{n} \frac{\|v\|_w}{k} \\
&\geq \liminf_{n\to\infty} \sum_{v\in A^{k\ell}\setminus\widetilde{Bad}(A^\ell,k,w,\varepsilon)} \frac{\|x[1\ldots nk\ell\|_v}{n} \frac{\|v\|_w}{k} \\
&\geq (1-\varepsilon)b^{-\ell} \liminf_{n\to\infty} \sum_{v\in A^{k\ell}\setminus\widetilde{Bad}(A^\ell,k,w,\varepsilon)} \frac{\|x[1\ldots nk\ell\|_v}{n} \\
&= (1-\varepsilon)b^{-\ell} \liminf_{n\to\infty} \left(1 - \sum_{v\in\widetilde{Bad}(A^\ell,k,w,\varepsilon)} \frac{\|x[1\ldots nk\ell\|_v}{n}\right) \\
&= (1-\varepsilon)b^{-\ell} \left(1 - \limsup_{n\to\infty} \sum_{v\in\widetilde{Bad}(A^\ell,k,w,\varepsilon)} \frac{\|x[1\ldots nk\ell]\|_v}{n}\right) \\
&\geq (1-\varepsilon)b^{-\ell} \left(1 - \sum_{v\in\widetilde{Bad}(A^\ell,k,w,\varepsilon)} \limsup_{n\to\infty} \frac{\|x[1\ldots nk\ell]\|_v}{n}\right) \\
&\geq (1-\varepsilon)b^{-\ell} \left(1 - \sum_{v\in\widetilde{Bad}(A^\ell,k,w,\varepsilon)} C \cdot b^{-k\ell}\right) \\
&\geq (1-\varepsilon)b^{-\ell}(1 - C\varepsilon).
\end{aligned}$$

Since this is true for all $\varepsilon > 0$,

$$\liminf_{n\to\infty} \frac{\|x[1\ldots nk\ell\|_w}{nk} \geq b^{-\ell}.$$

Finally, this last inequality is true for every $w \in A^\ell$; hence, by Lemma 7.3.6

$$\lim_{n\to\infty} \frac{\|x[1\dots n\ell]\|_w}{n} = b^{-\ell}.$$

The proof of implication 3 $\Rightarrow$ 1 is similar to 2 $\Rightarrow$ 1. Consider the set $Bad(A, w, k, \varepsilon)$ from Definition 7.3.3, the bound in Lemma 7.3.5, and the following fact. Fix w of length ℓ. Then for any n and k,

$$|x[1\dots n]|_w \leq \frac{1}{k} \sum_{v\in A^k} \sum_{r=0}^{k-1} \|x[1\dots n]\|_{v,r} \left(|v|_w + \ell - 1\right)$$

$$|x[1\dots n]|_w \geq \frac{1}{k-\ell+1} \sum_{v\in A^k} \sum_{r=0}^{k-1} \|x[1\dots n]\|_{v,r} |v|_w$$

Then,

$$\begin{aligned} \lim_{n\to\infty} \frac{|x[1\dots n]|_w}{n} &\leq \lim_{n\to\infty} \frac{1}{k} \sum_{v\in A^k} \sum_{r=0}^{k-1} \frac{\|x[1\dots n]\|_{v,r}}{n} \left(|v|_w + \ell - 1\right) \\ &= \lim_{n\to\infty} \frac{1}{k} \sum_{v\in A^k} \sum_{r=0}^{k-1} \frac{\|x[1\dots n]\|_{v,r}}{n} |v|_w. \end{aligned}$$

And

$$\begin{aligned} \lim_{n\to\infty} \frac{|x[1\dots n]|_w}{n} &\geq \lim_{n\to\infty} \frac{1}{k-\ell+1} \sum_{v\in A^k} \sum_{r=0}^{k-1} \frac{\|x[1\dots n]\|_{v,r}}{n} |v|_w \\ &\geq \lim_{n\to\infty} \frac{1}{k} \sum_{v\in A^k} \sum_{r=0}^{k-1} \frac{\|x[1\dots n]\|_{v,r}}{n} |v|_w \end{aligned}$$

Hence,

$$\lim_{n\to\infty} \frac{1}{k} \frac{|x[1\dots n]|_w}{n} = \lim_{n\to\infty} \sum_{v\in A^k} \sum_{r=0}^{k-1} \frac{\|x[1\dots n]\|_{v,r}}{n} |v|_w = \lim_{n\to\infty} \frac{1}{k} \sum_{v\in A^k} \frac{|x[1\dots n]|_v}{n} |v|_w.$$

7.5 Normality as Incompressibility by Finite Automata

The definition of normality can be expressed as a notion of incompressibility by finite automata with output also known as transducers. We consider *nondeterministic transducers*. We focus on transducers that operate in real time, that is, they process exactly one input alphabet symbol per transition. We start with the definition of a transducer (see Section 1.5.4 for the definition of automata without output).

Definition 7.5.1. A *nondeterministic transducer* is a tuple $\mathscr{T} = \langle Q, A, B, \delta, I, F \rangle$, where

- Q is a finite set of *states*,
- A and B are the input and output alphabets, respectively,
- $\delta \subset Q \times A \times B^* \times Q$ is a finite *transition* relation,
- $I \subseteq Q$ and $F \subseteq Q$ are the sets of *initial* and *final* states, respectively.

A transition of such a transducer is a tuple $\langle p, a, v, q \rangle$ which is written $p \xrightarrow{a|v} q$. A finite (respectively infinite) *run* is a finite (respectively infinite) sequence of consecutive transitions,

$$q_0 \xrightarrow{a_1|v_1} q_1 \xrightarrow{a_2|v_2} q_2 \cdots q_{n-1} \xrightarrow{a_n|v_n} q_n$$

A finite path is written $q_0 \xrightarrow{a_1 \cdots a_n | v_1 \cdots v_n} q_n$. An infinite path is *final* if the state q_n is final for infinitely many integers n. In that case, the infinite run is written $q_0 \xrightarrow{a_1 a_2 a_3 \cdots | v_1 v_2 v_3 \cdots} \infty$. An infinite run is accepting if it is final and furthermore its first state q_0 is initial. This is the classical Büchi acceptance condition. For two infinite words $x \in A^\omega$ and $y \in B^\omega$, we write $\mathscr{T}(x, y)$ whenever there is an accepting run $q_0 \xrightarrow{x|y} \infty$ in $\mathscr{T}$.

Definition 7.5.2. A transducer T is *bounded-to-one* if the function $y \mapsto |\{x : \mathscr{T}(x, y)\}|$ is bounded.

Definition 7.5.3. An infinite word $x = a_1 a_2 a_3 \cdots$ is *compressible* by a nondeterministic transducer if it has an accepting run $q_0 \xrightarrow{a_1|v_1} q_1 \xrightarrow{a_2|v_2} q_2 \xrightarrow{a_3|v_3} q_3 \cdots$ satisfying

$$\liminf_{n \to \infty} \frac{|v_1 v_2 \cdots v_n|}{n} \frac{\log |B|}{\log |A|} < 1.$$

It follows from the results in [175, 531] that the words which are not compressible by one-to-one deterministic transducers are exactly the normal words. A direct proof of this result appears in [56]. Extensions of this characterization for nondeterminisms and extra memory appear in [57, 131].

Theorem 7.5.4. *An infinite word is normal if and only if it not compressible by a bounded-to-one nondeterministic transducer.*

We first show that a non-normal word is compressible. We show a slightly stronger result since the transducer can be chosen deterministic and one-to-one.

Lemma 7.5.5. *A non-normal infinite word is compressible by a deterministic one-to-one transducer.*

Proof. Assume $x \in A^\omega$ is not normal. Let us show that x is compressible regardless of the choice of an output alphabet B. Since x is not normal, there is some word u_0 of length k such that

$$\lim_{n\to\infty} \frac{\|x[1\dots n]\|_{u_0}}{n/k} \neq \frac{1}{|A|^k}$$

meaning that the limit on the left side either does not exist or it does exist but it is different from $1/|A|^k$. There exists then an increasing sequence $(n_i)_{i\geq 0}$ of integers such that the limit $f_u = \lim_{i\to\infty} \|x[1\dots n_i]\|_u/(n_i/k)$ does exists for each word u of length k and furthermore $f_{u_0} \neq 1/|A|^k$. Note that $\sum_{u\in A^k} f_u = 1$. Let m be an integer to be fixed later. For each word $w \in A^{km}$, let f_w be defined by $f_w = \prod_{i=1}^m f_{u_i}$ where w is factorized $w = u_1 \cdots u_m$ with $|u_i| = k$ for each $1 \leq i \leq m$. Since $\sum_{w\in A^{km}} f_w = 1$, a word $v_w \in B^*$ can be associated with each word $w \in A^{km}$ such that $v_w \neq v_{w'}$ for $w \neq w'$, the set $\{v_w : w \in A^{mk}\}$ is prefix-free, and for each $w \in A^{km}$,

$$|v_w| \leq \lceil -\log f_w / \log |B| \rceil.$$

We claim that the words $(v_w)_{w\in A^{km}}$ can be used to construct a deterministic transducer $\mathcal{T}_m$ which compresses x for m large enough. The state set Q_m of $\mathcal{T}_m$ is the set $A^{<km}$ of words of length less than km. Its initial state is the empty word λ, and all states are final. Its set E_m of transitions is given by

$$E_m = \{w \xrightarrow{a|\lambda} wa : |wa| < km\} \cup \{w \xrightarrow{a|v_{wa}} \lambda : |wa| = km\}.$$

Let us denote by $\mathcal{T}_m(z)$ the output of the transducer $\mathcal{T}_m$ on some finite input word z. Suppose that the word z is factorized $z = w_1 \cdots w_n w'$ where $|w_i| = km$ for each $1 \leq i \leq n$ and $|w'| < km$. Note that $n = \lfloor |z|/km \rfloor$. Note also that the transducer $\mathcal{T}_m$ always comes back to its initial state λ after reading km symbols.

$$\begin{aligned}
|\mathcal{T}_m(z)| &= \sum_{i=1}^{n} |v_{w_i}| \\
&\leq \sum_{i=1}^{n} \lceil -\log f_{w_i} / \log |B| \rceil \\
&\leq \frac{|z|}{km} + \sum_{i=1}^{n} -\log f_{w_i} / \log |B| \\
&\leq \frac{|z|}{km} + \sum_{w\in A^{km}} \|z\|_w \frac{-\log f_w}{\log |B|} \\
&\leq \frac{|z|}{km} + \sum_{u\in A^{k}} \|z\|_u \frac{-\log f_u}{\log |B|}.
\end{aligned}$$

Applying this computation to the prefix $z = x[1..n]$ of x gives

$$\begin{aligned}\liminf_{n\to\infty} \frac{|\mathcal{T}_m(x[1..n]) \log |B|}{n \log |A|} &\le \lim_{i\to\infty} \frac{|\mathcal{T}_m(x[1..n_i]) \log |B|}{n_i \log |A|} \\ &\le \frac{\log |B|}{km \log |A|} + \frac{1}{k \log |A|} \sum_{u\in A^k} f_u(-\log f_u).\end{aligned}$$

Since at least one number f_u is not equal to $1/|A|^k$, the sum $\sum_{u\in A^k} f_u(-\log f_u)$ is strictly less than $k \log |A|$. For m chosen large enough, we obtain that $\mathcal{T}_m$ compresses x.

The following lemma is the key lemma to prove the converse.

Lemma 7.5.6. *Let ℓ be a positive integer, and let $u_1, u_2, u_3, \dots$ be words of length ℓ over the alphabet A such that $u_1u_2u_3\cdots$ is simply normal to word length ℓ. Let*

$$C_0 \xrightarrow{u_1|v_1} C_1 \xrightarrow{u_2|v_2} C_2 \xrightarrow{u_3|v_3} C_3 \cdots$$

be a run where each C_i is a configuration of some kind of transducer. Assume there is a real $\varepsilon > 0$ and a set $U \subseteq A^\ell$ of at least $(1-\varepsilon)|A|^\ell$ words such that $u_i \in U$ implies $|v_i| \ge \ell(1-\varepsilon)$. Then,

$$\liminf_{n\to\infty} \frac{|v_1v_2\cdots v_n|}{n\ell} \ge (1-\varepsilon)^3.$$

Proof. Assume words u_i as in the hypothesis. By definition of normality to word length ℓ, let n_0 be such that for every $u \in A^\ell$ and for every $n \ge n_0$,

$$|\{i : 1 \le i \le n, u_i = u\}| \ge n|A|^{-\ell}(1-\varepsilon).$$

Then, for every $n \ge n_0$,

$$\begin{aligned}|v_1v_2\cdots v_n| &= \sum_{i=1}^{n} |v_i| \\ &\ge \sum_{1\le i\le n, u_i\in U} |v_i| \\ &\ge \sum_{1\le i\le n, u_i\in U} \ell(1-\varepsilon) \\ &\ge n|A|^{-\ell}(1-\varepsilon) \sum_{u\in U} \ell(1-\varepsilon) \\ &\ge n|A|^{-\ell}(1-\varepsilon)(1-\varepsilon)|A|^\ell \ell(1-\varepsilon) \\ &\ge (1-\varepsilon)^3 n\ell.\end{aligned}$$

We now come back to the proof that normal words are not compressible by bounded-to-one transducers.

Proof. Fix a normal infinite word $x = a_1a_2a_3\cdots$, a real $\varepsilon > 0$, a bounded-to-one nondeterministic transducer $T = \langle Q, A, B, \delta, q_0, F\rangle$, and an accepting run $q_0 \xrightarrow{a_1|v_1} q_1 \xrightarrow{a_2|v_2} q_2 \xrightarrow{a_3|v_3} q_3 \cdots$. It suffices to show that there is ℓ and U such that Lemma 7.5.6 applies to this arbitrary choice of ε, T, and accepting run. For each word $u \in A^*$, let

$$h_u = \min\{|v| : \exists i, j, 0 \le i \le j, q_i \xrightarrow{u|v} q_j\}$$

be the minimum number of symbols that the processing of u can contribute to the output in the run we fixed. Let

$$U_\ell = \{u \in A^\ell : h_u \ge (1 - \varepsilon)\ell\}$$

be the set of words of length ℓ with relatively large contribution to the output. Let t be such that T is t-to-one. For each length ℓ, pair of states p, q that appear in the run, and for each word v, consider the set

$$U' = \{u \in A^\ell : p \xrightarrow{u|v} q\}.$$

Since p and q appear in the run, let $q_0 \xrightarrow{u_0|v_0} p$ be a prefix of the run and $q \xrightarrow{x_0|y_0} \infty$ be a suffix of the run. This implies $q \xrightarrow{x_0|y_0} \infty$ goes infinitely often through an accepting state. Thus, for different $u_1, u_2 \in U'$, there are accepting runs $q_0 \xrightarrow{u_0u_1x_0|v_0vy_0} \infty$ and $q_0 \xrightarrow{u_0u_2x_0|v_0vy_0} \infty$, from which it follows that $\mathcal{T}(u_0u_1x_0, v_0vy_0)$ and $\mathcal{T}(u_0u_2x_0, v_0vy_0)$. Therefore, by definition of t, $|U'| \le t$.

$$|\{u \in A^\ell : p \xrightarrow{u|v} q\}| \le t.$$

Thus,

$$|U_\ell| \ge |A|^\ell - |Q|^2 t |B|^{(1-\varepsilon)\ell+1}.$$

Fix ℓ such that $|U_\ell| > |A|^\ell(1 - \varepsilon)$ and apply Lemma 7.5.6 with $U = U_\ell$ to the considered run. This completes the proof.

7.6 Normality as Uniform Distribution Modulo 1

Let $(x_j)_{j\ge1}$ be a sequence of real numbers in the unit interval. The discrepancy of the N first elements is

$$D_N((x_j)_{j\ge1}) = \sup_{0\le u<v\le1} \left| \frac{|\{j : 1 \le j \le N \text{ and } u \le x_j \le v\}|}{N} - (v - u) \right|.$$

The sequence $(x_j)_{j\geq 1}$ is uniformly distributed in the unit interval if

$$\lim_{N\to\infty} D_N((x_j)_{j\geq 1}) = 0.$$

Schmidt [530] proved that for every sequence $(x_j)_{j\geq 1}$ of reals in the unit interval, there are infinitely many Ns such that

$$D_N((x_j)_{j\geq 1}) \geq \frac{\log N}{100\, N}.$$

There are sequences that achieve this lower bound, see [199].

Normality can be expressed in terms of uniform distribution modulo 1.

Theorem 7.6.1 (Wall 1949 [578]). *A real number x is normal to base b if and only if the sequence $(b^j x)_{j\geq 0}$ is uniformly distributed modulo* 1.

The discrepancy modulo 1 of the sequence $(b^j x)_{j\geq 0}$ gives the speed of convergence to normality to base b. Gál and Gál [236] and Philipp [480] proved that for almost all real numbers x, the discrepancy modulo 1 of the sequence $(b^j x)_{j\geq 0}$ is essentially the same and it obeys the law of iterated logarithm up to a constant factor that depends on b. Fukuyama [233] obtained the precise constant factor.

For a real number x, we write $\{x\} = x - \lfloor x \rfloor$ to denote the fractional part of x.

Theorem 7.6.2 (Fukuyama 2008 [233]). *For every real $\theta > 1$, there is a constant C_θ such that for almost all real numbers x (with respect to Lebesgue measure),*

$$\limsup_{N\to\infty} \frac{D_N(\{\theta^j x\}_{j\geq 0})\sqrt{N}}{\sqrt{\log\log N}} = C_\theta.$$

For instance, in case θ is an integer greater than or equal to 2,

$$C_\theta = \begin{cases} \sqrt{84/9}, & \text{if } \theta = 2 \\ \sqrt{2(\theta+1)/(\theta-1)}/2, & \text{if } \theta \text{ is odd} \\ \sqrt{2(\theta+1)\theta(\theta-2)/(\theta-1)^3}/2, & \text{if } \theta \geq 4 \text{ is even.} \end{cases}$$

It remains an open problem to establish the minimal discrepancy that can be achieved by a sequence $(\{b^j x\})_{j\geq 0}$ for some x.

The formulation of normality in terms of uniform distribution modulo 1 has been used in constructions of numbers that are normal to one base and not normal to another, where analytic tools come into play by way of Weyl's criterion of equidistribution [118, 364]. We give some references in Section 7.8.

7.7 Constructions of Numbers That Are Normal to a Given Base

Copeland and Erdős [166] generalized Champernowne's construction [141]. They show that for any increasing sequence of integers which does not grow too fast, the concatenation of its terms yields the expansion of a normal number. In particular, one can take the sequence of prime numbers. There are many other generalizations, such as [180, 435].

Other examples of normal numbers are defined by arithmetic constructions, the first ones are due to Stoneham [553] and Korobov [359]. For b, c be relatively prime integers greater than 1, the real numbers

$$\alpha_{b,c} = \sum_{n=1}^{\infty} \frac{1}{c^n b^{c^n}}$$

are normal to base b. Bailey and Borwein [31] showed that $\alpha_{2,3}$ is normal to base 2 but not to base 6. Noticeably, for any given integer base b, Levin [376] gives an arithmetic construction of a real number x, subtler than the series for $\alpha_{b,c}$, such that $D_N(\{b^n x\}_{n\geq 0})$is in $O((\log N)^2/N)$. This is the lowest discrepancy obtained so far, and it is close to the lower bound of $O(\log(N)/N)$ proved by Schmidt for arbitrary sequences (see Section 7.6 above). It is an open question whether there exists a real x for which $D_N(\{b^n x\}_{n\geq 0})$ reaches Schmidt's general lower bound.

Yet there is a very different kind of construction of expansions of normal numbers, based on combinatorics on words, specifically on de Bruijn words. This is due to Ugalde in [571].

In all the cases, the constructions have the form of an algorithm or can be turned into an algorithm. Recall that a real number x is computable if there is an algorithm that produces the expansion of x in some base, one digit after the other. The algorithm computes in linear time or has linear time complexity if it produces the first n digits in the expansion of x after performing a number of operations that is linear in n. Similarly, we consider polynomial, exponential, or hyper-exponential complexity. Algorithms with exponential complexity cannot run in human time, but algorithms with sub-exponential complexity can. In this monograph we analyze the computational complexity by counting the number of mathematical operations required to output the first k digits of the expansion of the computed number in a designated base. Thus, we do not count how many elementary operations are implied by each of the mathematical operations, which means that we neglect the computational cost of performing arithmetical operations with arbitrary precision.

In this section we present three constructions of real numbers that are insured to be normal to a given base. Since we care about the normality to just one base, we will just construct infinite words in a given alphabet. We first present the simplest possible construction à la Champernowne. Then we present Ugalde's construction, and we give a much simpler proof than the one in [571]. Finally we present a subtle construction of a normal word which has a self-similarity condition: the whole infinite word is identical to its subsequence at the even positions. This result is due to Becher, Carton, and Heiber (see [50, Theorem 4.2]).

7.7.1 À la Champernowne

Theorem 7.7.1. *Let A be an alphabet. Let w_j be the concatenation of all words over A of length j, in lexicographic order. The infinite word $w = w_1 w_2 w_3 \dots$ is normal to alphabet A.*

Proof. Let $w = w_1 w_2 w_3 \dots = a_1 a_2 \dots$ where each a_i is a symbol in A. Fix N and let n be such that

$$\sum_{j=1}^{n} j|A|^j \le N < \sum_{J-1}^{n+1} j|A|^j$$

Let u be a block of symbols in alphabet A. The occurrences of u in the prefix of $w[1..N]$ are divided into two classes: those that are fully contained in a single block of length i in some w_i and those that overlap several blocks.

$$\begin{aligned}
\frac{|a_1 a_2 \dots a_N|_u}{N} &\le \frac{|a_1 a_2 \dots a_{x_{n+1}}|_u}{n|A|^n} \\
&\le \frac{1}{n|A|^n}\left(\sum_{j=|u|}^{n+1} (j - |u| + 1)|A|^{j-|u|} + \sum_{j=1}^{n+1} (|u| - 1)|A|^j\right) \\
&\le \frac{(n+1)|A|^{-|u|}}{n|A|^n} \sum_{j=1}^{n+1} |A|^j + \frac{|u|}{n|A|^n} \sum_{j=1}^{n+1} |A|^j \\
&\le \frac{(n+1)}{n(|A|-1)}|A|^{-|u|} + \frac{|u||A|^2}{n(|A|-1)}.
\end{aligned}$$

The first term accounts for occurrences fully contained in a block and the second of for those that overlap several blocks. It follows that

$$\limsup_{N\to\infty} \frac{|a_1 a_2 \dots a_N|_u}{N} \le \frac{2}{|A|-1}|A|^{-|u|}.$$

By Lemma 7.4.1, w is normal to alphabet A.

The infinite word w can be computed very efficiently: the first N symbols can be produced in at most $O(N)$ elementary operations. It is also possible to produce just the N-th symbol of w in $O(\log N)$ many elementary operations.

7.7.2 *Infinite de Bruijn Words*

See [76] for a fine presentation and history of de Bruijn words.

Definition 7.7.2 ([182, 517]). A (noncyclic) *de Bruijn word* of order n over alphabet A is a word of length $|A|^n + n - 1$ such that every word of length n occurs in it exactly once.

Every de Bruijn word of order n over A with $|A| \geq 3$ can be extended to a de Bruijn word of order $n + 1$. Every de Bruijn word of order n over A with $|A| = 2$ can *not* be extended to order $n + 1$, but it can be extended to order $n + 2$. See [55] for a complete proof of this fact. This allows us to define infinite de Bruijn words, as follows.

Definition 7.7.3. An infinite de Bruijn word $w = a_1a_2\dots$ in an alphabet of at least three symbols is an infinite word such that, for every n, $a_1 \dots a_{|A|^n+n-1}$ is a de Bruijn word of order n. In case the alphabet has two symbols, an infinite de Bruijn word $w = a_1a_2\dots$ is such that, for every odd n, $a_1 \dots a_{|A|^n+n-1}$ is a de Bruijn word of order n.

Ugalde [571] was the first to prove that infinite de Bruijn words are normal.

Theorem 7.7.4. *Infinite de Bruijn words are normal.*

Proof. In case the alphabet A has two symbols, consider instead the words in the alphabet A' of four symbols obtained by the morphism mapping blocks two symbols in A to one symbol in A', and prove normality for alphabet A'.

Suppose that the alphabet A has at least 3 symbols. Let $x = a_1a_2\dots$ be an infinite de Bruijn word over A. Fix a word u of length ℓ and $n > |A|^\ell + \ell - 1$. Then u occurs in a de Bruijn word of order $n \geq \ell$ between $|A|^{n-\ell}$ and $|A|^{n-\ell} + n - \ell$ times. To see this, observe if u occurs at a position i, for some i such that $1 \leq i \leq |A|^n$, then position i is the beginning of an occurrence of a word of length n. There are exactly $|A|^{n-\ell}$ words of length n whose first ℓ symbols are u. In addition, there are exactly $n - \ell$ other positions in a de Bruijn word of order n at which a subword of length ℓ may start. Since x is infinite de Bruijn, by definition, for each n, $a_1 \dots a_{|A|^n+n-1}$ is a de Bruijn word or order n. Fix a position N, and let n be such that

$$|A|^n + n - 1 \leq N < |A|^{n+1} + n.$$

Then,

$$\frac{|a_1 \dots a_N|_u}{N} \leq \frac{|a_1 \dots a_{|A|^{n+1}+n}|_u}{|A|^n + n - 1} \leq \frac{|A|^{n+1-\ell} + n - \ell}{|A|^n + n - 1} \leq 2\,|A|^{-\ell+1}.$$

Thus,

$$\limsup_{N\to\infty} \frac{|a_1 \dots a_N|_u}{N} < 2\,|A|^{-\ell+1}.$$

By Lemma 7.4.1, using $C = 2\,|A|$, x is normal.

There is an obvious algorithm to compute an infinite de Bruijn word which, for each $n \geq 1$, extends a Hamiltonian cycle in a de Bruijn graph of order n to an Eulerian cycle in the same graph. This is done in time exponential in n. No efficient algorithm is known to compute the N-th symbol of an infinite de Bruijn word without computing the first N symbols.

7.7.3 A Normal and Self-Similar Word

For a given finite or infinite word $x = a_1a_2a_3\dots$ where each a_i is a symbol in alphabet A, define even$(x) = a_2a_4a_6\cdots$ and $odd(x) = a_1a_3a_5\cdots$. Thus, $x = \text{even}(x)$ means that $a_n = a_{2n}$ for all n.

Theorem 7.7.5 ([50, Theorem 4.2]). *There is a normal word x such that $x =$* even(x).

We construct a normal word $x = a_1a_2a_3\cdots$ over the alphabet $\{0, 1\}$ such that $a_{2n} = a_n$ for every n. The construction can be extended to an alphabet of size k to obtain a word $a_1a_2a_3\cdots$ such that $a_{kn} = a_n$ for each integer $n \geq 1$.

A finite word w is called *ℓ-perfect* for an integer $\ell \geq 1$, if $|w|$ is a multiple of ℓ and all words of length ℓ have the same number $|w|/(\ell 2^\ell)$ of aligned occurrences in w.

Lemma 7.7.6. *Let w be an ℓ-perfect word such that $|w|$ is a multiple of $\ell 2^{2\ell}$. Then, there exists a 2ℓ-perfect word z of length $2|w|$ such that* even$(z) = w$.

Proof. Since $|w|$ is a multiple of $\ell 2^{2\ell}$ and w is ℓ-perfect, for each word u of length ℓ, $\|w\|_u$ is a multiple of 2^ℓ. Consider a factorization of $w = w_1w_2\cdots w_r$ such that for each i, $|w_i| = \ell$. Thus, $r = |w|/\ell$. Since w is ℓ-perfect, for any word u of length ℓ, the set $\{i : w_i = u\}$ has cardinality $r/2^\ell$. Define z of length $2|w|$ as $z = z_1z_2\cdots z_r$ such that for each i, $|z_i| = 2\ell$, $even(z_i) = w_i$ and for all words u and u' of length ℓ, the set $\{i : z_i = u' \vee u\}$ has cardinality $r/2^{2\ell}$. This latter condition is achievable because, for each word u of length ℓ, the set $\{i : \text{even}(z_i) = u\}$ has cardinality $r/2^\ell$ which is a multiple of 2^ℓ, the number of possible words u'.

Corollary 7.7.7. *Let w be an ℓ-perfect word for some even integer ℓ. Then there exists an ℓ-perfect word z of length $2|w|$ such that* even$(z) = w$.

Proof. Since w is ℓ-perfect, it is also $\ell/2$-perfect. Furthermore, if u and v are words of length $\ell/2$ and ℓ, respectively, then $\|w\|_u = 2^{\ell/2+1}\|w\|_v$. Thus, the hypothesis of Lemma 7.7.6 is fulfilled with $\ell/2$.

Corollary 7.7.8. *There exist a sequence $(w_n)_{n\geq 1}$ of words and a sequence of positive integers $(\ell_n)_{n\geq 1}$ such that $|w_n| = 2^n$, $\mathrm{even}(w_{n+1}) = w_n$, w_n is ℓ_n-perfect and $(\ell_n)_{n\geq 1}$ is nondecreasing and unbounded. Furthermore, it can be assumed that $w_1 = 01$.*

Proof. We start with $w_1 = 01$, $\ell_1 = 1$, $w_2 = 1001$, and $\ell_2 = 1$. For each $n \geq 2$, if $\ell_n 2^{2\ell_n}$ divides $|w_n|$, then $\ell_{n+1} = 2\ell_n$ and w_{n+1} is obtained by Lemma 7.7.6. Otherwise, $\ell_{n+1} = \ell_n$ and w_{n+1} is obtained by Corollary 7.7.7. Note that the former case happens infinitely often, so $(\ell_n)_{n\geq 1}$ is unbounded. Also note that each ℓ_n is a power of 2.

Proof (of Theorem 7.7.5). Let $(w_n)_{n\geq 1}$ be a sequence given by Corollary 7.7.8. Let $x = 11w_1w_2w_3\cdots$ We first prove that x satisfies $x = \mathrm{even}(x)$. Note that $x[2^k + 1..2^{k+1}] = w_k$ for each $k \geq 1$ and $x[1..2^{k+1}] = 11w_1\cdots w_k$. The fact that $w_n = \mathrm{even}(w_{n+1})$ implies $x[2n] = x[n]$, for every $n \geq 3$. The cases for $n = 1$ and $n = 2$ hold because $x[1..4] = 1101$.

We prove that x is normal. Consider an arbitrary index n_0. By construction, w_{n_0} is ℓ_{n_0}-perfect, and for each $n \geq n_0$, w_n is also ℓ_{n_0}-perfect. For every word u of length ℓ_{n_0} and for every $n \geq n_0$,

$$\|x[1..2^{n+1}]\|_u \leq \|x[1..2^{n_0}]\|_u + \|w_{n_0}\dots w_n\|_u.$$

Then, for every N such that $2^n \leq N < 2^{n+1}$ and $n \geq n_0$,

$$\begin{aligned}
\frac{\|x[1..N]\|_u}{N/\ell_{n_0}} &\leq \frac{\|x[1..2^{n+1}]\|_u}{N/\ell_{n_0}} \\
&\leq \frac{\|x[1..2^{n_0}]\|_u + \|w_{n_0}\dots w_n\|_u}{N/\ell_{n_0}} \\
&\leq \frac{\|x[1..2^{n_0}]\|_u}{2^n/\ell_{n_0}} + \frac{\|w_{n_0}\dots w_n\|_u}{2^n/\ell_{n_0}} \\
&= \frac{\|x[1..2^{n_0}]\|_u}{2^n/\ell_{n_0}} + \frac{(2^{n_0} + \ldots + 2^n)/(\ell_{n_0}2^{\ell_{n_0}})}{2^n/\ell_{n_0}} \\
&< \frac{\|x[1..2^{n_0}]\|_u}{2^n/\ell_{n_0}} + \frac{2}{2^{\ell_{n_0}}}.
\end{aligned}$$

For large values of N and n such that $2^n \leq N < 2^{n+1}$, the expression $\|x[1..2^{n_0}]\|_u/(2^n/\ell_{n_0})$ becomes arbitrarily small. We obtain for every word u of length ℓ_{n_0},

$$\limsup_{N\to\infty} \frac{\|x[1..N]\|_u}{N/\ell_{n0}} \le 3\, 2^{-\ell_{n0}}.$$

Since the choice of ℓ_{n0} was arbitrary, the above inequality holds for each ℓ_n. Since $(\ell_n)_{n\ge1}$ is unbounded, the hypothesis of Lemma 7.4.1 is fulfilled, with $C = 3$, so we conclude that x is normal.

It is possible to compute a normal word x such that $x = even(x)$ in linear time.

7.8 Constructions of Absolutely Normal Numbers

The first constructions of absolutely normal numbers were given, independently, by Lebesgue [371] and Sierpiński [547], when the theory of computing was undeveloped. The numbers defined by these two constructions cannot be computed because they are just determined as the infimum of a set defined by infinite unions and intersections. The first example of a computable absolutely normal number was given by Turing [52, 570], and, unfortunately, it has doubly exponential time complexity. The computable reformulation of Sierpiński's construction [51] has also doubly exponential time complexity.

There are exponential algorithms that use analytic tools, such as Levin's construction [19, 375] of an absolutely normal number with fast convergence to normality and Schmidt's construction [529] of a number that is normal to all the bases in a prescribed set but not normal to the bases in the complement, see Theorem 7.9.3.

Some years ago, several efficient algorithms were published. Figueira and Nies gave in [222] an algorithm based on martingales with polynomial time complexity. Becher, Heiber, and Slaman [53] reworked Turing's strategy and obtained an algorithm with just above quadratic time complexity. Madritsch, Scheerer, and Tichy [397] adapted it and obtained an efficient algorithm to compute a number that is normal to all Pisot bases. Recently Lutz and Mayordomo [395] obtained an algorithm based on martingales with poly-logarithmic linear time complexity.

Another aspect in constructions of absolutely normal numbers is the speed of convergence to normality. Aistleitner et al. [8] constructed an absolutely normal real number x, so that for every integer b greater than or equal to 2 the discrepancy modulo 1 of the sequence $(b^n x)_{n\ge0}$ is strictly below that realized by almost all real numbers (see Section 7.6) The construction yields an exponential algorithm that achieves a discrepancy estimate lower than that in Levin's work [375]. According to Scheerer's analysis [525], currently there are no other known constructions achieving a smaller discrepancy. The problem of the existence of an absolutely normal number computable with polynomial complexity having fast rate of convergence to normality remains open.

We will present two algorithms, and we will analyze their computational complexity. We first need some notation.

If v is a block of digits in base b, I_v denotes b-ary interval

$$(.v, .v + b^{-|v|})$$

Definition 7.8.1. Let x be a real in the unit interval, and let x_b be its expansion in base b. We define

$$\Delta_N(x_b) = \max_{d\in\{1,\ldots,b\}} \left| \frac{|x_b[1\ldots N]|_d}{N} - \frac{1}{b} \right|.$$

If w is a finite block of digits in base b, we just write $\Delta(w)$ instead of $\Delta_{|w|}(w)$.

7.8.1 Turing's Construction of Absolutely Normal Numbers

Theorem 7.8.2 (Turing 1937? [52, 570]). *There is an algorithm that computes the expansion in base 2 of an absolutely normal number y in the unit interval.*

The construction is done by steps. We will use n as the step number, and we will define the following functions of n: N_n is the number of digits looked at step n, b_n is the largest base considered at step n, and ε_n is the maximum difference between the expected frequency of digits and the tolerated frequency of digits at step n. It is required that b_n be nondecreasing and unbounded and ε_n be nonincreasing and goes to zero. Many instantiations of these functions can work.

Definition 7.8.3. Define the following functions of n,

$$N_n = 2^{n_0+2n}, \text{where} n_0 = 11,$$

$$b_n = \lfloor \log N_n \rfloor$$

$$\varepsilon_n = 1/b_n.$$

Define the following sets of real numbers,

$$E_0 = (0,1), \text{ and for each } n$$

$$E_n = \bigcap_{b\in\{2,\ldots,b_n\}} \{x \in (0,1) : \Delta_{N_n}(x_b) < \varepsilon_n\}.$$

The value n_0 has been selected so that the forthcoming calculations are simple. Observe that for every n, $b_n \geq 2$. Thus, for each n the set E_n consists of all the real numbers whose expansion in the bases 2,3,..., b_n exhibit good frequencies of digits in the first N_n digits. We write μ for Lebesgue measure.

Proposition 7.8.4. *For each n, E_n is a finite union of open intervals with rational endpoints, $E_{n+1} \subset E_n$, and $\mu E_n > 1 - N_n^2$.*

Proof. The values of N_n and ε_n satisfy the hypotheses of Lemma 7.3.5 with digits in base b (i.e., let k be N_n, let ℓ be 1, and let ε be ε_n),

$$\mu\{x \in (0, 1) : \Delta_{N_n}(x_b) \geq \varepsilon_n\} < 2b^2 e^{-\varepsilon_n^2 b N_n/6}.$$

Then, for $b_n \leq \log N_n$, $\varepsilon \geq 1/\log N_n$ and $N_n > e^{10}$ can be checked that

$$\sum_{b=2}^{b_n} 2b^2 e^{-\varepsilon^2 b N_n/6} < 1/N_n^2.$$

Hence,

$$\mu E_n \geq 1 - \sum_{b=2}^{b_n} 2b^2 e^{-\varepsilon^2 b N_n/6} \geq 1 - 1/N_n^2.$$

Proposition 7.8.5. *The set $\bigcap_{n\geq 0} E_n$ has positive measure and consists just of absolutely normal numbers.*

Proof. From Proposition 7.8.4 follows that $\bigcap_{n\geq 0} E_n$ has positive measure. Suppose $x \in \bigcap_{n\geq 0} E_n$. Then, for every n, $x \in E_n$, so for each $b = 2, 3, \ldots, b_n$,

$$\Delta_{N_n}(x_b) \leq \varepsilon_n.$$

Let b be an arbitrary base, and let M be an arbitrary position. Let n be such that

$$N_n \leq M < N_{n+1}.$$

For each b smaller than b_n we have that for each digit d in $\{0, \ldots, b-1\}$,

$$\frac{|x_b[1\ldots M]|_d}{M} < \frac{|x_b[1\ldots N_{n+1}]|_d}{N_n} < \frac{N_{n+1}}{N_n}\left(\frac{1}{b} + \varepsilon_{n+1}\right) = 4\left(\frac{1}{b} + \varepsilon_{n+1}\right)$$

$$\frac{|x_b[1\ldots M]|_d}{M} > \frac{|x_b[1\ldots N_n]|_d}{N_{n+1}} > \frac{N_n}{N_{n+1}}\left(\frac{1}{b} - \varepsilon_n\right) = \frac{1}{4}\left(\frac{1}{b} - \varepsilon_n\right).$$

Since ε_n is decreasing in n and goes to 0, we conclude that for each base $b = 2, 3 \ldots$,

$$\limsup_{N\to\infty} \frac{|x_b[1\ldots N]|_d}{N} < 4\frac{1}{b}.$$

Using the morphism that maps digits in base b^ℓ to words in base b, this is equivalent to say that for each base b, for every length ℓ, and for every word u of length ℓ,

$$\limsup_{N\to\infty} \frac{\|x_b[1\ldots \ell N]\|_u}{N} < 4\frac{1}{b^\ell}.$$

By Theorem 7.4.1, x is normal to every base b, hence absolutely normal.

Turing's construction selects nested binary intervals $I_1, I_2, \ldots$ such that, for each n, $\mu I_n = 1/2^n$. Each interval I_{n+1} is either the left half or the right half of I_n. The base-2 expansion of the computed number y is denoted with the sequences $y_1, y_2, \ldots$ which is the trace of the left/right selection at each step. Recall Definition 7.8.3 where the sets E_n are defined, for every $n \geq 0$.

Initial step, $n = 0$. $I_0 = (0, 1)$, $E_0 = (0, 1)$.

Recursive step, $n > 1$. Assume that in the previous step we have computed I_{n-1}.
Let I_n^0 be left half of I_{n-1} and I_n^1 be right half of I_{n-1}.
If $\mu\left(I_n^0 \cap \bigcap_{j=0}^{n} E_j\right) > 1/N_n$ then let $I_n = I_n^0$ and $y_n = 0$.
Else let $I_n = I_n^1$ and $y_n = 1$.

Proof (of Theorem 7.8.2). From Algorithm 7.8.1 follows that the intervals $I_1, I_2, \ldots$ are nested, and for each n, $\mu I_n = 1/2^n$. To prove the correctness of the algorithm, we need to prove that the following condition is invariant along every step n of the algorithm:

$$\mu\left(I_n \cap \bigcap_{j=1}^{n} E_j\right) > 0.$$

We prove it by induction on n. Recall $N_n = 2^{n_0+2n}$.
Base case $n = 0$.

$$\mu(I_0 \cap E_0) = \mu((0,1)) > \frac{1}{N_0^2} = \frac{1}{2^{2n_0}}.$$

Inductive case, $n > 0$. Assume as inductive hypothesis that

$$\mu\left(I_n \cap \bigcap_{j=0}^{n} E_j)\right) > \frac{1}{N_n}.$$

We now show it holds for $n+1$. Recall $\mu E_n > 1 - 1/N_n^2$. Then,

$$\mu\left(I_n \cap \bigcap_{j=0}^{n+1} E_j\right) = \mu\left(I_n \cap \bigcap_{j=0}^{n} E_j \cap E_{n+1}\right) > \frac{1}{N_n} - \frac{1}{N_{n+1}^2} > \frac{2}{N_{n+1}}.$$

Since the algorithm chooses I_{n+1} among I_n^0 and I_n^1 ensuring $\mu(I_{n+1} \cap \bigcap_{j=0}^{n+1} E_j) > 1/N_{n+1}$, we conclude $\mu(I_{n+1} \cap \bigcap_{j=0}^{n+1} E_j) > 1/N_{n+1}$ as required.

Finally, since $(I_n)_{n\geq 0}$ is a nested sequence of intervals and $\mu(I_n \cap \bigcap_{j=0}^{n} E_j) > 0$, for every n, we obtain that

$$\bigcap_{n\geq 0} I_n = \bigcap_{n\geq 0}\left(I_n \cap \bigcap_{j=0}^{n} E_j \right).$$

contains a unique real number y. By Lemma 7.8.5, all the elements in $\bigcap_{j\geq 0} E_j$ are absolutely normal. This concludes the proof of Theorem 7.8.2.

We now bound the number of mathematical operations computed by the algorithm to output the first n digits of the expansion of the computed number in a designated base. We do not count how many elementary operations are implied by each of the mathematical operations, which means that we neglect the computational cost of performing arithmetical operations with arbitrary precision.

Proposition 7.8.6. *Turing's algorithm has double exponential time complexity.*

Proof. At step n the algorithm computes the set $I_{n-1} \cap E_n$ by computing first the set

$$I_{n-1} \cap E_n = \bigcap_{b\in\{2,\ldots,b_n\}} \{x \in I_{n-1} \cap E_{n-1} : \Delta_{N_n}(x_b) < \varepsilon_n\}$$

and choosing one of its halves. Then, the number of words to be examined to compute $I_n \cap E_n$ is

$$(b_n)^{N_n - N_{n-1} - (n-1)}.$$

Since $N_n = 2^{n_0+2n}$ and $b_n = \lfloor \log N_n \rfloor$, this number of words is in the order of

$$O\big((2n)^{2^{2n}}\big).$$

The examination of all these words requires $O\big((2n)^{2^{2n}}\big)$ mathematical operations. We conclude by noticing that using the set $I_n \cap E_n$ at step n the algorithm determines the $n-th$ binary digit of the computed number.

7.8.2 A Fast Construction of Absolutely Normal Numbers

We give a simplified version of the algorithm given by Becher, Heiber, and Slaman in [53].

Theorem 7.8.7. *There is an algorithm that computes an absolutely normal number x in nearly quadratic time completely: the first n digits in the expansion of x in base 2 are obtained by performing $O(n^2 \sqrt[4]{\log n})$ mathematical operations.*

The following two lemmas are not hard to prove.

Lemma 7.8.8 ([53, Lemma 3.1]). *Let u and v be blocks and let ε be a positive real number.*

1. *If $\Delta(u) < \varepsilon$ and $\Delta(v) < \varepsilon$ then $\Delta(uv) < \varepsilon$.*
2. *If $\Delta(u) < \varepsilon$, $v = a_1 \dots a_{|v|}$ and $|v|/|u| < \varepsilon$ then $\Delta(vu) < 2\varepsilon$, and for every ℓ such that $1 \le \ell \le |v|$, $\Delta(ua_1a_2 \dots a_\ell) < 2\varepsilon$.*

Lemma 7.8.9 (Lemma 3.4 [53]). *For any interval I and any base b, there is a b-ary subinterval J such that $\mu J \ge \mu I/(2b)$.*

The next two definitions are the core of the construction.

Definition 7.8.10. A t-sequence $\vec{\sigma}$ is a sequence of intervals $(\sigma_2, \dots, \sigma_t)$ such that for each base $b = 2, \dots, t$, σ_b is b-ary, for each base $b = 3, \dots, t$, $\sigma_b \subset \sigma_{b-1}$ and $\mu\sigma_b \ge \mu\sigma_{b-1}/(2b)$.

Observe that the definition implies $\mu\sigma_t \ge (\mu\sigma_2)/(2^t t!)$.

Definition 7.8.11. A t-sequence $\vec{\tau} = (\tau_2, \dots, \tau_t)$ *refines* a t'-sequence $\vec{\sigma} = (\sigma_2, \dots, \sigma_{t'})$ if $t' \le t$ and $\tau_b \subset \sigma_b$ for each $b = 2, \dots, t'$. A refinement has *discrepancy less than* ε if for each $b = 2, ..t'$, there are words u, v such that $\sigma_b = I_u$, $\tau_b = I_{uv}$, and $\Delta(v) < \varepsilon$.

We say that an interval is b-ary of *order* n if it is of the form

$$\left(\frac{a}{b^n}, \frac{a+1}{b^n}\right)$$

for some integer a such that $0 \le a < b^n$. If σ_b and τ_b are b-ary intervals, and $\tau_b \subseteq \sigma_b$, we say that the *relative order* of τ_b with respect to σ_b is the *order* of τ_b minus the *order* of σ_b.

Lemma 7.8.12. *Let t be an integer greater than or equal to 2, let t' be equal to t or to $t+1$, and let ε be a positive real less than $1/t$. Then, any t-sequence $\vec{\sigma} = (\sigma_2, \dots, \sigma_t)$ admits a refinement $\vec{\tau} = (\tau_2, \dots, \tau_{t'})$ with discrepancy less than ε. The relative order of τ_2 can be any integer greater than or equal to* $\max(6/\varepsilon, 24(\log_2 t)(\log(t!))/\varepsilon^2)$.

Proof. First assume $t' = t$. We must pick a t-sequence $(\tau_2, \dots, \tau_t)$ that refines $(\sigma_2, \dots, \sigma_t)$ in a zone of low discrepancy. This is possible because the measure of the zones of large discrepancy decreases at an exponential rate in the order of the interval. To prove the lemma, we need to determine the relative order N of τ_2 such that the measure of the union of the bad zones inside σ_2 for the bases $b = 2, \dots t$ is strictly less than the measure of the set all the possible t-ary subintervals τ_t of σ_2.

Let L be the largest binary subinterval in σ_t. Consider the partition of L in 2^N binary intervals τ_2 of equal length. For each τ_2, apply iteratively Lemma 7.8.9 to define $\tau_3, \ldots, \tau_{t_n}$. In this form, we have defined 2^N many t_n-sequences $(\tau_2, \ldots \tau_t)$. Let S be the union of the set of all possible intervals τ_t over these 2^N many t_n-sequences. Hence, by the definition of t-sequence,

$$\mu S \geq \mu L/(2^t t!).$$

By Lemma 7.8.9,

$$\mu L \geq \mu\sigma_t/4.$$

And by the definition of t-sequence again,

$$\mu\sigma_t \geq \mu\sigma_2/(2^t t!).$$

Combining inequalities we obtain,

$$\mu S \geq \mu\sigma_2/(2^t t!\; 4\; 2^t t!)$$

Now consider the bad zones inside σ_2. For each $b = 2, \ldots t$, for a length N and a real value ε, consider the following set of intervals of relative order $\lceil N/\log_2 b\rceil$ with respect to σ_2,

$$B_{b,\lceil N/\log_2 b\rceil,\varepsilon} = \bigcup_{\substack{u\in\{0,\ldots,b-1\}^{\lceil N/\log_2 b\rceil} \\ \Delta(u)\geq\varepsilon}} I_u.$$

Thus, the actual measure of the bad zones is

$$\mu\sigma_2\, \mu\Big(\bigcup_{b=2,..,t} \mu B_{b,\lceil N/\log_2 b\rceil,\varepsilon}\Big)$$

Then, N must be such that

$$\mu\sigma_2\, \mu\Big(\bigcup_{b=2,..,t} B_{b,\lceil N/\log_2 b\rceil,\varepsilon}\Big) < \mu S.$$

Using Lemma 7.3.5 on the left and the inequality above for μS on the right it suffices that N be greater than $6/\varepsilon$ and also N be such that

$$2t^2 \cdot e^{-\varepsilon^2 (N/3\log_2 t)} < \frac{1}{2^t t!}\frac{1}{4}\frac{1}{2^t t!}.$$

We can take N greater than or equal to $\max(6/\varepsilon, 24(\log_2 t)(\log(t!))/\varepsilon^2)$.

The case $t' = t + 1$ follows easily by taking first a t-sequence $\vec{\tau}$ refining $\vec{\sigma}$ with discrepancy less than ε. Definition 7.8.11 does not require any discrepancy considerations for τ_{t+1}. Take τ_{t+1} the largest $(t+1)$-ary subinterval of τ_t. By Lemma 7.8.9, $\mu\tau_{t+1} \geq (\mu\tau_t)/(2(t+1))$. This completes the proof of the lemma.

The algorithm considers three functions of the step number n: t_n is the maximum base to be considered at step n, ε_n is the maximum discrepancy tolerated at step n, and N_n is the number of digits in base 2 added at step n. It is required that t_n be increasing and ε_n be decreasing. Many instantiations of this functions can work.

The algorithm constructs $\vec{\sigma}_0, \vec{\sigma}_1, \vec{\sigma}_2, \ldots$ such that $\vec{\sigma}_0 = (0, 1)$, and for each $n \geq 1$, $\vec{\sigma}_n$ is t_n-sequence that refines $\vec{\sigma}_{n-1}$ with discrepancy ε_n and such that the order of $\sigma_{n,2}$ is N_n plus the order of $\sigma_{n-1,2}$.

Definition 7.8.13. Define the following functions of n,

$$\begin{aligned} t_n &= \max(2, \lfloor \sqrt[4]{\log n} \rfloor), \\ \varepsilon_n &= 1/t_n, \\ N_n &= \lfloor \log n \rfloor + n_{start}, \end{aligned}$$

where n_{start} is the minimum integer such that it validates the condition in Lemma 7.8.12. Thus, we require that for every positive n,

$$\lfloor \log n \rfloor + n_{start} \geq 6/\varepsilon_n \qquad \text{and}$$

$$\lfloor \log n \rfloor + n_{start} \geq 24(\log_2 t_n)(\log(t_n!))/\varepsilon_n^2.$$

Initial step, $n = 1$. $\vec{\sigma}_1 = (\sigma_2)$, with $\sigma_2 = (0, 1)$.

Recursive step, $n > 1$. Assume $\vec{\sigma}_{n-1} = (\sigma_2, \ldots, \sigma_{t_{n-1}})$. Take $\vec{\sigma}_n = (\tau_2, \ldots, \tau_{t_n})$ the leftmost t_n-sequence such that it is refinement of $\vec{\sigma}_{n-1}$ with discrepancy less than ε_n such that the relative order of τ_2 is N_n.

Proof (of Theorem 7.8.7). Consider Algorithm 7.8.2. The existence of the sequence $\vec{\sigma}_1, \vec{\sigma}_2, \ldots$ is guaranteed by Lemma 7.8.12. We have to prove that the real number x defined by the intersection of all the intervals in the sequence is absolutely normal. We pick a base b and show that x is simply normal to base b. Let $\tilde{\varepsilon} > 0$. Choose n_0 so that $t_{n_0} \geq b$ and $\varepsilon_{n_0} \leq \tilde{\varepsilon}/4$. At each step n after n_0 the expansion of x in base b was constructed by appending blocks u_n such that $\Delta(u_n) < \varepsilon_{n_0}$. Thus, by Lemma 7.8.8 (item 1) for any $n > n_0$,

$$\Delta(u_{n_0} \ldots u_n) < \varepsilon_{n_0}.$$

Applying Lemma 7.8.8 (item 2a), we obtain n_1 such that for any $n > n_1$

$$\Delta(u_1 \ldots u_n) < 2\varepsilon_{n_0}.$$

Let $N_n^{(b)}$ be the relative order of τ_b with respect to σ_b. By Lemma 7.8.9,

$$\frac{N_n}{\log_2 b} \leq N_n^{(b)} \leq \frac{N_n + 1}{\log_2 b} + 1.$$

Since $N_n = \lfloor \log n \rfloor + n_{start}$, N_n grows logarithmically and so does $N_n^{(b)}$ for each base b. Then, for n sufficiently large,

$$N_n^{(b)} \leq \frac{N_n + 1}{\log_2 b} + 1 \leq 2\varepsilon_{n_0} \sum_{j=1}^{n-1} \frac{N_j}{\log_2 b} \leq 2\varepsilon_{n_0} \sum_{j=1}^{n-1} N_j^{(b)}.$$

By Lemma 7.8.8 (item 2b), we conclude that for n sufficiently large, if $u_n = a_1 \ldots a_{|u_n|}$, then for every ℓ such that $1 \leq \ell \leq |u_n|$,

$$\Delta_\ell(u_1 \ldots u_{n-1} a_1 \ldots a_\ell) < 4\varepsilon_{n_0} < \tilde{\varepsilon}.$$

So, x is simply normal to base b for every $b \geq 2$.

We now analyze the computational complexity of the algorithm. Lemma 7.8.12 ensures the existence of the wanted t-sequence at each step n. To effectively find it, we proceed as follows. Divide the interval σ_2 into

$$2^{N_n}$$

equal binary intervals. In the worst case, for each of them, we need to check if it allocates a t_n-sequence $(\tau_2, \ldots, \tau_{t_n})$ that refines $(\sigma_2 \ldots, \sigma_{t_{n-1}})$ with discrepancy less than ε_n. Since we are just counting the number of mathematical operations ignoring the precision, at step n the algorithm performs

$$O(2^{N_n} t_n)$$

many mathematical operations. Since N_n is logarithmic in n and t_n is a rational power of $\log(n)$, we conclude that at step n the algorithm performs

$$O(n \sqrt[4]{\log n})$$

mathematical operations. Finally, in the first k steps, the algorithm will output at least k many digits of the binary expansion of the computed number having performed

$$O(k^2 \sqrt[4]{\log k})$$

many mathematical operations. This completes the proof of Theorem 7.8.7.

7.9 Normality, Non-normality, and Other Mathematical Properties

Recall that two positive integers are *multiplicatively dependent* if one is a rational power of the other. Then, 2 and 8 are dependent, but 2 and 6 are independent.

Theorem 7.9.1 (Maxfield 1953 [118]). *Let b and b' multiplicatively dependent. For any real number x, x is normal to base b if and only if x is normal to base b'.*

Theorem 7.9.2 (Cassels 1959 [135]). *Almost all real numbers in the middle third Cantor set (with respect to the uniform measure) are normal to every base which is not a power of 3.*

Theorem 7.9.3 (Schmidt 1961 [529]). *For any given set S of bases closed under multiplicative dependence, there are real numbers normal to every base in S and not normal to any base in its complement. Furthermore, there is a real x computable from S.*

Theorem 7.9.3 was improved in [58] to obtain lack of simple normality for the bases outside S instead of just lack of normality. Then Becher, Bugeaud, and Slaman [49] obtained the necessary and sufficient conditions on a set S for the existence of real numbers simply normal to every base in S and not simply normal to any base in its complement.

Theorem 7.9.4 (Becher, Bugeaud, and Slaman [49]). *Let S be a set of bases. There is a real x that is simply normal to exactly the elements in S if and only if*

1. *for each b, if b^k in S then b in S,*
2. *if infinitely many powers of b belong to S, then all powers of b belong to S.*

Moreover, the real x is computable from the set S. Furthermore, the set of real numbers that satisfy this condition has full Hausdorff dimension.

We end the section with references on the relation of normality and Diophantine approximations. The irrationality exponent m of a real number x reflects how well x can be approximated by rational numbers. Precisely, it is the supremum of the set of real numbers z for which the inequality

$$0 < \left| x - \frac{p}{q} \right| < \frac{1}{q^z}$$

is satisfied by an infinite number of integer pairs (p, q) with $q > 0$. Rational numbers have irrationality exponent equal to 1. Liouville numbers are those with infinite irrationality exponent. It follows from the fundamental work by [347] that almost all irrational numbers (with respect to Lebesgue measure) have irrationality exponent equal to 2. On the other hand, it follows from the theory of continued fractions that for every m greater than 2 or equal to infinity, there is a real number x with irrationality exponent equal to m.

Absolute normality places no restriction on irrationality exponents of irrational numbers. For every real number z greater than or equal to 2, there is an absolutely normal number with irrationality exponent equal to z. This existential result follows from Kaufman [338]. Bugeaud [117] showed there is an absolutely normal Liouville. In both cases, existence of such real numbers follows from the existence of a measure whose Fourier transform vanishes sufficiently quickly at infinity and which is supported by a subset of the real numbers with the appropriate irrationality exponent. Bugeaud's argument employs an adaptation of Kaufman's methods to the set of Liouville numbers due to Bluhm [92]. Becher, Heiber, and Slaman [54] exhibit a computable construction of an absolutely number Liouville number.

7.10 Selection

We consider the selection of symbols from an infinite word and define a word with the selected symbols. The general problem is which forms of selection preserve normality, that is, which families of functions f performing selection guarantee that $f(x)$ is normal when x is normal. Notice that if a selection procedure is allowed to read the symbol being decided, it would be possible to "select only zeroes" or yield similar schemes that do not preserve normality.

We consider three forms of selection. *Prefix selection* looks at just the prefix of length $i - 1$ to decide whether the symbol at position i is selected. *Suffix selection* looks at just the suffix starting at position $i+1$ to decide whether symbol at position i is selected. *Two-sided selection* looks at the prefix of length $i - 1$ and the suffix starting at position $i + 1$ to decide the selection of the symbol at position i. Prefix selection is the selection defined by Agafonov [6].

Let $x = a_0a_1a_2\cdots$ be an infinite word over alphabet A. Let $L \subseteq A^*$ be a set of finite words over A and $X \subseteq A^\omega$ a set of infinite words over A.

The word obtained by *prefix selection* of x by L is $x \upharpoonright L = a_{i_0}a_{i_1}a_{i_2}a_{i_3}\cdots$ where $i_0, i_1, i_2, \cdots$ is the enumeration in increasing order of all the integers i such that $a_0a_2\cdots a_{i-1} \in L$.

The word obtained by *suffix selection* of x by X is $x \upharpoonleft X = a_{i_0}a_{i_1}a_{i_2}a_{i_3}\cdots$ where $i_0, i_1, i_2, \cdots$ is the enumeration in increasing order of all the integers i such that $a_{i+1}a_{i+2}a_{i+3}\cdots \in X$.

Theorem 7.10.1 (Agafonov [6]). *If $x \in A^\omega$ is normal and $L \subset A^*$ is rational then $x \upharpoonright L$ is also normal.*

Before giving the proof of Theorem 7.10.1, we discuss some other results. Agafanov's theorem can be extended to suffix selection by replacing the rational set of finite words L by a rational set of infinite words X. The proof of this theorem is quite technical, so we do not give it here.

Theorem 7.10.2 ([57]). *If $x \in A^\omega$ is normal and $X \subset A^\omega$ is rational, then $x \upharpoonleft X$ is also normal.*

The prefix and suffix selections cannot be combined to preserve normality: in general, two-sided selection does not preserve normality. For instance, selecting all symbols surrounded by two symbols 1 in a normal word over $\{0, 1\}$ always destroys normality: the factor 11 occurs more frequently than the factor 00 in the resulting word.

We now give three lemmas to be used in the proof of Theorem 7.10.1.

Lemma 7.10.3. *For any set of finite words L, the function $x \mapsto \langle x \upharpoonright L, x \upharpoonright A^* \setminus L\rangle$ is one-to-one.*

Proof. Let $y_1 = x \upharpoonright L$ and $y_2 = x \upharpoonright A^* \setminus L$. By definition, y_1 contains some symbols of x, in the same relative order, and y_2 contains the complement, also in the same relative order. It is possible to reconstruct x by interleaving appropriately the symbols in y_1 and y_2. For each $i \geq 1$, the i-th symbol of x comes from y_1 if and only if the prefix of length i of x is in L. Thus, there is a unique x such that $y_1 = x \upharpoonright L$ and $y_2 = x \upharpoonright A^* \setminus L$.

A (deterministic) *two-output transducer* is like transducer, but it has two output tapes. Each of its transitions has the form $p \xrightarrow{a|v,w} q$ where a is the symbol read on the input tape and v and w are the words written to the first and the second output tape, respectively.

An infinite word $x = a_0a_1a_2\cdots$ is *compressible* by a two-output transducer if there is an accepting run $q_0 \xrightarrow{a_0|v_0,w_0} q_1 \xrightarrow{a_1|v_1,w_1} q_2 \xrightarrow{a_2|v_2,w_2} \cdots$ that satisfies

$$\liminf_{n\to\infty} \frac{(|v_0v_2\cdots v_n| + |w_0w_2\cdots w_n|)}{n+1}\frac{\log|B|}{\log|A|} < 1.$$

The following lemma states that an extra output tape does not help for compressing.

Lemma 7.10.4. *An infinite word is compressible by a bounded-to-one two-output transducer if and only if it is compressible by a bounded-to-one transducer.*

Proof. The "if" part is immediate by not using one of the output tapes.

Suppose that x is compressible by the bounded-to-one two-output transducer $\mathcal{T}_2$. We construct a transducer $\mathcal{T}_1$ with a single output tape which also compresses x. The main idea is to merge the two outputs into the single tape without losing the bounded-to-one assumption. Let m be an integer to be fixed later. The transducer $\mathcal{T}_1$ simulates $\mathcal{T}_2$ on the input and uses two buffers of size m to store the outputs made by $\mathcal{T}_2$. Whenever one of the two buffers is full and contains m symbols, its content is copied to the output tape of $\mathcal{T}_1$ with an additional symbol in front of it. This symbol is either 0 or 1 to indicate whether the m following symbols comes from the first or the second buffer. This trick preserves the bounded-to-one assumption. This additional symbol for each block of size m increases the length of the output by a factor $(m+1)/m$. For m large enough, the transducer $\mathcal{T}_1$ also compresses x.

Lemma 7.10.5. *Let $x = a_0a_1a_2\cdots$ be a normal word, and let $q_0 \xrightarrow{a_0} q_1 \xrightarrow{a_1} q_2 \xrightarrow{a_2} \cdots$ be a run in a deterministic automaton. If the state q is visited infinitely often then* $\liminf_{n\to\infty} |\{i \le n : q_i = q\}|/n > 0$.

Proof. Let $\mathscr{A}$ be a deterministic automaton. For a state p and a finite word w, the unique state q such that $p \xrightarrow{w} q$ is denoted $p \cdot w$.

Let $q = q_1, \ldots, q_n$ be the states occurring infinitely often in the run. For $1 \le i, j\ len$, let $u_{i,j}$ be a word such that $q_i \cdot u_{i,j} = q_j$. Let us define the sequence of words $(w_k)_{1\le k\le n}$ by $w_1 = \lambda$ and $w_{k+1} = w_k u_{i,1}$ where $q_{k+1} \cdot w_k = q_i$. By definition, $q_k \cdot w_k = q$, and thus the finite run $q_i \xrightarrow{w_n} q_i \cdot w_n$ visits the state q for each i since w_i is a prefix of w_n. Since the number of occurrences of w_n in x converges to $1/|A|^{|w_n|}$, the result holds.

Proof (of Theorem 7.10.1). Let x be a normal word. Let $L \subset A^*$ be a rational language. We suppose by constriction that $x \restriction L$ is not normal, and we show that x can be compressed, contradicting its normality.

Let $\mathscr{A}$ be a deterministic automaton accepting L. This automaton can be turned into a two-output transducer that outputs $x \restriction L$ and $x \restriction A^* \setminus L$ on its first and second output tapes, respectively. Each transition that leaves a final state copies its input symbol to the first output tape, and each transition that leaves a nonfinal state copies its input symbol to the second output tape. By hypothesis, $x \restriction L$ is not normal and therefore can be compressed by some deterministic transducer. Combining, these two transducers yield a two-output transducer that compresses x. This later result holds because, by Lemma 7.10.5, the states that select symbols from x are visited at least linearly often. Then, by Lemma 7.10.4, x can be compressed and is not normal.

Chapter 8
Normal Numbers and Symbolic Dynamics

Manfred Madritsch

Abstract The present chapter takes a dynamical point of view. The orbit of an element plays a central role in dynamics, and we can deduce several properties such as periodicity, uniqueness, randomness, etc. from the orbit. Starting with a description of the link between dynamical systems and numeration systems, we present the concept of normal and non-normal numbers providing different views on the dynamics of the system. Normal numbers are "normal" with respect to randomly chosen objects, whereas non-normal numbers and extreme variants thereof are examples of general objects from a topological point of view. In the following sections, we present how to obtain maximal randomness as well as constructing numbers with a given degree of chaos. Then we turn our attention to non-normal numbers. Since they are not completely random, we have to find a different measurement for analyzing their structure. The Hausdorff dimension will provide us with an interesting parameter in this context.

8.1 Introduction

In the present chapter, we want to take a different view on numeration systems and normality than the one developed in Chapter 3 and Chapter 7. Let $M = (M, d)$ be a compact metric space and $T\colon M \to M$ a continuous transformation. Then, the pair (M, T) is a *topological dynamical system*. For a point $x \in M$, one is interested in the *orbit* $\mathcal{O}(x)$ of x defined as

$$\mathcal{O}(x) = \overline{\{T^n(x)\colon n \in \mathbb{N}\}}.$$

In particular, we want to characterize elements having a dense orbit and those having a periodic orbit.

M. Madritsch (✉)
Institut Élie Cartan de Lorraine, Universtité de Lorraine, BP 70139, F-54506 Vandœuvre-Lès-Nancy Cedex, France
e-mail: manfred.madritsch@univ-lorraine.fr

V. Berthé, M. Rigo (eds.), *Sequences, Groups, and Number Theory*, Trends in Mathematics, https://doi.org/10.1007/978-3-319-69152-7_8

From the point of view of numeration systems, we may associate with each element $x \in M$ an "address," namely, its expansion. This description uses letters from an alphabet A (a finite or countable set) and associates an infinite word $\omega \in A^{\mathbb{N}}$ with each element $x \in M$. In order to define this address, we need a so-called topological partition $\mathscr{P}$. A *topological partition* $\mathscr{P}$ is a family of disjoint open subsets of M such that

$$\bigcup_{P \in \mathscr{P}} \overline{P} = M.$$

For the moment let us assume that $\mathscr{P}$ is finite. Then we associate with each element in the orbit $\mathscr{O}(x)$ the corresponding set $P_i \in \mathscr{P}$. Fixing a partition we may number the elements of $\mathscr{P}$ starting with 0, i.e., $\mathscr{P} = \{P_0, P_1, \dots, P_{N-1}\}$. Then we note for each element of the orbit of x the corresponding set P_i, with $0 \leq i < N$. In particular, for $k \geq 0$, we set $a_k = i$ if $T^k x \in P_i$. Then, for each element $x \in M$, we obtain an infinite word $\omega = a_1 a_2 a_3 \dots$ over the alphabet $A = \{0, 1, \dots, N-1\}$. Clearly this representation depends on the transformation T as well as on the partition $\mathscr{P}$.

Now we consider the portion of the set of all words over A that occur as representations. To this end we call a word $w = a_1 a_2 \dots a_n$ *allowed* if

$$\bigcap_{k=1}^{n} T^{-k}(P_{a_k}) \neq \emptyset.$$

Let L be the set of allowed words. Then L is a language and there is a unique shift space $X \subset A^{\mathbb{N}}$, whose language is L. We denote by S the shift that is induced by T on X. For the corresponding definitions in symbolic dynamics, see Section 1.7.

For a finite word $\mathbf{a} = a_1 a_2 \dots a_n$ we denote by $|\mathbf{a}| = n$ its length. Furthermore, let $L_n = L \cap A^n$ be the set of all words of length n in L. Then the *topological entropy* $h(X)$ of the symbolic dynamical system (X, S) measures the richness of its language, i.e., it measures the number of different blocks occurring in the expansion of x. It is defined as

$$h(X) = \lim_{n \to \infty} \frac{1}{n} \ln |L_n| .$$

Before going further we want to present the two main examples we have in mind: the N-adic and the β-expansions.

Example 8.1.1. Let $M = \mathbb{R}/\mathbb{Z}$ be the circle and $T: M \to M$ be defined by $T(x) = Nx - \lfloor Nx \rfloor$. We divide M into N subintervals $P_0, \dots, P_{N-1}$ of the form $P_i = (i/N, (i+1)/N)$, and let $A = \{0, \dots, N-1\}$. Then the underlying number system is the N-ary representation. Furthermore, it is easy to verify that the associated language L is the set of all words over A, so that the one-sided symbolic dynamical system $X_{\mathscr{P},T}$ is the full one-sided N-shift $A^{\mathbb{N}}$.

Our second example will be the main motivation behind this chapter. In particular, we will consider β-expansions, where $\beta > 1$ is not necessarily an integer. These systems are of special interest, since the underlying symbolic dynamical system is not the full shift. Among the first authors investigating these number systems were Parry [469] and Rényi [500]. For a more modern account on these number systems, we refer the interested reader to the book of Dajani and Kraaikamp [176]. See also Section 3.6.2.

Example 8.1.2. Let $\beta > 1$ be a real number, $M = \mathbb{R}/\mathbb{Z}$ the circle, and $T\colon [0,1) \to [0,1)$ be the transformation given by

$$T(x) = \beta x - \lfloor \beta x \rfloor.$$

The sets

$$P_i := \left(\frac{i}{\beta}, \frac{i+1}{\beta}\right) \quad (i = 0, \ldots, \lfloor \beta \rfloor - 1)$$

and

$$P_{\lfloor \beta \rfloor + 1} := \left(\frac{\lfloor \beta \rfloor}{\beta}, 1\right)$$

form a topological partition of M. The corresponding language is called the β-shift.

Every $x \in M$ yields a representation as an infinite word ω over the alphabet A. The converse needs not to be true as we will see in a moment. For an infinite word $\omega = a_0 a_1 a_2 \ldots \in X$ and a positive integer k, we denote by $\omega|k = a_0 a_1 \ldots a_{k-1}$ the truncation of ω to the first k letters. Similarly we denote by $\omega_i^j = a_i a_{i+1} \ldots a_j$ the factor from the ith to the jth letter of ω. Furthermore let X_k be the projection of X to the first k letters of its words.

For each $\mathbf{w} = w_1 \ldots w_k \in L$ we denote by $[\mathbf{w}] \subset X$ the *cylinder set* of all infinite words starting with the same letters as $\mathbf{w}$, i.e.,

$$\begin{aligned} [\mathbf{w}] &= \{a_1 a_2 a_3 \ldots \in X\colon a_1 = w_1, \ldots, a_k = w_k\} \\ &= \{\omega \in X\colon \omega|k = \mathbf{w}\}. \end{aligned}$$

Similarly for each $\omega \in X$ we define the cylinder set $D_n(\omega)$ of order n corresponding to ω in M by

$$D_n(\omega) := \bigcap_{k=0}^{n-1} T^{-k} P_{a_k} \subset M.$$

The reader may have observed that our representations have a little problem. Let us consider the decimal system (with $N = 10$). Looking at the orbit of $\frac{1}{10}$,

$\frac{13}{100}$, or $\frac{137}{1000}$, we may observe that after a number of steps, we land outside of any P_i. The reason is that, for example, $\frac{1}{10}$ lies on the edge of P_0 and P_1. On which side should we put it? Into P_0 or P_1? Recall that there are reals that do not have a unique expansion in the decimals. In particular, we have two expansion for $\frac{1}{10}$: $0.1000000\ldots$ and $0.0999999\ldots$. In the following paragraphs, we want to investigate the degree of randomness in their expansions. Since both are not very random, we may neglect them in our considerations.

In the more general case of a topological partition $\mathscr{P}$, this ambiguity in the expansion originates from the intersection of two parts $\overline{P_i} \cap \overline{P_j}$ with $i \neq j$. Using this observation we want to describe the set of elements $x \in M$ that have a unique expansion. To this end, let

$$U = \bigcup_{i=0}^{N-1} P_i,$$

be an open and dense ($\overline{U} = M$) set. By the disjointedness of the P_i, every $x \in U$ lies in a unique P_i. Now we want to guarantee this also after iteration of T. Thus for $n \geq 1$ we set

$$U_n = \bigcap_{k=0}^{n-1} T^{-k}(U).$$

This set is open and dense in M and contains all $x \in M$ that have a unique representation after n iterations of T. Note that the expansion of $x \in U_n$ needs not be unique after $m \neq n$ steps. Thus by the Baire category theorem, the set

$$U_\infty = \bigcap_{n=0}^{\infty} U_n \tag{8.1}$$

is dense. By construction every $x \in U_\infty$ has a unique expansion $\omega = a_1a_2a_3\ldots \in X$. However, different $x \in M$ may share the same expansion. Therefore, we suppose that, for any $\omega = a_1a_2a_3\ldots \in X$, the set $\bigcap_{n=0}^{\infty} \overline{D_k(a)}$ is a singleton set, which is the case in the examples above. Then define the map $\pi : X \to M$ by

$$\bigcap_{k=0}^{\infty} T^{-k} P_{a_k} = \{\pi(a)\}.$$

Together with our definition of U_∞, we have the usual feeling that for each expansion $a \in X$, there exists one element $\pi(a) \in M$, and for each element $x \in U_\infty$, there is exactly one expansion $a \in X$ such that $\pi(a) = x$. Moreover we have that the shift S on X acts such that the following diagram commutes:

$$\begin{array}{ccc} X & \xrightarrow{S} & X \\ \downarrow\pi & & \downarrow\pi \\ M & \xrightarrow{T} & M \end{array}$$

Since our considerations mostly depend on the expansion and not on the point it represents, we will focus on X in the following and identify M (or more precisely U_∞) as the image of X under π. In particular, we will see below that, for normal numbers, only their representation in X plays a role, whereas for non-normal numbers, we have that the set $M \setminus U_\infty$ is the union of nowhere dense sets and is therefore negligible (in the topological sense).

8.1.1 *Infinite Alphabet*

After working over a finite alphabet, we look for similar definitions over an infinite alphabet. We start with the continued fraction expansion as motivation and example, so to say a motivating example.

Let $\mathbb{I}$ denote the irrational numbers in the unit interval, i.e., $\mathbb{I} := [0, 1] \setminus \mathbb{Q}$. Then every $x \in \mathbb{I}$ can be represented in a unique way as an infinite simple continued fraction, namely,

$$x = [a_1(x), a_2(x), a_3(x), \ldots] = \cfrac{1}{a_1(x) + \cfrac{1}{a_2(x) + \cfrac{1}{a_3(x) + \ddots}}},$$

where $a_k(x) \in \mathbb{N}$ for $k \geq 1$.

As above, let M be a metric space and $T : M \to M$ be a continuous map. Let $\mathscr{P} := \{P_1, P_2, \ldots\}$ be a countable family of disjoint open sets. Then we call $\mathscr{P}$ a topological partition (of M) if M is the union of the closures $\overline{P_i}$ for $i \geq 1$, i.e.,

$$M = \bigcup_{i \in \mathbb{N}} \overline{P_i}.$$

For the rest of this section, let us assume that a dynamical system (M, T) together with an infinite topological partition $\mathscr{P}$ is given. As above, we take a closer look at the underlying symbolic dynamical system. Without loss of generality, we may denote by $\mathbb{N}$ the alphabet corresponding to the partition $\mathscr{P}$. Furthermore, let $\mathbb{N}^k$ be the set of words of length k and

$$\mathbb{N}^* = \bigcup_{k \in \mathbb{N}} \mathbb{N}^k$$

be the set of finite words. Finally we denote by $\mathbb{N}^{\mathbb{N}}$ the set of infinite words over $\mathbb{N}$.

As previously, for an infinite word $\omega = a_1a_2a_3\ldots \in \mathbb{N}^{\mathbb{N}}$ and a positive integer n, we denote by $\omega|n = a_1a_2\ldots a_n$ its truncation to the nth place. Furthermore for a given finite word $\omega \in \mathbb{N}^*$, we denote by $[\omega] \subset \mathbb{N}^{\mathbb{N}}$ the cylinder set consisting of all infinite words starting with the same letters as ω, i.e.,

$$[\omega] := \left\{\gamma \in \mathbb{N}^{\mathbb{N}} : \gamma|\,|\omega| = \omega\right\}.$$

Now we want to describe the shift space that is generated. Therefore we call an infinite word $\omega = a_1a_2a_3\ldots \in \mathbb{N}^{\mathbb{N}}$ allowed for $(\mathscr{P}, T)$ (or allowed for short) if

$$\bigcap_{k=1}^{\infty} T^{-k}\left(P_{a_k}\right) \neq \emptyset.$$

Let $L = L_{\mathscr{P},T}$ be the set of allowed words. Then L is a language, and there is a unique shift space $X = X_{\mathscr{P},T} \subseteq \mathbb{N}^{\mathbb{N}}$, whose language is L. We call $X \subseteq \mathbb{N}^{\mathbb{N}}$ the one-sided symbolic dynamical system corresponding to $(\mathscr{P}, T)$. Finally for each $\omega = a_1a_2a_3\ldots \in X$ and $n \geq 0$, we denote by $D_n(\omega)$ the cylinder set of order n corresponding to ω in M, i.e.,

$$D_n(\omega) := \bigcap_{k=0}^{n} T^{-k}(P_{a_k}) \subseteq M.$$

Now we can state the definition of an infinite Markov partition.

Definition 8.1.3. Let (M, T) be a dynamical system and $\mathscr{P} = \{P_1, P_2, P_3, \ldots\}$ be an infinite topological partition of M. Then we call $\mathscr{P}$ an infinite Markov partition if the generated shift space $X_{\mathscr{P},T}$ is of finite type and for every $\omega \in X_{\mathscr{P},T}$ the intersection $\bigcap_{n=0}^{\infty} \overline{D_n(\omega)}$ consists of exactly one point.

Note that in the case of a finite alphabet, we did not introduce the Markov partitions because this would mean that we have a shift of finite type as language. However, we want to consider the more general case of a shift with specification below. The lack of examples of shifts with specification over an infinite alphabet led us to the definition of Markov partitions in this case.

After introducing all the necessary ingredients, we want to link the introduced concept of infinite Markov partitions with the continued fraction expansion and Lüroth series (*cf.* Dajani and Kraaikamp [176]).

Example 8.1.4. The dynamical system $([0, 1], T)$, where T is the Gauss map

$$Tx = \begin{cases} \frac{1}{x} - \lfloor \frac{1}{x} \rfloor & \text{for } x \neq 0, \\ 0 & \text{for } x = 0, \end{cases}$$

together with the infinite topological partition

$$\mathscr{P} := \{(\tfrac{1}{2}, 1), (\tfrac{1}{3}, \tfrac{1}{2}), (\tfrac{1}{4}, \tfrac{1}{3}), \ldots\}$$

provides the continued fraction expansion.

In particular, if we set $x_1 := x$ and $x_{k+1} := Tx_k$ for $k \geq 1$, then $a_k(x) = \lfloor \frac{1}{x_k} \rfloor$ for $k \geq 1$ is the continued fraction expansion of x from the introduction.

There are several extensions of the continued fraction expansion like continued fractions from below, nearest integers continued fractions, α-continued fractions, Rosen continued fractions, and combinations of those. For criteria such that a given number is normal with respect to different continued fraction expansions, we refer the interested reader to Kraaikamp and Nakada [361].

Example 8.1.5. Let $T\colon [0, 1) \to [0, 1)$ be defined by

$$T(x) = \begin{cases} n(n+1)x - n, & x \in \left[\frac{1}{n+1}, \frac{1}{n}\right), \\ 0 & x = 0. \end{cases}$$

Then the pair $([0, 1), T)$ together with the infinite topological partition

$$\mathscr{P} := \{(\tfrac{1}{2}, 1), (\tfrac{1}{3}, \tfrac{1}{2}), (\tfrac{1}{4}, \tfrac{1}{3}), \ldots\}$$

provides the Lüroth series (*cf.* [394]).

Under some mild restrictions, one can replace the intervals $\left[(n+1)^{-1}, n^{-1}\right)$ by arbitrary ones in order to get the generalized Lüroth series (*cf.* Chapter 2.3 of [176]).

Before considering the elements of M having multiple expansions, we note that in contrast to the survey of Barat *et al.* [36], we did not use fibered systems for the definition of the dynamical system. The reason lies in the concrete treatment of the border in the case of Markov partitions. In particular, when considering these partitions, it is by definition clear that the sets P_i are all open sets, whereas this is a priori not specified in the case of fibered systems by Schweiger [533]. This plays a key role in the analysis of the one-to-one correspondence between the infinite word and the corresponding element of M below. By the definition of a Markov partition, we have that every $\omega \in X$ maps to a unique element $x \in M$. However, as above the converse need not be true. Let us consider the continued fraction expansion (Example 8.1.4). Then the rational $\frac{1}{4}$ has two expansions, namely, [4] and [3, 1]. One observes that this ambiguity originates from the intersections $\overline{P_i} \cap \overline{P_j}$ for $i \neq j$ (which means from the borders of $\overline{P_i}$). Thus we concentrate on the inner points, which provide us with an infinite and unique expansion. Let

$$U = \bigcup_{i=1}^{\infty} P_i,$$

which is an open and dense ($\overline{U} = M$) set. Then for each $n \geq 1$ the set

$$U_n = \bigcap_{k=0}^{n-1} T^{-k}(U),$$

is open and dense in M. Thus by the Baire category theorem, the set

$$U_\infty = \bigcap_{n=1}^{\infty} U_n \tag{8.2}$$

is dense. Since $M \setminus U_\infty$ is the countable union of nowhere dense sets, it suffices to show that a set is residual in U_∞ in order to show that in fact it is residual in M.

Since the definition of normal and thus non-normal numbers will involve the expansions of the elements in M, we need the map $\pi = \pi_{\mathscr{P},T} : X_{\mathscr{P},T} \to M$ defined by

$$\{\pi_{\mathscr{P},T}(\omega)\} = \bigcap_{n=1}^{\infty} \overline{D_n(\omega)}.$$

Since π is bijective on U_∞, we may call ω **the** symbolic expansion of x if $\pi_{\mathscr{P},T}(\omega) = x$. Thus in the following we will silently suppose that $x \in U_\infty$.

8.2 Normal Numbers

At the moment of writing, we find several properties which are seen as "normal." This ranges from normal subgroups, normal vectors over the normal degree to normal polytopes. Normal numbers are a concept introduced by Borel [99], who had a probabilistic view in mind. We will see below that non-normal numbers are "normal" from a topological point of view.

Let us choose x from the interval $[0, 1]$ at random. Then x has a decimal expansion of the form

$$x = \sum_{k \geq 1} a_k 10^{-k}, \tag{8.3}$$

where $a_k \in \{0, \ldots, 9\}$ are the digits. We would expect that the events $a_1 = 1$ and $a_1 = 2$ occur with the same probability, i.e., $\mathbb{P}(a_1 = 1) = \mathbb{P}(a_1 = 2)$. Moreover there is no reason why this should not also hold for any two blocks of digits of the same length. In particular, we expect that for each block of digits $b_1, \ldots, b_k \in \{0, 1, \ldots, 9\}$ of length k, we have the same probability

$$\mathbb{P}(a_1 = b_1, \ldots, a_k = b_k) = 10^{-k}$$

to see it in the decimal expansion of x.

We call x as in (8.3) simply normal in base 10 if for each $d = 0, \ldots, 9$ we have that

$$\lim_{n\to\infty} \frac{1}{n}\#\{0 \le i < n : a_i = d\} = \frac{1}{10}.$$

Furthermore we call x normal in base 10 if it is simply normal with respect to the bases $10, 10^2, 10^3, \ldots$ An equivalent definition would be that for each word $\mathbf{b} = b_1 \ldots b_k \in \{0, 1, \ldots, 9\}^k$ we have

$$\lim_{n\to\infty} \frac{1}{n}\#\{0 \le i < n : a_i = b_1, \ldots, a_{i+k-1} = b_k\} = \frac{1}{10^k}.$$

This notion obviously generalizes to any base $q \ge 2$. Borel [99] showed that almost all (with respect to the Lebesgue measure) real numbers are normal to all bases $q \ge 2$, that is, they are absolutely normal. On the one hand, we expect that the numbers $\sqrt{2}$, π, $\log 2$, etc. are all normal to all bases; however, almost nothing is known in this direction. On the other hand, we would like to construct such a number. The first construction of a normal number in base 10 is due to Champernowne [141], who showed that the number

$$0.1\,2\,3\,4\,5\,6\,7\,8\,9\,10\,11\,12\,13\,14\,15\,16\,17\,18\,19\,20\,\ldots$$

is normal in base 10. There are not many constructions known and even less on numbers, which are normal to all bases. In the present chapter we consider the normality from a more abstract point of view and refer the interested reader to Chapter 7 for more information on explicit constructions.

Now we want to transfer the definition of a normal number to symbolic dynamical systems. To this end let $\mathbf{b} = b_1 \ldots b_k \in L_k$ be a word of length k and $\omega = a_0a_1a_2 \ldots \in X$. Then we write

$$\mathrm{P}(\omega, \mathbf{b}, n) = \frac{|\{0 \le i < n : a_{i+1} = b_1, \ldots a_{i+k} = b_k\}|}{n}$$

for the frequency of occurrences of the word $\mathbf{b}$ in the first n letters of ω. Furthermore we collect all different words of length k in a vector. For $k \ge 1$ and $\omega = a_0a_1a_2 \in X$ we define

$$\mathrm{P}_k(\omega, n) = (\mathrm{P}(\omega, \mathbf{b}, n))_{\mathbf{b}\in L_k}\,.$$

This vector describes the distribution of the different blocks among the first n letters of a given infinite word $\omega \in X$.

These frequencies of occurrence can be seen as a measure μ on X. In particular, let $\mathbf{a} = (a_1, \ldots, a_k)$ be a block of digits and $f_{\mathbf{a}}$ its desired frequency of occurrences. Then we set $\mu([\mathbf{a}]) = f_{\mathbf{a}}$ and obtain a measure on the set of possible expansions. However, the $f_{\mathbf{a}}$ have to fulfill certain restrictions. In particular, let Δ_k be the simplex of all probability vectors, i.e.,

$$\Delta_k = \left\{ (p_{\mathbf{a}})_{\mathbf{a}\in L_k} : p_{\mathbf{a}} \geq 0, \sum_{\mathbf{a}} p_{\mathbf{a}} = 1 \right\}.$$

If we denote by $|\cdot|_1$ the 1-norm, then $(\Delta_k, |\cdot|_1)$ is a metric space. On the one hand, every possible vector of frequencies of blocks of length k belongs to Δ_k. On the other hand, we get certain restrictions by the structure of the sequences, namely, we read them in one direction. Looking at all possible ways we can extend a block of length 2, such as 35, to a block of length 3, we get that

$$\left| \sum_{i\in A} \mathrm{P}(\omega, i35, n) - \sum_{i\in A} \mathrm{P}(\omega, 35i, n) \right| \leq \frac{1}{n}$$

for all sequences $\omega \in X$. This implies that for each ω all but finitely many points in the sequence $(\mathrm{P}_3(\omega, n))_n$ are very close to the subsimplex

$$\Delta_3 \cap \left\{ (p_{\mathbf{a}})_{\mathbf{a}\in L_3} : \sum_{i\in A} p_{i35} = \sum_{i\in A} p_{35i} \right\}.$$

Looking at all possible blocks of length $k-1$, we get that Δ_k is not the "correct" object to consider. Rather we need to consider the subsimplex of shift invariant probability vectors S_k, i.e.,

$$\mathrm{S}_k := \left\{ (p_{\mathbf{a}})_{\mathbf{a}\in A^k} : p_{\mathbf{a}} \geq 0, \sum_{\mathbf{a}\in A^k} p_{\mathbf{a}} = 1, \sum_{i\in A} p_{i\mathbf{a}} = \sum_{i\in A} p_{\mathbf{a}i} \text{ for all } \mathbf{a} \in A^{k-1} \right\}.$$

All these shift invariant probability vectors can be extended to a shift invariant probability measure. Therefore we will only consider this kind of measures.

More generally, let μ be a probability measure on X. Then we call μ *shift invariant* if for each $A \subset X$ we have that $\mu(S^{-1}A) = \mu(A)$. Let $\mathcal{M}$ be the set of shift invariant probability measure on X. Then for each $k, n \geq 1$ and $\omega \in X$ the vector $\mathrm{P}_k(\omega, n)$ gives rise to a probability measure on X_k, which can be easily extended to X. Moreover, we call $\mu \in \mathcal{M}$ associated with $\omega \in X$ if there exists an infinite subset $F \subset \mathbb{N}$ such that for any $k \geq 1$ and any word $\mathbf{b} \in L_k$

$$\lim_{\substack{n\to\infty \\ n\in F}} \mathrm{P}(\omega, \mathbf{b}, n) = \mu([\mathbf{b}]).$$

Furthermore, we call ω a *generic point for* μ if we can take $F = \mathbb{N}$: then μ is the only measure associated with ω. A normal number (a randomly chosen number) should have maximum chaos in its expansion. Therefore we need to define (measure theoretic) entropy first. This kind of entropy grades different measures with respect to the uncertainty in the expansion of their generic points. For $\mu \in \mathcal{M}$ the *measure-theoretic entropy* of μ is defined as

$$h(\mu) = \lim_{n\to\infty} -\frac{1}{n} \sum_{\mathbf{a}\in L_n} \mu([\mathbf{a}]) \log (\mu ([\mathbf{a}])) .$$

In 1969 Goodwyn [259] showed that, for all measures on X, we have $h(\mu) \le h(X)$ (*cf.* Chapter 18 of Denker *et al.* [193]). This motivates the following definition. We call a measure $\mu \in \mathcal{M}$ a *measure of maximal entropy* (or *maximal measure*) if $h(\mu) = h(X)$. Then we call $\omega \in X$ *normal* if it is generic for a maximal measure μ.

Compare these definitions with the concept of the empirical measure. Let $x \in X$ be a point in our shift space. Then the empirical measure (of order n) is defined as

$$T_n(x) := \frac{1}{n} \sum_{k=0}^{n-1} \delta_{S^k(x)},$$

where δ_u denotes the Dirac measure. Clearly the sequence $\{T_n\}_n$ has limit points, which are shift invariant probability measures, and x is generic for each one of them. Therefore x is generic for a single measure if and only if the sequence $\{T_n\}_n$ has only one limit point.

A measure $\mu \in \mathcal{M}$ is called *ergodic* if $\mu(S^{-1}A) = \mu(A)$ for a set $A \subseteq X$ implies that $\mu(A) = 1$ or $\mu(A) = 0$. This means that the only invariant sets are either sets of full measure or negligible sets (in the measure theoretical sense). In all our examples, we have a unique maximal measure, and this has to be ergodic (*cf.* Section 8.3 of [579]). If there is only one maximal measure, then we call the dynamical system *intrinsically ergodic*.

The common point of all constructions, if it is for normal or non-normal words and sets, is the concatenation of words with prescribed distribution. This concatenation is not always possible. Recall the β-expansion (Example 8.1.2) with respect to the Golden Ratio $\phi = \frac{1+\sqrt{5}}{2}$. In this case we have $M = [0, 1]$ and $T(x) = \phi x - \lfloor \phi x \rfloor$. Furthermore the topological partition is given as

$$\mathscr{P} = \left\{ \left(0, \frac{1}{\phi}\right), \left(\frac{1}{\phi}, 1\right) \right\} .$$

Then we denote by (X, S) the corresponding symbolic dynamical system. Since $\phi = 1 + \frac{1}{\phi}$ we have that

$$T\left(\left(\frac{1}{\phi}, 1\right)\right) = \left(0, \frac{1}{\phi}\right)$$

and the language L of X consists of all words over the alphabet $\{0, 1\}$ with no factor 11. In particular $L = \{a_1 \dots a_k \in A^*: a_i \cdot a_{i+1} = 0 \text{ for } 1 \le i \le k - 1\}$.

Now clearly 1001 and 101 are elements of L; however, their concatenation 1001101 is not. We will circumvent this defect by using the specification property. In particular, we say that a language L has the *specification* with gap $g \ge 0$ if for any $\mathbf{a}, \mathbf{b} \in L$ there exists a $\mathbf{w} \in L$ with $|\mathbf{w}| \leq g$ such that $\mathbf{awb} \in L$. In this case we chose $\mathbf{w}$ for $\mathbf{a}$ and $\mathbf{b}$ and write $\mathbf{a} \odot \mathbf{b} = \mathbf{awb}$ for short.

The specification property was first described by Bowen [103] with respect to orbits of elements. In this context the specification property states the existence of a periodic element that stays near the orbits of given elements for a given time. In particular, let (M, T) be a topological dynamical system consisting of a compact metric space $M = (M, d)$ and a continuous map T from M onto itself. Then (M, T) has the specification property if for every $\varepsilon > 0$ there is an integer $N(\varepsilon)$ such that if $x_1, x_2, \ldots x_k$ are points in M and $A_i = [\![a_i, b_i]\!]$ are sets of consecutive integers with $a_i - b_{i-1} > N(\varepsilon)$ for $i = 2, 3, \ldots, k$ and if $p > b_k - a_1 + N(\varepsilon)$, then there exists a periodic point $x \in M$ with period p such that

$$d(T^j x, T^j x_i) < \varepsilon \quad \text{for } j \in A_i.$$

Since concatenation in general and the specification property in particular are very important for our constructions, we want to know if this is compatible with our notion of normality and intrinsic ergodicity.

Theorem 8.2.1 (Bowen [103]). *Let X be a shift. If X has the specification property, then X is intrinsically ergodic.*

This result of Bowen can be seen as optimal in the following sense. We will consider two weaker versions of the specification property for which we do not have intrinsic ergodicity anymore. The first one allows the length of the filling word to depend on the left word. A subshift X has *nonuniform specification property with gap function $g(n)$* if

- $g(n)$ is positive and nondecreasing
- $\frac{g(n)}{n} \to 0$
- For any words $w^{(1)}, w^{(2)}, \ldots, w^{(k)} \in L$, and for any integers $n_1, \ldots, n_{k-1}$, where $n_i \leq g(|w^{(i)}|)$ for all i, there exist words $v^{(1)} \in L_{n_1}, \ldots, v^{(k-1)} \in L_{n_{k-1}}$ so that the word $w^{(1)} v^{(1)} w^{(2)} v^{(2)} \ldots w^{(k-1)} v^{(k-1)} w^{(k)} \in L$.

Then clearly a subshift X has specification with gap g if it has nonuniform specification with the constant gap function $g(n) = g$. In a recent paper, however, Pavlov [472] could show that these systems are not intrinsically ergodic.

Theorem 8.2.2 (Pavlov [472]). *For any positive nondecreasing function $g(n)$ with $\liminf_{n\to\infty} \frac{g(n)}{\ln n} > 0$, there exists a subshift with nonuniform specification with gap function $g(n)$ with exactly two ergodic measures of maximal entropy, whose supports are disjoint.*

The second weaker form, which is essentially due to Pfister and Sullivan [479], allows to replace a certain number of letters in the words in order to perform concatenation. A subshift X has *almost specification with mistake function $g(n)$* if

- $g(n)$ is positive and nondecreasing
- $\frac{g(n)}{n} \to 0$

- For any words $w^{(1)}, w^{(2)}, \dots, w^{(k)} \in L$, there exist words $v^{(1)}, \dots, v^{(k)} \in L$ such that $\left|w^{(i)}\right| = \left|v^{(i)}\right|$ for $1 \le i \le k$, $w^{(i)}$ and $v^{(i)}$ differ on at most $g(\left|w^{(i)}\right|)$ letters for $1 \le i \le k$, and the concatenation $v^{(1)}v^{(2)} \dots v^{(k)}$ is in L.

In the very same paper, Pavlov [472] could also provide a counterexample for the existence of a unique ergodic measure for this weaker form of the specification property.

Theorem 8.2.3 (Pavlov [472]). *There exists a subshift with almost specification with mistake function $g(n) = 4$ with exactly two ergodic measures of maximal entropy, whose supports are disjoint.*

Thus both weaker forms do not lead to intrinsic ergodicity. The results on Besicovitch-Eggleston sets in this chapter hold true if one considers nonuniform specification (*cf.* [478]).

8.2.1 Infinite Alphabet

As in the introduction above, we want to shed some light on the differences if we have an infinite alphabet. We start with the continued fraction expansion (Example 8.1.4). For $k \ge 1$ a positive integer and a block $\mathbf{b} = b_1 \dots b_k \in \mathbb{N}^k$, we denote by $\Pi(x, \mathbf{b}, n)$ the number of occurrences of this block $\mathbf{b}$ among the first n digits of x, i.e.,

$$\Pi(x, \mathbf{b}, n) := \frac{\{0 \le i < n : a_{i+1}(x) = b_1, \dots, a_{i+k}(x) = b_k\}}{n}.$$

Furthermore we denote by

$$\Pi_k(x, n) = (\Pi(x, \mathbf{b}, n))_{\mathbf{b} \in \mathbb{N}^k}$$

the vector of frequencies $\Pi(x, \mathbf{b}, n)$ for all blocks $\mathbf{b}$ of length k.

For digits (blocks of length 1) a famous result of Lévy [377] states that for Lebesgue almost all $x \in \mathbb{I} = \mathbb{R} \setminus \mathbb{Q}$ we have

$$\Pi(x, b, n) \to \frac{1}{\log 2} \log \frac{(b+2)^2}{b(b+2)} \tag{8.4}$$

for all $b \in \mathbb{N}$. In analogy with normal numbers in q-adic number systems above, we call $x \in \mathbb{I}$ simple (continued fraction) normal if it satisfies (8.4) with $b \ge 1$.

We can extend this notion to (continued fraction) normal numbers by using the Gauss measure defined by

$$\mu(A) = \frac{1}{\log 2} \int_A \frac{1}{1+x} \mathrm{d}x,$$

where $A \subset [0, 1]$ is a Lebesgue measurable set. Then we call a number (continued fraction) normal if the asymptotic frequency of its block of digits is determined by the Gauss measure. Now an application of Birkhoff's ergodic theorem (*cf.* [87] or [176]) yields that almost all numbers are (continued fraction) normal with respect to the Lebesgue measure.

After defining the environment, we want to pull over the definitions of normal and non-normal numbers to the symbolic dynamical system. To this end let $\mathbf{b} \in \mathbb{N}^k$ be a block of letters of length k and $\omega = a_1 a_2 a_3 \ldots \in X$ be the symbolic representation of an element. Then we write

$$\mathrm{P}(\omega, \mathbf{b}, n) = |\{0 \le i < n : a_{i+1} = b_1, \ldots, a_{i+k} = b_k\}|$$

for the frequency of the block $\mathbf{b}$ among the first n letters of ω. In the same manner as above, let

$$\mathrm{P}_k(\omega, n) = (\mathrm{P}(\omega, \mathbf{b}), n)_{\mathbf{b} \in \mathbb{N}^k}$$

be the vector of all frequencies of blocks $\mathbf{b}$ of length k among the first n letters of ω.

Let μ be a given T-invariant probability measure on X and $\omega \in X$. Then we call the measure μ associated with ω if there exists an infinite subsequence F of $\mathbb{N}$ such that for any block $\mathbf{b} \in \Sigma^k$

$$\lim_{\substack{n \to \infty \\ n \in F}} \mathrm{P}(\omega, \mathbf{b}, n) = \mu([\mathbf{b}]).$$

Furthermore, we call ω a generic point for μ if we can take $F = \mathbb{N}$: then μ is the only measure associated with ω. If μ is the maximal measure, then we call ω normal.

An application of Birkhoff's ergodic theorem yields for μ being ergodic that almost all numbers $\omega \in X$ are generic for μ. In both Examples 8.1.4 and 8.1.5, we have that the system is intrinsically ergodic, which means that there exists a unique maximal ergodic measure μ (*cf.* Chapter 3.1.2 of [176]).

In the final section, we want to consider non-normal numbers with respect to the continued fraction expansion. Again we get a restriction on the possible limiting frequencies $\lim_{n\to\infty} \mathrm{P}_k(\omega, n)$. For each $k \ge 1$, we define the simplex of all probability vectors Δ_k by

$$\Delta_k = \left\{ (p_{\mathbf{i}})_{\mathbf{i} \in \mathbb{N}^k} : p_{\mathbf{i}} \ge 0, \sum_{\mathbf{i} \in \mathbb{N}^k} p_{\mathbf{i}} = 1 \right\}.$$

The set Δ_k together with the 1-norm $\|\cdot\|_1$ defined by

$$\|\mathbf{p}-\mathbf{q}\|_1 = \sum_{\mathbf{i}\in\mathbb{N}^k} |p_{\mathbf{i}} - q_{\mathbf{i}}| \,.$$

is a compact metric space.

On the one hand, we clearly have that any vector $\mathrm{P}_k(\omega, n)$ of frequencies of blocks of digits of length k belongs to Δ_k. On the other hand, any probability vector needs to be shift invariant (*cf.* Volkmann [575] or Olsen [457]). Therefore we define the subsimplex of shift invariant probability vectors S_k by

$$\mathrm{S}_k := \left\{ (p_{\mathbf{i}})_{\mathbf{i}\in\mathbb{N}^k} : p_{\mathbf{i}} \geq 0, \sum_{\mathbf{i}\in\mathbb{N}^k} p_{\mathbf{i}} = 1, \sum_{i\in\mathbb{N}} p_{\mathbf{i}i} = \sum_{i\in\mathbb{N}} p_{i\mathbf{i}} \text{ for all } \mathbf{i} \in \mathbb{N}^{k-1} \right\} .$$

8.3 Construction of the Maximal Measure

Let us summarize for the moment. We have a topological dynamical system (X, T) and a topological partition $\mathscr{P}$. The partition $\mathscr{P}$ and the transformation T induce a symbolic dynamical system (X, S). Furthermore we set $\mathscr{M}$ the set of all shift invariant measures. Finally we suppose that the symbolic dynamical system fulfills the specification property and then there is a unique measure $\mu \in \mathscr{M}$ that maximizes the entropy, i.e., $h(\mu) = h(X)$.

In the present section we want to show that the Champernowne word is generic for the unique maximal measure. In particular, we want to show a little bit more. First of all we generalize our point of view to so-called coded systems. Simply speaking we are concatenating words ("codes") instead of letters. Secondly we also consider the Copeland and Erdős [166] construction of normal numbers. In the Champernowne word, we consider a concatenation of *all* possible words, whereas in the Copeland and Erdős construction, we restrict ourselves to a sufficiently large portion. The proof is of combinatorial nature and uses the fact that the number of omitted blocks has only a negligible contribution to the frequency of occurrences of a single word.

Modifying the Champernowne word, we will show in the next section that we may construct a word that is generic with respect to any given shift invariant measure. This construction has to be seen as a proof of concept. It is not optimal, and more sophisticated analysis in special cases leads to a far better rate of convergence (*cf.* Vandehey [573]). As an open problem, it would be of interest to improve the number of repetitions in the case of shifts of finite type.

Let us start now with the definition of a coded system. We suppose that A is finite and let A^+ denote the semigroup generated by A under concatenation. A language X is called a *code* if $X \in A^+$ generates a free submonoid X^* of A^* and if each word $w \in X^+$ has a unique factorization of the form

$$w = w_1 \ldots w_\ell$$

with $w_i \in X$ for $1 \le i \le \ell$. Furthermore a code is a *prefix code* if no word $u \in X^*$ has a representation of the form $u = u_1 v$ where $u_1 \in X$ and $v \in A^+$. The set $W = W(X^*)$ of factors of all words $w \in X^*$ is called the language generated by the code X.

We denote by b_n the number of words of X of length n and similarly by c_n the number of words of X^* of length n. By convention we suppose that $b_0 = c_0 = 1$. Furthermore we call ρ_X the radius of convergence of the code X if it is the radius of convergence of the series $\sum_{n>0} b_n z^n$. Analogously the radius of convergence ρ_{X^*} of the code X^* is the radius of convergence of the series $\sum_{n>0} c_n z^n$.

For any language $X \subset A^*$, we denote by $W^\infty = W^\infty(X)$ the set of "infinite words" generated by X, i.e., the sequences $w^* = a_1 a_2 a_3 \ldots \in A^*$ of the form

$$w^* = a w_1 w_2 w_3 \ldots,$$

with $a \in A^*$ and $w_j \in X$ for $j = 1, 2, \ldots$. With each given language $L \in A^*$, we associate the symbolic dynamical system

$$S_L(W^\infty, \mathfrak{B}, T, \mathscr{M})$$

where $W^\infty = W^\infty(L)$; $\mathfrak{B}$ is the σ-algebra generated by all cylinders of $A^{\mathbb{N}}$, i.e., sets of the form

$$[w] = \{a_1 a_2 \in A^{\mathbb{N}} : a_1 \ldots a_n = w\}$$

for some $n \in \mathbb{N}$ and $w \in A^*$; T is the shift operator and $\mathscr{M}$ is the set of all T-invariant probability measure μ on $\mathfrak{B}$. We also write $\mu(w)$ for $\mu([w])$ for short.

By abuse of language, we say that the dynamical system S_L satisfies the specification property if the language L does.

Our main result in this section is the following.

Theorem 8.3.1 ([80, Corollaire]). *Let S be a measure theoretical system that satisfies the specification property. Then the sequence obtained by concatenating the words of length 1, then of length 2, then of length 3, and so on is generic for the maximal measure of S.*

The proof we present below is an extended version of the original proof of Bertrand-Mathis [80]. Before jumping right into the proof, we have to unravel the coded system and link the entropy of X with that of X^*. Let X be a prefix code and μ be an invariant probability measure on X^*. Then the mean length of X with respect to μ is the number

$$\ell(X, \mu) = \sum_{x \in X} |x| \, \mu(x).$$

The following result of Blanchard and Hansel [88] provides an upper bound for the entropy of a given probability measure.

Proposition 8.3.2 ([88, Proposition 2.15]). *Let X be a prefix code, μ be an invariant probability measure on X^*, and $h(\mu)$ its entropy. Then*

1. We have the inequality

$$h(\mu) \leq -\ell(X, \mu) \log \rho_{X^*}.$$

2. Equality holds if and only if the following two conditions are satisfied:

a.

$$\sum_{x \in X} \rho_{X^*}^{|x|} = 1,$$

b. μ is the Bernoulli probability measure on $X^{\mathbb{Z}}$ defined by $\mu([x]) = \rho_{X^*}^{|x|}$, for $x \in X$.

The second tool considers the relation of the radii of convergence ρ_X and ρ_{X^*}.

Proposition 8.3.3 ([88, Proposition 2.12]). *Let X be a prefix code and X^* the generated submonoid. Then*

1. we always have $0 \leq \rho_{X^*} \leq \rho_X$;

2. if $\rho_{X^*} < \rho_X$, then ρ_{X^*} is the unique solution of the equation

$$\sum_{x \in X} z^{|x|} = 1;$$

3. we have $\rho_{X^*} = \rho_X$ if and only if $\rho_X < +\infty$ and $\sum_{x \in X} \rho_X^{|x|} \leq 1$.

Now we unwrap the transformation S on the coded system. Let $\Omega \subset X^{\mathbb{N}} \times \mathbb{N}$ be the set of couples $((x_n)_{n \in \mathbb{N}}, i)$ such that $1 \leq i \leq |x_0|$. We define a transformation T of Ω onto itself by

$$T\left((x_n)_{n \geq 0}, i\right) = \begin{cases} ((x_n)_{n \geq 0}, i+1) & \text{if } i < |x_0|, \\ ((x_{n+1})_{n \geq 0}, 1) & \text{if } i = |x_0|. \end{cases}$$

The dynamical system (Ω, T) is called the associated unilateral tower of X. The set $X^{\mathbb{N}}$ may be identified with the grounding $(X^{\mathbb{N}}, 1)$ of the tower Ω. Using this identification every probability measure $\overline{\mu}$ of (Ω, T) induces for each $U \in \mathfrak{B}$ one on $X^{\mathbb{N}}$ via

$$\mu(U) = \frac{\overline{\mu}\left(U \times \{1\}\right)}{\overline{\mu}\left(X^{\mathbb{N}} \times \{1\}\right)}.$$

Remark that we have $\sum_{x \in X} |x| \, \overline{\mu}(c(x) \times \{1\}) = \overline{\mu}(\Omega) = 1$. Thus we get that

$$\ell(X, \mu) = \sum_{x \in X} |x| \, \mu(x) = \frac{1}{\overline{\mu}\left(X^* \times \{1\}\right)}$$

and therefore $\ell(X, \mu) < +\infty$.

On the contrary every probability measure μ on X^* such that $\ell(X,\mu) < +\infty$ is induced by a probability measure $\overline{\mu}$ on (Ω, T). Furthermore we have by the formula of Abramov (*cf.* Theorem 2 of Chapter 10 §6 of [167])

$$h(\overline{\mu})\ell(X,\mu) = h(\mu).$$

Finally we note that X has the discrete topology, X^* has the product topology, and Ω has a natural topology such that T is continuous. But Ω is only compact if X is finite.

We call a prefix code X positive recurrent if it satisfies the conditions

$$\sum_{x\in X} \rho_{X^*}^{|x|} = 1 \quad \text{and} \quad \sum_{x\in X} |x|\, \rho_{X^*}^{|x|} < +\infty.$$

Remark 8.3.4. Any prefix code X such that $\rho_{X^*} < \rho_X$ is in fact positive recurrent. We obtain $\sum_{x\in X} \rho_{X^*}^{|x|} = 1$ from Proposition 8.3.3 part 2. For the second condition, let ρ be such that $\rho_{X^*} < \rho < \rho_X$. Then for $|x|$ sufficiently large, $|x|\, \rho_{X^*}^{|x|} < \rho^{|x|}$, and therefore $\sum_{x\in X} \rho^{|x|} < +\infty$ implies $\sum_{x\in X} |x|\, \rho_{X^*}^{|x|} < +\infty$.

Proposition 8.3.5 ([88, Proposition 2.16]). *Let X be a prefix code and (Ω, T) the associated tower.*

1. *We have*

$$\sup_{\overline{\mu}} h(\overline{\mu}) = -\log \rho_{X^*},$$

where $\overline{\mu}$ runs through all probability measures of (Ω, T).

2. *There exists one and only one probability measure $\overline{\mu}$ on Ω such that $h(\overline{\mu}) = -\log \rho_{X^*}$ if an only if X is positive recurrent. In this case $\overline{\mu}$ is the unique probability measure on Ω inducing on X^* the Bernoulli probability measure defined by*

$$\mu(x) = \rho_{X^*}^{|x|}, \quad x \in X.$$

Let $f\colon \Omega \to A^{\mathbb{N}}$ be the projection defined such that

$$fT = Tf$$

$f((x_n)_{n\in\mathbb{N}}, i) = a_i$ where a_i is the ith letter of x_0. This function is continuous and we have the following.

Proposition 8.3.6 ([88, Proposition 2.17]). *Let $\overline{\mu}$ be an ergodic probability measure on Ω and $\overline{\mu} \circ f^{-1}$ its image under f in $A^{\mathbb{N}}$. Then*

$$h(\overline{\mu} \circ f^{-1}) = h(\overline{\mu}).$$

We note that $\overline{\mu}_X$ is only defined if X is positive recurrent, and therefore we may define generic points of the Champernowne type only if $\rho_{X^*} < \rho_X$.

There may be different codes that induce the same dynamical system. For example, if $A = \{0, 1\}$, then the codes $X = \{0, 1\}$ with $\rho_{X^*} = \frac{1}{2} < \rho_X = \infty$ and

$$Y = \{u_1, \ldots, u_n \in A^+ : \forall k \in \mathbb{N}, u_1, \ldots u_k \text{ contains more 0 than 1}$$

$$\text{and } u_1, \ldots, u_n \text{ contain as many 0 as 1}\}$$

with $\rho_{Y^*} = \frac{\sqrt{2}}{2} = \rho_Y$ both induce the system $(A^{\mathbb{N}}, T)$, but Y does not verify $\rho_{Y^*} < \rho_Y$ and is not even positive recurrent.

We prove the theorem following the steps of Bertrand-Mathis [80]. Let us start with the following.

Proposition 8.3.7. *Let X be a prefix code (satisfying $\rho_{X^*} < \rho_X$) over the alphabet A.*

Suppose that the gcd of the length of words in X equals $q \geq 1$. If we concatenate all messages of length q, then all of length 2q, then all of length 3q, and so on, then we obtain a sequence $(e_n)_{n \geq 0}$ of letters in a dynamical system associated with X, and the generated sequence is generic for a measure μ_X, which we call the Champernowne measure induced by X. The measure μ_X has entropy $-\log \rho_{X^*}$.

The proof of Proposition 8.3.7 relies on the following lemma.

Lemma 8.3.8. *Let X be a prefix code such that*

$$\rho_{X^*} < \rho_X$$

and let

$$\lambda = \frac{1}{b_1 \rho_{X^*} + 2b_2 (\rho_{X^*})^2 + \cdots + nb_n (\rho_{X^*})^n + \cdots}.$$

If the gcd of the lengths of the words of the code is 1, then there exist reals $d < 1/\rho_{X^*}$ and B such that

$$c_n = \frac{\lambda}{(\rho_{X^*})^n} + v_n, \quad \text{where } |v_n| < Bd^n.$$

If the gcd of the lengths of the words of the code is $q \neq 1$, then there exist reals $d < 1/\rho_{X^*}$ and B such that

$$c_{qn} = \frac{q\lambda}{(\rho_{X^*})} + v_{qn}, \quad \text{where } \left|v_{qn}\right| < Bd^{qn}$$

and $c_n = 0$ for $q \nmid n$.

Proof. Since $\rho_{X^*} < \rho_X$ we get with Proposition 8.3.3 that

$$\sum_{n>0} b_n (\rho_{X^*})^n = 1$$

and on the converse that if the only real solution of the equation $\sum_{n\geq 0} b_n z^n = 1$ is strictly less than ρ_X, then it must equal ρ_{X^*}.

Suppose that $\rho_{X^*} < \rho_X$. Since the b_n are nonnegative and the derivative $\sum_{n>0} nb_n z^{n-1}$ has no zero in ρ_{X^*}, we must have that ρ_{X^*} is a simple root of the equation $\sum_{n>0} b_n z^n - 1 = 0$.

Suppose that α is another solution to the equation. Then it must be greater in modulus. Because otherwise if $|\alpha| \leq \rho_{X^*}$, then

$$\sum_{n>0} b_n \alpha^n = \sum_{n>0} b_n (\rho_{X^*})^n \quad \Rightarrow \quad \forall n > 0\colon b_n \alpha^n = b_n (\rho_{X^*})^n ,$$

and therefore the gcd of the length of the words in X is either 1 and $\alpha = \rho_{X^*}$ or it is q and $\alpha = e^{2i\pi m/q} \rho_{X^*}$ is a solution of the equation for each integer m.

Recall that c_k is the number of words of X^* of length k. Then by definition $(c_k)_{k\geq 1}$ are the coefficients of the formal power series

$$\frac{1}{1 - \sum_{n>0} b_n Y^n} = 1 + \left(\sum_{n>0} b_n Y^n\right) + \left(\sum_{n>0} b_n Y^n\right)^2 + \cdots .$$

There exists a disk D with center 0 and radius strictly greater than ρ_{X^*} on which the function $1/\left(1 - \sum_{n\geq 0} b_n Y^n\right)$ is a meromorphic function. If the gcd of the lengths of the words of the code is 1, then (by calculating the residue in ρ_{X^*})

$$\frac{1}{1 - \sum_{n>0} b_n Y^n} = \frac{\lambda}{1 - (Y/\rho_{X^*})} + H(Y),$$

where

$$\lambda = \left(\sum_{n>0} nb_n \rho_{X^*}^n\right)^{-1}$$

and H is an analytic function on D. Thus

$$\frac{1}{1 - \sum_{n>0} b_n Y^n} = \sum_{n\geq 0} \left(\frac{\lambda}{\rho_{X^*}^n} + v_n\right) Y^n,$$

where v_n satisfies the required conditions.

The case of the gcd of the lengths of the words of the code being $q > 1$ reduces to the consideration of the code over A^q. Then the gcd is 1 and we follow the case above.

Proof (Proof of Proposition 8.3.7). The idea of construction of the generic point follows the one of Champernowne. We write

$$\theta = \frac{1}{\rho_{X^*}}$$

for short.

Let B_n be the block formed by the c_n words in C_n concatenated one after the other. The length of each B_k is kc_k. Now we calculate the length ℓ_n of the concatenation of the blocks $B_1B_2\cdots B_n$. By Lemma 8.3.8 we have

$$\ell_n = \sum_{k=1}^{n} kc_k = \sum_{k=1}^{n} \left(\lambda k\theta^k + kv_k\right).$$

We set

$$u_n = \sum_{k=1}^{n} kv_k.$$

Then

$$|u_n| \ll \sum_{k=1}^{n} kd^k = \frac{d(nd^{n+1} - (n+1)d^n + 1)}{(d-1)^2}$$

and there exists B' such that

$$|u_n| \leq B'nd^n.$$

Therefore

$$\ell_n = \frac{\lambda\theta}{(\theta-1)^2}(n\theta^{n+1} - (n+1)\theta^n + 1) + u_n.$$

Now we want to count how often a given message m of length p occurs in the generated message. We distinguish three cases:

1. The occurrence happens between two messages of length k and $k+1$, respectively. This means that m is on the edge of B_kB_{k+1}. Since this rarely happens, we will neglect this case.
2. The message m occurs on the interior of a message m_1 of length k in B_k and

$$m_1 = amb,$$

where a and b are also messages.

3. The message m occurs on the interior of a message m_1 of length k in B_k such that

$$m_1 = amb,$$

where a or b is not a message.

Lemma 8.3.9. *The frequency of occurrences as in case 2 of the message m of length p in the constructed word is equal to* λ/θ^p.

Proof. The number of occurrences of the message m in the block B_k as in case 2 is

$$\begin{aligned} f_k &= \sum_{i=0}^{k-p} c_i c_{k-p-i} = \sum_{i=0}^{k-p} \left(\lambda\theta^i + v_i\right)\left(\lambda\theta^{k-p-i} + v_{k-p-i}\right) \\ &= \sum_{i=0}^{k-p} \lambda^2\theta^{k-p} + 2\sum_{i=0}^{k-p} v_i\lambda\theta^{k-p-i} + \sum_{i=0}^{k-p} v_i v_{k-p-i} \\ &= x_k + y_k + z_k. \end{aligned}$$

The first part equals $x_k = (k-p+1)\lambda^2\theta^{k-p}$.

Since $|v_i| < Bd^i$, the second part may be estimated by

$$|y_k| < 2\sum_{i=0}^{k-p} B\lambda d^i\theta^{k-p-i} = 2B\lambda\theta^{k-p}\left(\sum_{i=0}^{k-p}\frac{d^i}{\theta^i}\right) \ll \lambda\theta^{k-p}\frac{1}{1-\left(\frac{d}{\theta}\right)},$$

where we used that $d < \theta$. Note that the implied constant neither depend on k nor on p.

Finally we obtain the following upper bound for the third part:

$$|z_k| \ll \sum_{i=0}^{k-p} d^i d^{k-p-i} \ll (k-p+1)d^{k-p},$$

where we may suppose that the implied constant neither depend on k nor on p.

Summing over all k, we obtain

$$g_n = \sum_{k=p}^{n} f_k = \sum_{k=p}^{n} (x_k + y_k + z_k)$$

with

$$\sum_{k=p}^{n} x_n = \sum_{k=p}^{n} (k-p+1)\lambda^2\theta^{k-p}$$

$$= \lambda^2 \frac{(n-p+1)\theta^{n+2-p} - (n-p+2)\theta^{n+1-p} + 1}{(\theta-1)^2}$$

$$\left| \sum_{k=p}^{n} y_k \right| \ll \sum_{k=p}^{n} \theta^{k-p} \ll \frac{\theta^{n+1-p} - 1}{\theta - 1}$$

$$\left| \sum_{k=p}^{n} z_k \right| \ll \sum_{k=p}^{n} (k-p+1) d^{k-p} = \frac{(n-p+1)d^{n-p+2} - (n-p+2)d^{n-p+1} + 1}{(d-1)^2}.$$

Dividing by the length ℓ_n, we observe that the frequency g_n/ℓ_n tends to λ/θ^p for $n \to \infty$ and therefore a message m as in case 2 occurs with asymptotic frequency $\lambda/\theta^{|m|}$.

For the moment we suppose that the gcd of the length of the words in X equals 1.

Lemma 8.3.10. *Let δ_u be the Dirac point measure in u. There exists a unique measure μ, which is a limit point of the sequence of measures*

$$\frac{1}{\ell_n} \sum_{k=1}^{\ell_n} \delta_{T^k((e_n)_{n\geq 0})}.$$

Proof. Let μ be a limit point of the sequence. For a given message m, let $K(m)$ be the set of messages m_1 such that there exist u and v in A^* but not in X^* such that $m_1 = umv$. Furthermore u has no representation of the form $u = m_2 u_1$, where m_2 is a message, and v has no representation of the form $v = v_1 m_3$, where m_3 is a message.

The message m occurs in the sequence $(e_n)_{n\geq 0}$ every time when a message of $K(m)$ occurs in case 2 (we still neglect the case 1) and

$$\mu(m) = \sum_{m_1 \in K(m)} a(m_1) \frac{\lambda}{\theta^{|m_1|}},$$

where m occurs $a(m_1)$ times in m_1. For a given word ω, let $H(\omega)$ be the set of messages m_4 such that $m_4 = u\omega v$ and u and v are no messages and where neither u has a representation of the form

$$u = m_5 u_1$$

nor v has one of the form

$$v = v_1 m_6$$

with m_5 and m_6 again messages. Then

$$\mu(\omega) = \sum_{m \in H(\omega)} b(m)\mu(m),$$

where $b(m)$ is the number of occurrences of ω in m. This shows that the sequence

$$\frac{1}{\ell_n} \sum_{k=1}^{\ell_n} \delta_{T^k((e_n)_{n\geq 0})}$$

only has one limit point $\mu = \mu_X$.

We show that μ_X is the only limit point of the sequence e. Suppose there exists another measure ν which is a limit point of the sequence of measures

$$\frac{1}{n} \sum_{k<n} \delta_{T^k(e)}.$$

Since ℓ_{n+1}/ℓ_n is bounded by 2θ, the measure ν is absolutely continuous with respect to μ_X, which is impossible. Therefore

$$\mu_X = \lim_{n\to\infty} \frac{1}{n} \sum_{k<n} \delta_{T^k(e)}.$$

It is clear that the constructed measure gives a positive measure to any message.

Now we turn our attention to the entropy of μ_X. Let $E_n = \{\bigcup[m] : m \in C_n\}$. Since there are about $\lambda\theta^n$ messages of length n and the measure of $[m]$ is greater than λ/θ^n, we get for n sufficiently large that

$$\mu(E_n) > \frac{\lambda^2}{2}.$$

For given $\delta > 0$ and $\varepsilon > 0$ using the Shannon-McMillan-Breiman theorem (*cf.* Theorem 5 of Chapter 10 §6 of [167]), we divide the words $(a_1 \dots a_n)$ in A^n into two classes. The set B of bad words satisfying

$$\sum_{a_1,\dots,a_n \in B} \mu_X([a_1 \dots a_n]) < \delta$$

and the set G of good words, where each $a_1 \dots a_n \in G$, satisfies

$$e^{-n(h(\mu_X)+\varepsilon)} < \mu([a_1 \dots a_n]) < e^{-n(h(\mu_X)-\varepsilon)}.$$

Since the measure of E_n is greater than $\lambda^2/2$, we get that

$$\frac{\lambda}{\theta^n} < \mu([a_1 \dots a_n]) < e^{-n(h(\mu_X)-\varepsilon)}.$$

Thus

$$h(\mu_X) \leq \log \theta.$$

Now let us suppose that $h(\mu_X) < \log \theta$ and find a contradiction. We set

$$\varepsilon = \frac{1}{2}(\log \theta - h(\mu_X)).$$

For δ sufficiently small and n sufficiently large, at least half of the words of E_n are in the set G. Let D_n be their union, then

$$e^{-n(h(\mu_X)+\varepsilon)} < \mu_X([a_1 \dots a_n]) \quad \text{if } [a_1 \dots a_n] \in D_n$$

$$e^{-n(\log \theta-\varepsilon)} < \mu_X([a_1 \dots a_n]) \quad \text{if } [a_1 \dots a_n] \notin D_n$$

Since D_n is the union of at least $\lambda\theta^n/4$ words, we get

$$\mu_X(D_n) > \frac{\lambda\theta^n}{4} e^{-n(\log \theta - \varepsilon)} > \frac{\lambda}{4}\frac{\theta^n}{\theta'^n} \quad \text{where } \log \theta' = \log \theta - \varepsilon.]$$

Since $\theta' < \theta$, we get that $\lim_{n\to\infty} \mu_X(D_n) = \infty$ which is absurd. Thus $h(\mu_X) = \log \theta$.

For the case of a gcd greater than one, we proceed as follows. Let X be a prefix code over A, and let $q \neq 1$ be the gcd of the length of the words in X. Now let Y be the prefix code over A^q whose words are the words of X seen as words over A^q. Then the gcd of the length of the words in Y is one and we have

$$\rho_Y = (\rho_X)^q \quad \text{and} \quad \rho_{Y^*} = (\rho_{X^*})^q.$$

If we concatenate the messages of length 1, then 2, then 3, and so on from Y^*, then we get a sequence $(f_n)_{n\geq 0}$ in $(A^q)^{\mathbb{N}}$. We may see this sequence as concatenating the messages of length q, then $2q$, then $3q$, and so on from X^* to construct a sequence $(e_n)_{n\geq 0}$ in $A^{\mathbb{N}}$. Since Y satisfies the requirements of Proposition 8.3.7, we get that it is generic for a measure μ_c^q on $(A^q)^{\mathbb{N}}$, with entropy $-q \log \rho_{X^*}$ with respect to the shift on $(A^q)^{\mathbb{N}}$. Therefore the sequence $(e_n)_{n\geq 0}$ is generic for a T-invariant measure on $A^{\mathbb{N}}$

$$\mu_c = \frac{1}{q}\left(\mu_c^q + T^{-1}\mu_c^q + \cdots + T^{q-1}\mu_c^q\right),$$

which has entropy $-\log \rho_{X^*}$.

Proposition 8.3.11. *Let X be a prefix code satisfying $\rho_{X^*} < \rho_X$. Let (Ω, T) be the associated tower, and let $(e_n)_{n\geq 0}$ be the sequence defined in Proposition 8.3.7.*

Then $(e_n)_{n\geq 0}$ is the image of a point $(\varepsilon_n)_{n\geq 0} \in \Omega$ under f, and the point $(\varepsilon_n)_{n\geq 0}$ is generic for a unique measure $\overline{\mu}_X$ with entropy $-\log \rho_{X^*}$ on the tower. The Champernowne measure μ_X is the image of $\overline{\mu}_X$ by f.

Proof. The c_n messages of C_n may be written (if we decompose them into words over X) as

$$(x_{1,n}^1, \ldots, x_{1,n}^{k_1}, x_{2,n}^1, \ldots, x_{2,n}^{k_2}, \ldots, x_{c_n,n}^1, \ldots, x_{c_n,n}^{k_{c_n}}),$$

where for each $1 \leq i \leq n$ and $1 \leq j \leq k_i$, $x_{i,n}^j$ is a word in X and for all i

$$\sum_{j=1}^{k_i} \left| x_{i,n}^j \right| = n.$$

The sequence $(\varepsilon_n)_{n\geq 0}$ defined by

$$\begin{aligned}(\varepsilon_0, \varepsilon_1, \varepsilon_2, \ldots) =& (x_{1,1}^1, 1)(x_{1,1}^1, 2) \cdots \left(x_{1,1}^1, \left|x_{1,1}^1\right|\right) (x_{1,1}^2, 1) \cdots \\ & \cdots \left(x_{c_1,1}^{k_{c_1}}, \left|x_{c_1,1}^{k_{c_1}}\right|\right) (x_{1,2}^1, 1) \cdots \\ & \cdots \left(x_{c_n,n}^{k_{c_n}}, \left|x_{c_n,n}^{k_{c_n}}\right|\right) (x_{1,n+1}^1, 1) \cdots\end{aligned}$$

is the preimage of the sequence $(e_n)_{n\geq 0}$ defined in Proposition 8.3.7, and we show as above that the point $(\varepsilon_n)_{n\geq 0}$ is generic for a measure ν, whose image by f is the measure μ_X. The entropy of ν is bounded from below by $-\log \rho_{X^*}$. Thus ν is the unique measure $\overline{\mu}$ with entropy $-\log \rho_{X^*}$ on the tower and $\mu_X = f(\overline{\mu})$.

Instead of directly computing the entropy, we could have shown that the restriction of ν to the base of the tower is with the exception of a factor the Bernoulli measure with entropy $-\log_{X^*}$. This implies $\nu = \overline{\mu}$ and with $\mu_X = f(\overline{\mu}_X)$ we get

$$h(\mu_X) = h(\overline{\mu}_X) = -\log \rho_{X^*}.$$

Champernowne's construction is based on the sequence of positive integers. He conjectured and Copeland and Erdős [166] later proved that the corresponding construction over the primes, i.e.,

$$0.2\,3\,5\,7\,11\,13\,17\,19\ldots$$

also yields a normal number. Moreover they showed that any strictly increasing sequence of integers not missing too many elements yields a normal number.

Theorem 8.3.12 ([166]). *Let $(a_n)_{n\geq 1}$ be a strictly increasing sequence of positive integers. If for each $\varepsilon > 0$ there exists $N_0(\varepsilon)$ such that for all $N > N_0(\varepsilon)$*

$$\#\{n: a_n \leq N\} > N^{1-\varepsilon}$$

holds, then the number

$$0.a_1\, a_2\, a_3\, a_4\, a_5\, \ldots$$

is normal in the base of expansion.

This construction was generalized to languages with specification by Bertrand-Mathis and Volkmann [81].

Theorem 8.3.13. *Let L be a language with specification and $(a_n)_{n\geq 1}$ a sequence of different elements of $W(L^*)$ with $|a_1| \leq |a_2| \leq \ldots$. If for all $\varepsilon > 0$ there exists $n_o(\varepsilon)$ such that for all $n \geq n_0$*

$$\#\{a_\nu: |a_n| \leq n\} > |L_n|^{1-\varepsilon}$$

holds, then the infinite word

$$a = a_1 a_2 a_3 \ldots \in W^\infty$$

is normal.

Remark 8.3.14. The proof is only for connecting languages. The difference with languages with specification is that here the gap always has the same size g. Otherwise said, for any pair $\mathbf{a}, \mathbf{b} \in L$ there exists $\mathbf{v}$ with $|\mathbf{v}| = g$ such that $\mathbf{avb}$. However, following the lines of the proof, it is easy to rewrite it for languages with specification.

The idea of the proof is that the average word has good distribution. In the spirit of Besicovitch [82], for $\varepsilon > 0$, k an integer, and ν a shift invariant measure, a word $w \in W$ is (ε, k, ν)-normal if, for every $\mathbf{b} \in L_k$,

$$\nu(\mathbf{b})(1-\varepsilon) < \frac{N(\mathbf{b}, w)}{|w|} < \nu(\mathbf{b})(1+\varepsilon).$$

Furthermore we say that w is (ε, k)-normal, if ν is the maximal measure. Now Bertrand-Mathis and Volkmann [81] show that the set $E_n(\varepsilon, k)$ of words of length n which are not (ε, k)-normal is small. Thus if we only eliminate a small portion of "good" words, we still have enough for the construction.

This whole idea breaks down, when it comes to the sequence of squares. Using a combinatorial argument, Besicovitch [82] succeeded to show that the number

$$0.1\,4\,9\,16\,25\,36\,49\,64\,\ldots$$

is normal in base 10. However, his combinatorial method gets more and more involved when it comes to cubes and higher powers. Furthermore it seems to be impossible to apply it to polynomials instead of monomials. The modern approach is to use Fourier analysis. For simplicity we consider the decimal case. For a word $\mathbf{b} = b_1 \dots b_k \in \{0, 1, \dots, 9\}^k$ we denote by $I_{\mathbf{b}}$ its indicator function, i.e.,

$$I_{\mathbf{b}} = \begin{cases} 1 & \text{if } x \in \left[\sum_{i=1}^k b_i 10^{-i}, \sum_{i=1}^k b_i 10^{-i} + 10^{-k}\right) \\ 0 & \text{otherwise.} \end{cases}$$

The function $I_{\mathbf{b}}$ detects whether the decimal expansion of x starts with the same digits as $\mathbf{b}$. Suppose that $\mathbf{w} = w_1 \dots w_\ell$ is a word of length ℓ over the alphabet $\{0, 1, \dots, 9\}$. We associate with $\mathbf{w}$ the positive integer $w = \sum_{i=1}^{\ell} w_i 10^{\ell-i}$ whose decimal expansion is exactly the word $\mathbf{w}$. Then $I_{\mathbf{b}}\left(wq^{-j}\right)$ with $k \leq j \leq \ell$ detects whether the word $\mathbf{b}$ occurs at position $\ell - j$ in $\mathbf{w}$. Thus

$$\#\{0 \leq i \leq \ell - k \colon w_{i+1} = b_1, \dots, w_{i+k} = b_k\} = \sum_{i=k}^{\ell} I_{\mathbf{b}}\left(\frac{w}{q^j}\right).$$

Now we consider the Fourier transform of $I_{\mathbf{b}}$. To this end it is convenient to interpret the indicator function as superposition of two rounding error functions φ (also called "saw-tooth function", because of its function graph) defined by $\varphi(x) = x - \lfloor x \rfloor + \frac{1}{2}$. Then we have

$$I_{\mathbf{b}}(x) = 10^{-k}\varphi\left(x - \frac{\sum_{i=1}^k b_i 10^{i-1} + 1}{10^k}\right) - \varphi\left(x - \frac{\sum_{i=1}^k b_i 10^{i-1}}{10^k}\right).$$

Vaaler [572] provided approximations to this functions by trigonometric functions. Using his upper and lower bounds, we get the following.

Theorem 8.3.15 ([572, Theorem 19]). *Let $I \subset [0, 1]$ be an interval and χ_I its indicator function. Then for each positive integer H, there exist coefficients $a_H(h)$ and C_h for $-H \leq h \leq H$ with $|a_H(h)| \leq 1$ and $|C_h| \leq 1$ such that the trigonometric polynomial*

$$\chi_{I,H}^*(t) = |I| + \frac{1}{\pi} \sum_{0<|h|\leq H} \frac{a_H(h)}{|h|} e(ht)$$

satisfies

$$|\chi_I(t) - \chi_{I,H}^*(t)| \leq \frac{1}{H+1} \sum_{|h|\leq H} C_h \left(1 - \frac{|h|}{H+1}\right) e(ht).$$

This transfers the problem into one on exponential sums.

For the estimation of the exponential sum, it is easier to fix the position first and sum over the different expansions. Our target of application are polynomials p. To this end we suppose that $\lim_{n\to\infty} p(n) = +\infty$. The case of $-\infty$ as limit is similar, and the case of a constant polynomial is completely uninteresting as it yields a periodic word. The important feature of polynomials p is that they are eventually growing, meaning that there exists $n_0 \in \mathbb{N}$ such that for all $n \geq n_0$ we have $p(n+1) > p(n)$. Now we group all n yielding an expansion of the same length: There exists ℓ_0 such that for $\ell \geq \ell_0$ we find n_ℓ such that $\lfloor \log_{10} p(n_\ell) \rfloor < \lfloor \log_{10} p(n_\ell + 1)$ and $\lfloor \log_{10} p(n) \rfloor = \ell - 1$ for all $n_{\ell-1} + 1 \leq n \leq n_\ell$. Note that the decimal expansion of a positive integer w has length ℓ if and only if $\lfloor \log_{10} w \rfloor = \ell - 1$. Then counting the number of occurrences of a given block $\mathbf{b}$ in the constructed word is transferred via an estimate of exponential sums of the form

$$\sum_{\ell=\ell_0}^{L} \sum_{j=k}^{\ell} \sum_{n=n_\ell+1}^{n_{\ell+1}} e\left(\frac{\nu p(n)}{q^j}\right).$$

The first result using this method is due to Davenport and Erdős [180] who applied this to polynomials $p \in \mathbb{Z}[X]$. Schiffer [527] extended this to polynomials $p \in \mathbb{Q}[X]$ such that $p(\mathbb{N}) \subset \mathbb{N}$. Furthermore Nakai and Shiokawa [435, 436] considered polynomial like functions p of the form $p(x) = \alpha_1 x^{\beta_1} + \cdots + \alpha_d x^{\beta_d}$ with $\alpha_i \in \mathbb{R}$ for $1 \leq i \leq d$ and $0 < \beta_1 < \cdots < \beta_d$. Finally we want to mention the construction of Madritsch, Thuswaldner, and Tichy [401] where p is an entire function of bounded logarithmic order, i.e.,

$$\limsup_{r\to\infty} \frac{\log\log \max_{|z|\leq r} p(z)}{\log\log r} < \frac{4}{3}.$$

In the last case, the eventually growing condition is replaced by longer and longer intervals where the function is bounded from below.

Similar constructions over the primes yield also normal numbers. We want to mention Nakai and Shiokawa [437] and Madritsch, Thuswaldner, and Tichy [401]. Finally a recent result by Scheerer [526] generalizes these results to β-expansions. However, in his proof he uses a combinatorial argument of the Besicovitch type.

8.4 Generic Sequences for Different Measures

After the construction of generic words for the maximal measure, we want to look at different measures. For example, suppose we want the following distribution of digits in the q-adic expansion. The letters $0, 1, 2, \ldots, q-2$ should appear with frequency 2^{-q}, and the last letter $q-1$ should appear with frequency $1-(q-1)2^{-q}$. For example in the binary expansion, we would have that 0 occurs with frequency $1/4$ and 1 with frequency $3/4$.

Now for a block of digits $a_1a_2\dots a_k$, we suppose that its frequency of occurrences is the product of the frequency of occurrences of the single digits $a_1, a_2 \dots, a_k$. Considering our example in the binary expansion, this means that the blocks 00, 01, 10, and 11 occur with the frequencies 1/16, 3/16, 3/16, and 9/16, respectively.

The construction itself consists of two parts. The first one is a characterization of useful sequences of words. Given such a sequence, we might construct a generic sequence for μ by repeated concatenation of the words in the sequence. This immediately leads to the second part: the construction of such a useful sequence of words. In both parts we have parameters, which allow us to adapt the construction to the specific dynamical system (like Lüroth-series, continued fraction, β-expansion, …).

We start by circumventing the issue of an infinite alphabet. Let μ be a shift invariant measure. A sequence (ν_i) of shift invariant measures is called an approximation scheme if ν_i converges weakly to μ. A word $\mathbf{b} \in L$ is μ-admissible if $\mu(\mathbf{b}) \neq 0$, i.e., the cylinder $[\mathbf{b}]$ is in the support of μ. Let $L_\mu \subset L$ denote the set of μ-admissible words, and let $L_{\mu,k}$ denote the set of μ-admissible words of length k.

Recall that a word $w = w_1 \dots w_n$ is called (ε, k, μ)-normal if for each word $\mathbf{b} \in \bigcup_{i=1}^{k} L_i$ we have that

$$\mu(\mathbf{b})(1-\varepsilon) < \frac{N(\mathbf{b}, w)}{|w|} < \mu(\mathbf{b})(1+\varepsilon).$$

Furthermore let $(\mathbf{w}_i)_{i\geq 1}$ be a sequence of finite words and $(\ell_i)_{i\geq 1}$ be a nondecreasing sequence of positive integers. Then we call the sequence of pairs $(\mathbf{w}_i, \ell_i)_{i\geq 1}$ μ-good with respect to the approximation scheme $(\nu_i)_{i\geq 1}$ if each $\mathbf{w}_i$ is $(\varepsilon_i, k_i, \nu_i)$-normal satisfying

$$\frac{1}{\varepsilon_{i-1} - \varepsilon_i} = o(|\mathbf{w}_i|); \tag{8.5}$$

$$\frac{\ell_{i-1}}{\ell_i} \cdot \frac{|\mathbf{w}_{i-1}|}{|\mathbf{w}_i|} = o(i^{-1}); \tag{8.6}$$

$$\frac{1}{\ell_i} \cdot \frac{|\mathbf{w}_{i+1}|}{|\mathbf{w}_i|} = o(1). \tag{8.7}$$

The main theorem in this section is the following.

Theorem 8.4.1 ([399, Main Theorem]). *Let A be an alphabet, $L \subset A^*$ be a language satisfying the specification property, and $W^\infty(L)$ be the set of sequences generated by L. Let T be a shift of $A^{\mathbb{N}}$ and μ be a shift invariant probability measure on $W^\infty(L)$. Let $(k_i)_{i\geq 1}$ be a sequence of positive integers, and for $i \geq 1$ let $\nu_i\colon A^{k_i} \to [0,1]$ be a shift invariant probability measure on A^{k_i} such that $(\nu_i)_{i\geq 1}$ is an approximation scheme for μ. Let $(\mathbf{w}_i)_{i\geq 1}$ be a sequence of finite words, and let*

$(\ell_i)_{i\geq 1}$ be a nondecreasing sequence of positive integers. Suppose that $(\mathbf{w}_i, \ell_i)_{i\geq 1}$ is μ-good with respect to $(\nu_i)_{i\geq 1}$, then for each integer $k \in [1, \limsup_{i\to\infty} k_i]$, the sequence $\omega = \mathbf{w}_1^{\odot \ell_1} \odot \mathbf{w}_2^{\odot \ell_2} \odot \cdots$ is μ-normal of order k. Moreover, if $\limsup_{i\to\infty} k_i = \infty$, *then ω is μ-normal*

Our construction is very similar to the Champernowne type construction of Bertrand-Mathis in Section 8.3. However, our goal is to construct a sequence which is generic for any given shift invariant measure and not necessarily the maximal one. As a consequence of this general case, our construction is not so efficient in that it uses many repetitions of words that have "good distribution" for the desired measure.

The Champernowne word has several advantages like

- every possible block occurs exactly once,
- two consecutive blocks are different,
- the blocks are ordered in size, meaning that first we have those of length 1, then of length 2, then of length 3, and so on.

On the contrary the Champernowne word is generic for the maximal measure. What if we want a different one? A different one means that there is at least one block that gets more weight. Let us return to our case of the binary expansion and the frequencies $1/4$ for 0 and $3/4$ for 1. Then an idea would be to simply repeat each block according to its weight. For example, the word

$$0\,1\,1\,1,$$

where we repeated 0 once and 1 three times, has the desired frequencies of length one. For length two, we look at the word

$$00\ 01\ 01\ 01\ 10\ 10\ 10\ 11\ 11\ 11\ 11\ 11\ 11\ 11\ 11\ 11,$$

where we repeated 00 once, 01 and 10 three times, and 11 nine times. In general, since the common denominator is always 4^k, we multiply the desired frequency by 4^k and obtain the number of times we have to repeat this word in the sequence.

This works for q-adic expansions. However, we want to obtain such a construction for general symbolic dynamical systems satisfying the specification property. In these systems we have to face two main issues: the possible infinite alphabet and the restriction of concatenation of words.

Recall our example with β-expansions (Example 8.1.2), and set $\beta = \phi$ the Golden Ratio. In this case we have that the word 11 is forbidden in the expansion. However, if we concatenate $\mathbf{a} = 1001$ and $\mathbf{b} = 1010$, which are admissible as such, that yields the word $\mathbf{ab} = 10011010$, which is not admissible. At this point we use the specification property in order to glue the words together yielding the admissible word $\mathbf{a} \odot \mathbf{b} = 100101010$.

For a fixed $\ell \geq 1$ we let $L_\ell = \{\mathbf{p}_1, \ldots, \mathbf{p}_{|L_\ell|}\}$ be an arbitrary ordering of the words of length ℓ. Furthermore let $m_\ell = \min\{\mu(\mathbf{b}) : \mathbf{b} \in L_\ell\}$ for $\ell \geq 1$ and M be an arbitrary large constant such that $M \geq \frac{1}{m_\ell}$.

Now we define a word $p_{b,\ell,M}$ that contains each $\mathbf{p}_i \in L_\ell$ with multiplicity near $M\mu(\mathbf{p})_i$. Thus

$$\mathbf{p}_{\ell,M} := \mathbf{p}_1^{\odot\lceil M\mu(\mathbf{p}_1)\rceil} \odot \mathbf{p}_2^{\odot\lceil M\mu(\mathbf{p}_2)\rceil} \odot \cdots \odot \mathbf{p}_{|L_\ell|}^{\odot\left\lceil M\mu\left(\mathbf{p}_{|L_\ell|}\right)\right\rceil}.$$

In order to be useful for our construction, we have to show the (ε, k, μ)-normality of the constructed word $\mathbf{p}_{\ell,M}$ for $k \leq \ell$, i.e.,

$$(1-\varepsilon)\mu(\mathbf{b}) \leq \frac{N(\mathbf{b}, \mathbf{p}_{b,\ell,M})}{|\mathbf{p}_{b,\ell,M}|} \leq (1+\varepsilon)\mu(\mathbf{b}). \tag{8.8}$$

Furthermore for the applications below, we need ε and k explicitly. We show the inequalities in (8.8) by proving upper and lower bounds for the length $|\mathbf{p}_{\ell,M}|$ as well as for the number of occurrences $N(\mathbf{b}, \mathbf{p}_{\ell,M})$ of a block $\mathbf{b}$.

Starting with the length of the constructed word, we get the following upper and lower bounds

$$M\ell \leq |\mathbf{p}_{\ell,M}| \leq (g+\ell)\left(M + |A|^\ell\right),$$

where A denotes the alphabet, which we suppose to be finite for the moment. For the lower bound we suppose that all words have length ℓ, whereas for the upper bound we always have to add the gab g. Moreover the lower bound for the number of elements originates from the fact that each word has to appear at least once, and in the upper bound we compensate the error from the ceiling function.

The lower and upper bounds of the number of occurrences of $\mathbf{b}$ in $\mathbf{p}_{\ell,M}$ are a little bit more work. In particular, as above in Section 8.3, we will distinguish several cases: the word $\mathbf{b}$ may occur

1. within $\mathbf{p}_i$,
2. between two equal words $\mathbf{p}_i \odot \mathbf{p}_i$ or
3. between two different words $\mathbf{p}_i \odot \mathbf{p}_{i+1}$.

For the lower bound, we only consider the case of occurrences with $\mathbf{p}_i$ yielding

$$N(\mathbf{b}, \mathbf{p}_{\ell,M}) \geq (\ell - k + 1)M\mu(\mathbf{b}).$$

A consideration of all three cases yields the upper bound

$$N(\mathbf{b}, \mathbf{p}_{\ell,M}) \leq (\ell+g)\left(M\mu(\mathbf{b}) + |A|^{2\ell+g-k}\right).$$

Putting these bounds together, we get that $\mathbf{p}_{\ell,M}$ is (ε, k, μ)-normal for

$$k \leq \ell \quad \text{and} \quad \varepsilon \leq \max\left(\frac{g+k-1}{\ell+g} + \frac{|A|^\ell}{M+|A|^\ell}, \frac{g}{\ell} + \frac{1}{m_k}\frac{|A|^{2\ell+g-k}}{M}\right). \tag{8.9}$$

This construction is far from optimal in the sense that even if you use it for the construction of a normal number in the decimal system, it uses way too many repetitions. In special cases Vandehey [573] reduced the number of copies necessary in the construction. However, it remains an open problem to reduce the number of copies in the case of β-expansions or more generally in the case of shifts of finite type.

In the following we want to apply this construction to β-expansions and continued fraction expansion. Note that further applications have been considered by Madritsch and Mance [399], who constructed normal numbers with respect to q-adic expansions with maximal and arbitrary measure and Lüroth series. Also the unfair coin as a special application in the binary system with not maximal measure has been considered.

We only have restrictions on the concatenation in the case of β-expansions; all other examples are in the full shift. It is easy to combine our construction for β-expansions and continued fractions in order to get constructions for α-continued fractions (*cf.* Nakada [434]) or Rosen-continued fractions [510], which have an infinite digit set with restrictions on the concatenation of words. For the relation of normal numbers with respect to different continued fraction expansions, we refer the interested reader to the paper of Kraaikamp and Nakada [361].

The main ingredient in all our constructions is the following lemma which is a consequence of Theorem 8.4.1 together with the construction of the words $\mathbf{p}_{\ell,M}$.

Lemma 8.4.2. *Let μ be a shift invariant probability measure and let $(\nu_i)_{i\geq 1}$ be an approximation scheme for μ. Suppose that $q_i \geq 2$, M_i and ℓ_i are sequences of positive integers such that*

$$M_i \geq (\min\{\mu(\mathbf{b}) : \mathbf{b} \in \mathscr{D}_{\nu_i,i}\})^{-1} \quad \textit{and} \quad q_i^{2i} = o(M_i) \tag{8.10}$$

and $(\mathbf{p}_{i,M_i}, \ell_i)$ is μ-good for the approximation scheme $(\nu_i)_{i\geq 1}$. Then the sequence $\omega = \mathbf{w}_1^{\odot \ell_1} \odot \mathbf{w}_2^{\odot \ell_2} \odot \cdots$ is μ-normal.

For the β expansion the size of the gap depends on the expansion of 1. To this end we denote by $\mathrm{d}_\beta(1) = b_1 \dots b_t (b_{t+1} \dots b_{t+p})^\ell$ the β-expansion of 1. If 1 has a finite expansion, then we set $p = 0$. We are looking for the longest possible sequence of zeroes occurring in the expansion of 1. As one easily checks, the longest occurs if $b_1 = \cdots = b_{t+p-1} = 0$ and $b_{t+p} \neq 0$. Thus we can set the gap size g to be

$$g = t + p.$$

We wish to minimize the length of a cylinder set defined by a word of length ℓ. Define

$$\phi_\beta(\ell) = \begin{cases} 1 \text{ if } 1 \leq \ell \leq t \\ r \text{ if } t + (r-2)p \leq \ell \leq t + (r-1)p \end{cases}.$$

Then the length of this interval is at least $\beta^{-(t+\phi_\beta(\ell)p)}$. We use the fact that $\mu_\beta(I) \geq (1 - 1/\beta)\lambda(I)$ and put

$$M_i = \max\left(\frac{\beta^{t+\phi_\beta(i)p}}{1-\frac{1}{\beta}}, \lceil\beta\rceil^{2i}\log i\right).$$

Put $\mathbf{w}_i = \mathbf{p}_{i,M_i}$ and $q_i = \lceil\beta\rceil$. Note that $\lim_{i\to\infty}\frac{\phi(i)}{i/p} = 1$, so for large i

$$(i+g)\lceil\beta\rceil^{2i}\log i \leq |\mathbf{w}_i| \leq (i+g)\left(\lceil\beta\rceil^{2i}\log i + \lceil\beta\rceil^{i}\right)$$

Thus, for large i

$$|\mathbf{w}_i| \approx i\lceil\beta\rceil^{2i}\log i.$$

Put $\ell_i = i^{2i}$ and the computation follows the same lines as above.

Now we turn our attention to the continued fraction expansion. For a word $\mathbf{b} = b_1 \ldots b_i$, let $\Delta_{\mathbf{b}}$ be the set of all real numbers in $(0, 1)$ whose first i digits of it's continued fraction expansions are equal to $\mathbf{b}$. Put

$$\mu(\mathbf{b}) = \frac{1}{\log 2}\int_{\Delta_{\mathbf{b}}}\frac{dx}{1+x}.$$

If there is an index n such that $b_n > i$, then let $\nu_i(\mathbf{b}) = 0$. Let $S = \{n : b_n = i\}$. For $i < 8$, set $\nu_i(\mathbf{b}) = \mu(\mathbf{b})$. For $i \geq 8$, if $S = \emptyset$, then let $\nu_i(\mathbf{b}) = \mu(\mathbf{b})$. If $S \neq \emptyset$, then let

$$\nu_i(\mathbf{b}) = \sum_{\mathbf{b}'}\mu(\mathbf{b}'),$$

where the sum is over all words $\mathbf{b}' = b_1' \ldots b_i'$ such that for each index n in S, $b_n' \geq i$.

Put $m_i = \min_{\mathbf{b}\in\mathscr{D}_{\nu_i},|\mathbf{b}|=i}\nu_i(\mathbf{b})$. We wish to find a lower bound for m_i. If $\mathbf{b} = b_1 \ldots b_k$, then let

$$\frac{p_k}{q_k} = \cfrac{1}{b_1 + \cfrac{1}{b_2 + \ddots + \cfrac{1}{b_k}}}.$$

It is well known that $\lambda(\Delta_{\mathbf{b}}) = \frac{1}{q_k(q_k+q_{k-1})}$ and $\mu(\mathbf{b}) > \frac{1}{2\log 2}\lambda(\Delta_{\mathbf{b}})$.

Thus, we may find a lower bound for m_i by minimizing $(q_i(q_i + q_{i-1}))^{-1}$ for words $\mathbf{b}$ in $\mathscr{D}_{\nu_i}$. The minimum will occur for $\mathbf{b} = ii\ldots i$. It is known that $q_n = iq_{n-1} + q_{n-2}$ if we set $q_0 = 1$ and $q_1 = i$. Set

$$r_1 = \frac{i + \sqrt{i^2+4}}{2}, \quad r_2 = \frac{i - \sqrt{i^2+4}}{2}.$$

Then

$$q_n = \frac{r_1^{n+1} - r_2^{n+1}}{\sqrt{i^2+4}}.$$

Thus,

$$\frac{1}{q_i(q_i+q_{i-1})} = \frac{i^2+4}{(r_1^{i+1} - r_2^{i+1})((r_1^{i+1}+r_1^i) - (r_2^{i+1} - r_2^i))} > \frac{\log 2}{i^{2i}} \text{ for } i \geq 8.$$

Thus, $m_i > \frac{1}{2\log 2}\left(\frac{\log 2}{i^{2i}}\right) = \frac{1}{2}i^{-2i}$. Let $M_i = 2i^{2i}\log i$, $g = 0$, $\mathbf{w}_i = \mathbf{p}_{i+1,i,M_i}$. Set $\ell_i = 0$ for $i < 8$ and $\ell_i = \lfloor i^2 \log i \rfloor$ for $i \geq 8$. Then for $i \geq 9$

$$\frac{\ell_{i-1}}{\ell_i}\frac{|\mathbf{w}_{i-1}|}{|\mathbf{w}_i|}i < \frac{2(i-1)^{2i-1} + i^{i-1}}{2i^{2i}} = \left(1 - \frac{1}{i}\right)^{2i}\frac{1}{i-1} + \frac{1}{2i^{i+1}} \to 0$$

and

$$\frac{|\mathbf{w}_{i+1}|}{\ell_i|\mathbf{w}_i|} \leq \frac{2(i+1)^{2i+3} + (i+2)^{i+1}}{i^2\log i \cdot 2i^{2i+1}} = \left(1 + \frac{1}{i}\right)^{2i}\frac{(i+1)^3}{i^3\log i} + o(i^{-i}) \to 0.$$

By 8.4.2 the number whose digits of its continued fraction expansions are formed by $\mathbf{w}_1^{\odot \ell_1} \odot \mathbf{w}_2^{\odot \ell_2} \odot \cdots$ is normal with respect to the continued fraction expansions.

8.5 Besicovitch-Eggleston Sets

After constructing generic sequences for different measures, we want to consider sets of non-normal numbers. We will distinguish different kinds of non-normality. First of all it suffices for a sequences to be not generic for the maximal measure in order to be a non-normal sequence.

In this vein we want to continue the idea from the above section and investigate sets of reals whose digital distribution follows a different measure. These sets are called Besicovitch-Eggleston sets. Besicovitch [82] considered binary expansions where 0 appears with probability p and 1 with probability $1-p$. Later Eggleston [210] generalized this to arbitrary q-adic expansions. For a given vector $(p_0, \ldots, p_{q-1})$ with $\sum_{d=0}^{q-1} p_d = 1$ the set $B(p_0, \ldots, p_{q-1})$ of sequences such that the digit d appears with probability p_d has Hausdorff dimension

$$\dim_H B(p_0, \ldots, p_{q-1}) = -\sum_{d=0}^{q-1} p_d \log p_d.$$

The connection with the entropy of the measure $\mu(d) = p_d$ for $0 \leq d < q - 1$ was investigated by Billingsley [85, 86] who showed that for a given measure μ over the full shift, the set $B(\mu)$ of all sequences which are generic for this measure has Hausdorff dimension

$$\dim_H B(\mu) = h(\mu).$$

In the following section, we want to show a generalization of this result to sets of measures over a symbolic dynamical system fulfilling the specification property.

For $\alpha \in \mathcal{M}$ we denote by $B(\alpha)$ the subset of $x \in X$, which are generic for α, i.e., all $x \in X$ such that for $k \geq 1$ and $\mathbf{w} \in L_k$

$$\lim_{n\to\infty} \mathrm{P}(x, \mathbf{w}, n) = \alpha(\mathbf{w}).$$

Then Pfister and Sullivan [478] proved the following.

Theorem 8.5.1 ([478]). *Let $\alpha \in \mathcal{M}$. Then*

$$\dim_H B(\alpha) = \frac{\mathrm{h}(\alpha)}{\mathrm{h}(X)}.$$

They even proved more. In particular, they investigated connected sets of measures and showed upper and lower bounds for their Hausdorff dimension (*cf.* [478]). The rest of this section is devoted to a presentation of their proof.

8.5.1 Reconstruction and Canonical Sequences

Before jumping right into the presentation of the steps of their proof, we want to shed some light on the general concept behind. In the preceding section, we concatenated words in order to obtain a certain measure. For the Besicovitch-Eggleston sets on the other hand, the main idea is to enclose the set with a larger and a smaller one where we keep control on the Hausdorff dimension. To this end it will be very handy to approximate a certain measure, which might be different from the maximal one.

Considering cylinder sets up to a maximal order only gives us some "finite" restriction, and we will get a cover (a larger set). For the smaller set we need a sequence of sets $\Gamma_n \in A^n$ such that the empirical measure of each word in Γ_n is very close to the desired measure. By playing with the parameters of this construction, we obtain a smaller set. In the end we need to show that the Hausdorff dimension of the cover and the inner set converge to the same value.

For a better understanding, we look at a concrete example. Consider the full shift over the alphabet $\{0, 1\}$, and let α be the measure corresponding to the distribution of the digits equal to $(\frac{2}{3}, \frac{1}{3})$. This means that $\frac{2}{3}$ of the digits are 0s and $\frac{1}{3}$ of the digits are 1s. An easy way of obtaining all words that satisfy this distribution (exactly) is for n being divisible by 3 to consider the set

$$\Gamma_n = \left\{ (a_1 \dots a_n) \in \{0,1\}^n \colon \frac{1}{n} \sum_{i=1}^{n} a_i = \frac{1}{3} \right\}.$$

For $n \geq 1$ let $\beta_n[\cdot|\Gamma_n]$ be the probability measures on L_n that put equal weight on the words in Γ_n and zero weight on the words outside Γ_n. Then for each subset $\Delta_n \subset L_n$ we have

$$\beta_n[\Delta_n|\Gamma_n] = \frac{\#(\Delta_n \cap \Gamma_n)}{\#\Gamma_n}.$$

Furthermore for $m < n$ every measure λ_n on L_n induces a measure λ_m on L_m by selecting the first m letters from each word of length n.

Definition 8.5.2. A sequence $\{\Gamma_n \in L_n \colon n \in \mathbb{N}\}$ is called a *reconstruction sequence* for α if each Γ_n is invariant under cyclic permutations and

$$\lim_{n\to\infty} \beta_m\left[\cdot|\Gamma_n\right] = \alpha_m$$

for each $m \in \mathbb{N}$.

We illustrate the concept of a reconstruction sequence for α by extending the sequence $\{\Gamma_n \subset L_n \colon n \in \mathbb{N}\}$ for the case $n/3$ not being an integer.

For the efficiency of the construction, we would like the sequence to grow as fast as possible. To this end we consider the following variant of the sets Γ_n from above:

$$\Gamma_n^{\delta} = \left\{ (a_1, \dots, a_n) \in L_n \colon \left| \frac{1}{n} \sum_{i=1}^{n} a_i - \frac{1}{3} \right| \leq \delta \right\}.$$

This sequence of sets grows faster than the original one, however, not to α. For $0 < \delta < \frac{1}{6}$ the sequence $\{\beta_m[\cdot|\Gamma_n^{\delta}]\}_n$ converges to the distribution $\left(\frac{2}{3} - \delta, \frac{1}{3} + \delta\right)$, and for $\delta \geq \frac{1}{6}$ we obtain the maximal measure $(\frac{1}{2}, \frac{1}{2})$ (*cf.* Chapter 4 of Lewis, Pfister, and Sullivan [379]). This shows the tendency of a system to maximize its entropy (principle of maximum entropy). Since the entropy equals $h(p) = -p \log p - (1 - p) \log(1 - p)$ where we denote by p the expected frequency of occurrences of 1s, we have for $0 < \delta < \frac{1}{6}$ that $h(\frac{1}{3} + \delta) > h(\frac{1}{3})$ and the entropy maximizes for $p = \frac{1}{2}$.

We deduce from these examples that a reconstruction sequence cannot grow too quickly. In fact, we have the following upper bound.

Lemma 8.5.3 ([378, Lemma 2.2]). *Let α be a shift invariant measure. If $\{\Gamma_n \subset L_n : n \in \mathbb{N}\}$ is a reconstruction sequence for α, then*

$$\limsup_{n\to\infty} \frac{1}{n} \log \#\Gamma_n \leq h(\alpha).$$

Now we turn our attention to another property of a sequence of sets of words. We call a sequence $\{\Gamma_n \subset L_n : n \in \mathbb{N}\}$ a supporting sequence for α if

$$\lim_{n\to\infty} \alpha_n[\Gamma_n] = 1$$

where α_n is the probability measure induced on L_n. The sequences with $\delta > 0$ are supporting sequences for the distribution $(\frac{2}{3}, \frac{1}{3})$; however, this is not the case for $\delta = 0$. Therefore a supporting sequence cannot grow too slowly. In the same spirit as above, we have the following lower bound.

Lemma 8.5.4 ([378, Lemma 2.1]). *Let α be a shift invariant measure. If $\{\Gamma_n \subset L_n : n \in \mathbb{N}\}$ is a supporting sequence for α, then*

$$\liminf_{n\to\infty} \frac{1}{n} \log \#\Gamma_n \geq h(\alpha).$$

These two lemmas support the following definition.

Definition 8.5.5. Let α be an invariant measure. A sequence $\{\Gamma_n \subset L_n : n \in \mathbb{N}\}$ has *entropic growth rate* for α if and only if

$$\lim_{n\to\infty} \frac{1}{n} \log \#\Gamma_n = h(\alpha).$$

Let us modify the sequence Γ_n from above in order to get a sequence with entropic growth rate. To this end we set

$$\Gamma_n' = \left\{ (a_1, \ldots, a_n) \in L_n : \left| \frac{1}{n} \sum_{i=1}^{n} a_i - \frac{1}{3} \right| \leq \frac{\log n}{\sqrt{n}} \right\}.$$

Using the conditional limit theorem (*cf.* [379]), we get that this sequence has the reconstruction property. Furthermore the central limit theorem tells us that it also has the supporting property.

We want to take a closer look at this construction. Using the frequency vector from the introduction, we may write the set as

$$\Gamma_n' := \mathrm{P}_1(\cdot, n)^{-1} F_n,$$

where F_n is the closed ball of radius $\log n / \sqrt{n}$ centered in the point $(\frac{2}{3}, \frac{1}{3})$. This general idea motivates the following definition.

Definition 8.5.6. Let $T_n : X \to \mathcal{M}$ be the cyclic empirical measure. We call a sequence $\{\Gamma_n \in L_n : n \in \mathbb{N}\}$ *canonical* for α if and only if

1. there exists a decreasing sequence $\{F_n\}$ of closed neighborhoods of α whose intersection is $\{\alpha\}$;
2. each set Γ_n is given by
$$\Gamma_n = X_n T_n^{-1} F_n,$$
where X_n is the projection on the first n letters of each word;
3. for all n sufficiently large, $\beta_n[\Gamma_n] > 0$.

In this case we say that the canonical sequence $\{\Gamma_n\}$ is based on the sequence $\{F_n\}$.

We note the following three properties. The first one tells us that the canonical sequences are those we are looking for.

Lemma 8.5.7 ([378, Lemma 2.3]). *Let α be a shift invariant measure. Every canonical sequence for α is a reconstruction sequence for α.*

Moreover there is a canonical sequence that has entropic growth rate.

Lemma 8.5.8 ([378, Lemma 2.4]). *Let α be a shift invariant measure. Then there exists a canonical sequence for α having entropic growth rate.*

Finally we have the following result if α is ergodic.

Lemma 8.5.9 ([378, Lemma 2.5]). *Let α be an ergodic shift invariant measure. Then there exists a canonical sequence for α which is a supporting sequence for α.*

Together they prove the following theorem.

Theorem 8.5.10. *Let α be a shift invariant measure. Then there exists a reconstruction sequence for α having entropic growth rate.*

If, in addition, α is ergodic, then the reconstruction sequence may be chosen so as to be a supporting sequence for α.

In the present section, we will use this theorem in the following form.

Corollary 8.5.11 ([478, Corollary 2.1]). *Let α be a shift invariant measure and X be a shift space satisfying the specification property. Then there exists a sequence $\{\Gamma_n \subset L_n : n \in \mathbb{N}\}$ with the following properties. Given $\varepsilon > 0$ and a neighborhood U of α, there exists $N(U, \varepsilon)$ such that for all $n \geq N(U, \varepsilon)$,*

$$\log |\Gamma_n| \geq n(h(\alpha) - \varepsilon) \quad \textit{and} \quad T_n(\omega) \in U, \ \forall \omega \in X_n^{-1}(\Gamma(n)).$$

8.5.2 A Cover

Recall that μ is the maximal probability measure on X. Suppose that there exists a continuous nonnegative function e_μ on L, and $C_\mu > 0$ such that

$$\langle e_\mu, \alpha \rangle := \int e_\mu \mathrm{d}\alpha \geq C_\mu \quad \forall \alpha \in \mathscr{M}. \tag{8.11}$$

Before we start with the proof, we first need some tools. A central one is the concept of local functions.

Definition 8.5.12. A function f is *local* if there exists $1 \leq i \leq j < \infty$ such that $f(\omega) = f(\eta)$ whenever $X_{[i,j]}(\omega) = X_{[i,j]}(\eta)$ where $X_{[i,j]}$ denotes the projection to the letters from i to j.

Let $\nu \in \mathscr{M}$ be a measure. Then we set

$$e_\nu(\omega) := -\log \nu(\omega_1|\omega_2, \omega_3, \ldots).$$

We have the following estimates around the function e_ν.

Lemma 8.5.13 ([478, Lemma 3.1]). *Let ν be the maximal measure. Then*

$$\limsup_{n\to\infty} \sup_{\omega\in\Sigma} \left| \frac{1}{n} \log \nu \left[\omega_1^n\right] + \langle e_\nu, T_n(\omega)\rangle \right| = 0. \tag{8.12}$$

Furthermore for each $\delta > 0$ there exist $m_\delta, N_\delta \in \mathbb{N}$, and $f_\delta \in \mathscr{F}_{m_\delta}$ such that for all $n \geq N_\delta$, for all $\omega \in \Sigma$, $|e_\nu(\omega) - f_\delta(\omega)| \leq \delta$ and

$$\left| \langle f_\delta, T_n(\omega)\rangle + \frac{1}{n} \log \nu \left[\omega_1^n\right] \right| < \delta. \tag{8.13}$$

As a corollary we obtain the following useful calculation.

Corollary 8.5.14. *For $\alpha \in \mathscr{M}$ we have $|\langle e_\nu, \alpha\rangle - \langle f_\delta, \alpha\rangle| \leq \delta$ and*

$$\lim_{n\to\infty} -\frac{1}{n} \sum_{\omega_1^n \in L_n} \alpha[\omega_1^n] \log \nu[\omega_1^n] = \langle e_\nu, \alpha\rangle.$$

We set the desired dimension to be

$$s_* := \frac{\mathrm{h}(\alpha)}{\mathrm{h}(X)}.$$

Then for the cover we may assume that

$$s_* < s < 1.$$

It is sufficient to show that there exists a collection of cylinder sets $\{B_n : n \in \mathbb{N}\}$ such that

$$B(\alpha) \subset \bigcup_n B_n \quad \text{and} \quad \sum_n \mu[B_n]^s < \varepsilon.$$

We define the closed neighborhoods

$$U_{M,\varepsilon} := \{\rho \in \mathscr{M} : \|\rho - \alpha\|_M \leq \varepsilon\},$$

such that

$$\bigcap_{M,\varepsilon} U_{M,\varepsilon} = \{\alpha\}.$$

Let $\delta > 0$. Since the entropy h is upper semicontinuous on $\mathscr{M}$, there exists a closed neighborhood, say, $F^\delta := U_{M,\varepsilon}$, such that $M \geq m_\delta$, and if $\rho \in F^\delta \cap \mathscr{M}$, then $h(\rho) \leq h(\alpha) + \delta$. Set

$$\Gamma_n^\delta := X_{n+M}\left(\{\omega \in L : T_n(\omega) \in F^\delta\}\right).$$

Now we cover $B(\alpha)$ by these sets and search an upper bound for

$$\sum_{\omega_1^{n+M} \in \Gamma_n^\delta} \mu\left([\omega_1^{n+M}]\right)^s.$$

Using Lemma 8.5.13 for $\omega \in L$ such that $X_{n+M}(\omega) \in \Gamma_n^\delta$, we have for $n \geq N_\delta$,

$$\left| \frac{1}{n+M} \log \mu\left[X_{n+M}(\omega)\right] + \langle f_\delta, T_{n+M}(\omega)\rangle \right| < \delta.$$

Thus

$$\begin{aligned} \log \sum_{\omega_1^{n+M} \in \Gamma_n^\delta} \mu\left([\omega_1^{n+M}]\right)^s &\leq \log\left|\Gamma_n^\delta\right| + s\left(\max_{\omega_1^{n+M} \in \Gamma_n^\delta} \log \mu\left([\omega_1^{n+M}]\right)\right) \\ &\leq \log\left|\Gamma_n^\delta\right| + s(n+M)(-\langle f_\delta, T_{n+M}(\omega)\rangle + \delta). \end{aligned}$$

For the estimation of the first term $\log\left|\Gamma_n^\delta\right|$, we use standard large deviation results on the empirical measure T_n over the probability space (Ω, β), where Ω is the full shift over A and β is the uniform probability: for any compact subset $K \subset \mathscr{M}$

$$\limsup_n \frac{1}{n} \log \beta\left(\{\omega \in \Omega : T_n(\omega) \in K\}\right) \leq \sup_{\alpha \in K}(h(\alpha) - \log r),$$

where we use the convention that if $K = \emptyset$, then the supremum is equal to $-\infty$. Applying this with $K = F^\delta$, the set $\{\omega: T_n(\omega) \in F^\delta\}$ is $\mathscr{F}_{n+M}$-measurable. Since

$$\beta\left(\{\omega \in \Omega: T_n(\omega) \in F^\delta\}\right) = \frac{\left|X_{n+M}(\{\omega \in \Omega: T_n(\omega) \in F^\delta\})\right|}{r^{n+M}},$$

we deduce for n large enough

$$\begin{aligned}\log\left|\Gamma_n^\delta\right| &\le \log\left|X_{n+M}(\{\omega \in \Omega: T_n(\omega) \in F^\delta\})\right| \\ &\le (n+M)\left(\sup_{\alpha\in F^\delta}(h(\alpha)+\delta)\right).\end{aligned}$$

Combining our choice of F^δ and Corollary 8.5.14, we obtain

$$\sup_{\rho\in F^\delta} h(\rho) \le h(\alpha) + \delta \le \frac{h(\alpha)}{h(X)}(\langle f_\delta, \alpha\rangle + \delta) + \delta \le \frac{h(\alpha)}{h(X)}\langle f_\delta, \alpha\rangle + \delta.$$

Therefore

$$\log \sum_{\omega_1^{n+M}\in\Gamma_n^\delta} \mu([\omega_1^{n+M}])^s \le (n+M)\left[\left(\frac{h(\alpha)}{h(X)} - s\right)\langle f_\delta, \alpha\rangle + 4\delta\right].$$

By Corollary 8.5.14 $\langle f_\delta, \alpha\rangle \ge \langle e_\mu, \alpha\rangle - \delta \ge C_\mu - \delta$. We choose $\delta > 0$, such that $C_\mu - \delta > 0$ and $\langle f_\delta, \alpha\rangle\left(\frac{h(\alpha)}{h(X)} - s\right) + 4\delta < 0$. For such δ, we find M such that the sum

$$\sum_{n\ge m}\sum_{\omega_1^{n+M}\in\Gamma_n^\delta} \mu\left([\omega_1^{n+M}]\right)^s$$

can be made arbitrarily small by taking m sufficiently large. If $\omega \in L$ is such that $\{T_n(\omega)\} \xrightarrow[n\to\infty]{} \alpha$, then ω is in

$$\bigcup_{n=m}\bigcup_{\omega_1^{n+M}\in\Gamma_n^\delta} X_{n+M}^{-1}\left\{\omega_1^{n+M}\right\}$$

for arbitrary large m. Thus, for arbitrary large m,

$$\{[\omega_1^{n+M}]: \omega_1^{n+M} \in \Gamma_n^\delta, n \ge m\}$$

is a cover of $B(\alpha)$.

8.5.3 The Lower Bound

Now we concentrate on the lower bound. We construct a large set $B \subset B(\alpha)$, and we show that each element of B satisfies the requirements. Finally we prove the lower bound by showing that B already has the desired dimension.

The central idea behind the construction of the set B is simple. We recursively construct a sequence of positive integers $\{\ell_k\}$ and a sequence of subsets $\Gamma_k \subset L_{\ell_k}$ in such a way that for ω, with $X_{\ell_k}(\omega) \in \Gamma_k$, the empirical measure $T_{\ell_k}(\omega)$ is in a close neighborhood of α and $\log |\Gamma_j|$ is close to $\ell_k h(\alpha)$. Using the specification property, we construct a sequence B_k of subsets of L such that each word of B_k is a prefix of a word in B_{k+1}. The set B is the limit of these sets:

$$B := \bigcap_k \bigcup_{\omega \in B_k} [\omega].$$

Let ℓ_1 be a positive integer. Then we define $B_1 := \Gamma_1 \subset L_{\ell_1}$ and set $b_1 = \ell_1$ to be the length of the elements of B_1. For each pair $w^1 \in B_1$ and $w^2 \in \Gamma_2$, there exists a word v^{12} of length $|v^{12}| \leq g$ such that $w^1 \odot w^2 = w^1 v^{12} w^2 \in L$. Thus we obtain for each $\omega^1 \in B_1$ a set $E'(w^1)$ of $|\Gamma_2|$ words having the same prefix w^1 but not the same length. Therefore we split the set $E'(w^1)$ in at most $g+1$ subsets, such that in each subset the words have the same length b_2. We denote by $E(w^1)$ a subset of maximal cardinality and define

$$B_2 := \bigcup_{w^1 \in B_1} E(w^1).$$

By our construction we get the following lower bound for the number of elements of B_2:

$$|B_2| \geq \frac{|B_1| \cdot |\Gamma_2|}{g+1}. \tag{8.14}$$

The set B_2 is called a *concat product* of B_1 and Γ_2. Since this product depends on the choice of the "gluing" words v^{12} in $w^1 v^{12} w^2$ and of $E(w^1)$, it is not unique. However, it is well defined. Iterating this procedure yields sets B_k consisting of words of length b_k as concat products of B_{k-1} and Γ_k. By construction we obtain the following.

Lemma 8.5.15 ([478, Lemma 4.1]). *Each $\omega \in B$ has a unique decomposition into*

$$\omega = w^1 v^{12} w^2 v^{23} w^3 \cdots \quad \text{with } w^k \in \Gamma_k,\ k \in \mathbb{N}.$$

Let $\omega, \tilde{\omega} \in B$. Then $\omega = \tilde{\omega}$ if and only if

$$\omega_{b_k - \ell_k + 1}^{b_k} = \tilde{\omega}_{b_k - \ell_k + 1}^{b_k}.$$

The following proposition summarizes the choices for the parameter of Pfister and Sullivan [478], and we refer the interested reader to their paper for details.

Proposition 8.5.16 ([478, Proposition 4.1]). *Let ε, $0 < \varepsilon < 1$ and let $\alpha \in \mathscr{M}$. Then, there exist a sequence of subsets $B_n \subset L$, $n \geq 1$, an increasing diverging sequence of integers ℓ_n, $n \geq 1$, and a decreasing sequence ε_n, $n \geq 1$, such that $\varepsilon_n \leq \varepsilon$ and $\lim_n \varepsilon_n = 0$ with the following properties.*

1. $\log |B_n| \geq \sum_{j=1}^{n} \ell_j(h(\alpha) - \varepsilon_j)$.
2. *Each word of B_n is a prefix of a word of B_{n+1}, and*

$$B := \bigcap_k \bigcup_{\omega \in B_k} [\omega] \subset B(\alpha).$$

After the construction of the set B, we turn our attention to actually showing that the constructed sequence yields a lower bound. This is trivial if $s_* = 0$ and we may assume that $0 < s_*$. The idea is to show for $0 < s < s_*$ that $\mathscr{C}^s(B) = \infty$. Since B is compact, it suffices to consider finite covers. Let $\mathscr{D}$ be such a finite cover of B. Each cylinder in $\mathscr{D}$ is labeled by a word in L, and by abuse of notation we also denote the set of these words by $\mathscr{D}$. By construction of the set B, we clearly have that two cylinders labeled by two different words, say w^1 and w^2, are either disjoint or one is a subset of the other. The latter case occurs if and only if one of the words is the prefix of the other. For each $n \in \mathbb{N}$ we set

$$y_n := \min\{\mu(w) \colon w \in B_n\},$$

where μ denotes the maximum measure. The main part of the proof of $\mathscr{C}^s(B) = \infty$ is established by the following lemma.

Lemma 8.5.17. *Let $0 < s \leq 1$. Let $\mathscr{D}$ be a cover of B, such that $\mu(w) < y_N$, for all $w \in \mathscr{D}$. Then, there exists $n \geq N$ such that*

$$\sum_{w \in \mathscr{D}} \mu(w)^s \geq \exp\left(\sum_{j=1}^{n} L_j(h(\alpha) - \varepsilon_j)\right) y_{n+1}^s.$$

Proof. Since $\mu(w) < y_N$, each $w \in \mathscr{D}$ is of the form $w = vu$, with $v \in B_N$. Moreover, any $v \in B_N$ appears as a prefix, since $\mathscr{D}$ is a cover of B. We write

$$\sum_{w \in \mathscr{D}} \mu(w)^s = \sum_{v \in B_N} \sum_{u \colon vu \in \mathscr{D}} \mu(vu)^s.$$

Let $v_* \in B_N$ be such that for all $v \in B_N$,

$$\sum_{u \colon vu \in \mathscr{D}} \mu(vu)^s \geq \sum_{u \colon v_* u \in \mathscr{D}} \mu(v_* u)^s.$$

Then, by Proposition 8.5.16,

$$\sum_{w\in\mathscr{D}}\mu(w)^s \geq |B_N| \sum_{u:v_*u\in\mathscr{D}} \mu(v_*u)^s \geq \exp\left(\sum_{j=1}^{N} L_j(h(\alpha)-\varepsilon_j)\right) \sum_{u:v_*u\in\mathscr{D}} \mu(v_*u)^s.$$

Either

$$\sum_{w\in\mathscr{D}}\mu(w)^s \geq \exp\left(\sum_{j=1}^{N} L_j(h(\alpha)-\varepsilon_j)\right) y_{N+1}^s$$

or

$$\sum_{w\in\mathscr{D}}\mu(w)^s < \exp\left(\sum_{j=1}^{N} L_j(h(\alpha)-\varepsilon_j)\right) y_{N+1}^s.$$

In the latter case, we have

$$\sum_{u:v_*u\in\mathscr{D}} \mu(v_*u)^s < y_{N+1}^s,$$

such that $\mu(v_*u) < y_{N+1}$ for all u. Let $P_{N+1} := \{v \in B_{N+1}: v_* \text{ is a prefix of } v\}$. For each u such that $v_*u \in \mathscr{D}$, we can write $v_*u = v'u'$, with $v' \in P_{N+1}$. Since $\mathscr{D}$ is a cover of B, all prefixes $v' \in P_{N+1}$ occur in the decompositions of $v_*u = v'u'$.

By Corollary 8.5.11, for fixed $\varepsilon > 0$ there exist sets $\Gamma(\alpha,n) \in L_n$ and $N(\alpha,\varepsilon) \geq N_\varepsilon$ such that for all $n \geq N(\alpha,\varepsilon)$,

$$\begin{aligned}
&|T_n(\omega)-\alpha|_{M(\varepsilon)} < \frac{\varepsilon}{3} \quad \forall\omega \in L, X_n(\omega) \in \Gamma(\alpha,n);\\
&\frac{j}{n} < \varepsilon;\\
&\log|\Gamma(\alpha,n)| > n\left(h(\alpha)+\frac{\log(j+1)}{n}-\varepsilon\right);\\
&\left|\langle e_\nu,\alpha\rangle + \frac{1}{n}\log\nu\left(\left[\omega_1^n\right]\right)\right| < \varepsilon \quad \forall\omega_1^n \in \Gamma(\alpha,n).
\end{aligned} \tag{8.15}$$

By the construction of B_{N+1} and (8.15), we have

$$|P_{N+1}| \geq \exp(L_{N+1}(h(\alpha)-\varepsilon_{N+1})).$$

Choose $v'_* \in P_{N+1}$ such that for all $v' \in P_{N+1}$,

$$\sum_{u':v'u'\in\mathscr{D}} \mu(v'u')^s \geq \sum_{u':v'_*u'\in\mathscr{D}} \mu(v'_*u')^s.$$

Therefore, we get

$$\sum_{w\in\mathscr{D}}\mu(w)^s \geq \exp\left(\sum_{j=1}^{N} L_j(h(\alpha)-\varepsilon_j)\right)\sum_{u:v_*u\in\mathscr{D}}\mu(v_*u)^s$$
$$= \exp\left(\sum_{j=1}^{N} L_j(h(\alpha)-\varepsilon_j)\right)\sum_{v'\in P_{N+1}}\sum_{u':v'u'\in\mathscr{D}}\mu(v'u')^s$$
$$\geq \exp\left(\sum_{j=1}^{N+1} L_j(h(\alpha)-\varepsilon_j)\right)\sum_{u':v'_*u'\in\mathscr{D}}\mu(v'_*u')^s.$$

We can repeat the above argument; since the cover $\mathscr{D}$ is finite, there exists $n \geq N$ such that

$$\sum_{w\in\mathscr{D}}\mu(w)^s \geq \exp\left(\sum_{j=1}^{N} L_j(h(\alpha)-\varepsilon_j)\right) y_{n+1}^s.$$

For the construction of the inner set, we need some information on the empirical measure if we concatenate two words. In particular, using (8.15) and the fact

$$\|T_n(\omega)-T_m(\omega)\| \leq \frac{2(n-m)}{n} \quad \text{if } n>m,$$

we get for $\omega \in B$ that

$$\|T_{b_m+k}(\omega) - \sum_{j=0}^{m-1}\frac{b_{j+1}-b_j}{b_m}\alpha\| \leq \frac{3k}{b_m} + 3\sum_{j=0}^{m-1}\frac{b_{j+1}-b_j}{b_m}\varepsilon_{j+1}. \tag{8.16}$$

Using Lemma 8.5.13 and (8.16) we have

$$\lim_n \sup_{w\in B_{n+1}}\left|\frac{1}{b_{n+1}}\log\mu(w) + \sum_{j=0}^{n}\frac{b_{j+1}-b_j}{b_{n+1}}\langle e_\mu,\alpha\rangle\right| = 0. \tag{8.17}$$

For $\delta > 0$ there exists N_δ such that for $n \geq N_\delta$ we deduce from (8.17), (8.14), (8.15), and the inequality $h(\alpha)/s_* \geq \langle e_\mu,\alpha\rangle \geq C_\mu$ that

$$\sum_{j=1}^{n} L_j(h(\alpha)-\varepsilon_j) + s\log y_{n+1} \geq \sum_{j=1}^{n} h(\alpha)\left(L_j - \frac{s}{s_*}(b_j-b_{j-1})\right)$$

$$- s(b_{n+1} - b_n)\langle e_\mu, \alpha \rangle - \sum_{j=1}^{n} L_j \varepsilon_j - \delta s b_{n+1}.$$

Taking $\delta > 0$ such that $s\delta < (s_* - s)C_\mu$ and noting that $\lim_n \sum_1^n L_j / b_{n+1} = 1$, we deduce

$$\liminf_n \exp\left(\sum_{j=1}^{n} L_j(h(\alpha) - \varepsilon_j)\right) y_{n+1}^s = \infty$$

and therefore $\mathscr{C}^s(B) = \infty$ by Lemma 8.5.17.

8.6 Extremely Non-normal Numbers

After considering the Hausdorff dimension for a given distribution, we would like to know it on the one hand for more extreme cases as well as over an infinite alphabet. First we stay with the finite alphabet in order to better understand these concepts.

8.6.1 *Finite Alphabet*

When thinking about non-normal numbers without limiting frequencies, one of the first ideas is to suppose that the limiting frequency does not exist for one digit, whereas it exists for another one. In this case we clearly have $|A| > 2$. Since otherwise if p denotes the asymptotic frequency of 0s, then $1 - p$ must be the asymptotic frequency of 1s. We call a sequence ω of digits over the alphabet $\{0, 1, \dots, q-1\}$ *particularly non-normal* if there exist two distinct digits $0 \leq d_1, d_2 \leq q - 1$ such that

$$\lim_{n\to\infty} \mathrm{P}(\omega, d_1, n) \text{ exist but } \lim_{n\to\infty} \mathrm{P}(\omega, d_2, n) \text{ does not exist.}$$

Albeverio *et al.* [10] could show that the set of particularly non-normal numbers has full Hausdorff dimension 1.

Now we could suppose a different result, if we want no limits for all digits. A sequence ω over the alphabet $\{0, 1, \dots, q-1\}$ is called *essentially non-normal* if for all digits $0 \leq d \leq q - 1$ the asymptotic frequency of occurrences of this digit does not exist, i.e.,

$$\lim_{n\to\infty} \mathrm{P}(\omega, d, n) \quad \text{does not exist for } 0 \leq d < q.$$

For the case of q-adic expansions, Albeverio *et al.* [10] could prove the following.

Theorem 8.6.1 ([10, Theorem 1]). *Let $(\mathscr{P}, \phi)$ be the N-ary representation of Example 8.1.1. Then the set of essentially non-normal numbers is residual.*

This result has been generalized to Markov partitions whose underlying language is the full shift by Madritsch [398]. For shifts fulfilling the specification property, we have the following more general theorem.

Theorem 8.6.2 ([400]). *Let $\mathscr{P} = \{P_0, \ldots, P_{N-1}\}$ be a topological partition for (M, ϕ). Suppose that*

- *$\bigcap_{n=0}^{\infty} \overline{D_n(\omega)}$ consists of exactly one point;*
- *$X_{\mathscr{P},T}$ fulfills the specification property;*
- *for all $i \in \Sigma$ there exist $\mathbf{q}_{i,1} = (q_{1,1}, \ldots q_{1,N-1}), \mathbf{q}_{i,2} = (q_{2,1}, \ldots q_{2,N-1}) \in S_1$ such that $|q_{1,i} - q_{2,i}| > 0$.*

Then the set of essentially non-normal numbers is residual.

Remark 8.6.3. The requirement that for each digit we need at least two possible distributions is sufficient in order to prevent that the underlying language is too simple. For example, we want to exclude the case of the shift over the alphabet $\{0, 1\}$ with forbidden words 00 and 11.

In the present section we want to consider an "extreme" case. Let $\mathrm{A}_k(\omega)$ be the set of accumulation points of the sequence of frequency vectors $(\mathrm{P}_k(\omega, n))_n$ with respect to the 1-norm $\|\cdot\|_1$, i.e., for $\omega \in X$ we set

$$\mathrm{A}_k(\omega) := \{\mathbf{p} \in \Delta_k : \mathbf{p} \text{ is an accumulation point of } (\mathrm{P}_k(\omega, n))_n\},$$

where Δ_k denotes the simplex of all shift invariant probabilities. Then we define S_k as union of all possible accumulation points, i.e.,

$$\mathrm{S}_k := \bigcup_{\omega \in X} \mathrm{A}_k(\omega).$$

We note that in the case of q-ary expansions, this definition leads to the shift invariant probability vectors (*cf.* Theorem 0 of Olsen [459]).

For any infinite word $\omega \in X$, we clearly have $\mathrm{A}_k(\omega) \subseteq S_k$. On the other hand, we call $\omega \in X$ *extremely non-k-normal* if the set of accumulation points of the sequence $(\mathrm{P}_k(\omega, n))_n$ (with respect to $\|\cdot\|_1$) equals S_k, i.e., $\mathrm{A}_k(\omega) = \mathrm{S}_k$. Furthermore, we call a number *extremely non-normal* if it is extremely non-k-normal for all $k \geq 1$.

The set of extremely non-normal numbers for the q-adic representation has been considered by Olsen [459].

Theorem 8.6.4 ([459, Theorem 1]). *Let $(\mathscr{P}, T)$ be the q-adic expansion of Example 8.1.1. Then the set of extremely non-normal numbers is residual in M.*

This result was generalized to iterated function systems by Baek and Olsen [30] and to finite Markov partitions by Madritsch [398].

We want to extend this notion to the Cesàro averages of the frequencies. To this end for a fixed block $\mathbf{b} = b_1 \ldots b_k \in L_k$, let

$$\mathrm{P}^{(0)}(\omega, \mathbf{b}, n) = \mathrm{P}(\omega, \mathbf{b}, n).$$

For $r \geq 1$ we recursively define

$$\mathrm{P}^{(r)}(\omega, \mathbf{b}, n) = \frac{\sum_{j=1}^{n} \mathrm{P}^{(r-1)}(\omega, \mathbf{b}, j)}{n}$$

to be the rth iterated Cesàro average of the frequency of the block of digits $\mathbf{b}$ under the first n digits. Furthermore, we define by

$$\mathrm{P}_k^{(r)}(\omega, n) := \left(\mathrm{P}^{(r)}(\omega, \mathbf{b}, n)\right)_{\mathbf{b} \in L_k}$$

the vector of rth iterated Cesàro averages. As above, we are interested in the accumulation points. Thus similar to above let $\mathrm{A}_k^{(r)}(\omega)$ denote the set of accumulation points of the sequence $(\mathrm{P}_k^{(r)}(\omega, n))_n$ with respect to $\|\cdot\|_1$, i.e.,

$$\mathrm{A}_k^{(r)}(\omega) := \left\{\mathbf{p} \in \Lambda_k : \mathbf{p} \text{ is an accumulation point of } (\mathrm{P}_k^{(r)}(\omega, n))_n\right\}.$$

Now we call a number *rth iterated Cesàro extremely non-k-normal* if the set of accumulation points is the full set, i.e., $\mathrm{A}_k^{(r)} = \mathrm{S}_k$.

For $r \geq 1$ and $k \geq 1$ we denote by $\mathbb{E}_k^{(r)}$ the set of rth iterated Cesàro extremely non-k-normal numbers of M. Furthermore, for $r \geq 1$ we denote by $\mathbb{E}^{(r)}$ the set of rth iterated Cesàro extremely non-normal numbers and by $\mathbb{E}$ the set of *completely Cesàro extremely non-normal* numbers, i.e.,

$$\mathbb{E} = \bigcap_k \mathbb{E}_k^{(r)} \quad \text{and} \quad \mathbb{E} = \bigcap_r \mathbb{E}^{(r)} = \bigcap_{r,k} \mathbb{E}_k^{(r)}.$$

As above, this has already been considered for the case of the q-ary expansion by Hyde *et al.* [307]. The general theorem for dynamical systems fulfilling the specification property is due to Petrykiewicz and the author.

Theorem 8.6.5. *Let k, r, and N be positive integers. Let $\mathscr{P} = \{P_0, \ldots, P_{N-1}\}$ be a topological partition for (M, ϕ). Suppose that $L_{\mathscr{P},T}$ fulfills the specification property. Then the set $\mathbb{E}_k^{(r)}$ is residual.*

Since the set of non-normal numbers is a countable intersection of sets $\mathbb{E}_k^{(r)}$, we get the following.

Corollary 8.6.6. *Let N be a positive integer and $\mathscr{P} = \{P_0, \ldots, P_{N-1}\}$ be a number system partition for (M, T). Suppose that $L_{\mathscr{P},T}$ fulfills the specification property. Then the sets $\mathbb{E}^{(r)}$ and $\mathbb{E}$ are residual.*

8.6.2 Infinite Alphabet

Now we turn our attention to continued fractions with maximal digital frequency oscillation. We use the definitions and notations of the corresponding Sections 8.1.1 and 8.2.1.

We call $x \in \mathbb{I} = [0,1] \setminus \mathbb{Q}$ extremely non-k-normal if each probability vector $\mathbf{p} \in \mathrm{S}_k$ is an accumulation point of the sequence of vectors of frequencies $(\Pi_k(x,n))_{n\in\mathbb{N}}$. Furthermore we call $x \in \mathbb{I}$ extremely non-normal if it is extremely non-k-normal for every $k \geq 1$. Then Olsen [457] could prove the following.

Theorem 8.6.7 ([457, Theorem 1]). *The set of extremely non-normal continued fractions is residual.*

We denote by $\mathrm{A}(x,\mathbf{b})$ the set of all accumulation points of the sequence $(\Pi(x,\mathbf{b},n))_{n\in\mathbb{N}}$. Furthermore we set

$$\mathrm{A}(\mathbf{b}) = \bigcup_{x\in\mathbb{I}} \mathrm{A}(x,\mathbf{b}).$$

Then the set of continued fractions with maximal frequency oscillation $\mathbb{F}$ is defined by

$$\mathbb{F} = \{x \in \mathbb{I} : \mathrm{A}(\mathbf{b}) = \mathrm{A}(x,\mathbf{b}) \text{ for all } \mathbf{b}\}.$$

Liao, Ma, and Wang [380] could show the following.

Theorem 8.6.8 ([380, Theorem 1.1]). *The set of continued fractions with maximal frequency oscillation is residual.*

Similar to the abovementioned paper of Hyde *et al.* [307], we extend our considerations to Cesàro averages of the frequencies. For a fixed block $\mathbf{b} = b_1 \dots b_k \in \mathbb{N}^k$, let

$$\mathrm{P}^{(0)}(\omega,\mathbf{b},n) = \mathrm{P}(\omega,\mathbf{b},n).$$

For $r \geq 1$ we recursively define

$$\mathrm{P}^{(r)}(\omega,\mathbf{b},n) = \frac{\sum_{j=1}^{n} \mathrm{P}^{(r-1)}(\omega,\mathbf{b},j)}{n}$$

to be the rth iterated Cesàro average of the frequency of the block of digits $\mathbf{b}$ under the first n digits. Furthermore we define by

$$\mathrm{P}_k^{(r)}(\omega,n) := \left(\mathrm{P}^{(r)}(\omega,\mathbf{b},n)\right)_{\mathbf{b}\in\mathbb{N}^k}$$

the vector of rth iterated Cesàro averages. As above, we are interested in the accumulation points. Let $\mathrm{A}_k^{(r)}(\omega)$ denote the set of accumulation points of the sequence $(\mathrm{P}_k^{(r)}(\omega,n))_n$ with respect to $\|\cdot\|_1$, i.e.,

$$\mathrm{A}_k^{(r)}(\omega) := \left\{\mathbf{p} \in \Delta_k : \mathbf{p} \text{ is an accumulation point of } (\mathrm{P}_k^{(r)}(\omega, n))_n\right\}.$$

We will denote the set of extremely non-k-normal numbers of M by $\mathbb{E}_k^{(0)}$. Similarly for $r \geq 1$ and $k \geq 1$ we denote by $\mathbb{E}_k^{(r)}$ the set of rth iterated Cesàro extremely non-k-normal numbers of M. Furthermore for $r \geq 1$ we denote by $\mathbb{E}^{(r)}$ the set of rth iterated Cesàro extremely non-normal numbers and by $\mathbb{E}$ the set of completely Cesàro extremely non-normal numbers, i.e.,

$$\mathbb{E}^{(r)} = \bigcap_k \mathbb{E}_k^{(r)} \quad \text{and} \quad \mathbb{E} = \bigcap_r \mathbb{E}^{(r)} = \bigcap_{r,k} \mathbb{E}_k^{(r)}.$$

Then our result is the following.

Theorem 8.6.9. *Let $k \geq 1$ and $r \geq 0$ be integers. Furthermore let $\mathscr{P} = \{P_1, P_2, \dots\}$ be an infinite Markov partition for (M, T). Suppose that the generated shift space $X_{\mathscr{P},T}$ is the one-sided full shift. Then the set $\mathbb{E}_k^{(r)}$ is residual.*

Remark 8.6.10. We note that the same holds true for $X_{\mathscr{P},T}$ being a one-sided shift of finite type. In fact the only change is a replacement of the definition of Z_n and of Lemma 8.6.12 (*cf.* Olsen [458] and Olsen and Winter [460]).

After considering extremely non-normal numbers, we want to turn our attention toward numbers with maximal oscillation frequency. Similarly to above, for $r \geq 0$, we denote by $\mathrm{A}^{(r)}(\omega, \mathbf{b})$ the set of all accumulation points of the sequence $\left(\mathrm{P}^{(r)}(\omega, \mathbf{b}, n)\right)_{n \in \mathbb{N}}$. Furthermore we set

$$\mathrm{A}^{(r)}(\mathbf{b}) = \bigcup_{\omega \in U_\infty} \mathrm{A}^{(r)}(\omega, \mathbf{b}).$$

Then the set of numbers with r-th iterated Cesàro maximal frequency oscillation $\mathbb{F}^{(r)}$ is defined by

$$\mathbb{F}^{(r)} = \left\{\omega \in U_\infty \colon \mathrm{A}^{(r)}(\mathbf{b}) = \mathrm{A}^{(r)}(\omega, \mathbf{b}) \text{ for all } \mathbf{b} \in \mathbb{N}^*\right\}.$$

Our result is a generalization of Theorem 1 of Liao, Ma, and Wang [380].

Theorem 8.6.11. *Let $r \geq 0$ be an integer, and let $\mathscr{P} = \{P_1, P_2, \dots\}$ be an infinite Markov partition for (M, T). Suppose that the generated shift space $X_{\mathscr{P},T}$ is the one-sided full shift. Then $\mathbb{F}^{(r)}$ is residual.*

In the subsequent sections, we will prove the two Theorems 8.6.9 and 8.6.11. We will start with a general section on properties of words which we need in the proof of Theorem 8.6.9 and which are interesting on their own. Then we will show Theorem 8.6.9. Finally in Section 8.6.5 we will prove Theorem 8.6.11 by showing that $\mathbb{E}^{(r)} \subset \mathbb{F}^{(r)}$.

8.6.3 Preliminaries on Words

First of all we want to reduce the infinite problem to a finite one. Thus instead of considering S_k as such, we concentrate on those probability vectors that only put weight on a finite set of digits. In particular, let

$$\mathrm{S}_{k,N} = \left\{ (p_{\mathbf{i}})_{\mathbf{i}\in\mathbb{N}^k} : \begin{array}{c} p_{\mathbf{i}} \geq 0, \sum_{\mathbf{i}\in\mathbb{N}^k} p_{\mathbf{i}} = 1, \sum_{i\in\mathbb{N}} p_{i\mathbf{i}} = \sum_{i\in\mathbb{N}} p_{\mathbf{i}i} \text{ for all } \mathbf{i} \in \mathbb{N}^{k-1} \\ p_{\mathbf{i}} = 0 \text{ for } \mathbf{i} \in \mathbb{N}^k \setminus \{1,\ldots,N\}^k \end{array} \right\} \tag{8.18}$$

be the set of shift invariant probability vectors, where only the first N digits are weighted. Furthermore let

$$\mathrm{S}_k^* = \bigcup_{N\geq 1} \mathrm{S}_{k,N} \tag{8.19}$$

be the union of all probability vectors over a finite alphabet.

Since S_k^* is a dense and separable subset of S_k, we may concentrate on a dense sequence $(\mathbf{q}_{k,m})_m$ in S_k^*. We fix $\mathbf{q} = \mathbf{q}_{k,m}$ throughout the rest of this section. Then $\mathbf{q} \in \mathrm{S}_{k,N}$ for some $N \geq 1$, such that $\mathbf{q}_{\mathbf{i}} = 0$ for $\mathbf{i} \in \mathbb{N}^k \setminus \{1,\ldots,N\}^k$. For $n \geq 1$ we put

$$Z_n = Z_n(\mathbf{q}, N, k) = \left\{ \omega \in \bigcup_{\ell\geq knN^k} \{1,\ldots,N\}^\ell \mid \|\mathrm{P}_k(\omega) - \mathbf{q}\|_1 \leq \frac{1}{n} \right\}.$$

Since $\mathbf{q}$, N, and k will be fixed, we may omit them throughout the rest of this section.

The main idea consists now in the construction of a word having the desired frequencies. In particular, for a given word ω, we want to show that we can add sufficiently many copies of any word from Z_n to get a word with the desired properties. To this end we first need that there is at least one word in Z_n, i.e., that Z_n is not empty.

Lemma 8.6.12 ([457, Lemma 2.4]). *For all $n \geq 1$, $\mathbf{q} \in \mathrm{S}_k^*$, $N \in \mathbb{N}$, and $k \in \mathbb{N}$, we have $Z_n(\mathbf{q}, N, k) \neq \emptyset$.*

Now we may construct our word by adding arbitrary many copies of an element of Z_n.

Lemma 8.6.13. *Let N, n, t be positive integers and $\mathbf{q} \in \mathrm{S}_{k,N}$. Furthermore let $\omega = \omega_1 \ldots \omega_t \in \mathbb{N}^t$ be a word of length t, and let $M = \max_{1\leq i\leq t} \omega_i$ be the maximal "digit" in ω. Then, for any $\gamma \in Z_n(\mathbf{q}, N, k)$ and any*

$$\ell \geq L := t + |\gamma| \max\left(n, \frac{t}{k} \max\left(1, \frac{M^k}{N^k}\right)\right) \tag{8.20}$$

we get that

$$\|\mathrm{P}_k(\omega\gamma^*,\ell)-\mathbf{q}\|_1 \le \frac{6}{n}.$$

Proof. We set $s := |\gamma|$ and

$$\sigma = \omega\gamma^*|\ell.$$

Furthermore we set q and $0 \le r < s$ such that $m = t + qs + r$. Since an occurrence can happen in ω, in γ, somewhere in between or at the end, for every $\mathbf{i} \in \mathbb{N}^k$ we clearly have that

$$\frac{qs}{\ell}\mathrm{P}(\gamma,\mathbf{i}) \le \mathrm{P}(\sigma,\mathbf{i}) \le \frac{qs\mathrm{P}(\gamma,\mathbf{i})}{\ell} + \frac{t+q(k-1)+r}{\ell}.$$

Now we concentrate on the occurrences in multiples of γ and show that we may neglect those outside of γ, i.e.,

$$\|\mathrm{P}_k(\sigma)-\mathbf{q}\|_1 \le \left\|\mathrm{P}_k(\sigma)-\frac{qs}{\ell}\mathrm{P}_k(\gamma)\right\|_1 + \left\|\frac{qs}{\ell}\mathrm{P}_k(\gamma)-\mathbf{q}\right\|_1.$$

We will estimate both parts separately. For the first one, we get that

$$\begin{aligned}
\left\|\mathrm{P}_k(\sigma)-\frac{qs}{\ell}\mathrm{P}_k(\gamma)\right\|_1 &= \sum_{\mathbf{i}\in\{1,\dots,N\}^k}\left|\mathrm{P}(\sigma,\mathbf{i})-\frac{qs}{\ell}\mathrm{P}(\gamma,\mathbf{i})\right| \\
&\quad + \sum_{\mathbf{i}\in\mathbb{N}^k\setminus\{1,\dots,N\}^k}\left|\mathrm{P}(\sigma,\mathbf{i})-\frac{qs}{\ell}\mathrm{P}(\gamma,\mathbf{i})\right| \\
&\le \sum_{\mathbf{i}\in\{1,\dots,N\}^k}\frac{t+qk+s}{\ell} + \sum_{\substack{\mathbf{i}\in\mathbb{N}^k\setminus\{1,\dots,N\}^k \\ \mathrm{P}(\omega,\mathbf{i})\neq 0}}\frac{t}{\ell} \\
&\le N^k\frac{t+qk}{qnkN^k} + \frac{1}{q} + M^k\frac{t}{qnkN^k} \\
&= \frac{1}{n} + (c+1)\frac{1}{q},
\end{aligned}$$

where we have used that $\ell \ge qs \ge qnkN^k$ and written

$$c = \frac{t}{nk}\left(1+\frac{M^k}{N^k}\right).$$

For the second part, we get that

$$\left\| \frac{qs}{\ell} \mathrm{P}_k(\gamma) - \mathbf{q} \right\|_1 \le \left\| \frac{qs}{\ell} \mathrm{P}_k(\gamma) - \mathrm{P}_k(\gamma) \right\|_1 + \| \mathrm{P}_k(\gamma) - \mathbf{q} \|_1$$
$$\le qs \left| \frac{1}{\ell} - \frac{1}{qs} \right| + \frac{1}{n}$$
$$\le \frac{t}{\ell} + \frac{1}{n} \le \frac{t}{qnkN^k} + \frac{1}{n}.$$

Putting these together yields

$$\| \mathrm{P}_k(\sigma) - \mathbf{q} \|_1 \le \frac{1}{n} + (c+1)\frac{1}{q} + \frac{t}{qnkN^k} + \frac{1}{n}.$$

By our assumptions on the size of ℓ in (8.20), this proves the lemma.

8.6.4 Proof of Theorem 8.6.9

The standard method of proof is to construct a subset E of $\mathbb{E}_k^{(r)}$ which is easier to handle and already residual. In our construction of the set E, we mainly follow the ideas of Hyde *et al.* [307]. We start by recursively defining the functions φ_m for $m \ge 1$ by $\varphi_1(x) = 2^x$ and $\varphi_m(x) = \varphi_1(\varphi_{m-1}(x))$ for $m \ge 2$. Furthermore we set $\mathbb{D} = (\mathbb{Q}^{(\mathbb{N}^k)} \cap \mathrm{S}_k^*)$. Since $\mathbb{D}$ is countable and dense in S_k^* and therefore dense in S_k, we may concentrate on the probability vectors $\mathbf{q} \in \mathbb{D}$.

Now we say that a sequence $(\mathbf{x}_n)_n$ in $\mathbb{R}^{(\mathbb{N}^k)}$ has property P if for all $\mathbf{q} \in \mathbb{D}$, $m \in \mathbb{N}$, $i \in \mathbb{N}$, and $\varepsilon > 0$, there exists a $j \in \mathbb{N}$ satisfying:

1. $j \ge i$,
2. $j/2^j < \varepsilon$,
3. if $j < n < \varphi_m(j)$ then $\|\mathbf{x}_n - \mathbf{q}\|_1 < \varepsilon$.

Then we define our set E to consist of all frequency vectors having property P, i.e.,

$$E = \{x \in U_\infty : (\mathrm{P}_k^{(0)}(x;n))_{n=1}^\infty \text{ has property } P\}.$$

We will proceed in three steps showing that

1. E is residual,
2. if $(\mathrm{P}^{(r)}(x;n))_{n=1}^\infty$ has property P, then also $(\mathrm{P}^{(r+1)}(x;n))_{n=1}^\infty$ has property P, and
3. $E \subseteq \mathbb{E}_k^{(r)}$.

Lemma 8.6.14. *The set E is residual.*

Proof. For fixed $h, m, i \in \mathbb{N}$ and $\mathbf{q} \in \mathbb{D}$, we say that a sequence $(\mathbf{x}_n)_n$ in $\mathbb{R}^{\mathbb{N}^k}$ has property $P_{h,m,\mathbf{q},i}$ if for every $\varepsilon > 1/h$, there exists $j \in \mathbb{N}$ satisfying:

1. $j \geq i$,
2. $j/2^j < \varepsilon$,
3. if $j < n < \varphi_m(2^j)$, then $\|\mathbf{x}_n - \mathbf{q}\|_1 < \varepsilon$.

Now let $E_{h,m,\mathbf{q},i}$ be the set of all points whose frequency vector satisfies property $P_{h,m,\mathbf{q},i}$, i.e.,

$$E_{h,m,\mathbf{q},i} := \left\{x \in U_\infty : \left(\mathrm{P}_k^{(0)}(x;n)\right)_{n=1}^{\infty} \text{ has property } P_{h,m,\mathbf{q},i}\right\}.$$

Obviously we have that

$$E = \bigcap_{h\in\mathbb{N}} \bigcap_{m\in\mathbb{N}} \bigcap_{\mathbf{q}\in\mathbb{D}} \bigcap_{i\in\mathbb{N}} E_{h,m,\mathbf{q},i}.$$

Thus it remains to show that $E_{h,m,\mathbf{q},i}$ is open and dense.

1. $E_{h,m,\mathbf{q},i}$ **is open.** Let $x \in E_{h,m,\mathbf{q},i}$, then there exists a $j \in \mathbb{N}$ such that $j \geq i$, $j/2^j < 1/h$, and if $j < n < \varphi_m(2^j)$, then

$$\left\|\mathrm{P}_k^{(1)}(x;n) - \mathbf{q}\right\|_1 < 1/h.$$

   Let $\omega \in X$ be such that $x = \pi(\omega)$ and set $t := \varphi_m(2^j)$. Since $D_t(\omega)$ is open, there exists a $\delta > 0$ such that the ball $B(x,\delta) \subseteq D_t(\omega)$. Furthermore, since by definition all $y \in D_t(\omega)$ have their first t digits the same as x, we get that

$$B(x,\delta) \subseteq D_t(\omega) \subseteq E_{h,m,\mathbf{q},i}.$$

2. $E_{h,m,\mathbf{q},i}$ **is dense.** Let $x \in U_\infty$ and $\delta > 0$. We must find $y \in B(x,\delta) \cap E_{h,m,\mathbf{q},i}$.

   Let $\omega \in X$ be such that $x = \pi(\omega)$. Since $\operatorname{diam}\overline{D_t(\omega)} \to 0$ for $t \to \infty$ and $x \in D_t(\omega)$ for $t \geq 1$, there exists a t such that $D_t(\omega) \subset B(x,\delta)$. Let $\sigma = \omega|t$ be the first t digits of x.

   Now, an application of Lemma 8.6.12 with $n = 6h$ yields that there exists a finite word γ such that

$$\|\mathrm{P}_k(\gamma) - \mathbf{q}\|_1 \leq \frac{1}{6h}.$$

   Let $\varepsilon \geq \frac{1}{h}$ and L be as in the statement of Lemma 8.6.13. Then we choose j such that

$$\frac{j}{2^j} < \varepsilon \quad \text{and} \quad j \geq \max(L,i).$$

   An application of Lemma 8.6.13 with $n = 6h$ then gives us that

$$\|\mathrm{P}_k(\sigma\gamma^*\lfloor j)) - \mathbf{q}\|_1 \leq \frac{6}{n} = \frac{1}{h} \leq \varepsilon.$$

Thus we choose $y \in D_j(\sigma\gamma^*)$. Then on the one hand, $y \in D_j(\sigma\gamma^*) \subset D_t(\omega) \subset B(x,\delta)$, and on the other hand, $y \in D_j(\sigma\gamma^*) \subset E_{h,m,\mathbf{q},i}$

It follows that E is the countable intersection of open and dense sets, and therefore E is residual in U_∞.

Lemma 8.6.15. *Let* $\omega \in X_{\mathcal{P},\phi}$. *If* $(\mathrm{P}^{(r)}(\omega,n))_{n=1}^{\infty}$ *has property P, then also* $(\mathrm{P}^{(r+1)}(\omega,n))_{n=1}^{\infty}$ *has property P.*

This is Lemma 2.2 of [307]. However, the proof is short so we present it here for completeness.

Proof. Let $\omega \in X_{\mathcal{P},\phi}$ be such that $(\mathrm{P}_k^{(r)}(\omega;n))_{n=1}^{\infty}$ has property P, and fix $\varepsilon > 0, \mathbf{q} \in \mathbb{D}$, $i \in \mathbb{N}$, and $m \in \mathbb{N}$. Since $(\mathrm{P}_k^{(r)}(\omega,n))_{n=1}^{\infty}$ has property P, there exists $j' \in \mathbb{N}$ with $j' \geq i$, $j'/2^{j'} < \varepsilon/3$, and such that for $j' < n < \varphi_{m+1}(2^{j'})$ we have that $\left\|\mathrm{P}_k^{(r)}(\omega,n) - \mathbf{q}\right\|_1 < \varepsilon/3$.

We set $j = 2^{j'}$ and show that $(\mathrm{P}_k^{(r+1)}(\omega,n))_{n=1}^{\infty}$ has property P with this j. For all $j < n < \varphi_m(2^j)$ (i.e., $2^{j'} < n < \varphi_{m+1}(2^{j'})$), we have

$$\begin{aligned}
&\left\|\mathrm{P}_k^{(r+1)}(\omega,n) - \mathbf{q}\right\|_1 \\
&= \left\|\frac{\mathrm{P}_k^{(r)}(\omega,1) + \mathrm{P}_k^{(r)}(\omega,2) + \cdots + \mathrm{P}_k^{(r)}(\omega,n)}{n} - \mathbf{q}\right\|_1 \\
&= \left\|\frac{\mathrm{P}_k^{(r)}(\omega,1) + \mathrm{P}_k^{(r)}(\omega,2) + \cdots + \mathrm{P}_k^{(r)}(\omega,j')}{n}\right. \\
&\quad \left. + \frac{\mathrm{P}_k^{(r)}(\omega,j'+1) + \mathrm{P}_k^{(r)}(\omega,j'+2) + \cdots + \mathrm{P}_k^{(r)}(\omega,n) - (n-j')\mathbf{q}}{n} - \frac{j'\mathbf{q}}{n}\right\|_1 \\
&\leq \frac{\left\|\mathrm{P}_k^{(r)}(\omega,1) + \mathrm{P}_k^{(r)}(\omega,2) + \cdots + \mathrm{P}_k^{(r)}(\omega,j')\right\|_1}{n} \\
&\quad + \frac{\left\|\mathrm{P}_k^{(r)}(\omega,j'+1) - \mathbf{q}\right\|_1 + \cdots + \left\|\mathrm{P}_k^{(r)}(\omega,n) - \mathbf{q}\right\|_1}{n} - \frac{\|j'\mathbf{q}\|_1}{n} \\
&\leq \frac{j'}{n} + \frac{\varepsilon}{3}\frac{n-j'}{n} + \frac{j'}{n} \leq \frac{j'}{2^{j'}} + \frac{\varepsilon}{3} + \frac{j'}{2^{j'}} \leq \frac{\varepsilon}{3} + \frac{\varepsilon}{3} + \frac{\varepsilon}{3} = \varepsilon.
\end{aligned}$$

Lemma 8.6.16. *The set E is a subset of* $\mathbb{E}_k^{(r)}$.

Proof. We will show that for any $x \in E$ we also have $x \in \mathbb{E}_k^{(r)}$. To this end, let $x \in E$ and $\omega \in X_{\mathcal{P},\phi}$ be the symbolic expansion of x, i.e., $x = \pi(\omega)$. Since $(\mathrm{P}_k^{(0)}(\omega,n))_n$ has property P, by iterating Lemma 8.6.15 we get that $(\mathrm{P}_k^{(r)}(\omega,n)$ has property P.

Thus it suffices to show that $\mathbf{p}$ is an accumulation point of $(\mathrm{P}_k^{(r)}(\omega, n))_n$ for any $\mathbf{p} \in \mathrm{S}_k$. Therefore we fix $h \in \mathbb{N}$ and, since $\mathbb{D}$ is dense in S_k, we find a $\mathbf{q} \in \mathbb{D}$ such that

$$\|\mathbf{p} - \mathbf{q}\|_1 < \frac{1}{h}.$$

Since $(\mathrm{P}_k^{(r)}(\omega, n))_n$ has property P for any $m \in \mathbb{N}$, we find $j \in \mathbb{N}$ with $j \geq h$ and such that if $j < n < \varphi_m(2^j)$ then $\left\|\mathrm{P}_k^{(r)}(\omega, n) - \mathbf{q}\right\|_1 < \frac{1}{h}$. Hence let n_h be any integer with $j < n_h < \varphi_m(2^j)$, then

$$\left\|\mathrm{P}_k^{(r)}(\omega, n_h) - \mathbf{q}\right\|_1 < \frac{1}{h}.$$

Thus each n_h in the sequence $(n_h)_h$ satisfies

$$\left\|\mathbf{p} - \mathrm{P}_k^{(r)}(\omega, n_h)\right\|_1 \leq \|\mathbf{p} - \mathbf{q}\|_1 + \left\|\mathrm{P}_k^{(r)}(\omega, n_h) - \mathbf{q}\right\|_1 < \frac{2}{h}.$$

Since $n_h > h$, we may extract an increasing subsequence $(n_{h_u})_u$ such that $\mathrm{P}_k^{(r)}(\omega, n_{h_u}) \to \mathbf{p}$ for $u \to \infty$. Thus $\mathbf{p}$ is an accumulation point of $\mathrm{P}_k^{(r)}(\omega, n)$, which proves the lemma.

Proof (Proof of Theorem 8.6.9). Since by Lemma 8.6.14 E is residual in U_∞ and by Lemma 8.6.16 E is a subset of $\mathbb{E}_k^{(r)}$, we get that $\mathbb{E}_k^{(r)}$ is residual in U_∞. Again we note that $M \setminus U_\infty$ is the countable union of nowhere dense sets and therefore $\mathbb{E}_k^{(r)}$ is also residual in M.

8.6.5 *Proof of Theorem 8.6.11*

Following the proof of Liao, Ma, and Wang [380], it suffices to show that

$$\mathbb{E}^{(r)} \subset \mathbb{F}^{(r)}.$$

First the following lemma provides us with a suitable definition of $\mathrm{A}^{(r)}(\omega, \mathbf{b})$.

Lemma 8.6.17. *Let $r \geq 0$ be an integer, $\omega \in X_{\mathscr{P},\phi}$, and $\mathbf{b} \in \mathbb{N}^*$. Then*

$$\mathrm{A}^{(r)}(\omega, \mathbf{b}) = \left[\liminf_{n\to\infty} P^{(r)}(\omega, \mathbf{b}, n), \limsup_{n\to\infty} P^{(r)}(\omega, \mathbf{b}, n)\right].$$

Proof. It suffices to show that the gaps between two consecutive frequencies tend to zero, i.e.,

$$\lim_{n\to\infty} \left(\mathrm{P}^{(r)}(\omega, \mathbf{b}, n+1) - \mathrm{P}^{(r)}(\omega, \mathbf{b}, n)\right) = 0.$$

For $r = 0$ direct upper and lower estimates for the number of occurrences yield

$$\left|\mathrm{P}^{(0)}(\omega, \mathbf{b}, n+1) - \mathrm{P}^{(0)}(\omega, \mathbf{b}, n)\right| \leq \frac{1}{n+1}.$$

Since, for $i, j \geq 1$ and $r \geq 0$, we have that

$$\left|\mathrm{P}^{(r)}(\omega, \mathbf{b}, i) - \mathrm{P}^{(r)}(\omega, \mathbf{b}, j)\right| \leq 1,$$

we get by the definition of $\mathrm{P}^{(r+1)}(\omega, \mathbf{b}, n)$ that

$$\left|\mathrm{P}^{(r+1)}(\omega, \mathbf{b}, n+1) - \mathrm{P}^{(r+1)}(\omega, \mathbf{b}, n)\right|$$

$$\leq \frac{\sum_{j=1}^{n}\left|\mathrm{P}^{(r)}(\omega, \mathbf{b}, n+1) - \mathrm{P}^{(r)}(\omega, \mathbf{b}, j)\right|}{(n+1)n} \leq \frac{1}{n+1}.$$

Let $\mathbf{b} = b_1 b_2 \ldots b_k \in \mathbb{N}^*$ be a word of length k. Then we denote by $\mathrm{per}(\mathbf{b})$ the basic period of $\mathbf{b}$, i.e.,

$$\mathrm{per}(\mathbf{b}) := \min\{p \leq k \colon b_{p+j} = b_j \text{ for } 1 \leq j \leq k - p\}.$$

Furthermore we call the factor $\widetilde{\mathbf{b}} := b_1 \ldots b_{\mathrm{per}(\mathbf{b})}$ the basic factor. Then we have the following.

Lemma 8.6.18. *Let $r \geq 0$ be an integer and $\mathbf{b} \in \mathbb{N}^*$ be a finite word with basic period p and basic factor $\widetilde{\mathbf{b}} = b_1 \ldots b_p$. Then, for each $n \geq 2$,*

$$\lim_{n\to\infty} P^{(r)}(\widetilde{\mathbf{b}}^*, \mathbf{b}, n) = \frac{1}{p}.$$

Proof. For $r = 0$ this is Lemma 2.2 of [380]. The case $r \geq 1$ follows, since

$$\lim_{n\to\infty} P^{(r)}(\widetilde{\mathbf{b}}^*, \mathbf{b}, n) = \lim_{n\to\infty} P^{(r-1)}(\widetilde{\mathbf{b}}^*, \mathbf{b}, n) = \cdots = \lim_{n\to\infty} P^{(0)}(\widetilde{\mathbf{b}}^*, \mathbf{b}, n) = \frac{1}{p}.$$

Now we have enough tools to state the proof of Theorem 8.6.11.

Proof (Proof of Theorem 8.6.11). Let $\omega \in \mathbb{E}^{(r)}$ and $\mathbf{b} = b_1 \ldots b_k \in \mathbb{N}^*$ be a finite word with basic period p. Then Lemma 8.6.17 and Lemma 8.6.18 imply that

$$\mathrm{A}^{(r)}(\mathbf{b}) = \left[0, \tfrac{1}{p}\right].$$

Therefore in order to prove that $\omega \in \mathbb{F}^{(r)}$, it suffices to show that 0 and $\frac{1}{p}$ are limit points of $(\mathrm{P}^{(r)}(\omega, \mathbf{b}, n))_{n\in\mathbb{N}}$. Furthermore, since $\omega \in \mathbb{E}$, for any $\varepsilon > 0$ and $\mathbf{q} \in \mathrm{S}_k$ we have $\left\|\mathrm{P}_k^{(r)}(\omega, n)) - \mathbf{q}\right\|_1$ for infinitely many n. Thus it suffices to find two suitable probability vectors $\mathbf{q}$ providing the limit points 0 and $\frac{1}{p}$ for $\left(\mathrm{P}^{(r)}(\omega, \mathbf{b}, n)\right)_{n\in\mathbb{N}}$.

- **0 is a limit point.** We chose a digit d which is bigger than any digit in $\mathbf{b}$, i.e., $d > \max\{w_i : 1 \le i \le k\}$. Then we define the probability vector $\mathbf{q} = (q_{\mathbf{i}})_{\mathbf{i}\in\mathbb{N}^k}$ by

$$q_{\mathbf{i}} = \begin{cases} 1 & \text{if } \mathbf{i} = \underbrace{d \dots d}_{k \text{ times}}, \\ 0 & \text{else.} \end{cases}$$

  We clearly have that $\mathbf{q} \in \mathrm{S}_k$. Since $\mathrm{P}^{(r)}(\omega, \mathbf{b}, n)) < \varepsilon$ infinitely often, we have that 0 is a limit point.

- $\frac{1}{p}$ **is a limit point.** We note that $\gamma = \mathbf{b}^\infty$ is a periodic point with minimal period p under the map ϕ. Let μ be the periodic orbit measure, which has mass $\frac{1}{p}$ at each of the points $\{\gamma, \phi\gamma, \dots, \phi^{p-1}\gamma\}$. Then μ is shift invariant and induces a shift invariant probability vector

$$\mathbf{q} = (q_{\mathbf{b}})_{\mathbf{b}\in\mathbb{N}^k} = (\mu(\mathbf{b}))_{\mathbf{b}\in\mathbb{N}^k}.$$

  Since $\mathbf{q} \in \mathrm{S}_k$ and $q_{\mathbf{b}} = \mu(\mathbf{b}) = \frac{1}{p}$, we have $\left|\mathrm{P}^{(r)}(\omega, \mathbf{b}, n) - \frac{1}{p}\right| < \varepsilon$ infinitely often. Therefore $\frac{1}{p}$ is also a limit point.

Chapter 9
About the Domino Problem for Subshifts on Groups

Nathalie Aubrun, Sebastián Barbieri, and Emmanuel Jeandel

Abstract From a classical point of view, the domino problem is the question of the existence of an algorithm which can decide whether a finite set of square tiles with colored edges can tile the plane, subject to the restriction that adjacent tiles share the same color along their adjacent edges. This question has already been settled in the negative by Berger in 1966; however, these tilings can be reinterpreted in dynamical terms using the formalism of subshifts of finite type, and hence the same question can be formulated for arbitrary finitely generated groups. In this chapter we present the state of the art concerning the domino problem in this extended framework. We also discuss different notions of effectiveness in subshifts defined over groups, that is, the ways in which these dynamical objects can be described through Turing machines.

9.1 Introduction

Symbolic dynamics is the study of a particular type of dynamical systems which are called shift spaces or subshifts. These systems can be defined as sets of colorings of a group G by a finite alphabet A which are closed under the product topology and invariant under the shift action induced by the group. These objects were first introduced as a tool to study dynamical systems through discretization in the work of Hadamard [279] and then largely popularized in the highly influential article by Morse and Hedlund [287] where they were studied not only as tools but as inherently interesting objects.

A fundamental property of subshifts is the fact that they can be defined in a purely combinatorial way. Namely, a set of colorings of a group G by a finite alphabet A is a subshift if and only if it can be defined as the set of configurations which avoid a list

---

N. Aubrun (✉) · S. Barbieri
LIP, ENS de Lyon, 46 allée d'Italie, F-69007 Lyon, France
e-mail: nathalie.aubrun@ens-lyon.fr; sebastian.barbieri@ens-lyon.fr

E. Jeandel
LORIA, Campus Scientifique, BP 239, F-54506 Vandœuvre-Lès-Nancy, France
e-mail: emmanuel.jeandel@loria.fr

V. Berthé, M. Rigo (eds.), *Sequences, Groups, and Number Theory*,
Trends in Mathematics, https://doi.org/10.1007/978-3-319-69152-7_9

of forbidden patterns. This motivates the notion of subshift of finite type (SFT), that is, the set of subshifts which can be defined through a finite number of forbidden patterns.

Subshifts of finite type are of high interest from a computational point of view since they can be described by a finite amount of information – a finite set of forbidden patterns that defines the subshift – and thus decidability and algorithmic questions arise naturally. For instance, given a finite set of forbidden patterns F, the simplest question one can formulate is the following: does the subshift X_F defined by F contain at least one configuration? The main goal of this chapter is to present the state of the art concerning that question.

The domino problem of a finitely generated group G (sometimes noted $\texttt{DP}(G)$ in the sequel) asks essentially the following: is there an algorithm that takes as input a coding of a finite set of forbidden patterns F and outputs `Yes` if the SFT defined by F is nonempty and `No` otherwise? In the particular case where the group G is $\mathbb{Z}$, this problem is decidable: one-dimensional SFTs can be represented by a finite graph [381], and the existence of a configuration in the SFT (i.e., an infinite word) is equivalent to the existence of an infinite path in the graph. The two-dimensional case is much more interesting, since SFTs lack a good graph representation as it exists in 1D, even if some generalizations exist [411, 440].

In the 2D case, the emptiness problem for SFTs is equivalent to the problem of tiling the plane with Wang tiles. A Wang tile is a unit square with a color on each side that cannot be rotated or reflected. The domino problem $\texttt{DP}(\mathbb{Z}^2)$ is the algorithmic question of whether a given finite set of Wang tiles can be arranged along a $\mathbb{Z}^2$-lattice in such a way that adjacent tiles have the same color on their adjacent edges. This model is just another way to express local constraints: a set of tilings by Wang tiles can be seen as an SFT – the finite set of tiles stands for the finite alphabet – with constraints on the adjacent tiles. Reciprocally and up to a local recoding of the alphabet, an SFT can be transformed into a set of tilings by Wang tiles.

Originally, the domino problem was formulated on $\mathbb{Z}^2$ by Wang [580] as a toy problem to study a fragment of first-order logic (FO). He conjectured that every nonempty SFT on $\mathbb{Z}^2$ admits a periodic configuration, which implies the decidability of domino problem on $\mathbb{Z}^2$. His conjecture was proven wrong by Berger [71] who both exhibited an aperiodic set of 104 Wang tiles and proved the undecidability of the domino problem. This construction, later simplified by Robinson [508], proceeds by reduction from the halting problem of Turing machines.

The domino problem on other structures than $\mathbb{Z}$ and $\mathbb{Z}^2$ has also been successfully investigated. Robinson did not manage to prove undecidability of the problem on the hyperbolic plane but obtained as a preliminary step the undecidability of the origin-constrained version [509] – in this weaker version, noted `OCDP` for origin-constrained domino problem, one asks whether there exists in the SFT a configuration with a given letter at the origin. The undecidability of the unconstrained problem on the hyperbolic plane was proven later by Kari [335] and can also be obtained from the construction of a hierarchical aperiodic tiling on the hyperbolic plane by Goodman-Strauss [258].

For finitely generated groups, the question was formulated as such only very recently. For now no characterization of the groups which have decidable DP is known, and the problem seems very difficult to solve. Nevertheless a sufficient condition for decidability of DP exists: virtually free groups have decidable domino problem. The fact that they are indeed the only groups where we know the domino problem is decidable motivates the following conjecture: a group has decidable domino problem if and only if it is virtually free. This is further motivated by the following result: if a group is not virtually free, it has a thick end [586] and then arbitrarily large grids as minors by Halin's theorem (see [196] for a recent proof). It should then be possible to somehow use these grids as computation zones – similarly to what is done in Robinson's tiling [508] – to encode Turing machine computations and use them to obtain the undecidability of DP. But the main problem is that even if we know that such grids exist, we do not know where they appear and even less how to code them using local rules. Recent preprints support the conjecture: to our knowledge, the only other results are that decidability of DP is a quasi-isometry invariant for finitely presented groups [157] – i.e., a geometric property of the group – and that the conjecture holds true for Baumslag-Solitar groups [24], polycyclic groups [309], and groups of the form $G_1 \times G_2$ [310].

To better understand SFTs, we can study them through the prism of projective subdynamics. This operation modifies the group G on which a subshift X is defined into one of its subgroups H: starting from a G-subshift X, the subshift $\pi_H(X)$ is defined as the set of configurations of X restricted to H. For instance one may consider the set of rows that appear in a $\mathbb{Z}^2$-subshift. What can be said about projective subdynamics of SFTs? Addressed this way, this question is unfortunately hard to solve, even for $\mathbb{Z}^2$-SFTs. No complete characterization is known, even if some partial results exist [473]. Nevertheless, if we allow the initial subshift to be sofic – the image of an SFT under a contiguous and shift-commuting map – we get a strong result, known as the simulation theorem. Initially proven by Hochman [294] for sofic $\mathbb{Z}^3$-subshifts, and then generalized to $\mathbb{Z}^2$-subshifts independently in [203] and [25], this result states that the class of projective subdynamics of sofic $\mathbb{Z}^3$- or $\mathbb{Z}^2$-subshifts coincides exactly with the class of effectively closed subshifts, i.e., subshifts that can be defined by a recursively enumerable set of forbidden patterns. This result motivates the study of these objects.

The chapter is organized as follows. Section 9.2 presents the standard background in dimension 2 and explains where undecidability comes from for the domino problem on $\mathbb{Z}^2$. The heart of the chapter is Section 9.3 that surveys all existing results about the decidability or undecidability of the domino problem for finitely generated groups. Finally Section 9.4 is a reflexion about the notion of effectively closed subshift on a finitely generated group. Three different notions of effectiveness are defined, studied, and compared.

9.2 Subshifts of Finite Type on $\mathbb{Z}^2$, Wang Tiles and the Domino Problem

A Wang tile is a unit square with a color on each side that cannot be rotated or reflected. In order to tile the plane, Wang tiles can be arranged side by side only if the colors on their adjacent sides match. With this model of tilings in hand, one may wonder whether a given finite set of Wang tiles can tile the entire plane or not. This problem is known as the domino problem and was originally formulated by Wang [580] as a toy problem to study the $\forall\exists\forall$ fragment of first-order logic. He conjectured that every finite set of Wang tiles that can tile the entire plane can also do it in a periodic way, which implies the decidability of domino problem. His conjecture was proven wrong by Berger [71, 72] who both exhibited an aperiodic set of 104 Wang tiles and proved the undecidability of the domino problem. The construction, later simplified by Robinson [508], consists in a tile set that forces all possible tilings to contain a hierarchy of arbitrarily big squares – the plane can be decomposed in squares of order n which are themselves obtained by gluing together the squares of order $n-1$ and so on; aperiodicity comes from the fact that every translation cannot leave all levels of the structure invariant. These squares are then used as computation zones to run one arbitrary Turing machine $\mathcal{M}$; the final set of tiles associated with the Turing machine $\mathcal{M}$ has the property of being nonempty if and only if the machine $\mathcal{M}$ halts on the empty entry. By reduction from the halting problem, we conclude the undecidability of the domino problem. Note that Wang tiles are also considered in Section 10.2 as as a major tool in order to study and understand the behavior of automaton (semi)groups.

In this section, we do not present Robinson's construction, but a proof due to Kari [334] that can be generalized to Baumslag-Solitar groups, as described in Section 9.3.4.3. This alternative proof uses an encoding of rational piecewise affine maps into Wang tiles, and undecidability of the domino problem follows from a reduction to the mortality problem for piecewise affine maps.

9.2.1 Definitions

A *Wang tile* is a 4-tuple $t = (t_N, t_W, t_S, t_E) \in C^4$ where C is a finite set. It represents a unit square whose edges are colored according to the tuple interpreting the letters N, S, W, E as north, south, west, and east, respectively. See Figure 9.1.

Fig. 9.1 If the set C is interpreted as a finite set of colors, a Wang tile defined by a tuple (t_N, t_W, t_S, t_E) of colors and can be represented as shown.

Fig. 9.2 A tileset τ and a partial valid tiling of the plane.

A set $\tau \subset C^4$ of Wang tiles is called a *tileset*. We say $x : \mathbb{Z}^2 \to \tau$ is a valid tiling of the plane by τ if and only if for every $(i,j) \in \mathbb{Z}^2$:

$$x(i,j)_N = x(i,j+1)_S \text{ and } x(i,j)_E = x(i+1,j)_W.$$

Said otherwise, a valid tiling is an assignment of tiles from τ to every position of $\mathbb{Z}^2$ such that adjacent Wang tiles share the same color over adjacent edges (Figure 9.2). See also Section 10.2 for more on Wang tilings.

A natural question which arises from this setting is the following: is there a finite procedure which takes as input a tileset τ and decides whether there exists a valid tiling of the plane? In the next section, we introduce the formal concepts needed to precisely state this question.

9.2.2 *Turing Machines and the Halting Problem*

In this section we give some classical definitions that can be found with more details in [550]. Turing machines were initially introduced by Alan Turing [569] as a mathematical model that would serve for representing computations made by a human being. It is commonly accepted that Turing machines exactly catch what human can compute. This constitutes the *Church-Turing thesis*, which is the hypothesis under which functions computable by Turing machines are exactly functions computable by a human being – with no memory nor time limitation.

Turing machines are similar to finite automata, except that they can use an unlimited memory with a read/write access. The memory is realized by an infinite tape divided into cells, each cell carrying a symbol chosen among a finite alphabet. At each step of the computation, the head – i.e., the finite automaton – of the Turing machine can read the content of the tape, and depending on the read symbol and its internal state, the head can do some of the following actions: modify the content of the tape, change its internal state, and move to a neighbor cell.

A *Turing machine* is a tuple $(Q, \Gamma, \Sigma, \sharp, \delta, q_0, q_a, q_r)$ where Q, Σ are finite sets and

- Q is the set of states,
- Γ is a finite alphabet, the tape alphabet,
- $\Sigma \subset \Gamma$ is the input alphabet,
- $\sharp \in \Gamma \setminus \Sigma$ is the blank symbol,
- $\delta : Q \times \Gamma \to Q \times \Gamma \times \{-1, 0, 1\}$ is the transition function,
- $q_0 \in Q$ is the initial state,
- q_a and q_r are the accepting and rejecting states, with $q_a \neq q_r$.

A *configuration of the Turing machine* $\mathcal{M}$ is a tuple (x, i, q) where $x \in \Sigma^{\mathbb{Z}}$ is an infinite tape, $i \in \mathbb{Z}$ is the position of the computation head, and q its state. If w is a finite word, we will write (w, i, q) for the configuration where the tape is filled with blank symbols, except on positions $0 \dots |w| - 1$ where the word w is written.

Given a configuration $C = (x, i, q)$, the machine $\mathcal{M}$ computes on C as follows. If $\delta(q, x) = (q', x', \epsilon)$, then the machines goes to configuration $(y, i + \epsilon, q')$ where $y_n = x_n$ for all $n \neq i$ and $y_i = x'$. When the Turing machine $\mathcal{M}$ can go from configuration C to C' in one step, we denote $C \xrightarrow{\mathcal{M}} C'$.

We say that the Turing machine $\mathcal{M}$ accepts (resp. rejects) an input word $w \in \Sigma^*$ if starting from configuration $(w, 0, q_0)$, the machine reaches the accepting state q_a (resp. rejecting state q_r) after a finite number of steps of computation. Given an input word $w \in \Sigma^*$, the machine $\mathcal{M}$ thus has three possible behaviors: it accepts, rejects, or runs infinitely (loops). If $\mathcal{M}$ does not run infinitely, it is said to halt on w.

Example 9.2.1. An example of Turing machine and the first steps of a computation starting with configuration $(\dots \sharp 10 \sharp \dots, 0, q_0)$.

$\delta(q,x)$	Symbol x		
	0	1	$\sharp$
State q: q_0	$(q_0,0,1)$	$(q_0,1,1)$	$(q_+,\sharp,-1)$
q_+	$(q_f,1,.)$	$(q_+,0,-1)$	$(q_f,1,.)$
q_f	$(q_f,0,.)$	$(q_f,1,.)$	$(q_f,\sharp,.)$

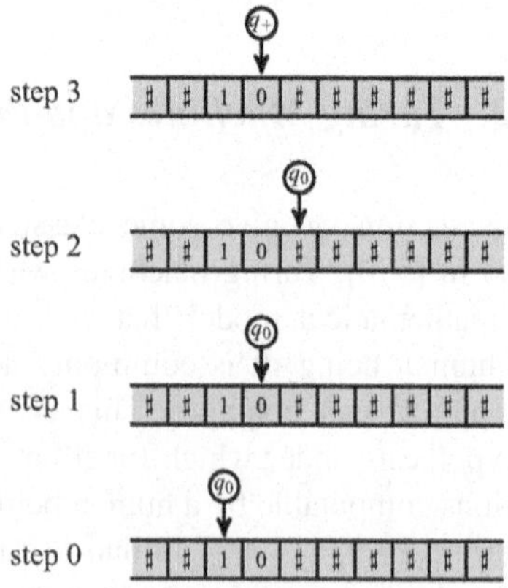

This Turing machine has the following behavior. If there is a finite number of symbols 0's and 1's around the computation head on the initial tape, then the machine adds 1 to the number coded in binary on the tape, and then halts. Otherwise, the machine never halts.

In our definition, the tape is fixed and the computation head moves, but Turing machines can equivalently be defined with a fixed computation head and a moving tape. This variant is called *moving tape Turing machine* [365], and configurations in this model are of the form (x, q) where $x \in \Sigma^{\mathbb{Z}}$ is an infinite tape and $q \in Q$ is the current state.

By definition, there are at most countably many Turing machines, and each given Turing machines can be encoded in a finite word, which is denoted $\langle \mathcal{M} \rangle$. More generally, we denote $\langle a, b \rangle$ the word that encodes the pair of objects (a, b) – objects can be words, Turing machines, or everything that possesses a finite description.

A language L is *decidable* if there exists a Turing machine $\mathcal{M}$ such that $\mathcal{M}$ accepts a word w if $w \in L$ and rejects w if $w \notin L$. A language L is *recursively enumerable* if there exists a Turing machine $\mathcal{M}$ such that $\mathcal{M}$ accepts a word w if and only if $w \in L$ – it can reject or loop otherwise.

A *decision problem* is a problem that takes an input that can be encoded into a finite word, and has a `Yes/No` answer. A decision problem is decidable if the language of encodings of its inputs with positive answer is decidable, undecidable otherwise.

A natural decision problem about Wang tiling is the so-called domino problem.

Definition 9.2.2. The *domino problem* is the decision problem that takes as input a finite tileset τ and outputs `Yes` if and only if there exists a valid tiling of the plane by τ.

Wang originally conjectured that if a set of Wang tiles can tile the plane, then they can always be arranged to do so periodically. Here by periodic tiling, we mean that the tiling can be constructed by repeating a rectangular pattern, where occurrences of this pattern are arranged on a sublattice of $\mathbb{Z}^2$. If this conjecture were true, then we could decide the domino problem by running in parallel the two following semi-algorithms – procedures that do not necessarily halt. The first semi-algorithm searches for a periodic rectangular pattern in the sense given above, by enumerating valid rectangular patterns by increasing size and checking if the sequences of colors that label the North and South edges (resp. West and East edges) match up to a cyclic permutation. The second semi-algorithm tries to tile bigger and bigger squares by a brute-force strategy. The first semi-algorithm halts if and only if there exists a periodic pattern; the second halts if and only if the set of tiles cannot tile the plane. Thus Wang's conjecture implies the decidability of the domino problem. In other words, the undecidability of the domino problem implies the existence of a set of Wang tiles that tiles the plane, but never in a periodic way – such sets of tiles are called aperiodic sets of tiles. Wang's conjecture was disproven by Berger [72], who proved the undecidability of the domino problem. It is noteworthy that his proof relies on the construction of an aperiodic set of tiles.

Theorem 9.2.3 (Berger, [71, 72]). *The domino problem is undecidable.*

A detailed proof of the undecidability of the domino problem will be given in Section 9.2.5.

Definition 9.2.4. The *halting problem* is the decision problem that takes as input a Turing machine $\mathscr{M}$ and an input word w and outputs `Yes` if and only if the machine $\mathscr{M}$ reaches a final state during its computation on w.

Theorem 9.2.5 (Turing, [569]). *The halting problem is undecidable.*

Proof. Suppose that the following language

$$HALT = \{\langle \mathscr{M}, w\rangle \mid \mathscr{M} \text{ halts on } w\}$$

is decidable. Then there exists a Turing machine $\mathscr{H}$ with the following behavior: $\mathscr{H}$ accepts the entry $\langle \mathscr{M}, w\rangle$ if $\mathscr{M}$ halts on w and rejects $\langle \mathscr{M}, w\rangle$ if $\mathscr{M}$ loops on w. We construct a Turing machine $\mathscr{N}$ that uses $\mathscr{H}$ as a subroutine. More precisely, on the input $\langle \mathscr{M}\rangle$, the machine $\mathscr{N}$ runs $\mathscr{H}$ on the input $\langle \mathscr{M}, \langle \mathscr{M}\rangle\rangle$, accepts if $\mathscr{H}$ rejects, and loops if $\mathscr{H}$ accepts. Running $\mathscr{N}$ on its own coding $\langle \mathscr{N}\rangle$ leads to a contradiction, since the machine $\mathscr{N}$ should both accept and loop! Thus the machine $\mathscr{H}$ cannot exist. □

In what precedes, Turing machines are seen as a device that can accept, reject, or loop on a given input. Another way to use this computational model is to consider the finite word written on the tape when a final state is reached as the output of the machine. A function $f : D \subseteq \{0, 1\}^* \to \{0, 1\}^*$ is *computable* if there exists a Turing machine $\mathscr{M}$ with the following behavior: if $w \in D$, then the machine accepts on entry w, and the tape contains $f(w)$ when the final state is reached.

9.2.3 Reductions

In what follows, $\overline{L}$ denotes the complement of the language L.

Reductions aim at comparing the computational difficulty of decision problems. In this section we present two of them which are meaningful for this chapter, Turing reduction and many-one reduction. In Section 9.4.3.2 a third one will be introduced, the enumeration reduction.

An *oracle Turing machine* is a couple $(\mathscr{M}, L)$, where $\mathscr{M}$ is a classical Turing machine with an additional state called the oracle state, and $L \subset \Gamma^*$ is a language called the oracle that can be queried in a single step of computation. When that machine $\mathscr{M}$ enters its oracle state, it can asks whether the word written on its tape belongs to L or not, and then evolves according to the answer. The oracle does not need to be a recursive or recursively enumerable language, so that the addition of an oracle may increase the computational power of the model. We will not explain in details how the oracle can be formalized as an extension of a classical Turing machine, but the reader can find details in [550].

Definition 9.2.6. The language L is *Turing reducible* to L', denoted $L \leq_T L'$, if there exists a Turing machine with oracle L' that computes L. We note $L \equiv_T L'$ if $L \leq_T L'$ and $L' \leq_T L$.

Example 9.2.7. One has that $HALT \leq_T \overline{HALT}$. Indeed, consider the Turing machine with oracle $\overline{HALT}$ that immediately requests the oracle on its input word, and then returns the negation of the oracle result. This machine accepts an input word $\langle \mathcal{M}, w\rangle$ if $\langle \mathcal{M}, w\rangle \notin \overline{HALT}$ and rejects otherwise. So it is a Turing machine with oracle $\overline{HALT}$ that computes $HALT$.

Definition 9.2.8. The language L is *many-one reducible* (also called *mapping reducible* in [550]) to L', denoted $L \leq_m L'$, if there exists a computable function f such that $x \in L$ if and only if $f(x) \in L'$ for every x. We note $L \equiv_m L'$ if $L \leq_m L'$ and $L' \leq_m L$.

Example 9.2.9. One has that $HALT \not\leq_m \overline{HALT}$. Indeed, suppose $HALT \leq_m \overline{HALT}$, that is to say there exists a computable function f such that $x \in HALT$ if and only if $f(x) \in \overline{HALT}$ for every x. We construct the Turing machine $\mathcal{M}_m$ as follows. On an input $\langle \mathcal{M}, w\rangle$, it first computes the word $f(\langle \mathcal{M}, w\rangle) = \langle \mathcal{M}', w'\rangle$. Then the machines simulates in parallel –one step for each simulation– the machine $\mathcal{M}$ on w and the machine $\mathcal{M}'$ on w'. If $\mathcal{M}$ halts on w, then the machine $\mathcal{M}_m$ accepts. If $\mathcal{M}'$ halts on w', the machine $\mathcal{M}_m$ rejects. One can check that $\mathcal{M}_m$ computes $HALT$, raising a contradiction.

One can show that many-one reducibility is stronger than Turing reducibility (see Exercise 9.5.1).

Definition 9.2.10. The *blank tape halting problem* is the decision problem that takes as input a Turing machine $\mathcal{M}$ and outputs `Yes` if and only if the machine $\mathcal{M}$ reaches a final state during its computation initiated on the empty tape

$$HALT_b = \{\langle \mathcal{M}\rangle \mid \mathcal{M} \text{ halts on the empty input}\}.$$

As an example of Turing reduction, we show the following.

Proposition 9.2.11 (Folklore). *The blank tape halting problem is undecidable.*

Proof. We prove that $HALT_b \leq_T HALT$, which is enough to prove that the blank tape halting problem is undecidable. The fundamental ingredient in this proof is that the encoding ($a \mapsto \langle a\rangle$ or $(a, b) \mapsto \langle a, b\rangle$) and decoding ($\langle a\rangle \mapsto a$ or $\langle a, b\rangle \mapsto (a, b)$) functions of finite objects are computable.

Let $\mathcal{M}_b$ be the Turing machine with oracle $HALT$ with the following behavior. On a given input word m, it first decodes m as a $\langle \mathcal{M}\rangle$ and then encodes $\langle \mathcal{M}, \varepsilon\rangle$ as a new word w that is now written on the tape (ε denotes the empty word). The machine now changes its state to enter the oracle state: if the word w written on the tape belongs to *HALT*, then the machine accepts; otherwise it rejects. Thus the language $HALT_b$ is computed by the machine $\mathcal{M}_b$ with oracle *HALT*. □

9.2.4 Domino Problem with Constrained Origin

As explained above, the behavior of a Turing machine only depends on local information (the state of the head and the content of the tape). Consequently, this is relatively easy to encode it inside a finite set of Wang tiles. We give a concrete encoding of the behavior of a given Turing machine $\mathcal{M}$ inside a finite tileset $\tau_{\mathcal{M}}$.

Definition 9.2.12. The *origin-constrained domino problem* is the decision problem that takes as input a finite tileset τ and a tile $t \in \tau$ and outputs `Yes` if and only if there exists a valid tiling of the plane by τ with the tile t at the origin.

Remark 9.2.13. Suppose the origin-constrained domino problem is decidable, and fix a finite tileset τ. Then it suffices to run the corresponding algorithm successively on the inputs (τ, t) for every tile $t \in \tau$ to get the decidability of the domino problem.

Theorem 9.2.14 (Kahr, Moore & Wang [322], Büchi [116]). *The origin-constrained domino problem is undecidable.*

Proof. Let $\mathcal{M}$ be a Turing machine. Consider the following tileset $\tau_{\mathcal{M}}$, where tiles are defined for every $a \in \Sigma$, for every $(q, a) \in Q \times \Sigma$ such that $\delta(q, a) = (q', a', .)$, for every $(r, a) \in Q \times \Sigma$ such that $\delta(r, a) = (r', a', 1)$, and for every $(s, a) \in Q \times \Sigma$ such that $\delta(s, a) = (s', a', -1)$:

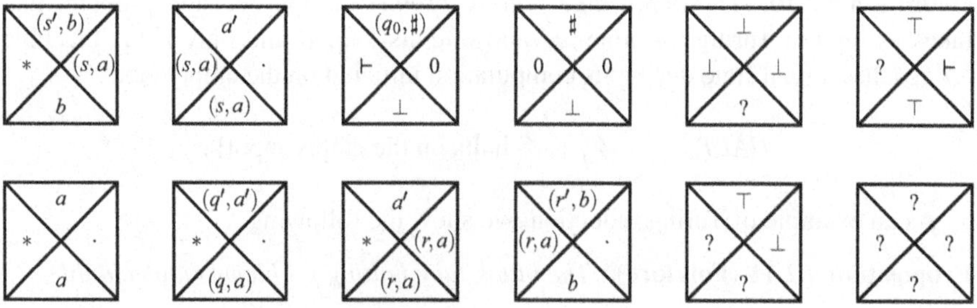

Impose that the tile $t_0 =$ 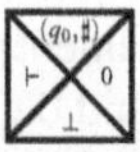

appears at the origin. Then there is exactly one way to fill in the first row and the half plane below it. Moreover, tiles located strictly to the left of the origin are also uniquely determined. Thus only the filling of upper right quadrant depends on the Turing machine $\mathcal{M}$.

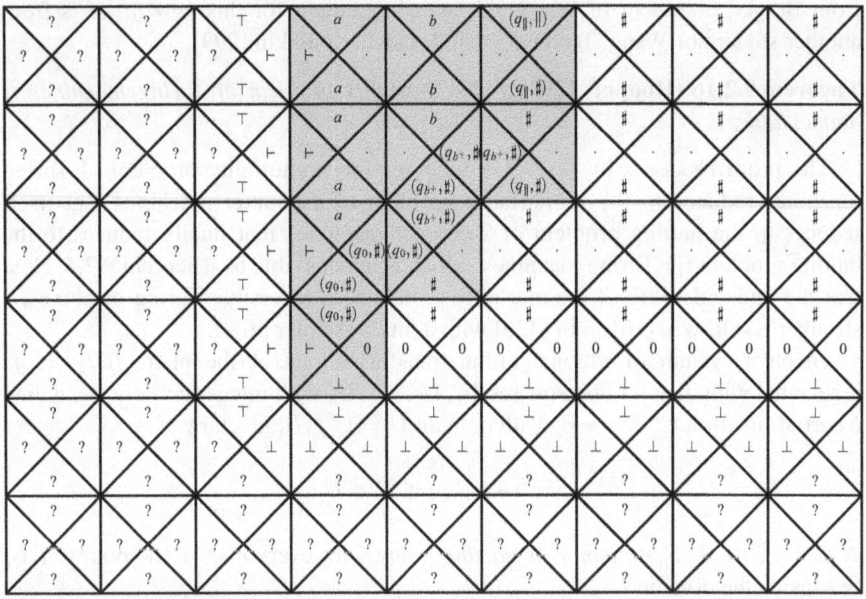

Suppose that the tiling presented above has been extended to a tiling until the ith row. Then on the top edge of row i, one can read the configuration $C_i = (w_0 \dots w_n \dots, j, q)$, the ith configuration of $\mathcal{M}$ on ε. So the tiling can be extended to row $i+1$ if and only if there exists a configuration $C_{i+1} = \mathit{Next}\ (C_i)$. Thus the computation of $\mathcal{M}$ on ε is infinite if and only if there exists a tiling by $\tau_{\mathcal{M}}$ with tile t_0 at the origin. Since the blank tape halting problem is undecidable, we conclude the origin-constrained domino problem is undecidable. □

9.2.5 *Domino Problem*

The undecidability of the domino problem was originally proven by Berger [72]. We present here an alternative proof, given by Kari [334], that has one main advantage for the purpose of this chapter: the construction can be adapted to other groups than $\mathbb{Z}^2$ (see Section 9.3.4.3).

Definition 9.2.15. The *mortality problem of Turing machines* is the decision problem that takes as input a deterministic Turing machine $\mathcal{M}$ with an halting state and outputs `Yes` if and only if there exists a non-halting configuration – configuration that never evolves into the halting state.

It is important to note that in this problem, the machine does not start from an initial configuration: the starting state and the starting tape are arbitrary. It is interesting to know that while the first proof of the undecidability of the domino problem comes from Berger, a student of Wang, the main ingredient for this new proof is from another student of Wang. Technical details can be found in [299].

Theorem 9.2.16 (Hooper, [299]). *The mortality problem of Turing machines is undecidable.*

The proof proceeds by several reductions: the immortality problem of Turing machines reduces to the immortality problem of 4-counter machines that itself reduces to the halting problem of 2-counter machines that finally reduces to the halting problem for Turing machines, which is undecidable by Theorem 9.2.5. Note that the undecidability of the mortality problem for reversible Turing machines, a stronger result, was proven in [336] with a much simpler proof.

Given a system of rational affine transformations of the plane $f_1, f_2, \ldots, f_n$ associated with disjoint unit squares $U_1, U_2, \ldots, U_n$ with integer corners, we define a partial function $f : \mathbb{R}^2 \to \mathbb{R}^2$ with domain $U = \cup_{i=1}^{n} U_i$ given by

$$\overrightarrow{x} \mapsto f_i(\overrightarrow{x}) \text{ if } \overrightarrow{x} \in U_i.$$

A point $\overrightarrow{x} \in \mathbb{R}^2$ is an *immortal starting point* if for every $n \in \mathbb{N}$, the point $f^n(\overrightarrow{x})$ lies inside the domain U.

Definition 9.2.17. The *mortality problem of piecewise affine maps* is the decision problem that takes as input a system of rational affine transformations of the plane $f_1, f_2, \ldots, f_n$ associated with disjoint unit squares $U_1, U_2, \ldots, U_n$ with integer corners and outputs `Yes` if and only if the system has an immortal starting point.

Theorem 9.2.18 ([334]). *The mortality problem of piecewise affine maps is undecidable.*

Proof. Given a Turing machine $\mathcal{M}$, we construct a system of piecewise affine maps that has an immortal starting point if and only if $\mathcal{M}$ has an immortal configuration. The construction presented here is the one from [334] and refers to [89, 355]. We assume that the machine $\mathcal{M}$ is a moving tape machine (see Section 9.2.2) and that its states and alphabet are $A = \{0, 1, \ldots, a-1\}$ and $Q = \{0, 1, \ldots, b-1\}$. The current configuration (x, q) of the machine will be coded by the two real numbers

$$\ell = \sum_{i=-1}^{-\infty} M^i x_i$$

and

$$r = Mq + \sum_{i=0}^{\infty} M^{-i} x_i,$$

where M is an integer such that $M > a$ and $M > b-1$. The integer $\lfloor r \rfloor = Mq + x_0$ is enough to determine the next configuration, and a transition of the Turing machine corresponds to an affine map with matrix

$$\begin{pmatrix} M & 0 \\ 0 & \frac{1}{M} \end{pmatrix}, \begin{pmatrix} 1 & 0 \\ 0 & 1 \end{pmatrix} \text{ or } \begin{pmatrix} \frac{1}{M} & 0 \\ 0 & M \end{pmatrix}$$

depending on the tape movement. The translation constant of the affine map is adjusted to code the change of state and the change of the symbol on the tape. For instance, the transition $\delta(q,a) = (q', a', 1)$ is coded by the affine transformation

$$\begin{pmatrix} \ell \\ r \end{pmatrix} \mapsto \begin{pmatrix} \frac{1}{M} & 0 \\ 0 & M \end{pmatrix} \begin{pmatrix} \ell \\ r \end{pmatrix} + \begin{pmatrix} a' \\ M(q' - a - Mq) \end{pmatrix}.$$

The domain of this affine map is the unit square with integer coordinates $[0,1] \times [Mq, Mq+1]$. With this procedure, we transform a Turing machine $\mathscr{M}$ into a finite set of rational affine transformations $f_1, \ldots, f_n$ and disjoint unit squares with integer coordinates $U_1, \ldots, U_n$. One can check that immortality is preserved under this transformation: the Turing machine $\mathscr{M}$ has an immortal configuration if and only if the system of affine maps $f_1, \ldots, f_n$ has an immortal point. From Theorem 9.2.16 we conclude that the immortality problem of piecewise affine maps is undecidable. □

Theorem 9.2.19. *The domino problem is undecidable.*

Proof. We present the proof due to Kari [334] that proceeds by reduction from the mortality problem of piecewise affine maps. Consider $f : \mathbb{R}^2 \to \mathbb{R}^2$ a rational affine map. We construct a finite set of Wang tiles τ_f whose colors are chosen in $\mathbb{R}^2$. We first give an idea of how the tileset is made and will explain further how colors are chosen to get only a finite number of tiles. The tile

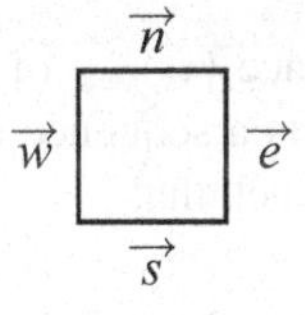

is said to compute the affine function f if

$$f(\vec{n}) + \vec{w} = \vec{s} + \vec{e}.$$

In other terms, $\vec{n}$ is the input on the top edge, and the output $\vec{s}$ is computed on the bottom edge. The computation is not exact: a carry $\vec{w}$ from the left edge is added to $f(\vec{n})$, and a carry $\vec{e}$ from the right edge is added to $\vec{s}$. Suppose now that a finite portion of a row is tiled with tiles that compute the function f as pictured below.

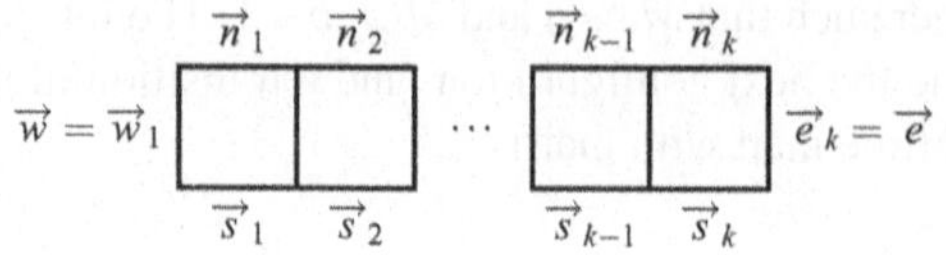

In the case where f is affine and since $e_i = w_{i+1}$ for $i = 1 \dots k-1$ by matching rules, it follows that

$$f\left(\frac{\vec{n}_1 + \cdots + \vec{n}_k}{k}\right) + \frac{1}{k}\vec{w} = \frac{\vec{s}_1 + \cdots + \vec{s}_k}{k} + \frac{1}{k}\vec{e}.$$

The carries will eventually vanish as the size of the finite portion tends to infinity, so that roughly speaking the average of the bottom labels will be the image by f of the average of the top labels.

Let f_i be the rational affine map with domain $U_i = [n, n+1] \times [m, m+1]$, given by

$$f_i(\vec{x}) = M\vec{x} + \vec{b}.$$

To describe the finite set of tiles that encodes f_i, we need some additional definitions. For $\vec{x} \in \mathbb{R}^2$ and $k \in \mathbb{Z}$, denote $A_k(\vec{x}) = \lfloor k\vec{x} \rfloor$, where $\lfloor (x, y) \rfloor = (\lfloor x \rfloor, \lfloor y \rfloor)$. Denote also

$$B_k(\vec{x}) = A_k(\vec{x}) - A_{k-1}(\vec{x}) = \lfloor k\vec{x} \rfloor - \lfloor (k-1)\vec{x} \rfloor.$$

If $\vec{x}$ is in the domain $U_i = [n, n+1] \times [m, m+1]$, one can check that $B_k(\vec{x}) \in \{(n, m), (n, m+1), (n+1, m), (n+1, m+1)\}$ for every $k \in \mathbb{Z}$. In other words, $\left(B_k(\vec{x})\right)_{k \in \mathbb{Z}}$ is a sequence of elements chosen in $\{(n, m), (n, m+1), (n+1, m), (n+1, m+1)\}$.

We say that a bi-infinite sequence $(x_k)_{k \in \mathbb{Z}}$ of i's and $(i+1)$'s represents a real number $x \in [i, i+1]$ if there exists a sequence of intervals $I_1 \subset I_2 \subset \cdots \subset \mathbb{Z}$ of increasing lengths $n_1 < n_2 < \dots$ such that

$$\lim_{k \to \infty} \frac{\sum_{j \in I_k} x_j}{n_k} = x,$$

that is to say there is an infinite sequence of intervals of increasing lengths whose averages converge to x. Note that if $(x_k)_{k \in \mathbb{Z}}$ is a representation of x, all the shifted sequences $(x_{\ell+k})_{k \in \mathbb{Z}}$ for every $\ell \in \mathbb{Z}$ are also representations of x. Note also that a sequence $(x_k)_{k \in \mathbb{Z}}$ can represent several distinct real numbers, since different interval sequences may converge to different points, and that by a compactness argument, every sequence $(x_k)_{k \in \mathbb{Z}}$ does represent at least one real number x.

Clearly the bi-infinite sequence $(B_k(\vec{x}))_{k \in \mathbb{Z}}$ is a representation of $\vec{x}$ in the sense defined above. It is called a balanced representation of $\vec{x}$.

The tileset τ_{f_i} corresponding to $f_i(\overrightarrow{x}) = M\overrightarrow{x} + \overrightarrow{b}$ consists of tiles

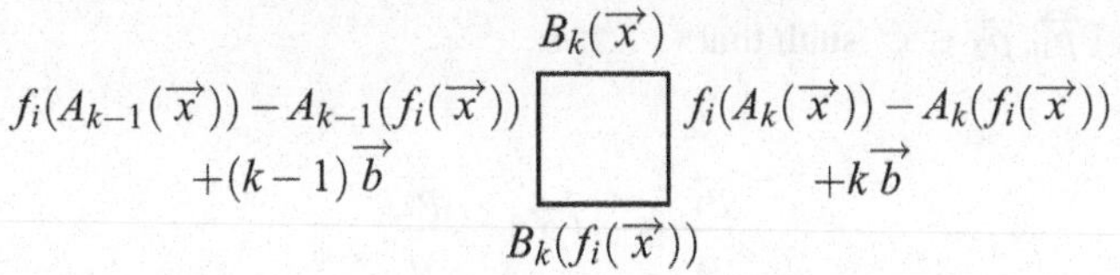

for every $k \in \mathbb{Z}$ and $\overrightarrow{x} \in U_i$. One can check that these tiles compute the function f_i.

$$
\begin{aligned}
\delta &= f_i(\overrightarrow{n}) + \overrightarrow{w} - \overrightarrow{s} - \overrightarrow{e} \\
&= f_i\left(B_k(\overrightarrow{x})\right) + f_i\left(A_{k-1}(\overrightarrow{x})\right) - A_{k-1}\left(f_i(\overrightarrow{x})\right) + (k-1)\overrightarrow{b} - B_k\left(f_i(\overrightarrow{x})\right) \\
&\quad - f_i\left(A_k(\overrightarrow{x})\right) + A_k\left(f_i(\overrightarrow{x})\right) - k\overrightarrow{b} \\
&= M\lfloor k\overrightarrow{x}\rfloor - M\lfloor (k-1)\overrightarrow{x}\rfloor + \overrightarrow{b} + M\lfloor (k-1)\overrightarrow{x}\rfloor + \overrightarrow{b} - \lfloor (k-1)f_i(\overrightarrow{x})\rfloor \\
&\quad + (k-1)\overrightarrow{b} - \lfloor kf_i(\overrightarrow{x})\rfloor + \lfloor (k-1)f_i(\overrightarrow{x})\rfloor - M\lfloor k\overrightarrow{x}\rfloor - \overrightarrow{b} + \lfloor kf_i(\overrightarrow{x})\rfloor - k\overrightarrow{b} \\
&= 0.
\end{aligned}
$$

Denote by A_i the finite alphabet used to color the edges of tiles in τ_{f_i}. Since the domain U_i is bounded, there are only finitely many possible values for the top and bottom colors in the tileset. The case of left and right colors is a little bit more subtle. Denote

$$c_k(\overrightarrow{x}) = f_i(A_k(\overrightarrow{x})) - A_k(f_i(\overrightarrow{x})) + k\overrightarrow{b}$$

the color on the right edge of the tile of the figure above, so that the left color is $c_{k-1}(\overrightarrow{x})$. By using the fact that

$$\overrightarrow{x} - \overrightarrow{1} \leq \lfloor \overrightarrow{x} \rfloor < \overrightarrow{x}$$

for every $\overrightarrow{x} \in \mathbb{R}^2$, where $\overrightarrow{1}$ denotes the vector $(1, 1)$, we get that

$$
\begin{aligned}
&M(k\overrightarrow{x} - \overrightarrow{1}) + \overrightarrow{b} - k(M\overrightarrow{x} + \overrightarrow{b}) + k\overrightarrow{b} \leq c_k(\overrightarrow{x}) \leq Mk\overrightarrow{x} \\
&+ \overrightarrow{b} - k(M\overrightarrow{x} + \overrightarrow{b}) + \overrightarrow{1} + k\overrightarrow{b} \\
&- M \cdot \overrightarrow{1} + \overrightarrow{b} \leq c_k(\overrightarrow{x}) \leq \overrightarrow{b} + \overrightarrow{1}.
\end{aligned}
$$

Since $\overrightarrow{b}$ is a rational vector and M has rational coefficients, by taking q the lcm of the denominators of all the rational numbers appearing in $\overrightarrow{b}$ and M, we get the existence of $\overrightarrow{p_1}, \overrightarrow{p_2} \in \mathbb{Z}^2$ such that

$$\frac{\overrightarrow{p_1}}{q} \leq c_k(\overrightarrow{x}) \leq \frac{\overrightarrow{p_2}}{q},$$

where $\overrightarrow{p_1}$ is chosen maximal and $\overrightarrow{p_2}$ minimal. And even better than that, it happens that all values $c_k(\overrightarrow{x})$ are in the finite set

$$\left\{\frac{\overrightarrow{p_1}}{q}, \frac{\overrightarrow{p_1}+(0,1)}{q}, \frac{\overrightarrow{p_1}+(1,0)}{q}, \frac{\overrightarrow{p_1}+\overrightarrow{1}}{q}, \ldots, \frac{\overrightarrow{p_2}}{q}\right\} \subset \mathbb{Q}$$

for every $k \in \mathbb{Z}$ and every $\overrightarrow{x} \in U_i$. Indeed, a careful observation of all rational numbers that appear inside the expression of $c_k(\overrightarrow{x})$ shows that it can be written as $\frac{\overrightarrow{p}}{q}$, and the fact that $\overrightarrow{p_1} \leq \overrightarrow{p} \leq \overrightarrow{p_2}$ directly follows from the definition of $\overrightarrow{p_1}$ and $\overrightarrow{p_2}$. So the tileset τ_{f_i} corresponding to f_i is finite. By definition of τ_{f_i}, for a given $\overrightarrow{x} \in U_i$, one can tile a row with τ_{f_i} such that the balanced representations of $\overrightarrow{x}$ and $f_i(\overrightarrow{x})$ appear on the top and bottom labels, respectively.

Suppose now that we have a system of rational affine maps $f_1, f_2, \ldots, f_n$ associated with unit squares $U_1, U_2, \ldots, U_n$ with integer corners. From each function f_i we construct a finite set of tiles τ_{f_i} that computes f_i as explained above, whose top colors $\overrightarrow{n}$ are in U_i and bottom colors $\overrightarrow{s}$ in $f_i(U_i)$. We use an additional marking on the tiles – for instance by adding the color $i \in \{1, \ldots, n\}$ to every color from A_i – so that a row can be tiled only with tiles constructed from the same f_i. We get a final finite tileset $\tau_f \subset \bigcup_{i=1}^{n} (A_i \times \{i\})^4$. It remains to prove that the tileset τ_f admits a tiling of the plane if and only if the system $f_1, f_2, \ldots, f_n$ has an immortal point.

Suppose that the system $f_1, f_2, \ldots, f_n$ has an immortal point x in one of the U_i. We construct the tiling $t \in \tau_f^{\mathbb{Z}^2}$ by assigning to every position $(k, j) \in \mathbb{Z}^2$ the tile

	$B_k\left(f^j(\overrightarrow{x})\right)$	
$f\left(A_{k-1}\left(f^j(\overrightarrow{x})\right)\right) - A_{k-1}(f^{j+1}(\overrightarrow{x})) + (k-1)\overrightarrow{b}$	$\square$	$f\left(A_k\left(f^j(\overrightarrow{x})\right)\right) - A_k(f^{j+1}(\overrightarrow{x})) + k\overrightarrow{b}$
	$B_k\left(f^{j+1}(\overrightarrow{x})\right)$	

which gives a valid tiling in X_{τ_f}. Reciprocally, suppose that τ_f admits a valid tiling of the plane $t \in \tau_f^{\mathbb{Z}^2}$. There is no reason that would force the sequences of top labels that appear on a given row to be the balanced representation of a number $x \in U$. Nevertheless, we can proceed by extraction to prove the existence of an immortal point for f. Consider the intervals $I_k = \{-k, \dots, k\}$ for all $k \in_{\mathbb{N}}$. Define vectors $(\overrightarrow{x_k})_{k\in\mathbb{N}}$ as follows:

$$\overrightarrow{x_k} = \frac{\sum_{i=-k}^{k} \overrightarrow{n}(t_{i,0})}{2k+1},$$

in other words we look at the mean of increasing sums of top labels of the first row in the tiling t. By definition of the tileset τ_f, we immediately get that there exists some $1 \leq i \leq n$ such that $\overrightarrow{x_k} \in U_i$ for every $k \in \mathbb{N}$. By compactness of U_i, we extract a sequence $\left(\overrightarrow{x_{\phi(k)}}\right)$ that converges to $\overrightarrow{x} \in U_i$. By continuity of each f_i, we can check that $\overrightarrow{x}$ is an immortal point for f. □

9.3 Subshifts of Finite Type on Finitely Generated Groups

9.3.1 Definitions

9.3.1.1 Group Presentations and the Word Problem

Let G be a group. For words $u, v \in G^*$ we write $u =_G v$ if after applying the group operation on each pair of contiguous symbols the same element of G is obtained on both sides.

Definition 9.3.1. Let G be a group and $F \subset G$. The group generated by F is the set

$$\langle F \rangle := \{g \in G \mid \exists u \in (F \cup F^{-1})^* \text{ such that } u =_G g\}.$$

It is clear that $\langle F \rangle$ is the smallest subgroup of G that contains F.

Definition 9.3.2. We say a group G is *finitely generated* if there exists a finite subset $\mathsf{S} \subset G$ such that $G = \langle \mathsf{S} \rangle$. Such a set S is called a set of *generators* for G. The rank of G is defined as the smallest cardinality of a set of generators for G.

Example 9.3.3. The group of complex numbers of the form $e^{2i\pi n\alpha}$ for $n \in \mathbb{Z}$ with multiplication as the operation is finitely generated with rank 1. Indeed, it is generated by $e^{2i\pi\alpha}$. Note that this group is infinite if and only if $\alpha \notin \mathbb{Q}$.

Example 9.3.4. The group $(\mathbb{Q}, +)$ of rational numbers with addition has infinite rank. Indeed, for any set of finite rational numbers $p_1/q_1, \dots, p_n/q_n$, the denominator of any element of $\langle p_1/q_1, \dots, p_n/q_n \rangle$ is bounded by $\prod_{i=1}^{n} q_i$. Therefore it cannot generate $\mathbb{Q}$.

By definition of $\langle \mathsf{S} \rangle$, each element of a finitely generated group can be seen as a word in $(\mathsf{S} \cup \mathsf{S}^{-1})^*$. From now on, we will use the convention that every set of generators contains its inverses to avoid writing $\mathsf{S} \cup \mathsf{S}^{-1}$.

Definition 9.3.5. Let G be a group and $\mathsf{S} \subset G$. The right *Cayley graph* of G with respect to S is the colored directed graph $\Gamma(G, \mathsf{S})$ whose vertex set is G and its set of arcs is given by $E = \bigcup_{s \in \mathsf{S}} E_s$ where E_s is the set of arcs colored by $s \in \mathsf{S}$ defined by $E_s := \{(g, gs) \mid g \in G\}$.

If S generates G then $\Gamma(G, \mathsf{S})$ is connected. For $g \in G$ we denote $|g|_{\mathsf{S}}$ the length of the shortest path from 1_G to g in $\Gamma(G, \mathsf{S})$. This induces a distance $d_{\mathsf{S}}(g, h) := |g^{-1}h|$. We denote the closed ball centered in $g \in G$ of radius r by $B_{\mathsf{S}}(g, r) = \{h \in G \mid d_{\mathsf{S}}(g, h) \leq r\}$.

Example 9.3.6. Consider the group $\mathbb{Z}^2$ endowed with coordinate-wise sum as the operation. Let $\mathsf{S} = \{(0, 1), (1, 0), (0, -1), (-1, 0)\}$ be the canonical set of generators. Then $\Gamma(\mathbb{Z}^2, \mathsf{S})$ is the bi-infinite grid, and $|(n_1, n_2)|_{\mathsf{S}} = |n_1| + |n_2|$ is the taxicab norm.

Definition 9.3.7. Let S be a set and consider a copy $\mathsf{S}^{-1} = \{s^{-1} \mid s \in \mathsf{S}\}$. We say a word in $(\mathsf{S} \cup \mathsf{S}^{-1})^*$ is *reduced* if it does not contain ss^{-1} or $s^{-1}s$ as subwords. Every word in can be reduced to an unique minimal word by successively eliminating every apparition of ss^{-1} or $s^{-1}s$.

Definition 9.3.8. The *free group* over S is defined as the group F_{S} of all reduced words in $(\mathsf{S} \cup \mathsf{S}^{-1})^*$ endowed with word concatenation followed by reduction as the operation.

A more combinatorial way to look at groups is using presentations. A *group presentation* is a pair (S, R) where S is a set and $\mathsf{R} \subset (\mathsf{S} \cup \mathsf{S}^{-1})^*$ is a set of words. Elements of S are called *generators* and words of R are called *relators*.

Definition 9.3.9. Let G be a group. We say (S, R) is a presentation of G if G is isomorphic to $\langle \mathsf{S} | \mathsf{R} \rangle$ where

$$\langle \mathsf{S} | \mathsf{R} \rangle = F_{\mathsf{S}} / N_{\mathsf{R}}.$$

Here F_{S} is the free group over S, and N_{R} is the conjugate closure of R, that is, $N_{\mathsf{R}} = \langle \{grg^{-1} \mid g \in F_{\mathsf{S}} \text{ and } r \in \mathsf{R}\} \rangle$.

In other words, $\langle \mathsf{S} | \mathsf{R} \rangle$ is the largest quotient of the free group over S such that every word in R is identified to the empty word.

Example 9.3.10. We have that $\mathbb{Z}^2 \cong \langle a, b \mid aba^{-1}b^{-1} \rangle$.

Definition 9.3.11. We say a group G is *recursively presented* if there exists a presentation (S, R) such that $G \cong \langle \mathsf{S} | \mathsf{R} \rangle$, S is recursive, and R is a recursively enumerable language. If there exists a presentation for G for which both S and R are finite, we say G is *finitely presented*.

Definition 9.3.12. The *word problem of a group G* with respect to a set of generators S is the language $\texttt{WP}(G, \mathsf{S}) = \{u \in \mathsf{S}^* \mid u =_G 1_G\}$.

Proposition 9.3.13. *Let* $\mathsf{S}_1, \mathsf{S}_2$ *be two finite sets of generators for* G*. Then* $\mathtt{WP}(G, \mathsf{S}_1)$ *is many-one equivalent to* $\mathtt{WP}(G, \mathsf{S}_2)$*.*

Proof. As $\langle \mathsf{S}_2 \rangle = G$, we have that each $s \in \mathsf{S}_1$ can be written as $u(s) \in \mathsf{S}_2^*$ such that $s =_G u(s)$. As S_1 is finite, the function which sends a word $s_0 \cdots s_k \in \mathsf{S}_1^*$ to $u(s_0) \cdots u(s_k) \in \mathsf{S}_2^*$ is total computable and $s_0 \cdots s_k = 1_G \iff u(s_0) \cdots u(s_k) = 1_G$.

In view of Proposition 9.3.13 in terms of computability, we can unambiguously speak about the *word problem* of a group G and denote it as $\mathtt{WP}(G)$.

Proposition 9.3.14. *A finitely generated group* G *is recursively presented if and only if* $\mathtt{WP}(G)$ *is recursively enumerable.*

Proof. If $\mathtt{WP}(G, \mathsf{S})$ is recursively enumerable, one can choose $(\mathsf{S}, \mathtt{WP}(G, \mathsf{S}))$ as a presentation for G. Conversely, as G is recursively presented, then $G \cong F_{\mathsf{S}}/N_{\mathsf{R}}$ for some recursively enumerable $\mathsf{R} \subset \mathsf{S}^*$. Given $u \in F_{\mathsf{S}}$ we have $u =_G 1_G \iff u \in N_{\mathsf{R}}$, therefore it suffices to be able to recognize this set. An algorithm which does this is the following: iteratively for each $n \in \mathbb{N}$ run for n steps the algorithm recognizing R on all words on S^* of length at most n. Let A_n be the list of accepted words so far. Build $B_n = \{w\ell w^{-1} \mid |w| < n, \ell \in B_n\}$ and $C_n = \{u \in B_n^* \mid |u| \leq n\}$. The set C_n approximates the conjugate closure of R. It is easy to see that every possible word in N_{R} appears in C_n for large enough n.

9.3.1.2 SFT on Finitely Generated Groups

Most of the definitions are analogous to the one-dimensional case. Let A be a finite alphabet. The set A^G endowed with the left group action $S : G \times A^G \to A^G$ given by $S^g(x)_h = x_{g^{-1}h}$ is a *full shift*. The elements $a \in A$ and $x \in A^G$ are called *symbols and* configurations, respectively. With the product of the discrete topology on A, the set of configurations A^G is a compact metric space that has the *cylinders* $[a]_g = \{x \in A^G | x_g = a\}$ *as a subbasis.* A support is a finite subset $F \subset G$. Given a support F, a *pattern with support* F is an element $p \in A^F$, and we write $\mathrm{supp}(p) = F$. We also denote the cylinder generated by p in position g as $[p]_g = \bigcap_{h \in F} [p_h]_{gh}$, and $[p] = [p]_{1_G}$.

Definition 9.3.15. A *subshift* is a subset $X \subset A^G$ which is closed and shift invariant, that is, $S(X) \subset X$. Equivalently a subshift is the set of configurations $X_{\mathscr{F}}$ defined by a set of forbidden patterns $\mathscr{F}$ as follows:

$$X_{\mathscr{F}} = A^G \setminus \bigcup_{p \in \mathscr{F}, g \in G} [p]_g.$$

Definition 9.3.16. The *language* of a subshift $L(X)$ is the set of patterns p that appear in a configuration of X, that is, $[p] \cap X \neq \emptyset$. In particular $L(A^G)$ is the set of all patterns.

Let $X \subset A^G$ and $Y \subset B^G$ be subshifts. A continuous map $\sigma : X \to Y$ such that $S_Y \circ \sigma = \sigma \circ S_X$ where S_X, S_Y are the shift actions on X and Y, respectively, is called a *morphism*. A well-known Theorem of Curtis, Lyndon, and Hedlund which can be found in full generality in [136] asserts that morphisms are equivalent to maps defined by local rules as follows: there exists a finite $F \subset G$ and $\Phi : A^F \to B$ such that $\forall x \in X : \sigma(x)_g = \Phi(S^{g^{-1}}(x)|_F)$. A surjective morphism is called a *factor map*, and we denote the existence of a factor map from X to Y by $X \twoheadrightarrow Y$. A bijective morphism is called a *conjugacy*, and the fact that two subshifts are conjugate is written $X \cong Y$.

Definition 9.3.17. A subshift $X \subset A^G$ is of *finite type* or SFT if there exist a finite set $\mathscr{F} \subset L(A^G)$ of forbidden patterns such that $X = X_{\mathscr{F}}$. A subshift is *sofic* if it is the image of an SFT via a factor map.

Definition 9.3.18. Let S be a set of generators for the group G. A subshift $X \subset A^G$ is said to be *nearest neighbor with respect to* S if there exists a set $\mathscr{F} \subset L(A_G)$ such that $X = X_{\mathscr{F}}$ and every pattern $p \in \mathscr{F}$ satisfies $\text{supp}(p) = \{1_G, s\}$ for some $s \in \mathsf{S}$. Such a set of forbidden patterns is also said to be *nearest neighbor*.

Nearest neighbor subshifts can be seen as colorings of the Cayley graph $\Gamma(G, \mathsf{S})$ such that for each edge (g, gs) the choices of color are restricted.

Example 9.3.19. The set $X = \{x \in A^G \mid \forall s \in \mathsf{S}, x_g \neq x_{gs}\}$ is a nearest neighbor subshift.

This notion also encompasses Wang tiles as studied in Section 9.2.1. The following example makes this explicit.

Example 9.3.20. Consider $\mathbb{Z}^2$ with the set of generators $\mathsf{S} = \{(1,0), (0,1)\}$. Given a set of Wang tiles τ, the set of all tilings of the plane by τ is a nearest neighbor subshift. Indeed, it corresponds to $X_{\mathscr{F}} \subset \tau^{\mathbb{Z}^2}$ where the patterns $p \in \mathscr{F}$ with support $\{(0,0), (1,0)\}$ (respectively, $\{(0,0), (1,0)\}$) are exactly those such that $(p_{(0,0)})_E \neq (p_{(1,0)})_W$ (respectively, $(p_{(0,0)})_N \neq (p_{(0,1)})_S$).

Every nearest neighbor subshift is of finite type, indeed, any set $\mathscr{F}$ satisfying the constrains satisfies $\#(\mathscr{F}) \leq \#(A)^{2\#(\mathsf{S})}$. The converse is false. For instance, the sequence of $\mathbb{Z}$-subshifts $\{X_n\}_{n \in \mathbb{N}}$ where $X_n \subset \{0,1\}^{\mathbb{Z}}$ is defined by $\mathscr{F}_n = \{1^n\}$ is a countable set of subshifts of finite type which satisfy that $1^{n-1} \in L(X_n) \setminus \bigcup_{m<n} L(X_m)$. Therefore an infinite number of them are forcefully not nearest neighbor. Nevertheless, every subshift of finite type is conjugate to a nearest neighbor subshift.

Before showing that result in generality, we illustrate informally in Figure 9.3 how this conjugacy works in the case we would like to turn a $\mathbb{Z}^2$-subshift into an equivalent set of Wang tiles. As the set of forbidden patterns is finite, there exists a big enough $n \in \mathbb{N}$ such that the support of every forbidden pattern is contained in $[0,n]^2$. Then one can construct the set of colorings of $[0,n]^2$ which do not contain forbidden patterns and turn each one of them into Wang tiles which through their adjacency colors force two contiguous patterns to overlap. This technique gives a

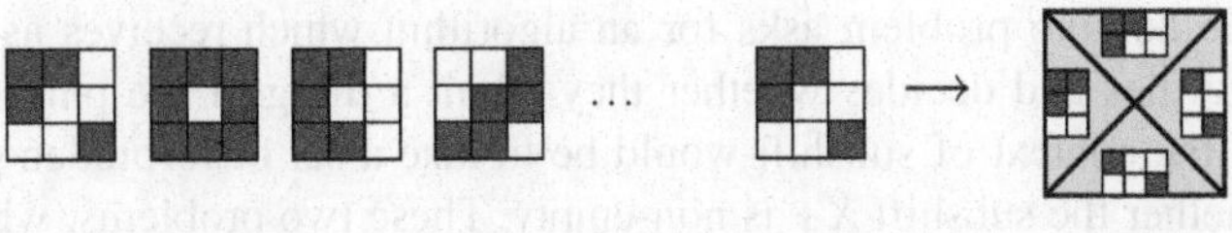

Fig. 9.3 In the left we see for $n = 2$ a set of patterns which do not contain forbidden subpatterns. In the right the transformation of one of these patterns into a Wang tile.

one to one correspondence between the set of valid tilings of the Wang tiles and the configurations in the original subshift which can be shown to be a conjugacy.

Proposition 9.3.21. *Every subshift of finite type is conjugate to a nearest neighbor subshift.*

Proof. Let $\mathcal{F}$ be a finite set of forbidden patterns defining $X_{\mathcal{F}} \subset A^G$ and let $N = \max_{p \in \mathcal{F}, g \in \mathrm{supp}(p)} |g|_{\mathsf{S}}$. We define the alphabet

$$B = \{\overline{p} \in A^{B_{\mathsf{S}}(1_G, N)} \mid \forall p \in \mathcal{F}, \overline{p}|_{\mathrm{supp}(p)} \neq p\}.$$

Consider the set of forbidden patterns $\mathcal{G}$ containing $q \in B^{\{1_G, s\}}$ if and only if $\exists g \in B_{\mathsf{S}}(1_G, N) \cap B_{\mathsf{S}}(s, N)$ such that $(q_{1_G})_g \neq (q_s)_{s^{-1}g}$. By definition $X_{\mathcal{G}}$ is nearest neighbor for S. We claim $X_{\mathcal{G}} \cong X_{\mathcal{F}}$. Indeed, consider the morphism $\sigma : X_{\mathcal{G}} \to X_{\mathcal{F}}$ given by $\sigma(y)_g = (y_g)_{1_G}$. Let y, z be two different configurations in $X_{\mathcal{G}}$. Modulo shifting these configurations we can suppose $y_{1_G} \neq z_{1_G}$, meaning there exists $h \in B_{\mathsf{S}}(1_G, N)$ such that $(y_{1_G})_h \neq (z_{1_G})_h$. Write $h =_G s_1, \cdots, s_n$ for some $n \leq N$ such that each $s_i \in \mathsf{S}$. The forbidden patterns of $\mathcal{G}$ force that $(y_{1_G})_h = (y_{s_1})_{s_1^{-1}h}$ and $(y_{s_1})_{s_1^{-1}h} = (y_{s_1 s_2})_{s_2^{-1}s_1^{-1}h}$ and so on we obtain:

$$(y_{1_G})_h = (y_{s_1,\cdots,s_n})_{s_n^{-1},\cdots,s_1^{-1}h} = (y_h)_{1_G}.$$

Similarly, $(z_{1_G})_h = (z_h)_{1_G}$, therefore $(y_h)_{1_G} \neq (z_h)_{1_G}$ and thus $\sigma(y)_h \neq \sigma(z)_h$ showing that σ is injective. Given $x \in X_{\mathcal{F}}$ we can define $y \in B^G$ given by $y_g = S^{g^{-1}}(x)|_{B_{\mathsf{S}}(1_G,N)}$. y satisfies $\sigma(y) = x$ and $y \in X_{\mathcal{G}}$ thus proving the surjectivity of σ. Hence σ is a conjugacy. □

We remark that the subshift $X_{\mathcal{G}}$ constructed in the previous proof is called the *higher-block shift* of X and denoted by $X^{[N]}$ in dimension one [381].

9.3.2 Domino Problem

From their representation with a finite automaton [381], the existence of a configuration in a $\mathbb{Z}$-SFT is equivalent to the existence of a cycle in a finite labeled graph, which is decidable.

In $\mathbb{Z}^2$ the domino problem asks for an algorithm which receives as an input a set of Wang tiles and decides whether they admit a tiling of the plane. A similar problem in the context of subshift would be to take a set of forbidden patterns $\mathcal{F}$ and ask whether the subshift $X_{\mathcal{F}}$ is non-empty. These two problems, while defined in different settings do inherently refer to the same objects and are both undecidable as shown in Section 9.2.4.

In general groups these problems become more complex to define for two reasons. From the side of the domino problem, we have to replace Wang tiles by nearest neighbor subshifts, which raises the question of which set of generators to use. From the side of the emptiness problem, we need a way to code the set of forbidden patterns such that a Turing machine can interpret them.

9.3.2.1 Definitions

We start this section by giving formal definitions for the domino problem and the emptiness problem, and then we prove that the decidability status of these two problems is the same and does not depend on the choice for the generating set of the group considered.

Definition 9.3.22. Let S be a fixed set of generators for a group G. The *domino problem with respect to* S is defined as the set $\mathtt{DP}(G, \mathsf{S})$ of codings of nearest neighbor for S sets of forbidden patterns $\mathcal{F}$ such that $X_{\mathcal{F}} \neq \emptyset$.

One way to formally code a nearest neighbor set of forbidden patterns is to identify each pattern as a triple in $A^2 \times S$ and identify A to a finite set of words in $\{0, 1\}^*$. We say that the domino problem with respect to S is *decidable* if $\mathtt{DP}(G, \mathsf{S})$ is a decidable language.

In order to define the emptiness problem, we first need to describe how to code general patterns.

Definition 9.3.23. Let G be a finitely generated group, $S \subset G$ a finite generating set, and A a finite alphabet. A *pattern coding* c is a finite set of tuples $c = \{(w_i, a_i)\}_{i\in I}$ where $w_i \in S^*$ and $a_i \in A$. Given a set $\mathscr{C}$ of pattern codings, we define the subshift $X_{\mathscr{C}}$ by:

$$X_{\mathscr{C}} = A^G \setminus \bigcup_{g\in G, c\in\mathscr{C}} \bigcap_{(w,a)\in c} [a]_{gw}$$

Note that if a pattern coding does not represent an actual pattern, that is, if two words representing the same group element get paired with different letters, then $\bigcap_{(w,a)\in c} [a]_{gw}$ is empty and the coding does not contribute at all in the formula above.

Definition 9.3.24. Let S be a fixed set of generators for a group G. The *emptiness problem with respect to* S is defined as the set $\mathtt{EP}(G, \mathsf{S})$ of sets of pattern codings $\mathscr{C}$ such that $X_{\mathscr{C}} \neq \emptyset$.

Using the same technique as in Proposition 9.3.13, we obtain that the computational properties of the emptiness problem are independent of the chosen set of generators. Therefore, analogously to the case of the word problem for groups, we can plainly speak about the emptiness problem for a given group $\texttt{EP}(G)$ and use an arbitrary set of generators.

Proposition 9.3.25. *For every pair* S, S' *of finite set of generators of* G*, we have that* $\texttt{EP}(G, \mathsf{S})$ *is many-one equivalent to* $\texttt{EP}(G, \mathsf{S}')$.

Proposition 9.3.26. *Let* S *be a finite set of generators of* G*. Then* $\texttt{DP}(G, \mathsf{S})$ *is many-one equivalent to* $\texttt{EP}(G, \mathsf{S})$.

Proof. Clearly $\texttt{DP}(G, \mathsf{S}) \leq_m \texttt{EP}(G, \mathsf{S})$ as any instance of $\texttt{DP}(G, \mathsf{S})$ is an instance of $\texttt{EP}(G, \mathsf{S})$. To prove the converse, we would like to use the conjugacy from Proposition 9.3.21, but the construction of the new alphabet might not be computable if the word problem of G is undecidable. We bypass this problem as follows. Given a set of pattern codings $\mathscr{C}$, we compute $N = \max_{c \in \mathscr{C}} \max_{(w_i, a_i) \in c} |w_i|$ and

$$B = \{b : \bigcup_{n \leq N} S^n \to A \mid \forall c \in \mathscr{C}, \exists (w, a) \in c : b_w \neq a\}$$

That is, the set of all colorings of words of length at most N such that no pattern coding from $\mathscr{C}$ appears. This set is computable, and the nearest neighbor set of forbidden patterns $\mathscr{G}$ containing $q \in B^{\{1_G, s\}}$ if and only if $\exists w \in \bigcup_{n \leq N-1} S^n$ such that $(q_{1_G})_{sw} \neq (q_s)_w$ also is. Therefore we obtain an instance of $\texttt{DP}(G, S)$.

If $X_{\mathscr{C}}$ is nonempty, we can construct $y \in X_{\mathscr{G}}$ by setting $(y_g)_w = x_{gh}$ where $h =_G w$ is the group element coded by w. It clearly does not contain any coding from $\mathscr{C}$ by definition. Conversely, analogously to the proof of Proposition 9.3.21, we obtain that for every $y \in X_{\mathscr{G}}$, $g \in B_{\mathsf{S}}(1_G, N)$, and $w \in \bigcup_{n \leq N} S^n$ such that $g =_G w$, then $(y_{1_G})_w = (y_g)_\epsilon$ where ϵ is the empty word. In particular we deduce that every symbol $b \in B$ appearing in a configuration must satisfy that $b_{w_1} = b_{w_2}$ for each $w_1 =_G w_2$. We can thus construct a configuration $x \in X_{\mathscr{C}}$ from $y \in X_{\mathscr{G}}$ defined as $x_g = (y_g)_\epsilon$. We conclude that $\texttt{EP}(G, \mathsf{S}) \leq_m \texttt{DP}(G, \mathsf{S})$ and thus $\texttt{DP}(G, \mathsf{S}) \equiv_m \texttt{EP}(G, \mathsf{S})$. □

Mixing the two previous propositions, we get:

$$\texttt{DP}(G, \mathsf{S}) \equiv_m \texttt{EP}(G, \mathsf{S}) \equiv_m \texttt{EP}(G, \mathsf{S}') \equiv_m \texttt{DP}(G, \mathsf{S}')$$

Corollary 9.3.27. *Let* S, S' *be a finite set of generators of* G*. Then* $\texttt{DP}(G, \mathsf{S})$ *is many-one equivalent to* $\texttt{DP}(G, \mathsf{S}')$.

We can therefore just speak plainly about the *domino problem* $\texttt{DP}(G)$ of a group G as the domino problem with respect to any set of generators. The results of this sections give us the liberty to treat the domino problem in any of the previous formats, that is, using any finite set of generators, and either with nearest neighbor forbidden patterns or with sets of pattern codings.

9.3.2.2 Basic Properties

Theorem 9.3.28. *For any group G then $\mathtt{WP}(G) \leq_m \overline{\mathtt{DP}(G)}$. In particular the domino problem is undecidable for any group with undecidable word problem.*

Proof. More precisely, we are going to show $\mathtt{WP}(G) \leq_m \overline{\mathtt{EP}(G)}$. Consider the alphabet $A = \{0, 1, 2\}$. Given $w \in \mathsf{S}^*$ an input of the word problem, we associate the set of pattern codings $\mathscr{C} = \{c_0, c_1, c_2\}$ where $c_i = \{(\epsilon, i), (w, i)\}$. This set $\mathscr{C}$ is clearly computable from w. In other words, the set $\mathscr{C}$ forces the symbol in each group element g to be different from the one in gw.

If $w \in \mathtt{WP}(G)$ then $w =_G 1_G$, therefore $[i]_\epsilon \cap [i]_w = [i]_{1_G}$ and so $X_{\mathscr{C}} \cap [i]_w = \emptyset$ for each $i \in \{0, 1, 2\}$. We deduce that

$$\emptyset = \bigcup_{i\in\{0,1,2\}} X_{\mathscr{C}} \cap [i]_w = X_{\mathscr{C}} \cap A^G = X_{\mathscr{C}} \text{ and thus } \mathscr{C} \in \overline{\mathtt{EP}(G)}.$$

In the case where $w \neq_G 1_G$, we show that $X_{\mathscr{C}} \neq \emptyset$. Indeed, let $w =_G g \in G$. As $g \neq 1_G$ then $\langle g \rangle$ is a nontrivial cyclic subgroup. So either $\langle g \rangle \cong \mathbb{Z}$ or $\langle g \rangle \cong \mathbb{Z}/n\mathbb{Z}$ for some $n \geq 2$. We construct $y \in A^{\langle g \rangle}$ differently for each case as follows: In the case $\langle g \rangle \cong \mathbb{Z}$ we set $y_{g^m} = m \mod 2$. In the case $\langle g \rangle \cong \mathbb{Z}/n\mathbb{Z}$, we distinguish again two cases; if n is even, then we set $y_{g^m} = m \mod 2$. Otherwise we just set $y_{g^m} = m \mod 2$ if $n \nmid m$, otherwise $y_{g^m} = 2$. One can verify that in each case $\forall h \in \langle g \rangle\ y \notin [i]_h \cap [i]_{hg}$.

Consider a set of left representatives L for the quotient $G/\langle g \rangle$. We can define $x \in A^G$ by $x_{\ell h} = y_h$ for every $\ell \in L$ and $h \in \langle g \rangle$. By definition we have for each ℓ, h, and $i \in A$, then $x \notin [i]_{\ell h} \cap [i]_{\ell h w}$ and thus $x \in X_{\mathscr{C}}$ and hence $X_{\mathscr{C}} \neq \emptyset$. □

Fix a group G, a finite generating set S, a finite alphabet, and a set of codings of nearest neighbor forbidden patterns $\mathscr{F}$. Suppose $X_{\mathscr{F}}$ is empty. Then by compactness, there exists a size N such that the ball of size $B_{\mathsf{S}}(1_G, N)$ fails to be colored without patterns from $\mathscr{F}$. So a naive procedure, which would consist in exhaustively searching for a valid coloring of balls of increasing size, will eventually stop because such a coloring does not exist for the ball of size N. The only restriction we need to perform such a procedure is that the group structure should be enumerable, which is formalized in the following proposition.

Proposition 9.3.29. *The domino problem is co-recursively enumerable for any recursively presented group.*

Proof. As G is recursively presented, there is a Turing machine $\mathscr{M}_1$ which on input $w \in \mathsf{S}^*$ returns `YES` if and only if $w =_G 1_G$. Consider the following algorithm $\mathscr{M}_2$:

1. Initialize $n \leftarrow 1$.
2. Do the following procedure:

- For each pair of words $u, v \in \mathsf{S}^*$ of length at most n. Run n steps $\mathcal{M}_1$ on entry uv^{-1}.
- For each $m \in \{1, \cdots, n\}$ construct the set X_m of functions $p : \bigcup_{k \leq m} \mathsf{S}^k \to A$ such that $p_u = p_v$ for each pair (u, v) where $\mathcal{M}$ answered YES and where no forbidden pattern appears.

3. If some X_m is empty, return YES. Otherwise do $n \leftarrow n + 1$ and go to 2.

The previous algorithm answers YES if and only if the instance of DP(G) generates an empty subshift. Indeed, if the subshift X is nonempty, then we can take $x \in X$ and define $p \in X_m$ as $p_w = x_w$. Conversely if $X \subset A^G$ is empty, there exists $N \in \mathbb{N}$ such that every $p \in A^{B_{\mathsf{S}}(1_G,N)}$ satisfies $X \cap [p] = \emptyset$. Otherwise we may choose $x_n \in [p_n] \cap X$ where $p_n \in A^{B_{\mathsf{S}}(1_G,n)}$ and any accumulation point of $\{x_n\}_{n \in \mathbb{N}}$ would be in X. Therefore it suffices to run the procedure for sufficient steps such that every pair (u, v) of length at most N such that $u =_G v$ is identified and $\mathcal{M}_2$ will forcefully obtain that $X_N = \emptyset$ and answer YES. □

In other words, Proposition 9.3.29 means that as soon as the group is recursively presented, the difficult part of the domino problem is to detect if a valid tiling exists.

9.3.3 Inheritance Properties

Proposition 9.3.30. *For every finitely generated $H \leq G$ we have* DP$(H) \leq_m$ DP(G).

Proof. Let S_H and S_G be sets of generators for H and G, respectively. As $H \leq G$ then $\mathsf{S}_G \cup \mathsf{S}_H$ also generates G. Any input of DP(H, S_H) is also an input of DP$(G, \mathsf{S}_H \cup \mathsf{S}_G)$. If the original input produces an empty subshift, then it also does so in the image as the subgroup $H \leq G$ admits no valid configuration. Conversely, if the original input admits a configuration, then it can be used to tile each lateral class G/H as in the proof of Theorem 9.3.28, and therefore the subshift produced by the image is also nonempty. □

From Theorem 9.2.19 and Proposition 9.3.30 we get:

Corollary 9.3.31. *If $\mathbb{Z}^2$ embeds into G then* DP(G) *is undecidable.*

Proposition 9.3.32. *For every finitely generated normal subgroup $H \trianglelefteq G$, we have* DP$(G/H) \leq_m$ DP(G).

Proof. Every quotient of a finitely generated group is finitely generated, so DP(G/H) is well defined. Let L be a set of representatives of G/H in G and let $\eta : G/H \to L$ be this identification. Consider finite sets $\mathsf{S}_{G/H}$, S_H of generators of G/H and H, respectively, and let $\mathsf{S}_L = \eta(\mathsf{S}_{G/H})$. We remark that if $f_1, f_2 \in G/H$, then $\eta(f_1 f_2) = \eta(f_1)\eta(f_2)h$ for some $h \in H$. In particular as every $g \in G$ can be written as $g = \ell h$ for some $\ell \in L$ and $h \in H$, we obtain that each $g \in G$ can be written as uv where $u \in \mathsf{S}_L$ and $v \in \mathsf{S}_H$. Therefore $\mathsf{S}_L \cup \mathsf{S}_H$ generate G.

Consider an instance $\mathscr{F}$ of $\mathrm{DP}(G/H, \mathrm{S}_{G/H})$. We construct an instance $\mathscr{G} = \mathscr{G}_1 \cup \mathscr{G}_2$ of $\mathrm{DP}(G, \mathrm{S}_L \cup \mathrm{S}_H)$ such that $X_\mathscr{G} = \emptyset$ if and only if $X_\mathscr{F} = \emptyset$. For each pattern with support $\{1_{G/H}, r\}$ with $r \in \mathrm{S}_{G/H}$ in the original instance, we add the same pattern with support $\{1_G, \eta(r)\}$ in $\mathscr{G}_1$. We construct $\mathscr{G}_2$ as the set of all patterns p with support $\{1_G, s\}$ for $s \in \mathrm{S}_H$ such that $p_{1_G} \neq p_s$. Namely, $\mathscr{G}_1$ copies the original rules in every quotient and $\mathscr{G}_2$ forces that $\forall y \in X_\mathscr{G}$ every configuration is invariant by translations by H. This construction of $\mathscr{G}$ is computable and gives an instance of $\mathrm{DP}(G, \mathrm{S}_L \cup \mathrm{S}_H)$.

Suppose $X_\mathscr{F} \subset A^{G/H}$ is nonempty. Then from $x \in X_\mathscr{F}$ we can construct $y \in A^G$ defined by $y_{\ell h} = x_{\eta^{-1}(\ell)}$. By definition we have that for each $g \in G$ and $s \in \mathrm{S}_H$ then $y_g = y_{gs}$ and so no pattern from $\mathscr{G}_2$ appears. Also, given $r \in \mathrm{S}_L$ we have

$$\begin{aligned} y_{\ell hr} &= y_{\ell r(r^{-1}hr)} \\ &= y_{(\ell r)h'} \text{ for some } h' \in H \text{ as } H \trianglelefteq G \\ &= y_{\ell' h'' h'} \text{ for some } h'' \in H \\ &= y_{\ell'} \end{aligned}$$

where $\ell' = \eta(\eta^{-1}(\ell)\eta^{-1}(r))$. Therefore $y_{\ell hr} = x_{\eta^{-1}(\ell)\eta^{-1}(r)}$ meaning that no patterns from $\mathscr{G}_1$ appear. Therefore $y \in X_\mathscr{G}$.

Conversely let $y \in X_\mathscr{G}$ and consider $x \in A^{G/H}$ defined by $x_g = y_{\eta(g)}$. Suppose a forbidden pattern with support $\{1_{G/H,r}\}$ from $\mathscr{F}$ appears in x in position g. Therefore $y_{\eta(gr)} = y_{\eta(g)\eta(r)h} = y_{\eta(g)\eta(r)}$ for some $h \in H$, and thus the same forbidden pattern appears in y with support $\{1_G, \eta(r)\}$. This implies that $x \in X_\mathscr{F} \neq \emptyset$. □

Proposition 9.3.33. *Let $H \leq G$ such that $[G : H] < \infty$. Then* $\mathrm{DP}(G) \equiv_m \mathrm{DP}(H)$.

Proof. The direction $\mathrm{DP}(H) \leq_m \mathrm{DP}(G)$ is direct from Proposition 9.3.30. Conversely, to prove $\mathrm{DP}(G) \leq_m \mathrm{DP}(H)$, we can suppose that $H \trianglelefteq G$. Indeed, if H is not normal, we can find $N \leq H$ such that $N \trianglelefteq G$ and $[G : N] < \infty$ (see Exercise 9.5.9). If we prove that $\mathrm{DP}(G) \leq_m \mathrm{DP}(N)$, we would have $\mathrm{DP}(G) \leq_m \mathrm{DP}(N) \leq_m \mathrm{DP}(H)$ and thus $\mathrm{DP}(G) \leq_m \mathrm{DP}(H)$.

Let $X \subset A^G$ be a subshift and R a set of representatives of the right lateral classes $G\backslash H$ which contains 1_G. We define the *R-higher power shift* of X as the set

$$X^{[R]} := \{y \in (A^R)^H \mid \exists x \in X, \forall (h, r) \in H \times R, (y_h)_r = x_{hr}\}.$$

The set $X^{[R]}$ is indeed an H-subshift and $X = \emptyset \iff X^{[R]} = \emptyset$. As every finite index subgroup of a finitely generated group is itself finitely generated (see Exercise 9.5.10), we can take a set of generators S_H for H and thus $\mathrm{S}_H \cup R$ is a finite set of generators for G. Let $D = \mathrm{S}_H \cup (RRR^{-1} \cap H)$ and $E = RDR^{-1}$. Note that as both R and S_H are finite then E also is. Furthermore as $1_G \in R$ then $\mathrm{S}_H \subset E$ and as $H \trianglelefteq G$ we have $E \subset H$, therefore $H = \langle E \rangle$. Given an instance $\mathscr{F}$ of $\mathrm{DP}(G, \mathrm{S}_H \cup R)$ with alphabet A, we are going to construct an instance $\mathscr{G}$ of $\mathrm{DP}(H, E)$ with alphabet A^R such that $X_\mathscr{G} = X_\mathscr{F}^{[R]}$ as follows: for every pattern $p \in \mathscr{F}$ with $\mathrm{supp}(p) = \{1_G, s\}$

with $s \in \mathsf{S}_H$ and $r \in R$, we put in $\mathscr{G}$ all the patterns q with support $\{1_H, rsr^{-1}\}$ such that $(q_{1_H})_r = p_{1_G}$ and $(q_{rsr^{-1}})_r = p_s$. This will take care of all patterns with support $\{1_G, s\}$ and $s \in \mathsf{S}_H$. For the remaining patterns let $(a, b) \in R^2$. By definition it is always possible to write $ab = \bar{h}c$ for some $c \in R$ and some $\bar{h} \in RRR^{-1} \cap H$. Now for every pattern $p \in \mathscr{F}$ with $\text{supp}(p) = \{1_G, b\}$ with $b \in R$ and $a \in R$, we let $\bar{h} \in RRR^{-1} \cap H$ such that $ab = \bar{h}c$ and we add to $\mathscr{G}$ all patterns q with support $\{1_H, a^{-1}\bar{h}\}$ such that $(q_{1_H})_a = p_{1_G}$ and $(q_{a^{-1}\bar{h}})_c = p_b$.

Some of the patterns in $\mathscr{G}$ defined above will have some trivial support, in this case we just consider that as a restriction on the alphabet A^R. Clearly as R is fixed beforehand, this construction can be computed from an instance of $\texttt{DP}(G, \mathsf{S}_H \cup R)$. We leave as an exercise to the reader to verify that $X_{\mathscr{G}} = X_{\mathscr{F}}^{[R]}$ and thus conclude that $\texttt{DP}(G) \leq_m \texttt{DP}(H)$. □

We say two groups G_1, G_2 are commensurable if they contain finite index subgroups $H_1 \leq G_1$ and $H_2 \leq G_2$ such that $H_1 \cong H_2$.

Corollary 9.3.34. *Let G_1, G_2 be two commensurable groups, then $\texttt{DP}(G_1) \equiv_m \texttt{DP}(G_2)$. Said otherwise, the domino problem is an invariant of commensurability.*

9.3.4 Classes of Groups

9.3.4.1 Virtually Free Groups

Proposition 9.3.35. *Let F be a free group of finite rank. Then $\texttt{DP}(F)$ is decidable.*

Proof. Let $n = \text{rank}(F)$ and let $S = \{s_1, \dots, s_n, s_1^{-1}, \cdots, s_n^{-1}\}$ be the set of free generators of F. Consider an instance $\mathscr{F}$ of $\texttt{DP}(F, S)$ over an alphabet A. We say a symbol $a \in A$ is *extensible with respect to* $B \subset A$ if for every $s \in S$ there exists $b \in B$ such that neither of the patterns p, q with supports $\text{supp}(p) = \{1_F, s\}$ and $\text{supp}(p) = \{1_F, s^{-1}\}$ defined by $p_{1_F} = a$, $p_s = b$, $q_{1_F} = b$, $q_{s^{-1}} = a$ belong to $\mathscr{F}$. Said otherwise, a is extensible with respect to B if for every direction $s \in S$ it's possible to put an a next to some $b \in B$ in position s without creating a forbidden pattern.

Consider the Turing machine $\mathscr{M}$ which receives an instance of $\texttt{DP}(F, S)$ and does the following:

1. Initialize $E \leftarrow A$.
2. Let E' be the subset of symbols of E which are extensible with respect to E.
3. If $E' \neq E$ assign $E \leftarrow E'$ and go to step 2.
4. If $E \neq \emptyset$ answer $\texttt{YES}$. Otherwise answer $\texttt{NO}$.

This procedure always ends in at most $|A|$ iterations of step 2. Clearly non-extensible symbols cannot appear in a configuration of $X_{\mathscr{F}}$. We deduce therefore that $X_{\mathscr{F}} \subset E^F$ at any step of the algorithm. This implies that if $\mathscr{M}$ answers $\texttt{NO}$ then indeed $X_{\mathscr{F}} = \emptyset$. If $\mathscr{M}$ answers $\texttt{YES}$, then E stabilizes into a nonempty set of extensible with respect to E symbols. Fix for every $a \in E$ a function $\varphi_a : S \to E$ which gives an symbol in E which can be put next to a in direction s. We define $x \in E^F$ inductively as follows: Fix $x_{1_G} = a \in E$. Suppose x is defined over all words $w \in S^*$ of length $|w| \leq n$. For each non-reducible word ws, we let $x_{ws} = \varphi_{(x_w)}(s)$. As the Cayley graph of F is a $2n$-regular infinite tree, this construction does not generate any forbidden patterns and hence $x \in X_{\mathcal{F}}$. □

Definition 9.3.36. Let $\mathscr{P}$ be a group property. A group is said to be virtually $\mathscr{P}$ if it contains a finite index subgroup which satisfies such property.

Integrating the previous proposition with Proposition 9.3.33, we obtain the following theorem.

Theorem 9.3.37. *Every virtually free group has decidable word problem.*

We would like to remark a nice application of Proposition 9.3.35. If G is finitely generated by some finite set S, it admits a presentation $G \cong \langle \mathsf{S}|\mathsf{R}\rangle = F_{\mathsf{S}}/N_{\mathsf{R}}$ where $N_{\mathsf{R}} \trianglelefteq F_{\mathsf{S}}$. As $\mathtt{DP}(F_{\mathsf{S}})$ is decidable, Proposition 9.3.32 implies that if N_{R} is finitely generated then $\mathtt{DP}(G)$ is decidable. If we put this together with the Nielsen-Schreier [402] theorem which states that every subgroup of a free group is itself free, we can write it in the following way.

Corollary 9.3.38. *Let* (S, R) *be a group presentation. If* $\mathtt{DP}(\langle \mathsf{S}|\mathsf{R}\rangle)$ *is undecidable, then the free group* N_{R} *generated by the conjugate closure of* R *has infinite rank.*

Example 9.3.39. Let $[a, b] = aba^{-1}b^{-1}$ denote the *commutator* of a and b. Let $\mathbb{Z}^2 \cong \langle a, b \mid [a, b]\rangle$. As $\mathtt{DP}(\mathbb{Z}^2)$ is undecidable, then $N_{[a,b]}$ has infinite rank. One can also easily verify that $N_{[a,b]} = [F, F] = \{[g, h] \mid g, h \in F\}$. This constitutes a new (and algebraic topology free) proof of the classical result stating that the commutator subgroup of a free group of rank 2 has infinite rank.

9.3.4.2 Polycyclic Groups

The class of polycyclic groups is one of the largest for which we can obtain a complete classification concerning the undecidability of the domino problem. This is achieved through the use of the properties proved in section 9.3.3. Polycyclic groups have indeed a lot of nice properties due to the fact that this is one of the largest classes of groups which is closed under subgroups and quotients, and that contain only finitely presented groups with decidable word problem. See [534] and [373] for more details on polycyclic groups.

Polycyclic groups are the solvable groups for which every subgroup is finitely generated. The best way to give examples of polycyclic groups is by the Auslander-Swan theorem:

Theorem 9.3.40. *Polycyclic groups are precisely solvable subgroups of* $GL_n(\mathbb{Z})$.

See [534, Chapter 5, Theorem 5] or [373, section 3.3] for a proof. By Tits Alternative [564], we therefore obtain that virtually polycyclic groups are precisely the subgroups of $GL_n(\mathbb{Z})$ that do not contain non-abelian free groups, or equivalently amenable subgroups of $GL_n(\mathbb{Z})$.

Polycyclic groups form a nice class of groups due to their many closure properties:

Proposition 9.3.41. *Quotients and subgroups of polycyclic groups are polycyclic. In particular, subgroups of polycyclic groups are always finitely generated.*

This property opens the possibility to do inductive proofs on polycyclic groups. This is done formally with the concept of the *Hirsch number*. The Hirsch number $h(G)$ of a polycyclic group G is the number of infinite factors in a series with cyclic or finite factors. The Hirsch number is always finite, and subgroups and quotients have a smaller Hirsch number than the group. More precisely:

Proposition 9.3.42.

- *If G_1 is a subgroup of G_2, then $h(G_1) \leq h(G_2)$.*
- *If H is a normal subgroup of G, then $h(G) = h(G/H) + h(H)$.*
- *$h(G) = 0$ if and only if G is finite.*
- *$h(G) = 1$ if and only if G is virtually $\mathbb{Z}$.*
- *$h(G) = 2$ if and only if G is virtually $\mathbb{Z}^2$.*

See in particular [534, Chapter 1.C].

We now are ready for the main theorem of this section.

Theorem 9.3.43. *Let G be a virtually polycyclic group. Then G has an undecidable domino problem if and only if G is not virtually cyclic.*

Proof. One direction is clear. By Proposition 9.3.33, it is sufficient to prove the result for polycyclic groups.

We prove the result by induction on the Hirsch number. The result is clear for Hirsch number $0, 1, 2$. Now let G be a group of Hirsch number no less than 3.

It is known that every polycyclic group admit a nontrivial normal free abelian subgroup [534, Chapter 1, lemma 8].

Let H be such a subgroup. If $H = \mathbb{Z}^n$ for some $n > 2$, then H has an undecidable domino problem, and therefore G also has an undecidable domino problem by Proposition 9.3.30. Otherwise $H = \mathbb{Z}$. Then G/H is a polycyclic subgroup of Hirsch number $h(G) - 1 \geq 2$ and therefore has an undecidable domino problem. We conclude again by Proposition 9.3.32 that G has an undecidable domino problem. □

To which extent this theorem can be extended is open. Of course any group that contains a polycyclic group of Hirsch number greater than 2 also has an undecidable domino problem.

A natural direction is extending the theorem to all finitely generated solvable groups. How to do this is unclear. First, there are finitely generated solvable groups with an undecidable word problem. Furthermore, some of them do not contain a copy of $\mathbb{Z}^2$ which means the previous method cannot work. Examples include the

Lamplighter groups and the Baumslag-Solitar groups. Baumslag-Solitar groups will be treated in the next section. Whether the Lamplighter group admits an undecidable domino problem remains open.

9.3.4.3 Baumslag-Solitar Groups

In this section we prove the undecidability of the domino problem on Baumslag-Solitar groups. Given two non-zero integers m and n, we define $\mathbf{BS}(m, n)$ the *Baumslag-Solitar group of order* (m, n) as the two-generators and one-relator group with presentation

$$\mathbf{BS}(m, n) =< a, b | a^m b = ba^n > .$$

In particular $\mathbf{BS}(1, 1)$ is isomorphic to $\mathbb{Z}^2$. Since $\mathbf{BS}(-m, -n)$ is isomorphic to $\mathbf{BS}(m, n)$, it is enough to consider groups with $m > 0$. For simplicity, we also assume that $n > 0$. The case $n < 0$ is analogous.

We first discuss $\Gamma_{m,n}$, the Cayley graph of $\mathbf{BS}(m, n)$ for $m, n > 0$. Since $\mathbf{BS}(m, n)$ has two generators, every vertex in its Cayley graph $\Gamma_{m,n}$ has in-degree and out-degree 2.

The level associated with g of the Cayley graph $\Gamma_{m,n}$ is the induced subgraph obtained by keeping only the vertices of the coset $g\langle a\rangle = \{g.a^k : k \in \mathbb{Z}\}$. We denote it by $\mathscr{L}_g$ and we say that the vertex g defines the level $\mathscr{L}_g$. The level $\mathscr{L}_{gb}$ is a predecessor of the level $\mathscr{L}_g$, while the latter is a successor of the former, for all group elements g. Note that each level has m predecessors and n successors.

Our tilings will be colorings of the edges of the Cayley graph $\Gamma_{m,n}$. The local constraint is given in terms of a set of allowed patterns on the edges

$$\varepsilon \longrightarrow a \longrightarrow a^2 \longrightarrow \ldots \longrightarrow a^m \longrightarrow a^m b = ba^n$$

and

$$\varepsilon \longrightarrow b \longrightarrow ba \longrightarrow ba^2 \longrightarrow \ldots \longrightarrow ba^n = a^m b,$$

see the left side of Figure 9.4 for the case $m = 3, n = 2$. For each group element g the pattern of this shape found at position g must be among the allowed patterns.

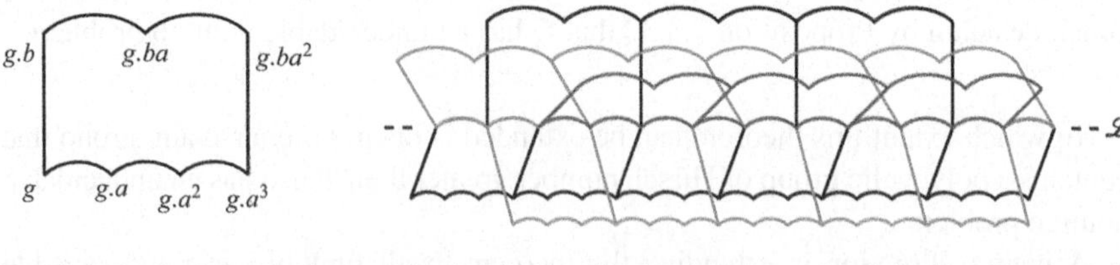

Fig. 9.4 On the left: the shape of the tiles in $\Gamma_{3,2}$. On the right: some levels in $\Gamma_{3,2}$. The level $\mathscr{L}$ has two successor levels (drawn below the level) and three predecessor levels (drawn above it).

The Cayley graph $\Gamma_{m,n}$ can be projected into the Euclidean plane by a function $\Phi_{m,n} : \mathbf{BS}(m,n) \to \mathbb{R}^2$, defined as follows. Let w be a finite word on the alphabet $A = \{a, b, a^{-1}, b^{-1}\}$. Then any element of $\mathbf{BS}(m,n)$ can be represented by such a word, but this representation is of course non unique. If x is a letter of A, we denote by $|w|_x$ the number of occurrences of x in the word w. We then define for $x \in A$ the contribution of x to w by $\| w \|_x = |w|_x - |w|_{x^{-1}}$.

Let $\psi_{m,n} : A^* \to \mathbb{R}$ be the function defined by induction on the length of the word by

$$\begin{cases} \psi_{m,n}(\varepsilon) = 0 \text{ where } \varepsilon \text{ is the empty word} \\ \psi_{m,n}(w.b) = \psi_{m,n}(w.b^{-1}) = \psi_{m,n}(w) \\ \psi_{m,n}(w.a) = \psi_{m,n}(w) + \left(\frac{m}{n}\right)^{\|w\|_b} \\ \psi_{m,n}(w.a^{-1}) = \psi_{m,n}(w) - \left(\frac{m}{n}\right)^{\|w\|_b} \end{cases}$$

Lemma 9.3.44. *For every $u, v \in A^*$ one has*

$$\psi_{m,n}(u.v) = \psi_{m,n}(u) + \left(\frac{m}{n}\right)^{\|u\|_b} \psi_{m,n}(v).$$

Proof. By induction on the length of v. □

We can now define the projection function $\Phi_{m,n} : \mathbf{BS}(m,n) \to \mathbb{R}^2$ which associates to every element g of $\mathbf{BS}(m,n)$ its coordinates on the Euclidean plane:

$$\Phi_{m,n}(g) = (\psi_{m,n}(w), \| w \|_{b^{-1}}),$$

where w is a word representing g. The following proposition states that the definition does not depend on the choice of w. Its proof is a simple application of Lemma 9.3.44.

Proposition 9.3.45. *The function $\Phi_{m,n}$ is well defined on* $\mathbf{BS}(m,n)$.

For every element $g \in \mathbf{BS}(m,n)$, define the shift of g as the first coordinate of $\Phi_{m,n}(g)$ that takes rational values and the height of g as the second coordinate of $\Phi_{m,n}(g)$ that takes integer values.

All elements belonging to the same level project on the same horizontal line, thus we can speak of the height of a level. The height is $\| w \|_{b^{-1}}$ for the words w that represent the elements of the level.

Remark 9.3.46. The function $\Phi_{m,n}$ is not injective. Let $m = 3$ and $n = 2$. Consider the word

$$\omega = bab^{-1}a^2ba^{-1}b^{-1}a^{-2}.$$

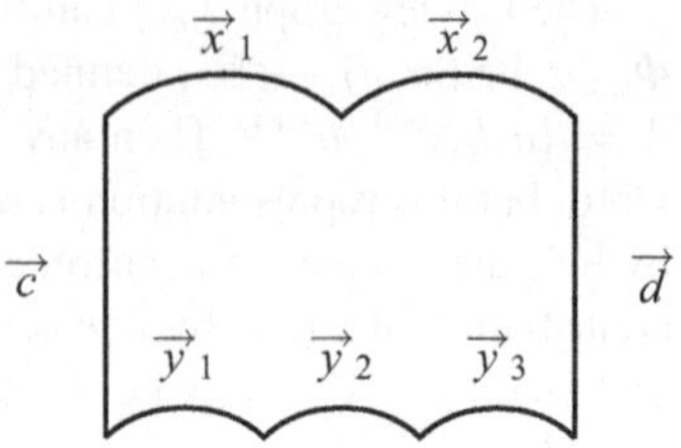

Fig. 9.5 The general form of tiles in **BS**(3, 2).

We have

$$\Phi_{3,2}(\omega) = \Phi_{3,2}(\varepsilon) = (0,0).$$

However, freely reduced words that do not contain $b^{-1}a^{km}b$ or $ba^{kn}nb^{-1}$ as subwords, for any integer k, cannot represent the identity in $\mathbf{BS}(m,n)$. Thus ω and ε represent different elements of the group. Moreover, Baumslag-Solitar groups are HNN-extensions of $\mathbb{Z}$, thus from Britton's lemma it follows that a finite subgroup of Baumslag-Solitar group is conjugate to a finite subgroup of $\mathbb{Z}$. Since ω is not the identity, it has infinite order in $\mathbf{BS}(3,2)$. We see that there is an infinite cyclic subgroup that is projected by $\Phi_{3,2}$ to point $(0,0)$. This will not be a problem in the sequel: the tile associated with an element $g \in \mathbf{BS}(m,n)$ will depend only on $\Phi_{m,n}(g)$.

Following the ideas from Section 9.2.5, we can construct a tile set such that if a level in a tiling of $\Gamma_{m,n}$ represents some $\vec{x} \in \mathbb{R}^2$, then its successor levels represent $f(\vec{x})$ where $f : \mathbb{R}^2 \to \mathbb{R}^2$ is a rational affine map. Going from one level to one of its successor level corresponds to one iteration of f, and a decrease of the height of the level by 1.

Consider the case $\mathbf{BS}(3,2)$. The tiles are of the form shown in Figure 9.5. We say that the tiles compute the function f if the relation

$$\frac{f(\vec{x}_1 + \vec{x}_2)}{2} + \vec{c} = \frac{\vec{y}_1 + \vec{y}_2 + \vec{y}_3}{3} + \vec{d}. \tag{9.1}$$

is satisfied.

Consider a sequence of k such tiles on some level, next to each other so that the left vertical edge of a tile is the same as the right edge of the previous tile. Averaging (9.1) over all k tiles yields then

$$f(\vec{x}) + \frac{\vec{c}_1}{k} = \vec{y} + \frac{\vec{d}_k}{k}$$

where $\vec{x}$ is the average of the labels on the segment of $2k$ edges on the previous level, $\vec{y}$ is the average of the labels on the corresponding segment of $3k$ edges on the level below, and $\vec{c}_1$ and $\vec{d}_k$ are the left and right vertical edges of the first and the last tile in the row, respectively. Letting k grow to infinity, we see that if the previous level represents some $\overrightarrow{x} \in \mathbb{R}^2$ then the next level necessarily represents $f(\overrightarrow{x})$, as required.

For all $g \in \mathbf{BS}(3,2)$ with $\Phi_{m,n}(g) = (\alpha, \beta)$ and for every $k \in \mathbb{Z}$, we define the translated balanced representation of $\overrightarrow{x}$ as the bi-infinite sequence $B^g(\overrightarrow{x}) = (B_k^g(\overrightarrow{x}))_{k \in \mathbb{Z}}$, where

$$B_k^g(\overrightarrow{x}) = \left\lfloor \left(\left(\frac{m}{n}\right)^\beta \alpha + k\right) \overrightarrow{x} \right\rfloor - \left\lfloor \left(\left(\frac{m}{n}\right)^\beta \alpha + (k-1)\right) \overrightarrow{x} \right\rfloor.$$

$$B_k^g(\overrightarrow{x}) = \left\lfloor \left(\left(\frac{m}{n}\right)^\beta \alpha + k\right) \overrightarrow{x} \right\rfloor - \left\lfloor \left(\left(\frac{m}{n}\right)^\beta \alpha + (k-1)\right) \overrightarrow{x} \right\rfloor.$$

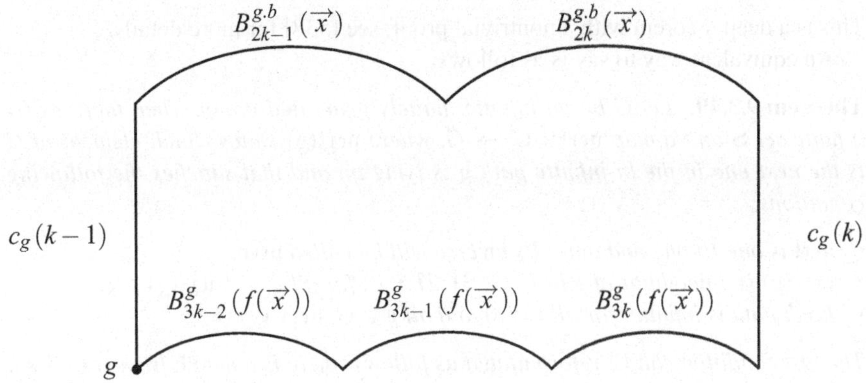

$$\text{with } c_g(k) = \tfrac{1}{2} f\left(\left\lfloor \left(\left(\tfrac{3}{2}\right)^{\beta-1} \alpha + 2k\right) \overrightarrow{x} \right\rfloor\right) - \tfrac{1}{3} \left\lfloor \left(\left(\tfrac{3}{2}\right)^{\beta} \alpha + 3k\right) f(\overrightarrow{x}) \right\rfloor + k\overrightarrow{b}$$

For the same reasons as those invoked in Section 9.2.5 (the domain U is bounded, and the function f_i has rational coefficients), any rational function $f : U \subset \mathbb{R}^2 \to \mathbb{R}^2$ can be encoded by a finite tile set, so that for a given $\overrightarrow{x} \in U$, one can tile a level of $\mathbf{BS}(3,2)$ such that the balanced representations of $\overrightarrow{x}$ and $f(\overrightarrow{x})$ appear on the top and bottom labels, respectively. We deduce the undecidability of the domino problem for $\mathbf{BS}(3,2)$. The proof for other Baumslag-Solitar groups is similar.

Theorem 9.3.47. *The domino problem is undecidable on Baumslag-Solitar groups.*

9.3.4.4 Groups $G_1 \times G_2$

In this section, we will prove the undecidability of the domino problem for groups that are direct products of infinite groups, i.e., groups of the form $G_1 \times G_2$, where G_1 and G_2 are infinite and finitely generated.

Of course, the only interesting case is when at least one of the two groups is an infinite torsion group, i.e., an infinite group where all elements are of finite order. Indeed, if it is not the case, then both G_1 and G_2 contain $\mathbb{Z}$ as a subgroup, and therefore $G_1 \times G_2$ contains $\mathbb{Z}^2$ as a subgroup and thus has an undecidable domino problem by Proposition 9.3.30.

However, even if G does not contain a copy of $\mathbb{Z}$, it is still true that the Cayley graph of G, as any infinite connected graph, contains infinitely many (undirected) paths. In fact, it can even been proven that some Cayley graph of G (for a suitable choice of generators) can be covered by such paths. This is the purpose of the following theorem:

Theorem 9.3.48 ([538]). *Let G be an infinite, finitely generated group. Then there exists a finite set S s.t. the Cayley graph $\Gamma(G, \mathsf{S})$ of G with S as generators can be covered by disjoint bi-infinite paths.*

This is a deep theorem with a nontrivial proof, see [538] for more details.

An equivalent way to say is as follows:

Theorem 9.3.49. *Let G be an infinite, finitely generated group. Then there exists a finite set S and a map $\mathfrak{next} : G \to G$, where $\mathfrak{next}(g)$ states which element of G is the next one in the bi-infinite path g is lying on and that satisfies the following conditions:*

- *$\mathfrak{next}$ is one-to-one and onto. Its inverse will be called $\mathfrak{prev}$.*
- *$\mathfrak{next}(g)$ is a neighbor of g in $\Gamma(G, \mathsf{S})$. That is, for all g, $g^{-1}\mathfrak{next}(g) \in \mathsf{S}$.*
- *Each path is infinite: for all $n > 0$ and all $g \in G$, $\mathfrak{next}^n(g) \neq g$.*

The last condition can be reformulated as follows: there is a map $\mathfrak{h}$ from G to $\mathbb{Z}$ s.t. $\mathfrak{h}(\mathfrak{next}(g)) = \mathfrak{h}(g) + 1$.

In all that follows, we suppose without loss of generality that S is symmetric, which means that additionally for all g, $g^{-1}\mathfrak{prev}(g) \in \mathsf{S}$.

Under these conditions, $\mathfrak{next}$ and $\mathfrak{prev}$ can be defined *locally*. Indeed, let $\mathfrak{n}(g) = g^{-1}\mathfrak{next}(g)$ and $\mathfrak{p}(g) = g^{-1}\mathfrak{prev}(g)$. Then $\mathfrak{n}$ and $\mathfrak{p}$ have values in the finite set S. Furthermore they satisfy the following two properties:

- If $\mathfrak{n}(g) = s$ then $\mathfrak{p}(gs) = s^{-1}$.
- If $\mathfrak{p}(g) = s$ then $\mathfrak{n}(gs) = s^{-1}$.

$\mathfrak{n}$ and $\mathfrak{p}$ can be interpreted in the Cayley graph $\Gamma(G, \mathsf{S})$. At each vertex g of the Cayley graph, two arrows are pointed by. One of them corresponds to $\mathfrak{n}(g)$, the other to $\mathfrak{p}(g)$. The first condition above states that if we start from some vertex g, follow the arrow pointed by $\mathfrak{n}$, and then at $g\mathfrak{n}(g)$ follow the arrow pointed by $\mathfrak{p}$ we come back to g.

Definition 9.3.50. A *valid pair* for (G, S) is a pair (n, p) of maps from G to S s.t.

- If $\mathsf{n}(g) = s$ then $\mathsf{p}(gs) = s^{-1}$.
- If $\mathsf{p}(g) = s$ then $\mathsf{n}(gs) = s^{-1}$.

By definition, $(\mathfrak{n}, \mathfrak{p})$ is a valid pair. The two following facts are obvious:

Proposition 9.3.51. *Let* (n, p) *be a valid pair for* (G, S). *Let* $\mathsf{next}(g) = g\mathsf{n}(g)$ *and* $\mathsf{prev}(g) = g\mathsf{p}(g)$. *Then* next *is one-to-one and onto with inverse* prev.

Proposition 9.3.52. *The set of all valid pairs for* (G, S) *is an SFT. More accurately, define:*

$$X_{G,\mathsf{S}} = \{x \in (\mathsf{S} \times \mathsf{S})^G | \forall s \in \mathsf{S}, [(x_g)_1 = s \Rightarrow (x_{gs})_2 = s^{-1}] \\ \wedge [(x_g)_2 = s \Rightarrow (x_{gs})_1 = s^{-1}]\}$$

Then $X_{G,\mathsf{S}}$ *is an SFT. Furthermore, if* $x \in X_{G,\mathsf{S}}$, *then the pair* (n, p) *defined by* $\mathsf{n}(g) = (x_g)_1$ *and* $\mathsf{p}(g) = (x_g)_2$ *is a valid pair. Conversely, if* (n, p) *is a valid pair, then the configuration* x *defined by* $x_g = (\mathsf{n}(g), \mathsf{p}(g))$ *is in* $X_{G,\mathsf{S}}$.

Note that it might be possible that next and prev are not cyclic. In fact, in a typical configuration of $X_{G,\mathsf{S}}$, it is quite likely that $\mathsf{next}^i(g) = g$ for some g and $i > 0$.

Intuitively, a configuration x of $X_{G,\mathsf{S}}$ corresponds therefore to a partition of the Cayley graph $\Gamma(G, \mathsf{S})$ into cycles and bi-infinite paths, where, at each position $g \in G$, we should look at x_g to read the information coding $\mathsf{n}(g)$ and $\mathsf{p}(g)$ to know what is the next and previous vertex in the cycle or path. In group terms, we could say intuitively that we have partitioned the vertices of the Cayley graph of G into copies of the (canonical) Cayley graphs of $\mathbb{Z}$ and/or $\mathbb{Z}/n\mathbb{Z}$ for some (possibly infinitely many) n. We know also that, if S is chosen to verify the conclusion of Theorem 9.3.49, then at least one configuration of $X_{G,\mathsf{S}}$ contains only bi-infinite paths and no cycles.

We are now almost ready to proceed to the proof. Let G_1 and G_2 be two infinite, finitely generated groups, and let $G = G_1 \times G_2$. Let S_1 and S_2 be two sets of generators for G_1 and G_2 that satisfy the conclusion of Theorem 9.3.49.

We then obtain two SFTs, X_{G_1,S_1} on G_1, and X_{G_2,S_2} on G_2. We extend X_{G_1,S_1} to an SFT on $G_1 \times G_2$ by extending periodically every configuration along G_2 (using the same idea as in Proposition 9.3.32) and proceed analogously with X_{G_2,S_2}. Taking the product of these two SFTs, we have now obtained an SFT X over the alphabet $\mathsf{S}_1 \times \mathsf{S}_1 \times \mathsf{S}_2 \times \mathsf{S}_2$ s.t.:

- If $(\mathsf{n}_1, \mathsf{p}_1)$ is a valid pair for (G_1, S_1) and $(\mathsf{n}_2, \mathsf{p}_2)$ a valid pair for (G_2, S_2), then the configuration x defined by $x_{(g_1,g_2)} = (\mathsf{n}_1(g_1), \mathsf{p}_1(g_1), \mathsf{n}_2(g_2), \mathsf{p}_2(g_2))$ is in X.
- Furthermore, every configuration of X is of this form.

Intuitively a configuration of X partitions the vertices of the Cayley graph $\Gamma(G, \mathsf{S}_1 \times \mathsf{S}_2)$ into copies of the (canonical) Cayley graphs of $\mathbb{Z} \times \mathbb{Z}$, $\mathbb{Z}/n\mathbb{Z} \times \mathbb{Z}$, $\mathbb{Z} \times \mathbb{Z}/n\mathbb{Z}$ and/or $\mathbb{Z}/m\mathbb{Z} \times \mathbb{Z}/n\mathbb{Z}$. We know also that as S_1 and S_2 were chosen to verify the conclusion of Theorem 9.3.49, some configuration of X contains only copies of $\mathbb{Z} \times \mathbb{Z}$.

We are now ready for the proof of the undecidability of the domino problem on $G_1 \times G_2$. Let Y be a nearest neighbor SFT on $\mathbb{Z}^2$ with alphabet A. Let $\mathcal{F}_1$ be the set of forbidden patterns of Y of support $\{(0,0),(1,0)\}$ and $\mathcal{F}_2$ the set of patterns of support $\{(0,0),(0,1)\}$. We can interpret $\mathcal{F}_1$ and $\mathcal{F}_2$ as subsets of A^2, so that $y \in Y$ if and only if for all i,j $(y_{(i,j)}, y_{(i+1,j)}) \notin \mathcal{F}_1$ and $(y_{(i,j)}, y_{(i,j+1)}) \notin \mathcal{F}_2$.

We can define a subshift Z on $G_1 \times G_2$ over the alphabet $(\mathsf{S}_1 \times \mathsf{S}_1 \times \mathsf{S}_2 \times \mathsf{S}_2 \times A)$ as follows:

$$Z = \left\{ z \in X \times A^{G_1 \times G_2} \middle| \forall g \in G_1 \times G_2 \begin{pmatrix} ((z_g)_5, (z_{g(z_g)_1})_5) \notin \mathcal{F}_1 \\ ((z_g)_5, (z_{g(z_g)_3})_5) \notin \mathcal{F}_2 \end{pmatrix} \right\}.$$

Basically, an element of $z \in Z$ is a configuration of X where each element g is additionally labeled with a letter from A. At each element g, we forbid the patterns of support $\{(0,0),(1,0)\}$ to appear in the direction indicated by $\mathsf{n}_1(g)$, and the patterns of support $\{(0,0),(0,1)\}$ to appear in the direction indicated by $\mathsf{n}_3(g)$.

Proposition 9.3.53. *Y is nonempty if and only if Z is nonempty.*

Proof. Suppose that Y is nonempty, and let y be an element of Y. Informally, we know some configuration of X partitions G into copies of $\mathbb{Z}^2$. We will thus use y in each copy of $\mathbb{Z} \times \mathbb{Z}$ to obtain our configuration of Z.

Formally, let $\mathfrak{next}_i, \mathfrak{prev}_i, \mathfrak{n}_i, \mathfrak{p}_i, h_i, i \in \{1,2\}$ the functions that come from Theorem 9.3.49. Consider the configuration x of X that correspond to the valid pairs $(\mathfrak{n}_1, \mathfrak{p}_1, \mathfrak{n}_2, \mathfrak{n}_2)$.

We now define z by $(z_g)_{1-4} = (x_g)$ and $(z_{(g_1,g_2)})_5 = y_{\mathfrak{h}_1(g_1),\mathfrak{h}_2(g_2)}$. Then it is clear that $z \in Z$. Indeed, let $g = (g_1, g_2) \in G_1 \times G_2$. Then $(z_{g(z_g)_1})_5 = (z_{g(\mathfrak{n}_1(g_1),1_{G_2})})_5 = (z_{\mathfrak{next}_1(g_1),g_2})_5$ so that $(z_5, (z_{(z_g)_1 g})_5) = (y_{\mathfrak{h}_1(g_1),\mathfrak{h}_2(g_2)}, y_{\mathfrak{h}_1(g_1)+1,\mathfrak{h}_2(g_2)}) \notin \mathcal{F}_1$ by definition of y. Similarly, $(z_5, (z_{(z_g)_1 g})_5) \notin \mathcal{F}_2$. Therefore $z \in Z$ and Z is nonempty.

Conversely, suppose Z is nonempty and let $z \in Z$. Informally, z partitions G into copies of $\mathbb{Z} \times \mathbb{Z}$ or bastardized versions of it of the form $\mathbb{Z}/n\mathbb{Z} \times \mathbb{Z}/m\mathbb{Z}$. We look at just one of this copy and use it to build our configuration on Y, possibly unfolding the copy if necessary.

Formally, let x be defined by $x_g = (z_g)_{1-4}$. Then $x \in X$ and therefore corresponds to two valid pairs $(\mathsf{n}_1, \mathsf{p}_2)$ and $(\mathsf{n}_2, \mathsf{p}_2)$. Let $\mathsf{next}_1, \mathsf{prev}_1$ and $\mathsf{next}_2, \mathsf{prev}_2$ be the associated functions. We now define $y \in A^{\mathbb{Z}^2}$ by $y_{(i,j)} = (z_{((\mathsf{next}_1)^i(1_G),(\mathsf{next}_2)^j(1_G))})_5$. Then $y \in Y$.

Indeed. Let $(i,j) \in \mathbb{Z}^2$. Let $g_1 = (\mathsf{next}_1)^i(1_G)$ and $g_2 = (\mathsf{next}_2)^j(1_G)$. Then

$$y_{(i+1,j)} = (z_{(\mathsf{next}_1)(g_1),g_2})_5 = (z_{(g_1 \mathsf{n}_1(g_1),g_2)})_5 = (z_{(g_1 z_{(g_1,g_2)})_1,g_2})_5$$

Therefore $(y_{(i,j)}, y_{(i+1,j)}) = ((z_{(g_1,g_2)})_5, (z_{((g_1 z_{(g_1,g_2)})_1,g_2)})_5) \notin \mathcal{F}_1$ by hypothesis on z. Similarly, $(y_{(i,j)}, y_{(i,j+1)}) \notin \mathcal{F}_2$. Therefore $y \in Y$ and Y is nonempty. □

Corollary 9.3.54. *Let G_1, G_2 be two infinite, finitely generated groups. Then $G_1 \times G_2$ has an undecidable domino problem.*

Proof. Forbidden words for Z can be built effectively from a set of forbidden words of a nearest neighbor $\mathbb{Z}^2$-subshift. □

Corollary 9.3.55. *The Grigorchuk group has an undecidable domino problem.*

Proof. The Grigorchuk group G is a well-known example of a torsion group. G contains a subgroup of finite index of the form $H = H_1 \times H_2$ with H_1, H_2 infinite [430]. $H_1 \times H_2$ has an undecidable domino problem by the previous propositions; therefore G has an undecidable domino problem by Proposition 9.3.33. □

9.3.5 Discussion

9.3.5.1 Muller & Schupp Theorem

In Section 9.3.4.1 we proved that virtually free groups have decidable domino problem. The proof directly gives an algorithm that decides the problem. It is noteworthy that this result can be obtained by a totally different argument. This comes from the combination of three facts: first, the domino problem can be expressed in Monadic Second Order logic (MSO) [33, 580]; second, a group is virtually free if and only if it has finite tree-width [431]; third, graphs with finite tree-width are exactly those with decidable MSO logic [366].

Proposition 9.3.56 ([366, 431]). *If G is virtually free, then G has decidable domino problem.*

The reasoning that leads to decidability of domino problem for virtually free groups using logic also suggests that this sufficient condition might also be necessary. This assumption comes from the following reasoning: if a group is not virtually free, then it has arbitrarily large grids as minors [507]. It should then be possible to somehow use these grids as computation zones – similarly to what is done in Robinson's tiling [508] – to encode Turing machine computations and conclude the undecidability of the domino problem. But the main issue is that even if we know that such grids exist, we do not know where they appear and even less how to make them appear inside a tiling by Wang tiles.

Conjecture 9.3.57. A finitely presented group has decidable domino problem if and only if it is virtually free.

9.3.5.2 Hyperbolic Groups

The theorems in the previous section indicate that, whenever a group contains (in some sense) $\mathbb{Z} \times \mathbb{Z}$ or any other nontrivial Baumslag-Solitar group, they automatically have undecidable domino problem.

A specific class of groups where this result cannot be applied is the class of hyperbolic groups. Indeed, hyperbolic groups do not contain any copy of $\mathbb{Z} \times \mathbb{Z}$ or any other nontrivial Baumslag-Solitar group. They also always have decidable word problem, so Theorem 9.3.28 cannot apply, and they are always finitely presented.

Furthermore, free groups are hyperbolic, so it is very tempting to think that they always have decidable domino problem. However, this is not true.

Indeed, by a theorem of Rips [506], there exist hyperbolic groups G and finitely generated normal subgroups H of G s.t. G/H is isomorphic to $\mathbb{Z}^2$. Therefore, they have an undecidable domino problem by Proposition 9.3.32. It is also tempting to think the idea used above for the domino problem in the Baumslag-Solitar group can be extended for hyperbolic surface groups.

While these two ideas give examples of hyperbolic groups with undecidable domino problem, how to prove the result for an arbitrary, not virtually free, hyperbolic group, remains an open problem.

9.3.5.3 Translation-Like and Quasi-Isometric Groups

Section 9.3.4.4 suggests that the undecidability of the domino problem is a geometric property: as soon as some Cayley graph of G contains a grid structure, G will have an undecidable domino problem. This is also hinted at in Section 9.3.5.1. There are a few theorems that indeed suggest, at least for recursively presented groups, that it is indeed the case. In order to study this idea, we need the notion of quasi-isometry.

The definitions we give here are only valid in the context of finitely generated groups, but these notions can be defined in the general case of metric spaces.

Definition 9.3.58. Let G_1, G_2 be two finitely generated groups and S_1 a set of generators for G_1.

A map $f : G_1 \to G_2$ is Lipschitz if there exists a finite set $\mathsf{S}_2 \in H$ s.t. for all $g \in G_1$, and all $s \in \mathsf{S}_1, f(g)^{-1}f(gs) \in \mathsf{S}_2$.

A map $f : G_1 \to G_1$ is at bounded distance from the identity if there exists a finite set F for all $g \in G_1, g^{-1}f(g) \in F$.

Compare the definition with Theorem 9.3.49. Geometrically, a Lipschitz map f sends adjacent vertices in the Cayley graph $\Gamma(G_1, \mathsf{S}_1)$ to adjacent (or identical) vertices in the Cayley graph $\Gamma(G_2, \mathsf{S}_2)$. It can be proven easily that the fact that f is Lipschitz does not depend on S_1, so that it is a property of the function and the group and not of the specific choice of generators.

Notice that if h is a homomorphism from G_1 to G_2, then h is a Lipschitz map by taking $\mathsf{S}_2 = h(\mathsf{S}_1)$.

Definition 9.3.59. G_1 and G_2 are quasi-isometric if and only if there exists maps $f : G_1 \to G_2, g : G_2 \to G_1$ such that:

- f, g are Lipschitz.
- g is a quasi-inverse for f: $f \circ g$ and $g \circ f$ are at bounded distance from the identity.

Quasi-isometry intuitively means that a pair of Cayley graphs of G_1 and G_2 look similar at large scale.

Definition 9.3.60 ([538, 584]). A right action of G on H is an infix operator $\star$: $H \times G \to H$ s.t. $h \star 1_G = h$ and $h \star (g_1 g_2) = (h \star g_1) \star g_2$ for all $g_1, g_2 \in G$ and all $h \in H$.

An action is free if $h \star g = h$ for some h implies $g = 1_G$.

An action is translation-like if it is free and for all $g \in G$, the map $h \to h \star g$ is at bounded distance from the identity.

It is easy to see that it is sufficient to prove the last property for a set of generators of G. Therefore the last property can be replaced by: for some (any) set S_1 of generators of G, there exist S_2 that generate H s.t. for all $s \in \mathsf{S}_1$ and all $h \in H$, $h^{-1}(h \star s) \in \mathsf{S}_2$. Compare the definition with Theorem 9.3.49 when $G = \mathbb{Z}$.

An intuitive way to understand translation-like actions is that it covers some Cayley graph of H by copies of some Cayley graph of G.

Theorem 9.3.61 ([157]). *Let G and H be quasi-isometric finitely presented groups. G has undecidable domino problem if and only if H has undecidable domino problem.*

Theorem 9.3.62 ([310]). *Let G be a finitely presented group with undecidable domino problem and H a finitely generated group. If G acts translation-like on H, then H has an undecidable domino problem.*

Both proofs are similar to the main proof of Section 9.3.4.4 in that they consist in seeing the Cayley graph of H as similar to the Cayley graph of G, or copies of the Cayley graph of G for the second theorem. The main difference is that we have replaced $\mathbb{Z}$, a free group, with an arbitrary *finitely presented* group. In the first theorem, they define a subshift of finite that codes all quasi-isometries that correspond to some finite sets $\mathsf{S}_1, \mathsf{S}_2$. In the second theorem, a subshift of finite type analogous to $X_{G,\mathsf{S}}$ from Corollary 9.3.54 will be defined that somehow codes all translation-like actions (and also non-free actions) that correspond to a function from a set of generators of G to a set of generators of H defined by the translation-like action. However, in this case the group G acting on H is not free; therefore the SFT needs also to code the relations of G (and hence needs G to be finitely presented for the construction to work).

9.4 Towards a Definition of Effective Subshifts on Groups

We start this section by presenting the notion of effectively closed subshift on $\mathbb{Z}^d$ for $d \geq 1$. This class of subshift appears naturally in dimension 1 as a generalization of sofic subshifts – those with a regular language, and thus whose complement of the language is also regular – as follows: effectively closed subshifts are those which can be defined by a recursively enumerable set of forbidden patterns. In addition to this fact, effectively closed subshifts also appear in a natural way when studying some subsystems of two-dimensional SFTs or sofic subshifts, called projective subdynamics. These links are explained in Section 9.4.1.

9.4.1 *Link Between $\mathbb{Z}$ and $\mathbb{Z}^2$*

To preserve the finiteness of the alphabet and thus compactness of the space of configurations, we use the projective subdynamics, to be defined in the Section 9.4.1.1. Note that this notion of projective subdynamics is different from the notion of subdynamics of a subshift defined by Hochman in [294].

9.4.1.1 Projective Subdynamics: Definition and Example

We define the *projective subdynamics* of a two-dimensional subshift $X \subseteq A^{\mathbb{Z}^2}$ as the set of rows that can appear inside configurations of X :

$$\pi(X) = \left\{ y \in A^{\mathbb{Z}} \mid \exists x \in X \text{ s.t. } x_{(i,0)} = y_i \text{ for every } i \in \mathbb{Z} \right\}.$$

The set of configurations $\pi(X)$ defined above is a subshift (see Exercise 9.5.2).

Example 9.4.1. Consider the two-dimensional SFT X on alphabet

$$A = \left\{a, a, b, b, c, c, c, \boxed{\uparrow}, \boxed{\nearrow}, \boxed{\nwarrow}, \boxed{\uparrow}, \boxed{\nearrow}, \boxed{\nwarrow}\right\}$$

and whose allowed patterns appear in the following configuration.

The idea is to ensure that on every row, patterns of the form $a^n b^n$ appear in a sea of c' – without preoccupying of the colors. Diagonal signals are sent from the leftmost *a*' and the rightmost *b*' – they are designated as so if they have a *c*' as a neighbor. These diagonal signals bump on other signals and on symbols *c*' when they reach one, and two signals can collide only at the middle of a pattern $a^n b^n$ – the middle is marked with up-arrows.

It is left as an exercise to prove that the projective subdynamics of the SFT X is not a sofic subshift (see Exercise 9.5.3).

9.4.1.2 Effectively Closed Subshifts on $\mathbb{Z}^d$

A subshift $X \subseteq A^{\mathbb{Z}}$ is *effectively closed* if there exists a recursively enumerable set of forbidden patterns that defines it. To defined effectively closed subshifts on $\mathbb{Z}^d$ for $d \geq 2$, take any recursive bijection Φ between $\mathbb{Z}^d$ and $\mathbb{Z}$ – such a bijection always exists. Then a pattern p with finite support $S \subset \mathbb{Z}^d$ can be coded by the finite set of tuples $\Phi(p) = \{(\Phi(i), a) \mid i \in S \text{ and } p_x = a\}$. The set $\Phi(p)$ is called the pattern coding of p. We then define an *effectively closed* subshift on $\mathbb{Z}^d$ as a subshift for which there exists a recursively enumerable set of pattern codings that defines it.

From that definition, it is easy to check that the class of effectively closed subshifts contains SFTs and sofic subshifts.

Proposition 9.4.2. *Subshifts of finite type and sofic subshifts are effectively closed.*

Nevertheless, the class of effectively closed subshifts is wider than the class of sofic subshifts, as the following example in dimension 1 shows.

Example 9.4.3. Let Y be the subshift defined as $\pi(X)$ where X is the two-dimensional SFT of Example 9.4.1. This subshift Y is not sofic, nevertheless it can be defined as the set of configurations that avoid the following patterns

$$\{ba, cb, ac\} \cup \{ca^m b^n c \mid m \neq n\}.$$

This set is obviously recursively enumerable, so that the subshift Y is effectively closed but not sofic.

The previous example cannot be adapted to dimension 2, since we lack an equivalent of graph representation of sofic subshifts in higher dimension. The difficulty lies in the fact that it is difficult to prove, in general, that a subshift is not sofic. There exist sufficient conditions for non-soficness, but no characterization exists. Nevertheless, there are explicit examples of effectively closed but non-sofic subshifts in dimension 2 (Figure 9.6). The proof of non-soficness uses a combinatorial argument also used in [567] that can be generalized to prove non-soficness of well-chosen subshifts on amenable groups [23].

Example 9.4.4. We define a two dimensional subshift X_{mirror}, called the *mirror shift*. It consists of all configurations over the alphabet $A = \{$□, ■, ■$\}$ which avoid the following patterns

Fig. 9.6 One configuration in the two-dimensional mirror subshift X_{mirror}.

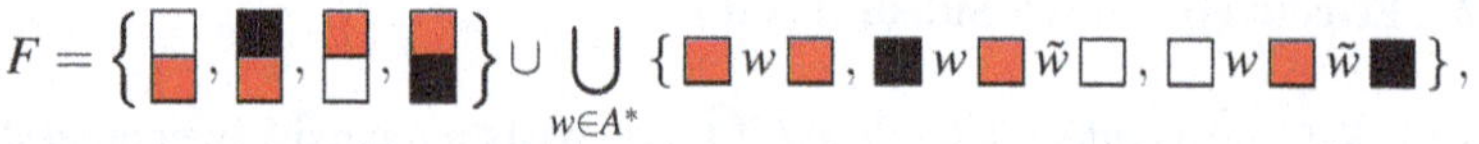

where $\tilde{w}$ denotes the mirror image of the word w, which is the word of length $|w|$ defined by $(\tilde{w})_i = w_{|w|-i+1}$ for all $1 \leq i \leq |w|$.

The mirror subshift X_{mirror} contains the $\mathbb{Z}^2$-fullshift $\{$□, ■$\}^{\mathbb{Z}^2}$ as a subsystem but also all configurations that respect the following conditions: a symbol 🟥 forces all symbols in the same column to be also 🟥 symbols; there is at most one column of 🟥 symbols; if a symbol 🟥 is present on a row, then □ and ■ symbols of this row are arranged symmetrically with respect to the 🟥 symbol.

The column of 🟥, if it appears in a configuration, behaves as a mirror towards the two half planes it defines, hence the name of the subshift. Obviously this subshift is effectively closed since the set of forbidden patterns F_{mirror} can be effectively enumerated, but one can prove it is not sofic by a direct combinatorial argument.

Proposition 9.4.5. *The mirror subshift $X_{mirror} \subset A^{\mathbb{Z}^2}$ is not sofic.*

Proof. Consider $S = \{(0, 1), (1, 0)\}$ and suppose that the mirror subshift is sofic on $\mathbb{Z}^2$, then there exists a S-nearest neighbor $\mathbb{Z}^2$-SFT $X \subset B^{\mathbb{Z}^2}$ on some finite alphabet B and a 1-block factor code $\phi : X \twoheadrightarrow X_{\text{mirror}}$.

Let n be a positive integer and define $\Lambda_n := [-n, n]^2$. Notice that $\overset{\circ}{\Lambda}_{n+1} = \Lambda_n$ and thus $\partial\Lambda_{n+1} = \Lambda_{n+1} \setminus \Lambda_n$. In $L_{\Lambda_n}(X_{\text{mirror}})$ there are exactly $2^{(2n+1)^2}$ different patterns that do not contain a 🟥. These patterns are images of patterns of X with support $[-n, n]^2$ under ϕ and are surrounded with a crown with support $\partial\Lambda_{n+1}$. There are at most $|B|^{4(2n+2)}$ different crowns.

Consider now all configurations $x \in X_{\text{mirror}}$ in which a mirror appears at the origin, that is to say $x_{(0,j)} =$ 🟥 for all $j \in \mathbb{Z}$. For n large enough one has $|B|^{4(2n+2)} < 2^{(2n+1)^2}$, consequently there exist two distinct patterns P_1 and P_2 with support Λ_n that appear, respectively, in configurations y_1 and y_2 of X_{mirror} – assume that y_1 and y_2 are such that $(x_1)|_{\Lambda_n+(n^2,0)} = P_1$ and $(x_2)|_{\Lambda_n+(n^2,0)} = P_2$ – and such that there exist two distinct configurations x_1, x_2 in the extension X of X_{mirror} with the same crown – $(x_1)|_{\partial\Lambda_{n+1}+(n^2,0)} = (x_2)|_{\partial\Lambda_{n+1}+(n^2,0)}$ – and such that $y_1 = \phi(x_1)$ and $y_2 = \phi(x_2)$. As X is the nearest neighbor, we can construct a new configuration $\widetilde{y} \in A^{\mathbb{Z}^2}$ defined by

$$\widetilde{y}_z = \begin{cases} (P_2)_{z-(n^2,0)}, \text{ if } z \in \Lambda_n + (n^2,0) \\ (y_1)_z \text{ otherwise,} \end{cases}$$

in other terms $\widetilde{y}$ is the same configuration as y_1 except that pattern P_1 have been replaced by pattern P_2. On the one hand, in configuration $\widetilde{y}$ a mirror appears at the origin, but since P_1 and P_2 have been chosen distinct $\widetilde{y} \notin X_{\text{mirror}}$. On the other hand the configuration $\widetilde{x} \in B^{\mathbb{Z}^2}$ defined by

$$\widetilde{x}_z = \begin{cases} (x_2)_{z-(n^2,0)}, \text{ if } z \in \Lambda_n + (n^2,0) \\ (x_1)_z \text{ otherwise,} \end{cases}$$

does not contain any forbidden pattern for X – that have been chosen nearest neighbor – and satisfies $\widetilde{y} = \phi(\widetilde{x})$, which proves that $\widetilde{y} \in X_{\text{mirror}}$ hence raising a contradiction (Figure 9.7). □

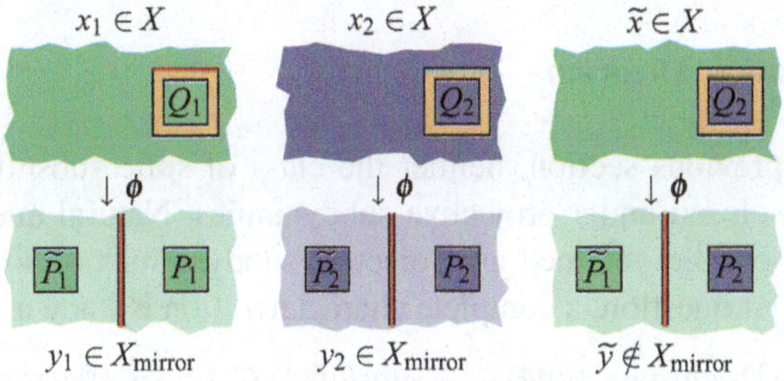

Fig. 9.7 Two configurations y_1 and y_2 in the mirror subshift X_{mirror} with a mirror at the origin, and that differ on $\Lambda_n+(n^2,0)$, but whose preimages in the nearest neighbor $\mathbb{Z}^2$-SFT extension X are the same on $\partial\Lambda_{n+1}$. If it were so, one could construct a configuration $\widetilde{y}$ – by replacing $(y_1)|_{\Lambda_n+(n^2,0)}$ by $(y_2)|_{\Lambda_n+(n^2,0)}$ in configuration y_1 – which belongs to the image $\phi(X)$ but does not belong to X_{mirror}. This proves X_{mirror} is not sofic.

Remark 9.4.6. One can define mirror subshifts in any dimension as the union of the $\mathbb{Z}^d$-fullshift $\{□, ■\}^{\mathbb{Z}^d}$ and the set of configurations $x \in A^{\mathbb{Z}^d}$ with the hyperplane $\{i\} \times \mathbb{Z}^{d-1}$ filled with ■ symbols for some $i \in \mathbb{Z}$, and such that $x|_{\{i+j\}\times\mathbb{Z}^{d-1}} = x|_{\{i-j\}\times\mathbb{Z}^{d-1}}$ for every $j \in \mathbb{Z}$. Then Proposition 9.4.5 can be generalized to any dimension.

We now list stability properties satisfied by effectively closed subshifts.

Proposition 9.4.7. *The class of effectively closed subshifts on $\mathbb{Z}^d$ is closed under finite intersection, finite union, and morphism.*

Proof. The key ingredient is that one can choose a maximal – for inclusion – and recursively enumerable set of forbidden patterns to define an effectively closed subshift X. Indeed, the complement of the language of an effectively closed subshift has this property: it can be recursively enumerated from any recursively enumerable set of forbidden patterns that defines X and is by definition maximal for inclusion.

If X_1 and X_2 are two effectively closed subshifts, then $X_1 \cap X_2$ (resp. $X_1 \cup X_2$) is defined by the set of forbidden patterns $\overline{\mathscr{L}_1} \cup \overline{\mathscr{L}_2}$ (resp. $\overline{\mathscr{L}_1} \cap \overline{\mathscr{L}_2}$). Since the union (resp. intersection) of two recursively enumerable languages is also recursively enumerable, effectively closed subshifts are closed under finite intersection (resp. finite union). Stability under morphisms comes from the fact that morphism are computable. □

Proposition 9.4.8. *The class of effectively closed subshifts on $\mathbb{Z}^d$ is closed under projective subdynamics.*

Proof. This stability result follows from the fact that projective subdynamics are special cases of factors of subactions, and by Theorem 3.1 and Proposition 3.3 of [294] which establish that symbolic factors and subactions preserve effectiveness. □

9.4.1.3 Simulation Theorem

As seen in the previous section, neither the class of sofic subshifts nor the class of SFTs are not closed under projective subdynamics. Natural questions are thus: which subshifts can be obtained as projective subdynamics of sofic subshifts? of SFTs? For the first question, a complete characterization is known.

Theorem 9.4.9 (Hochman, [294]). *A subshift $Y \subseteq B^{\mathbb{Z}^k}$ is effectively closed if and only if it the projective subdynamics of a sofic subshift $X \subseteq A^{\mathbb{Z}^{k+2}}$.*

The original construction by Hochman can be found in [294].

Theorem 9.4.10 (Aubrun & Sablik, [25], Durand, Romaschenko & Shen, [203]). *A subshift $Y \subseteq B^{\mathbb{Z}^k}$ is effectively closed if and only if it the projective subdynamics of a sofic subshift $X \subseteq A^{\mathbb{Z}^{k+1}}$.*

These two theorems can be used as a black box to prove soficness of some complex subshifts and lead to several applications. As a first example, consider the following result.

Theorem 9.4.11 (Myers, [432]). *There exist non-recursive two-dimensional SFTs, i.e., subshifts of finite type on some alphabet A such that none of their configurations can be described by a computable function $f : \mathbb{Z}^2 \to A$.*

This theorem was proven a few years after the publication of Robinson's proof of the undecidability of the domino problem. The proof is strongly inspired by Robinson's techniques and is thus very technical. The same result can be obtained as a direct application of the simulation theorem as follow. Consider a one-dimensional effectively closed subshift X with no computable configurations (see [139] for an explicit construction). The two-dimensional sofic subshift, obtained by the simulation theorem, that projects onto X has no computable configurations (otherwise X would also contain a computable configurations). Since inverses of factor maps are computable, we get a non-recursive two-dimensional SFT.

Consider now S-adic subshifts that are defined by a sequence of substitutions. In higher dimension and if the driving sequence is chosen to be computable, then S-adic subshifts are sofic [25]. This result generalizes theorems by Mozes [429] and Goodman-Strauss [257] on substitutive subshifts – when the driving sequence of substitutions is constant – to S-adic subshifts. The proofs of these two famous theorems are highly technical and difficult to handle. In contrast, the proof presented in [25] consists in a direct application of the simulation theorem, combined with a clever encoding of substitutions.

We list other examples of applications of the simulation theorem:

- In [516] the authors prove that tilings obtained by digitizing irrational vector spaces are aperiodic if and only if the digitized vector spaces are computable.
- In [312] it is proven that sets of periods for multidimensional SFTs are exactly sets of integers of the complexity class NP.
- In [34] an example of one-dimensional effectively closed subshift with a non-computable quasi-periodicity function is given.

9.4.2 *Effectiveness on Groups*

This section is devoted to the study of what kind of subshifts on groups can be defined using Turing machines. As opposed to what happens in $\mathbb{Z}^d$, a general group might present extra difficulties which are related to its word problem.

Before defining the main object of this section, we make a digression with regards to their name. In the literature a subshift defined by a recursively enumerable set of forbidden patterns is usually said to be *effective*. Nevertheless, the name has also been used to talk about subshifts X whose language $L(X)$ is decidable, see for example [192]. These two notions define different objects. To avoid confusion, we refer to these objects as *effectively closed subshifts*, as they would be called in computable analysis.

9.4.2.1 Definition and Basic Properties

For the definition of effectiveness in the context of subshift in a general group, we use Definition 9.3.23 of a pattern coding as the standard structure in which the information about forbidden patterns is given to a Turing machine.

Definition 9.4.12. We say a subshift $X \subset A^G$ is *effectively closed* if there exists a recursively enumerable set of pattern codings $\mathscr{C}$ such that $X = X_{\mathscr{C}}$.

Usually in computability theory the word *effective* is used for objects defined by a decidable set instead of just a recursively enumerable one. In this case the usage of the word is justified, as effectively closed subshifts coincide with those which are definable by a decidable set of forbidden pattern codings.

Proposition 9.4.13. *Let $X \subset A^G$ be an effectively closed subshift. Then there exists a decidable set of pattern codings $\mathscr{C}$ such that $X = X_{\mathscr{C}}$.*

Proof. Let $\mathscr{C}'$ a recursively enumerable set of pattern codings such that $X = X_{\mathscr{C}'}$. If $\mathscr{C}'$ is finite the result is trivial. Otherwise there exists a recursive enumeration $\mathscr{C}' = \{c_0, c_1, \dots\}$. For a pattern coding c we define its length as $|c| = \max_{(w,a)\in c} |w|$. For $n \in \mathbb{N}$ let $L_n = \max_{k \leq n} |c_k|$ and define $\mathscr{C}_n$ as the finite set of all pattern codings c which satisfy the following properties:

- Every $w \in S^*$ with $|w| \leq L_n$ appears in exactly one pair in c.
- $(w, a) \in c$ implies that $|w| \leq L_n$.
- If $(w, a) \in c_n$ then $(w, a) \in c$.

That is, $\mathscr{C}_n$ is the set of all pattern codings which are completions of c_n up to every word of length at most L_n in every possible way. Consider $\mathscr{C} = \bigcup_{n\in\mathbb{N}} \mathscr{C}_n$. It is easy to check that

$$\bigcup_{c\in\mathscr{C}_n} \bigcap_{(w,a)\in c} [a]_w = \bigcap_{(u,b)\in c_n} [b]_u.$$

Therefore $X_{\mathscr{C}'} = X_{\mathscr{C}}$. We claim $\mathscr{C}$ is decidable.

Consider the algorithm which does the following on input c: It initializes n to 0. Then it enters into the following loop: first it produces the pattern coding c_n. If $L_n > |c|$ it rejects the input. Otherwise it calculates the set $\mathscr{C}_n$. If $c \in \mathscr{C}_n$, then it accepts; otherwise it increases the value of n by 1.

As L_n is increasing and cannot stay in the same value indefinitely this algorithm eventually ends for every input. □

It is important to remark that the previous result just gives the existence of one decidable set of pattern codings which defines the subshift. Even in the case of $\mathbb{Z}$-subshifts there exist effectively closed subshifts whose language is undecidable. See Exercise 9.5.20.

Nevertheless, even with this result in hand, it is often more convenient to define effectively closed subshifts by a recursively enumerable set of forbidden pattern codings. In particular, if the group is recursively presented, this set can be chosen canonically as the set of pattern codings which represent the complement of the language.

Proposition 9.4.14. *Let $X \subset A^G$ be an effectively closed subshift. If G is recursively presented, then it is possible to choose $\mathscr{C}$ to be the recursively enumerable and maximal – for inclusion – set of pattern codings such that $X = X_{\mathscr{C}}$.*

Proof. A pattern coding c belongs to the maximal set $\mathscr{C}$ defining X if and only if $X \cap \bigcap_{(w,a)\in c}[a]_w = \emptyset$. Let $c \in \mathscr{C}$ and $\mathscr{C}'$ a recursively enumerable set such that $X = X_{\mathscr{C}'}$. Then:

$$\bigcap_{(w,a)\in c}[a]_w \subset \bigcup_{c'\in\mathscr{C}',g\in G}\ \bigcap_{(v,b)\in c'}[b]_{gv}.$$

By compactness we may extract a finite open cover indexed by c_i', g_i such that:

$$\bigcap_{(w,a)\in c}[a]_w \subset \bigcup_{i\leq n}\ \bigcap_{(v,b)\in c_i'}[b]_{g_iv} \qquad (\star)$$

Note that each of these g_i can be seen as a finite word in S^*. Now let T be the Turing machine which does iteratively for $n \in \mathbb{N}$ the following:

- Runs n steps the machine T_1 recognizing $\texttt{WP}(G)$ for every word in S^* of length smaller than n.
- Runs n steps the machine T_2 recognizing $\mathscr{C}'$ for every pattern coding defined on a subset of words of S^* of length smaller than n.
- Let $\sim_n$ be the equivalence relation for words in S^* of length smaller than n such that $u \sim_n v$ if uv^{-1} has been already accepted by T_1. Let $\mathscr{C}_n$ be the pattern codings already accepted by T_2. If every word in c has length smaller than n check if the following relation is true under $\sim_n$:

$$\bigcap_{(w,a)\in c}[a]_w \subset \bigcup_{c'\in\mathscr{C}_n,|u|\leq n}\ \bigcap_{(v,b)\in c'}[b]_{uv}$$

 If it is true, accept, otherwise increase n by 1 and continue.

Let m be the max of all $|w|$ such that $(w, a) \in c$, and $|w'|$ such that $(w', a') \in c_i'$ and all $|g_i|$. By definition, there exists an $N \in \mathbb{N}$ such that every c_i' for $i \leq n$ is accepted and every word representing 1_G of length smaller than $2m$ is accepted. This means that at stage N relation $(\star)$ is satisfied and T accepts c. If c is not in the maximal set, the machine never accepts. □

In what follows, we show that the class of effectively closed subshifts is closed under factors and projective subdynamics when the group is recursively presented.

Proposition 9.4.15. *For recursively presented groups the class of effectively closed subshifts is closed under factors.*

Proof. Let $X \subset A_X^G$ be an effectively closed subshift. As G is recursively presented, a recursively enumerable set of pattern codings $\mathscr{C}_X$ defining X can be chosen to be maximal by Proposition 9.4.14. Consider a factor code $\sigma : X \twoheadrightarrow Y$ defined by a local function $\Phi : A_X^F \to A_Y$. Let $f_1, \ldots, f_{|F|}$ be words in S^* representing the elements of F.

As σ is surjective, for each $a \in A_Y$ we have $|\Phi^{-1}(a)| > 0$. Therefore we can associate to a pair (w, a) a nonempty finite set of pattern codings

$$\mathscr{C}_{w,a} = \{(wf_i, p_{f_i})_{i=1,\ldots,|F|} \mid p \in \Phi^{-1}(a)\}.$$

That is, $\mathscr{C}_{w,a}$ is a finite set of pattern codings over A_X representing every possible preimage of a. For a pattern coding $c = (w_i, a_i)_{i \leq n}$ where $a_i \in A_Y$, we define

$$\mathscr{C}_c = \{ \bigcup_{(w,a)\in c} \widetilde{c}_{w,a} \mid \widetilde{c}_{w,a} \in \mathscr{C}_{w,a}\}.$$

That is, $\mathscr{C}_c$ is the finite set of pattern codings formed by choosing one possible preimage for each letter. Let $\mathscr{M}$ be the Turing machine which on entry c runs the machine recognizing $\mathscr{C}_X$ on every pattern coding in $\mathscr{C}_c$. If it accepts for every input, then $\mathscr{M}$ accepts c. Let $\mathscr{C}_Y$ be the set of pattern codings accepted by $\mathscr{M}$. We claim $Y = Y_{\mathscr{C}_Y}$.

Let $y \in Y_{\mathscr{C}_Y}$ and $n \in \mathbb{N}$. For each pattern coding c defined over all words of length at most n such that $y \in \bigcap_{(w,a)\in c}[a]_w$, there is a pattern coding $c_n \in \mathscr{C}_c$ which does not belong to $\mathscr{C}_X$. As $\mathscr{C}_X$ is maximal we have that $\bigcap_{(v,b)\in c_n}[b]_v \cap X \neq \emptyset$. Extracting a configuration x_n from $\bigcap_{(v,b)\in c_n}[b]_v \cap X$ we obtain a sequence $(x_n)_{n\in\mathbb{N}}$. By compactness there is a converging subsequence with limit $\widetilde{x} \in X$. By continuity of σ we have that $y = \sigma(\widetilde{x}) \in Y$. Conversely if $y \in Y$ there exists $x \in X$ such that $\sigma(x) = y$. Therefore for every finite $F' \subset G$ and pattern coding c such that $\bigcap_{(w,a)\in c}[a]_w = [y|_{F'}]$ there exists a pattern coding $\widetilde{c} \in \mathscr{C}_c$ such that $\bigcap_{(v,b)\in\widetilde{c}}[b]_v = [x|_{F'F}]$. Therefore, $c \notin \mathscr{C}_y$ and thus $y \in Y_{\mathscr{C}_Y}$. □

Definition 9.4.16. Let $H \leq G$ be a subgroup of G. Given a subshift $X \subset A^G$ the H-projective subdynamics of X is the subshift $\pi_H(X) \subset A^H$ defined as:

$$\pi_H(X) = \{x \in A^H \mid \exists y \in X, \forall h \in H, x_h = y_h\}$$

Proposition 9.4.17. *Let G be a recursively presented group and $H \leq G$ a finitely generated subgroup of G. If $X \subset A^G$ is effectively closed, then its H-projective subdynamics $\pi_H(X)$ is effectively closed.*

Proof. As H is finitely generated, there exists a finite set $\mathsf{S}' \subset H$ such that $\langle \mathsf{S}' \rangle = H$. As G is finitely generated by S, there exists a function $\gamma : \mathsf{S}' \to \mathsf{S}^*$ such that $s' =_G \gamma(s')$ (i.e., every element of S' can be written as a word in S^*). Extend the function γ to act by concatenation over words in S'^*.

As G is recursively presented, by Proposition 9.4.14, the set of pattern codings $\mathscr{C}_G$ defining X can be chosen to be maximal. Let $c = (w_i, a_i)_{i\in I}$ a pattern coding where $w_i \in \mathsf{S}'^*$ and define $\gamma(c) := (\gamma(w_i), a_i)_{i\in I}$. Let $\mathscr{M}$ be the Turing machine which on entry c runs the algorithm recognizing $\mathscr{C}_G$ on entry $\gamma(c)$ and accepts if and only if this machine accepts. Clearly $\mathscr{C}_H = \{c \mid \mathscr{M} \text{ accepts } c\}$ is recursively enumerable. Also, as $\mathscr{C}_G$ is a maximal set of pattern codings, then $c \in \mathscr{C}_H \iff \bigcap_{(w,a)\in\gamma(c)}[a]_w \cap X = \emptyset$. Therefore $\pi_H(X) = X_{\mathscr{C}_H}$. □

9.4.2.2 The Case of Non-recursively Presented Groups

In order to prove some of the properties of effectively closed subshifts, we have used the hypothesis that the group is recursively presented. What could go wrong if this is not the case? We aim to throw some light into this question.

The main problem we encounter is the failure of Proposition 9.4.14. Indeed, it even fails for the simplest example. The full shift is always effectively closed as the empty set of pattern codings defines it. Nevertheless if the alphabet A contains at least two symbols, then a maximal set of forbidden pattern codings contains in particular the coding $\{(\varepsilon, a), (w, b)\}$ for $b \neq a$ if and only if $w \in \texttt{WP}(G)$. If this set were recursively enumerable, one could use it to enumerate $\texttt{WP}(G)$, and thus G would be recursively presented.

Consider the case of a sofic subshift Y in a non-recursively presented group G. Any SFT extension $\sigma : X \twoheadrightarrow Y$ can be represented by a finite set of pattern codings for X and some fixed coding of the finite set F which defines the block code $\Phi : A_X^F \to A_Y$ which determines σ. A recursively enumerable set of pattern codings for Y would consist of a list of patterns –which in particular– do not appear in Y. This means that every pattern obtained by taking the preimage under Φ does not appear in X. Nevertheless, the previous argument implies that we cannot test this algorithmically in general. This is the reason why the proof of Proposition 9.4.15 does not extend to this case.

9.4.2.3 The One-or-Less Subshift $X_{\leq 1}$

Consider the subshift $X_{\leq 1} \subset \{0,1\}^G$ whose configurations contain at most one appearance of the letter 1.

$$X_{\leq 1} = \{x \in \{0,1\}^G \mid 1 \in \{x_g, x_h\} \Rightarrow g = h\}$$

We are going to show that even in the case of recursively presented groups this subshift can fail to be effectively closed.

Proposition 9.4.18. *If G is infinite, then $X_{\leq 1}$ is not an SFT.*

Proof. Suppose $X_{\leq 1} = X_{\mathscr{F}}$ for a finite $\mathscr{F}$ and let $F = \bigcup_{p\in\mathscr{F}} \operatorname{supp}(p)$, $U = \bigcup_{h\in F^{-1}} hF$ and note that $|U| < \infty$. As G is infinite, there exists $g \in G \setminus U$. Consider the configuration $x \in \{0,1\}^G$ which takes the value 1 in $\{1_G, g\}$ and 0 elsewhere. Clearly $x \notin [p]_h$ for every $h \in G$ and $p \in \mathscr{F}$ otherwise $\{1_G, g\} \subset hF$ implying that $hF \subset U$ and thus $g \in U$. Therefore $x \in X_{\mathscr{F}}$ but $x \notin X_{\leq 1}$. □

Proposition 9.4.19. *Let G be a recursively presented group. Then $X_{\leq 1}$ if effectively closed if and only if the word problem of G is decidable.*

Proof. If $\texttt{WP}(G)$ is decidable then $X_{\leq 1}$ is effectively closed. Indeed, an algorithm recognizing a maximal set of pattern codings $\mathscr{C}$ such that $X_{\leq 1} = X_{\mathscr{C}}$ is the following: on input c it considers every pair $(w_1, 1), (w_2, 1)$ in c and accept if and only if $w_1 w_2^{-1} \neq_G 1_G$ for a pair. Conversely, as G is recursively presented, the word problem is already recursively enumerable. It suffices to show it is co-recursively enumerable.

By Proposition 9.4.14 there exists a maximal set of forbidden pattern codings $\mathscr{C}$ with $X_{\leq 1} = X_{\mathscr{C}}$. Given $w \in S^*$, consider the pattern coding $c_w = \{(\epsilon, 1), (w, 1)\}$. Note that $w \neq_G 1_G \iff c_w \in \mathscr{C}$. Therefore the algorithm, which on entry $w \in S^*$ runs the algorithm recognizing $\mathscr{C}$ on entry c_w and accepts if and only if this one accepts, recognizes $S^* \setminus \texttt{WP}(G)$. Hence $\texttt{WP}(G)$ is co-recursively enumerable. □

Using Proposition 9.4.15 we obtain the following corollary.

Corollary 9.4.20. *If G is recursively presented and* $\texttt{WP}(G)$ *is undecidable, then* $X_{\leq 1}$ *is not a sofic subshift.*

9.4.3 Two Larger Notions of Effectiveness

As stated in Section 9.4.2, the classical notion of effectively closed $\mathbb{Z}^d$-subshifts extended to finitely generated groups fails to be completely satisfying for several reasons. First, the notion of effectively closed subshift is not directly related to the group G itself: we use pattern codings, which is in some sense a way to come back to dimension 1. Another way to formulate this reservation is that, unlike the case of $\mathbb{Z}$ where effectively closeness comes with a natural computational model – classical Turing machines – there is no similar correspondence for effectively closed subshifts on a finitely generated group G. And second, very simple subshifts like the one-or-less subshift $X_{\leq 1}$ are not effectively closed for recursively presented groups with undecidable word problem (Proposition 9.4.19): it would be natural to ask that this subshift is always effective, independently of the complexity of the word problem of the group.

In order to escape from these limitations, we study two different extensions of the notion of effectiveness which cover a larger countable class of subshifts. These notions are G-effectiveness and enumeration effectiveness.

The notion of G-effectiveness tackles the problems linked to the word problem of the group by adding the language $\texttt{WP}(G)$ as an oracle. The advantage of this class, besides repairing the previous problems related to the word problem of the group, is that the set of subshifts defined by it can be given a natural definition using modified Turing machines which use the group as a tape. This characterization is interesting in the sense that it allows explicit constructions to be made with the help of these machines. Some examples can be found in [23].

Then we show the limitations of the notion of G-effectiveness and propose an alternative definition of effectiveness for G-subshifts called enumeration effectiveness. This new notion, which is weaker than G-effectiveness, allows us to generalize Proposition 9.4.19 to any group – the recursive presentation hypothesis is no longer needed.

9.4.3.1 *G*-Effectiveness

Definition 9.4.21. A subshift $X \subset A^G$ is G-effectively closed if there is a set of pattern codings $\mathscr{C}$ such that $X = X_{\mathscr{C}}$, and $\mathscr{C}$ is recursively enumerable with oracle $\texttt{WP}(G)$.

Following from the results of this section, one can directly use the definition above to show that the following properties hold for any finitely generated group G.

1. If X is a G-effectively closed subshift, then a maximal set of pattern codings $\mathscr{C}$ such that $X = X_{\mathscr{C}}$ is recursively enumerable with oracle $\texttt{WP}(G)$.
2. The class of G-effectively closed subshifts is closed under finite intersections and unions.
3. The class of G-effectively closed subshifts is closed under factors.
4. Being G-effectively closed is a conjugacy invariant.
5. The class of G-effectively closed subshifts contains all sofic subshifts.
6. The class of G-effectively closed subshifts contains all effectively closed subshifts.
7. If $\texttt{WP}(G)$ is decidable, then every G-effectively closed subshift is effectively closed.
8. $X_{\leq 1}$ is a G-effectively closed subshift.

Nevertheless, this class fails to be stable under projective subdynamics.

Proposition 9.4.22. *Let G be a group which is not recursively presented. There exists a $(G \times \mathbb{Z})$-effectively closed subshift $X \subset A^{G\times\mathbb{Z}}$ such that its $\mathbb{Z}$-projective subdynamics is not $\mathbb{Z}$-effectively closed.*

Proof. Let $A = S \cup \{\star\}$. For $w \in S^*$, let p_w defined over the support $\{1_G\} \times \{0, \dots |w|+1\}$ such that $(p_w)_{(1_G,0)} = (p_w)_{(1_G,|w|+1)} = \star$ and for $j \in \{1, \dots, |w|\}$ then $(p_w)_{(1_G,j)} = w_j$. Let $X := X_{\mathscr{F}} \subset A^{G\times\mathbb{Z}}$ be defined by the set of forbidden patterns $\mathscr{F} = \{p_w \mid w \in \texttt{WP}(G)\}$. Clearly X is $(G \times \mathbb{Z})$-effectively closed. Every $\mathbb{Z}$-coset of a configuration $x \in X$ contains a bi-infinite sequence $y \in A^{\mathbb{Z}}$ such that either y contains at most one symbol $\star$ or every word appearing between two appearances of $\star$ represents 1_G in G.

We claim that $\pi_{\mathbb{Z}}(X)$ is not effectively closed. If it were, there would exist a maximal set of forbidden pattern codings which is recursively enumerable and defines $\pi_{\mathbb{Z}}(X)$. Therefore given $w \in S^*$ a machine could run the algorithm for the word $\star w \star$, and it would be accepted if and only if $w =_G 1_G$. This would imply that G is recursively presented. □

Although Proposition 9.4.22 is a theoretical drawback for the notion of G-effectiveness, it is noteworthy that this behavior only happens when the projective subdynamics is taken with respect to a group with strictly weaker word problem. If the projective subdynamics of a $(G \times \mathbb{Z})$-effectively closed subshift is taken with respect to G, the resulting subshift is indeed G-effectively closed.

The interest of this class is mainly due to the fact that they can be defined in a natural way using modified Turing machines. These objects which we call *G-machines* replace the bi-infinite tape by a Cayley graph $\Gamma(G, \mathsf{S})$ of the finitely generated group G.

Definition 9.4.23. A G-machine is a 6-tuple $(Q, \Sigma, \sqcup, q_0, Q_F, \delta)$ where Q is a finite set of states, Σ is a finite alphabet, $\sqcup \in \Sigma$ is the blank symbol, $q_0 \in Q$ is the initial state, $Q_F \subset Q$ is the set of accepting states, and $\delta : \Sigma \times Q \to \Sigma \times Q \times \mathsf{S}$ is the transition function.

G-machine T acts on the set $\Sigma^G \times Q$ as follows: let $(x, q) \in \Sigma^G \times Q$ and $\delta(x_{1_G}, q) = (a, q', s)$. Then $T(x, q) = (S_{s^{-1}}(\widetilde{x}), q')$ where $\widetilde{x}|_{1_G} = a$ and $\widetilde{x}|_{G\setminus\{1_G\}} = x|_{G\setminus\{1_G\}}$. Figure 9.8 illustrates this action when G is a free group. Here the head of the Turing machine is assumed to stay at a fixed position and the tape moves instead.

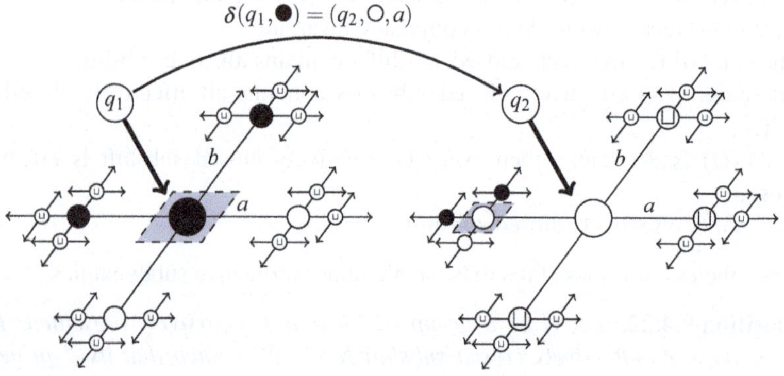

Fig. 9.8 A transition of a G-machine for the free group on two elements.

Let $F \subset G$ be a finite set and $p \in \Sigma^F$. Let $x^p \in \Sigma^G$ be the configuration such that $(x^p)|_F = p$ and $(x^p)|_{G\setminus F} \equiv \sqcup$. We say that T *accepts* p if there is $n \in \mathbb{N}$ such that $T^n(x^p, q_0) \in \Sigma^G \times Q_F$. $L \subset L(A^G)$ is *G-recursively enumerable* if there exists a G-machine T which accepts $p \in \Sigma_G^*$ if and only if $p \in L$. If both L and $L(A^G) \setminus L$ are G-recursively enumerable, we say L is *G-decidable*.

In [23] it is shown that these machines characterize G-effective subshifts. We say a set of patterns $L \subset L(A^G)$ is *closed by extensions* if for each $p_1, p_2 \in L(A^G)$ such that $p_1 \sqsubset p_2$ then $p_1 \in L \implies p_2 \in L$. Also, for a set of pattern codings $\mathscr{C}$, we denote by $\mathsf{p}(\mathscr{C})$ the set of patterns they define in the group G. Formally:

$$\mathsf{p}(\mathscr{C}) = \{p \in L(A^G) \mid \exists c \in \mathscr{C}, [p] = \bigcap_{(w,a)\in c} [a]_w\}$$

Theorem 9.4.24. *Let G be a finitely generated infinite group and $L \subset L(A^G)$ be a set of patterns. If L is G-recursively enumerable, then there exists a recursively enumerable with oracle $\mathtt{WP}(G)$ set of pattern codings $\mathscr{C}$ such that $L = \mathsf{p}(\mathscr{C})$. Conversely, if $\mathscr{C}$ a recursively enumerable with oracle $\mathtt{WP}(G)$ set of pattern codings. If $\mathsf{p}(\mathscr{C})$ is closed by extensions, then $\mathsf{p}(\mathscr{C})$ is G-recursively enumerable.*

Using the fact that the maximal set of forbidden patterns of a subshift is closed by extension, we obtain what follows.

Corollary 9.4.25. *A subshift $X \subset A^G$ is G-effectively closed if and only if there exists a G-recursively enumerable set $\mathscr{F} \subset L(A^F)$ such that $X = X_{\mathscr{F}}$.*

A big drawback of the class of G-effectively closed subshifts is that in general they do not admit a simulation theorem in the sense of Theorem 9.4.10. Specifically, $X_{\leq 1}$ is G-effective for each group, but it cannot be obtained as the projective subdynamics of a sofic subshift if G is finitely generated, recursively presented group but has undecidable word problem.

Proposition 9.4.26. *Let G be a finitely generated, recursively presented group with undecidable word problem and H a finitely generated group such that $\mathtt{WP}(H) \leq_m \mathtt{WP}(G)$. Then $X_{\leq 1}$ cannot be obtained as the G-projective subdynamics of a sofic $G \times H$-subshift.*

Proof. As G is recursively presented and $\mathtt{WP}(H) \leq_m \mathtt{WP}(G)$, then H is also recursively presented, and thus $G \times H$ is recursively presented. Applying Proposition 9.4.15 we get that sofic $G \times H$-subshifts are effectively closed. Hence, using Proposition 9.4.17 the G-projective subdynamics must also be effectively closed. We conclude as per the fact that $X_{\leq 1}$ is not effectively closed for recursively presented groups with undecidable word problem.

In particular, this means that if G is a finitely generated, recursively presented group with undecidable word problem, then there is no general simulation theorem for G-effective subshifts coming from sofic subshifts on $G \times \mathbb{Z}^d$ or even $G \times \cdots \times G$. It might therefore be interesting to weaken the notion of G-effectiveness, so that even if $X_{\leq 1}$ is no longer always in the class, a general simulation theorem is not proscribed.

9.4.3.2 Enumeration Effectiveness

If A and B are two sets of words in S, we say that A is *enumeration reducible* to B, denoted $A \leq_e B$, if there exists an algorithm that produces an enumeration of A from any enumeration of B. Formally, $A \leq_e B$ if there exists a partial computable function f that associates to each $\langle n, i \rangle$ a finite set $D_{n,i}$ such that $n \in A \iff \exists i, D_{n,i} \subseteq B$.

Remark 9.4.27. A set A is recursively enumerable if and only if $A \leq_e \emptyset$. Particularly, if A is recursively enumerable then $A \leq_e B$ for any set B.

From this remark, we get a characterization of effectively closed subshifts as those for which there exists a set of pattern codings $\mathscr{C}$ such that $X = X_{\mathscr{C}}$ and $\mathscr{C} \leq_e \emptyset$.

We will use a characterization of enumeration reducibility that can be found in [454, Exercise XIV.1.2] and in [535].

Proposition 9.4.28. *$A \leq_e B$ if and only if for every set C, if B is recursively enumerable with oracle C, then A is also recursively enumerable with oracle C.*

We can translate the notion of G-effectiveness in term of enumeration reducibility. If we denote $B \oplus C$ the set $\{(0,x) \mid x \in B\} \cup \{(1,x) \mid x \in C\}$, we get that A is recursively enumerable with oracle B if and only if $A \leq_e B \oplus \overline{B}$. As written in Definition 9.4.21, a subshift $X \subset A^G$ is G-effectively closed if there is a set of pattern codings $\mathscr{C}$ such that $X = X_{\mathscr{C}}$, and $\mathscr{C}$ is recursively enumerable with oracle $\texttt{WP}(G)$. We thus get the following characterization of G-effectively closed subshifts.

Proposition 9.4.29. *A G-subshift $X \subset A^G$ is G-effectively closed if there is a set of pattern codings $\mathscr{C}$ such that $X = X_{\mathscr{C}}$, and such that $\mathscr{C} \leq_e \texttt{WP}(G) \oplus \overline{\texttt{WP}(G)}$.*

Thus with G-effectiveness, forbidden patterns are produced from two enumerations: one enumeration of the words in S^* that represent the identity of the group (the word problem $\texttt{WP}(G)$), and one enumeration of the word in S^* that do not represent the identity of the group (the complement of the word problem $\overline{\texttt{WP}(G)}$). But are those two enumerations strictly necessary to produce forbidden patterns? In order to check whether a pattern coding is inconsistent, it suffices to check all pairs of words w, w', and if at some point two words happen to represent the same group element, check whether they are assigned different symbols. Only the enumeration of the word problem is needed for that. The notion of enumeration effectiveness is based on this observation.

Definition 9.4.30. A G-subshift $X \subset A^G$ is *G-enumeration effective* if there exists a set of pattern codings $\mathscr{C}$ such that $X = X_{\mathscr{C}}$ and $\mathscr{C} \leq_e \texttt{WP}(G)$.

We first compare the class of G-enumeration effective subshifts with the classes of effectively closed subshifts and G-effectively closed subshifts. From the characterizations of these classes with enumeration reduction, it follows immediately that G-enumeration effective subshifts are in between the two others. The diagram on Figure 9.9 summarizes the propositions listed below.

Proposition 9.4.31. *Let G be a finitely generated group and X an effectively closed subshift. Then X is G-enumeration effective.*

Proof. Since X is G-effectively closed, there exists a set of pattern codings $\mathscr{C}$ such that $X = X_{\mathscr{C}}$, and $\mathscr{C} \leq_e \emptyset$. A fortiori, we get that $\mathscr{C} \leq_e \texttt{WP}(G)$, thus X is G-enumeration effective. □

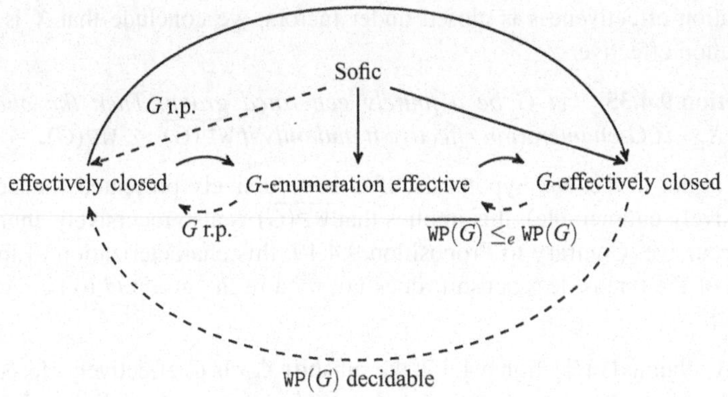

Fig. 9.9 Inclusion relations between different classes of G-subshifts for a finitely generated group G. Inclusion represented by a dashed arrow only holds for groups having the property labeling the arrow.

Proposition 9.4.32. *Let G be a finitely generated group and X a G-enumeration effective subshift. Then X is G-effectively closed.*

Proof. Since X is G-enumeration effective, there exists a set of pattern codings $\mathscr{C}$ such that $X = X_{\mathscr{C}}$ and $\mathscr{C} \leq_e \texttt{WP}(G)$. A fortiori, we get that $\mathscr{C} \leq_e \texttt{WP}(G) \oplus \overline{\texttt{WP}(G)}$; thus X is G-effectively closed. □

Conversely, if $\overline{\texttt{WP}(G)}$ is enumeration reducible to $\texttt{WP}(G)$, any set enumeration reducible to $\texttt{WP}(G)$ is also enumeration reducible to $\texttt{WP}(G) \oplus \overline{\texttt{WP}(G)}$, from which we deduce that groups with the property are exactly groups where G-enumeration effectiveness and G-effectiveness coincide.

Proposition 9.4.33. *Let G be a finitely generated such that $\overline{\texttt{WP}(G)} \leq_e \texttt{WP}(G)$ and X a G-effectively closed subshift. Then X is G-enumeration effective.*

Proposition 9.4.34. *G-enumeration effectiveness is closed under factors. In particular, if X is a sofic subshift, then it is G-enumeration effective.*

Proof. Recall that a finitely generated group is recursively presented if and only if $\texttt{WP}(G)$ is recursively enumerable. Let $\mathscr{C}$ be the list of forbidden pattern codings obtained in the proof of Proposition 9.4.15 (with an oracle to $\texttt{WP}(G)$ in this case). We have shown that $\mathscr{C}$ is recursively enumerable if $\texttt{WP}(G)$ is recursively enumerable.

Let C be an arbitrary language such that $\texttt{WP}(G)$ is recursively enumerable with oracle C, then obviously $\mathscr{C}$ is recursively enumerable with oracle C. By Proposition 9.4.28 this implies that $\mathscr{C} \leq_e \texttt{WP}(G)$. This shows that G-enumeration effectiveness is closed under factors.

Let Y be an SFT extension of X. Clearly Y is always effectively closed: take a finite list of forbidden patterns defining it hard code each pattern into a pattern coding. By Proposition 9.4.31 we have that Y is G-enumeration effective. As G-

enumeration effectiveness is closed under factors, we conclude that X is also G-enumeration effective. □

Proposition 9.4.35. *Let G be a finitely generated group. Then the one-or-less subshift $X_{\leq 1}$ is G-enumeration effective if and only if* $\overline{\mathtt{WP}(G)} \leq_e \mathtt{WP}(G)$.

This is not a vacuous hypothesis: if G is recursively presented (hence $\mathtt{WP}(G)$ is recursively enumerable), this implies that $\overline{\mathtt{WP}(G)}$ is also recursively enumerable, hence recursive. Contrary to Proposition 9.4.19, this characterization of the effectiveness of the one-or-less subshift does not require the group G to be recursively presented.

Proof. As claimed in Section 9.4.3.2, the subshift $X_{\leq 1}$ is G-effectively closed, which means that there is a set of pattern codings $\mathscr{C}$ such that $X = X_{\mathscr{C}}$, and such that $\mathscr{C} \leq_e \mathtt{WP}(G) \oplus \overline{\mathtt{WP}(G)}$ by Proposition 9.4.29. Suppose that $\overline{\mathtt{WP}(G)} \leq_e \mathtt{WP}(G)$. Let C be a set such that $\mathtt{WP}(G)$ is recursively enumerable with oracle C and denote $\mathscr{M}$ the Turing machine with oracle C that recognizes $\mathtt{WP}(G)$. We construct a new Turing machine $\mathscr{M}'$ with oracle C with the following behavior. Given a pattern coding c, $\mathscr{M}'$ in parallel simulates $\mathscr{M}_e$ to enumerate the set $\{0\} \times \mathtt{WP}(G)$ and recursively enumerates $\{1\} \times \overline{\mathtt{WP}(G)}$ from the latter enumeration. From this enumeration of $\mathtt{WP}(G) \oplus \overline{\mathtt{WP}(G)}$, the machine $\mathscr{M}'$ then produces an enumeration of $\mathscr{C}$. Thus $\mathscr{C} \leq_e \mathtt{WP}(G)$.

Reciprocally, first note that in an analogous way to Proposition 9.4.14, we can suppose that any G-enumeration effective subshift is given by a maximal set of forbidden pattern codings. Let $X_{\leq 1}$ be G-enumeration effective and $\mathscr{C} \leq_e \mathtt{WP}(G)$ be the maximal set of pattern codings defining $X_{\leq 1}$. As in the proof of Proposition 9.4.19, we can define $f : S^* \to \mathscr{C} \cup \overline{\mathscr{C}}$ where $f(w) = \{(\epsilon, 1), (w, 1)\}$. Clearly if $w \in \overline{\mathtt{WP}(G)}$ if and only if $f(w) \in \mathscr{C}$. Therefore $\overline{\mathtt{WP}(G)} \leq_m \mathscr{C} \leq_e \mathtt{WP}(G)$ which implies $\overline{\mathtt{WP}(G)} \leq_e \mathtt{WP}(G)$. □

9.4.3.3 Towards a Simulation Theorem

In the case of $\mathbb{Z}^d$-subshifts, the notion of effectively closed subshift is quite natural for two reasons. First it extends the notion of sofic $\mathbb{Z}$-subshift from the point of view of pattern exclusion: a sofic $\mathbb{Z}$-subshift has a regular language, while an effectively closed $\mathbb{Z}$-subshift has a recursively enumerable language. Hence sofic subshifts are effectively closed. Second the class of effectively closed subshifts is stable under projective subdynamics (see Section 9.4.1). A reasonable generalization of effectiveness to finitely generated groups should at least satisfy these two properties, with the hope that a simulation theorem may hold.

Clearly, the notion of effectiveness for a finitely generated group G given in Definition 9.4.12 is too restrictive: sofic subshifts are effectively closed for recursively presented groups (Proposition 9.4.15), but we do not know what happens for non-recursively presented groups. Moreover, this notion does not behave well with projective subdynamics (see Propositions 9.4.22). By Propositions 9.4.34 and 9.4.32, sofic subshifts are always G-enumeration effective, and thus G-effectively closed. Thus among the three definitions presented in this chapter, only G-effectiveness and G-enumeration effectiveness are likely to fulfill our requirements. Another argument that may be taken into account to choose between these two notion is the one-or-less subshift $X_{\leq 1}$. On the one hand, this subshift is always G-effectively closed (Section 9.3.4.1); on the other hand, it is G-enumeration effective only for finitely generated groups satisfying that $\overline{\texttt{WP}(G)} \leq_e \texttt{WP}(G)$.

What would be a general statement of a simulation theorem for a finitely generated group G? The two notions of G-enumeration effectiveness and G-effectiveness both depend on the group G. Apart from torsion groups and with no additional restriction, the operation of projective subdynamics may transform a G-subshift on a $\mathbb{Z}$-subshift, where the three notions of effectiveness coincide. So we should consider only projective subdynamics from a group to a subgroup with the same complexity of $\texttt{WP}$. The simulation theorems from [37, 294] suggest that adding the group $\mathbb{Z}^2$ to the group G – what is meant by *adding* has of course to be precise – makes possible a characterization of projective subdynamics of sofic subshifts on G equipped with $\mathbb{Z}^2$ as effective G-subshifts. In Section 9.3.4.4, the result of [538] is used to extract a grid structure from a direct product $G_1 \times G_2$ where G_1 and G_2 are both infinite. Hence the idea to study projective subdynamics of $G{\times}G{\times}G$-effectively closed subshifts. Unfortunately, Proposition 9.4.26 tells us that for recursively presented groups G with undecidable $\texttt{WP}$, some simple G-effectively closed subshifts cannot be realized as the projective subdynamics of a sofic $G \times H$-subshift, where H is any finitely generated group with $\texttt{WP}(H) \leq_m \texttt{WP}(G)$. So a simulation theorem is excluded for G-effectiveness. To the knowledge of the authors, the question remains open for G-enumeration effectiveness.

9.5 Exercises

Exercise 9.5.1. Show that if $L \leq_m L'$, then $L \leq_T L'$ **Hint**: Express $L \leq_m L'$ with a Turing machine with oracle L'.

Exercise 9.5.2. Let $X \subseteq A^{\mathbb{Z}^2}$ be a subshift. Show that

$$\pi(X) = \left\{y \in A^{\mathbb{Z}} \mid \exists x \in X \text{ s.t. } x_{(i,0)} = y_i \text{ for every } i \in \mathbb{Z}\right\}$$

is also a subshift.

Exercise 9.5.3. Prove that the projective subdynamics of the subshift defined in Example 9.4.1 is not sofic. **Hint**: Use the fact that the language $\{a^n b^n \mid n \in \mathbb{N}\}$ is not a regular language.

Exercise 9.5.4. Give an example of a recursively presented group which is not finitely presented.

Exercise 9.5.5. Show that the properties of being an SFT and being a sofic subshift are conjugacy invariants.

Exercise 9.5.6. Show that the domino problem is a group isomorphism invariant, that is, that if G is isomorphic to H then the domino problem of G is many-one equivalent to the domino problem of H.

Exercise 9.5.7. Find an example of a finitely generated group which contains a non-finitely generated subgroup. **Hint**: One possibility is to use Corollary 9.3.38.

Exercise 9.5.8. Let G be a finitely generated group and $H \trianglelefteq G$. Show that the quotient group G/H is also finitely generated.

Exercise 9.5.9. Let $H \leq G$ be a subgroup such that $[G : H] < \infty$. Show that there exists $N \leq H$ such that $N \trianglelefteq G$ and $[G : N] < \infty$. **Hint**: define N as the stabilizer of the action of G over the lateral classes G/H by left multiplication.

Exercise 9.5.10. Let G be a finitely generated group and $H \leq G$ a subgroup of finite index. Show that H is finitely generated. **Hint**: Let $L = \{\ell_1 \dots, \ell_n\}$ be a set of representatives of the left lateral classes containing 1_G. For each generator s of G write $s\ell_i = \ell_{i,s}h_{i,s}$ for some $\ell_{i,s} \in L$ and $h_{i,s} \in H$. Show that the set of all $h_{i,s}$ generates H.

Exercise 9.5.11. Fill in the details from Proposition 9.3.33. In particular, show that the subshift defined by $\mathcal{G}$ is indeed $X_{\mathcal{F}}^{[R]}$.

Exercise 9.5.12. Let G be the subgroup of $GL_2(\mathbb{Q})$ generated by the matrices $a = \begin{pmatrix} 1 & 1 \\ 0 & 1 \end{pmatrix}$ and $b = \begin{pmatrix} 2 & 0 \\ 0 & 1 \end{pmatrix}$. Show that every matrix in G is of the form $\begin{pmatrix} x & y \\ 0 & z \end{pmatrix}$ where x, y, z are dyadic rationals (rationals of the form $p/2^q$) and that all such matrices with $x = z = 1$ belong to G. Deduce that G contains a subgroup that is not finitely generated. Show that G is a Baumslag-Solitar group.

Exercise 9.5.13. Let $G = \mathbb{F}_2 = \langle a, b \rangle$ the free group generated by a and b. Consider the Cayley graph $\Gamma(G, \mathsf{S})$ for $\mathsf{S} = \{a, b, a^{-1}, b^{-1}\}$. Show that this Cayley graph can be covered by disjoints bi-infinite paths (in the sense of Theorems 9.3.48 and 9.3.49).

Exercise 9.5.14. Let $G = PSL_2(\mathbb{Z}) = \langle a, b | a^2, b^3 \rangle$. Consider the Cayley graph $\Gamma(G, \mathsf{S})$ for $\mathsf{S} = \{a, b, a^{-1}, b^{-1}\}$ depicted below.

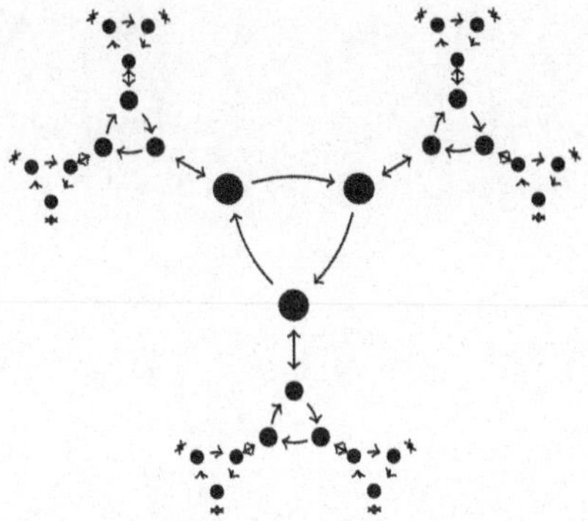

Show that this Cayley graph cannot be covered by disjoints bi-infinite paths (in the sense of Theorems 9.3.48 and 9.3.49). (This example shows that the choice of S is important for this theorem to hold.)

Exercise 9.5.15. Consider again $G = PSL_2(\mathbb{Z}) = \langle a, b | a^2, b^3 \rangle$. Show that $X_{G,\{a,b\}}$ is nonempty.

Exercise 9.5.16. Consider again $G = PSL_2(\mathbb{Z}) = \langle a, b | a^2, b^3 \rangle$. Draw the Cayley graph $\Gamma(G, \mathsf{S})$ for $\mathsf{S} = \{a, ab, a^{-1}, (ab)^{-1}\}$. Show that this Cayley graph can be covered by disjoints bi-infinite paths (in the sense of Theorems 9.3.48 and 9.3.49).

Exercise 9.5.17. A *weak valid pair* for (G, S) is a pair (n, p) from G to S s.t. $\mathsf{p}(gs) = s^{-1}$ where $s = \mathsf{n}(g)$. Give an example for $G = \mathbb{F}_2 = \langle a, b \rangle$ and $S = \{a, b\}$ of a weak valid pair that is not a valid pair.

Exercise 9.5.18. Show that if G is finite, any weak valid pair is a valid pair.

Exercise 9.5.19. Give the proofs of Propositions 9.3.51 and 9.3.52.

Exercise 9.5.20. Give an example of a $\mathbb{Z}$-subshift which is effectively closed but with an undecidable language. **Hint**: Consider the set of forbidden words $\{10^n1\}_{n \in L}$ for an appropriate set L.

Exercise 9.5.21. Show that the class of effectively closed subshifts is closed under finite intersections. Prove that the same result does not hold for countable intersections.

Exercise 9.5.22. Show that for recursively presented groups the class of effectively closed subshifts is closed under finite unions.

Chapter 10
Automaton (Semi)groups: Wang Tilings and Schreier Tries

Ines Klimann and Matthieu Picantin

Abstract Groups and semigroups generated by Mealy automata were formally introduced in the early 1960s. They revealed their full potential over the years, by contributing to important conjectures in group theory. In the current chapter, we intend to provide various combinatorial and dynamical tools to tackle some decision problems all related to some extent to the growth of automaton (semi)groups. In the first part, we consider Wang tilings as a major tool in order to study and understand the behavior of automaton (semi)groups. There are various ways to associate a Wang tileset with a given complete and deterministic Mealy automaton and various ways to interpret the induced Wang tilings. We describe some of these fruitful combinations, as well as some promising research opportunities. In the second part, we detail some toggle switch between a classical notion from group theory—Schreier graphs—and some properties of an automaton group about its growth or the growth of its monogenic subgroups. We focus on polynomial-activity automata and on reversible automata, which are somehow diametrically opposed families.

10.1 Introduction

Groups and semigroups generated by Mealy automata were formally introduced in the early 1960s (for details, see [123, 271] and references therein). They revealed their full potential over the years, by contributing to important conjectures in group theory.

We intend here to provide various combinatorial and dynamical tools to tackle some decision problems all related to some extent to the growth of automaton (semi)groups.

I. Klimann (✉) · M. Picantin
IRIF, UMR 8243, CNRS & Université Paris Diderot - Case 7014, F-750205 Paris Cedex 13, France
e-mail: klimann@irif.fr; picantin@irif.fr

V. Berthé, M. Rigo (eds.), *Sequences, Groups, and Number Theory*, Trends in Mathematics, https://doi.org/10.1007/978-3-319-69152-7_10

In Section 10.2, we consider Wang tilings as a major tool in order to study and understand the behavior of automaton (semi)groups. There are various ways to associate a Wang tileset with a given complete and deterministic Mealy automaton and various ways to interpret the induced Wang tilings. We describe some of these fruitful combinations, as well as some promising research opportunities.

In Section 10.3, we detail some toggle switch between a classical notion from group theory—Schreier graphs—and some properties of an automaton group about its growth or the growth of its monogenic subgroups. We focus on polynomial-activity automata and on reversible automata, which are somehow diametrically opposed families.

10.1.1 Mealy Automata

We first recall the formal definition of an automaton. A *(finite, deterministic, and complete) automaton* is a triple $\big(Q, \Sigma, \delta = (\delta_i\colon Q \to Q)_{i\in\Sigma}\big)$, where the *state set* Q and the *alphabet* Σ are nonempty finite sets and the δ_i are functions.

A *Mealy automaton* is a quadruple $(Q, \Sigma, \delta, \rho)$ such that (Q, Σ, δ) and (Σ, Q, ρ) are both automata. In other terms, a Mealy automaton is a complete, deterministic, letter-to-letter transducer with the same input and output alphabet. Its *size* is the cardinal of its state set.

The graphical representation of a Mealy automaton is standard; see Figures 10.1 and 10.2.

A Mealy automaton $\mathscr{A} = (Q, \Sigma, \delta, \rho)$ is *invertible* if each function ρ_x is a permutation of Σ and *reversible* if each function δ_i is a permutation of Q.

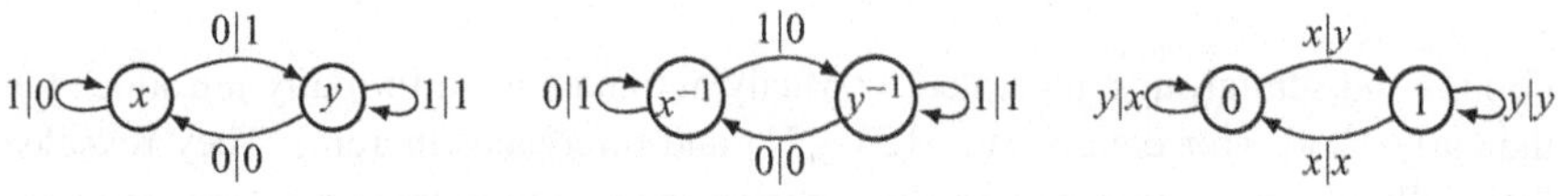

Fig. 10.1 The lamplighter automaton $\mathscr{L}$, its inverse automaton $\mathscr{L}^{-1}$, and its dual automaton $\mathfrak{d}\mathscr{L}$.

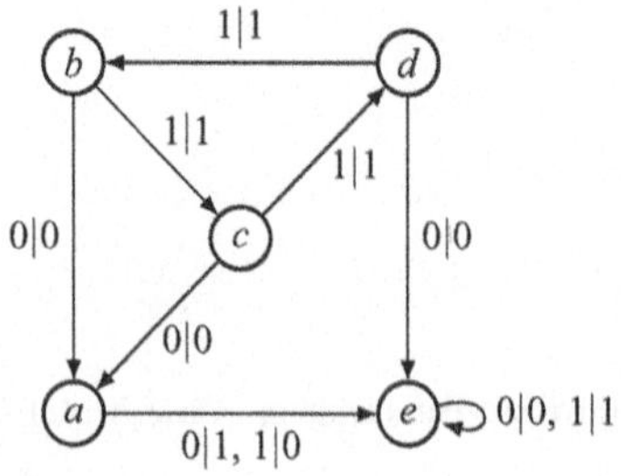

Fig. 10.2 The Grigorchuk automaton.

In the case where $\mathcal{A}$ is invertible, there is an explicit way to express the actions of the inverses functions by considering the *inverse automaton* $\mathcal{A}^{-1}$ having $Q^{-1} = \{x^{-1}, x \in Q\}$ as state set and a transition $x^{-1} \xrightarrow{j|i} y^{-1}$ whenever $x \xrightarrow{i|j} y$ is a transition in $\mathcal{A}$ (see Figure 10.1). In the case where $\mathcal{A}$ is reversible, its connected components are strongly connected.

In any Mealy automaton $\mathcal{A} = (Q, \Sigma, \delta, \rho)$, the sets Q and Σ play dual roles. So we may consider the *dual (Mealy) automaton* defined by $\mathfrak{d}\mathcal{A} = (\Sigma, Q, \rho, \delta)$, where we have the transition $i \xrightarrow{x|y} j$ whenever $x \xrightarrow{i|j} y$ is a transition in $\mathcal{A}$ (see Figure 10.1). Obviously, a Mealy automaton is reversible if and only if its dual is invertible.

An invertible Mealy automaton is *bireversible* if it is reversible (i.e., the input letters of the transitions act like permutations on the state set) and if so is its inverse (i.e., the output letters of the transitions act like permutations on the state set).

Whenever $\mathcal{A}$ is an invertible-reversible Mealy automaton, we can consider the letters and their inverses. By setting $\mathcal{A}' = \mathfrak{d}(\mathfrak{d}\mathcal{A} \sqcup (\mathfrak{d}\mathcal{A})^{-1})$, the (invertible-reversible) Mealy automaton $\widetilde{\mathcal{A}} = \mathcal{A}' \sqcup (\mathcal{A}')^{-1}$ is the extension of $\mathcal{A}$ with state set $Q \sqcup Q^{-1}$ and alphabet $\Sigma \sqcup \Sigma^{-1}$.

For any set Σ, we let Σ^+ denote the free semigroup over Σ (*resp.* Σ^* for the free monoid with unit $\mathtt{1}$) and call its elements Σ-*words*. We write $|w|$ for the length of a Σ-word w and ww' for the product of two Σ-words w and w'.

A state of a Mealy automaton can be seen as acting on the set Σ^* of finite words (equivalently on a regular rooted tree of arity $|\Sigma|$) or on the set Σ^ω of infinite words (equivalently on the boundary of the former tree).

10.1.2 Minimization and Nerode Classes

Let $\mathcal{A} = (Q, \Sigma, \delta, \rho)$ be a Mealy automaton.

The *Nerode equivalence* $\equiv$ *on* Q is the limit of the sequence of increasingly finer equivalences $(\equiv_k)_k$ recursively defined by:

$$\forall x, y \in Q, \qquad x \equiv_0 y \iff \rho_x = \rho_y ,$$
$$\forall k \geqslant 0,\ x \equiv_{k+1} y \iff \big(x \equiv_k y \quad \wedge \quad \forall i \in \Sigma,\ \delta_i(x) \equiv_k \delta_i(y)\big) .$$

Since the set Q is finite, this sequence is ultimately constant. For every element x in Q, we let $[x]$ denote the class of x *w.r.t.* the Nerode equivalence, called the *Nerode class* of x. Extending to the n-th power of $\mathcal{A}$, we let $[x]$ denote the Nerode class in Q^n of $x \in Q^n$.

Remark 10.1.1. The Nerode classes of a connected reversible Mealy automaton (i.e., a Mealy automaton with exactly one connected component) have the same cardinality.

The *minimization* of $\mathscr{A}$ is the Mealy automaton $\mathfrak{m}(\mathscr{A}) = (Q/\equiv, \Sigma, \tilde{\delta}, \tilde{\rho})$, where for every (x, i) in $Q \times \Sigma$, $\tilde{\delta}_i([x]) = [\delta_i(x)]$ and $\tilde{\rho}_{[x]} = \rho_x$. This definition is consistent with the standard minimization of "deterministic finite automata" where instead of considering the mappings $(\rho_x : \Sigma \to \Sigma)_x$, the computation is initiated by the separation between terminal and nonterminal states.

Two Mealy automata are *equivalent* if their minimizations are isomorphic as labeled graphs. A Mealy automaton is *minimal* if it is equivalent to its minimization.

A pair of dual Mealy automata is *reduced* if both automata are minimal. The $\mathfrak{md}$-*reduction* of a Mealy automaton, introduced in [9], consists in minimizing the automaton or its dual until the resulting pair of dual Mealy automata is reduced. It is well-defined: if both a Mealy automaton and its dual automaton are non-minimal, the reduction is confluent.

10.1.3 Automaton (Semi)groups

Let $\mathscr{A} = (Q, \Sigma, \delta, \rho)$ be a Mealy automaton. For each transition $x \xrightarrow{i|\rho_x(i)} \delta_i(x)$, we associate the *cross-transition* depicted in the following way:

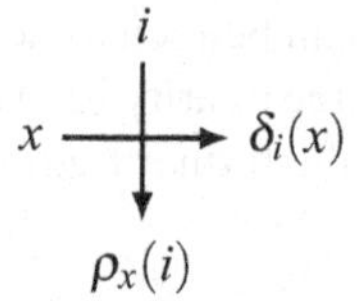

Each state $x \in Q$ defines a mapping from Σ^* into itself recursively defined by:

$$\forall i \in \Sigma,\ \forall s \in \Sigma^*, \qquad \rho_x(is) = \rho_x(i)\rho_{\delta_i(x)}(s)$$

that can also be depicted by a so-called *cross-diagram* obtained by gluing cross-transitions (see [9, 246]) representing the action of a word of states on a word of letters (or vice versa):

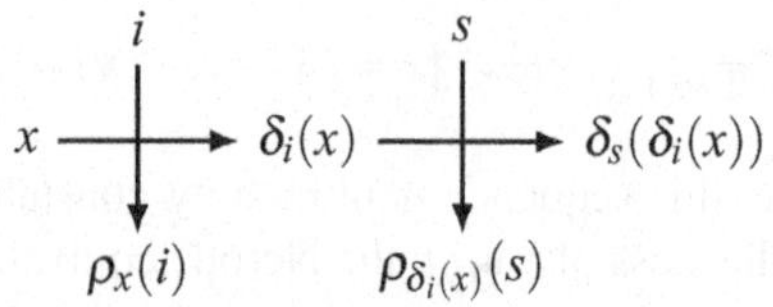

In a dual way, each letter $i \in \Sigma$ defines also an action on Q^*.

Both actions naturally extend to words, respectively, in Q^* and Σ^* with the following convention. The image of the empty word is itself. We define the function ρ_x *induced* by $x = x_1 \cdots x_n \in Q^n$, with $n > 0$, by setting, $\rho_x \colon \Sigma^* \to$

$\Sigma^*, \rho_x = \rho_{x_n} \circ \cdots \circ \rho_{x_1}$. We let $\delta_i : Q^* \to Q^*, i \in \Sigma$ denote dually the functions induced by the states of $\mathfrak{d}\mathscr{A}$. For $s = s_1 \cdots s_n \in \Sigma^n$ with $n > 0$, set $\delta_s : Q^* \to Q^*$, $\delta_s = \delta_{s_n} \circ \cdots \circ \delta_{s_1}$.

Alternatively, we consider the powers of $\mathscr{A}$: for $n > 0$, its *n-th power* $\mathscr{A}^n$ is the Mealy automaton

$$\mathscr{A}^n = \left(Q^n, \Sigma, (\delta_i : Q^n \to Q^n)_{i \in \Sigma}, (\rho_x : \Sigma \to \Sigma)_{x \in Q^n} \right) .$$

By convention, $\mathscr{A}^0$ is the trivial automaton on the alphabet Σ. Note that all the powers of a reversible Mealy automaton are reversible as well.

The semigroup of mappings from Σ^* to Σ^* generated by $\{\rho_x, x \in Q\}$ is called the *semigroup generated by* $\mathscr{A}$ and is denoted by $\langle \mathscr{A} \rangle_+$. When $\mathscr{A}$ is invertible, the functions induced by its states are permutations on words of the same length, and thus we may consider the group of mappings from Σ^* to Σ^* generated by $\{\rho_x, x \in Q\}$. This group is called the *group generated by* $\mathscr{A}$ and is denoted by $\langle \mathscr{A} \rangle$. Such a group is self-similar in the sense that for any element g of the group and any word $w \in \Sigma^*$, the unique mapping $g_{|w} : \Sigma^* \to \Sigma^*$ defined by

$$\forall u \in \Sigma^*, \; g(wu) = g(w) g_{|w}(u)$$

belongs to the group. See also Definition 11.2.15.

Two states of a Mealy automaton belong to the same Nerode class if and only if they represent the same element in the generated (semi)group, i.e., if and only if they induce the same action on Σ^*.

Remark 10.1.2. If two words of Q^* are equivalent, so are their images under the action of any element of $\langle \mathfrak{d}\mathscr{A} \rangle_+$.

Remark 10.1.3. If a state of a Mealy automaton induces the identity, so do all the states reachable from it. In particular, in a reversible connected component of a Mealy automaton, a state induces the identity if and only so do all of its states.

Remark 10.1.4. Let $\mathscr{A}$ and $\mathscr{B}$ be two reversible connected Mealy automata on the same alphabet Σ, and let x be some state of $\mathscr{A}$ and y be some state of $\mathscr{B}$. If x and y have the same action on Σ^*, then $\mathfrak{m}(\mathscr{A})$ and $\mathfrak{m}(\mathscr{B})$ are isomorphic; in particular they have the same size. Indeed the image of x in $\mathscr{A}$ by some word $s \in \Sigma^*$ and the image of y in $\mathscr{B}$ by this same word s have necessarily the same action on Σ^*, and $\mathscr{A}$ and $\mathscr{B}$ being strongly connected (because they are connected and reversible), for every state of $\mathscr{A}$ there is a state of $\mathscr{B}$ which acts similarly on Σ^* and vice versa.

Let us recall some known results from [9] and [351] (see also [178, 441, 521]) that will be used in our proofs.

Proposition 10.1.5. *An invertible-reversible Mealy automaton $\mathscr{A}$ and its extension $\widetilde{\mathscr{A}}$ generate isomorphic groups.*

Proposition 10.1.6. *An invertible Mealy automaton generates a finite group if and only if it generates a finite semigroup.*

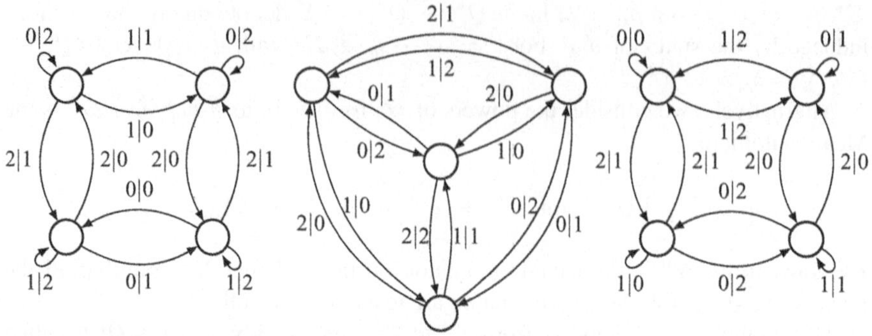

Fig. 10.3 Three non-$\mathfrak{md}$-trivial bireversible automata that generate finite groups. The inverse of the leftmost one provides a fourth minimal example.

Theorem 10.1.7. *A Mealy automaton generates a finite (semi)group if and only if so does its dual.*

Corollary 10.1.8. *A Mealy automaton generates a finite (semi)group if and only if so does its* $\mathfrak{md}$*-reduction.*

The trivial Mealy automaton generates the trivial (semi)group. If the $\mathfrak{md}$-reduction of a Mealy automaton $\mathscr{A}$ leads to the trivial Mealy automaton, $\mathscr{A}$ is said to be $\mathfrak{md}$-*trivial*. It is decidable whether a Mealy automaton is $\mathfrak{md}$-trivial. An $\mathfrak{md}$-trivial Mealy automaton generates a finite semigroup, but in general the converse is false [9].

The (bi)reversible Mealy automata seem to be especially sensitive to $\mathfrak{md}$-reduction. The four minimal examples of non-$\mathfrak{md}$-trivial bireversible Mealy automata generating finite groups are displayed on Figure 10.3 (see [9] and [483] for other examples).

Theorem 10.1.9. *Any 2-letter and/or 2-state bireversible Mealy automaton generates a finite group if and only if it is* $\mathfrak{md}$*-trivial.*

We shall see in Section 10.2.3 why and how we intend to generalize this fundamental result.

10.2 A Matter of Tilings

Wang tilings are a major tool in order to study and understand the behavior of automaton (semi)groups. Without even thinking about the potential generated algebraic structures, Mealy automata have earlier been associated with Wang tilings by K. Culik [170], by J. Kari [333], and ultimately by E. Jeandel and M. Rao [311] in their overwhelming and successful pursuit of small aperiodic Wang tilesets. Such Mealy automata need not to be either complete or deterministic, preventing to easily define automaton (semi)groups as in the current framework. Now there are

Fig. 10.4 A Wang tile.

various ways to associate a Wang tileset with a given complete and deterministic Mealy automaton and various ways to interpret the induced Wang tilings. We aim to describe some of these fruitful combinations, as well as some promising research opportunities.

Section 10.2.1 recalls basic definitions and undecidability results about Wang tilings. See also Section 9.2. Section 10.2.2 is devoted to the undecidability result by P. Gillibert of the finiteness problem for automaton semigroups and then sketches a bright connection between reset Mealy automata and some one-way cellular automata. Section 10.2.3 focuses on so-called helix graphs, a crucial notion capturing the whole dynamics by placing on a same footing the symmetric roles of the state set and the alphabet. Section 10.2.4 outlines an effective and natural approach to interpret any semigroup admitting a special language of greedy normal forms—based on rewriting systems and Wang tilings—as an automaton semigroup, namely, the semigroup generated by a Mealy automaton encoding the behavior of such a language of greedy normal forms under one-sided multiplication. In each of all these cases, the key notion is duality.

10.2.1 Background on Tilings

Named after H. Wang [580], a *Wang tile* is a unit square tile with a color on each edge: it is a quadruple $t = (t_w, t_s, t_e, t_n) \in C^4$ where C is a finite set of colors, as typically depicted in Figure 10.4. A *Wang tileset* is a finite set $\mathscr{T}$ of Wang tiles, and for each $t \in \mathscr{T}$ and $d \in \{n, s, e, w\}$, we put t_d for the color of the edge in the d-side. Given a Wang tileset $\mathscr{T}$, a *Wang tiling* of a subset P of $\mathbb{Z}^2$ is a map $f: P \to \mathscr{T}$. We say that such a Wang tiling f is *valid* whenever, with each point $(x, y) \in P$, f associates a tile $f(x, y)$ such that adjacent tiles share the same color on their common edge:

$$f(x,y)_n = f(x,y+1)_s, \qquad \text{for } (x,y) \in P \text{ and } (x,y+1) \in P,$$
$$f(x,y)_e = f(x+1,y)_w, \qquad \text{for } (x,y) \in P \text{ and } (x+1,y) \in P.$$

A simple compactness argument gives the following classical result.

Theorem 10.2.1. *For any Wang tileset $\mathscr{T}$, $\mathbb{Z}^2$ admits a valid Wang tiling for $\mathscr{T}$ if and only if so does each finite subset of $\mathbb{Z}^2$.*

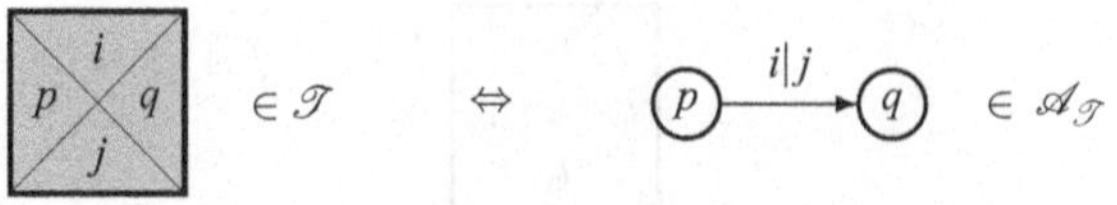

Fig. 10.5 The transducer $\mathcal{A}_{\mathcal{T}}$ associated with a Wang tileset $\mathcal{T}$ according to Culik-Kari.

In particular, if $\mathbb{Z}^2$ admits no valid Wang tiling, then there is a least integer $n \in \mathbb{N}$ such that the square $\{0, 1, \ldots, n\}^2$ admits no valid Wang tiling.

Following K. Culik [170] and J. Kari [333], any Wang tileset may be interpreted as a letter-to-letter transducer with the same input and output alphabet, according to the correspondence in Figure 10.5. Note that such a transducer may be neither complete nor deterministic (and has neither initial nor final states). Following [370] for instance, we say that a Wang tileset $\mathcal{T}$ is *cd-deterministic* with $(c, d) \in \mathcal{I} = \{(n, e), (s, e), (n, w), (s, w)\}$ if each tile $t \in \mathcal{T}$ is uniquely determined by its pair (t_c, t_d) of colors. Whenever $\mathcal{T}$ is *cd*-deterministic for each $(c, d) \in \mathcal{I}$, we say that $\mathcal{T}$ is *four-way deterministic*. The following lemma links properties of the Wang tileset $\mathcal{T}$ with properties of the associated transducer $\mathcal{A}_{\mathcal{T}}$.

Lemma 10.2.2. *Let $\mathcal{T}$ be a Wang tileset and $\mathcal{A}_{\mathcal{T}}$ be the associated transducer according to Culik-Kari. A necessary condition for $\mathcal{A}_{\mathcal{T}}$ to be a Mealy automaton is that $\mathcal{T}$ is nw-deterministic. In such a case, we have the following:*

- *$\mathcal{T}$ is ne-deterministic if and only if $\mathcal{A}_{\mathcal{T}}$ is reversible;*
- *$\mathcal{T}$ is sw-deterministic if and only if $\mathcal{A}_{\mathcal{T}}$ is invertible;*
- *$\mathcal{T}$ is 4-way deterministic if and only if $\mathcal{A}_{\mathcal{T}}$ is bireversible.*

This original Wang tiling viewpoint could provide a special insight to the dynamics of the Mealy automaton (see for instance [177]).

10.2.2 Finiteness and Order Problems

A second correspondence between Mealy automata and Wang tilings has allowed P. Gillibert [242] to prove that the finiteness and the order problems for automaton semigroups are undecidable (these decision problems are still open for automaton groups, see [271, Problems 7.2.1(a) and 7.2.1(b)]). P. Gillibert's proof relies on the local correspondence from *nw*-deterministic Wang tilesets to Mealy automata displayed in Figure 10.6, inspired by J. Kari's proof of the undecidability of the nilpotency problem for cellular automata [332].

The following results of R. Berger in [72] and of J. Kari in [332] illustrate how the existence of valid Wang tilings is hard to determine.

Theorem 10.2.3. *It is undecidable whether or not a Wang tileset admits a valid Wang tiling for the discrete plane $\mathbb{Z}^2$.*

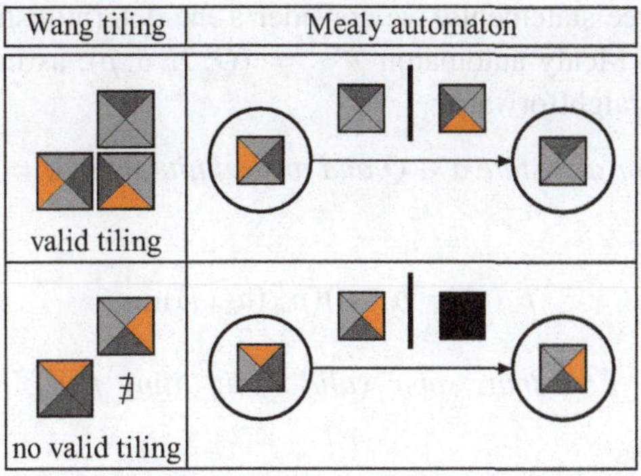

Fig. 10.6 The Mealy automaton $\mathscr{W}_{\mathscr{T}}$ associated with a *nw*-deterministic Wang tileset $\mathscr{T}$ according to Gillibert-Kari (see Definition 10.2.6).

Theorem 10.2.4. *It is undecidable whether or not a nw-deterministic Wang tileset admits a valid Wang tiling for the discrete plane* $\mathbb{Z}^2$.

Enhancing a result by K. Culik, J. K. Pachl, and S. Yu in [171], J. Kari proved the following in [332].

Theorem 10.2.5. *It is undecidable whether or not a one-dimensional cellular automaton is nilpotent.*

P. Gillibert adapted J. Kari's argument to address the finiteness and order problems for automaton semigroups. We have to be careful about a side effect: a cellular automaton acts on words indexed by $\mathbb{Z}$, while an automaton semigroup acts on words indexed by $\mathbb{N}$. According to P. Gillibert [242], we first define a Mealy automaton from a Wang tileset (Kari uses a similar construction to obtain a cellular automaton).

Definition 10.2.6. With any *nw*-deterministic Wang tileset $\mathscr{T}$, as illustrated in Figure 10.6, we associate the Mealy automaton $\mathscr{W}_{\mathscr{T}} = (Q, \Sigma, \delta, \rho)$ with $Q = \Sigma = \mathscr{T} \sqcup \{\blacksquare\}$, $\delta_b(a) = b$ for $(a, b) \in Q^2$, and

$$\rho_a(b) = \begin{cases} c & \text{for } a, b, c \in \mathscr{T} \text{ with } a_e = c_w \text{ and } c_n = b_s, \\ \blacksquare & \text{otherwise.} \end{cases}$$

The Mealy automaton $\mathscr{W}_{\mathscr{T}}$ associated with a *nw*-deterministic Wang tileset $\mathscr{T}$ should be understood in the following way. Any word over the state set can be seen as a word written over the tileset along some diagonal; the Mealy automaton transforms this word to the word written on the tiles along the diagonal right below. If it is impossible to put a tile at some place, then the *mistake* tile ■ is placed instead.

For the next three statements, we consider a *nw*-deterministic Wang tileset $\mathscr{T}$, with its associated Mealy automaton $\mathscr{W}_{\mathscr{T}} = (Q, \Sigma, \delta, \rho)$, as in Definition 10.2.6. The following is straightforward.

Lemma 10.2.7. *For any state $a \in Q$ and any infinite word $u = (u_k)_{k\in\mathbb{N}} \in \Sigma^\omega$, we have*

$$\rho_a(u) = \rho_a(u_0)(\rho_{u_k}(u_{k+1}))_{k\in\mathbb{N}}. \tag{10.1}$$

Lemma 10.2.8. *If $\mathbb{Z}^2$ admits some valid Wang tiling for $\mathscr{T}$, then $\langle\, \mathscr{W}_{\mathscr{T}} \,\rangle_+$ is infinite.*

Proof. Let $t\colon \mathbb{Z}^2 \to \mathscr{T}$ be some valid Wang tiling. Then we claim that the tile $\blacksquare$ induces an element of infinite order. The point is to show that, for any $n \in \mathbb{N}$, the word $w_n = (t(k+n,k))_{k\in\mathbb{N}}$ satisfies $\rho_{\blacksquare}^m(w_n) = \blacksquare^m w_{m+n}$ for every $m \in \mathbb{N}$, as illustrated in Figure 10.7. By very definition, for $(i,j) \in \mathbb{N}^2$, we have

$$\rho_{t(i,j)}(t(i+1,j+1)) = t(i+1,j)\,. \tag{10.2}$$

Given $n \in \mathbb{N}$, we find:

$$\begin{aligned} \rho_{\blacksquare}(w_n) &\overset{(10.1)}{=} \rho_{\blacksquare}(t(n,0))(\rho_{t(n+k,k)}(t(n+k+1,k+1)))_{k\in\mathbb{N}}\,, \\ &\overset{(10.2)}{=} \blacksquare(t(n+k+1,k))_{k\in\mathbb{N}}\,, \\ &= \blacksquare w_{n+1}\,. \end{aligned}$$

The result follows by induction. In particular, we deduce that the maps $\rho_{\blacksquare}^m$ are pairwise different, which means that the element $\rho_{\blacksquare}$ has infinite order and implies that the semigroup $\langle\, \mathscr{W}_{\mathscr{T}} \,\rangle_+$ is infinite. □

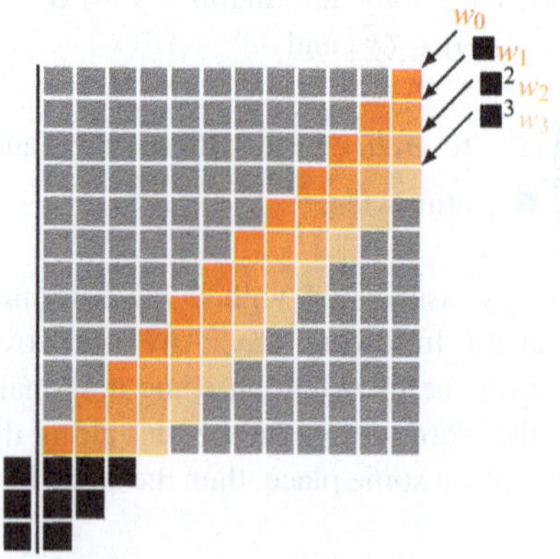

Fig. 10.7 The action of the tile $\blacksquare$ corresponds to an element of infinite order (proof of Lemma 10.2.8).

Proposition 10.2.9. *If $\mathbb{Z}^2$ admits no valid Wang tiling for $\mathscr{T}$, then $\langle \mathscr{W}_{\mathscr{T}} \rangle_+$ is finite.*

Proof. According to Theorem 10.2.1, there exists $n \in \mathbb{N}$ such that the set $\{0, 1, \ldots, n\}^2$ admits no valid Wang tiling for $\mathscr{T}$. Let fix $(p, q) \in \Sigma^n \times \Sigma^\omega$. As illustrated in Figure 10.8, we want to prove that any word $u \in Q^{2n}$ satisfies

$$\rho_u(pq) = \rho_u(p)\blacksquare^\omega . \tag{10.3}$$

Write $u = u_1 \cdots u_{2n}$ (those orange tiles on Figure 10.8), and set $\tau_0 = \mathrm{id}$ and

$$\tau_k = \rho_{u_1 u_2 \cdots u_k} = \rho_{u_k} \circ \rho_{u_{k-1}} \circ \cdots \circ \rho_{u_1} , \quad \text{for } 1 \le k \le 2n.$$

We have:

$$\rho_{u_{k+1}} \circ \tau_k = \tau_{k+1} , \quad \text{for } 0 \le k \le 2n-1. \tag{10.4}$$

Denoting by $f(i, j)$ the j-th letter of $\tau_i(pq)$ for $(i, j) \in \mathbb{N}^2$ with $0 \le i \le 2n$, we have:

$$\tau_i(pq) = (f(i, j))_{j \in \mathbb{N}} , \quad \text{for } 0 \le i \le 2n. \tag{10.5}$$

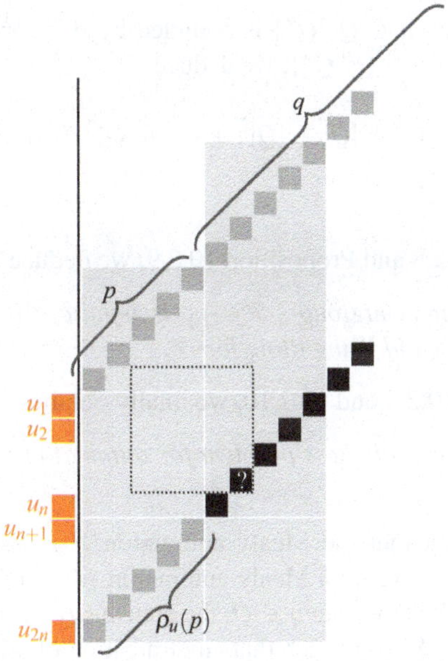

Fig. 10.8 If some tile (along the black diagonal) was not $\blacksquare$, the $n \times n$ square (with dotted line) could be tiled, inside the $2n$-width corridor (proof of Proposition 10.2.9).

For $0 \leq i < 2n$, we find:

$$\begin{aligned} f(i+1,j)_{j\in\mathbb{N}} &\overset{(10.5)}{=} \tau_{i+1}(pq)\,, \\ &\overset{(10.4)}{=} \rho_{u_{i+1}}(\tau_i(pq))\,, \\ &\overset{(10.5)}{=} \rho_{u_{i+1}}(f(i,j)_{j\in\mathbb{N}})\,, \\ &\overset{(10.1)}{=} \rho_{u_{i+1}}(f(i,0))(\rho_{f(i,j)}(f(i,j+1)))_{j\in\mathbb{N}}\,. \end{aligned}$$

We deduce

$$\rho_{f(i,j)}(f(i,j+1)) = f(i+1,j+1) \tag{10.6}$$

for $(i,j) \in \mathbb{N}^2$ with $0 \leq i < 2n$.

Now assume $f(2n, n+k) \neq \blacksquare$ for some $k \in \mathbb{N}$ (among the black diagonal on Figure 10.8). Applying inductively (10.6), we obtain $f(i+j, i+k) \neq \blacksquare$ for $0 \leq i,j \leq n$, which yields in particular a valid Wang tiling for some $n \times n$ square (with dotted line on Figure 10.8): this is a contradiction that allows to prove (10.3) as well.

Let $u = vw$ with $v \in Q^{2n}$ and $w \in Q^*$. We have

$$\rho_u(pq) = \rho_{vw}(pq) = \rho_w(\rho_v(pq)) \overset{(10.3)}{=} \rho_w(\rho_v(p)\blacksquare^\omega) = \rho_{vw}(p)\blacksquare^\omega = \rho_u(p)\blacksquare^\omega\,.$$

The cardinality of $\{\rho_u : u \in Q^{2n}Q^*\}$ is bounded by $|(\Sigma^n)^{(\Sigma^n)}|$. From $\langle\, \mathscr{W}_{\mathscr{T}} \,\rangle_+ = \{\rho_u : u \in Q^{<2n}\} \cup \{\rho_u : u \in Q^{2n}Q^*\}$, we deduce

$$|\langle\, \mathscr{W}_{\mathscr{T}} \,\rangle_+| \leq 1 + |Q| + |Q|^2 + \cdots + |Q|^{2n-1} + |(\Sigma^n)^{(\Sigma^n)}|\,.$$

Hence $\langle\, \mathscr{W}_{\mathscr{T}} \,\rangle_+$ is finite. □

Gathering Lemma 10.2.8 and Proposition 10.2.9, we deduce:

Theorem 10.2.10. *The semigroup $\langle\, \mathscr{W}_{\mathscr{T}} \,\rangle_+$ is infinite if and only if the discrete plane $\mathbb{Z}^2$ admits some valid Wang tiling for $\mathscr{T}$.*

Gathering Theorems 10.2.4 and 10.2.10, we finally obtain:

Corollary 10.2.11. *The finiteness problem for automaton semigroups is undecidable.*

It is worthwhile noting that the Mealy automaton $\mathscr{W}_{\mathscr{T}}$ associated with any *nw*-deterministic tileset $\mathscr{T}$ is reset: a Mealy automaton $\mathscr{A} = (Q, \Sigma, \delta, \rho)$ is said to be a *reset automaton* if, for any state $q \in Q$, the state $\delta_i(q)$ does not depend on q, that is, if there is a map $\varphi : \Sigma \to Q$ such that any transition of $\mathscr{A}$ has the form

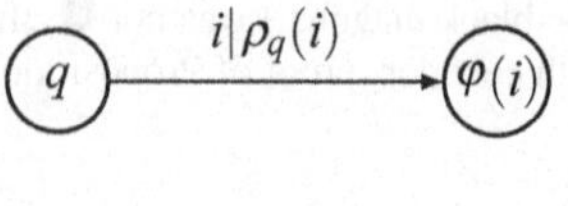

The very definition of $\mathscr{W}_{\mathscr{T}}$ induces also that its state set and its alphabet coincide. However, this feature is not specific in the framework of reset automata. Indeed, any reset automaton is equivalent to a reset automaton with $Q = \Sigma$ (and $\varphi = \mathrm{id}$).

The point now is that, by construction (with the mandatory adjunction of a tile ■), such reset Mealy automata $\mathscr{W}_{\mathscr{T}}$ are highly non-invertible, which seems to prevent to adapt the previous approach for automaton groups. Silva and Steinberg have studied groups generated by invertible reset automata [549]. They proved in particular that such a group is infinite if and only if any generator is of infinite order. However, the following problem remains open.

Problem 10.2.12. Is the finiteness problem for reset automaton groups decidable?

To conclude this section, we describe a link between the finiteness of reset automaton groups and the periodicity of one-way cellular automata. A *one-way cellular automaton* is a triple (Q, r, f) where Q is the finite state set, $r \in \mathbb{N}$ is the radius, and $f : Q^{r+1} \to Q$ is the local transition rule. A *configuration* of such an automaton is an element in $Q^{\mathbb{N}}$. The whole dynamics is described by the *global transition function* F defined by the local transition function f as

$$F(c)(k) = f(c(k), c(k+1), \dots, c(k+r))$$

for every configuration $c \in Q^{\mathbb{N}}$ and every $k \in \mathbb{N}$. Such a cellular automaton is said to be *periodic* if $F^p = \mathrm{id}$ holds for some integer $p > 0$.

Problem 10.2.13. Is the periodicity problem for one-way cellular automata decidable?

It must be recalled that J. Kari and N. Ollinger have shown that the periodicity problem for (reversible) cellular automata is undecidable [336]. One can restrict the study of the periodicity to those one-way cellular automata with radius 1 without loss of generality. Let (Q, f) denote the one-way cellular automaton $(Q, 1, f)$ with $f : Q^2 \to Q$.

Any such periodic cellular automaton (Q, f) has to preserve the value of the cell $c(0)$; hence, for any state $b \in Q$, the map $\rho_b : a \mapsto f(a, b)$ has to be a permutation. Such a cellular automaton is said to be *center-permutive*. The latter being a purely syntactic property, Problem 10.2.13 is equivalent to the following.

Problem 10.2.14. Is the periodicity problem for one-way center-permutive radius 1 cellular automata decidable?

As illustrated on Figure 10.9 and stated in [190], Problems 10.2.12 and 10.2.13 turn out to be a single one open problem.

Proposition 10.2.15. *For any family $\rho = (\rho_b)_{b \in \Sigma}$ of permutations of the alphabet Σ, the group generated by the Mealy automaton $(\Sigma, \Sigma, \mathrm{id}, \rho)$ is finite if and only if the cellular automaton $(\Sigma, (a, b) \mapsto \rho_b(a))$ is periodic.*

On the one hand, systematic experimentations on small reset Mealy automata (as well as randomly chosen large ones) seem to indicate that whenever a reset

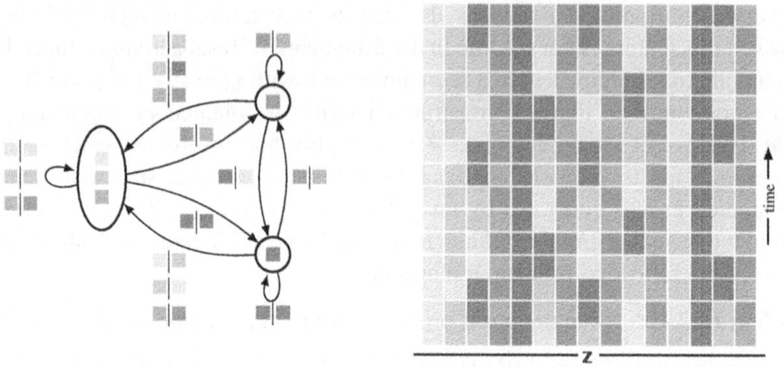

Fig. 10.9 Any reset Mealy automaton $(\Sigma, \Sigma, \mathrm{id}, \rho)$ (or its minimization, on the left) corresponds to a cellular automaton $(\Sigma, (a, b) \mapsto \rho_b(a))$ (with a fragment of a space-time diagram on the right), according to Proposition 10.2.15.

automaton group is finite, the semigroup generated by the dual automaton is very small. On the other hand, Delacourt and Ollinger have managed to inject some computations in one-way center-permutive cellular automata [190]. Tilting in opposite directions, these cooperating two points of view remain for now this crucial open question in some swing state.

10.2.3 Helix Graphs and Rigidity

The notion of a helix graph is a dynamical tool introduced in [9] and can be thought as some one-dimensional tiling.

Definition 10.2.16. The *helix graph* $\mathscr{H}_{n,k}$ of a Mealy automaton $\mathscr{A} = (Q, \Sigma, \delta, \rho)$ is defined to be the directed graph with nodes $Q^n \times \Sigma^k$ and arrows

$$\boxed{u, v} \longrightarrow \boxed{\delta_v(u), \rho_u(v)}$$

for all $(u, v) \in Q^n \times \Sigma^k$ (see Figure 10.10).

Merging together a Mealy automaton $\mathscr{A}$ and its dual $\partial\mathscr{A}$, their helix graphs allow to capture the whole dynamics by placing on a same footing the symmetric roles of the state set and the alphabet. This way, the two faces of the coin are coalesced into one.

Let us mention that a helix graph could be defined for any letter-to-letter transducer where input and output alphabets coincide. Such a transducer is a Mealy automaton if and only if from any vertex starts a unique edge.

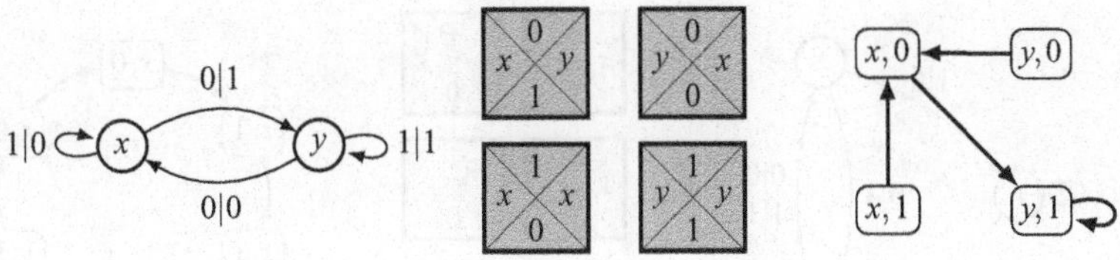

Fig. 10.10 The Mealy automaton $\mathscr{L}$ generating the lamplighter group, its associated Wang tileset $\mathscr{T}(\mathscr{L})$, and its helix graph $\mathscr{H}_{1,1}(\mathscr{L})$.

Proposition 10.2.17. *If the group generated by an invertible-reversible Mealy automaton is finite, then any of its helix graphs is a union of disjoint cycles.*

Proof. Let $\mathscr{A} = (Q, \Sigma, \delta, \rho)$ be an invertible-reversible Mealy automaton that generates a finite group. Theorem 10.1.7 implies that $\mathfrak{d}\mathscr{A}$ generates a finite group as well. Let $\mathscr{H}$ be the helix graph of $\mathscr{A}$ and s be the map from the finite set of vertices of $\mathscr{H}$ into itself that maps any vertex to its (unique) successor. The helix graph $\mathscr{H}$ is a union of disjoint cycles if and only if the map s is bijective, that is, if and only if the map s is surjective.

Let $x \in Q$ and $i \in \Sigma$. We have to show that the vertex (q, i) admits a unique predecessor in $\mathscr{H}$. There exist integers $m, n > 0$ satisfying $\rho_q^m = \rho_{q^m} = \mathrm{id}_{\langle \mathscr{A} \rangle}$ and $\delta_i^n = \delta_{i^n} = \mathrm{id}_{\langle \mathfrak{d}\mathscr{A} \rangle}$. This means that there is a transition $q^m \xrightarrow{i^n|i^n} q^m$ in the associated automaton of order (m, n). The corresponding cross-diagram is:

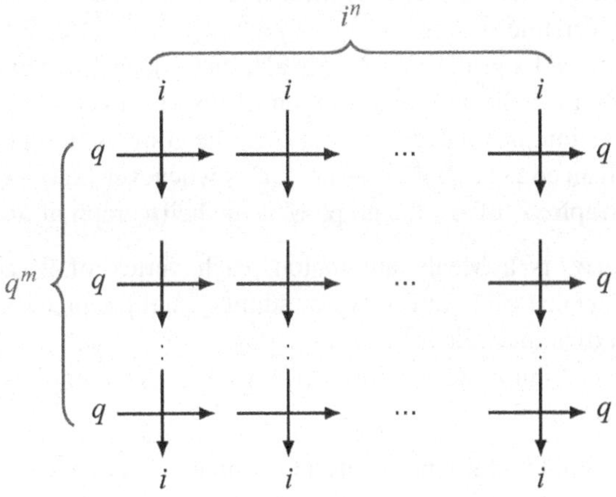

The most southeast cross gives a predecessor to the vertex (q, i). □

The condition of Proposition 10.2.17 is not sufficient: there are invertible-reversible Mealy automata whose helix graph is a union of disjoint cycles and that generate an infinite group, as for instance the Alešin automaton displayed on Figure 10.11. Proposition 10.2.18 characterizes the class of those invertible-reversible automata whose helix graph is a union of disjoint cycles.

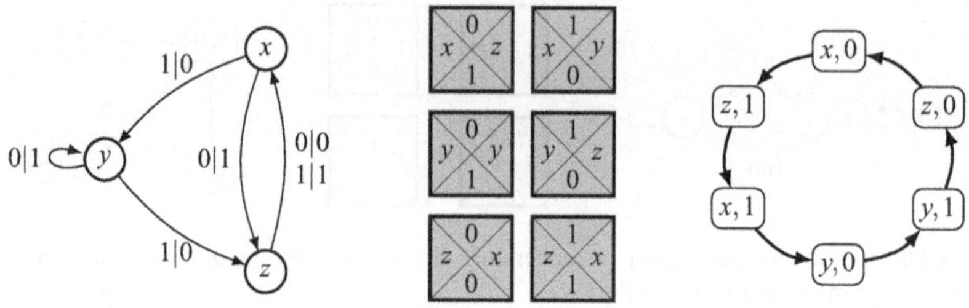

Fig. 10.11 Alešin automaton generates an infinite group, namely, the rank 3 free group. Its helix graph is a cycle.

Proposition 10.2.18. *Let $\mathscr{A}$ be an invertible-reversible Mealy automaton. The following are equivalent:*

1. *$\mathscr{A}$ is bireversible;*
2. *$\mathfrak{d}\mathfrak{i}\mathfrak{d}\mathfrak{i}\mathscr{A}$ is a Mealy automaton;*
3. *the helix graph of $\mathscr{A}$ is a union of disjoint cycles.*

Proof. ($1 \Rightarrow 2$) By hypothesis of bireversibility, $\mathscr{A}$ is invertible and $\mathfrak{i}\mathscr{A}$ is reversible. The latter means that $\mathfrak{d}\mathfrak{i}\mathscr{A}$ is invertible. Therefore $\mathfrak{i}\mathfrak{d}\mathfrak{i}\mathscr{A}$ is a Mealy automaton and so is its dual $\mathfrak{d}\mathfrak{i}\mathfrak{d}\mathfrak{i}\mathscr{A}$.

($2 \Rightarrow 1$) First, by hypothesis of invertibility, $\mathfrak{i}\mathscr{A}$ is a Mealy automaton, and so is its dual $\mathfrak{d}\mathfrak{i}\mathscr{A}$. Next, since $\mathfrak{d}\mathfrak{i}\mathfrak{d}\mathfrak{i}\mathscr{A}$ is assumed to be a Mealy automaton, so is its dual $\mathfrak{i}\mathfrak{d}\mathfrak{i}\mathscr{A}$. This means that $\mathfrak{d}\mathfrak{i}\mathscr{A}$ is invertible, that is, $\mathfrak{i}\mathscr{A}$ reversible. We deduce that $\mathscr{A}$ is bireversible.

($2 \Leftrightarrow 3$) In any helix graph (of a Mealy automaton), each vertex admits a unique successor. Such a helix graph is a union of disjoint cycles if and only if each vertex admits a unique predecessor. Let $\mathscr{G}$ be the graph with set of vertices $Q^{-1} \times \Sigma^{-1}$ and with an edge $(y^{-1}, j^{-1}) \to (x^{-1}, i^{-1})$ whenever $(x, i) \to (y, j)$ is an edge in the helix graph $\mathscr{H}$ of $\mathscr{A}$: the graph $\mathscr{G}$ is the helix graph of $\mathfrak{d}\mathfrak{i}\mathfrak{d}\mathfrak{i}\mathscr{A}$:

($\Rightarrow$) if $\mathfrak{d}\mathfrak{i}\mathfrak{d}\mathfrak{i}\mathscr{A}$ is a Mealy automaton, each vertex of $\mathscr{G}$ admits a unique successor, hence each vertex of $\mathscr{H}$ admits a unique predecessor, and $\mathscr{H}$ is a union of disjoint cycles;

($\Leftarrow$) if $\mathscr{H}$ is a union of disjoint cycles, so is $\mathscr{G}$. This implies that $\mathfrak{d}\mathfrak{i}\mathfrak{d}\mathfrak{i}\mathscr{A}$ is a Mealy automaton. □

We deduce a simple infiniteness criterion, which is very easy to check.

Corollary 10.2.19. *Any invertible-reversible Mealy automaton which is not bireversible generates an infinite group.*

We can also state the following characterization:

Theorem 10.2.20. *Let $\mathscr{A}$ be an invertible-reversible Mealy automaton. The group $\langle \mathscr{A} \rangle$ is finite if and only if there exists an integer C such that, for all k, ℓ, the helix graphs $\mathscr{H}_{k,\ell}$ of $\mathscr{A}$ are unions of disjoint cycles of lengths bounded by C.*

Note that such a characterization is not effective and does not directly lead to a decision procedure of finiteness.

Recall that, for any invertible-reversible automaton $\mathcal{A}$ with state set Q and alphabet Σ, we let $\widetilde{\mathcal{A}}$ denote the extension with state set $Q \sqcup Q^{-1}$ and alphabet $\Sigma \sqcup \Sigma^{-1}$.

Proof. Assume first that $\langle\, \mathcal{A} \,\rangle$ is finite: so is $\langle\, \widetilde{\mathcal{A}} \,\rangle$ by Proposition 10.1.5. Proposition 10.2.17 shows that the helix graphs of any level are unions of disjoint cycles. It remains to prove that the lengths of these cycles are uniformly bounded. By Theorem 10.1.7, the group $\langle\, \mathfrak{d}\widetilde{\mathcal{A}} \,\rangle$ is finite as well. Let $\mathscr{C}$ be a cycle in a helix graph of $\widetilde{\mathcal{A}}$ and let $(u, v) \in (Q \sqcup Q^{-1})^* \times (\Sigma \sqcup \Sigma^{-1})^*$ be a node of this cycle. Each node of $\mathscr{C}$ is of the form $(h(u), g(v))$, where g (*resp.* h) is an element of $\langle\, \widetilde{\mathcal{A}} \,\rangle$ (*resp.* $\langle\, \mathfrak{d}\widetilde{\mathcal{A}} \,\rangle$). Since the nodes are pairwise distinct, the length of the cycle $\mathscr{C}$ is at most $|\langle\, \widetilde{\mathcal{A}} \,\rangle| \times |\langle\, \mathfrak{d}\widetilde{\mathcal{A}} \,\rangle|$.

Let us prove the converse and assume that the group $\langle\, \mathcal{A} \,\rangle$ is infinite: so is $\langle\, \widetilde{\mathcal{A}} \,\rangle$ by Proposition 10.1.5. First, we argue that the orders of the elements of $\langle\, \widetilde{\mathcal{A}} \,\rangle$ are unbounded. Indeed, automata groups are residually finite by construction since they act faithfully on rooted locally finite trees. Moreover, it follows from Zelmanov's solution of the restricted Burnside problem [574, 593, 594] that any residually finite group with bounded torsion is finite. Since $\langle\, \widetilde{\mathcal{A}} \,\rangle$ is infinite, the orders of its elements are unbounded.

There exists either $x \in (Q \sqcup Q^{-1})^*$ such that the order of ρ_x is infinite or a sequence $(x_n)_{n \in \mathbb{N}} \subseteq (Q \sqcup Q^{-1})^*$ such that the sequence $(k_n)_n$ of orders of the elements ρ_{x_n} goes to infinity. We carry out the proof in the second case, the first one can be treated similarly. Let us concentrate on ρ_{x_n}, element of order k_n of $\langle\, \widetilde{\mathcal{A}} \,\rangle$. For all $1 \leq k < k_n$, there exists a word $u_k \in (\Sigma \sqcup \Sigma^{-1})^*$ satisfying $\rho_{x_n}^k(u_k) = u'_k \neq u_k$.

Say that a word $v \in (\Sigma \sqcup \Sigma^{-1})^*$ is *unital* if δ_v is the identity of $\langle\, \mathfrak{d}\widetilde{\mathcal{A}} \,\rangle$. Since $\langle\, \mathfrak{d}\widetilde{\mathcal{A}} \,\rangle$ is a group, the word u_k can be extended into a unital word $u_k v_k$. Set $w_n = u_1 v_1 \cdots u_{k_n-1} v_{k_n-1}$. By construction, we have $\rho_{x_n}(w_n) = u'_1 \cdots \neq w_n$. Since $u_1 v_1$ is unital, we also have:

$$\begin{aligned} \rho_{x_n}^2(w_n) &= \rho_{x_n}^2(u_1 v_1)\rho_{x_n}^2(u_2 v_2 \cdots u_{k_n-1} v_{k_n-1}) \\ &= \rho_{x_n}^2(u_1 v_1) u'_2 \cdots \neq w_n\,. \end{aligned}$$

In the same way, we prove that, for all $k < k_n$, we have $\rho_{x_n}^k(w_n) \neq w_n$.

In the helix graph of $\widetilde{\mathcal{A}}$ of level $(|x_n|, |w_n|)$, consider the cycle containing the node (x_n, w_n). Since w_n is unital, the successors of (x_n, w_n) on the cycle are $(x_n, \rho_{x_n}(w_n))$, $(x_n, \rho_{x_n}^2(w_n))$, ... Therefore, the cycle is of length k_n. Since k_n goes to infinity, the lengths of the cycles of the helix graphs of $\widetilde{\mathcal{A}}$ are not uniformly bounded. □

We finally mention a relevant perspective based on helix graphs. The original observation is that the cyclic helix graph of Alešin on Figure 10.11 happens to be what we call *rigid*: it admits a trivial symmetry group. We can simply illustrate the

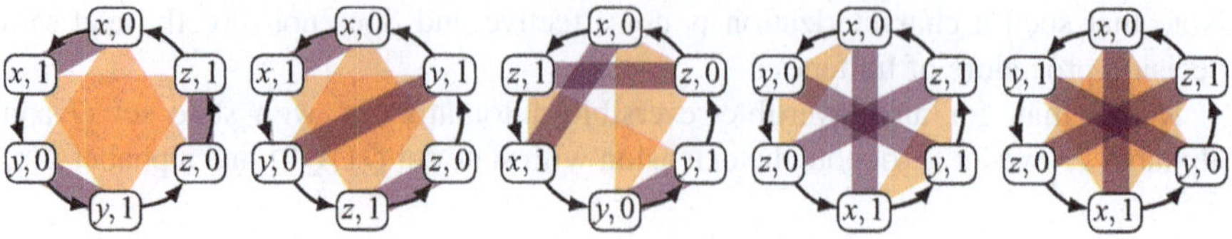

Fig. 10.12 The five helix graphs made of a unique cycle of size $2 \cdot 3 = 6$. Additional colors (orange for the letters *vs* violet for the states) emphasize the symmetries (or the absence of such ones). We give first the size s of the symmetry group of the helix graph together with the size e of the equivalence class of the corresponding automaton (with $s \times e = 2!\ 3!$) and then the generated group (only its finite or infinite nature matters here).

phenomenon on Figure 10.12 by comparing the latter (at the center) with the other four cyclic helix graphs with the same size.

This helix rigidity notion can be easily translated in terms of size of equivalence class to formulate the following claimed criterion:

Conjecture 10.2.21. Let $\mathscr{A}$ be a bireversible automaton. If the (nontrivial) $\mathfrak{m}\mathfrak{d}$-reduction of $\mathscr{A}$ admits an equivalence class of maximal size, then the group $\langle \mathscr{A} \rangle$ is infinite.

Note that Conjecture 10.2.21 is trivially true for 2-letter and/or 2-state bireversible automata by Theorem 10.1.9. Beyond, it would apply to more and more automata as suggesting by some experimentations:

States	Letters	$\mathfrak{m}\mathfrak{d}$-reduced	Rigid	%
2	5	190	154	81%
3	3	148	140	95%
4	3	6293	6117	97%

Furthermore, it seems that the lack of helix rigidity yields a conjugator such that the $\mathfrak{m}\mathfrak{d}$-reduction of the conjugated is smaller and so on. Based on the helix rigidity, we claim the following generalization of Corollary 10.1.8 and Theorem 10.1.9.

Conjecture 10.2.22. Let $\mathscr{A}$ be a bireversible automaton. The group $\langle \mathscr{A} \rangle$ is finite if and only if some conjugate of $\mathscr{A}$ is $\mathfrak{m}\mathfrak{d}$-trivial.

The latter would validate in particular the idea that $\mathfrak{m}\mathfrak{d}$-reduction allows to solve the finiteness problem for prime bireversible automata, that is, for those bireversible that admit no nontrivial decomposition. This notion of primality turns out to be recurrent in various contexts and seems to be especially relevant for the Burnside problem in Section 10.3.

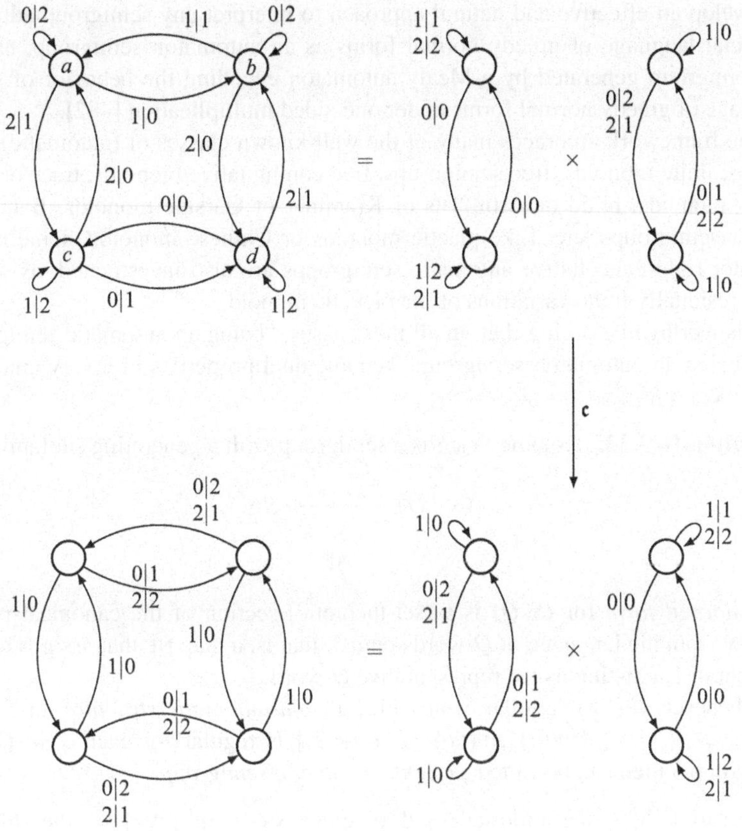

Fig. 10.13 An $\mathfrak{md}$-reduced bireversible automaton that admits an $\mathfrak{md}$-trivial conjugate.

Let us come back to the specimens from Figure 10.3. We have seen that they are $\mathfrak{md}$-reduced and generate finite groups. Each of their helix graphs admits a nontrivial symmetry; hence it is not rigid.

For instance, the leftmost one (on Figure 10.3) admits the symmetry $0 \leftrightarrow 1, a \leftrightarrow c, b \leftrightarrow d$. As shown on Figure 10.13, it admits a decomposition as a product of two components $\mathscr{C}_1$ and $\mathscr{C}_2$. The conjugate $\mathscr{C}_2 \times \mathscr{C}_1$ happens to be $\mathfrak{md}$-trivial. The statement of Conjecture 10.2.22 might be stronger, and we could use some notion of $\mathfrak{mdc}$-reduction that would alternate $\mathfrak{md}$-reduction and conjugacy.

10.2.4 Automat-ic-on Semigroups

Considering tilings as computations, they can be used to encode some rewriting systems, that, according to [184, 185], we call *quadratic normalizations*. This allow to develop an effective and natural approach to interpret any semigroup admitting a special language of greedy normal forms as an automaton semigroup, namely, the semigroup generated by a Mealy automaton encoding the behavior of such a language of greedy normal forms under one-sided multiplication [482].

The framework embraces many of the well-known classes of (automatic) semigroups: finite monoids, free semigroups, free commutative monoids, trace or divisibility monoids, braid or Artin-Tits or Krammer or Garside monoids, Baumslag-Solitar semigroups, etc. Like plactic monoids or Chinese monoids, some neither left- nor right-cancellative automatic semigroups are also investigated, as well as some residually finite variations of the bicyclic monoid.

It is worthwhile noting that, in all these cases, "being an automatic semigroup" and "being an automaton semigroup" become dual properties in a very automata-theoretical sense.

Definition 10.2.23. Assume that S is a semigroup with a generating subfamily Q.

$$\mathrm{EV} : Q^+ \longrightarrow S, \qquad \mathrm{NF} : S \longrightarrow Q^+$$

A *normal form* for (S, Q) is a (set-theoretic) section of the canonical projection EV from the language of Q-words onto S, that is, a map NF that assigns to each element of S a distinguished representative Q-word.

Whenever $\mathrm{NF}(S)$ is regular, it provides a *right-automatic structure* for S if the language $\mathscr{L}_q = \{\ (\mathrm{NF}(a), \mathrm{NF}(aq))\ :\ a \in S\ \}$ is regular for each $q \in Q$. The semigroup S then can be called a *(right-)automatic semigroup.*

We mention here the thorough and precious study in [296] of the different notions (right- or left-reading *vs* right- or left-multiplication) of automaticity for semigroups.

Remark 10.2.24. In his seminal work [215, Chapter 9], Thurston shows how the set of these different automata recognizing the multiplication—that is, recognizing the languages of those pairs of normal forms of elements differing by a right factor $q \in Q$—and the one recognizing the equality in Definition 10.2.23 can be replaced with advantage by a single letter-to-letter transducer (Definition 10.2.35) that computes the normal forms via iterated runs, each run both providing one brick of the final normal from and outputting a word still to be normalized.

One will often consider the associated normalization $\mathrm{N} = \mathrm{NF} \circ \mathrm{EV}$.

Definition 10.2.25. A *normalization* is a pair (Q, N), where Q is a set and N is a map from Q^+ to itself satisfying, for all Q-words u, v, w:

- $|\mathrm{N}(w)| = |w|$,
- $|w| = 1 \Rightarrow \mathrm{N}(w) = w$,
- $\mathrm{N}(u\,\mathrm{N}(w)\,v) = \mathrm{N}(u\,w\,v)$.

A Q-word w satisfying $\mathrm{N}(w) = w$ is called N-*normal*. If S is a semigroup (*resp.* a monoid), we say that (Q, N) is a *normalization for* S if S admits the presentation

$$\langle\ Q : \{w = \mathrm{N}(w) \mid w \in Q^{+}\}\ \rangle_{+} \quad (\textit{resp.}\ \langle\ Q : \{w = \mathrm{N}(w) \mid w \in Q^{*}\}\ \rangle_{+}^{1}\ ).$$

Following [186], we associate with every element $q \in Q$ a q-labeled edge and with a product the concatenation of the corresponding edges and represent equalities in the ambient semigroup using commutative diagrams, what we call here *square-diagram*: for instance, the following square illustrates an equality $q_1 q_2 = q_1' q_2'$.

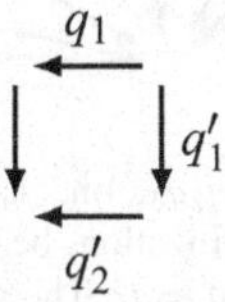

For a normalization (Q, N), we let $\overline{\mathrm{N}}$ denote the restriction of N to Q^2 and, for $i \geq 1$, by $\overline{\mathrm{N}}_i$ the (partial) map from Q^{+} to itself that consists in applying $\overline{\mathrm{N}}$ to the entries in positions i and $i+1$. For any finite sequence $i = i_1 \cdots i_n$ of positive integers, we write $\overline{\mathrm{N}}_i$ for the composite map $\overline{\mathrm{N}}_{i_n} \circ \cdots \circ \overline{\mathrm{N}}_{i_1}$ (so $\overline{\mathrm{N}}_{i_1}$ is applied first).

Definition 10.2.26. A normalization (Q, N) is *quadratic* if both following conditions hold:

- a Q-word w is N-normal if, and only if, so is every length-two factor of w;
- for every Q-word w, there exists a finite sequence i of positions, depending on w, such that $\mathrm{N}(w)$ is equal to $\overline{\mathrm{N}}_i(w)$.

Definition 10.2.27. As illustrated on Figure 10.14, with any quadratic normalization (Q, N) is associated its *breadth* (d, p) (called minimal left and right classes in [184, 185]) defined as:

$$d = \max_{(q_1,q_2,q_3)\in Q^3} \min\{\ \ell : \mathrm{N}(q_1q_2q_3) = \overline{\mathrm{N}}_{\underbrace{212\cdots}_{\text{length } \ell}}(q_1q_2q_3)\},$$

and

$$p = \max_{(q_1,q_2,q_3)\in Q^3} \min\{\ \ell : \mathrm{N}(q_1q_2q_3) = \overline{\mathrm{N}}_{\underbrace{121\cdots}_{\text{length } \ell}}(q_1q_2q_3)\}.$$

Such a breadth is ensured to be finite provided that Q is finite and then satisfies $|d - p| \leq 1$. For $d \leq 4$ and $p \leq 3$, (Q, N) is said to satisfy Condition ( ), corresponding to the so-called *domino rule* in [184–186] but with a different reading direction.

Remark 10.2.28. One can build quadratic normalizations with a (finite) breadth arbitrarily high (see [185]). A natural question would be to know what is the

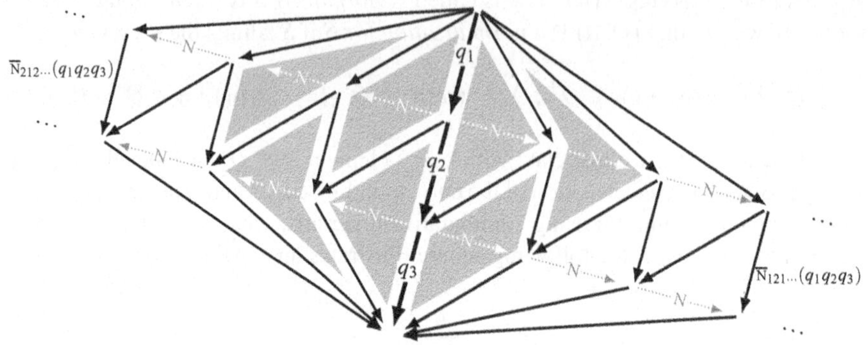

Fig. 10.14 From an initial Q-word $q_1q_2q_3$, one applies normalizations on the first and the second 2-factors alternatively up to stabilization, beginning either on the first 2-factor q_1q_2 (on the right-hand side here) or on the second q_2q_3. The gray zone corresponds to Condition ( ) as defined in Definition 10.2.27.

maximal breadth for a fixed size of Q. For instance, the semigroup

$$\mathbf{W} = \langle\, \mathtt{a}, \mathtt{b}, \mathtt{c} : \mathtt{aa} = \mathtt{cc} = \mathtt{bc}, \mathtt{ba} = \mathtt{cb} = \mathtt{ab}, \mathtt{bb} = \mathtt{ca} = \mathtt{ac} \,\rangle_+$$

admits a quadratic normalization with breadth $(11, 10)$, corresponding to the maximal breadth for $|Q| = 3$. Such a large breadth corresponds with a great height of the associated $\overline{N}$-graph as displayed on Figure 10.15. An easy general observation is the following: the larger the breadth, the higher the $\overline{N}$-graph, the most the associated semigroup approximates the rank 1 free semigroup. Here, as an ultimate example, $\mathbf{W}$ is precisely isomorphic to $\langle\, \mathtt{a} : \,\rangle_+$. Conversely, any quadratic normalization (Q, N) with a zero breadth corresponds to the rank $|Q|$ free semigroup $\langle\, Q : \,\rangle_+$ (except for $|Q| = 1$).

The first main result of [185] is an axiomatization of these quadratic normalizations satisfying Condition ( ) in terms of their restrictions to length-two words: any idempotent map $\overline{N}$ on Q^2 that satisfies $\overline{N}_{2121} = \overline{N}_{121} = \overline{N}_{1212}$ extends into a quadratic normalization (Q, N) satisfying Condition ( ). For larger breadths, a map on length-two words normalizing length-three words need not normalize words of greater length.

The second main result involves termination. Every quadratic normalization (Q, N) gives rise to a quadratic rewriting system, namely, the one with rules $w \longrightarrow \overline{N}(w)$ for $w \in Q^2$. By Definition 10.2.26, such a rewriting system is confluent and normalizing, meaning that, for every initial word, there exists a finite sequence of rewriting steps leading to a unique N-normal word, but its convergence, meaning that *any* sequence of rewriting steps is finite, is a quite different problem.

Theorem 10.2.29. *[185] If (Q, N) is a quadratic normalization satisfying Condition ( ), then the associated rewriting system is convergent.*

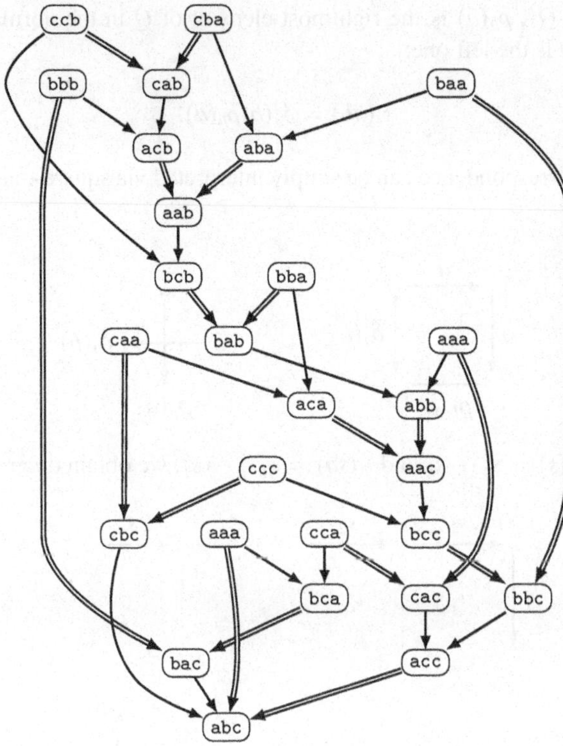

Fig. 10.15 The $\overline{\mathrm{N}}_{1,2}$-graph for the quadratic normalization associated with **W**: simple arrows correspond to $\overline{\mathrm{N}}_1$ and double arrows to $\overline{\mathrm{N}}_2$, while loops (on the sinks and on some sources) are simply omitted for better readability.

More precisely, every rewriting sequence starting from a word of Q^p has length at most $\frac{p(p-1)}{2}$ (*resp.* $2^p - p - 1$) in the case of a breadth $(3,3)$ (*resp.* either $(3,4)$ or $(4,3)$). Theorem 10.2.29 is essentially optimal since there exist non-convergent rewriting systems with breadth $(4,4)$.

The results of the current section rely on the special Condition (◆) outlined by Dehornoy and Guiraud (see [185]). However, none of their results (in particular Theorem 10.2.29 mentioned here for completeness) is neither applied nor required here. We want to emphasize that Condition (◆) appears as a common denominator for the different approaches.

All the ingredients are now in place to effectively and naturally interpret as an automaton monoid any automatic monoid admitting a special language of normal forms—namely, a quadratic normalization satisfying Condition (◆). The point is to construct a Mealy automaton encoding the behavior of its language of normal forms under one-sided multiplication.

Definition 10.2.30. Assume that S is a semigroup with a quadratic normalization (Q, N). We define the Mealy automaton $\mathscr{A}_{S,Q,\mathrm{N}} = (Q, Q, \delta, \rho)$ such that, for every $(a, b) \in Q^2$, $\rho_b(a)$ is the rightmost element of Q in the normal form $\mathrm{N}(ab)$ of ab and $\delta_a(b)$ is the left one:

$$\mathrm{N}(ab) = \delta_a(b)\rho_b(a).$$

The latter correspondence can be simply interpreted via square-diagram *vs* cross-diagram:

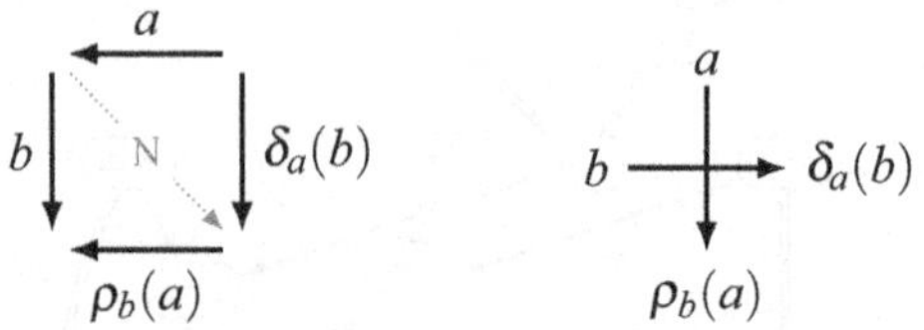

Then, for $\mathrm{N}(s) = s_n \cdots s_1$ and $\mathrm{N}(sq) = q_n s'_n \cdots s'_1$, we obtain diagrammatically:

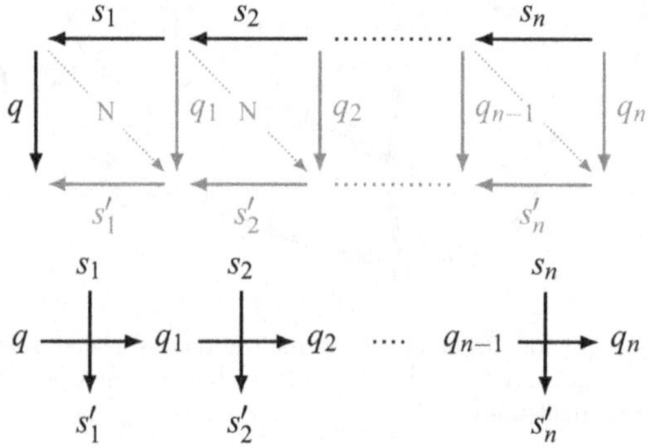

We choose on purpose to always draw a normalization square-diagram backward, such that it coincides with the associated cross-diagram. The function ρ_q induced by the state q maps any normal form (read backward) to the normal form of the right product by q (read backward).

We now aim to strike reasonable (most often optimal) hypotheses for a quadratic normalization (Q, N), associated with an original semigroup S to generate a semigroup $\langle \mathscr{A}_{S,Q,\mathrm{N}} \rangle_+$ that approximates S as sharply as possible. Since the generating sets coincide by Definition 10.2.30, we shall first focus on the case where S should be a quotient of $\langle \mathscr{A}_{S,Q,\mathrm{N}} \rangle_+$ (top-approximation), and next, on the case where $\langle \mathscr{A}_{S,Q,\mathrm{N}} \rangle_+$ should be a quotient of S (bottom-approximation).

Before establishing the top-approximation statement, we just recall that semigroups could appear much more difficult to handle, especially when it comes to automaticity (see [296]) or self-similarity (see [108, 109]). To any (not monoid) semigroup S with a quadratic normalization (Q, N), one obtains a monoid S^1 with a quadratic normalization (Q^1, N^1) by adjoining a unit $\mathtt{1}$ and then by setting $Q^1 = Q \sqcup \{\mathtt{1}\}$ and defining N^1 by $\mathrm{N}^1(w) = \mathrm{N}(w)$ and

$$\mathtt{N}^1(\mathtt{1}w) = \mathtt{N}^1(w\mathtt{1}) = \mathtt{1}\mathtt{N}(w) \qquad (\boxed{\mathtt{N}^1})$$

for $w \in Q^+$. The choice made for Condition ($\boxed{\mathtt{N}^1}$) becomes natural whenever we think of the (adjoined or not) unit $\mathtt{1}$ as some *dummy* element that simply ensures the length-preserving property for $\mathtt{N}^1$ (see Definition 10.2.25 and also [185, Section 2.2]).

Lemma 10.2.31. *Assume that S is a monoid with a quadratic normalization* $(Q, \mathtt{N})$ *satisfying Condition* ($\boxed{\mathtt{N}^1}$). *Then the Mealy automaton* $\mathscr{A}_{S,Q,\mathtt{N}}$ *generates a monoid of which S is a quotient.*

Proof. Let $S^1 = Q^*/\equiv_{\mathtt{N}^1}$ and $\mathscr{A}_{S^1,Q^1,\mathtt{N}^1} = (Q^1, Q^1, \delta, \rho)$ as in Definition 10.2.30. We have to prove that any relation in $\langle\, \mathscr{A}_{S^1,Q^1,\mathtt{N}^1} \,\rangle_+^1$ is a relation in S^1, thereby implying for all $u, v \in Q^*$:

$$\rho_u = \rho_v \implies u \equiv_{\mathtt{N}^1} v.$$

Let $\rho_{p_1} \cdots \rho_{p_k} = \rho_{q_1} \cdots \rho_{q_\ell}$ be some relation in $\langle\, \mathscr{A}_{S,Q,\mathtt{N}} \,\rangle_+^1$ with $p_i \in Q$ for $0 \leq i \leq k$ and $q_j \in Q$ for $0 \leq j \leq \ell$. Any word w over Q admits hence the same image under $\rho_{p_1} \cdots \rho_{p_k}$ and under $\rho_{q_1} \cdots \rho_{q_\ell}$. By taking $w = \mathtt{1}^\omega$ (or any sufficiently long power of $\mathtt{1}$, precisely any word from $\mathtt{1}^{\max(k,\ell)}\mathtt{1}^*$), such a common image corresponds to their normal forms by very definition of $\mathscr{A}_{S^1,Q^1,\mathtt{N}^1}$ (see Figure 10.16). Therefore, the resulting letter-wise equality $\mathtt{1}^{-\omega}\mathtt{N}(p_1 \cdots p_k) = \mathtt{1}^{-\omega}\mathtt{N}(q_1 \cdots q_\ell)$ (where $\mathtt{1}^{-\omega}$ denotes the left-infinite word $\cdots\mathtt{111}$) implies that the two corresponding Q-words $p_1 \cdots p_k$ and $q_1 \cdots q_\ell$ represent a same element in S^1 by definition of $\mathtt{N}^1$. □

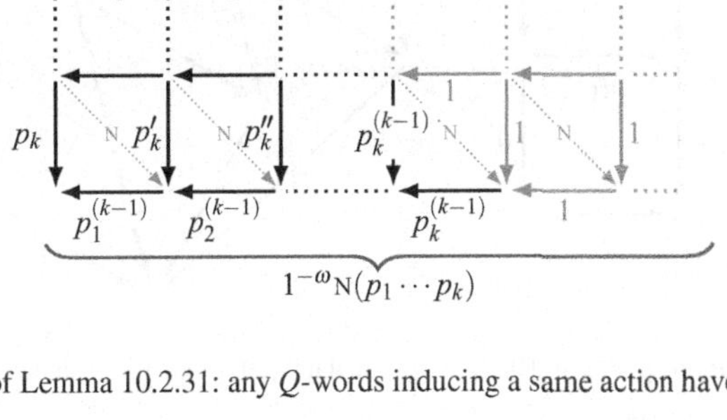

Fig. 10.16 Proof of Lemma 10.2.31: any Q-words inducing a same action have normal forms that coincide.

Although specific to a monoidal framework and then requiring the innocuous Condition ( ), the previous straightforward proof relies only on the definition of a quadratic normalization and on the well-fitted associated Mealy automaton (Definition 10.2.30). For the bottom-approximation statement, we consider an extra assumption, which happens to be necessary and sufficient.

Proposition 10.2.32. *Assume that S is a semigroup with a quadratic normalization (Q, N). If Condition ( ) is satisfied, then the Mealy automaton $\mathscr{A}_{S,Q,\mathrm{N}}$ generates a semigroup quotient of S. The converse holds provided that Condition ( ) is satisfied.*

Proof. Let $S = Q^+/\equiv_{\mathrm{N}}$ and $\mathscr{A}_{S,Q,\mathrm{N}} = (Q, Q, \delta, \rho)$ as in Definition 10.2.30.

($\Leftarrow$) Assume that Condition ( ) is satisfied and that there exists $(a, b, c, d) \in Q^4$ with $ab \equiv_{\mathrm{N}} cd$. We have to prove $\rho_{ab} = \rho_{cd}$. Without lost of generality, the word ab can be supposed to be N-normal, that is, $\mathrm{N}(ab) = \mathrm{N}(cd) = ab$ holds.

Let $u = qv \in Q^n$ for some $n > 0$ and $q \in Q$. We shall prove both $\rho_{ab}(u) = \rho_{cd}(v)$ (coordinate wise) and $\delta_u(ab) \equiv_{\mathrm{N}} \delta_v(cd)$ by induction on $n > 0$. For $n = 1$, we obtain the two square-diagrams on Figure 10.17 (left). With these notations, we have to prove $q_0'' = q_1''$ and $a'b' \equiv_{\mathrm{N}} c'd'$, the latter meaning $\mathrm{N}(a'b') = \mathrm{N}(c'd')$, that is, with the notations from Figure 10.17, the conjunction of $a'' = c''$ and $b'' = d''$. Now these three equalities hold whenever (Q, N) satisfies Condition ( ), as shown on Figure 10.17 (right).

This allows to proceed the induction and to prove the implication ($\Leftarrow$).

($\Rightarrow$) Consider an arbitrary length-three word over Q, say qcd. Let a, b denote the elements in Q satisfying $\mathrm{N}(cd) = ab$. By definition, we deduce $ab \equiv_{\mathrm{N}} cd$. This implies $\rho_{ab} = \rho_{cd}$ by hypothesis. In particular, the images of any word qv under ρ_{ab} and ρ_{cd} coincide: $\rho_{ab}(qv) = \rho_{cd}(qv)$, hence

$$\rho_{ab}(q) = q_0'' = q_1'' = \rho_{cd}(q)$$

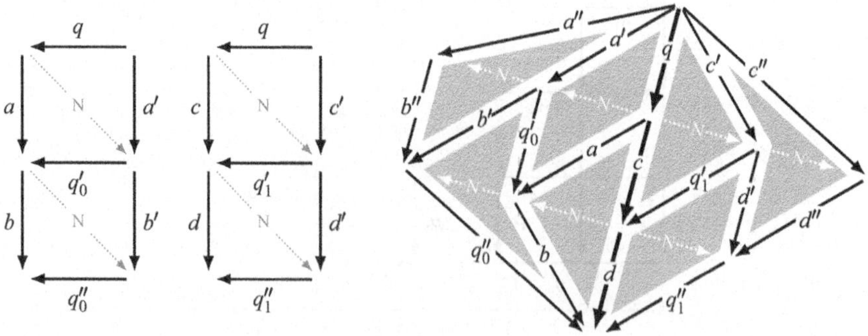

Fig. 10.17 Proof of Proposition 10.2.32: initial data (left) can be pasted into Condition ( ) (right).

and

$$\rho_{\delta_q(ab)}(v) = \rho_{a'b'}(v) = \rho_{c'd'}(v) = \rho_{\delta_q(cd)}(v)$$

(with notations of Figure 10.17). The last equality holds for any original word $v \in Q^*$ and implies $\rho_{a'b'} = \rho_{c'd'}$. Whenever, Condition ($\boxed{N}$) is satisfied, we deduce $\mathrm{N}(a'b') = \mathrm{N}(c'd')$ according to Lemma 10.2.31. We obtain

$$\overline{\mathrm{N}}_{121}(qcd) = \overline{\mathrm{N}}_{2121}(qcd).$$

Therefore (Q, N) satisfies Condition (◆). □

Gathering Lemma 10.2.31 and Proposition 10.2.32, we obtain the following result.

Theorem 10.2.33. *Assume that S is a monoid with a quadratic normalization (Q, N) satisfying Conditions ($\boxed{N}$) and (◆). Then the Mealy automaton $\mathscr{A}_{S,Q,\mathrm{N}}$ generates a monoid isomorphic to S.*

Proof. By construction, S and $\langle\ \mathscr{A}_{S,Q,\mathrm{N}}\ \rangle_+^1$ share a same generating subset Q. Now, any defining relation for S maps to a defining relation for $\langle\ \mathscr{A}_{S,Q,\mathrm{N}}\ \rangle_+^1$ by Proposition 10.2.32 and conversely by Lemma 10.2.31. □

Corollary 10.2.34. *Any monoid with a quadratic normalization satisfying Conditions ($\boxed{N}$) and (◆) is residually finite.*

To conclude this section, we come back to Remark 10.2.24 about the original transducer approach by Thurston.

Definition 10.2.35. With any quadratic normalization (Q, N) is associated its *Thurston transducer* defined as the Mealy automaton $\mathscr{T}_{Q,\mathrm{N}}$ with state set Q, alphabet Q, and transitions as follows:

Corollary 10.2.36. *Assume that S is a monoid with a quadratic normalization (Q, N) satisfying Conditions ($\boxed{N}$) and (◆). The Thurston transducer $\mathscr{T}_{Q,\mathrm{N}}$ and the Mealy automaton $\mathscr{A}_{S,Q,\mathrm{N}}$ being dual automaton, S possesses both the explicitly dual properties of automaticity and self-similarity.*

One of the simplest nontrivial examples is the following. Many others can be found in [482, 483]. The automatic monoid $\mathbf{M} = \langle\ \mathtt{a}, \mathtt{b} : \mathtt{ab} = \mathtt{a}\ \rangle_+^1$ admits a quadratic normalization (Q, N) with $Q = \{\mathtt{1}, \mathtt{a}, \mathtt{b}\}$, $\mathrm{N}(\mathtt{ab}) = \mathtt{1a}$, and $\mathrm{N}(xy) = xy$ for $(x, y) \in Q^2 \setminus \{(\mathtt{a}, \mathtt{b})\}$. The latter has width $(3, 3)$, hence satisfies Condition (◆). According to Theorem 10.2.33, $\mathbf{M}$ is therefore an automaton monoid. The corresponding Wang tileset and the Mealy automaton are displayed on Figure 10.18.

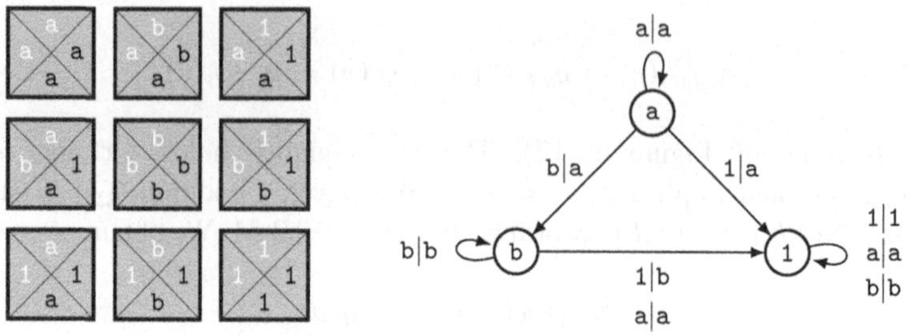

Fig. 10.18 The Wang tileset and the Mealy automaton associated with the monoid $\langle\, \mathtt{a}, \mathtt{b} : \mathtt{ab} = \mathtt{a} \,\rangle^1_+$.

10.3 A Matter of Orbits

In this section, we detail some toggle switch between a classical notion from group theory—Schreier graphs—and some properties of an automaton group about its growth or the growth of its monogenic subgroups. In the first part (Section 10.3.1), we see how automaton groups can provide examples for questions on Schreier graphs, and in the second part (Section 10.3.2), we see how Schreier graphs viewed in a meta structure called an orbit tree or a Schreier trie[1] can give answers to questions about these groups generated by invertible-reversible Mealy automata.

The first part considers essentially polynomial-activity automata which were introduced by S. Sidki [546], whereas the second part considers reversible automata. As we will see, these families of Mealy automata are somehow diametrically opposed.

10.3.1 Schreier Graphs and Polynomial-Activity Automata

There are many ways to define the Schreier graphs of a group acting on some set. Schreier graphs are essentially a generalization of Cayley graphs: let G be a group generated by S and acting on a set X, the vertices of its Schreier graph (depending on S) are the elements of X, and there is an edge $x \to y$ if y is the image of x under the action of some element of S. By considering the action of the group on itself by right-multiplication, this graph coincides with its Cayley graph. In this chapter, the considered group is always an automaton group which acts on the regular rooted tree, leading to the following definitions (see also Section 11.3):

[1]This latter denomination is introduced for the first time in this chapter, motivated in Section 10.3.2.

Definition 10.3.1. Let $\mathscr{A} = (Q, \Sigma, \delta, \rho)$ be a Mealy automaton. Let u be a finite or infinite word on Σ. The *Schreier graph* of $\langle \mathscr{A} \rangle$ *pinpointed* by u is the orbit of u under the action of $\langle \mathscr{A} \rangle$.

The *finite Schreier graph of level n* of $\langle \mathscr{A} \rangle$, $n \in \mathbb{N}$, is the union of the Schreier graphs pinpointed by the length n words, and the *infinite Schreier graph* of $\langle \mathscr{A} \rangle$ is the union of the Schreier graphs pinpointed by the infinite words.

Pinpointed infinite Schreier graphs were for instance used by R. Grigorchuk and V.V. Nekrashevych to prove the existence of amenable actions for non-amenable groups [272].

The *growth* of the infinite Schreier graph pinpointed by some infinite word w is the sequence $(\gamma_n(w))_{n\in\mathbb{N}}$, where $\gamma_n(w)$ is the number of vertices in the closed ball $B(w, n)$ of radius n centered at w:

$$B(w, n) = \{v \in \Sigma^* \mid \exists q \in Q^{\leq n}, \rho_q(w) = v\} .$$

The growth of a group (which can be seen as the growth of its Cayley graph) has been studied for a long time now (see an introduction to the growth problem in Section 10.3.4). In this section, we explore the growth of Schreier graphs of automaton groups, in the flavor of Bondarenko's article [98], focusing on polynomial-activity automata.

A Mealy automaton has *polynomial activity* if its nontrivial cycles are pairwise disjoint (a trivial cycle is a state with a loop inducing the identity). Its *degree* is m if the largest number of these cycles connected by a path is $m + 1$. A polynomial-activity automaton of degree -1 is *bounded*: any nontrivial cycles are disjoint and not connected by a directed path. The *degree* of a state of a Mealy automaton is the degree of the part of this automaton accessible from it. Note that the set of polynomial-activity automata of degree m forms a group for the usual product of Mealy automata.

The class of polynomial-activity automata is important within Mealy automata: it contains many interesting examples (let us mention the well-known automata of Grigorchuk and of Gupta-Sidki), and its subfamily of bounded automata has interesting deciding properties, for example, order and conjugacy [97], which are undecidable in the general framework of Mealy automata [243, 556].

We give here a property of Schreier graphs of polynomial-activity automata.

Theorem 10.3.2. *The infinite pinpointed Schreier graphs of a polynomial-activity automaton have subexponential growth.*

Sketch of the proof. Let $\mathscr{A} = (Q, \Sigma, \delta, \rho)$ be a polynomial-activity Mealy automaton of degree m. By passing at the right power of Σ, we can assume that $\mathscr{A}$ satisfies the following properties:

1. for every state q and every letter a, the state $\delta_a(q)$ either belongs to a cycle or has degree less than the degree of q;
2. for every state q in a nontrivial cycle, there exists a letter a such that $\delta_a(q) = q$.

If q is some state in some power $\mathcal{A}^n$ and $v \in \Sigma^*$ is a word, it is clear from the fact that $\mathcal{A}$ has polynomial activity of degree m that $\delta_v(q)$ is either a state in the same cycle as q or it is not and then its degree is less than the degree of q. In fact, the following property, called **(P)**, can be proven to be true using quite technical considerations (see [98]): there exists a constant C such that for any state q of $\mathcal{A}^n$ and any word $v \in \Sigma^*$ of length greater than $|v| \geq C(\log n)^{(m+1)}$, either $\delta_v(q) = p_1 \cdots p_n$ and each state p_i of $\mathcal{A}$ has degree at most $m-1$ or $\delta_v(q)$ belongs to a loop, that is, there exists a letter $a \in \Sigma$ such that $\delta_{va}(q) = \delta_v(q)$. We let k denote the least integer greater to $C(\log n)^{(m+1)}$.

The proof of the theorem is now by induction on the degree m. For $m = -1$, as the nontrivial cycles of the automaton are disjoint and not connected by a directed path, the number of vertices of each of its pinpointed infinite Schreier graphs is bounded.

Now, suppose that every pinpointed infinite Schreier graph of any polynomial-activity automaton of degree less than m has subexponential growth, less than or equal to $|\Sigma|^{C_1(\log n)^m}$ for some constant C_1. Let $\mathcal{A}$ be a polynomial-activity automaton of degree m and $w = a_1 a_2 \cdots \in \Sigma^\omega$ be an infinite word. Let us look at the growth of the Schreier graph pinpointed by w and in particular at the balls $B(w, n)$.

In what follows, we divide the word w in two factors: a length k prefix (where k is introduced above) whose contribution to the orbit of w is finite, and an infinite suffix $v = a_{k+1} a_{k+2} \cdots$ whose contribution to the orbit of w is shown to be subexponential through Property **(P)**.

Any word of $B(w, n)$ consists then of a length k prefix which is the image of $a_1 \cdots a_k$ by some element of $\langle \mathcal{A} \rangle$ and an infinite suffix which is the image of v by some element of

$$\mathcal{N}_{(n,k)} = \{\delta_{a_1 \cdots a_k}(q) \mid q \in Q^{\leq n}\} .$$

We have a first bound for the size of the considered ball:

$$|B(w, n)| \leq |\Sigma|^k \cdot |\mathcal{N}_{(n,k)}(v)| ,$$

where $\mathcal{N}_{(n,k)}(v)$ denotes the orbit of v under the action of $\mathcal{N}_{(n,k)}$.

Let $\mathcal{B}_n$ denote now the set of the states of $\mathcal{A}^{\leq n}$ of degree less than $m-1$: $\mathcal{B}_n$ satisfies the induction property, and so the size of $\mathcal{B}_n(v)$, the orbit of v under the action of $\mathcal{B}_n$, is bounded by $|\Sigma|^{C_1(\log n)^m}$. Furthermore, by Property **(P)**, for each $q \in \mathcal{N}_{(n,k)} \setminus \mathcal{B}_n$, there exists a letter a such that $\delta_a(q) = q$. Hence

$$\mathcal{N}_{(n,k)} \subseteq \bigcup_{a \in \Sigma} \mathcal{N}^a_{(n,k)} \cup \mathcal{B}_n ,$$

where $\mathcal{N}^a_{(n,k)}$ is the subset of $\mathcal{N}_{(n,k)}$ formed by the elements which are stabilized by the action induced by a.

We consider now the orbit of v under the action of $\mathscr{N}^a_{(n,k)}$ for some letter a.

For $v = a^\omega_{k+1}$, there exist some letter $b \in \Sigma$ and some state $p \in Q^{|q|}$ such that:

$$\rho_q(v) = \begin{cases} b^\omega & \text{for } q \in \mathscr{N}^{a_{k+1}}_{(n,k)} , \\ b\rho_p(a^\omega_{k+1}) & \text{otherwise} . \end{cases}$$

Hence

$$|\mathscr{N}^{a_{k+1}}_{(n,k)}(v)| \le |\Sigma| \quad \text{and} \quad |\mathscr{N}^{b}_{(n,k)}(v)| \le |\Sigma| \cdot \mathscr{B}_n(v) ,$$

for any letter $b \neq a_{k+1}$. For $v = a^\ell_{k+1} c v_1$ with $c \neq a_{k+1}$, we obtain similarly:

$$|\mathscr{N}^{a_{k+1}}_{(n,k)}(v)| \le |\Sigma|^2 \quad \text{and} \quad |\mathscr{N}^{b}_{(n,k)}(v)| \le |\Sigma| \cdot |\mathscr{B}_n(a^{\ell-1}_{k+1} c v_1)| ,$$

for any letter $b \neq a_{k+1}$. The conclusion follows. □

10.3.2 Schreier Tries and Reversible Automata

Until very recently, the Schreier graphs of an automaton (semi)group were seen individually, with no links between them. The notion of an *orbit tree* [353, 354] gives a new, more dynamical, vision of the whole set of finite Schreier graphs for a (semi)group generated by a reversible Mealy automaton. This notion stands on the connected components of the powers of a reversible Mealy automaton. In addition to the above remarks on these components (Remarks 10.1.1, 10.1.2, 10.1.3, and 10.1.4), let us add the following one:

Remark 10.3.3. It is known from [353] that a reversible automaton generates a finite semigroup if and only if the sizes of the connected components of its powers are uniformly bounded. It is straightforward to adapt the proof to show that a reversible automaton generates a finite semigroup if and only if the sizes of the minimizations of the connected components of its powers are uniformly bounded.

In the second part of this section, we will deal with labeled trees. There will be several possible label sets for these trees, but we need to set up some common terminology. All our trees are rooted, i.e., with a selected vertex called the *root*. We will visualize the trees traditionally as growing down from the root. A *path* is a (possibly infinite) sequence of adjacent edges without backtracking from top to bottom. A path is said to be *initial* if it starts at the root of the tree. A *branch* is an infinite initial path. The lead-off vertex of a nonempty path e is denoted by $\top(e)$ and its terminal vertex by $\bot(e)$ whenever the path is finite.

The *level of a vertex* is its distance to the root, and the *level of an edge* or *a path* is the level of its initial vertex.

The vertices of an orbit tree of a reversible Mealy automaton are the connected component of its powers, i.e., the finite Schreier graphs of its dual.

In fact, when the automaton is reversible, its powers have a particular form: each connected component of its $(n+1)$-th power, for some integer n, can be seen as several copies of some connected component of its n-th power. More precisely, we have the following property:

Property 10.3.4. [351] Let $\mathscr{A}$ be a reversible Mealy automaton with state set Q and n be a positive integer. The following links appear between the connected components of $\mathscr{A}^n$ and $\mathscr{A}^{n+1}$:

1. The length n prefixes of the states of a connected component of $\mathscr{A}^{n+1}$ belong to the same connected component in $\mathscr{A}^n$.
2. If u and v are two states of the same connected component in $\mathscr{A}^n$ ($u, v \in Q^n$), then any connected component of $\mathscr{A}^{n+1}$ contains as many states prefixed by u as states prefixed by v.

The *orbit tree* $t(\mathscr{A})$ of the dual of a Mealy automaton $\mathscr{A} = (Q, \Sigma, \delta, \rho)$ has vertices the connected components of the powers of $\mathscr{A}$, and the incidence relation built by adding an element of Q: for any nonnegative integer n, the connected component of a word $u \in Q^n$ is the parent of the connected component(s) of ux, for any $x \in Q$. This notion has been described in [537] for more general actions on trees and leads in this context to a graph. In the case of rooted trees (which is the only one we consider in this chapter), this graph is a tree as proved in [238]. To avoid the heaviness of saying "the orbit tree of the dual of the Mealy automaton," we introduce here a new terminology: if $\mathscr{A}$ is a Mealy automaton, we call the orbit tree of its dual its *Schreier trie*. Indeed, this tree is related to Schreier graphs as each of its levels is a finite Schreier graph of its dual, and it is a trie by looking at the states of the connected components labeling its vertices.

In [238], the vertices of a Schreier trie are labeled by the size of the connected component. Following [353], we use a different though equivalent labeling: if $\mathscr{C}$ is the parent of $\mathscr{D}$ in the orbit tree, we label the edge $\mathscr{C} \to \mathscr{D}$ by the ratio $\frac{|\mathscr{D}|}{|\mathscr{C}|}$ which is known to be an integer by Property 10.3.4. An example of the first levels of a Schreier trie is given in Fig. 10.19.

Each vertex of $t(\mathscr{A})$ is labeled by a connected automaton with state set in Q^n, where n is the level of this vertex in the tree. By a minor abuse, we can consider that each vertex is labeled by a finite language in Q^n or even by a word in Q^n.

Let u be a (possibly infinite) word over Q. The *path of* u in the Schreier trie $t(\mathscr{A})$ is the unique initial path going from the root through the connected components of the prefixes of u; u can be called *a representative* of this path (we can say equivalently that this path is *represented* by u or that the word u *represents* the path).

A branch of a Schreier trie is *active* if it has infinitely many coefficients greater than 1.

The Schreier trie $t(\mathscr{A})$ is built emphasizing the prefix relation. Nevertheless, once this tree obtained, it is interesting to highlight some paths that are suffix

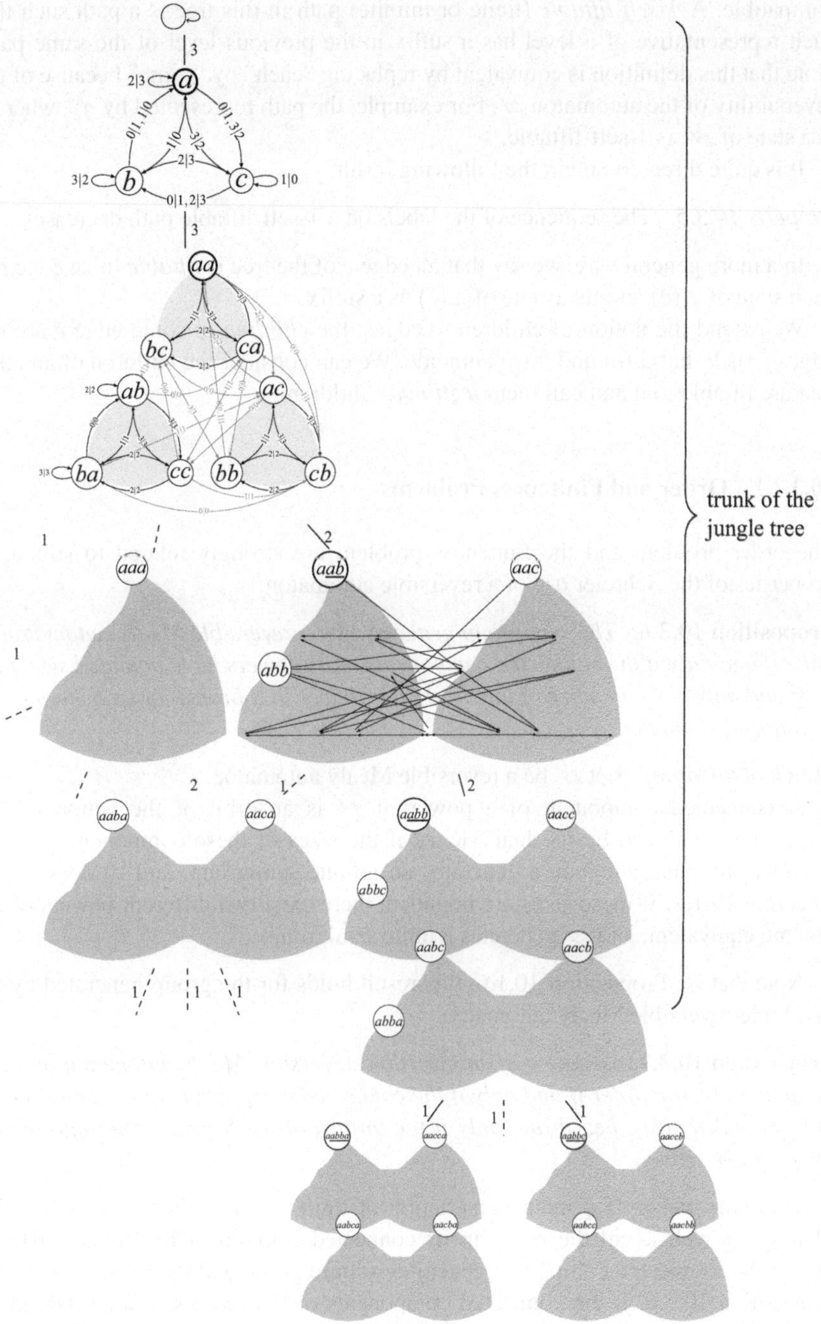

Fig. 10.19 An example of the first levels of a Schreier trie (all edges) and a jungle tree (plain edges). After the trunk the jungle tree consists in a regular binary tree (plain edges).

compatible. A *1-self-liftable* (finite or infinite) path in this tree is a path such that each representative of a level has a suffix in the previous level of the same path. Note that this definition is equivalent by replacing "each" by "some" because of the reversibility of the automaton $\mathscr{A}$. For example, the path represented by q^ω, where q is a state of $\mathscr{A}$, is 1-self-liftable.

It is quite direct to obtain the following result.

Property 10.3.5. The sequence of the labels on a 1-self-liftable path decreases.

In a more general way, we say that an edge e of the tree is *liftable* to an edge f if each state of $\bot(e)$ admits a state of $\bot(f)$ as a suffix.

We extend the notion of children to edges: the *children* of some edge e are the edges f such that $\bot(e)$ and $\top(f)$ coincide. We can consider the children of an edge that are liftable to it and call them *legitimate* children.

10.3.2.1 Order and Finiteness Problems

The order problem and the finiteness problem are strongly related to structural properties of the Schreier trie of a reversible automaton:

Proposition 10.3.6. *The semigroup generated by a reversible Mealy automaton is finite if and only if the connected components of its powers have bounded size, that is, if and only if the number of labels greater than 1 in a branch of its Schreier trie is uniformly bounded.*

Sketch of the proof. Let $\mathscr{A}$ be a reversible Mealy automaton.

A connected component of a power of $\mathscr{A}$ is an orbit of the action of the semigroup generated by its dual. Hence if the sizes of these components are not bounded, the dual automaton generates an infinite semigroup, and so does $\mathscr{A}$ by Theorem 10.1.7. If these sizes are bounded, there exist two different powers of $\mathscr{A}$ that are equivalent, and $\mathscr{A}$ generates a finite semigroup. □

Note that by Proposition 10.1.6, this result holds for the group generated by an invertible-reversible Mealy automaton.

Proposition 10.3.7. *A state q of an invertible-reversible Mealy automaton induces an action of finite order if and only if the connected components of the powers of q have bounded sizes, i.e., if and only if the branch of the Schreier trie represented by q^ω is not active.*

Sketch of the proof. If q induces an action of finite order n, then q^n acts like the identity, as well as all the states in its connected component by Remark 10.1.3. Hence this connected component generates a finite group and the result follows by Proposition 10.3.6. If the connected components of the powers of q are bounded, eventually two of them are isomorphic, with the powers of q labeling the same state, hence q induces an action of finite order. □

If j is a subtree of a Schreier trie, a *j-word* is a representative of one of its vertices. A *cyclic j-word* is a word whose all powers are representative of vertices of j.

10.3.3 The Burnside Problem

The Burnside problem is a famous, long-standing question in group theory. In 1902, W. Burnside asked if a finitely generated group whose all elements have finite order—henceforth called a *Burnside group*—is necessarily finite [122].

The question stayed open until E.S. Golod and I. Shafarevitch exhibit in 1964 an infinite group satisfying Burnside's conditions [255, 256], hence solving the general Burnside problem. In the early 1960s, V.M. Glushkov suggested using automata to attack the Burnside problem [247]. Later, S.V. Alešin [11] in 1972 and then R. Grigorchuk [266] in 1980 gave simple examples of automata generating infinite Burnside groups.

It is remarkable that all known examples of infinite Burnside automaton groups are generated by non reversible Mealy automata. It has been proven in fact that two specific subfamilies of invertible-reversible Mealy automata cannot generate infinite Burnside groups: non-bireversible automata [252] and connected automata of prime size [251]. Both proofs rely on the construction of a particular branch in the Schreier trie of the automaton.

Let $\mathscr{A} = (Q, \Sigma, \delta, \rho)$ be an invertible-reversible Mealy automaton and $t(\mathscr{A})$ be its Schreier trie.

10.3.3.1 When the Automaton Is Not Bireversible

In this case, the automaton $\mathscr{A}$ has at least one non-bireversible connected component, say $\mathscr{B}$, and all the 1-self-liftable branches of the Schreier trie of $\mathscr{B}$ are active (see Lemma 10.3.8 below), hence all the elements of the semigroup generated by $\mathscr{B}$ have infinite order by Proposition 10.3.7.

Lemma 10.3.8. *The 1-self-liftable branches of the Schreier trie of a connected invertible-reversible non-bireversible Mealy automaton are all active.*

Sketch of the proof. Because of the non-bireversibility of the automaton and of all of the connected components of its power, there exists no 1-self-liftable branch with a label 1. □

10.3.3.2 When the Automaton Is Connected of Prime Size

Of course, if the Schreier trie of this automaton has at least one active 1-self-liftable branch, the technique developed in the previous case can apply. But nothing ensures that the existence of such a branch. So we have to develop an alternative strategy in case $\mathscr{A}$ has no active 1-self-liftable branch. This strategy can be summarized into two steps:

Step 1 exhibit a subtree of the Schreier trie whose labels from some level on are 1: hence there is only a finite number of applications induced by the language of the labels of this subtree;

Step 2 prove that the action induced by a word of states has a uniform bounded power equivalent to the action induced by some word in the above language.

Then by E.I. Zelmanov's result [593, 594], the conclusion comes.

Assume that $t(\mathscr{A})$ has no active 1-self-liftable branch.

Jungle Trees

We first build the tree of step 1. This tree, called *jungle tree*, starts with a linear part whose labels decrease and eventually ends as a regular tree with all labels 1.

Definition 10.3.9. Let e be a finite initial 1-self-liftable path such that:

- the lowest (i.e., the last) edge of e has at least two legitimate children;
- each of its legitimate children has label 1.

The *jungle tree* $j(e)$ of e is the subtree of $t(\mathscr{A})$ build as follows:

- it contains the path e—its *trunk*;
- it contains the regular tree rooted by $\perp(e)$, and formed by all the edges which are descendant of $\perp(e)$ and liftable to the lowest edge of e.

The *arity* of this jungle tree is the number of legitimate children of $\perp(e)$. Since every legitimate child has label 1, it is also the label of the last edge of e.

Words in $\perp(e)$ are called *stems*. They have all the same length which is the length of the trunk of $j(e)$.

A tree is a *jungle tree* if it is the jungle tree of some finite initial 1-self-liftable path.

Note that (i) $t(\mathscr{A})$ has at least one jungle tree, since $\mathscr{A}$ has no active self-liftable branch by hypothesis; (ii) $t(\mathscr{A})$ has finitely many jungle trees.

Any jungle tree answers step 1. Let us look closer at the language of j-words, for some jungle tree j whose trunk has length n. In particular, the existence of cyclic j-words is ensured by the simple fact that any j-word of length $n \times (1 + |Q|^n)$ admits a cyclic j-word as a factor. Besides, every cyclic j-word induces an action of finite order, bounded by a uniform constant depending on j, by Proposition 10.3.7.

From now on, j denotes a jungle tree of $\mathscr{A}$, whose trunk has length n.

The set of stems of j has several interesting properties.

Proposition 10.3.10. *The relation over stems $u \sim v$ is defined as follows: there exists $s \in Q^*$ such that usv is a j-word and ρ_{us} acts like the identity on Σ^*, is an equivalence relation.*

Sketch of the proof. Because of the construction of the tree j, the j-words have a lot of good combinatorial properties (proved in [251]):

- any factor of a j-word is itself a j-word;
- if uv is a j-word, with $|v| \geq n$, what can follow uv in j is independent from u. In particular, if vw is also a j-word, then so is uvw;
- if $t, u, v \in Q^*$ are such that tuv is a j-word, then there exists $w \in Q^*$ such that $tuvwu$ is also a j-word (which means that if you are walking on a j-word and you have already seen some factor, you can find eventually this same factor); □

Proposition 10.3.11. *The set of states which appear as first letter of a stem in a $\sim$-class has a cardinal which is greater than or equal to 2 and divides the number of states of $\mathscr{A}$.*

Corollary 10.3.12. *If $\mathscr{A}$ has a prime size, all the states appear as first letter of a stem in a fixed $\sim$-class.*

The main tool of this section is the following one:

Proposition 10.3.13. *Let $\mathscr{A} = (Q, \Sigma, \delta, \rho)$ be a connected bireversible Mealy automaton of prime size, with no active self-liftable branch. Let j be a jungle tree of its Schreier trie, and u some (possibly empty) j-word. Then for any state $q \in Q$, there exists $w \in Q^*$ such that uwq is a j-word and ρ_w acts like the identity of Σ^*.*

Proof. Let s be a stem such that us is a j-word: there exists a stem x with first letter q in the $\sim$-class of s, from Corollary 10.3.12, i.e., there exists $v \in Q^*$ such that svx is a j-word and ρ_{sv} acts like the identity of Σ^*. Conclusion comes from the fact that what can follow a stem depend only on its length n suffix. □

We now have all elements to prove the main result of this section.

Theorem 10.3.14. *A connected invertible-reversible Mealy automaton of prime size cannot generate an infinite Burnside group.*

Proof. Let $\mathscr{A}$ be a connected invertible-reversible Mealy automaton of prime size. If $\mathscr{A}$ is not bireversible, apply step 1.

If $\mathscr{A}$ is bireversible and its Schreier trie has an active self-liftable branch, as seen above, the method of step 1 is still valid, and $\langle \mathscr{A} \rangle$ has an element of infinite order.

Therefore, we can assume that $\mathscr{A}$ is bireversible and its Schreier trie has no active self-liftable branch. Let us show that $\langle \mathscr{A} \rangle$ is finite. Let j be some jungle tree of $t(\mathscr{A})$. As in [353] we prove that for any word $u \in Q^*$, ρ_u has some uniformly bounded power which acts like some cyclic j-word.

Let $u \in Q^*$. We prove by induction that any prefix of u induces the same action as some j-word. It is obviously true for the empty prefix. Fix some $k < |u|$ and suppose that the prefix v of length k of u induces the same action as some j-word s. Let $x \in Q$ be the $(k+1)$-th letter of u. By Corollary 10.3.13, there exists a j-word w inducing the identity, such that swx is a j-word. But vx and swx induce the same action; the result follows. Hence we obtain a j-word $u^{(1)}$ inducing the same action as u.

By the very same process, we can construct, for any $i \in \mathbb{N}$, a j-word $u^{(i)}$ inducing the same action as u, such that $u^{(1)}u^{(2)} \dots u^{(i)}$ is a j-word. Since the set Q^n is finite, there exist $i < j$, $j - i \leq |Q|^n$, such that $u^{(i)}$ and $u^{(j)}$ have the same prefix of length

n. Take $v = u^{(i)}u^{(i+1)}\dots u^{(j-1)}$: v is a cyclic j-word and induces the same action as u^{j-i}. As seen before, the fact that v is a cyclic j-word implies that the order of its induced action ρ_v is bounded by a constant depending only on j, hence so does ρ_u (with a different constant, but still depending only on j). Consequently, every element of $\langle\, \mathcal{A} \,\rangle_+$ has a finite order, uniformly bounded by a constant, whence, as $\langle\, \mathcal{A} \,\rangle_+$ is residually finite, by Zelmanov's theorem [593, 594], $\langle\, \mathcal{A} \,\rangle_+$ is finite, which concludes the proof. □

10.3.4 Growth and Level-Transitivity

In this section, we give a negative answer to the Milnor problem on the existence of groups of intermediate growth for a very particular class of automaton groups: the ones generated by an invertible-reversible Mealy automaton whose Schreier trie has a unique branch.

This family of groups contain in particular automaton groups which are branch groups, one of the three classes into which the class of just infinite groups is naturally decomposed [42, 270].

10.3.4.1 Growth

We first recall some definitions concerning the notion of growth function for groups. See also Sections 11.3.1 and 11.4.

Let H be a semigroup generated by a finite set S. The *length* of an element g of the semigroup, denoted by $|g|$, is the length of its shortest decomposition:

$$|g| = \min\{n \mid \exists s_1, \dots, s_n \in S,\ g = s_1 \cdots s_n\}\,.$$

The *growth function* γ_H^S of the semigroup H with respect to the generating set S enumerates the elements of H with respect to their length:

$$\gamma_H^S(n) = |\{g \in H\,;\ |g| \leq n\}|\,.$$

The *growth functions* of a group are defined similarly by taking symmetrical generating sets.

The growth functions corresponding to two generating sets are equivalent [409], so we may define the *growth* of a group or a semigroup as the equivalence class of its growth functions. Hence, for example, a finite (semi)group has a bounded growth, while an infinite abelian (semi)group has a polynomial growth, and a non-abelian free (semi)group has an exponential growth.

It is quite easy to obtain groups of polynomial or exponential growth. Answering a question of J. Milnor [419], R. Grigorchuk gave the very first example of an automaton group of intermediate growth [269]: faster than any polynomial, slower than any exponential (see Grigorchuk automaton in Figure 10.2).

10.3.4.2 Level-Transitivity

The action of a (semi)group generated by an invertible Mealy automaton $\mathscr{A} = (Q, \Sigma, \delta, \rho)$ is *level-transitive* if its restriction to Σ^n has a unique orbit, for any n (this notion is equivalently called *spherically transitive* [271]). From a dual point of view, it means that the powers of the dual $\mathfrak{d}\mathscr{A}$ are connected, i.e., its Schreier trie has a unique branch.

The level-transitivity of an automaton semigroup has some influence on the growth of the semigroup generated by the dual automaton.

Theorem 10.3.15 ([352]). *The semigroup generated by an invertible-reversible Mealy automaton whose Schreier trie has a unique branch has exponential growth.*

Note that the exponential growth of the semigroup generated by an invertible Mealy automaton implies the exponential growth of the group generated by this same automaton.

The Nerode classes of two consecutive powers of its state set are linked in the following way:

Lemma 10.3.16. *Let $\mathscr{A} = (Q, \Sigma, \delta, \rho)$ be an invertible-reversible Mealy automaton whose Schreier trie has a unique branch. Let $(C_i)_{1\le i\le k}$ be the Nerode classes of Q^n for some n, and D be a Nerode class of Q^{n+1}. We have*

$$D = \bigcup_{q\in Q_D} C_{i_{q,D}}q \quad \text{and} \quad D = \bigcup_{q\in Q'_D} qC_{i'_{q,D}},$$

where $Q_D \subseteq Q$ and $Q'_D \subseteq Q$ have the same cardinality, and the $(i_{q,D})_{q\in Q_D}$ on the one hand and the $(i'_{q,D})_{q\in Q_D}$ on the other are pairwise distinct.

The automata $\mathfrak{m}(\mathscr{A}^n)$ and $\mathfrak{m}(\mathscr{A}^{n+1})$ have the same size if and only if $Q_D = Q'_D = Q$.

Theorem 10.3.15 can now be proven by observing that a relatively immediate consequence of Lemma 10.3.16 is that the sequence $(|\mathfrak{m}(\mathscr{A}^n)|)_{n\ge 0}$ increases strictly and exponentially.

The next theorem improves Theorem 10.3.15 for Mealy automata of prime size.

Theorem 10.3.17 ([351]). *The semigroup generated by an invertible-reversible Mealy automaton of prime size whose Schreier trie has a unique branch is free on the automaton state set.*

The idea is to bound the sizes of the Nerode classes in the powers of the Mealy automaton.

For the next three lemmas, let $\mathscr{A} = (A, \Sigma, \delta, \rho)$ be a reversible p-state Mealy automaton, p prime, whose Schreier trie is formed by a unique branch. By Proposition 10.3.6, $\mathscr{A}$ generates an infinite semigroup.

Lemma 10.3.18. *There cannot exist two equivalent words of different length in Q^*.*

Proof. For each m, $\mathscr{A}^m$ is connected, and so any two words of length m are mapped one onto the other by an element of $\langle \mathfrak{d}\mathscr{A} \rangle_+$.

Let u and v be two equivalent words of different lengths, say $|u| < |v|$. Every word of length $|v|$ is then equivalent to a word of length $|u|$: if w is of length $|v|$, then $w = \delta_t(v)$ for some $t \in \Sigma^*$, and, by Remark 10.1.2, w is equivalent to $\delta_t(u)$ of length $|u|$. By Remark 10.3.3, the semigroup $\langle \mathscr{A} \rangle_+$ is finite, which is impossible. □

Lemma 10.3.19. *All the Nerode classes of a given power Q^m have the same size, which happens to be a power of p.*

The proof of this lemma is direct from Remark 10.1.1.

Lemma 10.3.20. *There cannot exist two equivalent words of the same length in Q^*.*

Proof. Let u and v be two different equivalent words of the same length $n + 1$. Let us prove by induction on $m > n$ that $\mathfrak{m}(\mathscr{A}^m)$ has at most p^n states.

The automaton $\mathscr{A}^{n+1}$ has p^{n+1} states. The words u and v are in the same Nerode class: by Lemma 10.3.19, all Nerode classes of Q^{n+1} have at least p elements, and $\mathfrak{m}(\mathscr{A}^{n+1})$ has at most p^n states.

Suppose that $\mathfrak{m}(\mathscr{A}^m)$ has at most p^n states. Then, since all Nerode classes have the same size by Lemma 10.3.19, the induction hypothesis implies that they have at least p^{m-n} elements. Let us look at $[x_1^m]$: it contains

$$x_1^m, u_1, u_2, \ldots, u_{p^{m-n}-1} ,$$

which are pairwise distinct. Among these words, there is at least one whose suffix in x_1 is the shortest, say u_1 without loss of generality: $p^{m-n} > 1$ and x_1^m has the longest possible suffix in x_1. Hence $[x_1^{m+1}]$ contains the following pairwise distinct $p^{m-n} + 1$ words:

$$x_1^{m+1}, u_1x_1, u_2x_1, \ldots, u_{p^{m-n}-1}x_1, x_1u_1 .$$

By Lemma 10.3.19, $|[x_1^{m+1}]|$ is a power of p, so $|[x_1^{m+1}]| \geq p^{m+1-n}$. As all Nerode classes of Q^{m+1} have the same cardinality, we can conclude that $\mathfrak{m}(\mathscr{A}^{m+1})$ has at most $p^{m+1}/p^{m+1-n} = p^n$ elements, ending the induction.

Consequently, since there is only a finite number of different Mealy automata with up to p^n states, there exist $k < \ell$ such that $\mathfrak{m}(\mathscr{A}^k)$ and $\mathfrak{m}(\mathscr{A}^\ell)$ are equal up to state numbering. Hence the semigroup $\langle \mathscr{A} \rangle_+$ is finite, which is impossible. □

As a corollary of Lemmas 10.3.18 and 10.3.20, we can state the following proposition.

Proposition 10.3.21. *Let $\mathscr{A}$ be a reversible Mealy automaton of size p, with p prime. If the Schreier trie of $\mathscr{A}$ has a unique branch, then $\mathscr{A}$ generates a free semigroup of rank p, with the states of $\mathscr{A}$ being free generators of the semigroup.*

Chapter 11
Amenability of Groups and G-Sets

Laurent Bartholdi

Abstract This text surveys classical and recent results in the field of amenability of groups, from a combinatorial standpoint. It has served as the support of courses at the University of Göttingen and the École Normale Supérieure. The goals of the text are (1) to be as self-contained as possible, so as to serve as a good introduction for newcomers to the field; (2) to stress the use of combinatorial tools, in collaboration with functional analysis, probability, etc., with discrete groups in focus; (3) to consider from the beginning the more general notion of amenable *actions*; and (4) to describe recent classes of examples and in particular groups acting on Cantor sets and topological full groups.

11.1 Introduction

In 1929, John von Neumann introduced in [576] the notion of amenability of G-spaces. Fundamentally, he considers the following property of a group G acting on a set X: *The right G-set X is amenable if there exists a G-invariant mean on the power set of X*, namely, a function m: {subsets of X} $\to [0, 1]$ satisfying $m(A \sqcup B) = m(A) + m(B)$ and $m(X) = 1$ and $m(Ag) = m(A)$ for all $A, B \subseteq X$ and all $g \in G$.

Amenability may be thought of as a finiteness condition, since nonempty finite G-sets are amenable with $m(A) = \#A/\#X$; it may also be thought of as a fixed-point property: on a general G-set, there exists a G-invariant mean; on a compact G-set, there exists a G-invariant measure; on a convex compact G-set, there exists a G-fixed point, see Section 11.6; and on a G-measure space, there exists a G-invariant measurable family of means on the orbits, see Section 11.6.2.

Amenability may be defined for other objects such as graphs and random walks on sets. If X is a G-set and G is finitely generated, then X naturally has the structure

Supported by the "@raction" grant ANR-14-ACHN-0018-01.

L. Bartholdi (✉)
École Normale Supérieure, Paris, France

Mathematical Institute, Georg-August University of Göttingen, Bunsenstrasse, Göttingen, Germany
e-mail: laurent.bartholdi@gmail.com

V. Berthé, M. Rigo (eds.), *Sequences, Groups, and Number Theory*, Trends in Mathematics, https://doi.org/10.1007/978-3-319-69152-7_11

of a graph, with one edge from x to xs for every $x \in X, s \in S$. Amenability means, in the context of graphs, that there are finite subsets of X with arbitrarily small boundary with respect to their size. In terms of random walks, it means that there are finite subsets with arbitrarily small connectivity between the set and its complement and equivalently that the return probability of the random walk decreases subexponentially in time, see Section 11.8.

The definition may also be modified in another direction: rather than considering group actions, we may consider equivalence relations or more generally groupoids. The case we concentrate on is an equivalence relation with countable leaves on a standard measure space. The orbits of a countable group acting measurably naturally give rise to such an equivalence relation. This point of view is actually very valuable: quite different groups (*e.g.*, one with free subgroups, one without) may generate the same equivalence relation; see Section 11.6.2.

One of the virtues of the notion of amenability of G-sets is that there is a wealth of equivalent definitions; depending on context, one definition may be easier than another to check, and another may be more useful. In summary, the following will be shown, in the text, to be equivalent for a G-set X:

1. X is amenable; *i.e.*, there is a G-invariant mean on subsets of X;
2. There is a G-invariant normalized positive functional in $\ell^\infty(X)^*$, see Corollary 11.2.25;
3. For every bounded functions h_i on X and $g_i \in G$, the function $\sum_i h_i(1 - g_i)$ is nonnegative somewhere on X, see Theorem 11.2.29;
4. For every finite subset $S \subseteq G$ and every $\epsilon > 0$, there exists a finite subset $F \subseteq X$ with $\#(FS \setminus F) < \epsilon \# F$, see Theorem 11.3.23(5);
5. For every finite subset $S \subseteq G$, every $\epsilon > 0$, and every $p \in [1, \infty)$, there exists a positive function $\phi \in \ell^p(X)$ with $\|\phi s - \phi\| < \epsilon \|\phi\|$ for all $s \in S$, see Theorem 11.3.23(4);
6. Every convex compact set equipped with a G-equivariant map from X admits a fixed point, see Theorem 11.6.4;
7. Every compact set equipped with a G-equivariant map from X admits an invariant measure, see Theorem 11.6.7;
8. The isoperimetric constant (Definition 11.8.2) of every nondegenerate G-driven random walk on X vanishes, see Theorem 11.8.4(2);
9. The spectral radius (Definition 11.8.2) of every nondegenerate G-driven random walk on X is equal to 1, see Theorem 11.8.4(3);
10. For every field $\Bbbk$, every finite subset $S \subseteq G$, and every $\epsilon > 0$, there exists a finite-dimensional subspace $F \le \Bbbk X$ with $\dim(FS) < (1 + \epsilon) \dim F$, see Theorem 11.10.29.

In turn, *non*-amenability also amounts to existential statements: the following are equivalent:

1. X is *not* amenable;
2. There exists a "paradoxical decomposition" of X, namely, $X = Z_1 \sqcup \cdots \sqcup Z_m = Z_{m+1} \sqcup \cdots \sqcup Z_{m+n} = Z_1 g_1 \sqcup \cdots \sqcup Z_{m+n} g_{m+n}$ for some $Z_i \subset X$ and $g_i \in G$, see Theorem 11.5.14(2);

3. There exists a map $\phi: X \circlearrowleft$ and a finite subset $S \subseteq G$ with $\#\phi^{-1}(x) = 2$ and $\phi(x) \in xS$ for all $x \in X$, see Theorem 11.5.14(4);
4. There exists a free action of a non-amenable group H on X by G-wobbles, i.e., with the property that for every $h \in H$ there is a finite subset $S \subseteq G$ with $xh \in xS$ for all $x \in X$, see Theorem 11.5.15(3);
5. For every $n \in \mathbb{N}$ large enough and for every field $\Bbbk$, there exits an $n \times (n-1)$ matrix with entries in $\Bbbk G$ that gives an injective map $(\Bbbk X)^n \hookrightarrow (\Bbbk X)^{n-1}$, see Theorem 11.10.12.

Amenability has been given particular attention for groups themselves, seen as G-sets under right multiplication; see the next section. We stress that many results that exclusively concern groups (*e.g.*, the recent proofs that topological full groups are amenable) are actually proven using amenable G-sets in a fundamental manner. The reason is that a group is amenable if and only if it acts on an amenable G-set with amenable point stabilizers, see Proposition 11.2.26.

Quotients of amenable G-sets are again amenable; but sub-G-sets of amenable G-sets need not be amenable. A stronger notion will be developed in Section 11.9, that of *extensively amenable* G-sets. It has the fundamental property that if $\pi: X \twoheadrightarrow Y$ is a G-equivariant map between G-sets, then X is extensively amenable if and only if both Y and all $\pi^{-1}(y)$ are extensively amenable, the latter for the action of the stabilizer G_y.

We detail slightly the Day-Reiter characterization of amenability given above: the space $\ell^1(G)$ of summable functions on G is a Banach algebra under convolution, and $\ell^1(X)$ is a Banach $\ell^1(G)$-module. We denote by $\varpi(\ell^1 G)$ and $\varpi(\ell^1 X)$, respectively, the ideal and submodule of functions with 0 sum, and by $\ell^1_+(G)$ and $\ell^1_+(X)$ the cones of positive elements.

Then X_G is *amenable* if and only if for every $\epsilon > 0$ and every $g \in \varpi(\ell^1 G)$, there exists $f \in \ell^1_+(X)$ with $\|fg\| < \epsilon\|f\|$, see Proposition 11.3.25.

The quantifiers may be exchanged; we call X_G *laminable*[2] if for every $\epsilon > 0$ and every $f \in \varpi(\ell^1 X)$ there exists $g \in \ell^1_+(G)$ with $\|fg\| < \epsilon\|g\|$, see Theorem 11.8.20. It has the consequence that there exists a measure μ on G such that every μ-harmonic function on X is constant and equivalently that there are no nontrivial tail events for the random walk on X driven by μ.

In case $X = G_G$, these definitions are equivalent, but for G-sets the properties of being amenable or laminable are in general position.

11.1.1 Amenability of Groups

John von Neumann's purpose, in introducing amenability of G-spaces, was to understand better the group-theoretical nature of the Hausdorff-Banach-Tarski paradox. This paradox, due to Banach and Tarski [35] and based on Hausdorff's

[2]This has been considered recently by Kaimanovich under the name of "L-amenable actions."

work [286], states that a solid ball can be decomposed into five pieces, which when appropriately rotated and translated can be reassembled into two balls of same size as the original one. It could have been felt as a death blow to measure theory; it is now resolved by saying that the pieces are not measurable.

A group is called *amenable* if all nonempty G-sets are amenable; and it suffices to check that the regular G-set G_G is amenable, see Corollary 11.2.11.

Using the "paradoxical decompositions" criterion, it is easy to see that the free group $F_2 = \langle a, b \mid \rangle$ is not amenable: we exhibit a partition $F_2 = G_1 \sqcup \cdots \sqcup G_m \sqcup H_1 \sqcup \cdots \sqcup H_n$ and elements $g_1, \ldots, g_m, h_1, \ldots, h_n$ with $F_2 = G_1 g_1 \sqcup \cdots \sqcup G_m g_m = H_1 h_1 \sqcup \cdots \sqcup H_n h_n$ as follows. Set

$$\begin{aligned} G_1 &= \{\text{words whose reduced form ends in } a\} \cup \{1, a^{-1}, a^{-2}, \ldots\}, \\ G_2 &= \{\text{words whose reduced form ends in } a^{-1}\} \setminus \{a^{-1}, a^{-2}, \ldots\}, \\ H_1 &= \{\text{words whose reduced form ends in } b\}, \\ H_2 &= \{\text{words whose reduced form ends in } b^{-1}\}; \end{aligned}$$

then $F_2 = G_1 \sqcup G_2 \sqcup H_1 \sqcup H_2 = G_1 \sqcup G_2 a = H_1 \sqcup H_2 b$.

The group of rotations $SO_3(\mathbb{R})$ contains a free subgroup F_2 and even one that acts freely on the sphere S^2; so its orbits are all isomorphic to F_2. Choose a *transversal*: a subset $T \subset S^2$ intersecting every F_2-orbit in exactly one point. Consider then the sphere partition $S^2 = TG_1 \sqcup TG_2 \sqcup TH_1 \sqcup TH_2 = TG_1 \sqcup TG_1 a = TH_1 \sqcup TH_2 b$; this is the basis for the paradoxical Hausdorff-Banach-Tarski decomposition.

John von Neumann also noted that the class of amenable groups is closed under the following operations $(*)$: subgroups, quotients, extensions, and directed unions. It contains all finite and abelian groups. More generally, a criterion due to Følner, Theorem 11.3.23(5), shows that all groups in which every finite subset generates a group of subexponential word growth[3] are amenable. One may therefore define the following classes:

$$\begin{aligned} EG &= \text{the smallest class containing finite and abelian groups and closed under } (*), \\ SG &= \text{the smallest class containing groups of subexponential growth and closed} \\ &\quad\ \text{under}(*), \\ AG &= \text{the class of amenable groups}, \\ NF &= \text{the class of groups with no free subgroups}; \end{aligned}$$

and concrete examples show that all inclusions

$$EG \subsetneqq SG \subsetneqq AG \subsetneqq NF$$

[3]Namely, in which the number of elements expressible as a product of at most n generators grows subexponentially in n.

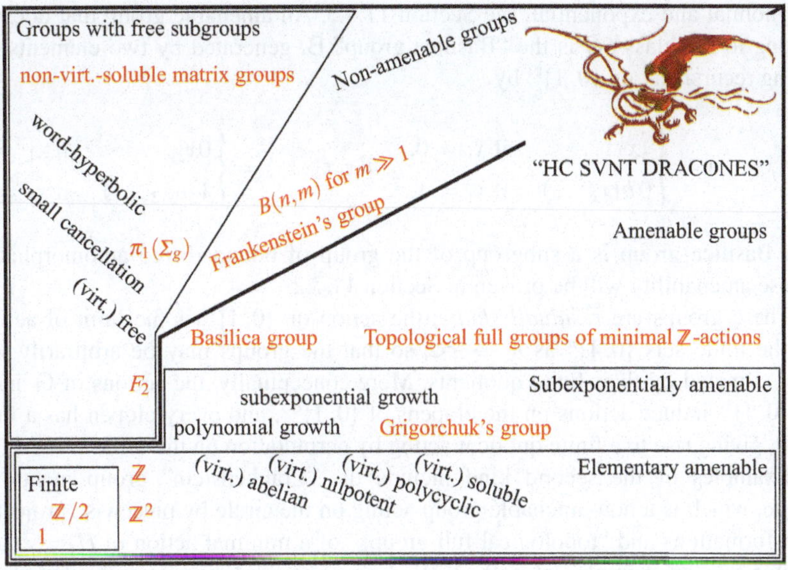

Fig. 11.1 The universe of groups

are strict: the "Grigorchuk group" G for the first inclusion, see Section 11.4.3; the group of "bounded tree automorphisms" for the second inclusion, see Section 11.7.2; and the "Frankenstein group" for the last one, see Section 11.7.3.

This text puts a strong emphasis on examples; they are essential to obtain a (however coarse) picture of the universe of discrete groups, see Figure 11.1. A fairly general framework contains a large number of important constructions: groups acting on Cantor sets. On the one hand, if we choose $X = \mathscr{A}^{\mathbb{N}}$ as model for the Cantor set, we have examples of groups defined by automatic transformations of X, namely, by actions of invertible transducers. On the other hand, we may fix a "manageable" group H acting on X and consider the group of self-homeomorphisms of X that are piecewise H.

Examples of the first kind may be constructed via their recursively defined actions on X. The Grigorchuk group G is the group acting on $\{0,1\}^{\mathbb{N}}$ and generated by four elements a, b, c, d defined by

$$a(x_0x_1\dots) = (1-x_0)x_1\dots, \qquad b(x_0x_1\dots) = \begin{cases} x_0\, a(x_1\dots) & \text{if } x_0 = 0,\\ x_0\, c(x_1\dots) & \text{if } x_0 = 1,\end{cases}$$

$$c(x_0x_1\dots) = \begin{cases} x_0\, a(x_1\dots) & \text{if } x_0 = 0,\\ x_0\, d(x_1\dots) & \text{if } x_0 = 1,\end{cases} \qquad d(x_0x_1\dots) = \begin{cases} x_0x_1\dots & \text{if } x_0 = 0,\\ x_0\, b(x_1\dots) & \text{if } x_0 = 1.\end{cases}$$

The Grigorchuk group gained prominence in group theory for being a finitely generated infinite torsion group and for having intermediate word growth between polynomial and exponential, see Section 11.4.3. An amenable group that does not belong to the class SG is the "Basilica group" B, generated by two elements a, b acting recursively on $\{0,1\}^{\mathbb{N}}$ by

$$a(x_1x_2\dots) = \begin{cases} 1x_2\dots & \text{if } x_1 = 0,\\ 0\,b(x_2\dots) & \text{if } x_1 = 1,\end{cases} \qquad b(x_1x_2\dots) = \begin{cases} 0x_2\dots & \text{if } x_1 = 0,\\ 1\,a(x_2\dots) & \text{if } x_1 = 1.\end{cases}$$

The Basilica group is a subgroup of the group of bounded tree automorphisms, whose amenability will be proven in Section 11.7.2.

These groups are *residually finite*: the action on $\{0,1\}^{\mathbb{N}}$ is the limit of actions on the finite sets $\{0,1\}^n$ as $n \to \infty$, so that the groups may be arbitrarily well approximated by their finite quotients. More conceptually, the actions of G and B on $\{0,1\}^{\mathbb{N}}$ induce actions on the clopens of $\{0,1\}^\infty$, and every clopen has a finite orbit, giving rise to a finite quotient acting by permutation on the orbit.

Examples of the second kind include the "Frankenstein" group mentioned above, which is a non-amenable group acting on the circle by piecewise projective transformations and "topological full groups" of a minimal action of $H = \mathbb{Z}$ on a Cantor set, for example, let $\sigma: 0 \mapsto 01, 1 \mapsto 0$ be the Fibonacci substitution and consider $H = \langle S \rangle$ the two-sided shift on the subset $X = \overline{\{S^n(\sigma^\infty(0)) \mid n \in \mathbb{Z}\}} \subset \{0,1\}^{\mathbb{Z}}$. Let G be the group of piecewise-H homeomorphisms of X. Then G' is an example of a simple, infinite, finitely generated, amenable group.

These groups' actions on the Cantor set exhibit behaviors at the exact opposite of G and B: the actions are *expansive*: the orbit of a clopen may be used to separate points in X.

Finally, we consider in Section 11.10 the adaptation of amenability to a linear setting: on the one hand, a natural notion of amenability of $\mathscr{A}$-modules for an associative algebra $\mathscr{A}$ and, on the other hand, a characterization of amenability by cellular automata.

11.1.2 Why This Text?

After John von Neumann's initial work in the late 1920s, amenability of groups has developed at great speed in the 1960s and then remained mostly dormant till the late 2000s, when a variety of new techniques and examples appeared. It seems now to be a good time to reread and rewrite the fundamentals of the field with these developments in mind.

I have done my best to include all the material I found digestible and to express it in the "best" generality, namely, the maximum generality that does not come at the price of arcane definitions or notation. Whenever possible, I included complete proofs of the results, so that the text may be used for a course as well as for a reference.

I have also striven to follow von Neumann's use of G-sets rather than groups; it seems to me that clarity is gained by separating the set X (with a right G-action) from the group G.

I have also, consciously, avoided any mention of amenability for topological groups. This notion is well developed for second-countable locally compact groups, see, *e.g.*, [62, 499], so I should justify its exclusion. I have felt that either the results stated for discrete groups extend more or less obviously to topological groups (and then there is no point in loading the notation with topology), or they don't extend, and then the additional effort would be a distraction from the main topic.

I have also devoted a fairly large part of the text to examples and, in particular, to groups defined by their action on a Cantor set, see the previous section. I have included exercises, with rankings *=just check the definitions, **=requires some thought, and ***=probably very difficult. Problems are like *** exercises but are questions rather than statements.

I have consulted a large number of sources and did my best to attribute to their original authors all results and fragments of proof that I have used. Apart from articles, these sources include notes from a course given by Nicolas Monod at EPFL in 2007 and from a course given by Anna Erschler and myself at ENS in 2016 and books in preparation by Kate Juschenko [317] and Gábor Pete [477]. I have also made abundant use of [137, 265], and [62, Chapter 5 and Appendix G].

I benefited from useful conversations with and remarks from Yves de Cornulier, Anna Erschler, Vadim Kaimanovich, Peter Kropholler, Yash Lodha, Nicolas Matte Bon, Nicolas Monod, Volodia Nekrashevych, and Romain Tessera. I thank all of them heartily.

11.1.3 Why Not This Text?

For lack of space, I have left out much material that I wanted to include. First and foremost, I have not touched at all at the boundary initiated by Furstenberg; the "size" of its boundary is an indication of the non-amenability of a G-set.

I have also left out much material related to quantitative invariants — drift, entropy, on- and off-diagonal probabilities of return of random walks, and their relation to other invariants such as growth and best-case distortion of embeddings in convex metric spaces such as Hilbert space. This topic is evolving rapidly, and I fear that my rendition would be immediately obsolete.

I would have preferred to write Section 11.6.2 in terms of groupoids, especially since groupoids appear anyways in Section 11.9.2. In the end, I have opted for directness at the cost of generality.

Finally, I put as much effort as I could into including applications and examples in the text; but I omitted the most important ones, *e.g.*, Margulis's work on lattices in semisimple Lie groups and percolation on graphs, feeling they would take us too far adrift.

11.1.4 Notation

We mainly use standard mathematical notation. We try to keep Latin capitals for sets, Latin lowercase for elements, and Greek for maps. A subset inclusion $A \subset B$ is strict, while $A \subseteq B$ means that A could equal B. The difference and symmetric difference of two sets A, B are, respectively, written $A \setminus B$ and $A \triangle B$. We denote by $\mathfrak{P}(X)$ the power set of X and by $\mathfrak{P}_f(X)$ the collection of finite subsets of X. Since it appears quite often in the context of amenability, we use $A \Subset B$ ("compactly contained") to mean that A is a finite subset of B.

We denote by A^X the set of maps $X \to A$, and by $A^{(X)}$ the restricted product of A, namely, the set of finitely supported maps $X \to A$. Under the operation of symmetric difference, $\mathfrak{P}(X)$ and $\mathfrak{P}_f(X)$ are, respectively, isomorphic to $(\mathbb{Z}/2)^X$ and $(\mathbb{Z}/2)^{(X)}$.

We denote by $\operatorname{Sym}(X)$ the group of finitely supported permutations of a set X and abbreviate $\operatorname{Sym}(n) = \operatorname{Sym}(\{1, \dots, n\})$. Groups and permutations always act on the right, and we denote by $X \leftarrow\!\!\!\!\!\circlearrowright G$ a set X equipped with a right G-action.

We denote by $\mathbb{1}_A$ the characteristic function of a set A, and also by $\mathbb{1}_{\mathscr{P}}$ the function that takes value 1 when property $\mathscr{P}$ holds and 0 otherwise.

Finally, we write $x \restriction S$ for various kinds of restriction of the object x to a set S.

11.2 Means and Amenability

Definition 11.2.1. Let X be a set. A *mean* on X is a function[4] $m: \mathfrak{P}(X) \to [0, 1]$ satisfying

$$m(X) = 1,$$
$$m(A \sqcup B) = m(A) + m(B) \text{ for all disjoint } A, B \subseteq X.$$

(This last property is often called *finite additivity*, as opposed to the σ-additivity property enjoyed by measures, in which countable unions are allowed.)

It easily follows from the definition that $m(\emptyset) = 0$; that $m(A) \leq m(B)$ if $A \subseteq B$; and that $m(A_1 \sqcup \cdots \sqcup A_k) = m(A_1) + \cdots + m(A_k)$ for pairwise disjoint $A_1, \dots, A_k$.

We denote by $\mathscr{M}(X)$ the set of means on X, with the usual topology on a set of functions, namely, a sequence $m_n \in \mathscr{M}(X)$ converges to m precisely if for every $\varepsilon > 0$ and every finite collection $A_1, \dots, A_k \subseteq X$ we have $|m_n(A_i) - m(A_i)| < \varepsilon$ for all $i \in \{1, \dots, k\}$ and all n large enough.

Observe that $\mathscr{M}$ is a covariant functor: if $f: X \to Y$, then we have a natural map $f_*: \mathscr{M}(X) \to \mathscr{M}(Y)$ given by

$$f_*(m): B \mapsto m(f^{-1}(B)) \qquad \text{for all } B \subseteq Y.$$

[4]By $\mathfrak{P}(X)$ we denote the power set of X, namely, the set of its subsets.

In particular, if a group G acts on X, then it also acts on $\mathscr{M}(X)$. For a right action $\cdot\colon X \times G \to X$, we have a right action on $\mathscr{M}(X)$ given by $(m \cdot g)(A) = m(A \cdot g^{-1})$ for all $A \subseteq X$.

Definition 11.2.2 (von Neumann [576]). Let G be a group and let $X \curvearrowleft G$ be a set on which G acts. The G-set X is *amenable* if there is a G-fixed element in $\mathscr{M}(X)$.

A group G is *amenable* if all nonempty right G-sets are amenable.

In other words, the G-set X is amenable if $\mathscr{M}(X)^G \neq \emptyset$, namely, if there exists a mean m on X such that $m(Ag) = m(A)$ for all $g \in G$ and all $A \subseteq X$.

11.2.1 First Examples

Proposition 11.2.3. *Every finite, nonempty G-set is amenable. More generally, every G-set with a finite orbit is amenable.*

Note that, trivially, the empty set is never amenable since a mean requires $m(\emptyset) = 0 \neq 1 = m(X)$.

Proof. Let xG be a finite G-orbit in the G-set X. Then $m(A) := \#(A \cap xG)/\#(xG)$ defines a G-invariant mean on X.

In particular, finite groups are amenable. We shall now see that, although amenable groups abound, extra logical tools are necessary to provide more examples.

Proposition 11.2.4. *The infinite cyclic group $\mathbb{Z}$ is amenable.*

Proof (False proof). Define $m \in \mathscr{M}(\mathbb{Z})$ by

$$m(A) = \lim_{n\to\infty} \frac{\#(A \cap \{1,2,\dots,n\})}{n}.$$

It is clear that $m(A)$ is contained in $[0,1]$, and the axioms of a mean are likewise easy to check. Finally, if g denote the positive generator of $\mathbb{Z}$,

$$\begin{aligned}|m(Ag) - m(A)| &= \lim_{n\to\infty} \frac{|\#(Ag \cap \{1,2,\dots,n\}) - \#(A \cap \{1,2,\dots,n\})|}{n}\\ &= \lim_{n\to\infty} \frac{|\#(A \cap \{0,1,\dots,n-1\}) - \#(A \cap \{1,2,\dots,n\})|}{n}\\ &= \lim_{n\to\infty} \frac{\#(A \cap \{0,n\})}{n} = 0.\end{aligned}$$

□

The problem in this proof, of course, is that the limit need not exist. Consider typically

$$A = \bigcup_{k\geq 0} \{2^k+1, 2^k+2, \dots, 2^k+2^{k-1}\} = \{2,3,5,6,9,10,11,12,17,\dots\}.$$

The arguments of the "limit" above oscillate between $2/3$ and $1/2$. To correct this proof, we make use of a logical axiom:

Definition 11.2.5. Let X be a set. A *filter* is a family $\mathfrak{F}$ of subsets of X, such that

1. $X \in \mathfrak{F}$ and $\emptyset \notin \mathfrak{F}$;
2. if $A \in \mathfrak{F}$ and $B \supseteq A$ then $B \in \mathfrak{F}$;
3. if $A, B \in \mathfrak{F}$ then $A \cap B \in \mathfrak{F}$.

An *ultrafilter* is a maximal filter (under inclusion). It therefore satisfies the extra condition:

4. if $A \subseteq X$, then either $A \in \mathfrak{F}$ or $X \setminus A \in \mathfrak{F}$.

For every $x \in X$, there is a *principal* ultrafilter $\mathfrak{F}_x = \{A \subseteq X \mid x \in A\}$.
The set of ultrafilters on X is called its *Stone-Čech compactification* and is written βX. Its topology is defined by declaring open, for every $Y \subseteq X$, the collection $\{\mathfrak{F} \in \beta X \mid Y \in \mathfrak{F}\} \cong \beta Y$.

Elements of a filter are thought of as "large." As a standard example, consider the "cofinite filter" on $\mathbb{N}$:

$$\mathfrak{F}_c = \{A \subseteq \mathbb{N} \mid \mathbb{N} \setminus A \text{ is finite}\}.$$

Using this notion, the standard definition of convergence in analysis can be phrased as follows: "a sequence (x_n) converges to x if for every $\epsilon > 0$ we have $\{n \in \mathbb{N} \mid \epsilon > |x_n - x|\} \in \mathfrak{F}_c$." More generally, for a filter $\mathfrak{F}$ on $\mathbb{N}$, we define *convergence with respect to* $\mathfrak{F}$ by

$$\lim_{\mathfrak{F}} x_n = x \quad \text{if and only if} \quad \forall \epsilon > 0 : \{n \in \mathbb{N} \mid \epsilon > |x_n - x|\} \in \mathfrak{F}.$$

A standard axiom asserts the existence of *non-principal* ultrafilters on every infinite set. In fact, Zorn's lemma implies that the cofinite filter $\mathfrak{F}_c$ is contained in an ultrafilter $\mathfrak{F}$. Using this axiom, βX is compact and in fact is universal in the sense that every map $X \to K$ with K compact Hausdorff factors uniquely through βX. We state this universal property in the following useful form sometimes called "Stone duality":

Lemma 11.2.6. *Let X be a set. The map $f \mapsto (\mathfrak{F} \mapsto \lim_{\mathfrak{F}} f)$ is an isometry between the spaces $\ell^\infty(X)$ of bounded functions on X and $\mathscr{C}(\beta X)$ of continuous functions on βX with supremum norm.*

In particular, if $\mathfrak{F}$ is an ultrafilter on $\mathbb{N}$, then every bounded sequence converges with respect to $\mathfrak{F}$.

Proof. We first prove that if $f: X \to \mathbb{R}$ is bounded and $\mathfrak{F}$ is an ultrafilter, then it has a well-defined limit with respect to $\mathfrak{F}$. Assume $f(x) \in [L_0, U_0]$ for all $x \in X$. For $i = 0, 1, \dots$ repeat the following.

1. Set $M_i = (L_i + U_i)/2$.
2. Define $A_i = \{x \in X \mid f(x) \in [L_i, M_i]\}$ and $B_i = \{x \in X \mid f(x) \in [M_i, U_i]\}$.
3. By induction, $A_i \cup B_i \in \mathfrak{F}$; so either $A_i \in \mathfrak{F}$ or $B_i \in \mathfrak{F}$. In the former case, set $(L_{i+1}, U_{i+1}) = (L_i, M_i)$ while in the latter case set $(L_{i+1}, U_{i+1}) = (M_i, U_i)$.

Then (L_i) is an increasing sequence, (U_i) is a decreasing sequence, and they both have the same limit; call that limit $f(\mathfrak{F})$.

We have extended f to βX. Let us show that this extension is continuous at every $\mathfrak{F}$: keeping the notation from the previous paragraph, for every $\epsilon > 0$, there is some i with $U_i - L_i < \epsilon$; so $\{x \in X \mid \epsilon > |f(x) - f(\mathfrak{F})|\} \supseteq A_i \cup B_i \in \mathfrak{F}$ and therefore $f(x) \to f(\mathfrak{F})$ when $x \to \mathfrak{F}$.

Finally the inverse map $\mathscr{C}(\beta X) \to \ell^\infty(X)$ is simply given by restriction to the discrete subspace $X \subseteq \beta X$.

Exercise 11.2.7 (*). Prove that the Stone-Čech compactification βX is homeomorphic to the set of continuous *algebra* homomorphisms $\ell^\infty(X) \to \mathbb{R}$, with the induced topology of $\ell^\infty(X)^*$.

Exercise 11.2.8 (*). Let $\mathfrak{F}$ be an ultrafilter on $\mathbb{N}$. Prove $\lim_{\mathfrak{F}}(x_n + y_n) = \lim_{\mathfrak{F}} x_n + \lim_{\mathfrak{F}} y_n$ when these last two limits exist.

Using a non-principal ultrafilter $\mathfrak{F}$ on $\mathbb{N}$, we may correct the "proof" that $\mathbb{Z}$ is amenable, by replacing "lim" by "$\lim_{\mathfrak{F}}$"; but in some sense we have done nothing except shuffling axioms around. Indeed, an ultrafilter $\mathfrak{F}$ on X is precisely the same thing as a $\{0, 1\}$-valued mean on X: given an ultrafilter $\mathfrak{F}$, we define a mean m on X by

$$m(A) = \begin{cases} 0 & \text{if } A \notin \mathfrak{F}, \\ 1 & \text{if } A \in \mathfrak{F}, \end{cases}$$

and given a mean m taking $\{0, 1\}$ values, we define a filter $\mathfrak{F} = \{A \subseteq X \mid m(A) = 1\}$; so the construction of complicated means is as hard as the construction of complicated filters.

Proposition 11.2.9. *The free group F_k is not amenable if $k \geq 2$.*

Proof. We reason by contradiction, assuming that the regular right F_k-set $F_k \curvearrowleft F_k$ is amenable. Assume that there were an invariant mean $m\colon \mathfrak{P}(F_k) \to [0, 1]$. In $F_k = \langle x_1, \dots, x_k \mid \rangle$, let A denote those elements whose reduced form ends by a nontrivial (positive or negative) power of x_1. Then clearly $F_k = A \cup Ax_1$, so

$$1 = m(F_k) \leq m(A) + m(Ax_1) = 2m(A).$$

On the other hand, $F_k \supseteq Ax_2^{-1} \sqcup A \sqcup Ax_2$, so

$$1 = m(F_k) \geq m(Ax_2^{-1}) + m(A) + m(Ax_2) = 3m(A).$$

These statements imply $1/2 \leq m(A) \leq 1/3$, a contradiction.

11.2.2 Elementary Properties

Proposition 11.2.10. *Let G, H be groups; let $X \curvearrowleft G$ and $Y \curvearrowleft H$ be, respectively, a G-set and an H-set; let $\phi\colon G \twoheadrightarrow H$ be a surjective homomorphism; and let $f\colon X \to Y$ be an equivariant map, namely, satisfying $f(xg) = f(x)\phi(g)$ for all $x \in X, g \in G$. If X is amenable, then Y is amenable.*

Proof. If $\mathscr{M}(X)^G \neq \emptyset$, then $f_*(\mathscr{M}(X)^G) = f_*(\mathscr{M}(X))^{\phi(G)} \subseteq \mathscr{M}(Y)^H$ so $\mathscr{M}(Y)^H \neq \emptyset$.

Corollary 11.2.11 ([264, Corollary 3.2]). *Let G be a group. Then G is amenable if and only if the right G-set G_G is amenable.*

Proof. Assume the right G-set $G \curvearrowleft G$ is amenable. For every nonempty G-set X, choose $x \in X$; then $g \mapsto xg$ is a G-equivariant map $G \to X$, so X is amenable by Proposition 11.2.10. The converse is obvious.

Thus amenability of a group is equivalent to amenability of the right regular action and also to amenability of all actions. We give another characterization:

Proposition 11.2.12. *Let G be a group. Then the following are equivalent:*

1. *the group G is amenable;*
2. *the G-set G_G is amenable;*
3. *G admits an amenable free action.*

Proof. In view of the previous corollary, it suffices to prove (3) $\Rightarrow$ (2). Let X be a free G-set, and choose a G-isomorphism $X \cong T \times G$. Let $m\colon \mathfrak{P}(X) \to [0, 1]$ be a G-invariant mean. Define a mean m' on G by $m'(A) = m(T \times A)$, and check that m' is G-invariant.

Exercise 11.2.13 (*). Let X, Y be G-sets. Then

1. $X \sqcup Y$ is amenable if and only if X or Y is amenable;
2. $X \times Y$ is amenable if and only if X and Y are amenable.

Proposition 11.2.10 says that quotients of amenable G-sets are amenable. Note however that subsets of amenable G-sets need not be amenable, the empty set being the extreme example. See Section 11.9 for a notion of amenability better suited to subsets and extensions of G-sets.

Definition 11.2.14 (Wreath product). We introduce a construction of groups that serve as important examples. Let A, G be groups and let X be a G-set. Their *(restricted) wreath product* is

$$A \wr_X G := A^{(X)} \rtimes G, \tag{11.1}$$

the semidirect product of the group of finitely supported maps $X \to A$ with G, under the action of G at the source. Elements of $A \wr_X G$ may be written as (f, g)

with $f: X \to A$ and $g \in G$; they multiply by $(f, g) \cdot (f', g') = (f \cdot (f'g^{-1}), gg')$ with $(f'g^{-1})(x) = f'(xg)$.

In case G acts faithfully on X, elements of $A \wr_X G$ may be thought of as "decorated permutations": permutations, say σ represented by a diagram with vertex set X and an arrow from x to $\sigma(x)$ and with a label in A on each arrow in such a manner that only finitely many labels are nontrivial. Decorated permutations are composed by concatenating their arrows and multiplying their labels.

The wreath product is associative, in the sense that if A, G, H are groups, X is a G-set and Y is an H-set, then $G \wr_Y H$ naturally acts on $X \times Y$ and $A \wr_{X \times Y} (G \wr_Y H) = (A \wr_X G) \wr_Y H$.

On the other hand, for groups A, G we write "$A \wr G$" for the wreath product $A \wr_G G$ with regular right action of G on itself, and that operation is *not* associative.

Definition 11.2.15 (Tree automorphisms). For a finite set $\mathscr{A}$, consider the set $X := \mathscr{A}^*$ of words over $\mathscr{A}$. This set is naturally the vertex set of a rooted tree $\mathscr{T}$; the root is the empty word, and there is an edge between $x_1 \cdots x_n$ and $x_1 \cdots x_n x_{n+1}$ for all $x_i \in \mathscr{A}$. The space $\mathscr{A}^{\mathbb{N}}$ corresponds to infinite paths in $\mathscr{T}$ and thus naturally describes the boundary of $\mathscr{T}$.

Let G be the group of graph automorphisms of $\mathscr{T}$: maps $\mathscr{A}^* \circlearrowleft$ that preserve the edge set. There is then a natural map $\pi: G \to \mathrm{Sym}(\mathscr{A})$ defined by restricting the action of G to the neighbors of the root; and $\ker(\pi)$ acts on the $\#\mathscr{A}$ disjoint trees hanging from the root, so it is isomorphic to $G^{\mathscr{A}}$. We therefore have a natural isomorphism

$$\Phi: G \longrightarrow G \wr_{\mathscr{A}} \mathrm{Sym}(\mathscr{A}). \tag{11.2}$$

A subgroup $H \leq G$ is called *self-similar* if the isomorphism (11.2) restricts to a homomorphism $\Phi: H \to H \wr_{\mathscr{A}} \mathrm{Sym}(\mathscr{A})$. In that case, elements of H may be defined recursively in terms of their image under Φ, and conversely such a recursive description defines uniquely an action on $\mathscr{T}$.

The Grigorchuk group G (see Section 11.4.3 or the Introduction) acts faithfully on the binary rooted tree $\mathscr{T}_2$ and as such is a subgroup of the automorphism group of $\mathscr{T}_2$. It is self-similar, and the generators $\{a, b, c, d\}$ of G may be written using decorated permutations as follows:

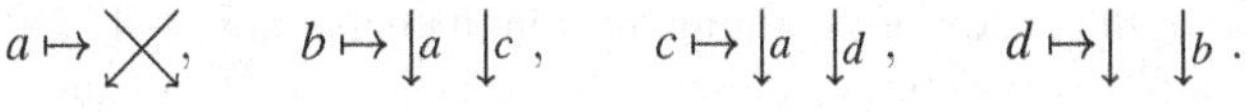

Example 11.2.16 (The "lamplighter group"). Consider $G = \mathbb{Z}$ acting on itself by translation, and $A = \mathbb{Z}/2$. The wreath product $W = A \wr G$ is called the "lamplighter group." The terminology is justified as follows: consider an bi-infinite street with a lamp at each integer location. The group G consists of invertible instructions for a person, the "lamplighter": either move up or down the street, or toggle the state of a lamp before him/her.

If we denote by a the operation of toggling the lamp at position 0 and by t the movement of the lamplighter one step up the street, then G is generated by $\{a, t\}$; and it admits as presentation

$$G = \langle a, t \mid [a, a^{t^k}] \text{ for all } k \in \mathbb{N} \rangle. \tag{11.3}$$

Exercise 11.2.17 (**). Let A be a simple group and let H be perfect. Let $G := H \wr_X A$ be their wreath product. Then G is perfect, and all normal subgroups of G are G or of the form N^X for a normal subgroup $N \triangleleft H$.

Example 11.2.18 (Monod-Popa [422]). There are groups $K \triangleleft H \triangleleft G$ such that the G-sets $K\backslash G$ and $H\backslash G$ are amenable, but the H-set $K\backslash H$ is not.

Choose indeed any non-amenable group Q, and set $G := Q \wr \mathbb{Z}$ and $H = Q^{(\mathbb{Z})}$ and $K = Q^{(\mathbb{N})}$.

The G-set $H\backslash G$ is clearly amenable, since the action of G factors through an action of $\mathbb{Z}$. To prove that $K\backslash G$ is amenable, it therefore suffices to find an H-invariant mean on $\ell^\infty(K\backslash G)$, and then apply Proposition 11.2.26. Let t denote the positive generator of $\mathbb{Z}$. For every $k \in \mathbb{N}$, define a mean m_k by $m_k(f) = f(Kt^k)$ for $f \in \ell^\infty(K\backslash G)$. This mean is invariant by the group K^{t^k}. Since $H = \bigcup_{k\in\mathbb{N}} K^{t^k}$, any weak limit of the m_k is an H-invariant mean.

On the other hand, $K\backslash H$ is just a restricted direct product of Q's, so it is not amenable by Proposition 11.2.12.

Exercise 11.2.19 (**). Give an amenable G-set such that none of its orbits are amenable.

Hint: Consider the "lamplighter group" $G = \langle a, t \rangle$, see Example 11.2.16, and the groups $G_n = \langle a, t \mid [a, a^{t^k}] \text{ for all } k = 1, \dots, n \rangle$. Consider the natural action of $F_2 = \langle a, t \mid \rangle$ on $X = \bigsqcup_{n\ge 0} G_n$, and show that (i) each G_n is non-amenable, (ii) the group G is amenable, and (iii) the action on X approximates arbitrarily well the action on G.

We return to the definition of means we started with; we shall see more criteria for amenability. Recall that $\mathscr{M}(X)$ denotes the set of means on X.

Lemma 11.2.20. *$\mathscr{M}(X)$ is compact.*

Proof. Since $\mathscr{M}(X)$ is a subset of $[0, 1]^{\mathfrak{P}(X)}$ which is compact by Tychonoff's theorem,[5] it suffices to show that $\mathscr{M}(X)$ is closed.

Now each of the conditions defining a mean, namely, $m(X) - 1 = 0$ and $m(A \cup B) - m(A) - m(B) = 0$, defines a closed subspace of $[0, 1]^{\mathfrak{P}(X)}$ because it is the zero set of a continuous map. The intersection of these closed subspaces is $\mathscr{M}(X)$ which is therefore closed.

Here are simple examples of means. For $x \in X$, define $\delta_x \in \mathscr{M}(X)$ by

$$\delta_x(A) = \begin{cases} 0 & \text{if } x \notin A, \\ 1 & \text{if } x \in A. \end{cases}$$

[5] We are using here, and throughout this chapter, the axiom of choice; see [341].

It is easy to see that the axioms of a mean are satisfied. We have thus obtained a map $\delta: X \to \mathscr{M}(X)$, which is clearly injective.

Lemma 11.2.21. *$\delta(X)$ is discrete*[6] *in $\mathscr{M}(X)$.*

Proof. Given $x \in X$, set

$$\mathscr{U} = \{m \in \mathscr{M}(X) \mid m(\{x\}) > 0\}.$$

□

Corollary 11.2.22. *If X is infinite, then $\delta(X)$ is not closed.*

Proof. Indeed, if $\delta(X)$ is closed in $\mathscr{M}(X)$, then it is compact; being furthermore discrete, it is finite; δ being injective, X itself is finite.

Recall that a subset K of a topological vector space is *convex* if for all $x, y \in K$ the segment $\{(1-t)x + ty \mid t \in [0,1]\}$ is contained in K; see Section 11.6 for more on convex sets. The *convex hull* of a subset S of a topological vector space is the intersection $\widehat{S}$ of all the closed convex subspaces containing S.

Lemma 11.2.23. *$\mathscr{M}(X)$ is convex.*

Proof. Consider means m_i and positive numbers t_i such that $\sum t_i = 1$. Then $\sum t_i m_i$ clearly satisfies the axioms of a mean.

For a set X and $p \in [1, \infty)$, we denote by $\ell^p(X)$ the Banach space of functions $\phi: X \to \mathbb{R}$ satisfying $\|\phi\|^p := \sum |\phi(x)|^p < \infty$ and by $\ell^\infty(X)$ the space of bounded functions with supremum norm. For $p \in [1, \infty]$ the space $\ell^p(X)$ carries a natural isometric G-action by $(\phi g)(x) = \phi(xg^{-1})$. Of particular interest is the space $\ell^1(X)$, and its subset

$$\mathscr{P}(X) = \{\mu \in \ell^1(X) \mid \mu \geq 0, \sum_{x \in X} \mu(x) = 1\}, \tag{11.4}$$

the space of *probability measures* on X. It is a convex subspace of $\ell^1(X)$, compact for the weak*-topology, and (for infinite X strictly) contained in $\mathscr{M}(X)$:

Proposition 11.2.24. *For a set X, consider the following subset of $\ell^\infty(X)^*$:*

$$\mathscr{B}(X) := \{m \in \ell^\infty(X)^* \mid m(f) \geq 0 \text{ whenever } f \geq 0,\ m(\mathbb{1}) = 1\}.$$

Then the map $\int: \mathscr{B}(X) \to \mathscr{M}(X)$ defined by

$$(\textstyle\int m)(A) := m(\mathbb{1}_A) \text{ with } \mathbb{1}_A \text{ the characteristic function of } A$$

is a homeomorphism, functorial in X.

[6] Recall that D is discrete in a topological space X if for every $x \in D$ there is an open set $\mathscr{U} \ni x$ with $D \cap \mathscr{U} = \{x\}$.

The subspace $\ell^1(X) \cap \mathscr{B}(X) \subset \ell^\infty(X)^*$ *corresponds via* $\int$ *to the convex hull* $\widehat{\delta(X)}$ *of* $\delta(X)$.

We recall that there is a natural nondegenerate pairing $\ell^1(X) \times \ell^\infty(X) \to \mathbb{R}$, given by $(f, g) \mapsto \sum f(x)g(x)$. For that pairing, $(\ell^1 X)^* = \ell^\infty(X)$; but $(\ell^\infty X)^*$ is much bigger than $\ell^1(X)$, as is clear from the proposition. In fact, $\ell^\infty(X)$ is in isometric bijection with the space of continuous functions on the Stone-Čech compactification βX of X, see Lemma 11.2.6, so

$$(\ell^\infty(X))^* = L^1(\beta X) \text{ the set of Borel measures on } \beta X. \tag{11.5}$$

Proof (Proof of Proposition 11.2.24). Let $\mathscr{S}$ be the set of *simple* functions on X, namely, the functions that take only finitely many values. Consider first $m \in \ell^\infty(X)^*$ with $m(\mathbb{1}_A) = 0$ for all $A \subseteq X$. Then m vanishes on $\mathscr{S}$ by linearity; and $\mathscr{S}$ is dense in $\ell^\infty(X)$, so $m = 0$. This proves that $\int$ is injective.

On the other hand, let $m\colon \mathfrak{P}(X) \to [0, 1]$ be a mean. For $f \in \mathscr{S}$, we have

$$\left(\int m\right)(f) = \sum_{v \in f(X)} v m(f^{-1}(v)).$$

We check that $\int m$ is a continuous function $\mathscr{S} \to \mathbb{R}$ for the ℓ^∞ norm on $\mathscr{S}$; indeed, for f, g simple functions on X,

$$\begin{aligned}
\left|\left(\int m\right)(f) - \left(\int m\right)(g)\right| &= \left|\sum_{v \in f(X), w \in g(X)} (v - w)\mu(f^{-1}(v) \cap g^{-1}(w))\right| \\
&\leq \sum_{v \in f(X), w \in g(X)} |v - w|\mu(f^{-1}(v) \cap g^{-1}(w)) \\
&\leq \|f - g\|_\infty \sum_{v \in f(X), w \in g(X)} \mu(f^{-1}(v) \cap g^{-1}(w)) \\
&\leq \|f - g\|_\infty.
\end{aligned}$$

Therefore, $\int m$ extends to a continuous function $\ell^\infty(X) \to \mathbb{R}$, which clearly belongs to $\mathscr{B}(X)$. Since $\mathscr{S}$ is dense, this extension is unique.

Finally recall that $\ell^1(X)$ embeds in $\ell^\infty(X)^*$ by $f \mapsto (f' \mapsto \sum_x f(x)f'(x))$. The element $f \in \ell^1(X)$ therefore corresponds to the affine combination $\sum f(x)\delta_x$ of Dirac means.

From now on, we will use interchangeably the notations $m \in \mathscr{M}(X)$ and $m \in (\ell^\infty(X))^*$; they correspond to each other via the proposition.

Corollary 11.2.25. *Let X be a G-set. Then X is amenable if and only if there exists a G-invariant positive functional in $\ell^\infty(X)^*$.* □

The fact that the "Dirac" means $\widehat{\delta(X)}$ constitute a small subset of $\mathscr{M}(X)$ may be confirmed as follows. Every mean $m \in \widehat{\delta(X)}$ enjoys an additional property, namely, σ-additivity: for disjoint $A_1, A_2, \dots$ we have

$$m(\bigcup A_i) = \sum m(A_i).$$

Consider now an invariant mean m on $\mathbb{Z}$, as given by Proposition 11.2.4. Assume for contradiction that m were σ-additive. Then either $m(\{0\}) = 0$, so $m(\{n\}) = 0$ for all $n \in \mathbb{Z}$ by $\mathbb{Z}$-invariance and $m(\mathbb{Z}) = 0$ by σ-additivity; or $m(\{0\}) = \epsilon > 0$ and $m(\{0, 1, \dots, n\}) > 1$ as soon as $n > 1/\epsilon$. In all cases we have reached a contradiction.

Proposition 11.2.26. *Let X be an amenable G-set such that all point stabilizers G_x are amenable. Then G itself is amenable.*

Proof. Thanks to Proposition 11.2.24, for all Y we view $\mathscr{M}(Y)$ as the set of normalized positive functionals $m \colon \ell^\infty(Y) \to \mathbb{R}$. Let us first define a map $\Phi \colon X \to \mathscr{M}(G)$.

Since every G_x is amenable, there exists for all $x \in X$ an invariant mean $m_x \in \mathscr{M}(G_x)^{G_x}$, which we extend via the inclusion $G_x \hookrightarrow G$ to mean still written $m_x \in \mathscr{M}(G)^{G_x}$. Choose for every G-orbit in X a point x, and set $\Phi(xg) = m_x g$ on that orbit. This is well defined: if $xg = xh$, then $hg^{-1} \in G_x$, so $m_x h = m_x hg^{-1}g = m_x g$. It follows automatically that Φ is G-equivariant.

By functoriality, Φ induces a G-equivariant map $\Phi_* \colon \mathscr{M}(X) \to \mathscr{M}(\mathscr{M}(G))$.

Now there is, for all Y, a functorial map $\beta \colon \mathscr{M}(\mathscr{M}(Y)) \to \mathscr{M}(Y)$ called the *barycentre*: it is given by

$$\Upsilon(m)(f) = m(n \mapsto n(f)) \text{ for } m \in \mathscr{M}(\mathscr{M}(Y)), f \in \ell^\infty(Y), n \in \mathscr{M}(Y). \qquad (11.6)$$

Composing, we get a map $\Upsilon \circ \Phi_* \colon \mathscr{M}(X) \to \mathscr{M}(G)$, which is still G-equivariant. Now since X is amenable $\mathscr{M}(X)^G$ is nonempty, so $\mathscr{M}(G)^G$ is also nonempty.

Corollary 11.2.27. *Let $1 \longrightarrow N \longrightarrow G \longrightarrow Q \longrightarrow 1$ be an exact sequence of groups. Then G is amenable if and only if both N and Q are amenable.*

Proof. If G is amenable, then its quotient Q is amenable by Proposition 11.2.10, and its subgroup H is amenable by Proposition 11.2.12, since it acts freely on the amenable G-set G.

Conversely, if N and Q are amenable, then the natural action of G on Q satisfies the hypotheses of Proposition 11.2.26.

Exercise 11.2.28 (*). Let G be a group. We might have called G *left-amenable* if there exists a *left-invariant* mean on G, namely, a mean $m \in \mathscr{M}(G)$ with $m(gA) = m(A)$ for all $g \in G, A \subseteq G$, and have called G *bi-amenable* if there exists a mean $m \in \mathscr{M}(G)$ with $m(gAh) = m(A)$ for all $g, h \in G, A \subseteq G$.

Prove that in fact G is amenable if and only if it is left-amenable, if and only if it is bi-amenable.

We conclude with yet another criterion, attributed[7] to Jacques Dixmier:

Theorem 11.2.29 (Følner [225, Theorem 4], Dixmier [197, Théorème 1]; see [263, Theorem 4.2]). *Let X be a G-set. Then X is amenable if and only if for any $h_1, \dots, h_n \in \ell^\infty(X)$ and any $g_1, \dots, g_n \in G$ the function*

$$H := \sum_{i=1}^{n} (h_i - h_i g_i) \qquad \textit{satisfies} \ \sup_{x \in X} H(x) \geq 0.$$

Proof. If X is amenable, then there is an invariant positive mean $m \in \ell^\infty(X)^*$; then for every function H as above, $m(H) = 0$ by invariance while $m(H) \leq \sup H$ by positivity.

On the other hand, if $\sup H \geq 0$ for all H as above, then an invariant mean may be constructed as follows: set

$$\tilde{m}(f) = \inf_{H \text{ as above}} \sup_X (f + H).$$

Clearly $\tilde{m}$ satisfies $\tilde{m}(\lambda f) = \lambda \tilde{m}(f)$ for $\lambda \geq 0$ and $\tilde{m}(fg) = \tilde{m}(f)$ for $g \in G$ and $\tilde{m}(\mathbb{1}) = 1$ and $\tilde{m}(f) \geq 0$ if $f \geq 0$; and $\tilde{m}(f + g) \leq \tilde{m}(f) + \tilde{m}(g)$ because if $\tilde{m}(f) \geq \sup_X(f + H) - \epsilon$ and $\tilde{m}(g) \geq \sup_X(g + K) - \epsilon$ then $\tilde{m}(f + g) \leq \sup_X(f + g + H + K) \leq \sup_X(f + H) + \sup_X(g + K) \leq \tilde{m}(f) + \tilde{m}(g) - 2\epsilon$. The Hahn-Banach theorem (see, *e.g.*, [515, Theorem 3.12]) implies the existence of a linear functional m with the same properties.

11.3 Følner, Day, and Reiter's Criteria

The following combinatorial criterion will be shown equivalent to amenability; it is sometimes the easiest path to prove a group's amenability. It was introduced by Erling Følner [226], though the idea of averaging over larger and larger finite sets to construct invariant means can be traced back at least to Ahlfors [7, Chapter III.25].

Definition 11.3.1. Let X be a G-set. We say that X satisfies *Følner's condition* if for all finite $S \Subset G$ and all $\epsilon > 0$, there is a finite subset $F \Subset X$ with

$$\#(FS \setminus F) < \epsilon \# F.$$

When we say that a group G satisfies Følner's condition, we mean it for the right G-set $X = G \looparrowleft G$.

[7] Erroneously!

For example, $\mathbb{Z}$ satisfies Følner's condition: given $\epsilon > 0$ and $S \subset \mathbb{Z}$ finite, find k such that $S \subseteq \{-k,\dots,k\}$. Let $\ell \in \mathbb{N}$ be such that $\ell > 2k/\epsilon$, and set $F = \{1,2,\dots,\ell\}$. Then $FS \setminus F \subseteq \{1-k,\dots,0,\ell+1,\dots,\ell+k\}$ has size at most $2k$, so $\#(FS \setminus F) < \epsilon \#F$.

Actually, the definition makes sense in a much more general context, that of *graphs*:

Definition 11.3.2. A directed graph (digraph) is a pair of sets $\mathscr{G} = (V,E)$ called *vertices* and *edges*, with maps $\pm: E \to V$ giving for each edge $e \in E$ its *head* $e^+ \in V$ and *tail* $e^- \in V$.

A graph $\mathscr{G} = (V,E)$ has *bounded valency* if there is a bound $K \in \mathbb{N}$ such that at every vertex $v \in V$ there are at most K incoming and outgoing edges, namely, if $\#\{e \in E \mid v = e^+\} \le K$ and $\#\{e \in E \mid v = e^-\} \le K$.

Consider a G-set X and a finite set $S \subset G$. The *Schreier graph* of X with respect to S is the graph with vertex set $V = X$ and edge set $E = X \times S$, with $(x,s)^- = x$ and $(x,s)^+ = xs$. In other words, there is an edge from x to xs for all $x \in X, s \in S$. If $X = G \looparrowleft G$, then the Schreier graph is usually called the *Cayley graph* of G. See also Definition 10.3.1.

Let (V,E) be a graph. For a subset $F \subseteq V$, its *boundary* is the set of edges connecting F to its complement, in formulæ

$$\partial F = \{e \in E \mid e^- \in F, e^+ \notin F\}.$$

Definition 11.3.3. A graph $\mathscr{G} = (V,E)$ satisfies *Følner's condition* if for all $\epsilon > 0$ there is a finite subset $F \Subset V$ with $\#\partial F < \epsilon \#F$.

Thus Følner's criterion asks for the existence of subgraphs of X with an arbitrarily small relative outer boundary. It is clear that a G-set X satisfies Følner's condition if and only if its Schreier graphs satisfy it for all choices of $S \Subset G$.

Lemma 11.3.4. *Let X be a G-set. Følner's condition is equivalent to: for all finite subsets $S \Subset G$ and all $\epsilon > 0$, there is a finite subset $F \Subset X$ with*

$$\#(Fs \setminus F) < \epsilon \#F \text{ for all } s \in S.$$

Proof. If $\#(FS \setminus F) < \epsilon \#F$, then in particular $\#(Fs \setminus F) < \epsilon \#F$ for all $s \in S$. Conversely, if $\#(Fs \setminus F) < \epsilon \#F / \#S$ for all $s \in S$ then $\#(FS \setminus F) < \epsilon \#F$.

Recall that a *directed set* is a partially ordered set $(\mathscr{N}, \le)$ with finite upper bounds, *i.e.*, for every $m,n \in \mathscr{N}$, there exists an element $\max\{m,n\} \in \mathscr{N}$ with $m,n \le \max\{m,n\}$. A *net* is a sequence indexed by a directed set. For $(x_n)_{n \in \mathscr{N}}$ a real-valued net, we write

$$\lim_{n\to\infty} x_n = x \quad \text{to mean} \quad \forall \epsilon > 0 : \exists n_0 \in \mathscr{N} : \forall n \ge n_0 : |x_n - x| < \epsilon, \tag{11.7}$$

as in usual calculus.

Exercise 11.3.5 (*). Let $\mathscr{N}$ be a nonempty net. Then $\{F \subseteq \mathscr{N} \mid \exists n_0 \in \mathscr{N} : n \geq n_0 \Rightarrow n \in F\}$ is a filter on $\mathscr{N}$, and the notions of convergence in (11.7) and in the filter coincide.

We have the following alternative definition of Følner's condition:

Lemma 11.3.6. *Let G be a group and let X be a G-set. Then X satisfies Følner's condition if and only if there exists a net $(F_n)_{n\in\mathscr{N}}$ of finite subsets of X with*

$$\lim_{n\to\infty} \frac{\#(F_n g \setminus F_n)}{\#F_n} = 0 \text{ for all } g \in G. \tag{11.8}$$

Proof. Assume (11.8), and let $S \Subset G, \epsilon > 0$ be given. For each $s \in S$, let $n(s) \in \mathscr{N}$ be such that $\#(F_n s \setminus F_n) < \epsilon \#F_n / \#S$ for all $n \geq n(s)$, and set $F = F_{\max\{n(s)\}}$; then $\#(FS \setminus F) \leq \sum_{s\in S} \#(Fs \setminus F) < \epsilon \#F$, so Følner's condition is satisfied.

Conversely, define $\mathscr{N} = \{(S, \epsilon) \mid S \Subset G \text{ finite}, \epsilon > 0\}$, ordered as follows: $(S, \epsilon) \leq (T, \delta)$ if $S \subseteq T$ and $\epsilon > \delta$; so $\max\{(S, \epsilon), (T, \delta)\} = (S \cup T, \min\{\epsilon, \delta\})$. For each $n = (S, \epsilon) \in \mathscr{N}$, choose a finite set $F_n \Subset X$ with $\#(FS \setminus F) < \epsilon \#F$. These satisfy (11.8).

In case G is finitely generated, we also have the following alternative definition:

Lemma 11.3.7. *Let G be finitely generated, say by a finite set S containing 1, and let X be a G-set. Then X satisfies Følner's condition if and only if for all $\epsilon > 0$ there is a finite subset $F \Subset X$ with*

$$\#(FS \setminus F) < \epsilon \#F.$$

Proof. One direction is obvious. In the other direction, let $S' \Subset G$ and $\epsilon' > 0$ be given. Since S generates G, there exists $k \in \mathbb{N}$ with $S' \subseteq S^k$. Set $\epsilon = \epsilon'/k$, and let $F \Subset X$ satisfy $\#(Fs \setminus F) < \epsilon \#F$ for all $s \in S$.

Consider $g \in S'$, and write it as $g = s_1 \dots s_k$ with $s_1, \dots, s_k \in S$. Then

$$Fg \setminus F = \bigsqcup_{j=1}^{k} Fs_j \cdots s_k \setminus Fs_{j+1} \cdots s_k,$$

so

$$\begin{aligned}\#(Fg \setminus F) &= \sum \#(Fs_j \cdots s_k \setminus Fs_{j+1} \cdots s_k)\\ &= \sum \#(Fs_j \setminus F)s_{j+1} \cdots s_k < k\epsilon \#F = \epsilon' \#F.\end{aligned}$$

We are done by Lemma 11.3.4.

We shall see in Theorem 11.3.23 that a G-space X satisfies Følner's criterion if and only if it is amenable. This can be used to prove (non-)amenability in numerous cases; for example,

Proposition 11.3.8. *A G-set $X \curvearrowleft G$ is amenable if and only if for every finitely generated subgroup $H \le G$ the H-set $X \curvearrowleft H$ is amenable.*

Proof. ($\Leftarrow$) Given $S \Subset G$ and $\epsilon > 0$, consider $H = \langle S \rangle$ and apply Følner's criterion.
($\Rightarrow$) Every G-invariant mean is also H-invariant.

Thus, for instance, the action of $\mathbb{Q}$ on $\mathbb{Q}/\mathbb{Z}$ is amenable, because every finitely generated subgroup of $\mathbb{Q}$ has a finite orbit on $\mathbb{Q}/\mathbb{Z}$. (We shall later see that all actions of $\mathbb{Q}$ are amenable.)

Example 11.3.9. The group of permutations $\mathrm{Sym}(\mathbb{N})$ of $\mathbb{N}$ with finite support is amenable; indeed every finite subset generates a finite group.

Example 11.3.10. The group of "bounded-displacement permutations of $\mathbb{Z}$"

$$G = W(\mathbb{Z}) = \{\tau\colon \mathbb{Z} \circlearrowleft \mid \sup_{n \in \mathbb{Z}} |\tau(n) - n| < \infty\}$$

acts amenably on $\mathbb{Z}$. Indeed given $S \subset G$ finite and $\epsilon > 0$, the maximum displacement of elements of S is bounded, say $\le k$; and then $\mathbb{Z} \curvearrowleft G$ satisfies Følner's condition with $F = \{0, \dots, \lceil k/\epsilon \rceil\}$.

Example 11.3.11. The "lamplighter group" G from Example 11.2.16 is amenable. Indeed elements of G may be written as pairs (f, m) with $f\colon \mathbb{Z} \to \mathbb{Z}/2$ and $m \in \mathbb{Z}$, and one may consider as Følner sets

$$F_n = \{(f, m) \mid \mathrm{support}(f) \subseteq [-n, n] \text{ and } m \in [-n, n]\}.$$

Example 11.3.12. The free group $F_k = \langle x_1, \dots, x_k \mid \rangle$ is amenable if and only if $k \le 1$, see Proposition 11.2.9. Indeed if $k \le 1$ then F_k is $\{1\}$ or $\mathbb{Z}$; while in general, choose $S = \{x_1^{\pm 1}, \dots, x_k^{\pm 1}\}$ and consider $F \Subset X$. In the Cayley graph of F_k, which is a $2k$-regular tree (see Figure 11.2 left), consider the subgraph spanned by

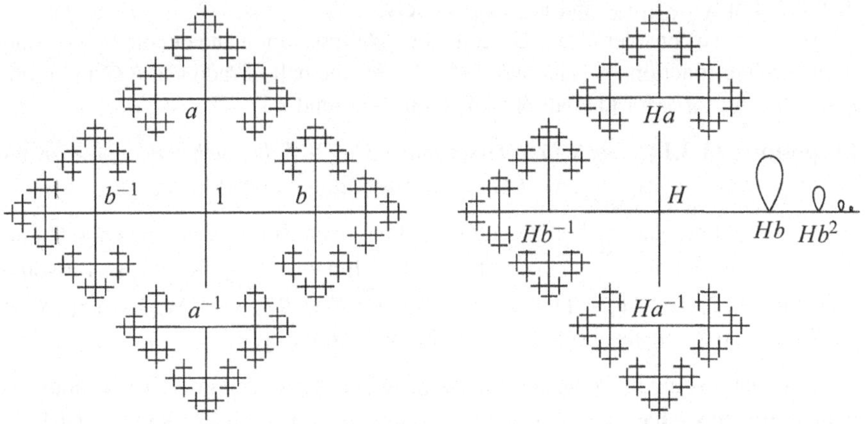

Fig. 11.2 The Cayley graph of the free group F_2, and the coset space of H (see Example 11.3.13)

F. It suffices to consider connected components of the graph once at a time; each connected component is a tree, with say v vertices and therefore $v-1$ edges. The sum of the vertex degrees within that tree is therefore $2v-2$, so the total number of edges pointing out of the component is at least $2kv-(2v-2) \geq (2k-2)v$; these edges point to distinct elements in $SF \setminus F$. Therefore, Følner's criterion is not satisfied as soon as $\epsilon < 2n-2$.

There are plenty of non-amenable groups with amenable actions and even faithful amenable actions; here is one.

Example 11.3.13. Consider $F_2 = \langle a, b \mid \rangle$ and its subgroup $H = \langle a^{b^n} : n \leq 0 \rangle$. Then F_2 acts naturally on the coset space $X := H\backslash F_2$, see Figure 11.2 right, and this action is amenable. Indeed with $S = \{a^{\pm 1}, b^{\pm 1}\}$ and $\epsilon > 0$ given, consider the set $F = \{H, Hb, \dots, Hb^n\}$ for $n > \epsilon^{-1}$. It satisfies Følner's criterion. Note that the action of F_2 on X is not free, but it is nevertheless faithful.

11.3.1 Growth of Sets

Let $X \looparrowleft G$ be a G-set, and consider $S \Subset G$ and $x_0 \in X$. The *orbit growth* of X is the function $v_{X,x_0,S}\colon \mathbb{N} \to \mathbb{N}$ given by

$$v_{X,x_0,S}(n) = \#\{x \in X \mid x = x_0 s_1 \cdots s_m \text{ for some } s_i \in S, m \leq n\}.$$

If G is finitely generated, then the orbit growth depends only mildly on the choice of S as soon as it generates G: if S' be another generating set of G, then there exists a constant $C > 0$ with $v_{X,x_0,S}(n) \leq v_{X,x_0,S'}(Cn)$ and $v_{X,x_0,S'}(n) \leq v_{X,x_0,S}(Cn)$. Similarly, if $x_0, x_0' \in X$ belong to the same G-orbit, then there exists a constant $C \in \mathbb{N}$ with $v_{X,x_0,S}(n) \leq v_{X,x_0',S}(n+C)$ and $v_{X,x_0',S}(n) \leq v_{X,x_0,S}(n+C)$. Therefore, the equivalence class of $v_{X,x_0,S}$ under linear transformations of its argument is independent of the choice of S if S generates G and of x_0 if X is transitive; it is denoted simply v_{X,x_0}, v_X, and v_{x_0}, respectively.

As usual, we consider G as a G-set under right translation and denote by $v_{G,S}$ and v_G its growth function. We also write $B_{G,S}(n)$ for the ball of radius n in G and more generally $B_{X,x_0,S}(n)$ for the ball of radius n in X around x_0.

Proposition 11.3.14. *Let X be a G-set and let $x_0 \in X$ be such that $v_{X,x_0,S}$ grows subexponentially for all $S \Subset G$. Then X satisfies Følner's condition.*

Proof. Let a finite subset $S \Subset G$ and $\epsilon > 0$ be given. Since X has subexponential growth, we have $\lim \sqrt[n]{v_{X,x_0,S}(n)} = 1$; therefore $\liminf \frac{v_{X,x_0,S}(n+1)}{v_{X,x_0,S}(n)} = 1$, so for some n we have $v_{X,x_0,S}(n+1) < (1+\epsilon)v_{X,x_0,S}(n)$. Set $F = B_{X,x_0,S}(n)$. We have $\#(FS) < (1+\epsilon)\#F$, so X satisfies Følner's condition by Lemma 11.3.7.

Note that it is unknown whether in every finitely generated group G of subexponential growth we have $\lim \frac{v_{G,S}(n+1)}{v_{G,S}(n)} = 1$; only the "lim inf" is known to equal 1.

One may study more quantitatively the Følner condition as follows: let X be a G-set and let S be a generating set for G. Define Føl: $\mathbb{N} \to \mathbb{N} \cup \{\infty\}$ by

$$\text{Føl}(n) = \inf\{\#F \mid F \Subset X, \#(F \triangle Fs) < \#F/n \text{ for all } s \in S\}. \tag{11.9}$$

Then $\text{Føl}(n) < \infty$ for all n precisely if X is amenable. A similar definition may be given for graphs, which we leave to the reader. Groups admit the following lower bound on Føl:

Proposition 11.3.15 (Coulhon-(Saloff-Coste) [168]). *Let $G = \langle S \rangle$ be a finitely generated group, with growth function $v_{G,S}(n)$. Then*

$$\text{Føl}(n) \geq \frac{1}{2} v_{G,S}(n) \textit{ for all } n \in \mathbb{N}.$$

Proof. We shall prove the following equivalent form: given $F \Subset G$, choose $n \in \mathbb{N}$ such that $v_{G,S}(n) \geq 2\#F$. We are required to find $s \in S$ with $\#(F \triangle Fs) \geq \#F/n$.

First, for all $x \in F$ we have $\#(xB_{G,S}(n) \setminus F) \geq \#F \geq \#(xB_{G,S}(n) \cap F)$, so

$$\sum_{g \in B_{G,S}(n)} \mathbb{1}_{xg \notin F} \geq v(n)/2 \geq \sum_{g \in B_{G,S}(n)} \mathbb{1}_{xg \in F},$$

$$\sum_{g \in B_{G,S}(n)} \sum_{x \in F} \mathbb{1}_{xg \notin F} \geq v(n)\#F/2,$$

so for some $g \in B_{G,S}(n)$, we have $\sum_{x \in F} \mathbb{1}_{xg \notin F} \geq \#F/2$, namely, $\#(F \triangle Fg) \geq \#F$. Write now $g = s_1 \cdots s_n$; then $F \triangle Fg = (F \triangle Fs_n) \triangle \cdots \triangle (Fs_2 \cdots s_n \triangle Fg) \subseteq (F \triangle Fs_n) \cup \cdots \cup (F \triangle Fs_1)s_2 \cdots s_n$. It follows that there exists some $k \in \{1, \ldots, n\}$ with $\#(F \triangle Fs_k) \geq \#F/n$.

On the other hand, we have an upper bound on Føl coming from balls: with $F = B_{G,S}(n)$ we have $F \triangle Fs \subseteq B_{G,S}(n+1)$, so $\#(F \triangle Fs) \leq v(n+1) - v(n)$, and therefore $\text{Føl}(v(n)/(v(n+1) - v(n))) \leq v(n)$. Assuming that v is the restriction to $\mathbb{N}$ of a differentiable function, we may seek a function f satisfying $f(1/\log(v)') = v$ to obtain an upper bound $\text{Føl}(n) \leq f(n)$. For example, if $v(n) \propto n^d$ then $f(n) \propto n^d$, and therefore the estimate given by Proposition 11.3.15 is at worst a constant off. The "1/2" in Proposition 11.3.15 cannot easily be eliminated: in a finite group, we shouldn't expect any good estimates for sets larger than half of the group.

Note also that we have $\text{Føl}(n) > n$ as soon as X is infinite, since then $\#(F \triangle Fs) \geq 1$ for all $F \Subset X$. No analogue of Proposition 11.3.15 may hold for G-sets in general:

Exercise 11.3.16 (**). Let X be a G-set for a finitely generated group G. Prove that $\text{Føl}(n)$ is linear (*i.e.*, $\text{Føl}(n) \leq Cn$ for some constant C) if and only if the Schreier graph of X has bounded cutsets, namely, there is a bound C' such that every finite set of vertices can be separated by removing at most C' vertices.

Exercise 11.3.17 (**). We saw in Exercise 11.2.28 that a group is "left-amenable" if and only if it is amenable. First prove directly that if a group admits sets that are almost invariant under right translation, then it admits sets that are almost invariant under left translation.

Next, prove that the infinite dihedral group $D_\infty = \langle a, b \mid a^2, b^2\rangle$ admits finite subsets that are almost right-invariant but far from left-invariant, namely, subsets $F_n \Subset D_\infty$ with $\#(F_n \triangle F_n g)/\#F_n \to 0$ for all $g \in D_\infty$ but $\#(F_n \triangle gF_n)/\#F_n \not\to 0$.

Give on the other hand a family of sets $F_n \Subset D_\infty$ with $\#(F_n \triangle gF_n h)/\#F_n \to 0$ for all $g, h \in D_\infty$.

We return to Definition 11.3.3. A connected graph $\mathscr{G} = (V, E)$ endows its set of vertices V with the structure of a metric space still written $\mathscr{G}$: the distance between two vertices is the minimal length of a path connecting them. Given two metric spaces (*e.g.*, connected graphs) X, Y, a map $f: X \to Y$ is *quasi-Lipschitz* if there is a constant C with

$$d(f(x), f(y)) \le C d(x, y) + C,$$

and f is a *quasi-isometry* if there is a quasi-Lipschitz map $g: Y \to X$ with $\sup_{x\in X} d(x, g(f(x))) < \infty$ and $\sup_{y\in Y} d(y, f(g(y))) < \infty$.

Exercise 11.3.18 (*). Let $\mathscr{G} = (V, E)$ be a graph, and let $\mathscr{G}' = (V', E')$ be its barycentric subdivision: $V' = V \sqcup E$ and $E' = E \times \{+, -\}$ with $(e, \pm)^{\pm} = e^{\pm}$ and $(e, \pm)^{\mp} = e$. Prove that $\mathscr{G}$ and $\mathscr{G}'$ are quasi-isometric.

Exercise 11.3.19 (*). Let G be a finitely generated group. Prove that all Cayley graphs of G with respect to finite generating sets are quasi-isometric; that all finite-index subgroups of G are have quasi-isometric Cayley graphs; and that all quotients of G by finite subgroups have quasi-isometric Schreier graphs.

Proposition 11.3.20. *Let $\mathscr{G} = (V, E)$ and $\mathscr{G}' = (V', E')$ be bounded-degree graphs, and let $f: \mathscr{G} \to \mathscr{G}'$ be quasi-Lipschitz with $\sup_{y\in V'} d(y, f(V)) < \infty$. If $\mathscr{G}$ is amenable then $\mathscr{G}'$ is amenable.*

In particular, if $\mathscr{G}, \mathscr{G}'$ are quasi-isometric, then $\mathscr{G}$ is amenable if and only if $\mathscr{G}'$ is amenable.

Proof. Let $\mathscr{G}'' = (V'', E'')$ be a graph. For $F \Subset V''$ and $k \in \mathbb{N}$, define

$$\partial^k(F) = \{(e_1, \dots, e_k) \mid e_i \in E'', e_i^+ = e_{i+1}^-, e_1^- \in F, e_k^+ \notin F\}.$$

Recall that $\mathscr{G}$ is amenable if and only if $\inf_{F\Subset V} \#\partial F/\#F = 0$. Equivalently, $\inf_{F\Subset V} \#\{e^+ \mid e \in \partial F\}/\#F = 0$. There exists a constant D such that, for every $F \Subset V$, we have $\{f(e^+) \mid e \in \partial F\} \subseteq \{e_D^+ \mid (e_1, \dots, e_D) \in \partial^D(f(F))\}$. Therefore, $\inf_{F\Subset V} \#\partial^D(f(F))/\#f(F) = 0$, and therefore $\inf_{F'\Subset V'} \#\partial(F')/\#F' = 0$.

Exercise 11.3.21 (*). Prove that if $\mathscr{G}, \mathscr{G}'$ are quasi-isometric graphs, then their Følner functions (11.9) are equivalent in the sense that $\text{Føl}_{\mathscr{G}}(n) \le C\text{Føl}_{\mathscr{G}'}(Cn + C) + C$ and conversely, for some constant C.

There are quasi-invariant groups with quite distinct algebraic properties, *e.g.*, $A\wr\mathbb{Z}$ and $B\wr\mathbb{Z}$ are quasi-isometric for all finite groups A, B of same cardinality. If A is Abelian but B is simple, then $A\wr\mathbb{Z}$ is metabelian and residually finite, but $B\wr\mathbb{Z}$ is neither. However, these groups are quasi-isometric (and both amenable).

11.3.2 Day's and Reiter's Criterion

Følner sets—finite subsets $F\Subset X$ that are almost invariant under translation—may be thought of as almost invariant characteristic functions.

Definition 11.3.22 (see [181, Theorem 1], [499, page 168]). Let X be a G-set. It satisfies the *Day-Reiter condition* for $p\ge 1$ if for every finite subset $S\Subset G$ and every $\epsilon>0$ there exists a positive function $\phi\in\ell^p(X)$ with $\|\phi s-\phi\|<\epsilon\|\phi\|$ for all $s\in S$.

Theorem 11.3.23. *Let X be a G-set. The following are equivalent:*

1. *X is amenable;*
2. *X satisfies the Day-Reiter condition for $p=1$;*
3. *X satisfies the Day-Reiter condition for some $p\in[1,\infty)$;*
4. *X satisfies the Day-Reiter condition for all $p\in[1,\infty)$;*
5. *X satisfies Følner's condition.*

Proof. (1) $\Rightarrow$ (2) Given $S\Subset G$ and $\epsilon>0$, consider the subset

$$K=\{\bigoplus_{s\in S}(\mu s-\mu)\mid \mu\in\mathscr{P}(X)\}\subset\ell^1(X)^S.$$

Since X is amenable, there exists a G-invariant functional $m\in\ell^\infty(X)^*$ by Corollary 11.2.25. Since $\ell^1(X)$ is weak*-dense in $\ell^\infty(X)^*$, there exists a net $(\mu_n)_{n\in\mathscr{N}}$ in $\ell^1(X)$ with $\mu_n\to\mu$ in the weak*-topology, so $\bigoplus_{s\in S}(\mu_n s-\mu_n)\in K$ converges to 0 in the weak*-topology on $\ell^1(X)^S$, so $\overline{K}^{\text{weak*}}\ni 0$. Since K is convex, its norm closure $\overline{K}$ also contains 0, by the Hahn-Banach theorem (see, *e.g.*, [515, Theorem 3.12]); so there exists $\mu\in\mathscr{P}(X)$ with $\|\mu s-\mu\|<\epsilon$ for all $s\in S$.

(2) $\Rightarrow$ (4) Let $\psi\in\ell^1(X)$ satisfy $\|\psi s-\psi\|<\epsilon\|\psi\|$ for all $s\in S$. Define $\phi(x):=\psi(x)^{1/p}$; then $\phi\in\ell^p(X)$ with $\|\phi\|_p=\|\psi\|^{1/p}$, and

$$\begin{aligned}\|\phi s-\phi\|_p^p&=\sum_{x\in X}|\phi(xs^{-1})-\phi(x)|^p=\sum_{x\in X}|\psi(xs^{-1})^{1/p}-\psi(x)^{1/p}|^p\\&\le\sum_{x\in X}|\psi(xs^{-1})-\psi(x)|\text{ because }|A^{1/p}-B^{1/p}|\le|A-B|^{1/p}\text{ for all }A,B\\&=\|\psi s-\psi\|<\epsilon\|\psi\|=\epsilon\|\phi\|_p^p.\end{aligned}$$

(4) $\Rightarrow$ (3) is obvious and so is (2) $\Rightarrow$ (3).

(3) $\Rightarrow$ (2) Let $\psi \in \ell^p(X)$ satisfy $\|\psi s - \psi\| < \epsilon\|\phi\|$ for all $s \in S$. Define $\phi(x) := \psi(x)^p$; then $\phi \in \ell^1(X)$ with $\|\phi\|_1 = \|\psi\|^p$, and

$$\begin{aligned}\|\phi s - \phi\| &= \sum_{x\in X} |\phi(xs^{-1}) - \phi(x)| = \sum_{x\in X} |\psi(xs^{-1})^p - \psi(x)^p|\\ &\le \sum_{x\in X} p|\psi(xs^{-1}) - \psi(x)| \max\{\psi(xs^{-1}), \psi(x)\}^{p-1} \text{ because } |X^p - Y^p|\\ &\le p|X-Y|\max\{X,Y\}^{p-1}\\ &\le p\Big(\sum_{x\in X} |\psi(xs^{-1}) - \psi(x)|^p\Big)^{1/p}\Big(\sum_{x\in X} |\psi(xs^{-1}) + \psi(x)|^p\Big)^{1-1/p}\\ &\quad\text{by Hölder's inequality}\\ &= p\|\psi s - \psi\|_p \|\psi s + \psi\|_p^{p-1} < p\epsilon\|\psi\|_p 2^{p-1}\|\psi\|_p^{p-1} = p2^{p-1}\epsilon\|\phi\|.\end{aligned}$$

(2) $\Rightarrow$ (5) Given $S \Subset G$ and $\epsilon > 0$, let $\phi \in \ell^1(X)$ be positive and satisfy $\|\phi s - \phi\| < \epsilon\|\phi\|$ for all $s \in S$. For all $r \in \mathbb{R}_+$, consider the set $F_r = \{x \in X \mid \phi(x) \ge r\}$. Then $\phi = \int \mathbb{1}_{F_r} dr$ and $\phi s = \int \mathbb{1}_{F_r s} dr$, so

$$\int (\#F_r s \triangle F_r) dr = \|\phi s - \phi\| < \epsilon\|\phi\| = \epsilon \int \#F_r dr;$$

therefore, there exists $r \in \mathbb{R}_+$ with $\#(F_r s \triangle F_r) < \epsilon \#F_r$, and X satisfies Følner's criterion by Lemma 11.3.4.

(5) $\Rightarrow$ (1) By Lemma 11.3.6, there exists a net $(F_n)_{n\in\mathscr{N}}$ with $\lim_{n\to\infty} \#(F_n g \setminus F_n)/\#F_n \to 0$ for all $g \in G$.

For each $n \in \mathscr{N}$, consider the "discrete" mean $\mu_n \in \mathscr{M}(X)$ defined by

$$\mu_n(A) = \frac{\#(A \cap F_n)}{\#F_n}.$$

Since $\mathscr{M}(X)$ is compact, the net $(\mu_n)_{n\in\mathscr{N}}$ has an accumulation point, say μ. We will show that μ is a G-invariant mean by a standard "$\delta/3$" argument.

Given $g \in G$ and $A \subseteq X$, we show $|\mu(A) - \mu(Ag)| < \delta$ for any $\delta > 0$. There is $n \in \mathscr{N}$ with

$$n > (\{g, g^{-1}\}, \delta/3), \quad |\mu_n(A) - \mu(A)| < \delta/3, \quad |\mu_n(Ag) - \mu(Ag)| < \delta/3,$$

because the μ_n converge pointwise to μ. Then

$$\begin{aligned}|\#(A \cap F_n) - \#(Ag \cap F_n)| &= |\#(A \cap F_n) - \#(A \cap F_n g^{-1})|\\ &\le \max\{\#(F_n \setminus F_n g^{-1}), \#(F_n g^{-1} \setminus F_n)\}\\ &\le \#(F_n\{g, g^{-1}\} \setminus F_n) < \epsilon \#F_n < \delta/3\#F_n,\end{aligned}$$

so $|\mu_n(A) - \mu_n(Ag)| < \delta/3$ and

$$|\mu(A) - \mu(Ag)| \le |\mu_n(A) - \mu(A)| + |\mu_n(A) - \mu_n(Ag)| + |\mu_n(Ag) - \mu(Ag)| < \delta.$$

Since this holds for all $\delta > 0$, we get $\mu(A) = \mu(Ag)$.

In fact, the "$\#(Fs \setminus F)/\#F \to 0$" in Følner's condition can be substantially weakened:

Proposition 11.3.24 (Gournay). *Let X be a G-set. Then X is amenable if and only if there is a constant $c < 1$ with the following property: for every finite subset $S \Subset X$, there is a finite subset $F \Subset X$ with $\#(Fs \setminus F) \le c\#F$ for all $s \in S$.*

Proof. ($\Rightarrow$) is obvious, by Lemma 11.3.4 and Theorem 11.3.23.

($\Leftarrow$) by the condition of the proposition, there exists a net $(F_n)_{n\in\mathscr{N}}$ of finite subsets of X (say indexed by $\mathfrak{P}_f(G)$) with $\limsup_{n\to\infty} \#(F_n g \setminus F_n)/\#F_n \le c$ for all $g \in G$. Let $\xi_n := \mathbb{1}_{F_n}/\sqrt{\#F_n} \in \ell^2(X)$ be the normalized characteristic function of F_n. We have

$$2 - 2\langle \xi_n, \xi_n g\rangle = \|\xi_n g - \xi_n\|_2^2 = \|\mathbb{1}_{F_n g} - \mathbb{1}_{F_n}\|_1/\#F_n = 2\#(F_n g \setminus F_n)/\#F_n,$$

so $\langle \xi_n, \xi_n g\rangle \ge 1 - c$ for all $n \gg 1$.

Choose now a non-principal ultrafilter $\mathfrak{F}$ on $\mathscr{N}$, and consider the ultraproduct space $\mathscr{H} := \ell^2(X)^{\mathfrak{F}}$: it is a Hilbert space, whose elements are equivalence classes of sequences $(\eta_n)_{n\in\mathscr{N}}$ with $\eta_n \in \ell^2(X)$ for all n and $\sum_{n\in\mathscr{N}} \|\eta_n\|^2 < \infty$, under the relation $(\eta_n) \sim (\eta'_n)$ if $\lim_{\mathfrak{F}} \|\eta_n - \eta'_n\| = 0$.

Write $\xi = (\xi_n) \in \mathscr{H}$, and let K denote the convex hull in $\mathscr{H}$ of $\{\xi g \mid g \in G\}$. We have $\langle \xi g, \xi\rangle \ge 1 - c$ for all $g \in G$, so $\langle \xi, \eta\rangle \ge 1 - c > 0$ for all $\eta \in K$, and in particular $0 \notin K$. Let ζ' be the element of K of minimal norm, and set $\zeta = \zeta'/\|\zeta'\|$, represented by a sequence $(\zeta_n)_{n\in\mathscr{N}}$ with $\zeta_n \in \ell^2(X)$ of norm 1. Since $\zeta g = \zeta$ by unicity of the element of minimal norm in K, we have $\|\zeta_n - \zeta_n g\| \to 0$ for all $g \in G$, so X is amenable by Theorem 11.3.23(3).

We finally present a result that puts as much symmetry between X and G as possible, with an eye toward the corresponding notion with the roles of G and X interchanged, see Theorem 11.8.20:

Proposition 11.3.25. *Let X be a G-set. Then X is amenable if and only if for every $\epsilon > 0$ and every $g \in \varpi(\ell^1 G)$ there exists a positive function $f \in \ell^1(X)$ with $\|fg\| < \epsilon\|f\|$.*

Proof. ($\Rightarrow$) Given $\epsilon > 0$ and $g = \sum_{x\in G} g_x x \in \ell^1 G$ with $\sum g_x = 0$, let $S \Subset G$ be such that $g' := g - \sum_{s\in S} g_s(s-1)$ satisfies $\|g'\| < \epsilon/2$. Since X is amenable, there exists $F \Subset X$ with $\#(Fs \setminus F) < \epsilon/4\|g\|\#F$ for all $s \in S$. Set $f := \mathbb{1}_F$. Then

$$\|fg\| \le \|fg'\| + \sum_{s\in S} \|g_s(fs - f)\| < \#F\|g'\| + 2\#(FS \setminus F)\|g\| < \epsilon\#F = \epsilon\|f\|.$$

($\Leftarrow$) Given $\epsilon > 0$ and $S \Subset G$, set $g = \sum_{s\in S} s - 1$, and let $f \in \ell^1(G)$ be a positive function satisfying $\|fg\| < \epsilon\|f\|/2$. Then

$$\epsilon\|f\| > 2\|fg\| \geq 2\Big\|\sum_{s\in S}\max(fs-f,0)\Big\| = \sum_{s\in S}2\|\max(fs-f,0)\| \geq \sum_{s\in S}\|fs-f\|.$$

□

11.3.3 Non-amenability

It may be interesting to consider weaker versions of amenability for groups, for instance, to consider groups admitting faithful amenable actions.

Denis Osin considers in [463] a class of "weakly amenable groups," which in our context are groups G admitting an amenable action X such that, for every finite $F \subset G$, there exists $x \in X$ with $\#(xF) = \#F$, namely, the orbit map $f \mapsto x \cdot f$ is injective on F. An example of a weakly amenable, non-amenable group is the Baumslag-Solitar group $\langle a, t \mid a^m t = ta^n\rangle$, for $m > n \geq 2$.

If a group G is not amenable, but all its proper subgroups are amenable, then G does not have any "interesting" amenable actions: by Proposition 11.2.26, every amenable action of G has a fixed point. This applies in particular to Tarski monsters [455], which are non-amenable torsion groups in which every proper subgroup is cyclic.

Definition 11.3.26 (Kazhdan, see [339] or [62]). A group G has *property (T)* if every unitary representation $G \to U(\mathscr{H})$ in a Hilbert space $\mathscr{H}$ with almost invariant vectors (in the sense that for every $\epsilon > 0$ and every finite $S \Subset G$ there exists nontrivial $x \in \mathscr{H}$ with $\|x - xs\| < \epsilon$ for all $s \in S$) has a nontrivial fixed vector.

If G is infinite, then Kazhdan's property (T) restricted to the unitary representation on $\ell^2(G)$ is thus precisely the negation of amenability: there are invariant vectors in $\ell^2(G)$ if and only if G is finite, and the existence of almost invariant vectors is the Day-Reiter condition for $p = 2$.

Thus an amenable group with property (T) is finite,[8] and more generally a group with property (T) does not have any "interesting" amenable actions: every amenable action has a finite orbit.

Glasner and Monod consider in [245] another group property, which they call *property (F)*: "every amenable action has a fixed point." They show that a free product of groups always has a faithful, transitive, amenable action unless one factor is (F) and the other is virtually (F). Thus, for example, $G * \mathbb{Z}$ is not amenable if $G \neq 1$ yet admits a faithful, transitive, amenable action.

[8]This was exploited in a fundamental manner by Margulis in [410] to prove his "normal subgroup theorem."

11.4 Growth of Groups

We cover here some classical material on asymptotic growth of groups. Recall from Section 11.3.1 that $v_{G,S}(n)$ denotes the number of elements in a group G that are expressible as products of at most n elements of S. The group G has *exponential growth* if $v_{G,S}(n) \geq B^n$ for some $B > 1$ and *subexponential growth* otherwise; it has *polynomial growth* if $v_{G,S}(n) \leq p(n)$ for some polynomial p; and it has *intermediate growth* if its growth is neither polynomial nor exponential. These properties are easily seen to be independent of the choice of generating set S.

11.4.1 Groups of Polynomial Growth

Groups of polynomial growth admit an elegant algebraic characterization. The "if" part is due to Hyman Bass [46] and independently Yves Guivarc'h [278], with an explicit computation of the growth degree of G, which is always an integer; the harder, "only if" part is due to Misha Gromov.

We recall some basic group theoretical terminology. For $\mathscr{P}$ a property of groups (abelian, ...), a group G is called *virtually* $\mathscr{P}$ if G admits a finite-index subgroup satisfying $\mathscr{P}$.[9]

A group G is *nilpotent* if there exists a constant c such that every $(c+1)$-fold iterated commutator $[g_0, [g_1, \dots, [g_{c-1}, g_c] \cdots]]$ vanishes in G; the minimal c is called the *nilpotency class* of G. A group G is *polycyclic* if it admits a sequence of subgroups $G = G_0 \rhd G_1 \rhd \cdots \rhd G_n = 1$ with G_k/G_{k+1} cyclic for all k. Finitely generated nilpotent groups are polycyclic.

Theorem 11.4.1 (Gromov [274]). *Let G be a finitely generated group. Then G has polynomial growth if and only if G is virtually nilpotent, namely, G has a finite-index nilpotent subgroup.*

Proof (Proof of Theorem 11.4.1, "if" direction). Let G_0 be a finite-index nilpotent subgroup of G. It suffices to prove that G_0 has polynomial growth, since then G will have polynomial growth of same degree as G_0. Denote by c the nilpotency class of G_0, so all $(c+1)$-fold iterated commutators vanish.

Let $(G_k)_{0\leq k\leq \ell}$ be a composition series for G, namely, a series of subgroups such that G_k/G_{k+1} is cyclic for all k; and for each k, let $x_k \in G_k$ be a lift of a generator of G_k/G_{k+1} so that $G_0 = \langle x_0, x_1, \dots x_{\ell-1}\rangle$.

We reason by induction on ℓ. If $\ell = 0$, or if G/G_1 is finite, we are done. Assume then $G_0/G_1 \cong \mathbb{Z}$, and by induction that the growth of G_1 is bounded by a polynomial, say of degree d.

[9] Much to the annoyance of finite group theorists, some people call finite groups "virtually trivial."

Consider $x \in G_0$, of the form $x = x_{i_1}^{\pm 1} \cdots x_{i_n}^{\pm 1}$. Write it in the form $x_0^e z$, with $e \in \mathbb{Z}$ and $z \in G_1$. This requires us to exchange past each other some letters x_0 and x_{i_j}, producing subexpressions $[x_{i_j}, x_0, \dots, x_0]$ along the process: indeed one has $Wx_0 = x_0 W[W, x_0]$ for any expression W.

There are at most n letters x_0 in x; each of them must be brought past at most n other letters, producing at most n^2 expressions $[x_i, x_0]$; each of these produces in turn at most n^3 expressions $[x_i, x_0, x_0]$; etc. We take as generating set S for G_1 all expressions of the form $[x_i, x_0, \dots, x_0]$ with $i \geq 1$ and length $\in \{1, \dots, c\}$. We have then expressed x by an integer $e \in \{-n, \dots, n\}$ and a word z of length at most $n + \cdots + n^c$ in these generators; so

$$v_{G_0, S\cup\{x_0\}}(n) \leq (2n+1) v_{G_1,S}(n + \cdots + n^c)$$

is bounded by a polynomial of degree $\leq cd + 1$.

We shall give at the end of Section 11.8.2 a sketch of the "only if" direction, via slowly growing harmonic functions.

Corollary 11.4.2. *Let G be a virtually nilpotent group. Then G is amenable.*

Proof. If G is virtually nilpotent, then every finitely generated subgroup of G is also virtually nilpotent, so by Theorem 11.4.1 has polynomial growth, so is amenable by Proposition 11.3.14.

11.4.2 Groups of Exponential Growth

At the other end of the growth spectrum, we find groups of *exponential growth*. In fact, as soon as a group has a non-abelian free subgroup, it has exponential growth; so a large class of groups, including all non-elementary hyperbolic groups [241], have exponential growth.

In the class of soluble groups, the growth of a group is either polynomial or exponential, as we shall see below. Recall that the *derived series* of a group G is the series of normal subgroups defined by $G^{(0)} = G$ and $G^{(i+1)} = [G^{(i)}, G^{(i)}]$ and that G is *soluble* if $G^{(n)} = 1$ for some n. The minimal such n is called the *derived length* of G.

Proposition 11.4.3 (see [418]*Lemma 1). *Let G be a finitely generated group of subexponential growth, and let $N \triangleleft G$ be a normal subgroup with $G/N \cong \mathbb{Z}$. Then N is also finitely generated.*

Proof. Let $S = \{x_1, \dots, x_d\}$ generate G, and let $x \in G$ generate G/N. Write each $x_i = x^{e_i} y_i$, with $y_i \in N$; so $G = \langle x, y_1, \dots, y_d \rangle$, and $N = \langle y_1, \dots, y_d \rangle^G$.

Consider further $N_i = \langle y_i^{x^n} \mid n \in \mathbb{Z} \rangle$, so that $N = \langle N_1, \dots, N_d \rangle$. It is sufficient to show that each N_i is finitely generated.

Write then $y = y_i$, and consider all expressions $x^{-1}y^{e_1}x^{-1}y^{e_2}\dots x^{-1}y^{e_n}x^n$, with all $e_j \in \{0,1\}$. There are 2^n such expressions, and their length is linear in n, so two must be equal in G because G has subexponential growth. Let

$$y^{e_1 x}\cdots y^{e_m x^m} = y^{f_1 x}\cdots y^{f_m x^m} \tag{11.10}$$

be such an equality in G, without loss of generality with $1 = e_m \neq f_m = 0$. It follows that y^{x^m} is in the group generated by $\{y^x,\dots,y^{x^{m-1}}\}$, so that $N_i = \langle y_i^{x^n} \mid n < m\rangle$. Now a similar argument, replacing x by x^{-1} in (11.10), shows that N_i is finitely generated.

Corollary 11.4.4. *Let G be a finitely generated group of subexponential growth, and let $N \triangleleft G$ be a normal subgroup such that G/N is virtually polycyclic. Then N is finitely generated.* □

Corollary 11.4.5 (Milnor [418]). *Let G be a finitely generated soluble group of subexponential growth. Then G is polycyclic.*

Proof. Consider the derived series $G^{(i)}$ of G; by assumption, $G^{(s+1)} = 1$ for some minimal $s \in \mathbb{N}$. Set $A = G^{(s)}$. We may assume, by induction, that G/A is polycyclic. By Corollary 11.4.4, the subgroup A is finitely generated and abelian, so is polycyclic too. It follows that G is polycyclic.

Lemma 11.4.6. *Let G be a finitely generated group that is an extension $N.Q$ of finitely generated virtually nilpotent groups. Then G is virtually soluble.*

Proof. Assume first that N is finite; we then claim that G is virtually nilpotent. Indeed the centralizer $Z_G(n)$ has finite index in G, so $Z = \bigcap_{n\ge 0} Z_G(n)$ has finite index in G. Then Z is a central extension of $Z \cap N$ by $Z/(Z \cap N)$, so is virtually nilpotent; and then so is G.

We turn to the general case. Let N_0 be a nilpotent subgroup of finite index in N. Up to replacing N_0 by $\bigcap_{[N:M]=[N:N_0]} M$, we may assume N_0 is characteristic in N and therefore normal in G. By the first paragraph, G/N_0 is virtually nilpotent, so G is virtually soluble.

We recall that a group is *noetherian* if all its subgroups are finitely generated; in other words, if every chain $H_1 < H_2 < \cdots$ of subgroups of G is finite.

Lemma 11.4.7. *A group G is polycyclic if and only if it is both soluble and noetherian.*

Proof. Note first that an abelian group is noetherian if and only if it is finitely generated: if finitely generated, it is of the form $\mathbb{Z}^d \times F$ for a finite abelian group F and is clearly noetherian.

If G is soluble and noetherian, then all quotients $G^{(i)}/G^{(i+1)}$ along its derived series are also noetherian, so finitely generated; the derived series may then be refined into a polycyclic series.

Conversely, an extension of noetherian groups is noetherian, so if G is polycyclic, then it is noetherian by induction.

This reduction to polycyclic groups brings us closer to groups of polynomial growth; the next step is the

Theorem 11.4.8 (Wolf). *Let G be a polycyclic group of subexponential growth. Then G is virtually nilpotent.*

Proof. Let $G = G_0 > G_1 > \cdots$ be a polycyclic series of minimal length. If $[G : G_1] < \infty$, proceed inductively with G_1. Assume therefore that $G/G_1 \cong \mathbb{Z} = \langle x \rangle$. By induction, there is a nilpotent subgroup $N \leq G_1$ of finite index. Furthermore, since G_1 is finitely generated by Proposition 11.4.3, we may suppose that N is characteristic in G_1, at the cost of intersecting it with its finitely many images under automorphisms of G_1; so we may assume $N \triangleleft G$. We have $N\langle x \rangle \leq G$ of finite index, and we replace G by $N\langle x \rangle$, to simplify notation.

We now seek a central series (N_k) in N, *i.e.*, a series with $N_0 = N$, all N_k normal in G, and $N_k/N_{k+1} \leq Z(N/N_{k+1})$; and we require that some nonzero power x^n centralizes N_k/N_{k+1} for all k. Then $\langle N, x^n \rangle$ will be the finite-index nilpotent subgroup of G we are after.

Among central series, choose one maximizing the number of k such that N_k/N_{k+1} is infinite; it exists because the number of factors is bounded by the Hirsch length of G. The torsion subgroup of N_k/N_{k+1} is characteristic, so insert it in the series between N_k and N_{k+1}. The resulting series is such that each quotient N_k/N_{k+1} is either finite or free abelian; and, in the latter case, if $M \triangleleft G$ and $N_{k+1} \leq M \leq N_k$, then either $N_{k+1} = M$ or N_k/M is finite.

If N_k/N_{k+1} is finite, then certainly some nonzero power of x will act trivially on it. We therefore consider $N_k/N_{k+1} \cong \mathbb{Z}^m$, and we study the $\mathbb{Q}[x]$-module $V := N_k/N_{k+1} \otimes \mathbb{Q} \cong \mathbb{Q}^m$.

The module V is irreducible; indeed, otherwise there would exist a proper, nontrivial invariant subspace $W < V$; then $M := \{x \in N_k \mid xN_{k+1} \in W\}$ is a normal subgroup of G, of infinite index in N_k, contradicting the maximality of the number of infinite factors in (N_k). We then use the

Lemma 11.4.9 (Schur). *Let V be an irreducible module. Then* $\mathrm{End}(V)$ *is a division ring.*

Proof. Let $\alpha \neq 0 \in \mathrm{End}(V)$ be an endomorphism; then $\ker(\alpha)$ and $\alpha(V)$ are invariant subspaces, so $\ker(\alpha) = 0$ and $\alpha(V) = V$; so α is invertible.

We see $x \in G$ as an endomorphism of V; by Lemma 11.4.9, the ring $\mathrm{End}(V)$ does not contain nilpotent elements, so x generates a field $\mathbb{Q}(x)$ within $\mathrm{End}(V)$. Since $\mathrm{End}(V)$ is finite-dimensional, x is algebraic. Since x preserves the lattice $N_k/N_{k+1} \subset V$, it is an algebraic integer. We now recall the classical

Lemma 11.4.10 (Kronecker). *Let τ be an algebraic number, all of whose conjugates have norm* 1. *Then τ is a root of unity.*

Proof. Let τ be algebraic of degree n, and consider some power $\sigma = \tau^N$. Then $\sigma \in \mathbb{Q}(\tau)$, and all conjugates of σ have norm 1, so the coefficients of the minimal polynomial of σ, which are symmetric functions of the conjugates of σ, have norm at most 2^n. It follows that there are finitely many such minimal polynomials, so $\sigma^N = \sigma^M$ for some $M > N$.

We are now ready to finish the proof. Either all conjugates of x (seen now as an algebraic number) have norm ≤ 1; and then x is a root of unity by Lemma 11.4.10, so x^n acts trivially for some $n > 0$; or there exists an embedding of $\mathbb{Q}(x)$ in $\mathbb{C}$ such that $|x| > 1$.

In that last case, we may replace x by a power of itself so that $|x| > 2$. Choose $y \in N_k \backslash N_{k+1}$, seen as a vector $v \neq 0 \in V$. Consider as in the proof of Proposition 11.4.3 all expressions $x^{-1}y^{e_1}x^{-1}y^{e_2}\dots x^{-1}y^{e_n}x^n$, with all $e_j \in \{0, 1\}$. There are 2^n such expressions, and their length is linear in n, so two must be equal in G because G has subexponential growth. This leads in V to the relation

$$(e_1 - f_1)x(v) + \cdots + (e_{n-1} - f_{n-1})x^{n-1}(v) + x^n(v) = 0,$$

so $(e_1 - f_1)x + \cdots + (e_{n-1} - f_{n-1})x^{n-1} + x^n = 0$, because only 0 is non-invertible in $\mathrm{End}(V)$. Now taking norms we get

$$|x|^n \leq (e_1 - f_1)|x| + \cdots + (e_{n-1} + f_{n-1})|x|^{n-1} \leq |x|\frac{|x|^{n-1} - 1}{|x| - 1} \leq |x|^n$$

using $|x| > 2$, a contradiction.

Corollary 11.4.11. *Let G be a virtually soluble finitely generated group. Then G has either polynomial of exponential growth, and has polynomial growth precisely when it is virtually nilpotent.* □

11.4.3 Groups of Intermediate Growth

The previous sections were aimed at showing that "most" groups have polynomial or exponential growth; John Milnor asked in 1968 whether there existed any groups of intermediate growth [419]. There can be no such examples among virtually soluble groups, as we saw above, nor among linear groups (subgroups of matrix groups over fields), by Tits' alternative [564].

Milnor's question has, however, a positive answer, which was given in the early 1980s by Slava Grigorchuk. We give here his example.

Set $\mathscr{A} = \{0, 1\}$, and consider the following group G acting recursively on the set $X := \mathscr{A}^{\mathbb{N}}$ of infinite sequences over $\mathscr{A}$. It is generated by four elements a, b, c, d defined by

$$(x_0x_1\cdots)a = (1 - x_0)x_1\cdots,$$

$$(x_0x_1\cdots)b = \begin{cases} x_0\cdots(1 - x_n)x_{n+1}\cdots & \text{if } x_0 = \cdots = x_{n-2} = 0 \neq x_{n-1}, n \not\equiv 0 \pmod 3 \\ x_0x_1\cdots & \text{else,} \end{cases}$$

$$(x_0x_1\cdots)c=\begin{cases}x_0\cdots(1-x_n)x_{n+1}\cdots & \text{if } x_0=\cdots=x_{n-2}=0\neq x_{n-1}, n\not\equiv 2 \pmod 3\\ x_0x_1\cdots & \text{else,}\end{cases}$$

$$(x_0x_1\cdots)d=\begin{cases}x_0\cdots(1-x_n)x_{n+1}\cdots & \text{if } x_0=\cdots=x_{n-2}=0\neq x_{n-1}, n\not\equiv 1 \pmod 3\\ x_0x_1\cdots & \text{else.}\end{cases}$$

This action is the limit of an action on finite sequences $\mathscr{A}^*$, which is the vertex set of the binary rooted tree, and G is self-similar, see Definition 11.2.15.

Theorem 11.4.12 (Grigorchuk [268]). *The group* G *has intermediate growth. More precisely, let* $\eta \approx 0.811$ *be the positive root of* $X^3+X^2+X-2=0$*; then*

$$\exp(n^{1/2}) \precsim v_{G,S}(n) \precsim \exp(n^{\log(2)/(\log(2)-\log(\eta))}).$$

We begin by a series of exercises deriving useful properties of G. Details may be found, *e.g.*, in [183, Chapter 8]. The self-similar structure of G is at the heart of all arguments; let us describe it again, starting from the action above.

There is an injective group homomorphism $\Phi\colon \mathsf{G} \to (\mathsf{G}\times\mathsf{G}) \rtimes C_2$, written $g \mapsto \langle\!\langle g_0, g_1\rangle\!\rangle \pi_g$ and defined as follows. If g permutes $0X$ and $1X$, then $\pi_g=\varepsilon\neq 1$, while if g preserves them setwise, then $\pi_g=1$. Then $g\pi_g^{-1}$ preserves $0X$ and $1X$, and for $i=0,1$ define a permutation g_i of X by $(x_0x_1\dots)g=(x_0\pi_g)\,(x_1\dots)g_{x_0}$. To see that the g_i belong to G, note that Φ is given on the generators by

$$\Phi\colon\begin{cases}a & \mapsto \langle\!\langle 1,1\rangle\!\rangle\varepsilon,\\ b & \mapsto \langle\!\langle a,c\rangle\!\rangle,\\ c & \mapsto \langle\!\langle a,d\rangle\!\rangle,\\ d & \mapsto \langle\!\langle 1,b\rangle\!\rangle.\end{cases}$$

Exercise 11.4.13 (*). Check in G the relations $a^2=b^2=c^2=d^2=bcd=(ad)^4=1$.

We fix once and for all the generating set $S=\{a,b,c,d\}$ of G. It follows from the exercise that every element of G may be written as a word of minimal length in the form $s_0as_1\cdots s_{n-1}as_n$ for some $s_0,s_n\in\{1,b,c,d\}$ and other $s_i\in\{b,c,d\}$.

We let $\eta\approx 0.811$ be the real root of $X^3+X^2+X-2=0$ and define a metric on G by setting

$$\|a\|=1-\eta^3,\quad \|b\|=\eta^3,\quad \|c\|=1-\eta^2,\quad \|d\|=1-\eta$$

and extending the metric to G by the triangle inequality: $\|g\|=\min\{\|s_1\|+\cdots+\|s_n\| \mid g=s_1\cdots s_n\}$.

Lemma 11.4.14. *If $\Phi(g) = \langle\!\langle g_0, g_1 \rangle\!\rangle \pi$, then $\|g_0\| + \|g_1\| \leq \eta(\|g\| + \|a\|)$.*

Proof. Consider $g \in \mathsf{G}$. Since $\|c\| + \|d\| \geq \|b\|$, etc., g may be written as a word of minimal norm in the form $s_0 a s_1 \cdots s_{n-1} a s_n$ for some $s_0, s_n \in \{1, b, c, d\}$ and other $s_i \in \{b, c, d\}$, using Exercise 11.4.13. Now among the s_i, each "b," taken with the "a" after it, contributes $\|b\| + \|a\| = 1$ to $\|g\|$ and contributes at most $\|a\| + \|c\| = \eta$ to $\|g_0\| + \|g_1\|$ because $\Phi(b) = \langle\!\langle a, c \rangle\!\rangle$. Similarly, each "$c$"+"$a$" contributes η to $\|g\|$ and at most η^2 to $\|g_0\| + \|g_1\|$, and each "d"+"a" contributes η^2 to $\|g\|$ and at most η^3 to $\|g_0\| + \|g_1\|$. Only the last s_n may not have an "a" after it. Summing all these inequalities proves the lemma.

Exercise 11.4.15 (**). Define $\sigma\colon \mathsf{G} \to \mathsf{G}$ by

$$\sigma\colon a \mapsto c^a, \quad b \mapsto d, \quad d \mapsto c, \quad c \mapsto b,$$

extended multiplicatively. Prove $\Phi(\sigma(g)) = \langle\!\langle \theta(g), g \rangle\!\rangle$ for all $g \in \mathsf{G}$, where $\theta(a) = d, \theta(b) = 1, \theta(c) = \theta(d) = a$ is a homomorphism to the finite group $\langle a, d \rangle \cong D_4$. Deduce that σ is well defined and is an injective endomorphism of G. For the usual word metric, prove that $|\sigma(g)| \leq 2|g| + 1$ for all $g \in \mathsf{G}$.

Proof (Proof of Theorem 11.4.12, see [38]). For the lower bound, consider the map (not quite a homomorphism!)

$$F\colon \mathsf{G} \times \mathsf{G} \to \mathsf{G}, \qquad (g_0, g_1) \mapsto \sigma(g_0)^a \sigma(g_1).$$

By the exercise, we have $\Phi(F(g_0, g_1)) = \langle\!\langle g_0\theta(g_1), \theta(g_0)g_1 \rangle\!\rangle$. Since $\#\theta(\mathsf{G}) = 8$ and Φ is injective, we have $\#\Phi^{-1}(g) = 8$ for all $g \in \mathsf{G}$. Also, $|\sigma(g)| \leq 2|g| + 1$ for the usual word metric, so $|F(g_0, g_1)| \leq 2|g_0| + 2|g_1| + 4$. Denoting by $B(n)$ the ball of radius n in G for the word metric, we have $F(B(n) \times B(n)) \subseteq B(4n + 4)$, so the growth function $v(n)$ of G satisfies $8v(n-2)^2 \leq v(4(n-2) + 4) \leq v(4n - 2)$. Iterating, we have $v(4^t n - 2) \geq 8^{2^t - 1} v(n-2)^{2^t}$, so $v(n) \geq 8^{\sqrt{n/8} - 1}$.

For the upper bound, we make use of the norm $\|\cdot\|$ and represent every $g \in \mathsf{G}$ by a finite rooted tree $R(g)$. Fix any constant $K > \|a\|/(\eta - 1)$. Given $g \in \mathsf{G}$, construct $R(g)$ as follows. If $\|g\| \leq K$, let $R(g)$ be the one-vertex tree with label g written at the root, which is also a leaf of the tree.

If $\|g\| > K$, compute $\Phi(g) = \langle\!\langle g_0, g_1 \rangle\!\rangle \pi$. Note $\|g_0\|, \|g_1\| < \|g\|$, and construct $R(g_0), R(g_1)$ recursively. Let then $R(g)$ be the tree with a root labeled π connected by two edges leading to the roots of $R(g_0)$ and $R(g_1)$, respectively. Since Φ is injective, the map R is injective, and it remains to count the number of trees of given size.

Up to replacing $\|g\|$ by $\max\{1, \|g\| - K\}$, we may assume that, in Lemma 11.4.14, we have $\|g_0\| + \|g_1\| \leq \eta\|g\|$ as soon as $\|g\|$ is large enough.

Let us denote by $\#R(g)$ the number of leaves of $R(g)$, and set $\alpha = \log 2/(\log 2 - \log\eta)$. We claim that there is a constant D such that $\#R(g) \le D\|g\|^\alpha$ for all $g \in \mathsf{G}$. This is certainly true if $\|g\|$ is small enough. For $\|g\| > K$, we proceed by induction:

$$\begin{aligned}\#R(g) &= \#R(g_0) + \#R(g_1) \le D(\|g_0\|^\alpha + \|g_1\|^\alpha)\\ &\le 2D\left(\frac{\|g_0\| + \|g_1\|}{2}\right)^\alpha \text{ by convexity of } X^\alpha\\ &\le 2D\|g\|^\alpha\left(\frac{\eta}{2}\right)^\alpha = D\|g\|^\alpha.\end{aligned}$$

We finally count the number of trees with n leaves. There are Catalan(n) such tree shapes; each of the $n-1$ non-leaf vertices has a label in $\{1, \varepsilon\}$, and each of the n leaf vertices has a label in $B(K)$. It follows that there are Catalan$(n)2^{n-1}B(K)^n \le E^n$ trees with at most n leaves, for some constant E; and then $v(n) \le E^{n^\alpha}$.

Exercise 11.4.16 (**). Prove that G is a torsion group.
Hint: Use Exercise 11.4.13, Lemma 11.4.14 and induction.

11.5 Paradoxical Decompositions

We consider again the general case of a group G acting on a set X and shall derive other characterizations of amenability, based on finite partitions of X.

Definition 11.5.1. A G-set X is *paradoxical* if there are partitions

$$X = Y_1 \sqcup \cdots \sqcup Y_m = Z_1 \sqcup \cdots \sqcup Z_n,$$

and $g_1, \ldots, g_m, h_1, \ldots, h_n \in G$, such that

$$X = Y_1 g_1 \sqcup \cdots \sqcup Y_m g_m \sqcup Z_1 h_1 \sqcup \cdots \sqcup Z_n h_n.$$

□

As a naive example, relax the condition that G be a group, and consider the monoid of affine transformations of $\mathbb{N}$. Then $\mathbb{N} = \mathbb{N}g_1 \sqcup \mathbb{N}h_1$ for $g_1(n) = 2n$, and $h_1(n) = 2n + 1$ defines a paradoxical decomposition.[10]

[10] This should not come as a surprise, since $\{g_1, h_1\}$ generate a free monoid.

Example 11.5.2. We return to Proposition 11.2.9. More precisely, now, consider $X = G = \langle x_1, x_2 \mid \rangle$ a free group of rank 2; and

$$Y_1 = \{\text{reduced words ending in } x_1\}, \qquad Y_2 = G \setminus Y_1,$$
$$Z_1 = \{\text{reduced words ending in } x_2\} \cup \{1, x_2^{-1}, x_2^{-2}, \dots\}, \qquad Z_2 = G \setminus Z_1;$$

then $G = Y_1 \sqcup Y_2 = Z_1 \sqcup Z_2 = Y_1 \sqcup Y_2 x_1^{-1} \sqcup Z_1 \sqcup Z_2 x_2^{-1}$.

11.5.1 Hausdorff's Paradox

John von Neumann had noted already in [576] that non-amenability of F_2 was at the heart of the Hausdorff-Banach-Tarski paradox. We first show:

Proposition 11.5.3. *The group* $\mathrm{SO}_3(\mathbb{R})$ *of rotations of the sphere contains a non-abelian free subgroup.*

Proof. There are many classical proofs of this fact. Consider, for example, the matrices

$$U = \begin{pmatrix} 0 & 1 & 0 \\ 1 & 0 & 0 \\ 0 & 0 & -1 \end{pmatrix}, \qquad V = \begin{pmatrix} 1 & 0 & 0 \\ 0 & -\frac{1}{2} & \frac{\sqrt{3}}{2} \\ 0 & -\frac{\sqrt{3}}{2} & -\frac{1}{2} \end{pmatrix}$$

in $\mathrm{SO}_3(\mathbb{R})$. They satisfy the relations $U^2 = V^3 = 1$, but no other, since in a product $W = U^{\varepsilon_1} V^{\pm 1} U \cdots V^{\pm 1} U^{\varepsilon_2}$ with $\varepsilon_1, \varepsilon_2 \in \{0, 1\}$ and n letters $V^{\pm 1}$ we have

$$W = \frac{1}{2^n} \begin{pmatrix} a_{1,1} & a_{1,2} & \sqrt{3} a_{1,3} \\ a_{2,1} & a_{2,2} & \sqrt{3} a_{2,3} \\ \sqrt{3} a_{3,1} & \sqrt{3} a_{3,2} & a_{3,3} \end{pmatrix}$$

with $a_{i,j} \in \mathbb{Z}$ and $a_{3,3}$ odd, as can be seen from computing $2^n W \bmod 2$; so $W \neq 1$. Then $\langle [U, V], [U, V^{-1}] \rangle$ is a free group of rank 2.

Here is another proof: $\mathrm{SO}_3(\mathbb{R})$ is the group of quaternions of norm 1. Let p be a prime $\equiv 1 \pmod 4$, and set

$$S = \{(a + bi + cj + dk)/\sqrt{p} \mid a \in 2\mathbb{N} + 1, b, c, d \in 2\mathbb{Z}, a^2 + b^2 + c^2 + d^2 = p\}.$$

It follows from Lagrange's theorem on sums of four squares that $\#S = p + 1$ and from the unique factorization of quaternions that S generates a free group of rank $(p + 1)/2$. See [305] for proofs of these facts.

The following paradox follows:

Theorem 11.5.4 (Hausdorff [286]). *There exists a partition of the sphere S^2, or of the ball B^3, in two pieces; and a further partition of each of these into, respectively, two and three pieces, in such a manner that these be reassembled, using only isometries of $\mathbb{R}^3$, into two spheres or balls, respectively.*

Proof. We first show the following: there is a countable subset $D \subset S^2$ such that one can decompose $S^2 \setminus D = P \sqcup Q$, and further decompose $P = P_1 \sqcup \cdots \sqcup P_m$ and $Q = Q_1 \sqcup \cdots \sqcup Q_n$, so that $S^2 \setminus D = P_1 g_1 \sqcup \cdots \sqcup P_m g_m = Q_1 h_1 \sqcup \cdots \sqcup Q_n h_n$.

Indeed, by Proposition 11.5.3, there is a free subgroup G of $\mathrm{SO}_3(\mathbb{R})$, acting on the sphere. Every nontrivial element of G acts as a rotation and therefore has two fixed points. Let D denote the collection of all fixed points of all nontrivial elements of G; clearly D is countable. The group G acts freely on $S^2 \setminus D$; let T be a choice of one point per orbit.[11] Let (Y_i, Z_j, g_i, h_j) be a paradoxical decomposition of G as in Definition 11.5.1. Set then $P_i = TY_i g_i^{-1}$ and $Q_j = TZ_j h_j^{-1}$ for $i = 1, \dots, m$ and $j = 1, \dots, n$.

Keeping the same notation, we now show that S^2 can be cut as $S^2 = U \sqcup V$, such that for an appropriate rotation ρ we have $\rho(U) \sqcup V = S^2 \setminus D$. Since D is countable, there is a direction $\mathbb{R}v \subset \mathbb{R}^3$ that does not intersect D. There are continuously many rotations ρ with axis $\mathbb{R}v$ and only countably many that satisfy $D \cap \rho^n(D) \neq \emptyset$ for some $n \neq 0$; let ρ be any other rotation. Set $U = \bigcup_{n \geq 0} \rho^n(D)$ and $V = S^2 \setminus U$; then $\rho(U) = U \setminus D$ and we are done.

These paradoxical decompositions can be combined (see Corollary 11.5.8 below for details), proving the statement for S^2.

The same argument works for all concentric spheres simultaneously and therefore for $B^3 \setminus \{0\}$. It remains to show that B^3 and $B^3 \setminus \{0\}$ can, respectively, be cut into isometric pieces. Let ρ be a rotation about $(\frac{1}{2}, 0, \mathbb{R})$ with angle 1 (in radians), and set $W = \{\rho^n(0) \mid n \in \mathbb{N}\}$. Then $\rho(W) = W \setminus \{0\}$, so $B^3 = W \sqcup (B^3 \setminus W)$ and $B^3 \setminus \{0\} = \rho(W) \sqcup (B^3 \setminus W)$.

11.5.2 Doubling Conditions

Let us restate paradoxical decompositions in a more sophisticated way.

Definition 11.5.5. Let a group G act on a set X. A *G-wobble* is a map $\phi \colon Y \to Z$ for two subsets $Y, Z \subseteq X$, such that there exists a finite decomposition $Y = Y_1 \sqcup \cdots \sqcup Y_n$ and elements $g_1, \dots, g_n \in G$ with $\phi(y) = y g_i$ whenever $y \in Y_i$.

[11] The axiom of choice is required here.

We define a preorder[12] on subsets of X by $Y \precsim Z$ if there exists an injective G-wobble $Y \to Z$ and an equivalence relation $Y \sim Z$ if there exists a bijective G-wobble $Y \to Z$; in that case, we say that Y and Z are *equidecomposable.*

Using that terminology, the G-set X is paradoxical if one may decompose $X = Y \sqcup Z$ with $Y \sim X \sim Z$.

Lemma 11.5.6. *The map $\phi: Y \to Z$ is a G-wobble if and only if there exists a finite subset $S \Subset G$ such that $\phi(y) \in yS$ for all $y \in Y$.*

Proof. If ϕ is a G-wobble, set $S = \{g_1, \dots, g_n\}$, and note $\phi(y) \in yS$ for all $y \in Y$.
Conversely, if $\phi(y) \in yS$ for all $y \in Y$, write $S = \{g_1, \dots, g_n\}$, and set

$$Y_n = \{y \in Y \mid \phi(y) = yg_n \text{ and } \phi(y) \neq yg_m \text{ for all } m < n\}.$$

□

Corollary 11.5.7. *The composition of G-wobbles is again a G-wobble, and the inverse of a bijective G-wobble is also a G-wobble.* □

It follows that the set of invertible G-wobbles is actually a group. If the space X is assumed compact and the pieces in the decomposition are open, then this group is known as the "topological full group" of G, see Section 11.9.2.

Corollary 11.5.8. *The relation $\precsim$ is a preorder, and $\sim$ is an equivalence relation.*

Proof. Consider injective G-wobbles $\phi: Y \to Z$ and $\psi: W \to Y$. By Lemma 11.5.6, there are $S, T \Subset G$ such that $\phi(y) \in yS$ and $\phi(w) \in wT$ for all $y \in Y, w \in W$. Then $\phi\psi(w) \in wTS$ for all $w \in W$, so $\phi\psi: W \to Z$ is an injective G-wobble, again by Lemma 11.5.6.

Theorem 11.5.9 (Cantor-Schröder-Bernstein [126]). *Let Y, Z be sets. If there exists an injection $\alpha: Y \to Z$ and an injection $\beta: Z \to Y$, then there exists a bijection $\gamma: Y \to Z$.*

Furthermore, γ may be chosen so that $\gamma(y) \in \{\alpha(y), \beta^{-1}(y)\}$ for all $y \in Y$.

Proof. Let $\alpha: Y \to Z$ and $\beta: Z \to Y$ be injective maps. Set $Y_0 = Y$ and $Z_0 = Z$; and, for $n \geq 1$, set $Y_n = \beta(Z_{n-1})$ and $Z_n = \alpha(Y_{n-1})$. Partition Y as follows:

$$U = \bigsqcup_{n \in \mathbb{N}} Y_{2n} \setminus Y_{2n+1}, \qquad V = \bigsqcup_{n \in \mathbb{N}} Y_{2n+1} \setminus Y_{2n+2}, \qquad W = \bigcap_{n \in \mathbb{N}} Y_n.$$

Define then $\gamma: Y \to Z$ as follows:

$$\gamma(y) = \begin{cases} \alpha(y) & \text{if } y \in U; \\ \beta^{-1}(y) & \text{if } y \in V \cup W. \end{cases}$$

[12] *That is*, a transitive, reflexive relation.

Therefore γ sends $Y_{2n} \setminus Y_{2n-1}$ to $Z_{2n+1} \setminus Z_{2n}$ and $Y_{2n+1} \setminus Y_{2n+2}$ to $Z_{2n} \setminus Z_{2n-1}$ while sending $\bigcap Y_n$ to $\bigcap Z_n$. It follows that γ is a bijection.

Corollary 11.5.10. *If $Y \precsim Z$ and $Z \precsim Y$, then $Y \sim Z$.*

Proof. Consider injective G-wobbles $\alpha: Y \to Z$ and $\beta: Z \to Y$. By Lemma 11.5.6, there are finite sets $S, T \Subset G$ such that $\alpha(y) \in yS$ and $\beta(z) \in zT$ for all $y \in Y, z \in Z$. Let $\gamma: Y \to Z$ be the bijection given by Theorem 11.5.9, with $\gamma(y) \in y(S \cup T^{-1})$. Then γ is a bijective G-wobble, again by Lemma 11.5.6.

We also need a little more terminology, coming from graph theory and following Definition 11.3.2:

Definition 11.5.11. A digraph (V, E) is *bipartite* if there is a decomposition $V = V^+ \sqcup V^-$ such that $e^+ \in V^+$ and $e^- \in V^-$ for every edge.

If V^+ and V^- are G-sets and are identified, the graph (V, E) is *bounded* if there exists a finite subset $S \Subset G$ with $e^+ \in e^-S$ for all $e \in E$.

An *m:n matching* in (V, E) is a subgraph $(V, \mathscr{M})$ with $\mathscr{M} \subset E$, such that for each $v \in V^+$ there are precisely n edges $e \in \mathscr{M}$ with $e^+ = v$, and for each $v \in V^-$ there are precisely m edges $e \in \mathscr{M}$ with $e^- = v$. We define similarly $m : (\le n)$ and $m : (\ge n)$ matchings.

If X is a G-set, a *bounded matching on X* is a matching in a bounded graph with vertex set $X \sqcup X$.

In particular, a $1:1$ matching is nothing but a bijection $V^- \to V^+$; and a bounded $1:1$ matching is a bijective G-wobble. A $1:(\le 1)$ matching is an injective map, and a $1:(\ge 0)$ matching is just a map.

Theorem 11.5.12 (Hall [281]-Hall-Rado [490]). *Let V, W be sets, and for each $v \in V$, let $E_v \subset W$ be a finite set. Assume that, for every finite subset $F \Subset V$,*

$$\text{the set } E_F := \bigcup_{v \in F} E_v \text{ contains at least } \#F \text{ elements.} \tag{11.11}$$

Then there exists an injection $e: V \to W$ with $e(v) \in E_v$ for all $v \in V$.

Proof. Assume first that (E_v) satisfies (11.11) and that $\#E_v \ge 2$ for some $v \in V$. We show that we may replace E_v by $E_v \setminus \{w\}$ for some $w \in E_v$ and still satisfy (11.11).

Indeed, consider $w_0 \ne w_1 \in E_v$, and assume that neither w_0 nor w_1 may be removed from E_v. Then there are $F_0, F_1 \Subset V$ and $N_i = E_{F_i} \cup (E_v \setminus \{w_i\}) \subseteq W$ for $i = 0, 1$ such that $\#N_i < \#(F_i \cup \{v\})$; *i.e.*, $\#N_i \le \#F_i$. Then

$$\begin{aligned}
\#F_0 + \#F_1 &\ge \#N_0 + \#N_1 = \#(N_0 \cup N_1) + \#(N_0 \cap N_1) \\
&\ge \#(E_{F_0 \cup F_1} \cup E_v) + \#(E_{F_0 \cap F_1}) \\
&\ge \#(F_0 \cup F_1) + 1 + \#(F_0 \cap F_1) = \#F_0 + \#F_1 + 1,
\end{aligned}$$

a contradiction. Then, inductively, we may suppose $\#E_v = 1$ for any given $v \in V$.

If V is finite, we are done by repeatedly replacing each E_v by a singleton; the injection is $v \mapsto w$ for the unique $w \in E_v$.

If V is countable, we may write $V = \{v_1, v_2, \dots\}$ and define recursively $E_v^0 = E_v$ for all $v \in V$, and, for $i, j > 0$,

$$E_{v_i}^j = \begin{cases} E_{v_i}^{j-1} & \text{if } j \neq i, \\ \text{the singleton coming from the above operation} & \text{if } j = i; \end{cases}$$

then the required injection is $v_i \mapsto w$ for the unique $w \in E_{v_i}^i$.

For general V, we need the help of an axiom. Order all systems (E_v') satisfying (11.11) by $(E_v') \leq (E_v'')$ if $E_v' \subseteq E_v''$ for all $v \in V$. By Zorn's lemma, $\{(E_v') \leq (E_v)\}$ admits a minimal element (E_v'). If $\#E_v' \geq 2$ for some $v \in V$, then by the above it could be made strictly smaller; therefore $\#E_v' = 1$ for all $v \in V$, and we again have an injection $V \to W$.

Note that if one drops the assumption that E_v is finite for all v, then there are counterexamples to the theorem, *e.g.*, $V = W = \mathbb{N}$, $E_0 = \mathbb{N}$ and $E_{n+1} = \{n\}$ for all $n \in \mathbb{N}$. For more details see [420].

Corollary 11.5.13. *Let (V, E) be a bipartite graph, and assume that for all $\varepsilon \in \{\pm 1\}$ and all finite subsets $F \subset V^\varepsilon$, the set*

$$\{v \in V^{-\varepsilon} \mid e^{-\varepsilon} = v, e^\varepsilon \in F \text{ for some } e \in E\}$$

is finite and contains at least $\#F$ elements. Then there exists a $1:1$ matching in (V, E).

Proof. By Theorem 11.5.12, there exists a subgraph of (V, E) defining an injection $V^- \to V^+$; and symmetrically there exists a subgraph of (V, E) defining an injection $V^+ \to V^-$. Applying Theorem 11.5.9, there exists a subgraph of (V, E) defining a bijection $V^- \to V^+$.

We are ready to prove the equivalence of our new notions:

Theorem 11.5.14. *Let X be a G-set. The following are equivalent:*

1. *X is paradoxical;*
2. *X is not amenable;*
3. *For any $m > n > 0$, there exists a bounded $m : n$ matching on X;*
4. *There exists a G-wobble $\phi: X \to X$ with $\#\phi^{-1}\{x\} = 2$ for all $x \in X$;*
5. *There exists a G-wobble $\phi: X \to X$ with $\#\phi^{-1}\{x\} \geq 2$ for all $x \in X$.*

Proof. (1) $\Rightarrow$ (2) Assume that there exists a G-invariant mean $\mu \in \mathscr{M}(X)$. Then

$$1 = \mu(X) = \sum_{i=1}^m \mu(Y_i g_i) + \sum_{j=1}^n \mu(Z_j h_j) = \sum_{i=1}^m \mu(Y_i) + \sum_{j=1}^n \mu(Z_j) = \mu(X) + \mu(X) = 2,$$

a contradiction.

(2) ⇒ (3) Assume that X does not satisfy Følner's condition, so there are $S \Subset G$ and $\epsilon > 0$ with $\#(FS) \geq (1+\epsilon)\#F$. Given $m > n > 0$, let $k \in \mathbb{N}$ be such that $(1+\epsilon)^k \geq m/n$.

Construct now the following bipartite graph: its vertex set is $V = X \times \{1,\dots,m\} \sqcup X \times \{\overline{1},\dots,\overline{n}\}$. There is an edge from (x,i) to $(xg,\overline{j})$ for all $g \in S^k$ and all $i \in \{1,\dots,m\}, j \in \{1,\dots,n\}$. Consider first a finite subset $F \Subset V^-$, and project it to $F' \subseteq X$. Then

$$\#(F'S^k \times \{\overline{1},\dots,\overline{n}\}) = n\#(F'S^k) \geq m\#F' \geq \#F,$$

and all these vertices are reached from F by edges in (V,E). Conversely, fix $g \in S^k$, and consider a finite subset $F \Subset V^+$. Because $m > n$, every $(x,\overline{i}) \in F$ is connected by an edge to $(xg^{-1},i) \in V^-$. Therefore, every finite $F \subset V^{\pm}$ has at least $\#F$ neighbors in $V^{\mp}$.

We now invoke the Hall-Rado theorem 11.5.12 to obtain a $1:1$ matching $(V,\mathscr{M})$, which we project to a bounded $m:n$ matching $(X \sqcup X, \mathscr{M})$ by setting $e^{\pm} = x$ whenever we had $e^{\pm} = (x,*)$ in $(V,\mathscr{M})$.

(3) ⇒ (4) Let $\mathscr{M}$ be a bounded $2:1$ matching on X. Given $x \in X$, there is a unique $e \in \mathscr{M}$ with $e^- = x$; set $\phi(x) = e^+$. This defines a G-wobble $\phi: X \to X$ with $\#\phi^{-1}(y) = 2$.

(4) ⇒ (5) is obvious.

(4) ⇒ (1) For each $x \in X$ choose $y_x \in X$ with $\phi(y_x) = y$; this is possible using the axiom of choice. Set $Y = \{y_x \mid x \in X\}$, and $Z = X \setminus Y$. We have $X = Y \sqcup Z$, and ϕ restricts to bijective G-wobbles $Y \to X$ and $Z \to X$, so $Y \sim X \sim Z$.

(5) ⇒ (2) Let $S \Subset G$ satisfy $\phi(x)S \ni x$ for all $x \in X$. Then, for any finite $F \Subset X$, we have $\phi^{-1}(F) \subseteq FS$ so $\#(FS) \geq 2\#F$.

If a group G contains a non-abelian free subgroup, then G is not amenable. The converse is not true, as we shall see in Section 11.7.3. However, the following weaker form of the converse holds:

Theorem 11.5.15 (see [584]). *Let X be a G-set. The following are equivalent:*

1. *X is not amenable;*
2. *There is a free action of the free group F_2 on X by bijective G-wobbles;*
3. *There is a free action of a non-amenable group on X by bijective G-wobbles.*

Proof. (1) ⇒ (2) Assume that X is non-amenable, so by Theorem 11.5.14 there exists a G-wobble $\phi: X \to X$ with $\#\phi^{-1}\{x\} = 2$ for all $x \in X$. Let $S \Subset G$ satisfy $\phi(x) \in xS$ for all $x \in X$.

View X as a directed graph T, with an edge from x to $\phi(x)$ for all x; and let U be the corresponding undirected graph. These graphs are 3-regular: in T every vertex has one outgoing and two incoming edges. Assume that there is a cycle in U. This cycle is necessarily oriented, for otherwise there would be two outgoing edges at a vertex. Furthermore, there cannot be two cycles in the same connected component of U: if there were two such cycles, consider a minimal path p joining them. At least one of p's extremities would be oriented away from its end, and again there would be two outgoing edges at a vertex.

It follows that all connected components of U are either 3-regular trees or cycles with 3-regular trees attached to them. Remove an edge from each cycle, creating in this manner either two vertices of degree 2 or one of degree 1. In all cases, at each vertex v of degree < 3, choose a ray ρ_v going to infinity consisting entirely of degree-3 vertices and shift the edges attached to ρ_v toward v along ρ_v so as to increase the degree of v. In this manner, we obtain a 3-regular forest U with vertex set X, with the following property: there exists a finite subset $S' \Subset G$ such that every edge of U, joining say x to y, satisfies $y \in xS'$. In fact, $S' = S \cup S^{-1}S^2$ will do.

Now label all edges of U with $\{a, b, c\}$ in such a manner that at every vertex all three colors appear exactly once on the incident edges. This is easy to do: on each connected component, label arbitrarily an edge; then at each extremity, label the two other incident edges by the two remaining symbols, and continue.

In this manner, every connected component of U becomes the Cayley graph of $H := \langle a, b, c \mid a^2, b^2, c^2\rangle$. In effect, we have defined an action of H on X by G-wobbles: the image of x under a, b, c, respectively, is the other extremity of the edge starting at x and labeled a, b, c, respectively. The group H contains a free subgroup of rank 2, namely, $\langle ab, bc\rangle$.

(2) $\Rightarrow$ (3) is obvious.

(3) $\Rightarrow$ (1) Assume that X admits a free action of a non-amenable group H by bijective G-wobbles; without loss of generality, H is finitely generated, say by a set T. Since H is not amenable and acts freely, it does not satisfy Følner's condition by Proposition 11.2.12, so there exists $\delta > 0$ such that $\#(FT) \geq \delta\#F$ for all $F \Subset X$.

Let $S \Subset G$ satisfy $xT \subset xS$ for all $x \in X$. In particular, $\#(FS) \geq \delta\#F$, so X does not satisfy Følner's condition.

Note that the proof becomes trivial in case $X = G \curvearrowleft G$, and G contains a non-abelian free subgroup; indeed the action of G itself is by G-wobbles.

It is possible to modify slightly this construction to make F_2 act transitively by G-wobbles, see [538].

11.6 Convex Sets and Fixed Points

We consider an abstract version of convex sets, introduced by Marshall Stone in [552] as sets with barycentric coordinates:

Definition 11.6.1. A *convex space* is a set K with an operation $[0, 1] \times K \times K \to K$ of taking convex combinations, written $(t, x, y) \mapsto t(x, y)$, satisfying the axioms

$$\begin{aligned} 0(x, y) &= x = t(x, x), \\ t(x, y) &= (1-t)(y, x), \qquad \text{for all } x, y, z \in K \text{ and } 0 \leq u \leq t \leq 1. \\ t(x, \tfrac{u}{t}(y, z)) &= \frac{t-u}{1-u}(x, u(y, z)), \end{aligned}$$

It is called *cancellative* if it furthermore satisfies the axiom

$$t(x, y) = t(x, z), t > 0 \Rightarrow y = z.$$

An *affine map* is a map $f: K \to L$ between convex spaces satisfying $t(f(x), f(y)) = t(x, y)$ for all $t \in [0, 1]$ and all $x, y \in K$.

Usual convex subsets of vector spaces are typical examples; if $K \subseteq V$ is convex, then $t(x, y) := (1 - t)x + ty$ gives K the structure of a convex space. There are other examples: for any set X, one may take $K = \mathfrak{P}(X)$ with $t(x, y) = x \cup y$ whenever $t \in (0, 1)$.

As another example, trees (and more generally $\mathbb{R}$-trees: geodesic metric spaces in which every triangle is isometric to a tripod) are convex spaces: for x, y in a tree, there is a unique geodesic from x to y, and $t(x, y)$ is defined as the point at distance $t\,d(x, y)$ from x along this geodesic. Unless the tree is a line segment, this convex space is not cancellative.

The set of closed balls in an ultrametric space,[13] with Hausdorff distance, is also an example of a convex space; it is actually isomorphic to the convex space associated with an $\mathbb{R}$-tree, see [302].

It turns out [552, Theorem 2] that those convex spaces that are embeddable in real vector spaces as convex subsets are precisely the cancellative ones.

A *topological convex space* is a convex space K with the structure of a topological space, such that the structure map $[0, 1] \times K \times K \to K$ is continuous. A *convex G-space* is a convex space on which a group G acts by affine maps. The *convex hull* of a subset $X \subseteq K$ of a convex space is the intersection $\widehat{X}$ of all convex subspaces of K containing X.

Exercise 11.6.2 (*). Convex spaces form a variety. Prove that the free convex space on $n + 1$ generators is isomorphic to the standard n-simplex $\{(x_0, \ldots, x_n) \in \mathbb{R}^{n+1} \mid x_i \geq 0, \sum x_i = 1\}$ and also to the convex hull of the basis vectors in $\mathbb{R}^{n+1}$. In particular, it is cancellative.

Definition 11.6.3. Let X, Y be G-sets. We say that Y is *X-markable* if there exists an equivariant G-map $X \to Y$.

Theorem 11.6.4. *Let X be a G-set. The following are equivalent:*

1. *X is amenable;*
2. *Every compact X-markable convex space admits a fixed point;*
3. *Every compact X-markable convex subset of a locally compact topological vector space admits a fixed point.*

Proof. (1) $\Rightarrow$ (2) By Lemma 11.3.6, there exists a net $(F_n)_{n \in \mathcal{N}}$ of Følner sets in X. Let K be a compact X-markable convex space, and let $\pi: X \to K$ be a G-equivariant map. For each $n \in \mathcal{N}$, set

[13] Namely, a metric space in which the ultratriangle inequality $d(x, z) \leq \max\{d(x, y), d(y, z)\}$ holds.

$$k_n := \sum_{x \in F_n} \frac{1}{\# F_n} \pi(x) \in K.$$

Then (k_n) is a net in K, so by compactness admits a cluster point, say k. The $k_n g$ have the same limit, so k is a fixed point.

(2) $\Rightarrow$ (3) is obvious.

(3) $\Rightarrow$ (1) Take $K = \mathscr{M}(X)$; it is compact by Lemma 11.2.20, X-marked by δ, convex by Lemma 11.2.23, and contained in the topological vector space $\ell^1(X)^*$ which is locally compact by the Banach-Alaoglu theorem [515, Theorem 3.15]. A fixed point is an invariant mean on X.

In particular, a group G is amenable if and only if every compact nonempty convex G-space admits a fixed point. We may thus show that amenability of G-sets is stable under amenable extensions:

Proposition 11.6.5. *Let X be a G-set, and let $N \triangleleft G$ be a normal subgroup with G/N amenable. Then $X \looparrowleft G$ is amenable if and only if $X \looparrowleft N$ is amenable.*

Proof. Let K be an X-markable convex compact space. The "if" direction is obvious, since every G-fixed point in K is N-fixed. Conversely, if $K^N \neq \emptyset$, then K^N is a nonempty convex compact space on which G/N acts and has a fixed point because G/N is amenable. Clearly $(K^N)^{G/N} = K^G$, so $X \looparrowleft G$ is amenable.

11.6.1 Measures

Consider a topological space X. We recall that $\mathscr{C}(X)$ denotes the space of continuous functions $X \to \mathbb{R}$ and that probability measures on X are identified with functionals $\lambda \in \mathscr{C}(X)^*$ such that $\lambda(\mathbb{1}) = 1$ and $\lambda(\phi) \geq 0$ if $\phi \geq 0$. One sometimes writes $\lambda(\phi) = \int \phi d\lambda$.

An important property of measures on subsets of vector spaces is that they have *barycentres*:

Lemma 11.6.6. *Let K be a nonempty convex compact subset of a locally compact topological vector space, and let $\mu \in \mathscr{C}(K)^*$ be a probability measure. Then there exists a unique $b \in K$ such that $\mu(\phi) = \phi(b)$ for all affine maps $\phi \in \mathscr{C}(K)$. We write $b = \int t d\mu(t)$ and call it the* barycentre *of μ.*

Proof. For any affine function $\phi: K \to \mathbb{R}$, set

$$K_\phi := \{x \in K \mid \mu(\phi) = \phi(x)\}.$$

It is clear that K_ϕ is convex and compact. Furthermore, it is nonempty; more generally, we will show that $K_{\phi_1} \cap \cdots \cap K_{\phi_n} \neq \emptyset$ for all affine $\phi_1, \ldots, \phi_n: K \to \mathbb{R}$.

Write $\phi = (\phi_1, \dots, \phi_n): K \to \mathbb{R}^n$. Define $L = \{\phi(x): x \in K\}$; this is a convex compact in $\mathbb{R}^n$. Define $p \in \mathbb{R}^n$ by $p_i = \mu(phi_i) = \int_K \phi_i d\mu$. We claim that p belongs to L; once this is shown, every $x \in K$ with $\phi(x) = p$ belongs to $K_{\phi_1} \cap \dots \cap K_{\phi_n}$, so the intersection is not empty.

We now show that, for any $q \notin L$, we have $p \neq q$. There exists then a hyperplane that separates q from L, namely, the nullspace of any affine map $\tau: \mathbb{R}^n \to \mathbb{R}$ with $\tau(q) < 0$ and $\tau(L) > 0$. In particular $\tau(\phi(x)) > 0$ for all $x \in K$, so by integrating $\tau(p) > 0$, and therefore $p \neq q$.

Set now $B = \bigcap_{\phi \text{ affine}} K_\phi$. It is nonempty by compactness of K, because any finite sub-intersection is nonempty.

Affine functions separate points[14] in K, so B contains a single point b.

Theorem 11.6.7. *Let X be a G-set. The following are equivalent:*

1. *X is amenable;*
2. *Every compact X-markable set admits an invariant probability measure.*

Proof. (1) $\Rightarrow$ (2) Let K be a compact G-set and let $\pi: X \to K$ be a G-equivariant map. Let $m \in \ell^\infty(X)^*$ be a G-invariant positive functional; then $m \circ \pi^*: \ell^\infty(K) \to \ell^\infty(X) \to \mathbb{R}$ is a G-invariant, positive functional on K, and its restriction to $\mathscr{C}(K)$ is an invariant probability measure on K.

(2) $\Rightarrow$ (1) Let K be a compact X-markable convex subset of a locally compact topological vector space, and let λ be an invariant probability measure on K. Then λ's barycentre, which exists by Lemma 11.6.6, is a fixed point in K, so X is amenable by Theorem 11.6.4(3) $\Rightarrow$ (1).

Exercise 11.6.8 (*). Reprove that the free group F_2 is not amenable as follows: write $F_2 = \langle a, b \mid \rangle$, and make it act on the circle $X = [0,1]/(0 \sim 1)$ by $xa = x^2$ and $xb = (x + 1/2) \mod 1$ for all $x \in [0,1]$. Show that the only a-invariant measure on X is δ_0 and that it is not b-invariant.

We proved in Corollary 11.4.2 that abelian groups are amenable. We may reprove it as follows:

Proposition 11.6.9 (Kakutani [326]-Markov [412]). *Let G be an abelian group. Then G is amenable.*

Proof. Let G act affinely on a convex compact K. For every $g \in G$ and every $n \geq 1$, define a continuous transformation $A_{n,g}: K \to K$ by

$$A_{n,g}(x) = \frac{1}{n} \sum_{i=0}^{n-1} xg^i.$$

Let $\mathscr{S}$ denote the monoid generated by $\{A_{n,g} \mid g \in G, n \geq 1\}$. We show that $\bigcap_{s \in \mathscr{S}} s(K)$ is not empty. Since K is compact, it suffices to show that every finite intersection $s_1(K) \cap \dots \cap s_k(K)$ is nonempty. To that end, set $t = s_1 \dots s_k$. We have

[14] Note that we use here the Hahn-Banach theorem, which requires certain logical axioms.

$$s_i(K) \subseteq s_i s_1 \dots \widehat{s_i} \dots s_k(K) = t(K),$$

because $\mathscr{S}$ is commutative. Therefore $s_1(K) \cap \dots s_k(K)$ contains $t(K)$, so it is not empty.

Pick now $x \in \bigcap_{s \in \mathscr{S}} s(K)$. To show that x is G-fixed, choose any affine function $\phi: K \to \mathbb{R}$, and any $g \in G$. For all n, write $x = A_{n,g}(y)$, and compute

$$\phi(x) - \phi(xg) = \frac{1}{n}(\phi(y) - \phi(yg^n)) \leq \frac{2}{n}\|\phi\|_\infty;$$

Since ϕ, g are fixed and n is arbitrary, we have $\phi(x) = \phi(xg)$ for all affine $\phi: K \to \mathbb{R}$, from which $x = xg$.

Furstenberg studied in [235] a condition at the exact opposite of amenability: a *boundary* for a group G is a compact G-space K which is minimal and such that every probability measure on K admits point measures in the closure of its G-orbit. By Theorem 11.6.7, if G is amenable then its only boundary is the point. See Section 11.11.1 for more details.

11.6.2 Amenability of Equivalence Relations

In the previous section, we gave conditions on a compact G-set to admit an invariant measure. Here, we assume that we are given a measure space on which a group acts by measure-class-preserving transformations.

In the abstract setting, we are given a set X, a σ-algebra $\mathfrak{M}$ of subsets of X, and a map $\lambda: \mathfrak{M} \to \mathbb{R}$.

To simplify the presentation, and focus on the interesting cases, we assume that (X, λ) is *σ-finite*, namely, X is the countable union of subsets of finite measure. In this case, it costs nothing to assume that λ is a *probability* measure, namely, $\lambda(X) = 1$. (Indeed, if $X = \bigsqcup_{n \in \mathbb{N}} X_n$ with $\lambda(X_n) < \infty$, define a new measure $\lambda'(A) = \sum_{n \in \mathbb{N}} 2^{-n}\lambda(A \cap X_n)/\lambda(X_n)$.) We will even assume that (X, λ) is a *standard probability space* [577], such as $([0, 1], \text{Lebesgue})$ or $(\{0, 1\}^{\mathbb{N}}, \text{Bernoulli})$; these spaces are isomorphic as measure spaces.

Let G be a group, and assume that G acts by measure-class-preserving transformations on (X, λ). Recall that this means that G acts on λ-null sets: if $A \subset X$ satisfies $\lambda(A) = 0$, then $(\lambda g)(A) = \lambda(Ag^{-1}) = 0$ for all $g \in G$. In other words, the measures λ and λg are absolutely continuous with respect to each other, and the Radon-Nikodym theorem [446] implies that there is an essentially unique measurable function $\partial(\lambda g)/\partial\lambda: X \to \mathbb{R}$ satisfying

$$\int_X f(xg)d\lambda(x) = \int_X f(x)\frac{\partial(\lambda g)}{\partial\lambda}d\lambda(x) \text{ for all } f \in L^1(X, \lambda).$$

If $(X, \lambda) = ([0,1], \text{Lebesgue})$ and $g: X \to X$ is differentiable, then $\partial(\lambda g)/\partial\lambda = dg/dx$, the usual derivative. The chain rule gives a "cocycle" identity

$$\frac{\partial(\lambda gh)}{\partial\lambda} = \frac{\partial(\lambda g)}{\partial\lambda} \cdot \left(\frac{\partial(\lambda h)}{\partial\lambda}^{g}\right).$$

In the extreme case (which is not the typical case we are interested in), the measure λ might be G-invariant: $\lambda(A) = \lambda(Ag)$ for all $A \subseteq X, g \in G$, and then the Radon-Nikodym derivative is constant $\equiv 1$.

To simplify the presentation and concentrate on the useful cases, we also restrict ourselves to a countable group G. Recall that an action is *essentially free* if λ-almost every point has a trivial stabilizer, namely, $\lambda(\{x \in X \mid G_x \neq 1\}) = 0$. More generally, everything is considered "up to measure 0": a group action, isomorphisms between actions, etc., only need to be defined on sets of full measure.

It will be useful to forget much about the group action and only remember its orbits. This is captured in the following definitions:

Definition 11.6.10. A *countable (respectively finite) measurable equivalence relation* on (X, λ) is a Borel equivalence relation $R \subseteq X \times X$ such that for every $x \in X$ the equivalence class $xR := \{y \in X \mid (x, y) \in R\}$ is countable (respectively finite) and such that for every measurable $A \subseteq X$ with $\lambda(A) = 0$ one has $\lambda(AR) = 0$.

The set R itself is treated as a measure space, with the counting measure on each equivalence class: for $E \subseteq R$, one sets $\mu(E) = \int_X \#\{y \in X \mid (x, y) \in E\} d\lambda(x)$.

A fundamental example is given by a measure-class-preserving action of a countable group G, as above: one sets $R_G = \{(x, y) \in X^2 \mid \exists g \in G \text{ with } xg = y\}$.

Definition 11.6.11. A countable measurable equivalence relation R on (X, λ) is *amenable* if there is a measurable invariant mean $m: X \to \mathscr{M}(R)$, written $x \mapsto m_x$, with $m_x \in \mathscr{M}(xR)$ for all $x \in X$. Here "measurable" means that for every $F \in L^\infty(X, \lambda)$, the map $x \mapsto m_x(F)$ is measurable, and "invariant" means that $m_x = m_y$ almost whenever $(x, y) \in R$.

By [161], a countable measurable equivalence R relation is amenable if and only if it is *hyperfinite*: R is the increasing union of countably many finite measurable equivalence relations; by [207], this in turn is equivalent to R being given by an action of $\mathbb{Z}$.

The following lemma rephrases amenability of equivalence relations as an analogue of the Day-Reiter criterion; we omit the proof which essentially follows that of Theorem 11.3.23; see [323]:

Lemma 11.6.12. *The equivalence relation R on (X, λ) is amenable if and only if there exists a system $(\phi_{x,n})_{x \in X, n \in \mathbb{N}}$ of measures, with $\phi_{x,n} \in \ell^1(xR)$, which is*

- measurable*: for all $n \in \mathbb{N}$ the function $(x, y) \mapsto \phi_{x,n}(y)$ is measurable on R,*
- asymptotically invariant*: $\|\phi_{x,n} - \phi_{y,n}\| \to 0$ for almost all $(x, y) \in R$.* □

Proposition 11.6.13. *If G is amenable and acts on (X, λ) by measure-class-preserving transformations, then G generates an amenable equivalence relation.*

Proof. Since G is amenable, there exists a sequence of almost invariant measures $\phi_n \in \ell^1(G)$, in the sense that $\|\phi_n - \phi_n g\| \to 0$ for all $g \in G$. Let R_G be the equivalence relation generated by G on X. For $x \in X$, set $\phi_{x,n} := x \cdot \phi_n$, the push-forward of ϕ_n along the orbit of x. Clearly $(\phi_{x,n})_{x \in X, n \in \mathbb{N}}$ is an asymptotically invariant system, and it is measurable since for all $n \in \mathbb{N}$ the level sets $\{(x,y) \in R \mid \phi_{x,n}(y) > a\}$ are the unions of the graphs of finitely many elements of G.

Note that the proposition does not admit a converse: for instance, if G is a discrete subgroup of a Lie group L and $P \leq L$ is soluble, then the action of G on $P\backslash L$ is amenable. Indeed the action of G on L is amenable: letting T be a measurable transversal of G in L, choose arbitrarily a measurable assignment $m\colon T \to \mathscr{M}(R_G)$ on the transversal, and extend it to L by translation. The map m may easily be required to be P-invariant, so it passes to the quotient $P\backslash L$.

Proposition 11.6.14. *If G acts essentially freely by measure-preserving transformations on the probability space (X, λ), and the generated equivalence relation R is amenable, then G is amenable.*

Proof. Given $f \in \ell^\infty(G)$, set

$$m(f) = \int_X m_x(xg \mapsto f(x))d\lambda(x). \square$$

It is possible for a non-amenable group to act essentially freely on a probability space:

Example 11.6.15. Let $F_k = \langle x_1, \dots, x_k \mid \rangle$ be a free group of rank k, and consider its *boundary* ∂F_k: it is the space of infinite reduced words over the generators of F_k,

$$\partial F_k = \{a_0 a_1 \cdots \in \{x_1^\pm, \dots, x_k^\pm\}^{\mathbb{N}} \mid a_i a_{i+1} \neq 1 \text{ for all } i \in \mathbb{N}\}.$$

The measure is equidistributed on cylinders: $\lambda(a_0 a_1 \dots a_n \{x_1^\pm, \dots, x_k^\pm\}^{\mathbb{N}}) = (2k)^{-1}(2k-1)^{1-n}$. The action of F_k on ∂F_k is by pre-catenation:

$$(a_0 a_1 \dots) \cdot x_i = \begin{cases} x_i a_0 a_1 \dots & \text{if } x_i a_0 \neq 1, \\ a_1 \dots & \text{if } x_i a_0 = 1. \end{cases}$$

Then the action of F_k on ∂F_k is essentially free and amenable, although F_k is not amenable.

Proof. For $1 \neq g = a_1 \dots a_n \in F_k$, its only fixed points in ∂F_k are g^∞ and $g^{-\infty}$; since F_k is countable and ∂F_k has the cardinality of the continuum, the action of F_k is free almost everywhere in ∂F_k.

For all $x = a_0 a_1 \cdots \in \partial F_k$, define probability measures $\mu_{x,n}$ on the orbit of x by

$$\mu_{x,n} = \frac{1}{n}\big(\delta_x + \delta_{xa_0} + \cdots + \delta_{xa_0 \cdots a_{n-1}}\big).$$

These measures converge weakly to a mean m_x on the orbit of x, and clearly m_x and m_{xg} have the same limit, since the sums defining $\mu_{x,n}$ and $\mu_{xg,n}$ agree on all but at most $|g|$ terms. Therefore, $m: X \to R_{F_k}$ is invariant, so R_{F_k} is amenable.

Consider a non-amenable group acting on (X, λ). So as to guarantee that the equivalence relation R_G be non-amenable, we may relax somewhat the condition that G preserve λ. We also assume that X is a compact topological space on which G acts by homeomorphisms. In fact, this is not a strong restriction: given an action of G on (X, λ) by measure-class-preserving transformations, we may always construct a compact topological G-space Y, with a measure μ on its Borel subsets, such that (X, λ) and (Y, μ) are isomorphic as G-measure spaces; see [59, Theorem 5.2.1].

We will call the action of G *indiscrete* if for every $\epsilon > 0$ and every neighborhood $\mathscr{U}$ of the diagonal in $X \times X$ there exists $g \neq 1 \in G$ with $\{(x, xg) \mid x \in X\} \subseteq \mathscr{U}$ and $\partial(\lambda g)/\partial\lambda \in (1-\epsilon, 1+\epsilon)$ almost everywhere.

The measurable-class-preserving action of G on X induces an action of G by isometries on the Banach space $L^1(X, \lambda)$ of integrable functions on X, by

$$(fg)(x) = \Big(\frac{\partial(\lambda g)}{\partial\lambda} f\Big)(xg^{-1}) \text{ for } f \in L^1(X, \lambda).$$

Lemma 11.6.16. *If we give G the topology of uniform convergence in its action on X, then the action of G on $L^1(X, \lambda)$ is continuous.*

Proof. Consider $f \in L^1(X, \lambda)$; we wish to show $fg \to f$ whenever $g \to 1$.

The closure of G in the homeomorphism group of X is second-countable locally compact; it therefore admits a Haar measure η. Let $K \subseteq \overline{G}$ be a compact with $\eta(K) = 1$, and let V be a compact neighborhood of 1 in $\overline{G}$. Since the Haar measure is invariant, we have

$$\|fg - f\| = \int_K \|fgh - fh\| d\eta.$$

Since f is measurable, there is for all $\epsilon > 0$ a continuous function $f': VK \to \mathbb{R}$ with $\int_{VK} \|fh - f'h\| d\eta < \epsilon$, and there is also a neighborhood W of 1 in V such that $\|f'gh - f'h\| < \epsilon$ for all $h \in K, g \in W$. Then $\|fg - f\| < 3\epsilon$ as soon as $g \in W$ by a standard "3δ" argument.

Proposition 11.6.17 (Monod; see [240, Théorème 2]). *Let G contain an indiscrete non-abelian free group acting essentially freely on a measure space (X, λ). Then G generates a non-amenable equivalence relation.*

Proof. It suffices to prove the claim with $G = \langle a, b \mid \rangle$ itself free. Let $A \subset G$ denote those elements whose reduced form starts with a nontrivial power of a, and define similarly B using b; so $G = A \sqcup B \sqcup \{1\}$.

Assume for contradiction that R_G is amenable, and let $m: X \to \mathscr{M}(R_G)$ be an invariant mean. Define measurable maps $u, v: X \to [0, 1]$ by

$$u(x) = m_x(xA), \qquad v(x) = m_x(xB).$$

Then $u + v = 1$ almost everywhere, and $0 \leq \sum_{n\in\mathbb{Z}} u(xb^n) \leq 1$ and $0 \leq \sum_{n\in\mathbb{Z}} v(xa^n) \leq 1$ almost everywhere, because the sets b^nA are all disjoint. In particular, if $v(x) > \frac{1}{2}$ then $v(xa^n) < \frac{1}{2}$ for all $n \neq 0$, so if $u(x) < \frac{1}{2}$ then $u(xa^n) > \frac{1}{2}$ for all $n \neq 0$. Define

$$P = \{x \in X \mid u(x) < \tfrac{1}{2}\}, \qquad Q = \{x \in X \mid u(x) > \tfrac{1}{2}\}.$$

Denote furthermore by $A' \subset A$ those elements of G that start and end with a nontrivial power of a, and by $B' \subset B$ those elements of G that start and end with a nontrivial power of b. Then $PA' \subseteq Q$, and $QB' \subseteq P$.

Since G is indiscrete, there exist $g_n \in G \setminus \{1\}$ with $g_n \to 1$ and $\partial(\lambda g_n)/\partial\lambda \to 1$ uniformly. Up to taking a subsequence, we may assume all g_n have the same first letter and the same last letter and have increasing lengths. Up to switching the roles of a and b, we may assume they all start with $a^{\pm 1}$. Up to replacing g_n by $g_n g_{n-1} g_n^{-1}$, we may assume they all belong to A'.

Since P is measurable, its characteristic function $\mathbb{1}_P$ is measurable and $\lambda(P) = \int_X \mathbb{1}_P d\lambda$. Then $\lambda(P\Delta Pg_n) = \int_X |\mathbb{1}_P - \mathbb{1}_{Pg_n}| d\lambda$; now $\mathbb{1}_{Pg_n} = \partial(\lambda g_n)/\partial\lambda \mathbb{1}_{P} g_n$ with $\partial(\lambda g_n)/\partial\lambda \to 1$, and by Lemma 11.6.16 $\mathbb{1}_P g_n \to \mathbb{1}_P$, so $\lambda(P\Delta Pg_n) \to 0$ as $n \to \infty$.

However, $Pg_n \subseteq Q \subseteq X\setminus P$ so $\lambda(P\Delta Pg_n) = 2\lambda(P)$; so $\lambda(P) = 0$. Next $\lambda(QB') \leq \lambda(P) = 0$ so $\lambda(Q) = 0$. It follows that $u = \frac{1}{2}$ almost everywhere, but this contradicts $0 \leq \sum_{n\in\mathbb{Z}} u(xb^n) \leq 1$.

Example 11.6.18. Let G be a countable indiscrete, non-soluble subgroup of $\mathrm{PSL}_2(\mathbb{R})$. Then G contains a non-discrete free group acting essentially freely on $X = \mathbb{P}^1(\mathbb{R})$. It follows that G generates a non-amenable equivalence relation on X.

Indeed, G contains an elliptic element of infinite order, namely, an element with $|\operatorname{trace}(g)| \in [-2, 2] \setminus 2\cos(\pi\mathbb{Q})$, see [314]. The group generated by some power of g and of a hyperbolic element not fixing g's fixed points is a non-discrete Schottky group.

Note that groups and equivalence relations are two special cases of *groupoids*, see Definition 11.9.17. There is a well-developed theory of amenability for groupoids with a measure on their space of units, see [323], and [22] for a full treatise.

11.7 Elementary Operations

We turn to a more systematic study of the class AG of amenable groups. John von Neumann already noted in [576] that AG is closed under the following operations:

Proposition 11.7.1. *Let G be a group.*

1. *Let $N \triangleleft G$ be a normal subgroup. If G is amenable, then G/N is amenable.*
2. *Let $H < G$ be a subgroup. If G is amenable, then H is amenable.*
3. *Let $N \triangleleft G$ be a normal subgroup. If N and G/N are amenable, then G is amenable.*

4. *Let $(G_n)_{n\in\mathcal{N}}$ be* directed *family of groups: $\mathcal{N}$ is a directed set, and for all $m<n$ there is a homomorphism $f_{mn}\colon G_m\to G_n$, with $f_{mn}f_{np}=f_{mp}$ whenever $m<n<p$. If G_n is amenable for all n, then $\varinjlim G_n$ is amenable.*

   In particular, if the G_n form a nested sequence of amenable groups, i.e., *$G_m\le G_n$ for $m<n$, then $\bigcup_{n\in\mathcal{N}} G_n$ is amenable.*

It is an amusing exercise to prove the proposition using a specific definition of amenability. Below we prove it using the fixed-point property of convex compact G-sets and give references to previous statements where other proofs were given.

Proof. 1. Proposition 11.2.10.

   For another proof, let G/N act on a nonempty convex compact K. Then in particular G acts on K, and since G is amenable, we have $K^G\neq\emptyset$ by Theorem 11.6.4. Then $K^{G/N}\neq\emptyset$ so G/N is amenable.
2. Proposition 11.2.12.

   For another proof, let H act on a nonempty convex compact K, and define

$$K^{G/H}=\{f\colon G\to K \mid f(xh)=f(x)h \text{ for all } x\in G, h\in H\}.$$

   Then $K^{G/H}$ is a convex compact G-set under the action $(f\cdot g)(x)=f(gx)$, so it admits a fixed point. This fixed point is a constant function, whose value is an H-fixed point in K.
3. Proposition 11.2.26.

   For another proof, let G act on a nonempty convex compact K. Since N is amenable, $K^N\neq\emptyset$. Since N is normal, G/N acts on K^N, and since G/N is amenable, $(K^N)^{G/N}\neq\emptyset$. But this last set is nothing but K^G.
4. Proposition 11.3.8.

   For another proof, write $G=\varinjlim G_n$, with natural homomorphisms $f_n\colon G_n\to G$ such that $f_m=f_{mn}f_n$ for all $m<n$. Let G act on a nonempty convex compact K. Then each G_n acts on K via f_n, and K^{G_n} is nonempty because G_n is amenable. Furthermore the K^{G_n} form a directed sequence of closed subsets of K: given $I\Subset\mathcal{N}$ finite, there is $n\in\mathcal{N}$ greater than I, so $\bigcap_{i\in I}K^{G_i}\supseteq K^{G_n}$ is not empty. By compactness, $\bigcap_{n\in\mathcal{N}}K^{G_n}=K^G\neq\emptyset$. □

We deduce immediately

Corollary 11.7.2. *A group G is amenable if and only if all its finitely generated subgroups are amenable.*

Indeed one direction follows from (2), the other from (4) with $\mathcal{N}$ the family of finite subsets of G, ordered by inclusion, and $G_n=\langle n\rangle$.

11.7.1 Elementary Amenable Groups

Finite groups are amenable; we saw in Corollary 11.4.2 and Proposition 11.6.9 that abelian groups are amenable and saw in Proposition 11.7.1 that the class of amenable groups is closed under extensions and colimits. Following Mahlon

Day [181], let us define the class of *elementary amenable groups*, *EG*. This is the smallest class of groups that contains finite and abelian groups and is closed under the four operations of Proposition 11.7.1: quotients, subgroups, extensions, and directed unions.

Example 11.7.3. Virtually soluble groups are in *EG*.

Indeed, they are obtained by a finite number of extensions using finite and abelian groups.

Example 11.7.4. For a set X, the group $\mathrm{Sym}(X)$ of finitely supported permutations is in *EG*.

Indeed, X is the union of its finite subsets, so $\mathrm{Sym}(X)$ is the directed limit of finite symmetric groups.

Example 11.7.5. Consider

$$G = \langle \dots, x_{-1}, x_0, x_1, \dots \mid \langle x_i, \dots, x_{i+k} \rangle^{(k)} \text{ for all } i \in \mathbb{Z}, k \in \mathbb{N} \rangle,$$

where $F^{(k)}$ denotes the kth term of the derived series of F. Then G is in *EG*.

Obviously the map $x_i \mapsto x_{i+1}$ extends to an automorphism of G; let $\widehat{G}$ denote the extension $G \rtimes \mathbb{Z}$ using this automorphism. Then $\widehat{G}$ also is in *EG*.

Indeed, $G = \bigcup_{k \in \mathbb{N}} \langle x_{-k}, \dots, x_k \rangle$, where each term is soluble. However, G itself is not soluble.

Example 11.7.6. This example is similar to 11.7.5 but more concrete. Consider formal symbols e_{mn} for all $m < n \in \mathbb{Z}$. The group M is the set of formal expressions $1 + \sum_{m<n} \alpha_{mn} e_{mn}$, with $\alpha_{mn} \in \mathbb{Z}$ and almost all 0; multiplication is defined by the formulas $e_{mn} e_{np} = e_{mp}$, all other products being 0. Then M is locally nilpotent, so it is in *EG*.

Extend then M by the automorphism $\sigma\colon e_{mn} \mapsto e_{m+1,n+1}$; the resulting group $\widehat{M} = M \rtimes \mathbb{Z}$ is again in *EG* and is finitely generated, by $1 + e_{12}$ and σ.

The class *EG* may be refined using transfinite induction. Let EG_0 denote the class of finite or abelian groups. For an ordinal α, let $EG_{\alpha+1}$ denote the class of extensions or directed unions of groups in EG_α; and for a limit ordinal α, set $EG_\alpha = \bigcup_{\beta<\alpha} EG_\beta$.

Lemma 11.7.7. *A group is elementary amenable if and only if it belongs to EG_α for some ordinal α.*

Proof. It suffices to see that the classes EG_α are closed under subgroups and quotients. This is clear for EG_0. If α is a successor, consider a subgroup $H \le G \in EG_\alpha$. Either $G = N.Q$ is an extension of groups in $EG_{\alpha-1}$; and then $H = (N \cap H).(H/N \cap H)$ with $H/N \cap H \le Q$; or $G = \bigcup G_i$, in which case $H = \bigcup (H \cap G_i)$; in both cases, $H \in EG_\alpha$ by induction. Consider next a quotient $\pi : G \twoheadrightarrow H$. Either $G = N.Q$, and $H = \pi(N).(H/\pi(N))$ with $Q \twoheadrightarrow H/\pi(N)$, or $G = \bigcup G_i$, in which case $H = \bigcup \pi(G_i)$; in both cases, $H \in EG_\alpha$ by induction.

If α is a limit ordinal, then each $G \in EG_\alpha$ actually belongs to EG_β for some $\beta < \alpha$, and there is nothing to do.

Example 11.7.8. Continuing Example 11.7.4, consider $H = \mathrm{Sym}(\mathbb{Z}) \rtimes \mathbb{Z}$, with $\mathbb{Z}$ acting on functions in $\mathrm{Sym}(\mathbb{Z})$ by shifting: $(n \cdot p)(x) = p(x - n)$. Then H is 2-generated, for example, by $(1, 2) \in \mathrm{Sym}(\mathbb{Z})$ and a generator of $\mathbb{Z}$.

Since $\mathrm{Sym}(\mathbb{Z})$ is a union of finite groups but is neither finite nor abelian, $\mathrm{Sym}(\mathbb{Z}) \in EG_1 \setminus EG_0$. Likewise, $H \in EG_2 \setminus EG_1$.

Example 11.7.5 is a bit more complicated. $F_k/F_k^{(k)}$ is soluble of class precisely k; so it belongs to $EG_{k-1} \setminus EG_{k-2}$. Therefore, $G \in EG_\omega$, but $G \notin EG_n$ for finite n. Similarly, $\widehat{G} \in EG_{\omega+1}$. The same holds for M and $\widehat{M}$ from Example 11.7.6.

Note also in Example 11.7.4 that the group of *all* permutations of $\mathbb{Z}$ is not amenable. Indeed it contains every countable group (seen as acting on itself); so if it were amenable, then by Proposition 11.7.1 every countable group would be amenable.

Recall that AG denotes the class of amenable groups. In [181], Mahlon Day asks whether the inclusion $EG \subseteq AG$ is strict; in other words, is there an amenable group that may not be obtained by repeated application of Proposition 11.7.1 starting with finite or abelian groups?

Theorem 11.7.9 (Chou [151, Theorems 2.3 and 3.2]). *Finitely generated torsion groups in EG are finite.*

No finitely generated group in EG has intermediate word growth.

The inequality $EG \neq AG$ follows, since there exist finitely generated infinite torsion groups (see [254] or Exercise 11.4.16) and groups of intermediate word growth, see Theorem 11.4.12 and [269].

Proof. The two statements are proven in the same manner, by transfinite induction. We only prove the second and leave the (easier) first one as an exercise. Let us show that if $G \in EG$ has subexponential word growth, then G is virtually nilpotent. Groups in EG_0 have polynomial growth and are therefore virtually nilpotent by Theorem 11.4.1. Consider next α a limit ordinal, and $G \in EG_\alpha$ a finitely generated group. We may assume that α is minimal, so in particular α is not a limit ordinal. Since G is finitely generated, we have $G = N.Q$ for $N, Q \in EG_{\alpha-1}$. By induction Q is virtually nilpotent, so in particular it is virtually polycyclic. By Corollary 11.4.4 the subgroup N is finitely generated, so is virtually nilpotent by induction. By Lemma 11.4.6, the group G is virtually soluble, and by Corollary 11.4.11 it has either polynomial or exponential growth.

11.7.2 Subexponentially Amenable Groups

In [137, Section 14], Tullio Ceccherini-Silberstein, Pierre de la Harpe and Slava Grigorchuk consider the class SG of *subexponentially amenable groups* as the smallest class containing groups of subexponential growth and closed under taking subgroups, quotients, extensions, and direct limits. We then have $EG \subsetneq SG \subseteq AG$, and we shall see promptly that the last inclusion is also strict.

We introduce a general construction of groups: let H be a permutation group on a set $\mathscr{A}$. We assume that the action is transitive and choose a point $0 \in \mathscr{A}$. Let us construct a self-similar group $\mathscr{M}(H)$ acting on the rooted tree $\mathscr{A}^*$, see Definition 11.2.15. The group $\mathscr{M}(H)$ is generated by two subgroups, written H and K and isomorphic, respectively, to H and to $H \wr H_0 = H^{(\mathscr{A}\setminus\{0\})} \rtimes H_0$. We first define the actions of H and K on the boundary $\mathscr{A}^{\mathbb{N}}$ of the tree. The action of $h \in H$ is on the first letter:

$$(a_0 a_1 \dots)h = (a_0 h)a_1 \dots .$$

The action of $(f, h) \in K$, with $f\colon \mathscr{A} \setminus \{a\} \to H$ finitely supported, fixes $a^{\mathbb{N}}$ and is as follows on its complement:

$$(a_0 a_1 \dots)(f, h) = 0 \dots 0(a_n h)(a_{n+1} f(a_n))a_{n+2} \dots \text{ with } n \text{ minimal such that } a_n \neq 0.$$

The self-similarity of $\mathscr{M}(H)$ is encoded by an injective homomorphism $\Phi\colon \mathscr{M}(H) \to \mathscr{M}(H) \wr_{\mathscr{A}} H = \mathscr{M}(H)^{\mathscr{A}} \rtimes H$, written $g \mapsto \langle\!\langle g_a \mid a \in \mathscr{A} \rangle\!\rangle \pi$ and defined as follows. Given $g \in \mathscr{M}(H)$, its image π in H is the natural action of g on $\{a\mathscr{A}^{\mathbb{N}} \mid a \in \mathscr{A}\} \cong \mathscr{A}$. The permutation g_a of $\mathscr{A}^{\mathbb{N}}$ is the composition $\mathscr{A}^{\mathbb{N}} \to a\mathscr{A}^{\mathbb{N}} \to (a\pi)\mathscr{A}^{\mathbb{N}} \to \mathscr{A}^{\mathbb{N}}$ of the maps $(w \mapsto aw)$, g and $((a\pi)w \mapsto w)$, respectively. On the generators of $\mathscr{M}(H)$, we have

$$\Phi(h) = \langle\!\langle 1 \mid a \in \mathscr{A} \rangle\!\rangle h, \qquad \Phi((f, h)) = \langle\!\langle f(a) \mid a \in \mathscr{A} \rangle\!\rangle h.$$

Proposition 11.7.10. *If H is perfect and 2-transitive, then Φ is an isomorphism.*

Proof. First, if H is 2-transitive, then $\mathscr{M}(H)$ is generated by three subgroups $H, H_0, \overline{H}$. Fix a letter $1 \in \mathscr{A}$; then H_0 consists of those $(1, h) \in K$, and $\overline{H}$ consists of those $(f, 1)$ where $f(a) = 1$ for all $a \neq 1$. To avoid confusions between these subgroups, we write $h, h_0, \overline{h}$ for respective elements of $H, H_0, \overline{H}$.

To prove that Φ is an isomorphism, it suffices to prove that $\langle\!\langle h, 1, \dots, 1 \rangle\!\rangle$, $\langle\!\langle h_0, 1, \dots, 1 \rangle\!\rangle$, and $\langle\!\langle \overline{h}, 1, \dots, 1 \rangle\!\rangle$ belong to $\Phi(\mathscr{M}(H))$ for all $h \in H, h_0 \in H_0, \overline{h} \in \overline{H}$.

First, choose $k \in H_0$ with $1k \neq 1$. For all $\overline{h}, \overline{h}' \in \overline{H}$, we have $\Phi([\overline{h}, (\overline{h}')^k]) = \langle\!\langle [\overline{h}, \overline{h}'], 1, \dots, 1 \rangle\!\rangle$; and since $\overline{H} \cong H$ is perfect, we get that the image of Φ contains $\overline{H} \times 1 \dots \times 1$. Consider next $h_0 \in H_0$; then $\Phi(h_0 h^{-1}) = \langle\!\langle h_0, 1, \dots, 1 \rangle\!\rangle$. Finally, $\Phi(\overline{h}) = \langle\!\langle \overline{h}, h, 1, \dots, 1 \rangle\!\rangle$ and $\langle\!\langle \overline{h}, 1, \dots, 1 \rangle\!\rangle$ belongs to the image of Φ, so $\langle\!\langle 1, h, 1, \dots, 1 \rangle\!\rangle$ also belongs to its image. Conjugating by an appropriate element of H, we see that $\langle\!\langle h, 1, \dots, 1 \rangle\!\rangle$ belongs to the image of Φ.

Theorem 11.7.11 ([43]; see [106] for the proof). *If H is finite, then the group $\mathscr{M}(H)$ is amenable.*

Proof. If $H \leq \widehat{H}$ as permutation groups then $\mathscr{M}(H) \leq \mathscr{M}(\widehat{H})$. It therefore does not reduce generality, in proving that $\mathscr{M}(H)$ is amenable, to consider H perfect and 2-transitive.

We consider $S = H \cup K$ as generating set for $\mathscr{M}(H)$. Let us define finite subsets $I_k \subseteq L_k$ of $\mathscr{M}(H)$ inductively as follows:

$$I_0 = K, \qquad L_0 = I_0 H,$$
$$I_k = H \cdot \Phi^{-1}(I_{k-1} \times L_{k-1}^{\mathscr{A}\setminus\{0\}}),$$
$$L_k = H \cdot \Phi^{-1}(L_{k-1}^{\mathscr{A}} \setminus (L_{k-1} \setminus I_{k-1})^{\mathscr{A}}).$$

Lemma 11.7.12. *For all $k \in \mathbb{N}$, we have $I_k K = I_k$ and $I_k H = L_k H = L_k$; therefore, $I_k S = L_k$.*

Proof. The claims are clear for $k = 0$. Also, $L_k H = L_k$ for all k. Consider $g \in I_k$ and $f \in K$, and write them $g = h\langle\!\langle g_a \mid a \in \mathscr{A} \rangle\!\rangle$ and $f = \langle\!\langle f_a \mid a \in \mathscr{A} \rangle\!\rangle h'$. Note $gf = a\langle\!\langle g_a f_a \mid a \in \mathscr{A} \rangle\!\rangle h'$. We have $f_a \in H$ for all $a \neq 0$, so $g_a f_a \in L_{k-1}$ for all $a \neq 0$; and $f_0 \in K$ so $g_0 f_0 \in I_{k-1}$.

Lemma 11.7.13. *Setting $\rho_k = \#I_k/\#L_k$, we have*

$$\rho_k = \frac{\rho_{k-1}}{1 - (1 - \rho_{k-1})^{\#\mathscr{A}}}.$$

Proof. Set $d = \#\mathscr{A}$. From the definition, we get $\#L_k = \#L_{k-1}^d \#H(1 - (1 - \rho_{k-1})^d)$ and $\#I_k = \#I_{k-1}\#L_{k-1}^{d-1}\#H$, so

$$\rho_k = \frac{\#I_k}{\#L_k} = \frac{\#I_{k-1}}{\#L_{k-1}(1 - (1 - \rho_{k-1})^d)}. \qquad \square$$

We are ready to prove that the sequence (I_k) is a Følner sequence. In view of Lemma 11.7.12, it suffices to prove $\rho_k \to 1$. Note $0 < \rho_{k-1} < \rho_k < 1$, so the sequence (ρ_k) has a limit, ρ. Then ρ satisfies $\rho = \rho/(1 - (1 - \rho)^d)$, so $\rho = 1$.

To prove that $\mathscr{M}(H)$ has exponential growth, we use a straightforward criterion:

Proposition 11.7.14. *Let a left-cancellative monoid $G = \langle S \rangle_+$ act on a set X; let there be a point $x \in X$ and disjoint subsets $Y_s \subseteq X \setminus \{x\}$ satisfying $xs \in Y_s$ and $Y_s S \subseteq Y_s$ for all $s \in S$. Then G is free on S, namely, $G \cong S^*$.*

Proof. Consider distinct words $u = u_1 \dots u_m, v = v_1 \dots v_n \in S^*$; we are to prove that they have distinct images in G. Since G is left-cancellative, we may assume either $m = 0$ or $u_1 \neq v_1$. In the first case, $xu = x \neq xv \in Y_{v_1}$, and in the second case, $Y_{u_1} \ni xu \neq xv \in Y_{v_1}$.

The proposition implies that $\mathscr{M}(H)$ has exponential growth for almost all H; it seems difficult to formulate a general result, so we content ourselves with an example:

Example 11.7.15. The group $\mathscr{M}(S_3)$ has exponential growth.

Proof. Write $\mathscr{A} = \{0, 1, 2\}$ and $S_3 = \langle(0, 1), (0, 2)\rangle$. In our notation, consider the elements $s = \overline{(0, 1)}(0, 1)$ and $t = \overline{(0, 2)}^{(1,2)_0}(0, 2)$. A quick calculation gives

$$\Phi(s) = \langle\!\langle s(0, 1), (0, 1), 1\rangle\!\rangle(0, 1), \qquad \Phi(t) = \langle\!\langle t(0, 2), 1, (0, 2)\rangle\!\rangle(0, 2),$$

Proposition 11.7.14 applies with $G = \langle s, t\rangle_+$ and $X = \mathscr{A}^{\mathbb{N}}$ and $x = 0^{\mathbb{N}}$ and $Y_s = \mathscr{A}^*10^{\mathbb{N}}$ and $Y_t = \mathscr{A}^*20^{\mathbb{N}}$.

The first construction of an amenable, not subexponentially amenable group appears in [45], with an explicit subgroup of (what was later defined to be) $\mathscr{M}(D_4)$.

Example 11.7.16. The group $\mathscr{M}(A_5)$ belongs to $AG \setminus SG$.

Proof. The group $G := \mathscr{M}(A_5)$ is amenable by Theorem 11.7.11. It contains $\mathscr{M}(S_3)$, *e.g.*, because the permutations $(0, 1)(3, 4)$ and $(0, 2)(3, 4)$ generate a copy of S_3 in A_5, so G has exponential growth by Example 11.7.15.

It remains to prove that G does not belong to SG, and we do this by transfinite induction, defining (just as we did for EG) the class SG_0 of groups of subexponential growth and for an ordinal α by letting SG_α denote those extensions and directed unions of groups in SG_β for $\beta < \alpha$.

By way of contradiction, let α be the minimal ordinal such that G belongs to SG_α. Since G is finitely generated, it is an extension of groups in SG_β for some $\beta < \alpha$. Now the only normal subgroups of G are 1 and the groups G_n in the series defined by $G_0 = G$ and $G_{n+1} = \Phi^{-1}(G_n^{\mathscr{A}} \times H)$; the argument is similar to that used to show that G is not in EG, see Exercise 11.2.17. In particular, every nontrivial normal subgroup of G maps onto G, so it cannot belong to SG_β for some $\beta < \alpha$.

11.7.3 Free Group Free Groups

For levity, in this section by "free group," we always mean "non-abelian free group." It follows from Proposition 11.7.1 that every group containing a free subgroup is itself not amenable; this covers surface groups, or more generally word-hyperbolic groups; free products of a group of size at least 2 with a group of size at least 3; and $SO_3(\mathbb{R})$; that last example is important in relation to the Banach-Tarski paradox (see Section 11.5.1).

Let us denote by NF the class of groups with no free subgroup. In [181], Mahlon Day asks whether the inclusion $AG \subseteq NF$ is an equality; in other words, does every non-amenable group contains a free subgroup?

This was made into a conjecture by Frederick Greenleaf [265, Page 9], attributed[15] to von Neumann. Ching Chou [151] proved $EG \neq NF$, while Alexander Ol'shanskiĭ [456] proved $AG \neq NF$, see also Sergei Adyan [5]. Indeed, they proved the much stronger result that the free Burnside groups

$$B(n,m) = \langle x_1, \dots, x_n \mid w^m \text{ for all words } w \text{ in } x_1^{\pm 1}, \dots, x_n^{\pm 1} \rangle \tag{11.12}$$

are non-amenable as soon as $n \geq 2$ and $m \geq 665$ are odd. These groups, of course, do not contain any non-*trivial* free subgroup.

The following examples of groups are called "Frankenstein groups," since (as their namesake) they have rather different properties than the groups they are built of:

Theorem 11.7.17 (Monod [421]). *Let* $\mathbb{A}$ *be a countable subring of* $\mathbb{R}$ *properly containing* $\mathbb{Z}$*; let* $P_{\mathbb{A}} \subseteq \mathbb{P}^1(\mathbb{R})$ *be the set of fixed points of hyperbolic elements in* $\mathrm{PSL}_2(\mathbb{A})$*, and let* $H(\mathbb{A})$ *be the group of self-homeomorphisms of* $\mathbb{P}^1(\mathbb{R})$ *that fix* ∞ *and are piecewise elements of* $\mathrm{PSL}_2(\mathbb{A})$ *with breakpoints in* $P_{\mathbb{A}}$*. Then* $H(\mathbb{A})$ *is a non-amenable free group free group.*

Proof. Since $\mathbb{A}$ properly contains $\mathbb{Z}$, it is dense in $\mathbb{R}$, so $\mathrm{PSL}_2(\mathbb{A})$ is a countable dense subgroup of $\mathrm{PSL}_2(\mathbb{R})$. It therefore generates a non-amenable equivalence relation on $\mathbb{P}^1(\mathbb{R})$, by Example 11.6.18.

Lemma 11.7.18 ([421, Proposition 9]). *For all* $p \in \mathbb{P}^1(\mathbb{R}) \setminus \{\infty\}$*, we have*

$$p \cdot \mathrm{PSL}_2(\mathbb{A}) \subseteq \{\infty\} \cup p \cdot H(\mathbb{A}).$$

Proof. Given $g \in \mathrm{PSL}_2(\mathbb{A})$ with $pg \neq \infty$, we seek $h \in H(\mathbb{A})$ with $ph = pg$. It will be made of two pieces, g near p and $z \mapsto z + r$ near ∞ for a suitable choice of $r \in \mathbb{A}$. Consider the quotient $q := g \cdot (z \mapsto z - r) \in \mathrm{PSL}_2(\mathbb{A})$; if q is hyperbolic, say with fixed points $\xi_\pm$, and $\{\xi_\pm\}$ separates p from ∞, then we may define h as g on the component of $\mathbb{P}^1(\mathbb{R}) \setminus \{\xi_\pm\}$ containing p and as $z \mapsto z + r$ on its complement. Now an easy calculation shows that q is hyperbolic for all $|r|$ large enough, and as $|r| \to \pm\infty$ one of the fixed points of q approaches ∞ and the other approaches ∞g, and as the sign of r changes the approach to ∞ is from opposite sides; so in all cases it is easy to find a suitable r.

Therefore, the equivalence relation generated by $H(\mathbb{A})$ is non-amenable, so $H(\mathbb{A})$ is itself non-amenable.

On the other hand, consider $f, g \in H(\mathbb{A})$. We claim that they do not generate a free group and more precisely that $\langle f, g \rangle$ either is metabelian or contains a subgroup isomorphic to $\mathbb{Z}^2$.

Let $1 \neq h \in \langle f, g \rangle''$ belong to the second derived subgroup, and intersect as few connected components of support(f) $\cup$ support(g) as possible — if no such h exists, we are already done. For every endpoint $p \in \partial(\text{support}(f) \cup \text{support}(g))$,

[15]Infelicitously!

the element h acts trivially in a neighborhood of p, because both f and g act as affine maps in a neighborhood of p; so the support of h is strictly contained in support(f) ∪ support(g). Since the dynamics of $\langle f, g\rangle$ has attracting elements in the neighborhood of p, there exists $k \in \langle f, g\rangle$ such that support(h) and support(h)k are disjoint; then $\langle h, h^k\rangle \cong \mathbb{Z}^2$.

Exercise 11.7.19 (***). Since $H(\mathbb{A})$ is not amenable, there is a free action of $PSL_2(\mathbb{A})$ on $\mathbb{R}$ by $H(\mathbb{A})$-wobbles. Construct explicitly such an action.

Hint: This is essentially what [382] does in computing the minimal number of pieces in a paradoxical decomposition of $H(\mathbb{A})$, but it's still highly non-explicit.

Thus we have $EG \subsetneqq SG \subsetneqq AG \subsetneqq NF$. The last inequality also holds for finitely generated groups — any finitely generated non-amenable subgroup of $H(\mathbb{A})$ will do. Lodha and Moore construct finitely *presented* examples in [383].

Problem 11.7.20. Is the group $H(\mathbb{Z})$ amenable?

The group $H(\mathbb{Z})$ is related to a famous group acting on the real line, consider Thompson's group F (see Problem 11.11.3), which we describe here.

Example 11.7.21. Let F be the group of self-homeomorphisms of $[0, 1]$ that are piecewise affine with slopes in $2^{\mathbb{Z}}$ and breakpoints in $\mathbb{Z}[\frac{1}{2}]$.

Conjugating F by Minkowski's "?" map, defined by $?(x) = \sum_{n\geq 0}(-1)^n 2^{-a_0-\cdots-a_n}$ if x's continued fraction expansion is $[a_0, a_1, \dots]$, one obtains a group of piecewise-$PSL_2(\mathbb{Z})$ homeomorphisms of the real line with rational breakpoints; it is easy to see that having rational breakpoints is equivalent to the maps being *diffeo*morphisms.

The same argument as that given in the proof of Theorem 11.7.17 shows that F is a free group free group.

The difference with $H(\mathbb{Z})$ is that breakpoints of maps in $H(\mathbb{Z})$ are in $\mathbb{P}_{\mathbb{Z}}$, which is disjoint from $\mathbb{Q}$. There are embeddings of F in $H(\mathbb{Z})$, so amenability of $H(\mathbb{Z})$ would imply that of F.

Yet another description of F is by an action on the Cantor set. For this, break the interval $[0, 1]$ open at every dyadic rational; one obtains in this manner a Cantor set, modeled on $\{0, 1\}^{\mathbb{N}}$ by the usual binary expansion of real numbers, except that one does not identify $a_1 \dots a_n 01^\infty$ with $a_1 \dots a_n 10^\infty$. The action of F is then by lexicographical order-preserving maps that are piecewise of the form $a_1 \dots a_n v \mapsto b_1 \dots b_k v$ for a collection of words $(a_1 \dots a_n, b_1 \dots b_k)$ and every $v \in \{0, 1\}^{\mathbb{N}}$. The group F is finitely generated, by the elements $x_0\colon 00v \mapsto 0v, 01v \mapsto 10v, 1v \mapsto 11v$ and $x_1\colon 0v \mapsto 0v, 1v \mapsto 1x_0(v)$, and is even finitely presented. See [125] for a detailed survey of F.

11.8 Random Walks

We now turn to other criteria for amenability, expressed in terms of random walks. For a thorough treatment of random walks, consult the book [588]; we content ourselves with the subset most relevant to amenability. One is given a space X, and a random walker W moving at random in X. There is thus a random process $W \in X \rightsquigarrow S(W) \in X$, describing a *single step* of the random walk. One asks for the distribution W_n of the random walker after a large number n of iterations of S.

More formally, we are given *one-step* transition probabilities $p_1(x, y) = \mathbb{P}(W_n = x | W_{n-1} = y)$ of moving to x for a particle lying at y; they satisfy $p_1(x, y) \geq 0$ and $\sum_{x\in X} p_1(x, y) = 1$ for all $y \in X$. We define iteratively $p_n(x, y) = \sum_{z\in X} p_{n-1}(x, z)p_1(z, y)$ and then ask for asymptotic properties of p_n.

Here are two fundamental examples. First, if X is a graph with finite degree, set $p_1(x, y) = 1/\deg(y)$ if x, y are neighbors, and $p_1(x, y) = 0$ otherwise. This is called the *simple random walk* (SRW) on the graph X.

Another fundamental example is given by a group G, a right G-set X, and a probability measure μ on G, namely, a map $\mu: G \to [0, 1]$ with $\sum_{g\in G} \mu(g) = 1$ as in (11.4). The random walk is then defined by

$$p_1(x, y) = \sum_{g\in G, x=yg} \mu(g). \tag{11.13}$$

It is called the *random walk driven by* μ. The measure μ is called *symmetric* if $\mu(g) = \mu(g^{-1})$ for all $g \in G$ and is called *adapted* if its support generates G qua semigroup.

These two examples coincide in case $G = \langle S\rangle$ is finitely generated and the driving measure μ is equidistributed on S; one considers then SRW on the Schreier graph of the action of G on X.

A random walk p on a set X is *reversible* if there exists a function $s: X \to (0, \infty)$ satisfying $s(x)p_1(x, y) = s(y)p_1(y, x)$ for all $x, y \in X$. SRW is reversible on undirected graphs, with $s(x) = \deg(x)$, and if μ is symmetric, then the random walk driven by μ is reversible with $s(x) \equiv 1$.

11.8.1 Spectral Radius

We shall prove a criterion, due to Harry Kesten in the case of groups and Dodziuk-Kendall and Gerl in the case of graphs, relating the spectral radius of the linear operator associated with p to amenability. It first appeared in [345]. Let p be a reversible random walk on a set X, for simplicity assumed symmetric throughout this subsection. Set $E = \{(x, y) \in X^2 \mid p_1(x, y) > 0\}$. We introduce two Hilbert spaces:

$$\ell_0^2 = \{f: X \to \mathbb{R} \mid \langle f, f\rangle < \infty\},$$

$$\ell_1^2 = \{g: E \to \mathbb{R} \mid g(x, y) = -g(y, x), \langle g, g\rangle < \infty\}$$

with scalar products

$$\langle f, f'\rangle = \sum_{x\in X} \overline{f(x)}f'(x) \text{ and } \langle g, g'\rangle = \frac{1}{2}\sum_{x,y\in X} p_1(x, y)\overline{g(x, y)}g'(x, y).$$

Elements of ℓ_1^2 are naturally extended to functions on X^2 which vanish on $X^2 \setminus E$.

One step of the random walk p induces a linear operator T on ℓ_0^2 given by

$$(Tf)(x) = \sum_{y \in X} p_1(x, y) f(y).$$

Writing δ_x for the function taking value 1 at $x \in X$ and 0 elsewhere, we then have $p_n(x, y) = (T^n \delta_y)(x)$. We also define operators d, d^* between ℓ_0^2 and ℓ_1^2 by

$$d\colon \ell_0^2 \to \ell_1^2, \qquad (df)(x, y) = f(x) - f(y),$$

$$d^*\colon \ell_1^2 \to \ell_0^2, \qquad (d^*g)(x) = \sum_{y \in X} p_1(x, y) g(x, y).$$

Lemma 11.8.1. *T is a self-adjoint operator on ℓ_0^2 of norm at most* 1. *The operator d^* is the adjoint of d, and $T = 1 - d^* d$.*

Proof. The first claim follows from the second. For $f \in \ell_0^2$ and $g \in \ell_1^2$, we compute

$$\begin{aligned}
\langle df, g \rangle &= \frac{1}{2} \sum_{(x,y) \in E} p_1(x, y) (\overline{f(x)} - \overline{f(y)}) g(x, y) \\
&= \frac{1}{2} \sum_{x \in X} \overline{f(x)} \sum_{y \in X} p_1(x, y)(g(x, y) - g(y, x)) \\
&= \sum_{x \in X} \overline{f(x)} (d^* g)(x) = \langle f, d^* g \rangle,
\end{aligned}$$

and

$$(1 - d^* d) f(x) = f(x) - \sum_{y \in X} p_1(x, y)(f(x) - f(y)) = \sum_{y \in X} p_1(x, y) f(y).$$

□

The following definitions are more commonly given in the context of graphs; our more general setting coincides with it if p is the simple random walk:

Definition 11.8.2. Let p be a random walk on a set X. The *isoperimetric constant* of p is

$$\iota(p) = \inf_{F \Subset X} \frac{p_1(F, X \setminus F)}{\#F} = \inf_{F \Subset X} \frac{\sum_{x \in F, y \in X \setminus F} p_1(x, y)}{\#F}.$$

The *spectral radius* of p is the spectral radius—or, equivalently, the norm—of the operator T.

The following inequalities relating spectral radius and isoperimetric constant appear, with different notation and normalization, in [84]:

Proposition 11.8.3. *Let p be a symmetric random walk on a set X. Then the isoperimetric constant ι and spectral radius ρ of p are related by*

$$\iota^2 + \rho^2 \le 1 \le \iota + \rho.$$

Proof. We begin by the second inequality. For $\epsilon > 0$, let $F \Subset X$ satisfy $p_1(F, X \setminus F)/\#F < \iota + \epsilon$. Let $\phi \in \ell_0^2$ denote the characteristic function of F. Then $\|\phi\|^2 = \#F$, and

$$\|d\phi\|^2 = \frac{1}{2} \sum_{(x,y)\in E} p_1(x,y)(\phi(x) - \phi(y))^2 = \sum_{x\in F, y\in X\setminus F} p_1(x,y) < (\iota + \epsilon)\|\phi\|^2;$$

then $(\rho + \iota + \epsilon)\|\phi\|^2 > \langle \phi, T\phi\rangle + \|d\phi\|^2 = \langle \phi, (1 - d^*d)\phi\rangle + \|d\phi\|^2 = \|\phi\|^2$. The conclusion $\rho + \iota \ge 1$ follows under $\epsilon \to 0$.

In the other direction, consider for finite $F \Subset X$ the projection $\pi_F \colon \ell_0^2 \circlearrowleft$ defined by $(\pi_F f)(x) = f(x)$ if $x \in F$ and 0 otherwise, and set $T_F := \pi_F T \pi_F$. The operator T_F is self-adjoint and converges strongly to T as F increases, so the spectral radius of T_F converges to ρ. For $\epsilon > 0$, let F be such that the spectral radius ρ_F of T_F is larger than $\rho - \epsilon$. Since T_F has nonnegative entries, its eigenvalue ρ_F is simple and has a nonnegative eigenvector ϕ, by the Perron-Frobenius theorem. We extend ϕ by 0 into an element of ℓ_0^2 and normalize it so that $\|\phi\| = 1$. Set then

$$A := \frac{1}{2} \sum_{(x,y)\in E} p_1(x,y)|\phi(x)^2 - \phi(y)^2|,$$

and compute

$$\begin{aligned} A^2 &= \Big(\frac{1}{2} \sum_{(x,y)\in E} p_1(x,y)|\phi(x) + \phi(y)| \cdot |\phi(x) - \phi(y)|\Big)^2 \\ &\le \frac{1}{2} \sum_{(x,y)\in E} p_1(x,y)(\phi(x) + \phi(y))^2 \cdot \frac{1}{2} \sum_{(x,y)\in E} p_1(x,y)(\phi(x) - \phi(y))^2 \\ &= (\|\phi\|^2 + \langle \phi, T_F\phi\rangle)(\|\phi\|^2 - \langle \phi, T_F\phi\rangle) = (1 + \rho_F)(1 - \rho_F), \end{aligned}$$

because $\sum_{(x,y)\in E} p_1(x,y)\phi(x)\phi(y) = \sum_{x\in F} \phi(x) \sum_{y\in X} p_1(x,y)\phi(y) = \langle \phi, T_F\phi\rangle$.

On the other hand, let $0 < s_1 < s_2 < \cdots < s_n$ denote the finitely many values that ϕ takes, and define, for $k = 1, \dots, n$,

$$F_k = \{x \in X \mid \phi(x) \ge s_k\},$$

with the additional conventions $s_0 = 0$ and $F_{k+1} = \emptyset$. Then

$$A = \frac{1}{2}\sum_{(x,y)\in E} p_1(x,y)|\phi(x)^2 - \phi(y)^2| = \sum_{k=1}^{n}\sum_{x\in F_k, y\notin F_k} p_1(x,y)(s_k^2 - s_{k-1}^2)$$

$$\geq \sum_{k=1}^{n} \iota\#F_k(s_k^2 - s_{k-1}^2) = \iota\sum_{k=1}^{n}(\#F_k - \#F_{k+1})s_k^2 = \iota\|\phi\| = \iota.$$

Combining, we get

$$(1-(\rho-\epsilon)^2) \geq 1 - \rho_F^2 \geq A^2 \geq \iota^2;$$

and the conclusion $\rho^2 + \iota^2 \leq 1$ follows under $\epsilon \to 0$.

This section's main result is the following characterization of amenable G-sets:

Theorem 11.8.4. *Let μ be a symmetric, adapted probability measure on a group G, let X be a G-set, and let p be the random walk on X driven by μ. Then the following are equivalent:*

1. *X is amenable;*
2. *$\iota(p) = 0$;*
3. *$\rho(p) = 1$.*

Proof. (1) $\Rightarrow$ (2) Assume first that X is amenable, and let $\epsilon > 0$ be given. Let $S \Subset G$ satisfy $\mu(S) > 1 - \epsilon/2$. Let $F \Subset X$ satisfy $\#(FS \setminus F) < \epsilon\#F/2$. Then

$$\sum_{x\in F, y\notin F} p_1(x,y) \leq \sum_{x\in F, y\in FS\setminus F} \mu(\{s\in S \mid y = xs\}) + \sum_{x\in F, g\in G\setminus S} \mu(g) \leq \epsilon\#F,$$

so $\iota(p) \leq \epsilon$ for all $\epsilon > 0$. (Note that we have not used the assumption that μ is adapted here.)

(2) $\Rightarrow$ (1) Let $\epsilon > 0$ and a finite subset S of G be given. By assumption, there exist $n \in \mathbb{N}$ and $\delta > 0$ such that $\mu^n(s) \geq \delta$ for all $s \in S$. Let F be a finite subset of X such that $\sum_{x\in F, y\notin F} p_n(x,y) < \delta\epsilon\#F$. Then $\#(FS \setminus F) < \epsilon\#F$, so X is amenable by Følner's criterion, Theorem 11.3.23(5) $\Rightarrow$ (1).

The equivalence (2) $\Leftrightarrow$ (3) is given by Proposition 11.8.3.

The spectral radius of the random walk has a direct interpretation in terms of probabilities of return of the random walk, at least when we restrict to *transitive* random walks: random walks with the property that for any two $x, y \in X$, there exists $n \in \mathbb{N}$ such that $p_n(x,y) > 0$ (not to be confused with random walks invariant under a transitive group action!). Let us make the following temporary

Definition 11.8.5. The *spectral radius* of the random walk p based at x is

$$\rho(p,x) := \limsup_{n\to\infty} \sqrt[n]{p_n(x,x)}.$$

Lemma 11.8.6 (Fekete). *Let $N \in \mathbb{N}$ be given, and let $\alpha\colon \{N, N+1, \dots\} \to \mathbb{R}$ be a subadditive function,* i.e., *a function satisfying $\alpha(m+n) \le \alpha(m) + \alpha(n)$. Then*

$$\lim_{n\to\infty} \frac{\alpha(n)}{n} = \inf_{n>0} \frac{\alpha(n)}{n};$$

in particular $\alpha(n)/n$ either converges or diverges to $-\infty$.

Proof. Consider any $a \ge N$, and write every $k \ge N$ as $k = qa + r$ with $q \in \mathbb{N}$ and $r \in \{N, N+1, \dots, N+a-1\}$. Then, for $k \ge N$,

$$\frac{\alpha(k)}{k} \le \frac{q\alpha(a) + \alpha(r)}{qa + r} \le \frac{\alpha(a)}{a} + \frac{\alpha(r)}{k};$$

letting $k \to \infty$, we get $\limsup_{n\to\infty} \alpha(k)/k \le \alpha(a)/a$ for every $a \ge N$; so $\limsup_{n\to\infty} \alpha(k)/k = \inf_{a\in\mathbb{N}} \alpha(a)/a$ converges or diverges to $-\infty$.

The "limsup" in the definition of the spectral radius is in fact a limit and is independent of the starting and endpoints:

Proposition 11.8.7. *Assume p is transitive. Then*

$$\rho(p,z) = \limsup_{n\to\infty} \sqrt[n]{p_n(x,y)} = \lim_{n\to\infty} \sqrt[2n]{p_{2n}(x,x)} \textit{ for all } x,y,z \in X.$$

Proof. For the first claim, consider more generally $w,x,y,z \in X$. There are $\ell \in \mathbb{N}$ such that $p_\ell(x,w) > 0$; and $m \in \mathbb{N}$ such that $p_m(z,y) > 0$. Since $p_{n+\ell+m}(x,y) \ge p_\ell(x,w)p_n(w,z)p_m(z,y)$ for all $n \in \mathbb{N}$, we have

$$\limsup_{n\to\infty} \sqrt[n]{p_n(x,y)} \ge \limsup_{n\to\infty} \sqrt[n-\ell-m]{p_\ell(x,w)p_m(z,y)} \sqrt[n-\ell-m]{p_n(w,z)} = \limsup_{n\to\infty} \sqrt[n]{p_n(w,z)}.$$

Applying it to $(w,x,y,z) = (z,x,y,z)$ and (x,z,z,y), respectively, gives the claim.

It is then clear that $\limsup_{n\to\infty} \sqrt[2n]{p_{2n}(x,x)} \le \rho(p,x)$; but conversely $p_{2n}(x,x) \ge p_n(x,x)^2$, so $\limsup_{n\to\infty} \sqrt[2n]{p_{2n}(x,x)} \ge \limsup_{n\to\infty} \sqrt[n]{p_n(x,x)} = \rho(p,x)$.

Now $p_{2r+2s}(x,x) \ge p_{2r}(x,x)p_{2s}(x,x)$ for all $r,s \in \mathbb{N}$. Setting $\alpha(r) = -\log p_{2r}(x,x)$, we get $\alpha(r+s) \le \alpha(r) + \alpha(s)$; furthermore, because p is transitive, $\alpha(r)$ is defined for all r large enough, and $\alpha(r) \ge 0$ because $p_{2r}(x,x) \le 1$. By Lemma 11.8.6, $\alpha(r)/r$ converges, whence $\sqrt[2n]{p_{2n}(x,x)}$ converges.

Proposition 11.8.8. *Let p be symmetric and transitive. Then the spectral radius of p is equal to the norm of T acting on $\ell^2(X)$.*

Proof. Let us write $\|T\|$ the operator norm of T on $\ell^2(X)$. First, by Proposition 11.8.7,

$$\rho(p,x) = \lim_{n\to\infty} \sqrt[2n]{p_{2n}(x,x)} = \lim_{n\to\infty} \sqrt[2n]{\langle T^{2n}\delta_x, \delta_x\rangle} \le \sqrt[2n]{\|T^{2n}\|} \le \|T\|.$$

Next, consider $f \in \mathbb{C}X$. By Cauchy-Schwartz's inequality, for all $m \in \mathbb{N}$ we have

$$\langle T^{m+1}f, T^{m+1}f\rangle = \langle T^m f, T^{m+2}f\rangle \le \|T^m f\| \cdot \|T^{m+2}f\|;$$

so $\|T^{m+1}f\|/\|T^m f\|$ is increasing, with limit $\lim_{n\to\infty} \sqrt[n]{\|T^n f\|}$. Now

$$\begin{aligned}\lim_{n\to\infty} \sqrt[n]{\|T^n f\|} &= \lim_{n\to\infty} \sqrt[2n]{\langle T^n f, T^n f\rangle} = \lim_{n\to\infty} \sqrt[2n]{\langle T^{2n} f, f\rangle} \\ &= \lim_{n\to\infty} \sqrt[2n]{\sum_{x,y\in\text{support}(f)} p_{2n}(x,y)f(x)f(y)} = \rho(p,x),\end{aligned}$$

because the sum is finite. Taking $m = 0$, we obtain $\|Tf\|/\|f\| \le \rho(p,x)$ for all $f \in \mathbb{C}X$; and since $\mathbb{C}X$ is dense in $\ell^2(X)$, we have $\|T\| \le \rho(p,x)$.

The probabilities of return, in the case of SRW, have a straightforward interpretation in terms of paths: say we consider a k-regular graph X with basepoint $*$. Then there are k^n paths of length n starting at $*$, and among these asymptotically $\rho(p)^n$ will end at $*$. Therefore, non-amenable graphs are characterized as those graphs in which exponentially few paths are closed.

Example 11.8.9. Let us look first at an amenable example: $X = \mathbb{Z}$ and $p_1(x, x \pm 1) = \frac{1}{2}$; this is SRW on the line. We write $p_n(x,y)$ for the probability that a particle starting at y reaches x at time n, that is, the probability that $T^n(x) = y$. The simple formula

$$p_n(x,y) = \begin{cases} \frac{1}{2^n}\binom{n}{\frac{n+x-y}{2}} & \text{if } n+x-y \equiv 0 \pmod 2, \\ 0 & \text{else} \end{cases}$$

is easily justified as follows: at each step, one chooses $+1$ or -1 with equal probabilities; at time n we then made 2^n choices. If $(n+x-y)/2$ of these are $+1$ and $(n-x+y)/2$ are -1, then we end up at $x + (n+x-y)/2 - (n-x+y)/2 = y$.

In particular, if n is even, we have $p_n(x,x) = 2^{-n}\binom{n}{n/2}$, so by Stirling's formula $n! \propto \sqrt{2\pi n}(n/e)^n$ we get

$$p_n(x,x) \propto \sqrt{\frac{2}{\pi n}}.$$

Example 11.8.10. Consider the free group F_d, whose Cayley graph is a $2d$-regular tree $\mathscr{T}$. Fix an edge of this tree, *e.g.*, between 1 and x_1, let A_n denote the number of closed paths in $\mathscr{T}$ based at 1 and by B_n the number of closed paths in $\mathscr{T}$, based at 1, that do not cross the fixed edge. Consider the generating series $A(z) = \sum A_n z^n$ and $B(z) = \sum B_n z^n$. Then $A(z) = 1/(1 - 2dz^2 B(z))$, because every closed path factors uniquely as a product of closed paths that reach 1 only at their endpoints; and $B(z) = 1/(1-(2d-1)z^2 B(z))$ for the same reason; so

$$A(z) = \frac{1-d+d\sqrt{1-4(2d-1)z^2}}{1-4d^2z^2}$$

and $A_n \propto (8d-4)^{n/2}$ and $p_n(1,1) \propto (8d-4)^{n/2}/(2d)^n$. Therefore, SRW on F_d has spectral radius $\rho(p) = \sqrt{2d-1}/d$.

The isoperimetric constant of SRW may also easily be computed. A connected, finite subset F of $\mathscr{T}$ has $\#F$ vertices and is connected to $2n\#F$ edges, of which $2(\#F-1)$ point back to F, so $\sum_{x\in F, y\notin F} p_1(x,y) = ((2n-2)\#F+2)/2n$. The isoperimetric constant is therefore $\iota(p) = 1-1/n$.

Exercise 11.8.11 (**). Compute the isoperimetric constant of SRW on the surface group $\Sigma_g = \langle a_1, b_1, \dots, a_g, b_g \mid [a_1,b_1]\cdots[a_g,b_g] = 1\rangle$.

Hint: its Cayley graph is a tiling of hyperbolic plane by $4g$-gons, meeting $4g$ per vertex. Use Euler characteristic.

Note that it is substantially harder to compute the spectral radius of SRW; only estimates are known, proportional to $\sqrt{g}$, see [261] for the best bounds.

It is sometimes easier to count *reduced* paths in graphs, rather than general paths. Formally, this may be expressed as follows: let $G = \langle S \cup S^{-1}\rangle$ be a finitely generated group, and write $S^{\pm} = S \cup S^{-1}$. There is a natural map $\pi\colon F_S \to G$ induced by the inclusion $S \hookrightarrow G$. The spectral radius of SRW on G is

$$\rho = \lim_{n\to\infty} \frac{\sqrt[n]{\#\{w \in (S^{\pm})^n \mid w =_G 1\}}}{\#S^{\pm}} \in [0,1].$$

The *cogrowth* of G is

$$\gamma = \lim_{n\to\infty} \sqrt[n]{\#w \in F_S \mid \pi(w) = 1\#(S\cup S^{-1})} \in [1, \#S^{\pm}-1].$$

Theorem 11.8.12 ([267]; see also [39, 158, 557, 587]). *The parameters γ, ρ are related by the equation*

$$\rho = \frac{\gamma + (\#S^{\pm}-1)/\gamma}{\#S^{\pm}} \text{ if } \gamma > 1.$$

In particular, G is amenable if and only if $\gamma = \#S^{\pm}-1$.

Proof. The most direct proof is combinatorial. Define formal matrices B, C indexed by G with power series coëfficients by

$$B(z)_{g,h} = \sum_{w\in F_S : g\pi(w)=h} z^{|w|}, \qquad C(z)_{g,h} = \sum_{w\in(S^{\pm})^* : gw =_G h} z^{|w|}.$$

Set for convenience $q := \#S^{\pm} - 1$. We shall prove the formal relationship

$$\frac{B(z)}{1-z^2} = \frac{C(z/(1+qz^2))}{1+qz^2} \tag{11.14}$$

from which the claim of the theorem follows. Define the adjacency matrix

$$A_{g,h} = \sum_{s \in S^{\pm}: gs=h} 1;$$

then $C(z) = 1/(1-zA)$. If for all $s \in S^{\pm}$ we define

$$B_s(z)_{g,h} = \sum_{w \in F_S \setminus \{1\}: w_1 = s, g\pi(w)=h} z^{|w|}$$

then

$$B(z) = 1 + \sum_{s \in S^{\pm}} B_s(z), \qquad B_s(z) = sz(B(z) - B_{s^{-1}}(z))$$

which solve to $B_s(z) = (1-z^2)^{-1}(sz - z^2)B(z)$ and therefore to

$$\frac{1+qz^2}{1-z^2}B(z) = 1 + \sum_{s \in S^{\pm}} \frac{z}{1-z^2} sB(z) = 1 + \frac{z}{1-z^2}AB(z);$$

so $(1+qz^2)/(1-z^2) \cdot B(z) = 1/(1 - z/(1+qz^2)A)$, which is equivalent to (11.14).

It is also known that $\rho \geq \sqrt{\#S^{\pm} - 1}$, with equality if and only if $G \cong F_S$, see [470].

11.8.2 Harmonic Functions

We shall obtain, in this subsection, yet another characterization of amenability in terms of bounded harmonic functions.

Definition 11.8.13. Let p be a random walk on a set X. A *harmonic function* is a function $f: X \to \mathbb{R}$ satisfying

$$f(x) = \sum_{y \in X} p_1(y,x) f(y).$$

In other words, f is a *martingale*: along a trajectory (W_n) of a random walk, the expectation of $f(W_n)$ given $W_0, \dots, W_{n-1}$ is $f(W_{n-1})$.

A random walk is called *Liouville* if the only bounded harmonic functions are the constants.

If X is a G-set and p is the random walk driven by a measure μ on the group G, we say that (X, μ) is Liouville when the corresponding random walk is Liouville.

Bounded harmonic functions are fundamental in understanding long-term behavior of random walks. The space of trajectories of a random walk on X started at $* \in X$ is the probability space $(X^{\mathbb{N}}, \nu)$, in which the trajectory $(W_0, W_1, \dots)$ has probability $\nu(W_0, W_1, \dots) = \prod_{n\geq 0} \mu(\{g \in G \mid W_n g = W_{n+1}\})$ when $W_0 = *$. An *asymptotic event* on $(X^{\mathbb{N}}, \nu)$ is a measurable subset of $X^{\mathbb{N}}$ that is invariant under the shift map of $X^{\mathbb{N}}$. Given a asymptotic event E, we define a bounded function f on X by $f(x) = \nu(E \cap \{W_n = x\})/\nu(\{W_n = x\})$, with n chosen large enough so that the denominator does not vanish; and check that it is harmonic by conditioning on the first step of the random walk; conversely, given a bounded harmonic function f the limit $f(W_n)$ almost surely exists along trajectories, by Doob's martingale convergence theorem, so $E_{[a,b]} = \{(W_0, W_1, \dots) \mid \lim f(W_n) \in [a, b]\}$ is an asymptotic event. In summary, a random walk is Liouville if and only if there are no nontrivial asymptotic events.

Let us continue with the example of SRW on $\mathbb{Z}$: a harmonic function satisfies $f(x-1) + f(x+1) = 2f(x)$, so f is affine. In particular, SRW on $\mathbb{Z}$ is Liouville.

Let us consider next the example of SRW, started at the identity, on the Cayley graph of $F_2 = \langle a, b \mid \rangle$, which is a tree. The random walk (W_n) escapes at speed $1/2$ toward the boundary of the tree, since at every position except the origin it has three ways of moving one step farther and one way of moving one step closer; so in particular almost surely $W_n \neq 1$ for all n large enough. Let $A \subset F_2$ denote those elements whose reduced form starts with a, and define

$$f(g) = \mathbb{P}(W_n \in g^{-1}A \text{ for all } n \text{ large enough}).$$

In words, $f(g)$ is the probability that a random walk started at g escapes to the boundary of the tree within A. It is clear that f is bounded, and it is seen to be harmonic by conditioning on the first step of the random walk. More succinctly, "the random walk eventually escapes in A" is a nontrivial asymptotic event. Therefore, SRW on a regular tree is not Liouville.

Exercise 11.8.14 (*). Let (X, μ) and (Y, ν) be Liouville random walks. Prove that $(X \times Y, \mu \times \nu)$ is Liouville.

Let us recal some properties of measures and random walks. The set $\ell^1(G)$ of summable functions on G is a Banach $*$-algebra, for the *convolution product*

$$(\mu\nu)(g) = \sum_{g=hk} \mu(h)\nu(k) \text{ for } \mu, \nu \in \ell^1(G). \tag{11.15}$$

We denote by $\check{\mu}$ the adjoint of μ, defined by $\check{\mu}(g) = \mu(g^{-1})$. If X is a G-set, then $\ell^p(X)$ is an $\ell^1(G)$-module for all $p \in [1, \infty]$, under

$$(f\mu)(x) = \sum_{x=yh} f(y)\mu(h) \text{ for } f \in \ell^p(X), \mu \in \ell^1(G).$$

If a random walk is driven by a measure μ, then from (11.13) we get $Tf = f\mu$. With our notation, a function $f \in \ell^\infty(X)$ is harmonic for the random walk driven by a measure μ if and only if $f\check{\mu} = f$.

The Liouville property is fundamentally associated with a measure or a random walk. It has a counterpart which solely depends on the space and is a variant of amenability with switched quantifiers (see Proposition 11.3.25):

Definition 11.8.15. A G-set X is called *laminable*[16] if for every $\epsilon > 0$ and every $f \in \varpi(\ell^1 X)$ there exists a positive function $g \in \ell^1(G)$ with $\|fg\| < \epsilon\|g\|$.

Proposition 11.8.16. *Let G be a group, viewed as a right G-set G_G. Then G_G is amenable if and only if G_G is laminable.*

Let G be an amenable group, and let $X \looparrowleft G$ be a G-set. Then X is laminable if and only if it is transitive or empty.

Proof. By Proposition 11.3.25, G_G is amenable if for every $\epsilon > 0$ and every $0 \neq g \in \varpi(\ell^1 G)$ there exists a positive function $f \in \ell^1(X)$ with $\|fg\| < \epsilon\|f\|\,\|g\|$; equivalently, $\|\check{g}\check{f}\| < \epsilon\|g\|\,\|f\|$, which is the definition of laminability of G_G.

For the second statement: if there is more than one G-orbit on X, choose x, y in different orbits; then $\|(\delta_x - \delta_y)g\| = 2\|g\|$ for all positive $g \in \ell^1(G)$. Conversely, given $\epsilon > 0$ and $f \in \varpi(\ell^1 X)$, choose $x \in X$ and $h \in \varpi(\ell^1 G)$ with $f = \delta_x h$. Since G is amenable, there is a positive function $g \in \ell^1(G)$ with $\|g\check{h}\| < \epsilon\|g\|$; so $\|f\check{g}\| = \|xh\check{g}\| = \|g\check{h}\| < \epsilon\|g\|$, and X is laminable.

The following easy proposition is an analogue to Proposition 11.2.10:

Proposition 11.8.17. *Let G, H be groups; let $X \looparrowleft G$ and $Y \looparrowleft H$ be, respectively, a G-set and an H-set; let $\phi: G \twoheadrightarrow H$ be a homomorphism; and let $f: X \to Y$ be a surjective equivariant map, namely, satisfying $f(xg) = f(x)\phi(g)$ for all $x \in X, g \in G$. If X is laminable, then Y is laminable.*

Proof. Given $\epsilon > 0$ and $e \in \varpi(\ell^1 Y)$, there is $e' \in \varpi(\ell^1 X)$ with $e' = e \circ f$, because f is surjective; then there is a positive function $g \in \ell^1(G)$ with $\|e'g\| < \epsilon\|g\|$, because X is laminable; then $\|e\phi(g)\| \leq \|e'g\| < \epsilon\|g\| = \epsilon\phi(g)$, so Y is laminable.

Corollary 11.8.18. *Let $X \looparrowleft G$ be a G-set and let $H \leq G$ be a subgroup. If H is amenable and transitive, then X is laminable.* □

Lemma 11.8.19. *If $X \looparrowleft G$ is laminable, then for every $x \in X$, every finite subset $S \Subset X$, and every $\epsilon > 0$, there exists a positive function $g \in \ell^1(G)$ with*

$$\|\delta_s g - \delta_x g\| < \epsilon\|g\| \text{ for all } s \in S.$$

Furthermore g may be supposed to be of finite support.

[16] This is a contraction of "Liouville" and "amenable."

Proof. Consider $f = \sum_{s\in S}\delta_s - \#S\delta_x$. Since X is laminable, there is for every $\epsilon > 0$ a positive function $g \in \ell^1(G)$ with $\|fg\| < \epsilon\|g\|/2$. Then

$$\epsilon\|g\| > 2\|fg\| \geq 2\Big\|\sum_{s\in S}\max(\delta_s g - \delta_x g, 0)\Big\| = \sum_{s\in S} 2\|\max(\delta_s g - \delta_x g, 0)\|$$
$$\geq \sum_{s\in S}\|\delta_s g - \delta_x g\|.$$

Using density of finitely supported functions in $\ell^1(G)$ gives the last claim.

The main result of this section is:

Theorem 11.8.20. *Let X be a G-set. The following are equivalent:*

1. *X is laminable;*
2. *There exists a symmetric measure μ with support equal to G such that (X,μ) is Liouville;*
3. *There exists a measure μ on G such that (X,μ) is Liouville.*

Corollary 11.8.21 (Kaimanovich-Vershik [325, Theorem 4.4]). *Let G be a countable group. Then G is amenable if and only if there exists a measure μ* (ad lib. *symmetric, with full support) such that (G,μ) is Liouville.* □

Note that there exist amenable non-laminable G-sets, such as Example 11.3.13, and non-amenable graphs for which SRW is Liouville, see [69] or [67, Chapter 13]. At the extreme, note that the empty set is laminable but not amenable, and the disjoint union of two points is amenable but not laminable. Here is a slightly less contrived example:

Example 11.8.22 (Kaimanovich). Consider the binary rooted tree with vertex set $\{0,1\}^*$ and an edge between $a_1\dots a_n$ and $a_1\dots a_{n+1}$ for all $a_i\in\{0,1\}$. Fix a function $f\colon\mathbb{N}\to\mathbb{N}$ satisfying $f(n) < n$ for all $n\in\mathbb{N}$, and put also an edge between $a_1\dots a_n$ and $a_1\dots\widehat{a_{f(n)}}\dots a_n$ for all $a_i\in\{0,1\}$. Finally add some loops at the root so as to make the graph 6-regular; we have constructed a graph $\mathcal{G}$, with a natural action of F_6 once the edges are appropriately labeled. We consider SRW on $\mathcal{G}$.

On the one hand, $\mathcal{G}$ is not amenable, for example, because SRW drifts away from the root at speed $(4-2)/6 = 1/3$ or because the isoperimetric inequality in $\mathcal{G}$ is at least as bad as in a binary tree.

On the other hand, if f grows slowly enough, then SRW on $\mathcal{G}$ is Liouville; indeed SRW converges to the boundary of the binary tree, represented by binary sequences $\{0,1\}^{\mathbb{N}}$, and it suffices to show that there are no asymptotic events on this boundary. If f is such that $f^{-1}(n)$ is infinite for all $n\in\mathbb{N}$, then each cöordinate in $\{0,1\}^{\mathbb{N}}$ is randomized infinitely often by the walk when it follows the f-edges, so there is no nonconstant measurable function on the space of trajectories.

For the remainder of the section, we assume the hypotheses of the theorem: a countable group G and a transitive G-set X are fixed. We also assume that all measures μ under consideration satisfy $\mu(1) > 0$ and call such μ *lazy*. This is harmless: a function f is harmonic for μ if and only if it is harmonic for $q\mu + (1-q)\delta_1$ whenever $q \in (0, 1]$.

Lemma 11.8.23. *Let μ be a lazy measure on G. Then there exists a sequence $(\epsilon_n) \to 0$, depending only on $\mu(1)$, such that, for all $f \in \ell^\infty(X)$,*

$$\|f\mu^n - f\mu^{n+1}\| \le \epsilon_n \|f\|.$$

Proof. Since $\|f\mu^n - f\mu^{n+1}\|_\infty \le \|f\|_\infty \cdot \|\mu^n - \mu^{n+1}\|_1$, it suffices to prove $\|\mu^n - \mu^{n+1}\|_1 \to 0$. Set $q = \mu(1)$; we assume $q \in (0, 1)$. Define a measure λ on $\mathbb{N}$ by $\lambda(0) = q, \lambda(1) = p = 1 - q$, and let ν be the probability measure on G such that $\mu = q\delta_1 + p\nu$. Then

$$\mu^n(g) = (q\delta_1 + p\nu)^n(g) = \sum_{i=0}^{n} \lambda^n(i)\nu^i(g),$$

$$\text{so } \|\mu^n - \mu^{n+1}\| = \left\| \sum_{i=0}^{n+1} (\lambda^n(i) - \lambda^{n+1}(i))\nu^i \right\|.$$

Since $\lambda^n(i) - \lambda^{n+1}(i) = \lambda^n(i) - q\lambda^n(i) - p\lambda^n(i-1) = p(\lambda^n(i) - \lambda^n(i-1))$, it suffices to prove $\sum_{i=0}^{n+1} |\lambda^n(i) - \lambda^n(i-1)| \to 0$. Remembering $\lambda^n(i) = \binom{n}{i} p^i q^{n-i}$, the argument of the absolute value is positive for $i < pn$ and negative for $i > pn$, so $\sum_{i=0}^{n+1} |\lambda^n(i) - \lambda^n(i-1)| \le \lambda^n(\lfloor pn \rfloor) + \lambda^n(\lceil pn \rceil) \to 0$.

Corollary 11.8.24. *For every bounded sequence of functions (F_n) in $\ell^\infty(X)$, every pointwise accumulation point of the sequence $(F_n\check{\mu}^n)$ is harmonic.* □

Proposition 11.8.25 ([194, Théorème 4]). *Let μ be lazy and adapted. Then (X, μ) is Liouville if and only if*

$$\textit{for all } x, y \in X: \quad \|\delta_x\mu^n - \delta_y\mu^n\|_1 \to 0 \textit{ as } n \to \infty.$$

Proof. Let first $f \in \ell^\infty(X)$ be harmonic. Then for all $n \in \mathbb{N}$

$$|f(x) - f(y)| = \left| \sum_{z \in X} f(z) \Big(\sum_{z = xh} \mu^n(h) - \sum_{z = yh} \mu^n(h) \Big) \right|$$
$$\le \|f\|_\infty \cdot \sum_{z \in X} \big| (\delta_x\mu^n)(z) - (\delta_y\mu^n)(z) \big| = \|f\|_\infty \cdot \|\delta_x\mu^n - \delta_y\mu^n\| \to 0,$$

so f is constant.

Conversely, assume that there exist $x, y \in X$ and a sequence (n_i) such that $\|\delta_x\mu^{n_i} - \delta_y\mu^{n_i}\| \ge 4c > 0$ for all $i \in \mathbb{N}$. Set $V_i := \{z \in X \mid (\delta_x\mu^{n_i})(z) > (\delta_y\mu^{n_i})(z)\}$; then

$$\sum_{z \in V_i} (\delta_x\mu^{n_i})(z) - (\delta_y\mu^{n_i})(z) \ge 2c.$$

Set then $W_i := \{z \in X \mid (\delta_x\mu^{n_i})(z) \geq (1+c)(\delta_y\mu^{n_i})(z)\}$; then $(\delta_x\mu^{n_i})(W_i) \geq c$, for otherwise one would have $\sum_{z\in V_i}(\delta_x\mu^{n_i})(z) - (\delta_y\mu^{n_i})(z) = \sum_{z\in W_i}(\dots) + \sum_{z\in V_i\setminus W_i}(\dots) < c + c = 2c$. Set finally $f_i = \mathbb{1}_{W_i}\check{\mu}^{n_i}$. Note then

$$f_i(y) = \sum_{z=yh} \mathbb{1}_{W_i}(z)\mu^{n_i}(h) = \sum_{z\in W_i}(\delta_y\mu^{n_i})(z) \leq \sum_{z\in W_i}(\delta_x\mu^{n_i})(z)/(1+c) = f_i(x)/(1+c),$$

and similarly $f_i(x) = (\delta_x\mu^{n_i})(W_i) \geq c$, so any accumulation point of the f_i is harmonic by Corollary 11.8.24, bounded and nonconstant.

Proof (Proof of Theorem 11.8.20, see [325, Theorem 4.3]). (1) $\Rightarrow$ (2) We assume throughout that X is transitive and therefore countable. Fix a basepoint $x_0 \in X$, and let $\{x_0, x_1, \dots\}$ be an enumeration of X. Choose two sequences $(t_i)_{i\in\mathbb{N}}$ and $(\epsilon_i)_{i\in\mathbb{N}}$ of positive real numbers with $\sum t_i = 1$ and $\lim \epsilon_i = 0$. Let (n_i) be a sequence of integers with $(t_1+\dots+t_{i-1})^{n_i} < \epsilon_i$ for all i. Since X is laminable, by Lemma 11.8.19, there exists for every i a positive function $\alpha_i \in \ell^1(G)$, normalized by $\|\alpha_i\| = 1$ and supported on a finite set (say F_i), with

$$\|(\delta_s - \delta_{x_0})\alpha_i\| < \epsilon_i \text{ for all } s \in \{x_0, \dots, x_i\} \cdot (\{1\} \cup F_1 \cup \dots \cup F_{i-1})^{n_i}.$$

Let us set $\mu = \sum_{i\in\mathbb{N}} t_i\alpha_i$. To prove that (G,μ) is Liouville, it suffices, by Proposition 11.8.25, to prove that $\|\delta_x\mu^n - \delta_{x_0}\mu^n\| \to 0$ for all $x \in X$. Say $x = x_\ell$; we claim that $\|\delta_x\mu^{n_\ell} - \delta_{x_0}\mu^{n_\ell}\| < 4\epsilon_\ell$, and this is sufficient to conclude the proof. For convenience let us write $n_\ell = n$, and expand

$$\mu^n = \sum\sum\nolimits_{k_1,\dots,k_n} t_{k_1}\cdots t_{k_n}\alpha_{k_1}\cdots\alpha_{k_n}. \tag{11.16}$$

We subdivide the sum (11.16) into two summands, ν_1 on which all $k_i < \ell$ and $\nu_2 = \mu^n - \nu_1$. First, $\|\nu_1\| = \sum_{k_i<\ell} t_{k_1}\cdots t_{k_n} = (t_1 + \dots + t_{\ell-1})^{n_\ell} < \epsilon_\ell$, so $\|\delta_x\nu_1 - \delta_{x_0}\nu_1\| < 2\epsilon_\ell$. Secondly, consider a summand $\theta = \alpha_{k_1}\cdots\alpha_{k_n}$ appearing in ν_2; by hypothesis $k_i \geq \ell$ for some i, which we choose minimal. The summand then has the form $\theta_1\alpha_{k_i}\theta_2$. The supports of $\delta_x\theta_1$ and of $\delta_{x_0}\theta_1$ are by hypothesis contained in $\{x, x_0\} \cdot (\{1\} \cup F_1 \cup \dots \cup F_{\ell-1})^{n_\ell}$, so $\|\delta_x\theta_1\alpha_{k_i} - \delta_{x_0}\alpha_{k_i}\| < \epsilon_\ell$ and $\|\delta_{x_0}\theta_1\alpha_{k_i} - \delta_{x_0}\alpha_{k_i}\| < \epsilon_\ell$. Consequently, $\|\delta_x\theta - \delta_{x_0}\theta\| < 2\epsilon_\ell$, so $\|\delta_x\nu_2 - \delta_{x_0}\nu_2\| < 2\epsilon_\ell$ and finally $\|\delta_x\mu^n - \delta_{x_0}\mu^n\| < 4\epsilon_\ell$ as required.

(2) $\Rightarrow$ (3) is obvious.

(3) $\Rightarrow$ (1) By Proposition 11.8.25, the sequence $(\mu^n)_{n\in\mathbb{N}}$ is asymptotically invariant. Consider $\epsilon > 0$ and $f \in \varpi(\ell^1 X)$. There is then a subset $S \Subset X \times X$ such that $f = f' + \sum_{(x,y)\in S} \delta_x - \delta_y$ with $\|f'\| < \epsilon/2$; we have $\|f\mu^n\| \leq \sum_{(x,y)\in S} \|\delta_x\mu^n - \delta_y\mu^n\| + \|f'\mu^n\|$, and for n large enough, each $\delta_x\mu^n - \delta_y\mu^n$ has norm at most $\epsilon/2\#S$, from which $\|f\mu^n\| < \epsilon = \epsilon\|\mu^n\|$, and X is laminable.

Exercise 11.8.26 (, [325]§6.5).** Let G be a group and let μ be a finitely supported probability measure on G. Prove that (G,μ) is Liouville if and only if $(G,\check{\mu})$ is Liouville.

On the other hand, give a group G and a probability measure μ on G such that (G,μ) is Liouville but $(G,\check{\mu})$ is not Liouville.

Hint for the second part: take $G = \mathbb{Z}/2 \wr \mathbb{Z}$, the "Lamplighter group" from Example 11.2.16. Choose a positive sequence $(\epsilon_n)_{n\ge 0}$ with $\sum \epsilon_n = \frac{1}{2}$ and $\sum n\epsilon_n = \infty$. Write elements of G as pairs (f,m) with $f\colon \mathbb{Z} \to \mathbb{Z}/2$ and $m \in \mathbb{Z}$, and define $\mu\colon G \to [0,1]$ by

$$\mu(0,1) = \frac{3}{8}, \qquad \mu(0,-1) = \frac{1}{8},$$

$$\mu(f,0) = \begin{cases} \frac{\epsilon_n}{2^n} & \text{if } \{n\} \subseteq \text{support}(f) \subseteq \{0,\dots,n\} \text{ for some } n \in \mathbb{N},\\ 0 & \text{else.}\end{cases}$$

11.9 Extensive Amenability

We introduce now a property stronger than amenability for G-sets, a property that behaves better with respect to extensions of G-sets (whence the name). This section is based on [320].

Definition 11.9.1. Let X be a set; recall that $\mathfrak{P}_f(X)$ denotes the collection of finite subsets of X. An *ideal* in $\mathfrak{P}_f(X)$ is a subset, for some $x \in X$, of the form $\{E \Subset X \mid x \in E\}$.[17]

Let X be a G-set. It is *extensively amenable* if there exists a G-invariant mean m on $\mathfrak{P}_f(X)$ giving weight 1 to every ideal.

It follows immediately from the definition that $m(\{\emptyset\}) = 0$ if $X \neq \emptyset$ and that for every $E \Subset X$ we have $m(\{F \Subset X \mid E \subseteq F\}) = 1$.

Recall that $\mathfrak{P}_f(X)$ is an abelian group under symmetric difference Δ and is naturally isomorphic to $(\mathbb{Z}/2)^{(X)}$ under the map $E \mapsto \mathbb{1}_E$. Recall also from (11.1) that the *wreath product* $\mathbb{Z}/2 \wr_X G$ is the semidirect product $G \ltimes (\mathbb{Z}/2)^{(X)}$, with G acting on $(\mathbb{Z}/2)^{(X)}$ by permuting its factors.

Lemma 11.9.2. *If G is amenable, then all G-sets are extensively amenable. Every extensively amenable nonempty G-set is amenable.*

Proof. Let G be amenable and let X be a G-set. Consider the set K of means on $\mathfrak{P}_f(X)$ giving full weight to every ideal. Clearly K is a convex compact subset of $\ell^\infty(\mathfrak{P}_f(X))^*$ and is nonempty because it contains any cluster point of $(\delta_E)_{E\Subset X}$. Since G is amenable, there exists a fixed point in K, so X is extensively amenable.

[17] It is really the ideal generated by $\{x\}$ in the semigroup $(\mathfrak{P}_f(X), \cup)$.

Let next $X \curvearrowleft G$ be extensively amenable, and let m be an invariant mean in $\ell^\infty(\mathfrak{P}_f(X) \setminus \{\emptyset\})^*$. Define a mean on X by

$$\ell^\infty(X) \ni f \mapsto m\Big(E \mapsto \frac{1}{\#E}\sum_{x\in E} f(x)\Big),$$

and note that it is G-invariant because m is.

Lemma 11.9.3. *Let X be a G-set. Then the following are equivalent:*

1. *X is extensively amenable;*
2. *For every finitely generated subgroup H of G and every H-orbit $Y \subseteq X$, the H-set Y is extensively amenable;*
3. *For every finitely generated subgroup H of G and every $x_0 \in X$, there is an H-invariant mean on $\mathfrak{P}_f(x_0 H)$ that gives nonzero weight to $\{E \Subset x_0 H \mid x_0 \in E\}$;*
4. *There is a G-invariant mean on $\mathfrak{P}_f(X)$ that gives nonzero weight to $\{E \Subset X \mid x_0 \in E\}$ for all $x_0 \in X$.*

Proof. (1) $\Rightarrow$ (4) by definition.

(4) $\Rightarrow$ (3) There is a natural map $\ell^\infty(\mathfrak{P}_f(x_0 H)) \to \ell^\infty(\mathfrak{P}_f(X))$ given by $f \mapsto f(- \cap x_0 H)$, inducing an H-equivariant map $\mathscr{M}(\mathfrak{P}_f(X)) \to \mathscr{M}(\mathfrak{P}_f(x_0 H))$.

(3) $\Rightarrow$ (2) Let $Y = x_0 H$ be an H-orbit, and let m_0 be an H-invariant mean on $\mathfrak{P}_f(Y)$ that gives positive weight to $\{A \Subset Y \mid x_0 \in A\}$. As in Theorem 11.3.23, the mean m_0 may be approximated by a net p_n of probability measures on $\mathfrak{P}_f(Y)$: these are maps $\mathfrak{P}_f(Y) \to [0,1]$ with total mass 1. Define now for every $k \in \mathbb{N}$ new probability measures on $\mathfrak{P}_f(Y)$ by

$$p_{n,k}(E) = \sum_{E_1\cup\cdots\cup E_k = E} p_n(E_1)\cdots p_n(E_k).$$

Let m be an cluster point of the $p_{n,k}$ as $n, k \to \infty$; then m is an H-invariant mean on $\mathfrak{P}_f(Y)$, and we check that it gives mass 1 to the ideal $S := \{E \Subset Y \mid x_0 \in E\}$ and therefore also to every ideal because H acts transitively on Y and m is H-invariant: since $m_0(S) > 0$, there exists $\delta < 1$ such that $p_n(S) > 1 - \delta$ for all n large enough, and then $p_{n,k}(S) > 1 - \delta^k$ so at the cluster point $m(S) = 1$.

(2) $\Rightarrow$ (1) For every finitely generated subgroup H of G and every finite union $Y = Y_1 \cup \cdots \cup Y_n$ of H-orbits, choose for $i = 1, \dots, n$ an H-invariant mean m_i on $\mathfrak{P}_f(Y_i)$, and construct a mean $m_{H,Y}$ on $\mathfrak{P}_f(X)$ by $m_{H,Y}(S) = m_1(\{E \cap Y_1 \mid E \in S\}) \cdots m_n(\{E \cap Y_n \mid E \in S\})$. Clearly $m_{H,Y}$ is H-invariant and gives full weight to ideals in $\mathfrak{P}_f(Y)$. Order the pairs (H, Y) by inclusion, and consider a cluster point of the net $(m_{H,Y})$. It is G-invariant and gives full weight to ideals in $\mathfrak{P}_f(X)$.

Note that Lemma 11.9.3(2) implies in particular that extensively amenable sets are *hereditarily amenable*: every subgroup acting on every orbit is amenable. We obtain in this manner an abundance of amenable actions that are not extensively amenable. For instance, consider Example 11.3.13 of an amenable action of $F_2 = \langle a, b \mid \rangle$, and the subgroup $K = \langle a^{b^{-1}}, a^{b^{-2}} \rangle$. Then K is a free group of rank 2, and the K-orbit Y of 1 in X is free, so Y is not an amenable K-set, and therefore X is not extensively amenable. We shall see in Example 11.9.20 a hereditarily amenable G-set that is not extensively amenable.

We come to the justification of the terminology "extensive amenability": the analogue of Corollary 11.2.27 for G-sets.

Proposition 11.9.4. *Let G be a group acting on two sets X, Y, and let $q: X \to Y$ be G-equivariant. If Y is extensively amenable and if for every $y \in Y$ the G_y-set $q^{-1}(y)$ is an extensively amenable, then X is extensively amenable. The converse holds if q is onto.*

Proof. The proof follows closely that of Proposition 11.2.26; see [320, Proposition 2.4] for details. Assume that $q^{-1}(y)$ is extensively amenable for all $y \in Y$, and let m_y be a G_y-invariant mean giving full weight to ideals. By making one choice per G-orbit, we may also assume that $m_{y'}$ is the push-forward by g of m_y whenever $y' = yg$. Extend every m_y to a mean on $\mathfrak{P}_f(X)$; then (m_y) is a G-equivariant map $Y \to \mathscr{M}(\mathfrak{P}_f(X))$.

For every $F = \{y_1, \dots, y_n\} \Subset Y$, we set

$$m_F(S) = m_{y_1}(\{E \cap q^{-1}(y_1) \mid E \in S\}) \cdots m_{y_n}(E \cap q^{-1}(y_n)),$$

and note that m_F gives full weight to every ideal of the form $\{E \Subset X \mid x \in E\}$ for some $x \in q^{-1}(F)$. The map $F \mapsto m_F$ defines a G-equivariant map $\mathfrak{P}_f(Y) \to \mathscr{M}(\mathfrak{P}_f(X))$. Composing with the barycentre Υ as in (11.6), we obtain a G-equivariant map $m_*: \mathscr{M}(\mathfrak{P}_f(Y)) \to \mathscr{M}(\mathfrak{P}_f(X))$.

Assume now that Y is extensively amenable, and let n be a G-invariant mean on $\mathfrak{P}_f(Y)$ giving full weight to ideals. Set $m := m_*(n)$; then m is a G-invariant mean on $\mathfrak{P}_f(X)$ giving full weight to ideals, so X is extensively amenable.

Assume finally that q is onto and that X is extensively amenable. By Lemma 11.9.3, the G_y-subset $q^{-1}(y)$ of X is extensively amenable for all $y \in Y$. Let m be a mean on $\mathfrak{P}_f(X)$ giving full weight to ideals, and define a mean n on $\mathfrak{P}_f(Y)$ by $n(S) = m(\{E \Subset X \mid q(E) \in S\})$. Given $y \in Y$, choose $x \in q^{-1}(y)$, and note

$$n(\{F \Subset Y \mid y \in F\}) = m(\{E \Subset X \mid y \in q(E)\}) \geq m(\{E \Subset X \mid x \in E\}) = 1.$$

□

In particular, let $K \leq H \leq G$ be groups. Then $K \backslash G$ is an extensively amenable G-set if and only if both $K \backslash H$ and $H \backslash G$ are extensively amenable. This is in contrast with Example 11.2.18, where the corresponding property is shown *not* to hold for amenability of sets.

The following proposition relates Definition 11.9.1 to the original definition; we begin by introducing some vocabulary. Let **A** denote the category of group actions:

its objects are pairs $X \looparrowleft G$ of a set X and an action of G on X, and a morphism $(X \looparrowleft G) \to (Y \looparrowleft H)$ is a pair of maps $(f\colon X \to Y, \phi\colon G \to H)$ intertwining the actions on X and Y, namely, satisfying $f(x)\phi(g) = f(xg)$ for all $x \in X, g \in G$. We denote by **AA** and **EA** the subcategories of amenable, respectively, extensively amenable actions.

We are interested in functors F: {finite sets, injections} $\to$ **AA**, written $F(X) = F_0(X) \looparrowleft F_1(X)$ for a group $F_1(X)$ and an $F_1(X)$-set $F_0(X)$. Since amenable actions are closed under directed unions, and every set is the directed union of its finite subsets, we get by continuity a functor still written F: {sets, injections} $\to$ **AA**, called an *amenable functor*. If furthermore F takes values in **EA** then we call it an *extensively amenable functor*. We call the functor F *tight* if the map $F_0(X \setminus \{x\}) \to F_0(X)$ is never onto.

We already saw some examples of tight functors: for any amenable group A, the functor $X \mapsto A^{(X)} \looparrowleft A^{(X)}$ since $A^{(X)}$ is the directed union of its amenable subgroups A^E over all $E \Subset X$; the functor $X \mapsto \mathrm{Sym}(X) \looparrowleft \mathrm{Sym}(X)$, by the same reasoning (see Example 11.7.4); and the functor $X \mapsto X \looparrowleft \mathrm{Sym}(X)$. Note that if X is a G-set, then $F_0(X)$ and $F_1(X)$ inherit G-actions by functoriality.

Proposition 11.9.5 ([320, Theorem 3.14]). *Let F be a functor as above, and let X be a G-set. If X is extensively amenable and F is amenable, then $F_0(X) \looparrowleft (G \ltimes F_1(X))$ is amenable, and if furthermore F is extensively amenable, then $F_0(X) \looparrowleft (G \ltimes F_1(X))$ is extensively amenable.*

Conversely, if F is tight and $F_0(X) \looparrowleft (G \ltimes F_1(X))$ is amenable, then X is extensively amenable.

Proof. Assume first that F is amenable. For every $E \Subset X$ let $m_E \in \mathcal{M}(F_0(E))^{F_1(E)}$ be an invariant mean, and extend it functorially to a mean still written $m_E \in \mathcal{M}(F_0(X))^{F_1(E)}$. By choosing once m_E per cardinality class of subsets of X, we may ensure that we have $f_*(m_E) = m_{E'}$ for every bijection $f\colon E \to E'$. We obtain in this manner a G-equivariant map $\mathfrak{P}_f(X) \to \mathcal{M}(F_0(X))$, and therefore, composing with the barycentre Υ as in (11.6), a map $\mathcal{M}(\mathfrak{P}_f(X))^G \to \mathcal{M}(F_0(X))^G$.

By assumption, there exists $m_0 \in \mathcal{M}(\mathfrak{P}_f(X))^G$ giving full mass to ideals; let m be the image of m_0 under the above map. Clearly m is a G-invariant mean on $F_0(X)$. It is also $F(A)$-invariant for every $A \Subset X$: one may restrict m_0 to $\{E \Subset X \mid A \subseteq E\}$ and still obtain a mean. Every m_E is $F_1(E)$-invariant, so it is in particular $F(A)$-invariant, and therefore m is also $F(A)$-invariant. In summary, m is $G \ltimes F_1(X)$-invariant, so $F_0(X)$ is an amenable $G \ltimes F_1(X)$-set.

For the converse, define a G-equivariant map support: $F_0(X) \to \mathfrak{P}_f(X)$ by

$$\mathrm{support}(x) = \bigcap\{E \Subset X \mid x \in \mathrm{image}(F_0(E) \to F_0(X))\}.$$

Assume that $F_0(X)$ is an amenable $G \ltimes F_1(X)$-set, and let m_0 be a G-invariant mean on $F_0(X)$. Let m the push-forward of m_0 via support; it is a G-invariant mean on $\mathfrak{P}_f(X)$. Choose $x_0 \in X$. By definition, $m(\{E \Subset X \mid x_0 \in E\}) = m_0(S)$ for the ideal

$$\begin{aligned} S &= \{x \in F_0(X) \mid x_0 \in \operatorname{support}(x)\} \\ &= \bigcap_{x_0 \notin E \Subset X} (F_0(X) \setminus \operatorname{image}(F_0(E) \to F_0(X))) \\ &= F_0(X) \setminus F_0(X \setminus \{x_0\}). \end{aligned}$$

Since F is tight, $S \neq \emptyset$. Furthermore, m_0 is $F_1(X)$-invariant, so $m_0(S) > 0$. We conclude by Lemma 11.9.3 that X is extensively amenable.

Finally, to prove that $F_0(X)$ is an extensively amenable $G \ltimes F_1(X)$-set whenever F is an extensively amenable functor, we apply the converse just proven to the functor $H(X) = (\mathbb{Z}/2)^{(X)} \looparrowleft (\mathbb{Z}/2)^{(X)}$. For every X, we know from the first part of the proof that $H_0(F_0(X))$ is an amenable $(G \ltimes F_1(X)) \ltimes H_1(F_0(X))$-set, since we assumed $F_1(X) \looparrowleft G \ltimes F_1(X)$ is extensively amenable. Therefore, the functor $X \mapsto H_0(F_0(X)) \looparrowleft (F_1(X) \ltimes H_1(F_0(X)))$ is amenable, and yet again the second part of the proof allows us to deduce that $F_0(X) \looparrowleft G \ltimes F_1(X)$ is extensively amenable.

A fundamental application of Proposition 11.9.5 is the following.

Corollary 11.9.6. *Let H be a subgroup of $G \ltimes F(X)$ for some extensively amenable G-set X. If $H \cap (G \times 1)$ is amenable, then H is amenable too.*

Proof. By Proposition 11.9.5, $F(X)$ is extensively amenable, so by Lemma 11.9.3 the H-orbit $1 \cdot H \subseteq X$ is an extensively amenable H-set and is therefore amenable by Lemma 11.9.2. The stabilizers in this action are conjugate to $H \cap (G \times 1)$, which is amenable by assumption, so H is amenable by Proposition 11.2.26.

There is also a connection between extensive amenability and laminability, see Definition 11.8.15: by Corollary 11.8.18, if $X \looparrowleft G$ is extensively amenable then $F(X) \looparrowleft G \ltimes F(X)$ is laminable.

In the next section, we shall see a sufficient condition for an action to be extensively amenable, and in Example 11.9.15 an application to interval exchange transformations.

We finish this section by a very brief summary of the "only if" part of a proof of Theorem 11.4.1 due to Kleiner [350] and simplified by Tao; we include it here because it combines amenability and the study of (now unbounded) harmonic functions.

Let G be a group of polynomial growth; we are to show that G has a nilpotent subgroup of finite index. We may of course assume that G is infinite, and by induction on the growth degree, it suffices to show that G has a finite-index subgroup mapping onto $\mathbb{Z}$. For that purpose, it suffices to show that G has an infinite image in some virtually soluble group. By [543], every amenable finitely generated subgroup of $\mathrm{GL}_n(\mathbb{C})$ is virtually soluble, and G is amenable by Proposition 11.3.14, so it suffices to construct a representation $G \to \mathrm{GL}_n(\mathbb{C})$ with infinite image. The proof uses the following arguments:

Lemma 11.9.7. *Let G be a countably infinite amenable group. Then there exists an action of G on a Hilbert space $\mathscr{H}$ with no fixed points.*

Proof. Consider $\mathscr{H} = \ell^2(\mathbb{N} \times G)$, the space of square-summable functions $(f_1, f_2, \dots)$ in $\ell^2(G)$. There is a natural, diagonal action of G on $\mathscr{H}$ by right translation. This action has a fixed point 0, but we can construct an affine action without fixed point as follows.

Let $(F_n)_{n\in\mathbb{N}}$ be a Følner sequence in G, and define $h = (\mathbb{1}_{F_n}/\sqrt{\#F_n})_{n\in\mathbb{N}}$. Then $h \notin \mathscr{H}$, but $h - hg \in \mathscr{H}$ for all $g \in G$, using the almost invariance of (F_n). We let G act on $\mathscr{H}$ by $f \cdot g = fg + h - hg$, namely, we move the fixed point to h.

The main result, whose proof we omit, is the following control on the growth of harmonic functions. It follows easily from Gromov's theorem, but Bruce Kleiner gave a direct and elementary proof of it:

Lemma 11.9.8. *Let G be a group of polynomial growth, and let μ be a measure on G. Then for every $d \in \mathbb{N}$, the vector space of harmonic maps $u: G \to \mathbb{R}$ of growth degree at most d (namely, for which there is a constant C with $|u(g)| \le C|g|^d$ for all $g \in G$) is finite-dimensional.* □

The proof of Theorem 11.4.1 is then finished: a group G of polynomial growth is amenable, so by Lemma 11.9.7 it has an affine, fixed-point-free action on a Hilbert space $\mathscr{H}$. Let μ be SRW on G, and define

$$E: \mathscr{H} \to \mathbb{R}_+, \qquad v \mapsto \frac{1}{2}\sum_{s\in S} \mu(s)\|vs - v\|^2.$$

Since $\mathscr{H}$ has no fixed point, $E(v) > 0$ for all $v \in \mathscr{H}$. Let us assume that $E(v)$ attains its minimum—this can be achieved by considering a sequence of better and better approximations to a minimum in an ultrapower of $\mathscr{H}$—and call its minimum h. One directly sees from $\partial E(v)/\partial v|_h = 0$ that h is μ-harmonic, and it is not constant. Then $V := \{\langle h|v\rangle \mid v \in \mathscr{H}\}$ is a vector space of Lipschitz harmonic maps, so it is finite-dimensional by Lemma 11.9.8, and G's action on V has infinite image because V has no nonzero G-fixed point.

11.9.1 Recurrent Actions

We saw in Proposition 11.3.14 that actions on subexponentially growing spaces are amenable and in Theorem 11.8.4 that random walks on graphs in which the probability of return to the origin decays subexponentially give amenable actions. We see here that stronger conditions—quadratic growth, recurrent random walks—produce extensively amenable actions.

Let $p_1: X \times X \to [0,1]$ be a random walk on a set X. It is *recurrent* at $x \in X$ if $\sum_{n\ge 0} p_n(x,x) = \infty$, namely, if a random walk started at x is expected to return infinitely often to x and equivalently if it is certain to return to x. It is *transient* if it is not recurrent.

We computed in Example 11.8.9 that the probability of return in to the origin in n steps of SRW on $\mathbb{Z}$ is $\propto n^{-1/2}$; so the probability of return to the origin on $\mathbb{Z}^d$ is $\propto n^{-d/2}$. It follows that SRW on $\mathbb{Z}^d$ is recurrent precisely for $d \le 2$.

Lemma 11.9.9. *The random walk p is recurrent if and only if for every $x \in X$, there exists a sequence of functions (a_n) in $\ell^2(X)$ with $a_n(x) = 1$ and $\|a_n - Ta_n\| \to 0$, for T the associated random walk operator.*

Proof. For a function $\phi \in \ell^2(X)$, define its *Dirichlet norm* as $D(\phi) = \|d\phi\|^2 = \frac{1}{2}\sum_{x,y\in X}(f(x) - f(y))^2 p_1(x,y)$. The claim is equivalent to requiring the existence of functions $a_n \in \ell^2(X)$ with $a_n(x) = 1$ and arbitrarily small Dirichlet norm. If X is finite, there is nothing to do, as the functions $a_n \equiv 1$ have $D(a_n) = 0$.

Choose $x \in X$. Assume first that p is transient, so that $G(y) := \sum_{n\ge 0} p_n(x,y)$ is well defined. Then for all $\phi \in \ell^2(X)$, we have

$$\langle d\phi, dG\rangle = \langle \phi, d^* dG\rangle = \phi(x),$$

and $|\langle d\phi, dG\rangle|^2 \le D(\phi)D(g)$, so $D(\phi) \ge \phi(1)/D(g)$ is bounded away from 0.

Assume next that p is recurrent. For every $n \in \mathbb{N}$, set $G_n(y) = \sum_{m=0}^{n} p_m(x,y)$ and $a_n(y) = G_n(y)/G_n(x)$. Since by assumption $G_n(x) \to \infty$, the functions $a_n(y)$ satisfy the requirement. □

For random walks with finite range, the following criterion due to Nash-Williams is very useful. Let p be a transitive random walk on a set X, and let $x \in X$ be a basepoint. A *slow constriction* of X is a family $\{x\} = V_0 \subset V_1 \subset \cdots$ of finite subsets of X, such that $\bigcup V_n = X$ and $p_1(V_m, V_n) = 0$ whenever $|m - n| \ge 2$ and $\sum_{n\ge 0} p_1(V_n, V_{n+1})^{-1} = \infty$. A *refinement* of p is the random walk on a set obtained by subdividing arbitrarily each transition $p_1(x,y)$ by inserting midpoints along it.

Theorem 11.9.10 (Nash-Williams [439]). *Let p be a transitive random walk on a set X. Then p is recurrent if and only if it has a refinement admitting a slow constriction.*

The result applies to $\mathbb{Z}^d$ for $d \le 2$: the sets V_n may be chosen as $\{-n, \dots, n\}^d$. We only prove the "only if" direction, which is the important direction for us.

Proof (First proof of Theorem 11.9.10, "only if" direction). Given a constriction (V_n), set $c_n = 1/p_1(V_n, V_{n+1})$, and define an associated random walk q on $\mathbb{N}$ by $q_1(n, n+1) = c_n/(c_n + c_{n-1})$ and $q_1(n, n-1) = c_{n-1}/(c_n + c_{n-1})$. It is easy to check $\sum_n q_n(1,1) = \infty$ if the constriction is slow. □

We shall give another proof of the "only if" direction, based on Lemma 11.9.9; we begin by a simple

Lemma 11.9.11. *Let $\sum_i v_i$ be a positive, divergent series. Then there exist $\lambda_{i,n} \ge 0$ such that $\sum_i \lambda_{i,n} v_i = 1$ for all n and $\sum_i \lambda_{i,n}^2 v_i \searrow 0$ as $n \to \infty$.*

Proof. Let $\alpha_n = 1+1/n$ be a decreasing sequence converging to 1. Group the terms in $\sum v_i$ into blocks $w_1 + w_2 + \cdots$ such that $w_i \geq 1$ for all i. Set

$$\lambda_{i,n} = \frac{\alpha_n - 1}{w_k \alpha_n^{k-1}} \text{ if } v_i \text{ belongs to the block } w_k.$$

□

Proof (Second proof of Theorem 11.9.10, "only if" direction). Let $(V_i)_{i\geq 1}$ be a slow constriction of X with basepoint x, and set $v_i := 1/p_1(V_i, V_{i+1})$. Apply the lemma to the divergent series $\sum v_i$, and define maps $a_n : X \to [0,1]$ by

$$a_n(y) = 1 - \sum_i \lambda_{i,n} v_i \mathbb{1}_{y \notin V_i}.$$

Then a_n has finite support, so in particular it belongs to $\ell^2(X)$; $a_n(x) = 1$ because $x \in V_i$ for all i; and $\|a_n - a_n g\|_2 \to 0$ for all $g \in G$ because $\sum \lambda_{i,n}^2 v_i \to 0$.

The main result of this section is the following. We will prove it in two different manners and, in fact in this manner, recover the "if" direction of Theorem 11.9.10:

Theorem 11.9.12. *If X is a G-set with an adapted recurrent random walk, then X is extensively amenable.*

We begin with some preparation for the proof. Let μ be a symmetric, adapted measure on a group G, and let X be a G-set. For a basepoint $x \in X$ and a trajectory $x, xg_1, xg_1g_2, \ldots$ of the random walk on X, the corresponding length-n *inverted orbit* is the random subset

$$O_n = \{x, xg_n, xg_{n-1}g_n, \ldots, xg_1 \cdots g_n\}.$$

If X is transitive, then $\#O_n$ depends only mildly on the choice of x.

Proposition 11.9.13. *Let X be a transitive G-set and let μ be a symmetric, adapted probability measure on G. Then X is extensively amenable if and only if*

$$\lim_{n\to\infty} \frac{-1}{n} \log \mathbb{E}(2^{-\#O_n}) = 0. \tag{11.17}$$

Proof. Thanks to Proposition 11.9.5, it is enough to prove that (11.17) is equivalent to the amenability of the $G \ltimes (\mathbb{Z}/2)^{(X)}$-set $(\mathbb{Z}/2)^{(X)}$. Choose a basepoint $x \in X$, and consider on $G \ltimes (\mathbb{Z}/2)^{(X)}$ the probability distribution $\nu := \frac{1}{2}(1+\delta_x) * \mu * \frac{1}{2}(1+\delta_x)$, called the "switch-walk-switch" measure: in the action on $(\mathbb{Z}/2)^{(X)}$, it amounts to randomizing the current copy of $\mathbb{Z}/2$, moving to another position in X, and randomizing the new copy of $\mathbb{Z}/2$. By Kesten's Theorem 11.8.4, amenability of the action on $(\mathbb{Z}/2)^{(X)}$ is equivalent to subexponential decay of return probabilities of a random walk $(f_0 = 1, f_1, f_1 f_2, \ldots)$ on $(\mathbb{Z}/2)^{(X)}$, namely, to

$\lim_{n\to\infty} \frac{-1}{n} \log \mathbb{P}(f_1 \cdots f_n = 1) = 0$. Now the support of $f_1 \cdots f_n$ is contained in O_n: writing each $f_i = \delta_x^{\epsilon_i} g_i \delta_x^{\zeta_i}$ with $g_i \in G$, we get $f_1 \cdots f_n = g_1 \cdots g_n \delta_{xg_1\cdots g_n}^{\epsilon_1} \delta_{xg_2\cdots g_n}^{\zeta_1+\epsilon_2} \cdots \delta_x^{\zeta_n}$; and f_n randomizes every copy of $\mathbb{Z}/2$ indexed by O_n, so $\mathbb{P}(f_n = 1) = \mathbb{E}(2^{-\#O_n})$.

Lemma 11.9.14. *Let p be a transitive random walk on a G-set X driven by a symmetric probability measure μ. Then X is recurrent if and only if $\lim \frac{1}{n}\mathbb{E}(\#O_n) = 0$.*

Proof. Choose a basepoint $x \in X$ for the random walk $(x = x_0, x_1, \dots)$, and define the random variable $\Theta = \min\{n \geq 1 \mid x_n = x\}$. Then

$$\begin{aligned}\mathbb{E}(\#O_{n+1} - \#O_n) &= \mathbb{P}(xg_{n+1} \notin O_n)\\ &= \mathbb{P}(xg_{n+1} \notin \{x, xg_n, xg_{n-1}g_n, \dots, xg_1 \cdots g_n\})\\ &= \mathbb{P}(\{xg_{n+1}, xg_{n+1}g_n^{-1}, xg_{n+1}g_n^{-1}g_{n-1}^{-1}, \cdots, xg_{n+1}g_n^{-1} \cdots g_1^{-1}\} \not\ni x)\\ &= \mathbb{P}(\Theta > n+1),\end{aligned}$$

because the random walk with increments $g_{n+1}, g_n^{-1}, \dots, g_1^{-1}$ has the same law as μ^n. Therefore, $\mathbb{E}(\#O_n)/n \to \mathbb{P}(\Theta = \infty)$, which vanishes if and only if X is recurrent.

Proof (First proof of Theorem 11.9.12). We may assume, by Lemma 11.9.3, that X is transitive. Let p be an adapted, transitive, recurrent random walk on X. By Lemma 11.9.14, we have $\frac{1}{n}\mathbb{E}(\#O_n) \to 0$, so by convexity

$$\frac{-1}{n} \log \mathbb{E}(2^{-\#O_n}) \leq \frac{1}{n}\mathbb{E}(\#O_n) \log 2 \to 0,$$

so X is extensively amenable by Proposition 11.9.13.

Proof (Second proof of Theorem 11.9.12). Let $x \in X$ be arbitrary. We start, using Lemma 11.9.9, with a sequence of functions (a_n) in $\ell^2(X)$ satisfying $a_n(x) = 1$ and $\lim \|a_n - a_n g\| = 0$ for all $g \in G$. (This is also the outcome of the second proof of Theorem 11.9.10.) We construct then maps $b_n : \mathfrak{P}_f(X) \to [0, 1]$ by

$$b_n(E) = \prod_{y \in E} a_n(y).$$

They are finitely supported and therefore may be viewed in $\ell^2(\mathfrak{P}_f(X))$. It remains to check that they are almost invariant under the action of $\mathbb{Z}/2 \wr_X G$. Assuming that X is transitive, this last group is generated by $\delta_x : X \to \mathbb{Z}/2$ and G. We have $b_n \delta_x = b_n$, because $b_n(E) = b_n(E \triangle \{x\})$.

The spaces $\ell^2(\mathfrak{P}_f(X))$ and $\bigotimes_X \ell^2(\mathbb{C}^2)$ are isometric; the isometry is the obvious one mapping δ_E to $\bigotimes_{x \in X} \delta_{x \in E}$, if we take $\{\delta_{\text{false}}, \delta_{\text{true}}\}$ as basis of $\ell^2(\mathbb{C}^2)$. We compute

$$\|b_n\|^2 = \langle b_n, b_n \rangle = \prod_{y \in X} (1^2 + a_n(y)^2),$$

and for $g \in G$ we similarly have $\langle b_n, b_n g^{-1}\rangle = \prod_{y\in X}(1 + a_n(y)a_n(yg))$, so

$$\underbrace{\left(\frac{\langle b_n, b_n\rangle}{\langle b_n, b_n g^{-1}\rangle}\right)^2}_{A} = \prod_{y\in X}\underbrace{\frac{(1 + a_n(y)^2)(1 + a_n(yg)^2)}{(1 + a_n(y)a_n(yg))^2}}_{B}.$$

Taking logarithms, and using the approximation $\log(t) \le t - 1$,

$$0 \le 2\log(A) \le \sum_{y\in X}\log(B) \le \sum_{y\in X}\frac{(a_n(y) - a_n(yg))^2}{(1 + a_n(y)a_n(yg))^2} \le \|a_n - a_n g^{-1}\|^2 \to 0,$$

so $\langle b_n, b_n\rangle / \langle b_n, b_n g^{-1}\rangle \to 1$ and therefore $\|b_n - b_n g\| \to 0$.

Example 11.9.15. An *interval exchange* is a piecewise translation self-map of the circle. More precisely, it is a right-continuous map $g\colon \mathbb{R}/\mathbb{Z} \circlearrowleft$ such that $\sphericalangle(g) := \{g(x) - x \mid x \in \mathbb{R}/\mathbb{Z}\}$ is finite.

The rotation $x \mapsto x + \alpha$ is an extreme example of interval exchange.[18] The interval exchange transformations naturally form a group IET acting on $\mathbb{R}/\mathbb{Z}$; and every countable subgroup $G \le \mathsf{IET}$ can be made to act on the Cantor set by letting $\mathscr{D}$ be the union of the G-orbits of discontinuity points of G (or of 0 if all elements of G are rotations) and replacing $\mathbb{R}/\mathbb{Z}$ by

$$X := (\mathbb{R}/\mathbb{Z} \setminus \mathfrak{D}) \cup (\mathfrak{D} \times \{+, -\}),$$

namely, by opening up the circle at every point of $\mathfrak{D}$; see [340, Section 5].

Little is known on the group IET; in particular, it is not known whether it contains non-abelian free groups or whether it is amenable. We prove:

Theorem 11.9.16 ([320, Theorem 5.1]). *Let $\Lambda \le \mathbb{R}/\mathbb{Z}$ be a finitely generated subgroup with free rank at most* 2*, namely,* $\dim(\Lambda \otimes \mathbb{Q}) \le 2$. *Then*

$$\mathsf{IET}(\Lambda) := \{g \in \mathsf{IET} \mid \sphericalangle(g) \subseteq \Lambda\}$$

is an amenable subgroup of IET.

Proof. We first prove that the action of $\mathsf{IET}(\Lambda)$ on $\mathbb{R}/\mathbb{Z}$ is extensively amenable. Choose a finite generating set for Λ; then the Cayley graph of Λ is quasi-isometric to $\mathbb{Z}^d$ for $d \le 2$ and in particular is recurrent. Let $G = \langle S\rangle$ be a finitely generated subgroup of $\mathsf{IET}(\Lambda)$. For $x \in \mathbb{R}/\mathbb{Z}$, the orbit xG injects into Λ under the map $y \mapsto y - x$, and this map is Lipschitz with Lipschitz constant $\max_{s\in S}\max_{\lambda\in\sphericalangle(s)}\|\lambda\|$, so the

[18] The name "interval exchange" comes from opening up the circle into an interval $[0, 1]$; the rotation on the circle may be viewed as an exchange of two intervals $[0, 1 - \alpha] \mapsto [\alpha, 1], [1 - \alpha, 1] \mapsto [0, \alpha]$.

Schreier graph of xG is recurrent. Theorem 11.9.12 implies that xG is an extensively amenable G-set, so $\mathbb{R}/\mathbb{Z}$ is an extensively amenable $\mathsf{IET}(\Lambda)$-set by Lemma 11.9.3.

We wish to apply Corollary 11.9.6 to $X = \mathbb{R}/\mathbb{Z}$ and $G = \mathsf{IET}(\Lambda)$ and $F(X) = X \nleftarrow \mathrm{Sym}(X)$. Given an interval exchange map $g \in \mathsf{IET}$, let $\tilde{g}$ be the unique left-continuous self-map of $\mathbb{R}/\mathbb{Z}$ that coincides with g except at its discontinuity points, and let $\tau_g = g^{-1}\tilde{g} \in \mathrm{Sym}(\mathbb{R}/\mathbb{Z})$ be the corresponding permutation of the discontinuity points of g^{-1}. We have a cocycle identity $\tau_{gh} = \tau_g^h \tau_h$, so the map

$$\iota: \begin{cases} \mathsf{IET} & \to \mathsf{IET} \ltimes \mathrm{Sym}(\mathbb{R}/\mathbb{Z}) \\ g & \mapsto (g, \tau_g) \end{cases}$$

is an embedding. Observe that $\tau_g = 1$ if and only if g is continuous, namely, is a rotation. Therefore, $\iota(\mathsf{IET}(\Lambda)) \cap (\mathsf{IET}(\Lambda) \times 1) = \Lambda$ consists of rotations, so it is amenable. We deduce by Corollary 11.9.6 that $\mathsf{IET}(\Lambda)$ is amenable.

11.9.2 Topological Full Groups

We apply the results from the previous sections to exhibit a wide variety of amenable groups.

We begin by a fundamental construction. Let G be group acting on a compact set X. The associated *topological full group* is the group $[[G, X]]$ of piecewise-G homeomorphisms of X:

$$[[G, X]] = \{\phi: X \circlearrowleft \mid \exists \nu: X \to G \text{ continuous with } \phi(x) = x\nu(x) \text{ for all } x\}.$$

Note that ν takes finitely many values since it is a map from a compact set to a discrete set. If we suppose X discrete rather than compact, then $[[G, X]]$ becomes the group of bijective G-wobbles of X that we saw in Section 11.5.2. The connection is even more direct: let $x \in X$ be such that its orbit xG is dense in X. Then $[[G, X]]$ acts faithfully on the orbit xG by G-wobbles.

The natural setting for the definition of the topological full group is that of *groupoids of germs*. We recall the basic notions:

Definition 11.9.17. A *groupoid* is a set $\mathfrak{G}$ with source and range maps $s, r: \mathfrak{G} \circlearrowleft$, with an associative multiplication $\gamma_1\gamma_2$ defined whenever $r(\gamma_1) = s(\gamma_2)$, and with an everywhere-defined inverse satisfying $\gamma\gamma^{-1} = s(\gamma) = r(\gamma^{-1})$. Its *set of units* is the subset $\mathfrak{G}_0$ of elements of the form $\gamma\gamma^{-1}$. The groupoid $\mathfrak{G}$ is called *topological* if $\mathfrak{G}$ is a topological space and the multiplication and inverse maps are continuous. Note that for every $x \in \mathfrak{G}_0$, the subset $\mathfrak{G}_x := \{\gamma \in \mathfrak{G} \mid s(\gamma) = r(\gamma) = x\}$ is a group, called the *isotropy group* of $\mathfrak{G}$ at x.

A fundamental example is given by a G-set X: the associated groupoid is $X \times G$ as a set, with $s(x, g) = x$ and $r(x, g) = xg$ and $(x, g)(xg, h) = (x, gh)$ and $(x, g)^{-1} = (xg, g^{-1})$. One writes this groupoid as $X \rtimes G$ and calls in the *action groupoid* of $X \looparrowleft G$.

Another example is given by the groupoid of germs, see Section 11.7.3. Let $X \rtimes G$ be an action groupoid, and declare $(x, g) \sim (y, h)$ when $x = y$ and there exists an open neighborhood of x on which g and h agree. The set of equivalence classes $\mathfrak{G}$ is called the *groupoid of germs* of $X \rtimes G$.

Definition 11.9.18. Let $\mathfrak{G}$ be a groupoid of germs, and let $\mathfrak{G}_0$ be its space of units. A *bisection* is a subset F of $\mathfrak{G}$ such that $s, r: F \to \mathfrak{G}_0$ are homeomorphisms. Note in particular that bisections are open and closed. Bisections may be composed and inverted, qua subsets of $\mathfrak{G}$. The *full group* $[[\mathfrak{G}]]$ of a groupoid $\mathfrak{G}$ is the group of its bisections.

Note that the topological full group of the groupoid of germs of the action of a group G coincides with the earlier definition of topological full group. It is more convenient to consider the full group of a groupoid of germs, because it is defined only in terms of local homeomorphisms and not of the global action of a group.

Theorem 11.9.19 ([321, Theorem 11]). *Let X be a G-topological space, let $\mathfrak{G}$ denote the groupoid of germs of X, and let $\mathfrak{H}$ be a groupoid of germs of homeomorphisms of X. Assume that*

1. *G is finitely generated;*
2. *At every $x \in X$ the group of germs $\mathfrak{G}_x$ is amenable;*
3. *For every $g \in G$, there are only finitely many $x \in X$ such that $(x, g) \notin \mathfrak{H}$, and then for each of these x, the action of G on xG is extensively amenable;*
4. *The topological full group $[[\mathfrak{H}]]$ is amenable.*

Then G is amenable, and if X is compact, then $[[\mathfrak{G}]]$ is amenable too.

Proof. Let P be the space of "finitely supported sections of $\mathfrak{H}\backslash\mathfrak{G}$": the quotient $\mathfrak{H}\backslash\mathfrak{G}$ is the set of equivalence classes in $\mathfrak{G}$ under $\gamma \sim \delta\gamma$ for all $\gamma \in \mathfrak{G}, \delta \in \mathfrak{H}$, and

$$P = \{\phi: X \to \mathfrak{H}\backslash\mathfrak{G} \text{ finitely supported } \mid s(\phi(x)) \in x\mathfrak{H} \text{ for all } x \in X\}.$$

There is a natural action of G on P, by $(\phi g)(x) = \phi(x) \cdot (t(\phi(x)), g)$.

We claim that P is an amenable G-set. For this, note first that there are only finitely many G-orbits in X at which at element of P can possibly be nontrivial: let S be a finite generating set for G; then for every $s \in S$ there is a finite subset $\Sigma_s \subseteq X$ at which $(x, s) \notin \mathfrak{H}$, so if $(x, g) \notin \mathfrak{H}$ for some $g = s_1 \dots s_n$, then $(xs_1 \dots s_{i-1}, s_i) \notin \mathfrak{H}$ for some i, and therefore $x \in \Sigma_{s_i} G$ for some i.

The G-set P is naturally the direct product, with diagonal action, of its restrictions to the finitely many G-orbits in X at which P can possibly be nontrivial. We therefore restrict ourselves to a single G-orbit $Y \subseteq X$, and the corresponding image $P_Y = \{\phi: Y \to \mathfrak{H}\backslash G\}$ of P.

Let us choose, for every $y,z \in Y$, an element $f_{y,z} \in \mathfrak{G}$ with $s(f_{y,z}) = y$ and $r(f_{y,z}) = z$, taking $f_{z,y} = f_{y,z}^{-1}$ and $f_{y,y} = 1$. Choose also a basepoint $x \in Y$. We have a "twisted" embedding $\iota\colon G \to \mathfrak{G}_x \wr_Y G$ given by $g \mapsto ((y \mapsto f_{x,y}(y,g)f_{yg,x}), g)$. Note that P_Y is isomorphic, qua G-set, to $\bigsqcup_Y \mathfrak{H}_x \backslash \mathfrak{G}_x$ with natural action of $\iota(G)$.

Now since $\mathfrak{G}_x$ is amenable, we have a functor {finite sets} $\to$ {amenable groups} given by $E \mapsto \mathfrak{G}_x^{(E)}$; since X and therefore Y are extensively amenable, Proposition 11.9.5 implies that $\bigsqcup_Y \mathfrak{G}_x$ is an amenable $\mathfrak{G}_x \wr_Y G$-set, and *a fortiori* so is its quotient P.

We next prove that the stabilizer G_ϕ of every $\phi \in P$ is amenable. Let $\{v_1, \dots, v_n\}$ be the support of ϕ, and set $K = G_\phi \cap G_{v_1} \cap \cdots \cap G_{v_n}$. We have a natural homomorphism $K \to \mathfrak{G}_{v_1} \times \cdots \times \mathfrak{G}_{v_n}$ to an amenable group, whose kernel is contained in $[[\mathfrak{H}]]$; so K is amenable. Iteratively applying Proposition 11.2.26 proves that $G_\phi \cap G_{v_1} \cap \cdots \cap G_{v_i}$ is amenable for all $i = n, n-1, \dots, 0$.

We apply once more Proposition 11.2.26 to deduce that G is amenable. Finally, the full group $[[\mathfrak{G}]]$ is the union of groups generated by finite sets of bisections, to which the theorem applies, so $[[\mathfrak{G}]]$ itself is amenable.

Example 11.9.20 ([320, Theorem 6.1]). Consider the "Frankenstein group" $H(\mathbb{A})$ from Theorem 11.7.17. Then the action of $H(\mathbb{A})$ on $\mathbb{R}$ is hereditarily amenable but is not extensively amenable.

Indeed, consider first $H \le H(\mathbb{A})$ and any $x \in \mathbb{R}$, and set $m := \inf(xH) \in \mathbb{R} \cup \{\infty\}$, as at the end of the proof of Theorem 11.7.17. Every element of H'' acts trivially in a neighborhood of m. Consider a sequence (x_n) in $\mathbb{R}$ converging to m; then any cluster point of the sequence of measures (δ_{x_n}) is an H''-invariant mean on xH. Since H/H'' is amenable, there is also an H-invariant mean on xH.

On the other hand, since $H(\mathbb{A})$ is not amenable, there exists a non-amenable finitely generated subgroup $G \le H(\mathbb{A})$, and Theorem 11.9.19 should *not* apply to G with $\mathfrak{H}$ the groupoid of germs of the action of $\mathrm{PSL}_2(\mathbb{R})$ on $\mathbb{R} \cup \{\infty\}$. However, the first condition is satisfied by assumption, the second one is satisfied because the group of germs at $x \in \mathbb{R}$ is at most $\mathrm{Affine}(\mathbb{R}) \times \mathrm{Affine}(\mathbb{R})$, and the fourth one is satisfied because projective transformations are analytic, so their germs coincide with point stabilizers, namely, with $\mathrm{Affine}(\mathbb{R})$. Therefore, the third condition fails, so there exists $x \in \mathbb{R}$ such that the action of G on xG is not extensively amenable.

We now specialize the results to X, a Cantor set, and more precisely the Cantor set of paths in a specific kind of graph:

Definition 11.9.21 ([104]; see [204]). A *Bratteli diagram* is a directed graph $\mathscr{D} = (V, E)$ along with decompositions $V = \bigsqcup_{i \ge 0} V_i$ and $E = \bigsqcup_{i \ge 1} E_i$ in nonempty finite subsets, such that $e^- \in V_{i-1}$ and $e^+ \in V_i$ for all $e \in E_i$. For $v \in V$ we denote by X_v the set of paths starting at V_0 and ending at v; by $X_n = \bigcup_{v \in V_n} X_v$ the set of paths of length n starting at V_0; and by X the set of infinite paths starting at V_0.

If for any $n \gg m$ there exists a path from every vertex in V_m to every vertex in V_n, the diagram is called *simple*.

For $e = (e_1, \dots, e_n) \in X_n$, we denote by eX the set of paths beginning with e; it is a basic open set for the topology on X, which turns X into a compact, totally disconnected space. If $\mathscr{D}$ is simple then X has no isolated points, so it is a Cantor set.

For two paths $e, f \in X_v$ for some $v \in V_n$, we define a homeomorphism $T_{e,f}: eX \to fX$ by

$$T_{e,f}(e, e_{n+1}, \dots) = (f, e_{n+1}, \dots) \text{ for all } e_i \in E_i.$$

Denote by $\mathfrak{T}$ the groupoid of germs of all homeomorphisms of $T_{e,f}$. It coincides with the *tail equivalence groupoid* of $\mathscr{D}$:

$$\mathfrak{T} = \{(e,f) \in X \times X \mid e = (e_i)_{i\geq 1}, f = (f_i)_{i\geq 1}, \text{ and } e_i = f_i \text{ for all } i \text{ large enough}\},$$

with the obvious groupoid structure $s(e,f) = e$, $r(e,f) = f$, and $(e,f) \cdot (f,g) = (e,g)$. The topology on $\mathfrak{T}$ has as basic open sets $\{\text{germs of } T_{e,f}\}$.

Let us describe the topological full group $[[\mathfrak{T}]]$. Every $g \in [[\mathfrak{T}]]$ acts locally like $T_{e,f}$ for some $v \in X_n$ and some $e, f \in X_v$; since X is compact, there exists a common $n(g) \in \mathbb{N}$, assumed minimal, for all these local actions. Write $[[\mathfrak{T}]]_n = \{g \in [[\mathfrak{T}]] \mid n(g) \leq n\}$; then $[[\mathfrak{T}]]_n$ is a group and is in fact isomorphic to $\prod_{v\in V_n} \mathrm{Sym}(X_v) \leq \mathrm{Sym}(X_n)$, since every $g \in [[\mathfrak{T}]]_n$ is uniquely determined by the rule $(e, e_{n+1}, \dots)^g = (e^g, e_{n+1}, \dots)$. It follows that $[[\mathfrak{T}]] = \bigcup_{n\geq 0}[[\mathfrak{T}]]_n$ is a locally finite group.

Definition 11.9.22 ([321]). Consider a homeomorphism $a: X \circlearrowleft$. For $v \in V_n$ denote by $\alpha_a(v)$ the number of paths $e \in X_v$ such that $a\restriction eX$ does not coincide with a transformation of the form $T_{e,f}$ for some $f \in X_v$. The homeomorphism a is called of *bounded type* if $\|a\| := \sup_{v\in V} \alpha_a(v)$ is finite and there are only finitely many points $x \in X$ at which the germ (a, x) does not belong to $\mathfrak{T}$.

It is easy to see that the set of bounded-type self-homeomorphisms of X forms a group. The following result produces a wide variety of amenable groups:

Theorem 11.9.23 ([321, Theorem 16]). *Let $\mathscr{D}$ be a Bratteli diagram, and let G be a group of homeomorphisms of bounded type of X. If the groupoid of germs of G has amenable isotropy groups, then G is amenable.*

Proof. We may assume without loss of generality that G is finitely generated. We apply Theorem 11.9.19 with $\mathfrak{H} = \mathfrak{T}$; since $[[\mathfrak{T}]]$ is locally finite, it is amenable. The only condition to check is that the action of G on X is extensively amenable; we prove that it is recurrent and apply Theorem 11.9.12.

Consider therefore an orbit xG of G and a finite generating set S of G. We will in fact prove that the simple random walk on xG admits a slow constriction and apply Theorem 11.9.10.

The Schreier graph of the orbit $xG \subset X$ is an S-labeled graph. In it, remove all edges $y \to ys$ such that the germ (y, s) does not belong to $\mathfrak{T}$. By assumption, only finitely many edges were removed, so the resulting graph has finitely many connected components; let $P \subseteq xG$ be a choice of one point per connected component. We have covered xG by finitely many $\mathfrak{T}$-orbits. For $e = (e_i)_{i\geq 1} \in P$ consider

$$F_{n,e} = \{(a_1, a_2, \dots, a_n, e_{n+1}, \dots) \in xG \mid a_1 \in E_1, \dots, a_n \in E_n\},$$

and set $F_n = \bigcup_{e \in P} F_{n,e}$. The F_n are finite subsets of xG, and $xG = \bigcup F_n$. For $e \in P, s \in S$, there are at most $\alpha_s(e_n^+)$ paths $f \in F_{n,e}$ with $fs \notin F_{n,e}$; so $\#(F_n S \setminus F_n) \leq \#P \cdot \#S \cdot \max_{s \in S} \|s\|$ are bounded. Furthermore the $F_n S \setminus F_n$ may be assumed disjoint by passing to a subsequence.

Definition 11.9.24 ([204, Definition 6.3.2]). A *Bratteli-Vershik diagram* is a Bratteli diagram $\mathscr{D} = (V, E)$ together with a partial order $\leq$ on E such that e, f are comparable if and only if $e^+ = f^+$. For every $v \in V$, there is an induced linear order on X_v: if $e = (e_1, \dots, e_n), f = (f_1, \dots, f_n) \in X_v$, then $e \leq f$ if and only if $e_i \leq f_i, e_{i+1} = f_{i+1}, \dots, e_n = f_n$ for some $i \in \{1, \dots, n\}$. We let $X^{\max}$ denote those $e = (e_1, \dots) \in X$ such that $(e_1, \dots, e_n)$ is maximal for all $n \in \mathbb{N}$, define $X^{\min}$ similarly, and say $\mathscr{D}$ is *properly ordered* if $\#X^{\max} = \#X^{\min} = 1$.

The *adic transformation* of a properly ordered Bratteli-Vershik diagram $(\mathscr{D}, \leq)$ is the self-homeomorphism $a \colon X \circlearrowleft$ defined as follows. If $e = (e_1, \dots) \in X$ is such that $(e_1, \dots, e_n)$ is not maximal in $X_{e_n^+}$ for some $n \in \mathbb{N}$, then $e^a := (f_1, \dots, f_n, e_{n+1}, \dots)$. Otherwise, e is the unique maximal path in X, and e^a is defined to be the unique minimal path in X.

If $\mathscr{D}$ is simple, then a is a minimal transformation of X. Bratteli-Vershik diagrams encode *all* minimal homeomorphisms of Cantor sets:

Theorem 11.9.25 ([289], see [204, Theorem 6.4.6]). *Every minimal homeomorphism of the Cantor set is topologically conjugate to the adic transformation of a properly ordered simple Bratteli-Vershik diagram.* □

(The idea of the proof is to choose a decreasing sequence $(C_n)_{n \geq 0}$ of clopen sets, shrinking down to a base point $\{x\}$, and to consider the associated "Kakutani-Rokhlin tower": the largest collection of iterated images of C_n under the homeomorphism that are disjoint. These translates of C_n make up the nth level of the Bratteli-Vershik diagram.)

Corollary 11.9.26 ([318]). *Let a be a minimal homeomorphism of a Cantor set X. Then the topological full group $[[\langle a \rangle, X]]$ is amenable.*

Proof. Using Theorem 11.9.25, we may assume a is the adic transformation of a Bratteli-Vershik diagram. It follows directly that $\alpha_a(v) = 1$ for every $v \in V$ and that the germs of a belong to $\mathfrak{T}$ for all points $x \in X \setminus X^{\max}$. No power of a has fixed points so their germs are all trivial.

Here are some typical examples of minimal $\mathbb{Z}$-actions on a Cantor set, to which Corollary 11.9.26 applies to produce amenable groups:

Example 11.9.27. Consider an irrational $\alpha \in (0, 1)$ and the transformation $x \mapsto x + \alpha$ on $\mathbb{R}/\mathbb{Z}$. It is minimal, since $\mathbb{Z} + \mathbb{Z}\alpha$ is dense in $\mathbb{R}$. We can replace $\mathbb{R}/\mathbb{Z}$ by a Cantor set as follows: set

$$X_\alpha := (\mathbb{R} \setminus \mathbb{Z}\alpha \sqcup (\mathbb{Z}\alpha \times \{+, -\}))/\mathbb{Z},$$

namely, replace every point $x \in \mathbb{Z}\alpha \subset \mathbb{R}/\mathbb{Z}$ by a pair $x^{\pm}$. Give X_α the cyclic order induced from the circle and $x^- < x^+$ and its associated topology. Then X_α is a Cantor set, and $x \mapsto x+\alpha$ is a minimal transformation of X_α; see Example 11.9.15.

As another example, consider the substitution $a \mapsto ab, b \mapsto a$ on $\{a,b\}^*$ and let $x \in \{a,b\}^{\mathbb{Z}}$ denote a fixed point of the substitution; for example, with "$\underline{a}$" denoting the position of the 0th letter, $x = \lim(\underline{a}b, ab\underline{a}ab, aba\underline{a}baba, \dots)$. Set $X = \overline{x\mathbb{Z}}$. Then the action of $\mathbb{Z}$ by shift on X is minimal.

In fact, this example coincides with the first one if one takes $\alpha = (\sqrt{5}-1)/2$ the golden ratio and $x = 0^+$, decomposes $X_\alpha = [0^+, \alpha^-] \cup [\alpha^+, 1^-]$, defines $\pi : X_\alpha \to \{a,b\}$ by $\pi(x) = a$ if $x \in [0^+, \alpha^-]$ and $\pi(x) = b$ if $x \in [\alpha^+, 1^-]$, and puts X_α in bijection with X via the map $x \mapsto (n \mapsto \pi(x+n))$.

The encoding of this example as a Bratteli diagram $\mathscr{D}$ is as follows:

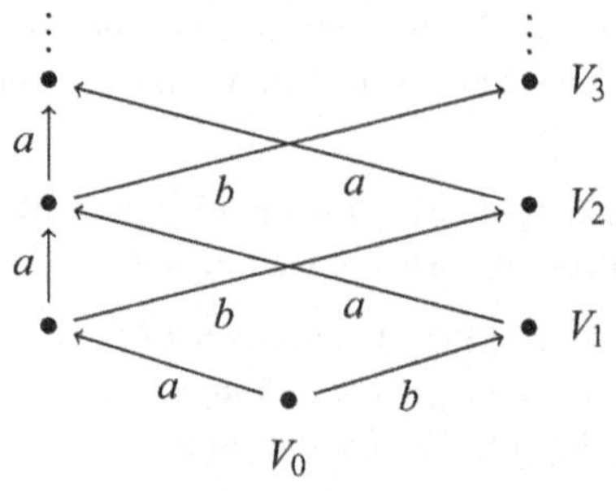

where now a point $x \in X_\alpha$ is encoded by the path in $\mathscr{D}$ with labels $(\pi(\tilde{x}/\alpha^n))_{n\geq 1}$ for the unique representative $\tilde{x}$ of x in $[0,1]$.

We next quote some results from [442] to exhibit some properties of the topological full groups $[[G,X]]$ constructed above.

Definition 11.9.28. Let $\mathfrak{G}$ be a groupoid. A *multisection* of degree d is a collection M of d^2 nonempty, disjoint bisections $\{F_{i,j}\}_{i,j=1,\dots,d}$ of $\mathfrak{G}$ such that $F_{i,i} \subseteq \mathfrak{G}_0$ and $F_{i,j}F_{j,k} = F_{i,k}$ for all $i,j,k \in \{1,\dots,d\}$.

For $\pi \in \mathrm{Sym}(d)$, we denote by M_π the element of $[[\mathfrak{G}]]$ that maps x to xF_{i,i^π} if $x \in F_{i,i}$ and fixes $\mathfrak{G}_0 \setminus \bigcup_{i=1}^d F_{i,i}$, and by $\mathrm{Alt}(M)$ the subgroup $\{M_\pi \mid \pi \in \mathrm{Alt}(d)\}$ of $[[\mathfrak{G}]]$. Finally, we denote by $\mathrm{Alt}(\mathfrak{G})$ the subgroup of $[[\mathfrak{G}]]$ generated by $\mathrm{Alt}(M)$ for all multisections M of $\mathfrak{G}$.

Proposition 11.9.29 ([442, Theorem 4.1]). *Let $\mathfrak{G}$ be a minimal groupoid of germs. Then every nontrivial subgroup of* $[[\mathfrak{G}]]$ *normalized by* $\mathrm{Alt}(\mathfrak{G})$ *contains* $\mathrm{Alt}(\mathfrak{G})$. *In particular,* $\mathrm{Alt}(\mathfrak{G})$ *is simple and is contained in every nontrivial normal subgroup of* $[[\mathfrak{G}]]$. □

(Note that the minimality assumption is always necessary: if $\mathfrak{G}$ does not act minimally, then let $Y \neq \mathfrak{G}_0$ be a closure of an orbit; then there is a natural quotient map $[[\mathfrak{G}]] \to [[\mathfrak{G} \upharpoonright Y]]$, proving that $[[\mathfrak{G}]]$ is not simple.)

We call a groupoid $\mathfrak{G}$ *compactly generated* if there exists a compact subset S of $\mathfrak{G}$ that generates it. This is, for example, the case if $\mathfrak{G}$ is the action groupoid of a finitely generated group G acting on a compact set (in which case one bisection per generator of G suffices to generate $\mathfrak{G}$).

Let $\mathfrak{G}$ be a compactly generated groupoid, say by $S \subseteq \mathfrak{G}$. We call $\mathfrak{G}$ *expansive* if there exists a finite cover $\mathscr{S}$ of S by bisections such that $\bigcup_{n\geq 0} \mathscr{S}^n$ generates the topology on $\mathfrak{G}$; so in particular for every $x \neq y \in \mathfrak{G}_0$, there exists a bisection $F \in \mathscr{S}^n$ with $x \in s(F) \not\ni y$.

Proposition 11.9.30 ([442, Theorem 5.6]). *If $\mathfrak{G}$ is compactly generated and expansive, then* $\mathrm{Alt}(\mathfrak{G})$ *is finitely generated.* □

Example 11.9.31. Consider the $\mathbb{Z}$-action from Example 11.9.27 for α the golden ratio. We claim that the group $G = [[\mathbb{Z}, X]]'$ is infinite, amenable, finitely generated and simple.

Amenability of G was proven in Corollary 11.9.26. Let $\mathfrak{G}$ be the groupoid of the action of $\mathbb{Z} = \langle a \rangle$ on X. It is minimal, so $\mathrm{Alt}(\mathfrak{G})$ is simple by Proposition 11.9.29; and it is easy to check $\mathrm{Alt}(\mathfrak{G}) = [[\mathfrak{G}]]'$. The groupoid $\mathfrak{G}$ is compactly generated, say by $S = X \cup \{(x, xa^{\pm 1}) \mid x \in X\}$. Finally $X \subset \{0,1\}^{\mathbb{Z}}$ is a subshift, so $\mathfrak{G}$ is expansive: the cover of S by X by $\{\{x \in X \mid x_0 = 0\} \cup \{x \in X \mid x_0 = 1\} \cup \{(x, xa) \mid x \in X\} \cup \{(x, xa^{-1}) \mid x \in X\}\}$ generates the topology on X and therefore on $\mathfrak{G}$.

Finally, we end with examples of topological full groups of non-minimal $\mathbb{Z}$-actions and of minimal $\mathbb{Z}^2$-actions which are *not* amenable, showing that Corollary 11.9.26 does not generalize without extra conditions:

Example 11.9.32 (Geodesic flow). Consider a free group F_k and the space X of geodesic maps $a\colon \mathbb{Z} \to F_k$ into the Cayley graph of F_k, namely, of bi-infinite geodesic rays. The $\mathbb{Z}$-action is by shifting: $\sigma(a) = (i \mapsto a_{i+1})$. The space X is a Cantor set and may be identified with $\{a \in \{x_1^{\pm}, \ldots, x_k^{\pm}\}^{\mathbb{Z}} \mid a_i a_{i+1} \neq 1 \text{ for all } i \in \mathbb{Z}\}$. For $a \in X$ and $j \in \{1, \ldots, k\}$, define

$$a \cdot x_j = \begin{cases} \sigma(a) & \text{if } a_0 = x_j, \\ \sigma^{-1}(a) & \text{if } a_{-1} = x_j^{-1}, \\ a & \text{otherwise.} \end{cases}$$

This defines a piecewise-$\mathbb{Z}$ action of F_k on X, which is easily seen to be faithful: for a nontrivial reduced word $w \in F_k$, extend w arbitrarily but non-periodically to a bi-infinite geodesic a containing w at positions $\{0, \ldots, |w| - 1\}$; then $a \cdot w = \sigma^{|w|}(a) \neq a$.

We may modify the example above by letting $C_2 * C_2 * C_2$ rather than F_k act on the space of geodesics of its Cayley graph and then embed that system into a minimal $\mathbb{Z}^2$-action, as follows:

Example 11.9.33 ([212]). Consider the space X of proper colorings of the edges of the standard two-dimensional grid by $\mathscr{A} = \{A, B, C, D, E, F\}$. There is a natural action of $\mathbb{Z}^2$ on X by translations.

To each $a \in \mathscr{A}$ corresponds a continuous involution $a: X \circlearrowleft$, defined as follows. For $\sigma \in X$, if there is an edge between $(0,0)$ and one of its neighbors v with color a, then $\sigma \cdot a := \sigma \cdot v$; otherwise $\sigma \cdot a := \sigma$. These involutions clearly belong to $[[\mathbb{Z}^2, X]]$.

We shall exhibit a minimal nonempty closed $\mathbb{Z}^2$-invariant subset Y of X on which $\mathbb{Z}^2$ acts freely and $H := \langle A, B, C \mid A^2, B^2, C^2 \rangle$ acts faithfully as subgroup of $[[\mathbb{Z}^2, X]]$; since H contains free subgroups, we will have proved that $[[\mathbb{Z}^2, Y]]$ may contain free subgroups (and therefore be non-amenable) for minimal, free $\mathbb{Z}^2$-spaces Y.

We create a specific coloring of the grid, namely, an element $\sigma \in X$, as follows: first, color every horizontal line of the grid alternately with E and F. Enumerate $H = \{w_0, w_1, \dots\}$. For all $x \in \mathbb{N}$, write $x = 2^i x'$ with x' odd, and color the vertical lines $\{x\} \times \mathbb{R}$ and $\{-x\} \times \mathbb{R}$ by the infinite word $(w_i D)^\infty$. Set $Y = \overline{\sigma \mathbb{Z}^2}$.

Every finite patch of $\sigma \restriction S$ repeats infinitely, and moreover there exists $n(S)$ such that every ball of radius $n(S)$ in the grid contains a copy of $\sigma \restriction S$. It follows (see [260]) that Y is minimal, that $\mathbb{Z}^2$ acts freely on Y because σ is aperiodic, and that every $\tau \in Y$ also uniformly contains copies of every patch.

Consider now $w \neq 1 \in \langle A, B, C \rangle$, and let τ be a translate of σ in which wD reads vertically at the origin. Then τw reads D vertically at the origin, so $\tau w \neq \tau$, and therefore w acts nontrivially.

11.10 Cellular Automata and Amenable Algebras

Von Neumann defined[19] *cellular automata* as creatures built out of infinitely many finite-state devices arranged on the nodes of $\mathbb{Z}^2$ or $\mathbb{Z}^3$, each device being capable of interaction with its immediate neighbors. Algebraically, we consider the natural generalization to creatures living on the vertices of a Cayley graph. We shall see that some fundamental properties of the automaton are characterized by amenability of the underlying graph.

Definition 11.10.1. Let G be a group. A finite *cellular automaton* on G is a G-equivariant continuous map $\Theta: \mathscr{A}^G \circlearrowleft$, where $\mathscr{A}$, the *state set*, is a finite set, and G acts on $\mathscr{A}^G$ by left translation: $(xg)(h) = x(gh)$ for $x \in \mathscr{A}^G$ and $g, h \in G$. Elements of $\mathscr{A}^G$ are called *configurations*.

A *linear cellular automaton* is defined similarly, except that $\mathscr{A}$ is rather required to be a finite-dimensional vector space, and Θ is required to be linear.

Note that usually G is infinite; much of the theory holds trivially if G is finite. The map Θ computes the one-step evolution of the automaton; its continuity implies that the evolution of a site depends only on a finite neighborhood, and its G-equivariance implies that all sites evolve with the same rule.

[19] It seems that von Neumann never published his work on cellular automata—see [121] for history of the subject.

Lemma 11.10.2 (Lyndon-Curtis-Hedlund). *A map $\Theta\colon \mathscr{A}^G \circlearrowleft$ is a cellular automaton if and only if there exists a finite subset $S \Subset G$ and a map $\theta\colon \mathscr{A}^S \to \mathscr{A}$ such that*

$$\Theta(x)(g) = \theta(s \mapsto x(gs))$$

for all $x \in \mathscr{A}^G$. The minimal such S is called the memory set *of Θ.*

Proof. Such a map Θ is continuous in the product topology if and only if $\Theta(x)(1)$ depends only on the restriction of x to S for some finite S.

A classical example of cellular automaton is Conway's *Game of Life*. It is defined by $G = \mathbb{Z}^2$ and $\mathscr{A} = \{\mathsf{alive}, \mathsf{dead}\}$, and by the following local rule θ as in Lemma 11.10.2: $S = \{-1, 0, 1\} \times \{-1, 0, 1\}$, and $\theta(x)$ depends only on $x(0,0)$ and on the number of alive cells among its eight neighbors:

$$\theta(x)(0,0) = \begin{cases} \mathsf{alive} & \text{if } x(0,0) \text{ is alive and two or three of its neighbors are alive,} \\ \mathsf{alive} & \text{if } x(0,0) \text{ is dead and exactly three of its neighbors are alive,} \\ \mathsf{dead} & \text{in all other cases, from loneliness or overpopulation.} \end{cases}$$

For example, here is the evolution of a piece of the plane; we represent alive in black and dead in white:

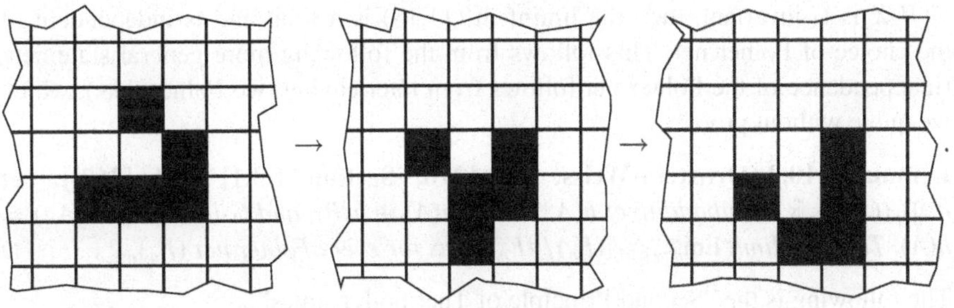

Note that the last configuration is the first one, transformed by $(x, y) \mapsto (1 - y, -x)$, so the pattern moves by a sliding reflection along the $x + y = 0$ direction.

Some properties have been singled out in attempts to understand the global, long-term behavior of cellular automata: a cellular automaton Θ may

have "Gardens of Eden" (GOE) if the map Θ is not surjective, the biblical metaphor expressing the notion of paradise lost forever. Note$\mathscr{A}^G$, so if Θ is not surjective, then there exists a finite subset $F \Subset G$ such that the projection of $\Theta(\mathscr{A}^G)$ to $\mathscr{A}^F$ is not onto;

have "Mutually Erasable Patterns" (MEP) if Θ fails in a strong way to be injective: there are configurations $x \neq y$ which nevertheless agree at all but finitely many places and such that $\Theta(x) = \Theta(y)$. The opposite is sometimes called *pre-injectivity*;

preserve the Bernoulli measure; open sets of the form $\mathscr{O}_{g,q} = \{x \in \mathscr{A}^G \mid x(g) = q\}$ are declared to have measure $\beta(\mathscr{O}_{g,q}) = 1/\#\mathscr{A}$, and one may ask whether $\beta(M) = \beta(\Theta^{-1}(M))$ for every measurable $M \subseteq \mathscr{A}^G$.

For example, it is clear that the Game of Life has Mutually Erasable Patterns, because of the "loneliness" clause:

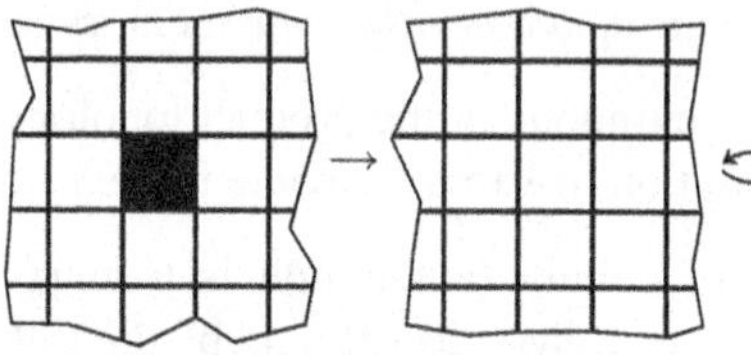

but it is less clear that there are also Gardens of Eden (there are some; the smallest known one is specified by $\#F = 92$ cells).

Before addressing the question of relating the GOE and MEP properties, we introduce one more tool: *entropy*. Assume that the group G is amenable, and let (F_n) be a Følner net in G, which exists by Lemma 11.3.6 and Theorem 11.3.23. For subsets $X \subseteq \mathscr{A}^G$ and $S \subseteq G$, we let $X \restriction S$ denote the projection of X to $\mathscr{A}^S$. We set

$$h(X) = \liminf_n \frac{\log(\#X \restriction F_n)}{\#F_n}. \tag{11.18}$$

If X is G-invariant, then the liminf in (11.18) is a limit and is independent of the choice of Følner net. This follows from the following more general statement (independence of the Følner net follows from interleaving two Følner nets), which we quote without proof:

Lemma 11.10.3 (Ornstein-Weiss, see [276, Section 1.3.1] and [362]). *Let* $h\colon \mathfrak{P}_f(G) \to \mathbb{R}$ *be subadditive:* $h(A \cup B) \leq h(A) + h(B)$, *and G-invariant:* $h(Ag) = h(A)$. *Then the limit* $\lim_{n\to\infty} h(F_n)/\#F_n$ *exists for every Følner net* $(F_n)_{n\in\mathscr{N}}$. □

The following is the "Second Principle of Thermodynamics":

Lemma 11.10.4. *For every cellular automaton Θ and every G-invariant $X \subseteq \mathscr{A}^G$, we have $h(\Theta(X)) \leq h(X)$.*

Proof. Let $S \Subset G$ be a memory set for G. For every finite $F \Subset G$, consider $E \subset G$ such that $ES \subseteq F$; then $\Theta(x)\restriction E$ depends only on $x\restriction F$. Therefore, $\#(\Theta(X)\restriction E) \leq \#(X\restriction F)$, so $\#(\Theta(X)\restriction F) \leq \#(X\restriction F)\#\mathscr{A}^{\#F-\#E}$. Take now $F = F_nS$ and $E = F_n$ for a net of Følner sets, and apply the definition from (11.18).

Finally, given a measure ν on $\mathscr{A}^G$, we may define a *measured entropy* as follows: for $S \subseteq G$ and $y \in \mathscr{A}^S$, denote by $\mathscr{O}_y$ the open set $\{x \in \mathscr{A}^G \mid x\downharpoonleft S = y\downharpoonleft S\}$; and for $X \subseteq \mathscr{A}^G$ set

$$h_\nu(X) = \liminf \frac{-\sum_{y\in X\downharpoonleft F_n} \nu(\mathscr{O}_y)\log \nu(\mathscr{O}_y)}{\#F_n}.$$

Note that $\beta(\mathscr{O}_y) = 1/\#\mathscr{A}^{\#S}$ if $y \in \mathscr{A}^S$, so the measured entropy coincides with (11.18) if $\nu = \beta$.

We are ready to state the main result, called the "Garden of Eden theorem." It was first proven for $G = \mathbb{Z}^d$ by Moore [423, the (1) $\Rightarrow$ (2) direction], Myhill [433, the (2) $\Rightarrow$ (1) direction], and Hedlund [288, the (1) $\Leftrightarrow$ (3) equivalence]:

Theorem 11.10.5 ([138, 415]). *Let G be an amenable group, and let Θ be a cellular automaton. Then the following are equivalent:*

1. *Θ has Gardens of Eden;*
2. *Θ has Mutually Erasable Patterns;*
3. *Θ does not preserve Bernoulli measure β;*
4. *$h(\Theta(\mathscr{A}^G)) < \log\#\mathscr{A}$.*

Remark 11.10.6. The same theorem holds for linear cellular automata (except that I do not know an analogue of Bernoulli measure), with the entropy replaced in the last statement by *mean dimension*:

$$\operatorname{mdim}(X) = \liminf_n \frac{\dim(\#X\downharpoonleft F_n)}{\#F_n}.$$

Proof. Throughout the proof, we let S denote the memory set of Θ.

(1) $\Rightarrow$ (4) If there exists a GOE, then there exists $F \Subset G$ with $\Theta(\mathscr{A}^G)\downharpoonleft F \neq \mathscr{A}^F$, so

$$h(\Theta(\mathscr{A}^F)) \le \frac{\log\#\Theta(\mathscr{A}^G)\downharpoonleft F}{\#F} < \log\#\mathscr{A}.$$

(4) $\Rightarrow$ (1) If $h(\Theta(\mathscr{A}^G)) < \log\#\mathscr{A}$, then there exists $F \Subset G$ with $\Theta(\mathscr{A}^G)\downharpoonleft F \neq \mathscr{A}^F$, and a GOE exists in $\mathscr{A}^F \setminus \Theta(\mathscr{A}^G)\downharpoonleft F$.

(2) $\Rightarrow$ (4) If $y \neq z$ are MEP, which differ on F and agree elsewhere, set $E = FS$ and let $T \subset G$ be maximal such that $Et_1 \cap Et_2 = \emptyset$ for all $t_1 \neq t_2 \in T$; note that T intersects every translate of $E^{-1}E$. Define

$$Z = \{x \in \mathscr{A}^G \mid x\downharpoonleft Et \neq y\downharpoonleft Et \text{ for all } t \in T\},$$

and compute $h(\Theta(\mathscr{A}^G)) = h(\Theta(Z)) \le h(Z) < \log\#\mathscr{A}$; the first equality follows since given in $x \in \Theta(\mathscr{A}^G)$, say $x = \Theta(w)$, one may replace in w every occurrence of $y\restriction Et$ by $z\restriction Et$ so as to obtain a $y\restriction Et$-free configuration, which therefore belongs to Z and has the same image as x under Θ; the second inequality follows from Lemma 11.10.4; and the last inequality because there are forbidden patterns $y\restriction Et$ in Z, with "density" at least $1/\#(E^{-1}E)$.

(4) $\Rightarrow$ (2) If $h(\Theta(\mathscr{A}^G)) < \log\#\mathscr{A}$, there exists F_n with $\log\#(\Theta(\mathscr{A}^G)\restriction F_nS)/\#F_n < \log\#A$, because $\#F_nS$ may be made arbitrarily close to $\#F_n$ for n large enough. Therefore, by the pigeonhole principle, there exist $y \neq z \in \mathscr{A}^G$ with $y\restriction(G\setminus F_n) = z\restriction(G\setminus F_n)$ and $\Theta(y) = \Theta(z)$.

(1) $\Rightarrow$ (3) This is always true: if Θ has GOE, then there exists a nonempty open set $\mathscr{U}$ in $\mathscr{A}^G\setminus\Theta(\mathscr{A}^G)$; then $\beta(\mathscr{U}) \neq 0$ while $\beta(\Theta^{-1}(\mathscr{U})) = 0$.

(3) $\Rightarrow$ (1) Define

$$K = \{\nu \text{ probability measure on } \mathscr{A}^G \mid \beta = \Theta_*\nu\}.$$

Note that K is convex and compact, no admits a G-fixed point because G is amenable. Consider $\nu \in K^G$. Then $\phi\colon(\mathscr{A}^G,\nu) \to (\mathscr{A}^G,\beta)$ is a factor map because Θ is onto, so $h_\nu(\mathscr{A}^G) \geq h_\beta(\mathscr{A}^G)$. However, β is the unique measure of maximal entropy,[20] so $\nu = \beta$ and therefore $\beta = \Theta_*\beta$.

It turns out that Theorem 11.10.5 is essentially optimal, and yields characterizations of amenable groups:

Theorem 11.10.7 ([41, 44]). *Let G be a non-amenable group. Then there exist*

1. *cellular automata (ad lib linear) that admit Mutually Erasable Patterns but no Gardens of Eden;*
2. *cellular automata (ad lib linear) that admit Gardens of Eden but no Mutually Erasable Patterns;*
3. *cellular automata that do not preserve Bernoulli measure but have no Gardens of Eden.*

In fact, we shall prove Theorem 11.10.7 for finite fields, answering at the same time the classical and linear questions. Let Θ be a linear cellular automaton; then $\mathscr{A} = \Bbbk^n$ for some field $\Bbbk$ and some integer n, and there exists an $n\times n$ matrix $\mathbf{M}$ over $\Bbbk G$ such that $\Theta(x) = x\mathbf{M}$ for all $x \in \mathscr{A}^G$. Conversely, every such matrix defines a linear cellular automaton.

The ring $\Bbbk G$ admits an anti-involution $*$, defined on its basis G by $g^* = g^{-1}$ and extended by linearity. This involution extends to an anti-involution on square matrices by $(\mathbf{M}^*)_{i,j} = (\mathbf{M}_{j,i})^*$, and $\mathbf{M}^*$ is called the *adjoint* of $\mathbf{M}$.

We put on $\mathscr{A}$ the natural scalar product $\langle x,y\rangle = \sum_{i=1}^n x_iy_i$. Consider the vector space $\mathscr{A}G = \bigoplus_{g\in G}\mathscr{A}$. Then $\mathscr{A}^G$ may be naturally identified with the dual of $\mathscr{A}G$, under the nondegenerate pairing $\langle x,y\rangle = \sum_{g\in G}\langle x(g),y(g)\rangle$ for $x \in \mathscr{A}G$ and $y \in \mathscr{A}^G$.

[20]One says that the G-action is *intrinsically ergodic*.

Exercise 11.10.8 (*). Prove that $\mathbf{M}^*$ is the adjoint with respect to this pairing; namely, $\langle x\mathbf{M}, y\rangle = \langle x, y\mathbf{M}^*\rangle$ for all $x \in \mathscr{A}G, y \in \mathscr{A}^G$.

We put a topology on $\mathscr{A}^G$ by declaring that, for every finite $S \Subset G$ and every vector space $V \le \mathscr{A}^S$, the subset $\{x \in \mathscr{A}^G \mid x\restriction S \in V\}$ is closed. With this topology, $\mathscr{A}^G$ is compact (but not Hausdorff). Nevertheless,

Lemma 11.10.9. *If Θ is a cellular automaton, then $\Theta(\mathscr{A}^G)$ is closed.*

Proof. Let S be the memory of Θ. Consider y in the closure of $\Theta(\mathscr{A}^G)$. Then for every $F \Subset G$, the affine space $L_F = \{x \in \mathscr{A}^{FS} \mid \Theta(x)\restriction F = y\restriction F\}$ is finite-dimensional and nonempty, and if $F \subseteq F'$ then $L_{F'}\restriction FS \subseteq L_F$; so $\{L_{F'}\restriction FS \mid F' \supseteq F\}$ is a nested sequence of nonempty affine spaces and in particular stabilizes at a nonempty affine space J_F. We still have restriction maps $J_{F'} \to J_F$ for all $F \subseteq F'$, which are easily seen to be surjective. Then $\varprojlim_{F\Subset G} J_F$ is nonempty and contains all preimages of y.

The following proposition extends to the infinite-dimensional setting the classical statement that the image of a matrix is the orthogonal of the nullspace of its transpose:

Proposition 11.10.10 ([565]). *Let $\mathbf{M}$ be an $n \times n$ matrix over $\Bbbk G$, let $\mathbf{M}^*$ be its adjoint, and set $\mathscr{A} = \Bbbk^n$. Then right multiplication by $\mathbf{M}$ is injective on $\mathscr{A}G$ if and only if right multiplication by $\mathbf{M}^*$ is surjective on $\mathscr{A}^G$.*

Proof. Assume first that right multiplication by $\mathbf{M}$ is not injective, and consider a nontrivial element $c \in \mathscr{A}G$ with $c\mathbf{M} = 0$. We claim that for every $y \in (\mathscr{A}^G)\mathbf{M}^*$, we have $\langle c, y\rangle = 0$. Say $y = z\mathbf{M}^*$; then the claim follows from the computation

$$\langle c, y\rangle = \langle c, z\mathbf{M}^*\rangle = \langle c\mathbf{M}, z\rangle = \langle 0, z\rangle = 0.$$

Since $\langle -, -\rangle$ is nondegenerate, this implies that y cannot range over all of $\mathscr{A}^G$, so right multiplication by $\mathbf{M}^*$ is not surjective.

Conversely, suppose that right multiplication by $\mathbf{M}$ is not surjective. Since $\mathscr{A}^G\mathbf{M}$ is closed, there exists an open set in its complement; so there exists a finite subset $S \Subset G$ and a proper subspace $V \lneq \mathscr{A}^S$ such that, for every $c \in \mathscr{A}^G\mathbf{M}$, its projection $c\restriction S$ belongs to V. Since $\mathscr{A}^S$ is finite-dimensional, there exists a linear form y on $\mathscr{A}^S$ that vanishes on V. Note that y, qua element of $(\mathscr{A}^S)^*$, is canonically identified with an element of $\mathscr{A}^S$ and therefore with an element of $\mathscr{A}G$. We claim $y\mathbf{M}^* = 0$, proving that right multiplication by $\mathbf{M}^*$ is not injective. This follows from the following computation: consider an arbitrary $c \in \mathscr{A}^G$. Then

$$\langle y\mathbf{M}^*, c\rangle = \langle y, c\mathbf{M}\rangle = 0.$$

Since $\langle -, -\rangle$ is nondegenerate and $c \in V^G$ is arbitrary, this forces $y\mathbf{M}^* = 0$.

Before embarking on the main step of the proof of Theorem 11.10.7, we give a simple example of a cellular automaton that is pre-injective but not surjective:

Example 11.10.11 (Muller, see [396, page 55]). Consider the free product of cyclic groups $G = \langle a, b, c | a^2, b^2, c^2 \rangle$. Fix a field $\Bbbk$, and set $\mathscr{A} := \Bbbk^2$. Define the linear cellular automaton $\Theta\colon \mathscr{A}^G \circlearrowleft$ by

$$\Theta(x) = x \cdot \begin{pmatrix} a+b & 0 \\ b+c & 0 \end{pmatrix}.$$

It is obvious that Θ is not surjective: its image is $(\Bbbk \times 0)^G$. To show that it is pre-injective, consider x a nonzero configuration with finite support, and let $F \Subset G$ denote its support. Let $f \in F$ be an element of maximal length; then at least two among fa, fb, fc will be reached precisely once as products of the form $F \cdot \{a, b, c\}$. Write $x(f) = (\alpha, \beta) \neq (0, 0)$; then at least two among the equations

$$\Theta(x)(fa) = \alpha, \qquad \Theta(x)(fb) = \alpha + \beta, \qquad \Theta(x)(fc) = \beta$$

hold, and this is enough to force $\Theta(x) \neq 0$.

In the general case of a non-amenable group $G = \langle S \rangle$, we may not claim that there exist two elements reached exactly once from an arbitrary finite set F under right S-multiplication; but we shall see that there exists "many" elements reached "not too many" times, in the sense that there exists $f \in F$ with $\sum_{s \in S} 1/\#\{t \in S \mid fs \in Ft\} > 1$; and this will suffice to construct a pre-injective, non-surjective cellular automaton. We phrase our result as the following algebraic criterion, which may be of independent interest:

Theorem 11.10.12. *Let $X \curvearrowleft G$ be a right G-set, and let $\Bbbk$ be a field. Then X is non-amenable if and only if there exists $n \in \mathbb{N}$ and an $n \times (n-1)$ matrix with entries in $\Bbbk G$ that gives an injective map $(\Bbbk X)^n \hookrightarrow (\Bbbk X)^{n-1}$.*

Obviously if there exists such an n, then every larger n is also suitable; and we shall see in the proof that n depends only on $X \curvearrowleft G$ and the cardinality of $\Bbbk$. We begin the proof by a combinatorial lemma:

Lemma 11.10.13. *Let n be an integer. Then there exists a set Y and a family of subsets $Z_1, \dots, Z_n$ of Y such that, for all $I \subseteq \{1, \dots, n\}$ and all $i \in I$, we have*

$$\#\Big(Z_i \setminus \bigcup_{j \in I \setminus \{i\}} Z_j\Big) \geq \frac{\#Y}{(1 + \log n)\#I}. \tag{11.19}$$

Furthermore, if $n \geq 2$ then we may require $Z_1 \cup \cdots \cup Z_n \neq Y$.

Proof. We denote by $\mathrm{Sym}(n)$ the symmetric group on n letters. Define

$$Y := \frac{\{1, \dots, n\} \times \mathrm{Sym}(n)}{(i, \sigma) \sim (j, \sigma) \text{ if } i \text{ and } j \text{ belong to the same cycle of } \sigma};$$

in other words, Y is the set of cycles of elements of $\mathrm{Sym}(n)$. Let Z_i be the natural image of $\{i\} \times \mathrm{Sym}(n)$ in the quotient Y.

First, there are $(i-1)!$ cycles of length i in $\mathrm{Sym}(i)$, given by all cyclic orderings of $\{1,\dots,i\}$; so there are $\binom{n}{i}(i-1)!$ cycles of length i in $\mathrm{Sym}(n)$, and they can be completed in $(n-i)!$ ways to a permutation of $\mathrm{Sym}(n)$; so

$$\#Y = \sum_{i=1}^{n} \binom{n}{i}(i-1)!(n-i)! = \sum_{i=1}^{n} \frac{n!}{i} \le (1+\log n)n! \tag{11.20}$$

since $1 + 1/2 + \cdots + 1/n \le 1 + \log n$ for all n.

Next, consider $I \subseteq \{1,\dots,n\}$ and $i \in I$, and set $Z_{i,I} := Z_i \setminus \bigcup_{j\in I\setminus\{i\}} Z_j$. Then $Z_{i,I} = \{(i,\sigma) \mid (i,\sigma) \nsim (j,\sigma) \text{ for all } j \in I \setminus \{i\}\}$. Summing over all possibilities for the length-$(j+1)$ cycle $(i,t_1,\dots,t_j)$ of σ intersecting I in $\{i\}$, we get

$$\begin{aligned}\#Z_{i,I} &= \sum_{j=0}^{n-\#I} \binom{n-\#I}{j} j!(n-j-1)!\\ &= \sum_{k:=n-j=\#I}^{n} (n-\#I)!(\#I-1)!\binom{k-1}{k-\#I} \\ &= (n-\#I)!(\#I-1)!\binom{n}{n-\#I} = \frac{n!}{\#I}.\end{aligned} \tag{11.21}$$

Combining (11.20) and (11.21), we get

$$\#Z_{i,I} = \frac{n!}{\#I} = \frac{(1+\log n)n!}{(1+\log n)\#I} \ge \frac{\#Y}{(1+\log n)\#I}.$$

Finally, if $n \ge 2$ then (11.19) may be improved to $\#Y \le (0.9 + \log n)n!$; for even larger n one could get to $\#Y \le (0.57721\cdots + \log n)n!$. Since clearly $\#Y/(0.9+\log n) \ge (\#Y+1)/(1+\log n)$, one may simply replace Y by $Y \sqcup \{\cdot\}$.

Proof (Proof of Theorem 11.10.12). Assume first that X is amenable, and let $\mathbf{M}\colon (\Bbbk X)^n \to (\Bbbk X)^{n-1}$ be an injective map. Let $S \Subset G$ be such that $\mathbf{M}$'s entries belong to $\Bbbk S$. Since X is amenable, there is $F \Subset X$ with $\#(FS) < \frac{n}{n-1}\#F$. Consider then the restriction $\mathbf{M}\colon (\Bbbk F)^n \hookrightarrow (\Bbbk FS)^{n-1}$. Since the dimension of the range is greater than that of the source, it has a nontrivial kernel, so $\mathbf{M}$ is not injective.

Assume now that X is not amenable, so there exists by Theorem 11.3.23 a finite subset $S_0 \subset G$ and $\epsilon > 0$ with $\#(FS_0) \ge (1+\epsilon)\#F$ for all finite $F \subset X$. We then have $\#(FS_0^k) \ge (1+\epsilon)^k\#F$ for all $k \in \mathbb{N}$. Let k be large enough so that $(1+\epsilon)^k > 1 + k\log\#S_0$, and set $S := S_0^k$ and $n := \#S$. We will seek $\mathbf{M}$ supported in $\mathbb{K}S$. We have

$$\begin{aligned}\#(FS) &\ge (1+\epsilon)^k\#F > (1+k\log\#S_0)\#F\\ &\ge (1+\log n)\#F \text{ for all finite } F \subset G.\end{aligned} \tag{11.22}$$

Apply Lemma 11.10.13 to this n, and identify $\{1,\dots,n\}$ with S to obtain a set Y and subsets Z_s for all $s\in S$. We have $\bigcup_{s\in S} Z_s \subsetneq Y$ and

$$\#\Big(Z_s \setminus \bigcup_{t\in T\setminus\{s\}} Z_t\Big) \geq \frac{\#Y}{(1+\log n)\#T} \text{ for all } s\in T\subseteq S.$$

In case $\Bbbk$ is a finite field, we shall perhaps have to replace it by a finite extension $\mathbb{K}$; this will produce an injective map $(\mathbb{K}X)^n \to (\mathbb{K}X)^{n-1}$ and therefore by restriction of scalars an injective map $(\Bbbk X)^{n\dim_\Bbbk \mathbb{K}} \to (\Bbbk X)^{(n-1)\dim_\Bbbk \mathbb{K}} \to (\Bbbk X)^{n\dim_\Bbbk \mathbb{K}-1}$.

We shall specify soon how large the finite extension $\mathbb{K}$ of $\Bbbk$ should be. Under that future assumption, set $\mathscr{A} := \mathbb{K}Y$. For each $s\in S$, we shall construct a linear map $\alpha_s\colon \mathscr{A} \to \mathbb{K}Z_s \subset \mathscr{A}$; for this, we introduce the following notation: for $T\ni s$ denote by $\alpha_{s,T}\colon \mathscr{A} \to \mathbb{K}Z_{s,T}$ the composition of α_s with the coördinate projection $\mathscr{A} \to \mathbb{K}Z_{s,T}$. We wish to impose the condition that, whenever $\{T_s \mid s\in S\}$ is a family of subsets of S with $\sum_{s\in S}\#Z_{s,T_s} \geq \#Y$, we have

$$\bigcap_{s\in S}\ker(\alpha_{s,T_s}) = 0. \tag{11.23}$$

As a first step, we treat each α_s as a $\#Z_s\times\#Y$ matrix with variables as coëfficients, by considering only its rows indexed by $Z_s\subset Y$; and we treat each $\alpha_{s,T}$ as a $\#Z_{s,T}\times\#Y$ submatrix of α_s. The space of all $(\alpha_s)_{s\in S}$ therefore consists of $N := \#Y\sum_{s\in S}\#Z_s$ variables, so it is an affine space of dimension N.

Equation (11.23) amounts to the condition on these variables that all matrices obtained by stacking vertically a collection of α_{s,T_s}'s have full rank as soon as $\sum_{s\in S}\#Z_{s,T_s} \geq \#Y$. The complement of these conditions is an algebraic subvariety of $\mathbb{K}^N$, given by a finite union of hypersurfaces of the form "$\det(\cdots) = 0$." Crucially, the equations of these hypersurfaces are defined over $\mathbb{Z}$ and in particular are independent of the field $\mathbb{K}$. Therefore, as soon as $\mathbb{K}$ is large enough, there exist points that belong to none of these hypersurfaces; and any such point gives a solution to (11.23).

Define now the matrix $\mathbf{M}$ with coëfficients in $\mathbb{K}G$ by

$$\mathbf{M} = \sum_{s\in S}\alpha_s s. \tag{11.24}$$

It maps $\mathbb{K}X^n = \mathbb{K}Y\otimes\mathbb{K}X$ to $\mathbb{K}X^{n-1} = \mathbb{K}(Y\setminus\{\cdot\})\otimes\mathbb{K}X$ as required, since we assumed $\bigcup_{s\in S}Z_s \subseteq Y\setminus\{\cdot\}$. To show that $\mathbf{M}$ is injective, consider $u\in\mathbb{K}X^n$ nontrivial, and let $\emptyset\neq F\Subset X$ denote its support. Define $\rho\colon FS\to(0,1]$ by $\rho(x) := 1/\#\{s\in S \mid x\in Fs\}$. Now

$$\sum_{f\in F}\Big(\sum_{s\in S}\rho(fs)\Big) = \sum_{x\in FS}\sum_{s\in S: x\in Fs}\rho(x) = \sum_{x\in FS}1 = \#(FS),$$

so there exists $f \in F$ with $\sum_{s\in S} \rho(fs) \geq \#(FS)/\#F \geq 1 + \log n$ by (11.22). For every $s \in S$, set $T_s := \{t \in S \mid fs \in Ft\}$, so $\#T_s = 1/\rho(fs)$. We obtain

$$\sum_{s\in S} \#Z_{s,T_s} \geq \sum_{s\in S} \frac{\#Y}{(1+\log n)\#T_s} \text{ by Lemma 11.10.13}$$
$$= \sum_{s\in S} \frac{\#Y\rho(fs)}{1+\log n} \geq \#Y,$$

so by (11.23) the map $\mathscr{A} \ni a \mapsto (\alpha_{s,T_s}(a))_{s\in S}$ is injective. Set $v := u\mathbf{M}$. Since by assumption $u(f) \neq 0$, we get $(\alpha_{s,T_s}(u(f)))_{s\in S} \neq 0$, namely, there exists $s \in S$ with $\alpha_{s,T_s}(u(f)) \neq 0$. Now $v(fs) \downharpoonleft Z_{s,T_s} = \alpha_{s,T_s}(u(f))$ by (11.24), so $v \neq 0$ and we have proven that $\mathbf{M}$ is injective.

Proof (Proof of Theorem 11.10.7). We start by (2). Apply Theorem 11.10.12 to $\Bbbk = \mathbb{F}_2$, and let $\mathbf{M}$ be the $n \times (n-1)$ resulting matrix over $\Bbbk G$. Set $\mathscr{A} = \Bbbk^n$, and extend $\mathbf{M}$ to an $n \times n$ matrix by adding a column on 0's to its right. Then $\Theta \colon \mathscr{A}^G \circlearrowleft$ given by $\Theta(x) = x\mathbf{M}$ is a G-equivariant endomorphism of $\mathscr{A}^G$, is pre-injective because $\mathbf{M}$ is injective on $\mathscr{A}G$, and is not surjective because no configuration in its image has a nontrivial last coördinate.

Right multiplication by $\mathbf{M}^*$ on $\mathscr{A}^G$ is surjective and not pre-injective by Proposition 11.10.10, so this answers (1).

Finally, let $y \in \mathscr{A}^S$, for some $S \Subset G$, be such that $\mathscr{O}_y$ is a Garden of Eden for $\mathbf{M}$. Then $\mathscr{O}_y\mathbf{M}^* = 0$, so $\mathbf{M}^*$ does not preserve Bernoulli measure, answering (3).

11.10.1 Goldie Rings

We saw in the last section that linear cellular automata are closely related to group rings. We give now a characterization of amenability of groups in terms of ring theory. We recommend [471] as a reference for group rings.

Definition 11.10.14. Let R be a ring. It is *semiprime* if $aRa \neq 0$ whenever $a \in R \setminus \{0\}$. An element $a \in R$ is *regular* if $xay \neq 0$ whenever $x, y \in R \setminus \{0\}$, and the ring R is a *domain* if $xy \neq 0$ whenever $x, y \in R \setminus \{0\}$. The *right annihilator* of $a \in R$ is $\{x \in R \mid ax = 0\}$ and is a right ideal in R.

The ring R is *Goldie* if (1) there is no infinite ascending chain of right annihilators in R and (2) there is no infinite direct sum of nonzero right ideals in R.

Clearly R is a domain if and only if all its nonzero elements are regular; annihilators of regular elements are trivial; and all domains are semiprime.

These definitions may be difficult to digest, but they have strong consequences for the structure of R, see [169] and Goldie's theorem below. In terms of their ideal structure, the simplest rings are *skew fields*, in which all nonzero elements are invertible. Next best are *Artinian rings*, which do not admit infinite descending chains of ideals. Finitely generated modules over Artinian rings have a well-defined notion of dimension, namely, the maximal length of a composition series.

Øystein Ore studied in [461] when a ring R may be imbedded in a ring in which all regular elements of R become invertible. Let us denote by R^* the set of regular elements in R. A naive attempt is to consider expressions of the form as^{-1} with $a, s \in R$ and s regular; then to multiply them, one must rewrite $as^{-1}bt^{-1} = ab'(s')^{-1}t^{-1} = (ab')(ts')^{-1}$, and to add them, one must rewrite $as^{-1} + bt^{-1} = (at' + bs')(st')^{-1}$. In all cases, it is sufficient that R satisfy the following property, called *Ore's condition:*

$$\text{for all } a, s \in R \text{ with } s \text{ regular, there exist } b, t \in R \text{ with } t \text{ regular and } sb = at,$$

namely, every pair of elements a, s admits a common "right multiple" $at = sb$. The ring

$$R(R^*)^{-1} := \{as^{-1} \mid a \in R, s \in R^*\}/\langle as^{-1} = at(st)^{-1} \text{ for all } a \in R, s, t \in R^*\rangle$$

is called R's *classical ring of fractions*. It naturally contains R as the subring $\{a1^{-1}\}$. If R is a domain, then $R(R^*)^{-1}$ is a skew field.

Theorem 11.10.15 (Goldie [253]). *Let R be a semiprime Goldie ring. Then R satisfies Ore's condition, and its classical ring of fractions is Artinian.* □

Let $R \subseteq S$ be a subring of a ring. The ring S is called *flat* over R if for every exact sequence $0 \to A \to B \to C \to 0$ of R-modules the corresponding sequence $0 \to A \otimes_R S \to B \otimes_R S \to C \otimes_R S \to 0$ of S-modules is exact.

Exercise 11.10.16 (**). For R a domain, show that $S := R(R^*)^{-1}$ is flat.

Hint: there is an equational criterion for flatness—S is flat if and only if every R-linear relation $\sum r_i x_i = 0$, with $r_i \in R$ and $x_i \in S$, "follows from linear relations in R," in the following sense: the equation in matrix form $\mathbf{r}^T\mathbf{x} = 0$, with $\mathbf{r} \in R^n$ and $\mathbf{x} \in S^n$, implies equations $\mathbf{r}^T\mathbf{B} = 0$ and $\mathbf{x} = \mathbf{B}\mathbf{y}$ for some $n \times m$ matrix $\mathbf{B}$ over R and some $\mathbf{y} \in S^m$; see [368, 4.24(2)].

Using Ore's condition, apply this criterion by expressing in a R-linear relation $\sum r_i x_i = 0$ every $x_i = a_i s^{-1}$ for $a_i \in R$ and a common denominator $s \in R^*$.

Let now G be a group, let $\Bbbk$ be a field, and consider the group ring $\Bbbk G$. It is the $\Bbbk$-vector space with basis G, and multiplication extended multilinearly from the multiplication in G. Is is well understood when the group ring $\Bbbk G$ is semiprime:

Theorem 11.10.17 (Passman, see [471, Theorems 2.12 and 2.13]). *If $\Bbbk$ has characteristic 0, then $\Bbbk G$ is semiprime for all G. If $\Bbbk$ has characteristic $p > 0$, then $\Bbbk G$ is semiprime if and only if G has no finite normal subgroup of order divisible by p.* □

Exercise 11.10.18 (*). If G is non-amenable, then it has a non-amenable quotient $\overline{G}$ whose group ring $\Bbbk\overline{G}$ is semiprime for all $\Bbbk$.

Theorem 11.10.19 (Tamari [558], Kielak [44], Kropholler). *Let $\Bbbk$ be a field and let G be group such that $\Bbbk G$ is Goldie and semiprime. Then G is amenable.*

Furthermore, if $\Bbbk G$ *is a domain,*[21] *then* $\Bbbk G$ *satisfies Ore's condition if and only if* G *is amenable.*

Proof. Assume first that G is amenable and that $\Bbbk G$ is a domain, and let $a, s \in \Bbbk G$ be given. Let $S \Subset G$ contain the supports of a and s. Since G is amenable, there exists $F \Subset G$ with $\#(FS) < 2\#F$, by Følner's Theorem 11.3.23. Consider $b, t \in \Bbbk G$ as unknowns in $\Bbbk F$. The equation $sb = at$ which they must satisfy is linear in their coëfficients, and there are more variables ($2\#F$) than constraints ($\#(FS)$), so there exists a nontrivial solution, in which $t \neq 0$ if $s \neq 0$; so Ore's condition is satisfied.

Assume next that G is not amenable. By Theorem 11.10.12, there exists an $n \times (n-1)$ matrix $\mathbf{M}$ over $\Bbbk G$ such that multiplication by $\mathbf{M}$ is an injective map $(\Bbbk G)^n \to (\Bbbk G)^{n-1}$, namely, we have an exact sequence of free $\Bbbk G$-modules

$$0 \longrightarrow (\Bbbk G)^{dn} \longrightarrow (\Bbbk G)^{d(n-1)}. \tag{11.25}$$

Suppose now for contradiction that $\Bbbk G$ is a semiprime Goldie ring, and let S be its classical ring of fractions, which exists and is Artinian by Theorem 11.10.15. By Exercise 11.10.16, the ring S is flat over $\Bbbk$, so tensoring (11.25) with S we obtain an exact sequence

$$0 \longrightarrow S^{dn} \longrightarrow S^{d(n-1)}$$

which is impossible for reasons of composition length.

11.10.2 Amenable Banach Algebras

We concentrated, in this text, on amenability of groups. The topic of amenability of associative algebras has been developed in various directions; although the different definitions are in general inequivalent, we stress here the connections between amenability of a group (or a set) and that of an associated algebra (or module). In this section, all algebras are over $\mathbb{C}$, including the $\ell^p(X)$.

Let $\mathscr{A}$ be a Banach algebra, and let V be a Banach bimodule: a Banach space V endowed with commuting actions $V \widehat{\otimes} \mathscr{A} \to V$ and $\mathscr{A} \widehat{\otimes} V \to V$. Recall that a *derivation* is a map $\delta \colon \mathscr{A} \to V$ satisfying $\delta(ab) = a\delta(b) + \delta(a)b$, and a derivation δ is *inner* if it is of the form $\delta(a) = av - va$ for some $v \in V$. The dual V^* of a Banach bimodule is again a Banach bimodule, for the adjoint actions $(g \cdot \phi \cdot h)(x) = \phi(h^{-1}xg^{-1})$.

Definition 11.10.20. The Banach $\mathscr{A}$-module V is *amenable* if all bounded derivations of $\mathscr{A}$ into V are inner. More pedantically: the Hochschild cohomology group $H^1(\mathscr{A}, V)$ is trivial.

[21] Conjecturally (see [327] and [328, Problem 6]), $\Bbbk G$ is a domain if and only if G is torsion-free.

The algebra $\mathscr{A}$ itself is called amenable if all $H^1(\mathscr{A}, V^*) = 0$ for all Banach bimodules V.

Exercise 11.10.21 (, see Johnson [313, Proposition 5.1]).** Prove that the tensor product of amenable Banach algebras is amenable.

This definition seems quite distinct from everything we have seen in the context of groups and G-sets; yet it applies to the Banach algebra $\ell^1(G)$ introduced in (11.15). For a set X, denote by $\ell^\infty(X)_0^*$ those functionals $\Phi\colon \ell^\infty(X) \to \mathbb{C}$ such that $\Phi(\mathbb{1}_X) = 0$.

Theorem 11.10.22 ([313, Theorem 2.5]). *Let G be a group. Then the following are equivalent:*

1. *G is amenable;*
2. *$\ell^1(G)$ is amenable;*
3. *the Banach $\ell^1(G)$-module $\ell^\infty(G)_0^*$ is amenable.*

Proof. We begin by remarking that the bimodule structure on V can be modified into a right module structure: let $\overline{V}$ be V qua Banach space, with actions $g \cdot \overline{v} \cdot h = \overline{h^{-1}vh}$ for $g, h \in G$; in other words, the left action becomes trivial while the right action is by conjugation. A derivation $\delta\colon \ell^1(G) \to V$ gives rise to a "crossed homomorphism" $\eta\colon \ell^1(G) \to \overline{V}$, defined by $\eta(g) = \overline{g^{-1}\delta(g)}$. It satisfies $\eta(gh) = \eta(g)h + \eta(h)$. Inner derivations give rise to crossed homomorphisms of the form $\eta(g) = v - vg$ for some $v \in \overline{V}$. For the rest of the proof, we replace V by $\overline{V}$.

(1) $\Rightarrow$ (2) Let $m\colon \ell^\infty(G) \to \mathbb{C}$ be a mean on G. Given a Banach module V and a crossed homomorphism $\eta\colon \ell^1(G) \to V^*$, define $v \in V^*$ by

$$v(f) = m(g \mapsto \eta(g)(f)) \text{ for all } f \in V.$$

Compute then, for $h \in G$,

$$\begin{aligned}(vh)(f) &= v(fh^{-1}) = m(g \mapsto \eta(g)(fh^{-1})) = m\big(g \mapsto (\eta(g)h)(f)\big) \\ &= m\big(g \mapsto (\eta(gh) - \eta(h))(f)\big) = (v - \eta(h))(f),\end{aligned}$$

so $\eta(h) = v - vh$.

(2) $\Rightarrow$ (3) is obvious.

(3) $\Rightarrow$ (1) More generally, if X is a G-set and $\ell^\infty(X)_0^*$ is amenable, then X is amenable: choose $\Phi \in \ell^\infty(X)^*$ with $\Phi(\mathbb{1}_X) = 1$, and set $\eta(g) := \Phi - \Phi g$. Then $\eta\colon \ell^1(G) \to \ell^\infty(X)_0^*$ is a crossed homomorphism, so since $\ell^\infty(X)_0^*$ is amenable, there exists $\Psi \in \ell^\infty(X)_0^*$ with $\Psi - \Psi g = \Phi - \Phi g$, namely, $(\Phi - \Psi)g = \Phi - \Psi$. Then $\Phi - \Psi\colon \ell^\infty(X) \to \mathbb{C}$ is a G-invariant functional on X.

Furthermore, using (11.5), $\Phi - \Psi$ may be viewed as a measure on the Stone-Čech compactification βX; its normalized absolute value is a positive measure and therefore a G-invariant mean on X.

As a corollary, we may deduce that $\ell^1(G)$ is amenable if and only if its augmentation ideal has *approximate identities*, though we prefer to give a direct proof. Recall that an approximate identity in a Banach algebra $\mathscr{A}$ is a bounded net (e_n) in $\mathscr{A}$ with $e_n a \to a$ for all $a \in \mathscr{A}$ and that the *augmentation ideal* $\varpi(\ell^1 G)$ is $\{f \in \ell^1(G) \mid \sum_{g\in G} f(g) = 0\}$.

Lemma 11.10.23. *Let $\mathscr{A}$ be a Banach algebra with approximate identities, and let $f_1, \dots, f_N \in \mathscr{A}$ and $\epsilon > 0$ be given. Then there exists $e \in \mathscr{A}$ with $\|f_i - ef_i\| < \epsilon$ for all $i = 1, \dots, N$.*

Proof. Let $K = \sup \|e_n\|$ be a bound on the norms of approximate identities in $\mathscr{A}$. For $N = 0$ there is nothing to do. If $N \geq 1$, find by induction $e' \in \mathscr{A}$ satisfying $\|f_i - e'f_i\| < \epsilon/(1 + K)$ for all $i < N$, and let $e'' \in \mathscr{A}$ satisfy $\|(f_N - e'f_N) - e''(f_N - e'f_N)\| < \epsilon$. Set $e := e' + e'' - e''e'$, and check.

Theorem 11.10.24. *Let G be a group. Then G is amenable if and only if $\varpi(\ell^1 G)$ has approximate identities.*

Proof. ($\Rightarrow$) Given $f \in \varpi(\ell^1 G)$ and $\epsilon > 0$, let $S \Subset G$ be such that $\sum_{g\in G\setminus S} |f(g)| < \epsilon/2$. Since G is amenable, there exists $h \in \ell^1(G)$ with $h \geq 0$ and $\|h\| = 1$ and $\|h - hs\| < \epsilon/2$ for all $s \in S$; so $\|hf\| < \epsilon$. Set $e := 1 - h$; then $\|e\| \leq 2$, and $\|f - ef\| = \|hf\| < \epsilon$.

($\Leftarrow$) Let $S = \{s_1, \dots, s_n\} \Subset G$ and $\epsilon > 0$ be given, and apply Lemma 11.10.23 with $f_i = 1 - s_i$ to obtain $e \in \mathscr{A}$ satisfying $\|1 - s - e(1 - s)\| < \epsilon$ for all $s \in S$; set $g := 1 - e$ to rewrite this as $\|g - gs\| < \epsilon$. Finally set $h(x) = |g(x)|/\|g\|$ for all $x \in G$; we have obtained $h \geq 0$ and $\|h\| = 1$ and $\|h - hs\| < \epsilon$, so G is amenable by Theorem 11.3.23(2).

We recall without proof Cohen's factorization theorem:

Lemma 11.10.25 (Cohen [159]). *Let $\mathscr{A}$ be a Banach algebra with approximate identities, and consider $z \in \mathscr{A}$. Then for every $\epsilon > 0$ there exists $x, y \in \mathscr{A}$ with $z = xy$ and $\|z - y\| < \epsilon$.* □

For instance, it follows that if G is an amenable group, then $\varpi(\ell^1 G)^2 = \varpi(\ell^1 G)$. Amenability and the Liouville property are tightly related to the ideal structure of $\ell^1(G)$. The following is in fact a reformulation of Theorem 11.8.21.

Theorem 11.10.26 (Willis [585]). *Let G be a group and let X be a G-set. For a probability measure μ on G, let*

$$\ell^1_\mu(X) := \overline{\{f - f\mu \mid f \in \ell^1(X)\}}$$

denote the closed submodule of $\ell^1(X)$ generated by $1-\mu$, and write $\varpi(\ell^1 X) = \{f \in \ell^1(X) \mid \sum_{g\in G} f(g) = 0\}$. Then (X, μ) is Liouville if and only if $\ell^1_\mu(X) = \varpi(\ell^1 X)$.

In particular, G is amenable if and only if $\{\ell^1_\nu(G) \mid \mu \in \mathscr{P}(G)\}$ has a unique maximal element, which is $\varpi(\ell^1 G)$.

Proof. Assume first that (X, μ) is Liouville, and consider an arbitrary $f \in \varpi(\ell^1 X)$. By Proposition 11.8.25, we have $\|f\mu^n\| \to 0$, so $f - f\mu^n \to f$, and $f - f\mu^n = f(1 + \mu + \dots + \mu^{n-1})(1 - \mu) \in \ell^1_\mu(X)$, so $f \in \ell^1_\mu(X)$.

Conversely, if μ is such that $\ell^1_\mu(X) = \varpi(\ell^1 X)$, then given $f \in \varpi(\ell^1 X)$, we may for every $\epsilon > 0$ find $g \in \ell^1(X)$ with $\|f - g(1-\mu)\| < \epsilon$; then $\|f \cdot \frac{1}{n}\sum_{i=0}^{n-1} \mu^i\| \approx \|g(1-\mu^n)/n\| \to 0$, so $f\mu^n \to 0$. By Proposition 11.8.25, the random walk (X, μ) is Liouville.

By Theorem 11.8.21, G is amenable if and only if there exists a Liouville measure on G.

It remains to prove that if $\ell^1_\mu(X)$ is the unique maximal element in $\{\ell^1_\nu(X) \mid \nu \in \mathscr{P}(G)\}$, then $\ell^1_\mu(X) = \varpi(\ell^1 X)$. For this, let f belong to $\varpi(\ell^1 X)$ and write $f = g + ih$ with g, h real. Furthermore, write $g = g^+ - g^-$ and $h = h^+ - h^-$ for positive $g^\pm, h^\pm$, and set $c = \sum_{x\in X} g^+(x) = \sum_{x \in X} g^-(x)$ and $d = \sum_{x\in X} h^+(x) = \sum_{x\in X} h^-(x)$. Then

$$f = c(1 - g^+/c) + (-c)(1 - g^-/c) + (id)(1 - h^+/d) + (-id)/(1 - h^-/d),$$

and each term belongs to some $\ell^1_\nu(X)$ and therefore to $\ell^1_\mu(X)$ because $\ell^1_\mu(X)$ is maximal; so $f \in \ell^1_\mu(X)$.

Exercise 11.10.27 (, see Johnson [313, Proposition 5.1]).** Let $\mathscr{A}$ be an amenable algebra, and let $J \lhd \mathscr{A}$ be a closed ideal. Prove that if J and $\mathscr{A}/J$ are amenable, then $\mathscr{A}$ is amenable. Conversely, if $\mathscr{A}$ is amenable then $\mathscr{A}/J$ is amenable, and if J has approximate identities then it is amenable.

11.10.3 Amenable Algebras

We now turn to the group algebra $\Bbbk G$ for a field $\Bbbk$. Note that we do not make any assumption on the field, which could be finite.

Definition 11.10.28. Let $\mathscr{A}$ be an associative algebra, and let V be an $\mathscr{A}$-module. We call V *amenable* if for every finite-dimensional subspace $S \leq \mathscr{A}$ and every $\epsilon > 0$ there exists a finite-dimensional subspace $F \leq V$ with

$$\dim(FS) < (1 + \epsilon)\dim(F).$$

The algebra $\mathscr{A}$ itself is called *amenable* if all nonzero $\mathscr{A}$-modules are amenable.[22]

We note in passing that if $\mathscr{A}$ is finitely generated, then the "S" in Definition 11.10.28 may be fixed once and for all to be a generating subspace of $\mathscr{A}$.

Theorem 11.10.29 ([40]). *Let G be a group and let X be a G-set. Then $\Bbbk X$ is an amenable $\Bbbk G$-module if and only if X is amenable.*

[22] Some people defined amenability of algebras—erroneously, in my opinion—as mere amenability of the regular right module.

Proof (Proof, after [273, Section 3.6]). (⇒) Consider the set $\mathscr{O}(X)$ of orders on X; it is a closed subspace of $\{0,1\}^{X\times X}$, so it is compact. It is also the inverse limit of $\mathscr{O}(F)$ over all $F \Subset X$.

Let Π denote the group of all bijections of X. There exists a unique Π-invariant probability measure on $\mathscr{O}(X)$, which may be defined as the inverse limit of the uniform probability measures on $\mathscr{O}(F)$ over $F \Subset X$. For an order $\le \in \mathscr{O}(X)$, consider

$$\Phi^{\le}\colon \begin{cases} \{\text{finite-dim'l subspaces of } \Bbbk X\} & \to \{\text{finite subsets of } X\} \\ W & \mapsto \{\min^{\le}(\text{support}(w)) \mid w \in W \setminus \{0\}\}, \end{cases}$$

and let $m_W^{\le} := \mathbb{1}_{\Phi^{\le}(W)}$ be the corresponding characteristic function in $\ell^1(X)$. We clearly have

$$\|m_W^{\le}\| = \dim W, \qquad W_1 \le W_2 \Rightarrow m_{W_1}^{\le} m_{W_2}^{\le} \text{ pointwise.} \tag{11.26}$$

Define then $m_W := \int_{\mathscr{O}(X)} m_W^{\le} d\lambda(\le)$, and observe that (11.26) still holds for m_W. Furthermore, the map $W \mapsto m_W$ is Π-equivariant, so in particular it is G-equivariant; and (11.26) further implies $\|m_{W_2} - m_{W_1}\| = \dim W_2 - \dim W_1$ whenever $W_1 \le W_2$.

Now given $S \Subset G$ finite and $\epsilon > 0$, there exists $W \le \Bbbk X$ with $\dim(W + Ws) < (1+\epsilon)\dim W$ for all $s \in S$, because $\Bbbk X$ is amenable. Thus $\|m_{W+Ws} - m_W\| < \epsilon \dim W$, and similarly $\|m_{W+Ws} - m_{Ws}\| < \epsilon \dim W$, so

$$\|m_W - m_{Ws}\| = \|m_W - m_W s\| < 2\epsilon\|m_W\|,$$

and G is amenable by Theorem 11.3.23(2).

(⇐) Let a finite-dimensional subspace S of $\Bbbk G$ and $\epsilon > 0$ be given. There is a finite subset $T \Subset G$ with $S \le \Bbbk T$, so because X is amenable, there is $F \Subset X$ with $\#(FT) < (1+\epsilon)\#F$. Set $E := \Bbbk F$; then

$$\dim(ES) \le \dim((\Bbbk F)(\Bbbk T)) \le \#(FT) < (1+\epsilon)\#F = (1+\epsilon)\dim E.$$

□

Note that, although G_G is amenable if and only if $\Bbbk G_{\Bbbk G}$ is amenable, the growth of almost invariant subsets and subspaces may behave quite differently. In Example 11.3.11, we saw Følner sets F_n for the "lamplighter group" G, and we may convince ourselves that they are optimal, so G's Følner function, see (11.9), satisfies $\text{Føl}(n) = n2^n$. On the other hand,

$$W_n = \Bbbk\Big\{ \sum_{\text{support}(f)\subseteq[-n,n]} (f,m) \,\Big|\, m \in [-n,n] \Big\}$$

are subspaces of $\Bbbk G$ of dimension $2n+1$ with $\dim(W_n+W_ns)/\dim W_n = \#(F_n\cup F_ns)/\#F_n$, so the "linear Følner function" of G grows linearly.

The following is an analogue, for linear spaces, of the space ℓ^1 of summable functions on a set. Let V be a vector space. Consider the free $\mathbb{Z}$-module with basis $\{[A] \mid A\le V \text{ a finite-dimensional subspace}\}$, and let $\ell^1(V,\mathbb{Z})$ be its quotient under the relations $[A]+[B]=[A\cap B]+[A+B]$ for all $A,B\le V$. Note that every $x\in\ell^1(V,\mathbb{Z})$ may be represented as $x=\sum_i[X_i^+]-\sum_j[X_j^-]$. Define a metric on $\ell^1(V,\mathbb{Z})$ by

$$d(x,y)=\|x-y\|,\quad \|x\|=\inf\{\sum_i\dim(X_i)+\sum_j\dim(X_j^-)\mid x=\sum_i[X_i^+]-\sum_j[X_j^-]\}.$$

Lemma 11.10.30. *Let $\mathscr{A}$ be an algebra generated by a set B of invertible elements, and let V be an $\mathscr{A}$-module. Then V is amenable if and only if for every $S\Subset B$ and every $\epsilon>0$ there exists $f\in\ell^1(V,\mathbb{N})$ with $\|f-fs\|<\epsilon\|f\|$ for all $s\in S$.*

Proof. If V is amenable, then for every $S\Subset B$ and every $\epsilon>0$, there exists $F\le V$ finite-dimensional with $\dim(F+FS)<(1+\epsilon)\dim F$; so in particular $\dim(F+Fs)<(1+\epsilon)\dim F$ for all $s\in S$; since $\dim(Fs)=\dim F$ because s is invertible, we get $\dim(F\cap Fs)>(1-\epsilon)\dim F$ so $f:=[F]$ satisfies $\|f-fs\|<2\epsilon\|f\|$.

Conversely, given $S\Subset B$ and $f\in\ell^1(V,\mathbb{N})$ with $\|f-fs\|<\epsilon\|f\|$ for all $s\in S$, we have $\sum_{s\in S}\|f-fs\|<\epsilon\#S\|f\|$. There is a unique expression $f=[X_0]+\dots+[X_n]$ with $X_0\le\dots\le X_n\le V$; so there exists $i\in\{0,\dots,n\}$ with $\sum_{s\in S}\|[X_i]-[X_i]s\|<\epsilon\#S\|[X_i]\|$, and therefore $\sum_{s\in S}\dim(X_i+X_is)<(1+\epsilon\#S)\dim X_i$, so $\dim(X_i+X_iS)<(1+\epsilon\#S)\dim X_i$. We are done since $S\Subset B$ was arbitrary and B generates $\mathscr{A}$.

Corollary 11.10.31. *Let $\mathscr{A}$ be a group ring. Then $\mathscr{A}$ is amenable if and only if the regular right module $\mathscr{A} \looparrowleft \mathscr{A}$ is amenable.*

Proof. Consider $\mathscr{A}=\Bbbk G$ a group ring. If $\mathscr{A}$ is amenable, then obviously the regular module $\mathscr{A}_{\mathscr{A}}$ is amenable.

Conversely, if $\mathscr{A}_{\mathscr{A}}$ is amenable, then G_G is amenable by Theorem 11.10.29. Let V be a nonzero $\mathscr{A}$-module, and consider $v\in V\setminus\{0\}$. By Theorem 11.3.23(5) for every $S\Subset G$ and every $\epsilon>0$, there exists a subset $F\Subset G$ with $\#(F\triangle Fs)<\epsilon\#F$. Consider $x:=\sum_{f\in F}[vf]\in\ell^1(V,\mathbb{N})$, and note $\|x-xs\|<\epsilon\|x\|$. Thus $\mathscr{A}$ is amenable by Lemma 11.10.30.

Problem 11.10.32 (Gromov). Let G be a group. If the $\mathbb{R}G$-module

$$\mathscr{C}_0(G)=\{f\colon G\to\mathbb{R}\mid \inf_{F\Subset G}\sup(f\restriction G\setminus F)=0\}$$

is amenable, does it follow that G is amenable?

11.11 Further Work and Open Problems

For lack of space, some important and interesting topics have been omitted from this text. Here are a few of the most significant ones, with very brief descriptions.

11.11.1 Boundary Theory

Harry Furstenberg initiated a deep theory of "boundaries" for random walks. Given a random walk on a set X, say driven by a measure μ on a group G, a boundary is a measure space (Y, ν) with a measurable map from the orbit space $(X^{\mathbb{N}}, \mu^{\mathbb{N}}) \to Y$ that quotients through asymptotic equivalence, namely, if $(x_0, x_1, \dots)$ and $(x'_0, x'_1, \dots)$ differ in only finitely many positions, then their images are the same in Y.

There is a universal such space, written $\partial(X, \mu)$ and called the *Poisson boundary* of (X, μ), such that all boundaries are quotients of $\partial(X, \mu)$. This space, as a measure space, may be characterized by the identity

$$L^1(\partial(X, \mu), \nu) = \ell^1(X)/\ell^1_\mu(X),$$

see Theorem 11.10.26. The Poisson boundary is reduced to a point if and only if (X, μ) is Liouville.

In fact, it is better to view $\partial(X, \mu)$ as a measure space with a family of measures ν_x, one for each starting point $x \in X$ of the random walk, and satisfying $\sum_{g \in G} \mu(g)\nu_{xg} = \nu_x$ for all $x \in X$. One then has a "Poisson formula" for harmonic functions on X: if $f \in \ell^\infty(X)$ is harmonic, then there exists an integrable function $\hat{f}$ on $\partial(X, \mu)$ such that

$$f(x) = \int_{\partial(X,\mu)} \hat{f}(\xi) d\nu_x(\xi).$$

There is another construction of $\partial(X, \mu)$ based on $\ell^\infty(X)$ rather than $\ell^1(X)$: the subspace $h^\infty(X) \leq \ell^\infty(X)$ of harmonic functions is a commutative Banach algebra, under the product

$$(f_1 \cdot f_2)(x) = \lim_{n\to\infty} \sum_{g \in G} f_1(xg) f_2(xg) \mu^n(g).$$

The spectrum of $h^\infty(X)$, namely, the set of algebra homomorphisms $h^\infty(X) \to \mathbb{C}$, is naturally a measure space and is isomorphic to $\partial(X, \mu)$. The function $\hat{f}$ is the Gelfand transform of f, given by $\hat{f}(\xi) = \xi(f)$.

The Poisson boundary is naturally defined as a measure space and is directly connected to the space of bounded harmonic functions; but other notions of boundary have been considered, for example, the space of *positive* harmonic functions, leading to the *Martin boundary* which is a well-defined topological space; for a natural measure, it becomes measure-isomorphic to the Poisson boundary.

Shmuel Glasner considers in [244] "strongly amenable" groups: they are groups all of whose proximal actions on a compact space have a fixed point; see the comments at the end of Section 11.6.1. Recall that an action of G on a compact Hausdorff space X is *proximal* if for every $x, y \in X$, there exists a net (g_n) of elements of G such that $\lim_n xg_n = \lim_n yg_n$.

For details, we refer to the original articles [234, 235], the classical [325], and the survey [218].

11.11.2 Consequences

Little has been said about the uses of amenability. On the one hand, it plays a major role in the study of Lie groups and their lattices; for example, Margulis's "normal subgroup theorem" states that a normal subgroup of a lattice in a higher-rank semisimple Lie group is either finite or finite index [410]. Ruling out finite-index subgroups, the strategy is to show that such a group is amenable and has property (T).

Dave Witte Morris uses amenability, and Poincaré's recurrence theorem, to prove in [424] that all finitely generated amenable groups that act on the real line have homomorphisms onto $\mathbb{Z}$.

Benjamini and Schramm consider in [68] *percolation* on Cayley graphs. One fixes $p \in (0, 1)$ and a finitely generated group $G = \langle S \rangle$; call $\mathcal{G}$ the corresponding Cayley graph. Then every vertex $v \in \mathcal{G}$ is made independently at random "open" with probability p (and "closed" with probability $1 - p$). "Open clusters" are connected components of the subgraph of $\mathcal{G}$ spanned by open vertices. We define *critical probabilities*

$$p_c = \sup\{p \in (0, 1) \mid \text{the open cluster containing 1 is almost surely finite}\},$$
$$p_u = \inf\{p \in (0, 1) \mid \text{there is almost surely a single infinite open cluster}\}.$$

They conjecture that $p_c < 1$ for all G which are not virtually cyclic; this is known for all groups of polynomial or exponential growth and for all groups containing subgroups of the form $A \times B$ with A, B infinite, finitely generated groups.

They also conjecture that $p_c < p_u$ holds precisely when G is not amenable; see [280] for a survey of known results.

11.11.3 Ergodic Theory

One of the standard tools of ergodic theory is the "Rokhlin-Kakutani lemma": let $T: X \circlearrowleft$ be an invertible, measure-preserving transformation of a measure space (X, μ) that is *aperiodic* in the sense that almost all points have infinite orbits. Then for every $n \in \mathbb{N}$ and every $\epsilon > 0$, there exists a measurable subset $E \subseteq X$ such that $E, T(E), \dots, T^{n-1}(E)$ are all disjoint with $\mu(E \sqcup \cdots \sqcup T^{n-1}(E)) > 1 - \epsilon$.

It may be understood as the following statement. Given $S, T: X \circlearrowleft$, define their distance as $d(S, T) = \mu(\{x \in X \mid S(x) \neq T(x)\})$. Then for every $n \in \mathbb{N}, \epsilon > 0$, there exists S of period n with $d(S, T) < \epsilon$. In other words, $\mathbb{Z}$ may be approximated arbitrarily closely by $\mathbb{Z}/n$.

The Rokhlin lemma is essential in reducing ergodic theory problems to combinatorial ones. For example, it serves to prove that two Bernoulli shifts (the shift on $\mathscr{A}^{\mathbb{Z}}$ for a given probability measure on $\mathscr{A}$) are isomorphic if and only if they have the same entropy.

Ornstein and Weiss generalize in [462] the Rokhlin lemma to some amenable groups; see also [581]. Let G be a group; we say that a subset $F \Subset G$ *tiles* G if G is a disjoint union of translates of F; namely, if there exists a subset $C \subseteq G$ with $G = \bigsqcup_{c \in C} Fc$. They prove:

Theorem 11.11.1. *Let G be amenable, and let $F \Subset G$ be a finite subset. Then F tiles G if and only if for every free measure-preserving action of G on a probability space (X, μ) and every $\epsilon > 0$ there is a measurable subset $E \subseteq X$ such that $\{Ef \mid f \in F\}$ are all disjoint and $\mu(EF) > 1 - \epsilon$.*

In [583], Weiss calls G *monotileable* if it admits arbitrarily large tiles. He proves that amenable, residually finite groups are monotileable; more precisely, in Følner's definition of amenability, it may be assumed that the Følner sets tile G. For example, $\mathbb{Z}$ is tiled by sets of the form $\{-n, \dots, n\}$ which form an exhausting sequence of Følner sets and are also transversals for the subgroups $(2n+1)\mathbb{Z}$.

Let us denote by MG the class of monotileable groups; then MG contains all residually amenable groups and is closed under taking extensions, quotients, subgroups, and directed unions [151, Section 4].

It is at the present (2017) unknown whether *every* group is monotileable and whether $AG \subseteq MG$. It is also unknown whether if a group G belongs to $MG \cap AG$, then G may be tiled by Følner sets.

11.11.4 C^*- and von Neumann Algebras

We considered, in §11.10, amenability of groups in their relation to group rings and Banach algebras. Given a group G, other algebras, closed under weaker topologies, may be considered, in particular C^* and von Neumann algebras. We restrict, here, to countable groups.

Given a unitary representation $\pi: G \to U(\mathscr{H})$ on a Hilbert space $\mathscr{H}$, one considers the norm closure $C^*_\pi(G)$ of $\pi(\mathbb{C}G)$ in the Banach algebra $B(\mathscr{H})$ of bounded operators on $\mathscr{H}$. In particular, if π is the regular representation on $\mathscr{H} = \ell^2(G)$, one obtains the *reduced* C^*-algebra $C^*_\lambda(G)$.

There is also a *maximal* C^*-algebra $C^*_{\max}(G)$, defined as the completion of $\mathbb{C}G$ with respect to the norm $\|x\| = \sup_\pi \|x\|_\pi$, where π ranges over all unitary representations of G on Hilbert spaces. We have natural maps

$$\mathbb{C}G \hookrightarrow \ell^1(G) \hookrightarrow C^*_{\max}(G) \twoheadrightarrow C^*_\lambda(G).$$

Amenability of G is characterized by the fact that the last map $C^*_{\max}(G) \twoheadrightarrow C^*_\lambda(G)$ is an isomorphism, as shown by Hulanicki [303].

The *von Neumann algebra* $W^*(G)$ is, by contrast, defined as the *weak* closure of $\mathbb{C}G$ in $B(\ell^2(G))$. Amenability of G is characterized by the fact that $W^*(G)$ is *hyperfinite*: $W^*(G)$ contains as a dense subalgebra the union of its finite-dimensional subalgebras, as shown by Connes [160].

11.11.5 Numerical Invariants

Recall that the *entropy* of a probability measure μ on a countable set X is defined as

$$H(\mu) = -\sum_{x\in X} \mu(x) \log \mu(x), \text{ where as usual } 0 \log(0) = 0.$$

The Liouville property can, in some favorable cases, be detected by a single numerical invariant, its *entropy* or its *drift*. Given a random walk p on a set X, starting at $x \in X$, its *entropy growth* is the function $h(n) := H(p_n(x, -))$ computing the entropy of distribution of the random walker after n steps. If furthermore X is a metric space, the *drift growth* of p is the function $\ell(n) := \sum_{y\in X} p_n(x, y)d(x, y)$ estimating the expected distance from the random walker to the origin after n steps.

Let us assume for simplicity that X admits a transitive group action so that the functions h, ℓ, v are independent of the choice of x. A celebrated criterion by Avez [26] (for finitely supported μ) and Derriennic [195] and Kaimanovich-Vershik [325] (in the general case) shows that if $H(\mu) < \infty$, then (X, μ) is Liouville if and only if h is sublinear. Moreover, the volume, entropy, and drift growth are related by the inequality

$$\lim_{n\to\infty} \frac{h(n)}{n} \le \left(\lim_{n\to\infty} \frac{\log v(n)}{n}\right)\left(\lim_{n\to\infty} \frac{\ell(n)}{n}\right).$$

Finer estimates relate these functions $\log v, \ell, h$, in particular if all are sublinear; additionally, the probability of return $p(n) = -\log p_n(x,x)$ and the ℓ^p *distortion*

$$d_p(n) = \sup_{\Phi: G\to\ell^p \text{ 1-Lipschitz}} \inf\{\|\Phi(g)-\Phi(h)\| \mid d(g,h)\geq n\}$$

are all related by various inequalities; see [262, 438, 474].

11.11.6 Sofic Groups

The class of sofic groups is a common extension of amenable and residually finite groups. We refer to [275, 582] for its introduction. The definition may be seen as a variant of Følner's criterion:

Definition 11.11.2. Let G be a group. It is *sofic* if for every finite subset $S \Subset G$ and every $\epsilon > 0$, there exists a finite set F and a mapping $\pi: S \to \mathrm{Sym}(F)$ such that

$$\text{if } s,t,st \in S \text{ then } \#\{f \in F \mid f\pi(s)\pi(t) \neq f\pi(st)\} < \epsilon\#F,$$
$$\text{if } s \neq t \in S \text{ then } \#\{f \in F \mid f\pi(s) = f\pi(t)\} < \epsilon\#F.$$

Two cases are clear: if G is residually finite, then for every $S \Subset G$ there exists a homomorphism $\rho: G \to F$ to a finite group that is injective on S; define then $f\pi(s) = f\rho(s)$ for all $s \in S, f \in F$, showing that G is sofic. If on the other hand G is amenable, then for every $S \Subset G$ and every $\epsilon > 0$, there exists $F \Subset G$ with $\#(FS \setminus F) < \epsilon\#F$; define then $f\pi(s) = fs$ if $fs \in F$, and extend the partial map $\pi(s): F \dashrightarrow F$ arbitrarily into a permutation, showing that G is sofic.

Remarkably, there is at the present time (2017) no known example of a non-sofic group.

11.11.7 Is This Group Amenable?

We list here some examples of groups for which it is not known whether they are amenable or not. These problems are probably very hard.

Problem 11.11.3 (Geoghegan). Is Thompson's group F amenable?

Recall that F is the group of piecewise-linear homeomorphisms of $[0,1]$, with slopes in $2^{\mathbb{Z}}$ and breakpoints in $\mathbb{Z}[\frac{1}{2}]$; see [125] and Example 11.7.21.

There have been numerous attempts at answering Problem 11.11.3, too many to cite them all; a promising direction appears in [560]. Kaimanovich proves in [324] that, for every finitely supported measure μ on F, the orbit $(\frac{1}{2}F, \mu)$ is not Liouville; however Juschenko and Zhang prove in [319] that $\frac{1}{2}F$ *is* laminable.

There is a group that is related to F and acts on the circle $[0,1]/(0 \sim 1)$: it satisfies the same definition as F, namely, the group T of piecewise-linear self-homeomorphisms with slopes in $2^{\mathbb{Z}}$ and breakpoints in $\mathbb{Z}[\frac{1}{2}]/\mathbb{Z}$. Its amenable subgroup $\mathbb{Z}[\frac{1}{2}]/\mathbb{Z}$ acts transitively on the orbit $0T$, so $0T \looparrowleft T$ is laminable by Corollary 11.8.18.

Problem 11.11.4 (Nekrashevych). Are all contracting self-similar groups amenable?

Recall that a self-similar group is a group G generated by invertible transducers; it acts on $\mathscr{A}^{\mathbb{N}}$ and may be given by a map $\phi\colon G \to G \wr_{\mathscr{A}} \mathrm{Sym}(\mathscr{A})$, as in (11.2). It is *contracting* if there is a proper metric on G and constants $\lambda < 1, C$ such that whenever $\phi(g) = \langle\!\langle g_1, \ldots, g_{\#\mathscr{A}} \rangle\!\rangle \pi$, we have $\|g_i\| < \lambda\|g\| + C$. See [441].

Problem 11.11.5 (Folklore, often attributed to Katok). Is the group of interval exchange transformations amenable? Does it contain non-abelian free subgroups?

A partial, positive result appears in Example 11.9.15. It would suffice, following the strategy in that example (see [320, Proposition 5.3]), to prove that the group of $\mathbb{Z}^d$-wobbles $W(\mathbb{Z}^d)$ acts extensively amenably on $\mathbb{Z}^d$ for all $d \in \mathbb{N}$; at present (2017), this is known only for $d \leq 2$, see Theorem 11.9.16.

References

1. Adamczewski, B., Bell, J.P.: An analogue of Cobham's theorem for fractals. Trans. Am. Math. Soc. **363**(8), 4421–4442 (2011)
2. Adamczewski, B., Bell, J.P.: A problem around Mahler functions. Ann. Sc. Norm. Super. Pisa. (2013, to appear). ArXiv:1303.2019
3. Adamczewski, B., Bugeaud, Y.: On the complexity of algebraic numbers. I. Expansions in integer bases. Ann. Math. (2) **165**(2), 547–565 (2007)
4. Adamczewski, B., Faverjon, C.: Méthode de Mahler: relations linéaires, transcendance et application aux nombres automatiques. Proc. Lond. Math. Soc. **115**, 55–90 (2017)
5. Adyan, S.I.: Random walks on free periodic groups. Izv. Akad. Nauk SSSR Ser. Mat. **46**(6), 1139–1149, 1343 (1982)
6. Agafonov, V.N.: Normal sequences and finite automata. Sov. Math. Dokl. **9**, 324–325 (1968)
7. Ahlfors, L.: Zur theorie der überlagerungsflächen. Acta Math. **65**(1), 157–194 (1935)
8. Aistleitner, C., Becher, V., Scheerer, A.M., Slaman, T.: On the construction of absolutely normal numbers. Acta Arith. (to appear), arXiv:1702.04072
9. Akhavi, A., Klimann, I., Lombardy, S., Mairesse, J., Picantin, M.: On the finiteness problem for automaton (semi)groups. Int. J. Algebra Comput. **22**(6), 1–26 (2012)
10. Albeverio, S., Pratsiovytyi, M., Torbin, G.: Singular probability distributions and fractal properties of sets of real numbers defined by the asymptotic frequencies of their s-adic digits. Ukraïn. Mat. Zh. **57**(9), 1163–1170 (2005)
11. Alešin, S.: Finite automata and the Burnside problem for periodic groups. Mat. Zametki **11**, 319–328 (1972)
12. Allouche, J.-P., Rampersad, N., Shallit, J.: Periodicity, repetitions, and orbits of an automatic sequence. Theor. Comput. Sci. **410**(30–32), 2795–2803 (2009)
13. Allouche, J.-P., Scheicher, K., Tichy, R.F.: Regular maps in generalized number systems. Math. Slovaca **50**, 41–58 (2000)
14. Allouche, J.-P., Shallit, J.: Automatic Sequences: Theory, Applications, Generalizations. Cambridge University Press, Cambridge (2003)
15. Allouche, J.-P., Shallit, J.: The ring of k-regular sequences. II. Theor. Comput. Sci. **307**(1), 3–29 (2003)
16. Allouche, J.-P., Shallit, J.O.: The ring of k-regular sequences. In: Choffrut, C., Lengauer, T. (eds.) STACS 90, Proceedings of the 7th Symposium on Theoretical Aspects of Computer Science. Lecture Notes in Computer Science, vol. 415, pp. 12–23. Springer, Berlin (1990)
17. Allouche, J.-P., Shallit, J.O.: The ring of k-regular sequences. Theor. Comput. Sci. **98**, 163–197 (1992)

V. Berthé, M. Rigo (eds.), *Sequences, Groups, and Number Theory*,
Trends in Mathematics, https://doi.org/10.1007/978-3-319-69152-7

18. Allouche, J.-P., Shallit, J.O.: The ubiquitous Prouhet-Thue-Morse sequence. In: Ding, C., Helleseth, T., Niederreiter, H. (eds.) Sequences and Their Applications, Proceedings of SETA '98, pp. 1–16. Springer, London (1999)
19. Alvarez, N., Becher, V.: M. Levin's construction of absolutely normal numbers with very low discrepancy. Math. Comput. (to appear)
20. Amitsur, A.S., Levitzki, J.: Minimal identities for algebras. Proc. Am. Math. Soc. **1**, 449–463 (1950)
21. Amitsur, S.A., Small, L.W.: Affine algebras with polynomial identities. Rend. Circ. Mat. Palermo (2) Suppl. **31**, 9–43 (1993). Recent developments in the theory of algebras with polynomial identities (Palermo, 1992)
22. Anantharaman-Delaroche, C., Renault, J.: Amenable groupoids. Monographies de L'Enseignement Mathématique, vol. 36. L'Enseignement Mathématique, Geneva (2000)
23. Aubrun, N., Barbieri, S., Sablik, M.: A notion of effectiveness for subshifts on finitely generated groups. Theor. Comput. Sci. **661**, 35–55 (2017)
24. Aubrun, N., Kari, J.: Tiling problems on Baumslag-Solitar groups. In: MCU'13, pp. 35–46 (2013)
25. Aubrun, N., Sablik, M.: Multidimensional effective S-adic systems are sofic. Uniform Distribution Theory **9**(2), 7–29 (2014)
26. Avez, A.: Entropie des groupes de type fini. C. R. Acad. Sci. Paris Sér. A-B **275**, A1363–A1366 (1972)
27. Badkobeh, G., Crochemore, M.: Finite-repetition threshold for infinite ternary words. In: WORDS. EPTCS, vol. 63, pp. 37–43 (2011)
28. Badkobeh, G., Crochemore, M.: Fewest repetitions in infinite binary words. RAIRO Theor. Inform. Appl. **46**(1), 17–31 (2012)
29. Badkobeh, G., Crochemore, M., Rao, M.: Finite repetition threshold for large alphabets. RAIRO Theor. Inform. Appl. **48**(4), 419–430 (2014)
30. Baek, I.S., Olsen, L.: Baire category and extremely non-normal points of invariant sets of IFS's. Discrete Contin. Dyn. Syst. **27**(3), 935–943 (2010)
31. Bailey, D.H., Borwein, J.M.: Nonnormality of stoneham constants. Ramanujan J. **29**(1), 409–422 (2012)
32. Baker, K.A., McNulty, G.F., Taylor, W.: Growth problems for avoidable words. Theor. Comput. Sci. **69**(3), 319–345 (1989)
33. Ballier, A., Jeandel, E.: Tilings and model theory. First Symposium on Cellular Automata Journées Automates Cellulaires (2008)
34. Ballier, A., Jeandel, E.: Computing (or not) quasi-periodicity functions of tilings. In: Second Symposium on Cellular Automata "Journées Automates Cellulaires", JAC 2010, Turku, Finland, December 15–17, 2010. Proceedings, pp. 54–64 (2010)
35. Banach, S., Tarski, A.: Sur la décomposition des ensembles de points en parties respectivement congruentes. Fundam. Math. **6**, 244–277 (1924)
36. Barat, G., Berthé, V., Liardet, P., Thuswaldner, J.: Dynamical directions in numeration. Ann. Inst. Fourier (Grenoble) **56**(7), 1987–2092 (2006)
37. Barbieri, S., Sablik, M.: A generalization of the simulation theorem for semidirect products. Ergodic Theory Dyn. Syst. (to appear)
38. Bartholdi, L.: The growth of Grigorchuk's torsion group. Int. Math. Res. Not. **20**, 1049–1054 (1998)
39. Bartholdi, L.: Counting paths in graphs. Enseign. Math. (2) **45**(1–2), 83–131 (1999)
40. Bartholdi, L.: On amenability of group algebras, I. Isr. J. Math. **168**, 153–165 (2008)
41. Bartholdi, L.: Gardens of Eden and amenability on cellular automata. J. Eur. Math. Soc. **12**(1), 241–248 (2010)
42. Bartholdi, L., Grigorchuk, R.I., Šunić, Z.: Handbook of Algebra, vol. 3, chap. Branch groups, pp. 989–1112. Elsevier BV (2003)
43. Bartholdi, L., Kaimanovich, V.A., Nekrashevych, V.V.: On amenability of automata groups. Duke Math. J. **154**(3), 575–598 (2010)

44. Bartholdi, L., Kielak, D.: Amenability of groups is characterized by Myhill's theorem (2016). ArXiv:1605.09133
45. Bartholdi, L., Virág, B.: Amenability via random walks. Duke Math. J. **130**(1), 39–56 (2005)
46. Bass, H.: The degree of polynomial growth of finitely generated nilpotent groups. Proc. Lond. Math. Soc. (3) **25**, 603–614 (1972)
47. Baumgartner, J.E.: A short proof of Hindman's theorem. J. Comb. Theory Ser. A **17**, 384–386 (1974)
48. Bean, D.R., Ehrenfeucht, A., McNulty, G.: Avoidable patterns in strings of symbols. Pac. J. Math. **85**, 261–294 (1979)
49. Becher, V., Bugeaud, Y., Slaman, T.: On simply normal numbers to different bases. Math. Ann. **364**(1), 125–150 (2016)
50. Becher, V., Carton, O., Heiber, P.A.: Finite-state independence. Theory Comput. Syst. (to appear)
51. Becher, V., Figueira, S.: An example of a computable absolutely normal number. Theor. Comput. Sci. **270**, 947–958 (2002)
52. Becher, V., Figueira, S., Picchi, R.: Turing's unpublished algorithm for normal numbers. Theor. Comput. Sci. **377**, 126–138 (2007)
53. Becher, V., Heiber, P., Slaman, T.: A polynomial-time algorithm for computing absolutely normal numbers. Inf. Comput. **232**, 1–9 (2013)
54. Becher, V., Heiber, P., Slaman, T.: A computable absolutely normal Liouville number. Math. Comput. **84**(296), 2939–2952 (2015)
55. Becher, V., Heiber, P.A.: On extending de Bruijn sequences. Inf. Process. Lett. **111**(18), 930–932 (2011)
56. Becher, V., Heiber, P.A.: Normal numbers and finite automata. Theor. Comput. Sci. **477**, 109–116 (2013)
57. Becher, V., Heiber, P.A., Carton, O.: Normality and automata. J. Comput. Syst. Sci. **81**, 1592–1613 (2015)
58. Becher, V., Slaman, T.A.: On the normality of numbers to different bases. J. Lond. Math. Soc. **90**(2), 472–494 (2014)
59. Becker, H., Kechris, A.S.: The descriptive set theory of Polish group actions. London Mathematical Society Lecture Note Series, vol. 232. Cambridge University Press, Cambridge (1996)
60. Becker, P.G.: Effective measures for algebraic independence of the values of Mahler type functions. Acta Arith. **58**(3), 239–250 (1991)
61. Becker, P.G.: k-regular power series and Mahler-type functional equations. J. Number Theory **49**(3), 269–286 (1994)
62. Bekka, M.E.B., de la Harpe, P., Valette, A.: Kazhdan's property (T). New Mathematical Monographs, vol. 11. Cambridge University Press, Cambridge (2008)
63. Bell, J.P., Bugeaud, Y., Coons, M.: Diophantine approximation of Mahler numbers. Proc. Lond. Math. Soc. (3) **110**(5), 1157–1206 (2015)
64. Bell, J.P., Coons, M.: Transcendence tests for Mahler functions. Proc. Am. Math. Soc. **145**(3), 1061–1070 (2017)
65. Bell, J.P., Coons, M., Hare, K.G.: The minimal growth of a k-regular sequence. Bull. Aust. Math. Soc. **90**(2), 195–203 (2014)
66. Bell, J.P., Coons, M., Rowland, E.: The rational-transcendental dichotomy of Mahler functions. J. Integer Seq. **16**(2), 11 (2013). Article 13.2.10
67. Benjamini, I.: Coarse Geometry and Randomness. Lecture Notes in Mathematics, vol. 2100. Springer, Cham (2013). Lecture notes from the 41st Probability Summer School held in Saint-Flour, 2011
68. Benjamini, I., Schramm, O.: Percolation beyond $\mathbf{z}^d$, many questions and a few answers. Electron. Commun. Probab. **1**, no. 8, 71–82 (1996)
69. Benjamini, I., Kozma, G.: Nonamenable Liouville graphs (2010). ArXiv:1010.3365
70. Berend, D., Frougny, C.: Computability by finite automata and Pisot bases. Math. Syst. Theory **27**, 275–282 (1994)

71. Berger, R.: The undecidability of the Domino Problem. Ph.D. thesis, Harvard University (1964)
72. Berger, R.: The undecidability of the domino problem. Mem. Am. Math. Soc. **66**, 72 (1966)
73. Bernardino, A., Pacheco, R., Silva, M.: Coloring factors of substitutive infinite words (2016). ArXiv:1605.09343
74. Berstel, J.: Axel Thue's papers on repetitions in words: a translation. In: Monographies du LaCIM, vol. 11, pp. 65–80. LaCIM (1992)
75. Berstel, J., Karhumäki, J.: Combinatorics on words—a tutorial. Bull. Eur. Assoc. Theor. Comput. Sci. **79**, 178–228 (2003)
76. Berstel, J., Perrin, D.: The origins of combinatorics on words. Eur. J. Comb. **28**, 996–1022 (2007)
77. Berstel, J., Reutenauer, C.: Noncommutative rational series with applications. Encyclopedia of Mathematics and Its Applications, vol. 137. Cambridge University Press, Cambridge (2011)
78. Berthé, V., Rigo, M. (eds.): Combinatorics, automata and number theory. Encyclopedia of Mathematics and Its Applications, vol. 135. Cambridge University Press, Cambridge (2010)
79. Berthé, V., Rigo, M. (eds.): Combinatorics, words and symbolic dynamics. Encyclopedia of Mathematics and Its Applications, vol. 159. Cambridge University Press, Cambridge (2016)
80. Bertrand-Mathis, A.: Points génériques de Champernowne sur certains systèmes codes; application aux θ-shifts. Ergodic Theory Dyn. Syst. **8**(1), 35–51 (1988)
81. Bertrand-Mathis, A., Volkmann, B.: On (ϵ, k)-normal words in connecting dynamical systems. Monatsh. Math. **107**(4), 267–279 (1989)
82. Besicovitch, A.S.: The asymptotic distribution of the numerals in the decimal representation of the squares of the natural numbers. Math. Z. **39**, 146–156 (1934)
83. Bézivin, J.P.: Sur une classe d'équations fonctionnelles non linéaires. Funkcial. Ekvac. **37**(2), 263–271 (1994)
84. Biggs, N.L., Mohar, B., Shawe-Taylor, J.: The spectral radius of infinite graphs. Bull. Lond. Math. Soc. **20**(2), 116–120 (1988)
85. Billingsley, P.: Hausdorff dimension in probability theory. Ill. J. Math. **4**, 187–209 (1960)
86. Billingsley, P.: Hausdorff dimension in probability theory. II. Ill. J. Math. **5**, 291–298 (1961)
87. Billingsley, P.: Ergodic Theory and Information. Wiley, New York (1965)
88. Blanchard, F., Hansel, G.: Systèmes codés. Theor. Comput. Sci. **44**(1), 17–49 (1986)
89. Blondel, V.D., Bournez, O., Koiran, P., Papadimitriou, C., Tsitsiklis, J.N.: Deciding stability and mortality of piecewise affine dynamical systems. Theor. Comput. Sci. **255**(1–2), 687–696 (2001). Article dans revue scientifique avec comité de lecture
90. Blondel, V.D., Nesterov, Y., Theys, J.: On the accuracy of the ellipsoid norm approximation of the joint spectral radius. Linear Algebra Appl. **394**, 91–107 (2005)
91. Blondel, V.D., Theys, J., Vladimirov, A.A.: An elementary counterexample to the finiteness conjecture. SIAM J. Matrix Anal. Appl. **24**(4), 963–970 (electronic) (2003)
92. Bluhm, C.: Liouville numbers, Rajchman measures, and small Cantor sets. Proc. Am. Math. Soc. **128**(9), 2637–2640 (2000)
93. Boigelot, B., Brusten, J.: A generalization of Cobham's theorem to automata over real numbers. Theor. Comput. Sci. **410**(18), 1694–1703 (2009)
94. Boigelot, B., Brusten, J., Bruyère, V.: On the sets of real numbers recognized by finite automata in multiple bases. Log. Methods Comput. Sci. **6**, 1–17 (2010)
95. Boigelot, B., Brusten, J., Leroux, J.: A generalization of Semenov's theorem to automata over real numbers. In: Automated Deduction—CADE-22. Lecture Notes in Computer Science, vol. 5663, pp. 469–484. Springer, Berlin (2009)
96. Boigelot, B., Rassart, S., Wolper, P.: On the expressiveness of real and integer arithmetic automata (extended abstract). In: ICALP. Lecture Notes in Computer Science, vol. 1443, pp. 152–163. Springer, Berlin (1998)
97. Bondarenko, I., Bondarenko, N., Sidki, S., Zapata, F.: On the conjugacy problem for finite-state automorphisms of regular rooted trees. Groups Geom. Dyn. **7**(2), 323–355 (2013)

98. Bondarenko, I.V.: Growth of Schreier graphs of automaton groups. Math. Ann. **354**(2), 765–785 (2012)
99. Borel, É.: Les probabilités dénombrables et leurs applications arithmétiques. Rendiconti Circ. Mat. Palermo **27**, 247–271 (1909)
100. Borel, É.: Sur les chiffres décimaux $\sqrt{2}$ et divers problémes de probabilités en chaîne. C. R. Acad. Sci. Paris **230**, 591–593 (1950)
101. Borwein, J., Bailey, D.: Mathematics by Experiment, Plausible Reasoning in the 21st Century, 2nd edn. A. K. Peters, Ltd, Wellesley, MA (2008)
102. Bousch, T., Mairesse, J.: Asymptotic height optimization for topical IFS, Tetris heaps, and the finiteness conjecture. J. Am. Math. Soc. **15**(1), 77–111 (electronic) (2002)
103. Bowen, R.: Periodic points and measures for Axiom *A* diffeomorphisms. Trans. Am. Math. Soc. **154**, 377–397 (1971)
104. Bratteli, O.: Inductive limits of finite dimensional C^*-algebras. Trans. Am. Math. Soc. **171**, 195–234 (1972)
105. Braun, A.: The nilpotency of the radical in a finitely generated PI ring. J. Algebra **89**(2), 375–396 (1984)
106. Brieussel, J.: Folner sets of alternate directed groups. Ann. Inst. Fourier (Grenoble) **64**(3), 1109–1130 (2014)
107. Brlek, S.: Enumeration of factors in the Thue-Morse word. Discrete Appl. Math. **24**, 83–96 (1989)
108. Brough, T., Cain, A.J.: Automaton semigroup constructions. Semigroup Forum **90**(3), 763–774 (2015)
109. Brough, T., Cain, A.J.: Automaton semigroups: new constructions results and examples of non-automaton semigroups. Theor. Comput. Sci. **674**, 1–15 (2017)
110. Brown, T.C.: Colorings of the factors of a word (2006). Preprint, Department of Mathematics, Simon Fraser University, Canada
111. Brusten, J.: On the sets of real vectors recognized by finite automata in multiple bases. PhD thesis, University of Liège (2011)
112. Bruyère, V.: Entiers et automates finis (1985). Mémoire de fin d'études, Université de Mons
113. Bruyère, V., Hansel, G.: Bertrand numeration systems and recognizability. Theor. Comput. Sci. **181**, 17–43 (1997)
114. Bruyère, V., Hansel, G., Michaux, C., Villemaire, R.: Logic and *p*-recognizable sets of integers. Bull. Belg. Math. Soc. **1**, 191–238 (1994). Corrigendum, Bull. Belg. Math. Soc. **1** (1994), 577
115. Büchi, J.R.: Weak second-order arithmetic and finite automata. Zeitschrift für mathematische Logik und Grundlagen der Mathematik **6**, 66–92 (1960). Reprinted in Mac Lane, S., Siefkes, D. (eds.) The Collected Works of J. Richard Büchi. Springer, 1990, pp. 398–424
116. Büchi, J.R.: Turing-Machines and the Entscheidungsproblem. Math. Ann. **148**(3), 201–213 (1962)
117. Bugeaud, Y.: Nombres de Liouville et nombres normaux. C. R. Acad. Sci. Paris **335**(2), 117–120 (2002)
118. Bugeaud, Y.: Distribution modulo one and Diophantine approximation. Cambridge Tracts in Mathematics, vol. 193. Cambridge University Press, Cambridge (2012)
119. Bugeaud, Y.: Expansions of algebraic numbers. In: Four Faces of Number Theory. EMS Series of Lectures in Mathematics, pp. 31–75. European Mathematical Society, Zürich (2015)
120. Bundschuh, P.: Algebraic independence of infinite products and their derivatives. In: Number Theory and Related Fields. Springer Proceedings in Mathematics and Statistics, vol. 43, pp. 143–156. Springer, New York (2013)
121. Burks, A.W. (ed.): Essays on Cellular Automata. University of Illinois Press, Urbana, IL (1970)
122. Burnside, W.: On an unsettled question in the theory of discontinuous groups. Q. J. Math. **33**, 230–238 (1902)
123. Cain, A.: Automaton semigroups. Theor. Comput. Sci. **410**, 5022–5038 (2009)

124. Calkin, N.J., Wilf, H.S.: Binary partitions of integers and Stern-Brocot-like trees. unpublished pp. updated version August 5, 2009, 19 pp. (1998)
125. Cannon, J.W., Floyd, W.J., Parry, W.R.: Introductory notes on Richard Thompson's groups. Enseign. Math. (2) **42**(3–4), 215–256 (1996)
126. Cantor, G.: Mitteilungen zur lehre vom transfiniten. Zeitschrift fr Philosophie und philosophische Kritik **91**, 81–125 (1887)
127. Carpi, A.: On abelian power-free morphisms. Int. J. Algebra Comput. **03**(02), 151–167 (1993)
128. Carpi, A.: On the number of abelian square-free words on four letters. Discret. Appl. Math. **81**(1–3), 155–167 (1998)
129. Carpi, A., D'Alonzo, V.: On factors of synchronized sequences. Theor. Comput. Sci. **411**(44–46), 3932–3937 (2010)
130. Carpi, A., Maggi, C.: On synchronized sequences and their separators. Theor. Inform. Appl. **35**(6), 513–524 (2002) (2001)
131. Carton, O., Heiber, P.A.: Normality and two-way automata. Inf. Comput.**241**, 264–276 (2015)
132. Cassaigne, J.: Motifs évitables et régularité dans les mots. Ph.D. thesis, Université Paris VI (1994)
133. Cassaigne, J., Currie, J.D., Schaeffer, L., Shallit, J.: Avoiding three consecutive blocks of the same size and same sum. J. ACM **61**(2), 10:1–10:17 (2014)
134. Cassaigne, J., Nicolas, F.: Quelques propriétés des mots substitutifs. Bull. Belg. Math. Soc. Simon Stevin **10**(suppl.), 661–676 (2003)
135. Cassels, J.W.S.: On a problem of Steinhaus about normal numbers. Colloq. Math. **7**, 95–101 (1959)
136. Ceccherini-Silberstein, T., Coornaert, M.: Cellular Automata and Groups. Springer Monographs in Mathematics. Springer, New York (2010) https://books.google.cl/books?id=N-LSFFaHTKwC
137. Ceccherini-Silberstein, T.G., Grigorchuk, R.I., de la Harpe, P.: Amenability and paradoxical decompositions for pseudogroups and discrete metric spaces. Trudy Mat. Inst. Steklov. **224**(Algebra. Topol. Differ. Uravn. i ikh Prilozh.), 68–111 (1999)
138. Ceccherini-Silberstein, T.G., Machì, A., Scarabotti, F.: Amenable groups and cellular automata. Ann. Inst. Fourier (Grenoble) **49**(2), 673–685 (1999)
139. Cenzer, D., Dashti, S.A., King, J.L.F.: Computable symbolic dynamics. Math. Log. Q. **54**(5), 460–469 (2008)
140. Chalopin, J., Ochem, P.: Dejean's conjecture and letter frequency. RAIRO Theor. Inform. Appl. **42**(3), 477–480 (2008)
141. Champernowne, D.G.: The construction of decimals normal in the scale of ten. J. Lond. Math. Soc. **s1–8**, 254–260 (1933)
142. Chan, D.H.Y., Hare, K.G.: A multi-dimensional analogue of Cobham's theorem for fractals. Proc. Am. Math. Soc. **142**(2), 449–456 (2014)
143. Charlier, É., Kärki, T., Rigo, M.: Multidimensional generalized automatic sequences and shape-symmetric morphic words. Discret. Math. **310**(6–7), 1238–1252 (2010)
144. Charlier, É., Lacroix, A., Rampersad, N.: Multi-dimensional sets recognizable in all abstract numeration systems. RAIRO Theor. Inform. Appl. **46**(1), 51–65 (2012)
145. Charlier, É., Leroy, J., Rigo, M.: An analogue of Cobham's theorem for graph directed iterated function systems. Adv. Math. **280**, 86–120 (2015)
146. Charlier, É., Leroy, J., Rigo, M.: Asymptotic properties of free monoid morphisms. Linear Algebra Appl. **500**, 119–148 (2016)
147. Charlier, É., Rampersad, N.: The growth function of S-recognizable sets. Theor. Comput. Sci. **412**(39), 5400–5408 (2011)
148. Charlier, É., Rampersad, N., Shallit, J.: Enumeration and decidable properties of automatic sequences. Int. J. Found. Comput. Sci. **23**(5), 1035–1066 (2012)
149. Charlier, É., Rigo, M., Steiner, W.: Abstract numeration systems on bounded languages and multiplication by a constant. Integers **8**, A35, 19 (2008)
150. Choffrut, C., Karhumäki, J.: Combinatorics of words. In: Rozenberg, G., Salomaa, A. (eds.) Handbook of Formal Languages, vol. 1, pp. 329–438. Springer, New York (1997)

151. Chou, C.: Elementary amenable groups. Ill. J. Math. **24**(3), 396–407 (1980)
152. Clark, R.J.: Avoidable formulas in combinatorics on words. Ph.D. thesis, University of California, Los Angeles (2001)
153. Clark, R.J.: The existence of a pattern which is 5-avoidable but 4-unavoidable. Int. J. Algebra Comput. **16**(02), 351–367 (2006)
154. Cobham, A.: On the Hartmanis-Stearns problem for a class of tag machines. In: IEEE Conference Record of 1968 Ninth Annual Symposium on Switching and Automata Theory, pp. 51–60 (1968). Also appeared as IBM Research Technical Report RC-2178, August 23 1968
155. Cobham, A.: On the base-dependence of sets of numbers recognizable by finite automata. Math. Syst. Theory **3**, 186–192 (1969)
156. Cobham, A.: Uniform tag sequences. Math. Syst. Theory **6**, 164–192 (1972)
157. Cohen, D.B.: The large scale geometry of strongly aperiodic subshifts of finite type. Adv. Math. **308**, 599–626 (2017)
158. Cohen, J.M.: Cogrowth and amenability of discrete groups. J. Funct. Anal. **48**(3), 301–309 (1982)
159. Cohen, P.J.: Factorization in group algebras. Duke Math. J. **26**, 199–205 (1959)
160. Connes, A.: Classification of injective factors cases ii_1, ii_∞, iii_λ, $\lambda \neq 1$. Ann. Math. (2) **104**(1), 73–115 (1976)
161. Connes, A., Feldman, J., Weiss, B.: An amenable equivalence relation is generated by a single transformation. Ergodic Theory Dyn. Syst. **1**(4), 431–450 (1981)
162. Conway, J.H.: On Numbers and Games. Academic Press, New York (1976)
163. Coons, M.: Regular sequences and the joint spectral radius. Int. J. Found. Comput. Sci. **28**(2), 135–140 (2017)
164. Coons, M.: Zero order estimate for Mahler functions. N. Z. J. Math. **46**, 83–88 (2016)
165. Coons, M., Tyler, J.: The maximal order of Stern's diatomic sequence. Mosc. J. Comb. Number Theory **4**(3), 3–14 (2014)
166. Copeland, A.H., Erdős, P.: Note on normal numbers. Bull. Am. Math. Soc. **52**, 857–860 (1946)
167. Cornfeld, I.P., Fomin, S.V., Sinaĭ, Y.G.: Ergodic theory. Springer, New York (1982). Translated from the Russian by A.B. Sosinskiĭ
168. Coulhon, T., Saloff-Coste, L.: Isopérimétrie pour les groupes et les variétés. Rev. Mat. Iberoamericana **9**(2), 293–314 (1993)
169. Coutinho, S.C., McConnell, J.C.: The quest for quotient rings (of noncommutative noetherian rings). Am. Math. Mon. **110**(4), 298–313 (2003)
170. Culik II, K.: An aperiodic set of 13 Wang tiles. Discret. Math. **160**, 245–251 (1996)
171. Culik II, K., Pachl, J.K., Yu, S.: On the limit sets of cellular automata. SIAM J. Comput. **18**(4), 831–842 (1989)
172. Currie, J.D.: The number of binary words avoiding abelian fourth powers grows exponentially. Theor. Comput. Sci. **319**(1), 441–446 (2004)
173. Currie, J.D., Rampersad, N.: Fixed points avoiding abelian k-powers. J. Comb. Theory Ser. A **119**(5), 942–948 (2012)
174. Currie, J.D., Visentin, T.I.: Long binary patterns are abelian 2-avoidable. Theor. Comput. Sci. **409**(3), 432–437 (2008)
175. Dai, J., Lathrop, J., Lutz, J., Mayordomo, E.: Finite-state dimension. Theor. Comput. Sci. **310**, 1–33 (2004)
176. Dajani, K., Kraaikamp, C.: Ergodic Theory of Numbers. Carus Mathematical Monographs, vol. 29. Mathematical Association of America, Washington, DC (2002)
177. D'Angeli, D., Godin, T., Klimann, I., Picantin, M., Rodaro, E.: Boundary action of automaton groups without singular points and Wang tilings (2016). ArXiv:1604.07736
178. D'Angeli, D., Rodaro, E.: A geometric approach to (semi)-groups defined by automata via dual transducers. Geom. Dedicata **174-1**, 375–400 (2015)
179. Daubechies, I., Lagarias, J.C.: Sets of matrices all infinite products of which converge. Linear Algebra Appl. **161**, 227–263 (1992)

180. Davenport, H., Erdős, P.: Note on normal decimals. Can. J. Math. **4**, 58–63 (1952)
181. Day, M.M.: Amenable semigroups. Ill. J. Math. **1**, 509–544 (1957)
182. de Bruijn, N.G.: A combinatorial problem. Proc. Konin. Neder. Akad. Wet. **49**, 758–764 (1946)
183. de la Harpe, P.: Topics in Geometric Group Theory. University of Chicago Press, Chicago, IL (2000)
184. Dehornoy, P.: Garside and quadratic normalisation: a survey. In: 19th International Conference on Developments in Language Theory (DLT 2015). Lecture Notes in Computer Science, vol. 9168, pp. 14–45 (2015)
185. Dehornoy, P., Guiraud, Y.: Quadratic normalization in monoids. Int. J. Algebra Comput. **26**(5), 935–972 (2016)
186. Dehornoy, P., et al.: Foundations of Garside theory. Eur. Math. Soc. Tracts in Mathematics, vol. 22 (2015) http://www.math.unicaen.fr/ garside/Garside.pdf
187. Dejean, F.: Sur un théorème de Thue. J. Comb. Theory Ser. A **13**(1), 90–99 (1972)
188. Dekking, F.M.: Strongly non-repetitive sequences and progression-free sets. J. Comb. Theory Ser. A **27**(2), 181–185 (1979)
189. Dekking, F.M., Mendès France, M., Poorten, A.J.v.d.: Folds! Math. Intelligencer **4**, 130–138, 173–181, 190–195 (1982). Erratum, **5** (1983), 5
190. Delacourt, M., Ollinger, N.: Permutive one-way cellular automata and the finiteness problem for automaton groups. In: 13th Conference on Computability in Europe (CiE 2017). Lecture Notes in Computer Science, vol. 10307, pp. 234–245 (2017)
191. Delange, H.: Sur la fonction sommatoire de la fonction "somme des chiffres". Enseign. Math. **21**, 31–47 (1975)
192. Delvenne, J.C., Kůrka, P., Blondel, V.D.: Computational Universality in Symbolic Dynamical Systems, pp. 104–115. Springer, Berlin/Heidelberg (2005)
193. Denker, M., Grillenberger, C., Sigmund, K.: Ergodic Theory on Compact Spaces. Lecture Notes in Mathematics, vol. 527. Springer, Berlin (1976)
194. Derriennic, Y.: Lois "zéro ou deux" pour les processus de Markov. Applications aux marches aléatoires. Ann. Inst. H. Poincaré Sect. B (N.S.) **12**(2), 111–129 (1976)
195. Derriennic, Y.: Quelques applications du théorème ergodique sous-additif. In: Conference on Random Walks, Kleebach, 1979. Astérisque, vol. 74, pp. 183–201, 4. Soc. Math. France, Paris (1980)
196. Diestel, R.: A short proof of Halin's grid theorem. Abh. Math. Semin. Univ. Hambg. **74**, 237–242 (2004)
197. Dixmier, J.: Les moyennes invariantes dans les semi-groups et leurs applications. Acta Sci. Math. Szeged **12**(Leopoldo Fejer et Frederico Riesz LXX annos natis dedicatus, Pars A), 213–227 (1950)
198. Downey, R.G., Hirschfeldt, D.: Algorithmic randomness and complexity. Theory and Applications of Computability, vol. xxvi, 855 p. Springer, New York, NY (2010)
199. Drmota, M., Tichy, R.F.: Sequences, Discrepancies, and Applications. Lecture Notes in Mathematics, vol. 1651. Springer, Berlin (1997)
200. Dumas, P.: Récurrences mahlériennes, suites automatiques, études asymptotiques. Institut National de Recherche en Informatique et en Automatique (INRIA), Rocquencourt (1993). Thèse, Université de Bordeaux I, Talence, 1993
201. Dumas, P.: Joint spectral radius, dilation equations, and asymptotic behavior of radix-rational sequences. Linear Algebra Appl. **438**(5), 2107–2126 (2013)
202. Dumas, P.: Asymptotic expansions for linear homogeneous divide-and-conquer recurrences: algebraic and analytic approaches collated. Theor. Comput. Sci. **548**, 25–53 (2014)
203. Durand, B., Romashchenko, A., Shen, A.: Effective Closed Subshifts in 1D Can Be Implemented in 2D, pp. 208–226. Springer, Berlin/Heidelberg (2010)
204. Durand, F.: Combinatorics on Bratteli diagrams and dynamical systems. In: Combinatorics, Automata and Number Theory. Encyclopedia of Mathematics and its Applications, vol. 135, pp. 324–372. Cambridge University Press, Cambridge (2010)

205. Durand, F., Host, B., Skau, C.: Substitutional dynamical systems, Bratteli diagrams and dimension groups. Ergodic Theory Dyn. Syst. **19**(4), 953–993 (1999)
206. Durand, F., Rigo, M.: On Cobham's theorem. In: Handbook of Automata. European Mathematical Society Publishing House (in press)
207. Dye, H.A.: On groups of measure preserving transformation. I. Am. J. Math. **81**, 119–159 (1959)
208. Edgar, G.: Measure, Topology, and Fractal Geometry, 2nd edn. Undergraduate Texts in Mathematics. Springer, New York (2008)
209. Edlin, A.E.: The number of binary cube-free words of length up to 47 and their numerical analysis. J. Differ. Equ. Appl. **5**(4–5), 353–354 (1999)
210. Eggleston, H.G.: The fractional dimension of a set defined by decimal properties. Q. J. Math. Oxford Ser. **20**, 31–36 (1949)
211. Eilenberg, S.: Automata, Languages, and Machines, vol. A. Academic Press, New York (1974)
212. Elek, G., Monod, N.: On the topological full group of a minimal cantor $\mathbf{Z}^2$-system. Proc. Am. Math. Soc. **141**(10), 3549–3552 (2013)
213. Elekes, M., Keleti, T., Máthé, A.: Self-similar and self-affine sets: measure of the intersection of two copies. Ergodic Theory Dyn. Syst. **30**(2), 399–440 (2010)
214. Entringer, R.C., Jackson, D.E., Schatz, J.A.: On nonrepetitive sequences. J. Comb. Theory Ser. A **16**(2), 159–164 (1974)
215. Epstein, D.B.A., Cannon, J.W., Holt, D.F., Levy, S.V.F., Paterson, M.S., Thurston, W.P.: Word Processing in Groups. Jones and Bartlett Publishers, Boston, MA (1992)
216. Erdős, P.: Some unsolved problems. Mich. Math. J. **4**(3), 291–300 (1957)
217. Erdős, P.: Some unsolved problems. Magyar Tud. Akad. Mat. Kutató Int. Közl. **6**, 221–254 (1961)
218. Erschler, A.G.: Poisson-furstenberg boundaries, large-scale geometry and growth of groups. In: Proc. ICM Hyderabad, India, vol. II, pp. 681–704 (2010)
219. Evdokimov, A.A.: Strongly asymmetric sequences generated by a finite number of symbols. Dokl. Akad. Nauk SSSR **179**, 1268–1271 (1968)
220. Feng, D.J., Wang, Y.: On the structures of generating iterated function systems of Cantor sets. Adv. Math. **222**(6), 1964–1981 (2009)
221. Ferrante, J., Rackoff, C.: A decision procedure for the first order theory of real addition with order. SIAM J. Comput. **4**, 69–76 (1975)
222. Figueira, S., Nies, A.: Feasible analysis, randomness, and base invariance. Theory Comput. Syst. **56**, 439–464 (2015)
223. Fiorenzi, F., Ochem, P., Vaslet, E.: Bounds for the generalized repetition threshold. Theor. Comput. Sci. **412**(27), 2955–2963 (2011)
224. Flajolet, P., Sedgewick, R.: Analytic Combinatorics. Cambridge University Press, Cambridge (2009)
225. Følner, E.: Note on a generalization of a theorem of Bogolioùboff. Math. Scand. **2**, 224–226 (1954)
226. Følner, E.: On groups with full Banach mean value. Math. Scand. **3**, 243–254 (1955)
227. Fraenkel, A.S.: Systems of numeration. Am. Math. Mon. **92**, 105–114 (1985)
228. Fraenkel, A.S., Simpson, R.J.: How many squares must a binary sequence contain? Electron. J. Comb. **2**, R2, 9pp. (1995)
229. Frougny, C.: Representations of numbers and finite automata. Math. Syst. Theory **25**, 37–60 (1992)
230. Frougny, C., Pelantova, E.: Beta-representations of 0 and Pisot numbers. J. Théor. Nombres Bordeaux (in press)
231. Frougny, C., Sakarovitch, J.: Number representation and finite automata. In: Combinatorics, Automata and Number Theory. Encyclopedia of Mathematics and Its Applications, vol. 135, pp. 34–107. Cambridge University Press, Cambridge (2010)
232. Frougny, C., Solomyak, B.: On representation of integers in linear numeration systems. In: Pollicott, M., Schmidt, K. (eds.) Ergodic Theory of $\mathbb{Z}^d$ Actions (Warwick, 1993–1994).

London Mathematical Society Lecture Note Series, vol. 228, pp. 345–368. Cambridge University Press, Cambridge (1996)
233. Fukuyama, K.: The law of the iterated logarithm for discrepancies of $\{\theta^n x\}$. Acta Math. Hung. **118**(1), 155–170 (2008)
234. Furstenberg, H.: A poisson formula for semi-simple lie groups. Ann. Math. (2) **77**, 335–386 (1963)
235. Furstenberg, H.: Boundary theory and stochastic processes on homogeneous spaces. In: Harmonic Analysis on Homogeneous Spaces. Proceedings of Symposia in Pure Mathematics, vol. XXVI, Williams Coll., Williamstown, MA, 1972, pp. 193–229. American Mathematical Society, Providence, RI (1973)
236. Gál, S., Gál, L.: The discrepancy of the sequence $\{(2^n x)\}$. Koninklijke Nederlandse Akademie van Wetenschappen Proceedings. Seres A 67 = Indagationes Mathematicae **26**, 129–143 (1964)
237. Gamard, G., Ochem, P., Richomme, G., Séébold, P.: Avoidability of circular formulas (2016). ArXiv:1610.04439
238. Gawron, P.W., Nekrashevych, V.V., Sushchansky, V.I.: Conjugation in tree automorphism groups. Int. J. Algebra Comput. **11**(5), 529–547 (2001)
239. Georgiadis, L., Goldberg, A.V., Tarjan, R.E., Werneck, R.F.: An Experimental Study of Minimum Mean Cycle Algorithms, pp. 1–13. SIAM (2009)
240. Ghys, É., Carrière, Y.: Relations d'équivalence moyennables sur les groupes de Lie. C. R. Acad. Sci. Paris Sér. I Math. **300**(19), 677–680 (1985)
241. Ghys, É., de la Harpe, P.: Sur les groupes hyperboliques d'après Mikhael Gromov. Progress in Mathematics, vol. 83. Birkhäuser, Boston, MA (1990). Papers from the Swiss Seminar on Hyperbolic Groups held in Bern, 1988
242. Gillibert, P.: The finiteness problem for automaton semigroups is undecidable. Int. J. Algebra Comput. **24**(1), 1–9 (2014)
243. Gillibert, P.: Simulating Turing machines with invertible Mealy automata (2017, in preparation)
244. Glasner, S.: Proximal Flows. Lecture Notes in Mathematics, vol. 517. Springer, Berlin/New York (1976)
245. Glasner, Y., Monod, N.: Amenable actions, free products and a fixed point property. Bull. Lond. Math. Soc. **39**(1), 138–150 (2007)
246. Glasner, Y., Mozes, S.: Automata and square complexes. Geom. Dedicata **111**, 43–64 (2005)
247. Gluškov, V.: Abstract theory of automata. Uspehi Mat. Nauk **16**(5), 3–62 (1961)
248. Goč, D., Henshall, D., Shallit, J.: Automatic theorem-proving in combinatorics on words. Int. J. Found. Comput. Sci. **24**(6), 781–798 (2013)
249. Goč, D., Mousavi, H., Schaeffer, L., Shallit, J.: A new approach to the paperfolding sequences. In: Evolving Computability. Lecture Notes in Computer Science, vol. 9136, pp. 34–43. Springer, Cham (2015)
250. Goč, D., Mousavi, H., Shallit, J.: On the number of unbordered factors. In: Language and Automata Theory and Applications. Lecture Notes in Computer Science, vol. 7810, pp. 299–310. Springer, Heidelberg (2013)
251. Godin, T., Klimann, I.: Connected reversible Mealy automata of prime size cannot generate infinite Burnside groups. In: 41st International Symposium on Mathematical Foundations of Computer Science (MFCS 2016). LIPIcs, vol. 58, pp. 44:1–44:14 (2016)
252. Godin, T., Klimann, I., Picantin, M.: On torsion-free semigroups generated by invertible reversible Mealy automata. In: 9th International Conference on Language and Automata Theory and Applications (LATA). Lecture Notes in Computer Science, vol. 8977, pp. 328–339 (2015)
253. Goldie, A.W.: Semi-prime rings with maximum condition. Proc. Lond. Math. Soc. (3) **10**, 201–220 (1960)
254. Golod, E.S.: On nil-algebras and finitely approximable p-groups. Izv. Akad. Nauk SSSR Ser. Mat. **28**, 273–276 (1964). English translation: Am. Math. Soc. Transl. **48**, 108–111 (1965)

255. Golod, E.S.: On nil-algebras and finitely residual groups. Izv. Akad. Nauk SSSR. Ser. Mat. **28**, 273–276 (1964)
256. Golod, E.S., Shafarevich, I.: On the class field tower. Izv. Akad. Nauk SSSR Ser. Mat. **28**, 261–272 (1964)
257. Goodman-Strauss, C.: Matching rules and substitution tilings. Ann. Math. **147**(1), 181–223 (1998)
258. Goodman-Strauss, C.: A hierarchical strongly aperiodic set of tiles in the hyperbolic plane. Theor. Comput. Sci. **411**(7), 1085–1093 (2010)
259. Goodwyn, L.W.: Topological entropy bounds measure-theoretic entropy. Proc. Am. Math. Soc. **23**, 679–688 (1969)
260. Gottschalk, W.H.: Almost periodic points with respect to transformation semi-groups. Ann. Math. (2) **47**, 762–766 (1946)
261. Gouëzel, S.: A numerical lower bound for the spectral radius of random walks on surface groups. Comb. Probab. Comput. **24**(6), 838–856 (2015)
262. Gournay, A.: The Liouville property via Hilbertian compression (2014). ArXiv:1403.1195
263. Gournay, A.: Amenability criteria and critical probabilities in percolation. Expo. Math. **33**(1), 108–115 (2015)
264. Greenleaf, F.P.: Amenable actions of locally compact groups. J. Funct. Anal .**4**, 295–315 (1969)
265. Greenleaf, F.P.: Invariant Means on Topological Groups and Their Applications. Van Nostrand Mathematical Studies, No. 16. Van Nostrand, New York (1969)
266. Grigorchuk, R.I.: On Burnside's problem on periodic groups. Funktsional. Anal. i Prilozhen. **14**(1), 53–54 (1980)
267. Grigorchuk, R.I.: Symmetrical random walks on discrete groups. In: Multicomponent Random Systems, pp. 285–325. Dekker, New York (1980)
268. Grigorchuk, R.I.: On the Milnor problem of group growth. Dokl. Akad. Nauk SSSR **271**(1), 30–33 (1983)
269. Grigorchuk, R.I.: Degrees of growth of finitely generated groups and the theory of invariant means. Izv. Akad. Nauk SSSR Ser. Mat. **48**(5), 939–985 (1984)
270. Grigorchuk, R.I.: New Horizons in pro-p Groups, chap. Just Infinite Branch Groups, pp. 121–179. Birkhäuser, Boston, MA (2000)
271. Grigorchuk, R.I., Nekrashevich, V.V., Sushchanskiĭ, V.I.: Automata, dynamical systems, and groups. Tr. Mat. Inst. Steklova **231**, 134–214 (2000)
272. Grigorchuk, R.I., Nekrashevych, V.V.: Amenable actions of non-amenable groups. J. Math. Sci. **140**(3), 391–397 (2007)
273. Gromov, M.: Entropy and isoperimetry for linear and non-linear group actions. Groups Geom. Dyn. **2**(4), 499–593 (2008)
274. Gromov, M.L.: Groups of polynomial growth and expanding maps. Inst. Hautes Études Sci. Publ. Math. **53**, 53–73 (1981)
275. Gromov, M.L.: Endomorphisms of symbolic algebraic varieties. J. Eur. Math. Soc. (JEMS) **1**(2), 109–197 (1999)
276. Gromov, M.L.: Topological invariants of dynamical systems and spaces of holomorphic maps. I. Math. Phys. Anal. Geom. **2**(4), 323–415 (1999)
277. Guégan, G., Ochem, P.: A short proof that shuffle squares are 7-avoidable. RAIRO Theor. Inform. Appl. **50**(1), 101–103 (2016)
278. Guivarc′h, Y.: Groupes de Lie à croissance polynomiale. C. R. Acad. Sci. Paris Sér. A-B **271**, A237–A239 (1970)
279. Hadamard, J.: Théorème sur les séries entières. Acta Math. **22**, 55–63 (1899)
280. Häggström, O., Jonasson, J.: Uniqueness and non-uniqueness in percolation theory. Probab. Surv. **3**, 289–344 (2006)
281. Hall, P.: On representatives of subsets. J. Lond. Math. Soc. **10**, 26–30 (1935)
282. Hansel, G.: A simple proof of the Skolem-Mahler-Lech theorem. In: Brauer, W. (ed.) Proceedings of the 12th International Conference on Automata, Languages, and Programming (ICALP). Lecture Notes in Computer Science, vol. 194, pp. 244–249. Springer, Berlin (1985)

283. Hardy, G.H., Wright, E.M.: An Introduction to the Theory of Numbers, 5th edn. Oxford University Press, Oxford (1985)
284. Hare, K.G., Morris, I.D., Sidorov, N., Theys, J.: An explicit counterexample to the Lagarias-Wang finiteness conjecture. Adv. Math. **226**(6), 4667–4701 (2011)
285. Hartmanis, J., Stearns, R.E.: On the computational complexity of algorithms. Trans. Am. Math. Soc. **117**, 285–306 (1965)
286. Hausdorff, F.: Bemerkung über den Inhalt von Punktmengen. Math. Ann. **75**(3), 428–433 (1914)
287. Hedlund, G., Morse, M.: Symbolic dynamics. Am. J. Math. **60**(4), 815–866 (1938)
288. Hedlund, G.A.: Endormorphisms and automorphisms of the shift dynamical system. Math. Syst. Theory **3**, 320–375 (1969)
289. Herman, R.H., Putnam, I.F., Skau, C.F.: Ordered Bratteli diagrams, dimension groups and topological dynamics. Int. J. Math. **3**(6), 827–864 (1992)
290. Hindman, N.: Finite sums of sequences within cells of a partition of $\mathbb{N}$. J. Comb. Theory. Ser. A **17**, 1–11 (1974)
291. Hindman, N.: Partitions and sums of integers with repetition. J. Comb. Theory. Ser. A **27**, 19–32 (1979)
292. Hindman, N., Leader, I., Strauss, D.: Pairwise sums in colourings of the reals. Abh. Math. Semin. Univ. Hambg. 1–13 (2016)
293. Hindman, N., Strauss, D.: Algebra in the Stone-Čech compactification: theory and applications. Walter de Gruyter, Berlin (2012)
294. Hochman, M.: On the dynamics and recursive properties of multidimensional symbolic systems. Invent. Math. **176**(1), 131–167 (2009)
295. Hochman, M., Meyerovitch, T.: A characterization of the entropies of multidimensional shifts of finite type. Ann. Math. (2) **171**(3), 2011–2038 (2010)
296. Hoffmann, M.: Automatic semigroups. Ph.D. thesis, Univ Leicester (2001)
297. Hollander, M.: Greedy numeration systems and regularity. Theory Comput. Syst. **31**, 111–133 (1998)
298. Honkala, J.: On the simplification of infinite morphic words. Theor. Comput. Sci. **410**(8-10), 997–1000 (2009)
299. Hooper, P.K.: The undecidability of the Turing machine immortality problem. J. Symb. Log. **31**(2), 219–234 (1966)
300. Hopcroft, J.E., Motwani, R., Ullman, J.D.: Introduction to Automata Theory, Languages, and Computation, 3rd edn. Addison-Wesley, Boston (2006)
301. Hopcroft, J.E., Ullman, J.D.: Introduction to Automata Theory, Languages, and Computation. Addison-Wesley, Boston (1979)
302. Hughes, B.: Trees and ultrametric spaces: a categorical equivalence. Adv. Math. **189**(1), 148–191 (2004)
303. Hulanicki, A.: Means and Følner condition on locally compact groups. Stud. Math. **27**, 87–104 (1966)
304. Huova, M., Karhumki, J., Saarela, A.: Problems in between words and abelian words: k-abelian avoidability. Theor. Comput. Sci. **454**, 172–177 (2012)
305. Hurwitz, A.: Vorlesungen über die Zahlentheorie der Quaternionen. Springer, Berlin (1919)
306. Hutchinson, J.E.: Fractals and self-similarity. Indiana Univ. Math. J. **30**(5), 713–747 (1981)
307. Hyde, J., Laschos, V., Olsen, L., Petrykiewicz, I., Shaw, A.: Iterated Cesàro averages, frequencies of digits, and Baire category. Acta Arith. **144**(3), 287–293 (2010)
308. Ilie, L., Ochem, P., Shallit, J.: A generalization of repetition threshold. Theor. Comput. Sci. **345**(2), 359–369 (2005)
309. Jeandel, E.: Aperiodic subshifts on polycyclic groups (2015). ArXiv:1510.02360
310. Jeandel, E.: Translation-like actions and aperiodic subshifts on groups (2015). ArXiv:1508.06419
311. Jeandel, E., Rao, M.: An aperiodic set of 11 Wang tiles (2015). ArXiv:1506.06492
312. Jeandel, E., Vanier, P.: Characterizations of periods of multi-dimensional shifts. Ergodic Theory Dyn. Syst. **35**(2), 431–460 (2015)

313. Johnson, B.E.: Cohomology in Banach Algebras. American Mathematical Society, Providence, RI (1972). Memoirs of the American Mathematical Society, No. 127
314. Jørgensen, T.: A note on subgroups of $sl(2, C)$. Q. J. Math. Oxf. Ser. (2) **28**(110), 209–211 (1977)
315. Jungers, R.: The Joint Spectral Radius, Theory and Applications. Lecture Notes in Control and Information Sciences, vol. 385. Springer, Berlin (2009)
316. Jungers, R.M., Blondel, V.D.: On the finiteness property for rational matrices. Linear Algebra Appl. **428**(10), 2283–2295 (2008)
317. Juschenko, K.: Amenability of discrete groups by examples (2015). Book draft
318. Juschenko, K., Monod, N.: Cantor systems, piecewise translations and simple amenable groups. Ann. Math. (2) **178**(2), 775–787 (2013)
319. Juschenko, K., Zhang, T.: Infinitely supported Liouville measures of Schreier graphs (2016). ArXiv:1608.03554
320. Juschenko, K., Matte Bon, N., Monod, N., de la Salle, M.: Extensive amenability and an application to interval exchanges (2015). ArXiv:1503.04977
321. Juschenko, K., Nekrashevych, V., de la Salle, M.: Extensions of amenable groups by recurrent groupoids. Invent. Math. **206**(3), 837–867 (2016)
322. Kahr, A., Moore, E.F., Wang, H.: Entscheidungsproblem reduced to the $\forall\exists\forall$ case. Proc. Natl. Acad. Sci. U. S. A. **48**(3), 365–377 (1962)
323. Kaimanovich, V.A.: Amenability, hyperfiniteness, and isoperimetric inequalities. C. R. Acad. Sci. Paris Sér. I Math. **325**(9), 999–1004 (1997)
324. Kaimanovich, V.A.: Thompson's group f is not Liouville (2016). ArXiv:1602.02971
325. Kaimanovich, V.A., Vershik, A.M.: Random walks on discrete groups: boundary and entropy. Ann. Probab. **11**(3), 457–490 (1983)
326. Kakutani, S.: Two fixed-point theorems concerning bicompact convex sets. Proc. Imp. Acad. **14**(7), 242–245 (1938)
327. Kaplansky, I.: Problems in the theory of rings. Report of a conference on linear algebras, June, 1956, pp. 1–3. National Academy of Sciences-National Research Council, Washington, Publ. 502 (1957)
328. Kaplansky, I.: "Problems in the theory of rings" revisited. Am. Math. Mon. **77**, 445–454 (1970)
329. Karhumäki, J.: Generalized Parikh mappings and homomorphisms. Inf. Control **47**(3), 155–165 (1980)
330. Karhumäki, J., Saarela, A., Zamboni, L.Q.: On a generalization of Abelian equivalence and complexity of infinite words. J. Comb. Theory Ser. A **120**(8), 2189–2206 (2013)
331. Karhumäki, J., Shallit, J.: Polynomial versus exponential growth in repetition-free binary words. J. Comb. Theory Ser. A **105**(2), 335–347 (2004)
332. Kari, J.: The nilpotency problem of one-dimensional cellular automata. SIAM J. Comput. **21**(3), 571–586 (1992)
333. Kari, J.: A small aperiodic set of Wang tiles. Discret. Math. **160**, 259–264 (1996)
334. Kari, J.: The tiling problem revisited. In: MCU, pp. 72–79 (2007)
335. Kari, J.: On the undecidability of the tiling problem. In: Current Trends in Theory and Practice of Computer Science (SOFSEM), pp. 74–82 (2008)
336. Kari, J., Ollinger, N.: Periodicity and immortality in reversible computing. In: 33rd International Symposium on Mathematical Foundations of Computer Science (MFCS 2008). LNCS, vol. 5162, pp. 419–430 (2008)
337. Karp, R.: A characterization of the minimum mean cycle in a digraph. Discret. Math. **23**, 309–311 (1978)
338. Kaufman, R.: On the theorem of Jarník and Besicovitch. Acta Arith. **39**(3), 265–267 (1981)
339. Každan, D.A.: On the connection of the dual space of a group with the structure of its closed subgroups. Funkcional. Anal. i Priložen. **1**, 71–74 (1967)
340. Keane, M.: Interval exchange transformations. Math. Z. **141**, 25–31 (1975)
341. Kelley, J.L.: The Tychonoff product theorem implies the axiom of choice. Fundam. Math. **37**, 75–76 (1950)

342. Keränen, V.: Abelian squares are avoidable on 4 letters. In: ICALP, pp. 41–52 (1992)
343. Keränen, V.: New abelian square-free DT0L-languages over 4 letters. Manuscript (2003)
344. Keränen, V.: A powerful abelian square-free substitution over 4 letters. Theor. Comput. Sci. **410**(38), 3893–3900 (2009)
345. Kesten, H.: Symmetric random walks on groups. Trans. Am. Math. Soc. **92**, 336–354 (1959)
346. Khalyavin, A.: The minimal density of a letter in an infinite ternary square-free word is 883/3215. J. Integer Seq. **10**, Art. 07.6.5 (2007)
347. Khintchine, A.: Einige sätze über kettenbrüche, mit anwendungen auf die theorie der diophantischen approximationen. Math. Ann. **92**(1), 115–125 (1924)
348. Kitchens, B.P.: Symbolic dynamics, one-sided, two-sided and countable state Markov shifts. Universitext. Springer, Berlin (1998)
349. Kleene, S.C.: Representation of events in nerve nets and finite automata. In: Automata Studies, pp. 3–42. Princeton University Press, Princeton (1956)
350. Kleiner, B.: A new proof of Gromov's theorem on groups of polynomial growth. J. Am. Math. Soc. **23**(3), 815–829 (2010)
351. Klimann, I.: Automaton semigroups: the two-state case. Theory Comput. Syst. 1–17 (2014)
352. Klimann, I.: On level-transitivity and exponential growth. Semigroup Forum, pp. 1–7 (2016)
353. Klimann, I., Picantin, M., Savchuk, D.: A connected 3-state reversible Mealy automaton cannot generate an infinite burnside group. In: 19th International Conference on Developments in Language Theory (DLT). Lecture Notes in Computer Science, vol. 9168, pp. 313–325 (2015)
354. Klimann, I., Picantin, M., Savchuk, D.: Orbit automata as a new tool to attack the order problem in automaton groups. J. Algebra **445**, 433–457 (2016)
355. Koiran, P., Cosnard, M., Garzon, M.: Computability with low-dimensional dynamical systems. Theor. Comput. Sci. **132**(1), 113–128 (1994)
356. Kolpakov, R.: Efficient lower bounds on the number of repetition-free words. J. Integer Seq. **10**, Art. 07.3.2 (2007)
357. Kolpakov, R., Kucherov, G., Tarannikov, Y.: On repetition-free binary words of minimal density. Theor. Comput. Sci. **218**, 161–175 (1999)
358. Kolpakov, R., Rao, M.: On the number of Dejean words over alphabets of 5, 6, 7, 8, 9 and 10 letters. Theor. Comput. Sci. **412**(46), 6507–6516 (2011)
359. Korobov, A.N.: Continued fractions of certain normal numbers. Mat. Z. **47**(2), 28–33 (1990, in Russian). English translation in Math. Notes Acad. Sci. USSR **47**, 128–132 (1990)
360. Kozyakin, V.S.: A dynamical systems construction of a counterexample to the finiteness conjecture. In: Proceedings of the 44th IEEE Conference on Decision and Control, European Control Conference, pp. 2338–2343 (2005)
361. Kraaikamp, C., Nakada, H.: On normal numbers for continued fractions. Ergodic Theory Dyn. Syst. **20**(5), 1405–1421 (2000)
362. Krieger, F.: Le lemme d'Ornstein-Weiss d'après Gromov. In: Dynamics, Ergodic Theory, and Geometry. Mathematical Sciences Research Institute Publications, vol. 54, pp. 99–111. Cambridge University Press, Cambridge (2007)
363. Kucherov, G., Ochem, P., Rao, M.: How many square occurrences must a binary sequence contain? Electron. J. Comb. **10**(1), R12 (2003)
364. Kuipers, L., Niederreiter, H.: Uniform distribution of sequences. Dover, Mineola (2006)
365. Kurka, P.: On topological dynamics of Turing machines. Theor. Comput. Sci. **174**, 203–216 (1997)
366. Kuske, D., Lohrey, M.: Logical aspects of Cayley-graphs: the group case. Ann. Pure Appl. Logic **131**(1–3), 263 – 286 (2005)
367. Lagarias, J.C., Wang, Y.: The finiteness conjecture for the generalized spectral radius of a set of matrices. Linear Algebra Appl. **214**, 17–42 (1995)
368. Lam, T.Y.: Lectures on Modules and Rings. Graduate Texts in Mathematics, vol. 189. Springer, New York (1999)
369. Lansing, J.: Distribution of values of the binomial coefficients and the Stern sequence. J. Integer Seq. **16**(3), Article 13.3.7, 10 (2013)

370. Le Gloannec, B.: The 4-way deterministic periodic domino problem is undecidable. HAL:00985482 (2014)
371. Lebesgue, H.: Sur certains démonstrations d'existence. Bull. Soc. Math. France **45**, 127–132 (1917)
372. Lecomte, P.B.A., Rigo, M.: Numeration systems on a regular language. Theory Comput. Syst. **34**, 27–44 (2001)
373. Lennox, J.C., Robinson, D.J.: The Theory of Infinite Soluble Groups. Oxford Mathematical Monographs. Oxford Science Publications, Oxford (2004)
374. Leroux, J.: A polynomial time Presburger criterion and synthesis for number decision diagrams. In: Proceedings of the 20th IEEE Symposium on Logic in Computer Science (LICS 2005), 26–29 June 2005, Chicago, IL, USA, pp. 147–156. IEEE Computer Society (2005)
375. Levin, M.: On absolutely normal numbers. Vestnik Moscov. Univ. ser. I, Mat-Meh **1**, 31–37, 87 (1979). English translation in Moscow Univ. Math. Bull. 34(1), 32–39 (1979)
376. Levin, M.: On the discrepancy estimate of normal numbers. Acta Arith. Warzawa **88**, 99–111 (1999)
377. Levy, P.: Sur les lois de probabilité dont dependent les quotients complets et incomplets d'une fraction continue. Bull. Soc. Math. France **57**, 178–194 (1929)
378. Lewis, J.T., Pfister, C.E., Russell, R.P., Sullivan, W.G.: Reconstruction sequences and equipartition measures: an examination of the asymptotic equipartition property. IEEE Trans. Inform. Theory **43**(6), 1935–1947 (1997)
379. Lewis, J.T., Pfister, C.E., Sullivan, W.G.: Large deviations and the thermodynamic formalism: a new proof of the equivalence of ensembles. In: Fannes, M., Maes, C., Verbeure, A. (eds.) On Three Levels: Micro-, Meso-, and Macro-Approaches in Physics, pp. 183–192. Springer US, Boston, MA (1994)
380. Liao, L., Ma, J., Wang, B.: Dimension of some non-normal continued fraction sets. Math. Proc. Camb. Philos. Soc. **145**(1), 215–225 (2008)
381. Lind, D., Marcus, B.: An Introduction to Symbolic Dynamics and Coding. Cambridge University Press, Cambridge (1995)
382. Lodha, Y.: An upper bound for the Tarski numbers of non amenable groups of piecewise projective homeomorphisms (2016). ArXiv:1604.05997
383. Lodha, Y., Moore, J.T.: A nonamenable finitely presented group of piecewise projective homeomorphisms. Groups Geom. Dyn. **10**(1), 177–200 (2016)
384. Löding, C.: Efficient minimization of deterministic weak ω-automata. Inform. Process. Lett. **79**(3), 105–109 (2001)
385. Lothaire, M.: Combinatorics on Words. Encyclopedia of Mathematics and Its Applications, vol. 17. Addison-Wesley, Reading (1983)
386. Lothaire, M.: Algebraic Combinatorics on Words. Encyclopedia of Mathematics and Its Applications, vol. 90. Cambridge University Press, Cambridge (2002)
387. Lothaire, M.: Applied Combinatorics on Words. Encyclopedia of Mathematics and Its Applications, vol. 105. Cambridge University Press, Cambridge (2005)
388. Loxton, J.H.: A method of Mahler in transcendence theory and some of its applications. Bull. Aust. Math. Soc. **29**(1), 127–136 (1984)
389. Loxton, J.H., van der Poorten, A.J.: Arithmetic properties of automata: regular sequences. J. Reine Angew. Math. **392**, 57–69 (1988)
390. de Luca, A., Pribavkina, E., Zamboni, L.: A coloring problem for infinite words. J. Comb. Theory. Ser. A **125**, 306–332 (2014)
391. de Luca, A., Varricchio, S.: On the factors of the Thue-Morse word on three symbols. Inf. Process. Lett. **27**, 281–285 (1988)
392. de Luca, A., Zamboni, L.: On prefixal factorisations of words. Eur. J. Comb. **52**, 59–73 (2016)
393. de Luca, A., Zamboni, L.: On some variations of coloring problems of infinite words. J. Comb. Theory. Ser. A **137**, 166–178 (2016)
394. Lüroth, J.: Ueber eine eindeutige Entwickelung von Zahlen in eine unendliche Reihe. Math. Ann. **21**(3), 411–423 (1883)

395. Lutz, J., Mayordomo, E.: Computing absolutely normal numbers in nearly linear time (2016). ArXiv:1611.05911
396. Machì, A., Mignosi, F.: Garden of Eden configurations for cellular automata on Cayley graphs of groups. SIAM J. Discret. Math. **6**(1), 44–56 (1993)
397. Madritsch, M., Scheerer, A.M., Tichy, R.: Computable absolutely Pisot normal numbers. Acta Aritmetica (2017, to appear). arXiv:1610.06388
398. Madritsch, M.G.: Non-normal numbers with respect to Markov partitions. Discret. Contin. Dyn. Syst. **34**(2), 663–676 (2014)
399. Madritsch, M.G., Mance, B.: Construction of μ-normal sequences. Monatsh. Math. **179**(2), 259–280 (2016)
400. Madritsch, M.G., Petrykiewicz, I.: Non-normal numbers in dynamical systems fulfilling the specification property. Discret. Contin. Dyn. Syst. **34**(11), 4751–4764 (2014)
401. Madritsch, M.G., Thuswaldner, J.M., Tichy, R.F.: Normality of numbers generated by the values of entire functions. J. Number Theory **128**(5), 1127–1145 (2008)
402. Magnus, W., Karrass, A., Solitar, D.: Combinatorial Group Theory, 2nd rev. edn. Dover, Mineola (1976)
403. Mahler, K.: Arithmetische Eigenschaften der Lösungen einer Klasse von Funktionalgleichungen. Math. Ann. **101**(1), 342–366 (1929)
404. Mahler, K.: Arithmetische Eigenschaften einer Klasse transzendental-transzendenter Funktionen. Math. Z. **32**(1), 545–585 (1930)
405. Mahler, K.: Uber das Verschwinden von Potenzreihen mehrerer Veränderlichen in speziellen Punktfolgen. Math. Ann. **103**(1), 573–587 (1930)
406. Mahler, K.: An unsolved problem on the powers of $3/2$. J. Aust. Math. Soc. **8**, 313–321 (1968)
407. Mahler, K.: Remarks on a paper by W. Schwarz. J. Number Theory **1**, 512–521 (1969)
408. Maler, O., Staiger, L.: On syntactic congruences for ω-languages. Theor. Comput. Sci. **183**(1), 93–112 (1997)
409. Mann, A.: How Groups Grow. Lecture Note Series, vol. 395. London Mathematical Society (2012)
410. Margulis, G.A.: Discrete subgroups of semisimple Lie groups. Ergebnisse der Mathematik und ihrer Grenzgebiete (3), vol. 17. Springer, Berlin (1991)
411. Markley, N.G., Paul, M.E.: Matrix subshifts for $\mathbb{Z}^\nu$ symbolic dynamics. Proc. Lond. Math. Soc. **3**(43), 251–272 (1981)
412. Markov, A.A.: Quelques théorèmes sur les ensembles abeliens. C. R. (Dokl.) Acad. Sci. URSS, n. Ser. **1936**(1), 311–313 (1936)
413. Marsault, V., Sakarovitch, J.: Ultimate periodicity of b-recognisable sets: a quasilinear procedure. In: Developments in Language Theory. Lecture Notes in Computer Science, vol. 7907, pp. 362–373. Springer, Heidelberg (2013)
414. McNaughton, R., Zalcstein, Y.: The Burnside problem for semigroups. J. Algebra **34**, 292–299 (1975)
415. Meyerovitch, T.: Finite entropy for multidimensional cellular automata. Ergodic Theory Dyn. Syst. **28**(4), 1243–1260 (2008)
416. Milchior, A.: Büchi automata recognizing sets of reals definable in first-order logic with addition and order (2016). ArXiv:1610.06027
417. Milliken, K.: Ramsey's theorem with sums or unions. J. Comb. Theory Ser. A **18**, 276–290 (1975)
418. Milnor, J.: Growth of finitely generated solvable groups. J. Differ. Geom. **2**, 447–449 (1968)
419. Milnor, J.: Problem 5603. Am. Math. Mon. **75**(6), 685–686 (1968)
420. Mirsky, L.: Transversal Theory. An Account of Some Aspects of Combinatorial Mathematics. Mathematics in Science and Engineering, vol. 75. Academic Press, New York (1971)
421. Monod, N.: Groups of piecewise projective homeomorphisms. Proc. Natl. Acad. Sci. U. S. A. **110**(12), 4524–4527 (2013)
422. Monod, N., Popa, S.: On co-amenability for groups and von Neumann algebras. C. R. Math. Acad. Sci. Soc. R. Can. **25**(3), 82–87 (2003)

423. Moore, E.F.: Machine models of self-reproduction. In: Mathematical Problems in the Biological Sciences. Proceedings of the Symposium in Applied Mathematics, vol. XIV, pp. 17–33. American Mathematical Society, Providence, RI (1962)
424. Morris, D.W.: Amenable groups that act on the line. Algebr. Geom. Topol. **6**, 2509–2518 (2006)
425. Moshe, Y.: On some questions regarding *k*-regular and *k*-context-free sequences. Theor. Comput. Sci. **400**(1-3), 62–69 (2008)
426. Mousavi, H.: Automatic theorem proving in Walnut (2016). ArXiv:1603.06017
427. Mousavi, H., Schaeffer, L., Shallit, J.: Decision algorithms for Fibonacci-automatic words, I: Basic results. RAIRO Theor. Inform. Appl. **50**(1), 39–66 (2016)
428. Mousavi, H., Shallit, J.: Mechanical proofs of properties of the Tribonacci word. In: Combinatorics on Words. Lecture Notes in Computer Science, vol. 9304, pp. 170–190. Springer, Cham (2015)
429. Mozes, S.: Tilings, substitutions systems and dynamical systems generated by them. J. Anal. Math. **53**, 139–186 (1989)
430. Muchnik, R., Pak, I.: Percolation on Grigorchuk Groups. Commun. Algebra **29**(2), 661–671 (2001)
431. Muller, D.E., Schupp, P.E.: The theory of ends, pushdown automata, and second-order logic. Theor. Comput. Sci. **37**, 51–75 (1985)
432. Myers, D.: Non recursive tilings of the plane II. J. Symb. Log. **39**(2), 286–294 (1974)
433. Myhill, J.: The converse of Moore's Garden-of-Eden theorem. Proc. Am. Math. Soc. **14**, 685–686 (1963)
434. Nakada, H.: Metrical theory for a class of continued fraction transformations and their natural extensions. Tokyo J. Math. **4**(2), 399–426 (1981)
435. Nakai, Y., Shiokawa, I.: A class of normal numbers. Jpn. J. Math. (N.S.) **16**(1), 17–29 (1990)
436. Nakai, Y., Shiokawa, I.: Discrepancy estimates for a class of normal numbers. Acta Arith. **62**(3), 271–284 (1992)
437. Nakai, Y., Shiokawa, I.: Normality of numbers generated by the values of polynomials at primes. Acta Arith. **81**, 345–356 (1997)
438. Naor, A., Peres, Y.: Embeddings of discrete groups and the speed of random walks. Int. Math. Res. Not. IMRN pp. Art. ID rnn 076, 34 (2008)
439. Nash-Williams, C.S.J.A.: Random walk and electric currents in networks. Proc. Camb. Philos. Soc. **55**, 181–194 (1959)
440. Nasu, M.: Textile Systems for Endomorphisms and Automorphisms of the Shift. Memoirs of the American Mathematical Society, vol. 114. American Mathematical Society, Providence (1995)
441. Nekrashevych, V.V.: Self-Similar Groups. Mathematical Surveys and Monographs, vol. 117. American Mathematical Society, Providence, RI (2005)
442. Nekrashevych, V.V.: Simple groups of dynamical origin (2015). ArXiv:1511.08241
443. Nesterenko, Y.V.: Estimate of the orders of the zeroes of functions of a certain class, and their application in the theory of transcendental numbers. Izv. Akad. Nauk SSSR Ser. Mat. **41**(2), 253–284, 477 (1977)
444. Nesterenko, Y.V.: Algebraic independence of algebraic powers of algebraic numbers. Mat. Sb. (N.S.) **123(165)**(4), 435–459 (1984)
445. Nikishin, E.M., Sorokin, V.N.: Rational approximations and orthogonality. Translations of Mathematical Monographs, vol. 92. American Mathematical Society, Providence, RI (1991). Translated from the Russian by Ralph P. Boas
446. Nikodym, O.: Sur une généralisation des intégrales de M. J. Radon. Fundam. Math. **15**, 358 (1930)
447. Nishioka, K.: Algebraic independence measures of the values of Mahler functions. J. Reine Angew. Math. **420**, 203–214 (1991)
448. Niven, I.: Irrational Numbers. MAA (1963)
449. Ochem, P.: A generator of morphisms for infinite words. RAIRO Theor. Inform. Appl. **40**, 427–441 (2006)

450. Ochem, P.: Letter frequency in infinite repetition-free words. Theor. Comput. Sci. **380**(3), 388–392 (2007)
451. Ochem, P.: Doubled patterns are 3-avoidable. Electron. J. Comb. **23**(1) (2016)
452. Ochem, P., Reix, T.: Upper bound on the number of ternary square-free words. In: Workshop on Words and Automata (2006)
453. Ochem, P., Rosenfeld, M.: Avoidability of formulas with two variables (2016). ArXiv:1606.03955
454. Odifreddi, P.: Chapter {XIV} enumeration degrees. In: Classical Recursion Theory. Studies in Logic and the Foundations of Mathematics, vol. 143, pp. 827–861. Elsevier, Amsterdam (1999)
455. Ol′shanskiĭ, A. Y.: Infinite groups with cyclic subgroups. Dokl. Akad. Nauk SSSR **245**(4), 785–787 (1979)
456. Ol′shanskiĭ, A.Y.: On the question of the existence of an invariant mean on a group. Uspekhi Mat. Nauk **35**(4(214)), 199–200 (1980)
457. Olsen, L.: Extremely non-normal continued fractions. Acta Arith. **108**(2), 191–202 (2003)
458. Olsen, L.: Multifractal analysis of divergence points of deformed measure theoretical Birkhoff averages. J. Math. Pures Appl. (9) **82**(12), 1591–1649 (2003)
459. Olsen, L.: Extremely non-normal numbers. Math. Proc. Camb. Philos. Soc. **137**(1), 43–53 (2004)
460. Olsen, L., Winter, S.: Multifractal analysis of divergence points of deformed measure theoretical Birkhoff averages. II. Non-linearity, divergence points and Banach space valued spectra. Bull. Sci. Math. **131**(6), 518–558 (2007)
461. Ore, Ø.: Linear equations in non-commutative fields. Ann. Math. (2) **32**(3), 463–477 (1931)
462. Ornstein, D.S., Weiss, B.: Entropy and isomorphism theorems for actions of amenable groups. J. Anal. Math. **48**, 1–141 (1987)
463. Osin, D.V.: Weakly amenable groups. In: Computational and Statistical Group Theory (Las Vegas, NV/Hoboken, NJ, 2001). Contemporary Mathematics, vol. 298, pp. 105–113. American Mathematical Society, Providence, RI (2002)
464. Owings, J.: Problem e2494. Am. Math. Mon. **81** (1974)
465. Ochem P., Rao M.: Minimum frequencies of occurrences of squares and letters in infinite words. In: Mons Days of Theoretical Computer Science. Mons, August 27–30 (2008)
466. Pansiot, J.J.: Hiérarchie et fermeture de certaines classes de tag-systèmes. Acta Inform. **20**, 179–196 (1983)
467. Pansiot, J.J.: Complexité des facteurs des mots infinis engendrés par morphismes itérés. In: Paredaens, J. (ed.) Proceedings of the 11th International Conference on Automata, Languages, and Programming (ICALP). Lecture Notes in Computer Science, vol. 172, pp. 380–389. Springer, Berlin (1984)
468. Pansiot, J.J.: Subword complexities and iteration. Bull. Eur. Assoc. Theor. Comput. Sci. **26**, 55–62 (1985)
469. Parry, W.: On the β-expansions of real numbers. Acta Math. Acad. Sci. Hung. **11**, 401–416 (1960)
470. Paschke, W.L.: Lower bound for the norm of a vertex-transitive graph. Math. Z. **213**(2), 225–239 (1993)
471. Passman, D.S.: The Algebraic Structure of Group Rings. Pure and Applied Mathematics. Wiley-Interscience, New York (1977)
472. Pavlov, R.: On intrinsic ergodicity and weakenings of the specification property. Adv. Math. **295**, 250–270 (2016)
473. Pavlov, R., Schraudner, M.: Classification of sofic projective subdynamics of multidimensional shifts of finite type. Trans. Am. Math. Soc. **367**, 3371–3421 (2015)
474. Peres, Y., Zheng, T.: On groups, slow heat kernel decay yields Liouville property and sharp entropy bounds (2016). ArXiv:1609.05174
475. Perrin, D.: Symbolic dynamics and finite automata. In: Wiedermann, J., Hájek, P. (eds.) Proceedings of the 20th Symposium, Mathematical Foundations of Computer Science 1995. Lecture Notes in Computer Science, vol. 969, pp. 94–104. Springer, Berlin (1995)

476. Perrin, D., Pin, J.E.: Infinite Words. Automata, Semigroups, Logic and Games. Elsevier/Academic Press, Amsterdam (2004)
477. Pete, G.: Probability and geometry on groups (2015). Book in progress
478. Pfister, C.E., Sullivan, W.G.: Billingsley dimension on shift spaces. Nonlinearity **16**(2), 661–682 (2003)
479. Pfister, C.E., Sullivan, W.G.: On the topological entropy of saturated sets. Ergodic Theory Dyn. Syst. **27**(3), 929–956 (2007)
480. Philipp, W.: Limit theorems for lacunary series and uniform distribution mod 1. Acta Arith. **26**(3), 241–251 (1975)
481. Piatetski-Shapiro, I.I.: On the distribution of the fractional parts of the exponential function. Moskov. Gos. Ped. Inst. Uč. Zap. **108**, 317–322 (1957)
482. Picantin, M.: Automatic semigroups vs automaton semigroups (2016). ArXiv:1609.09364
483. Picantin, M.: Automates, (semi)groupes, dualités. Habilitation à diriger des recherches, Univ. Paris Diderot (2017)
484. Pirillo, G.: A proof of Shirshov's theorem. Adv. Math. **124**(1), 94–99 (1996)
485. Pirillo, G., Varricchio, S.: On uniformly repetitive semigroups. Semigroup Forum **49**(1), 125–129 (1994)
486. Pleasants, P.A.B.: Non-repetitive sequences. Math. Proc. Camb. Philos. Soc. **68**, 267–274 (1970)
487. Pytheas Fogg, N.: In: Berthé, V., Ferenczi, S., Mauduit, C., Siegel, A. (eds.) Substitutions in Dynamics, Arithmetics and Combinatorics. Lecture Notes in Mathematics, vol. 1794. Springer, Berlin (2002)
488. Queffélec, M.: Substitution Dynamical Systems—Spectral Analysis. Lecture Notes in Mathematics, vol. 1294. Springer, Berlin (1987)
489. Rabin, M.O., Scott, D.: Finite automata and their decision problems. IBM J. Res. Dev. **3**, 115–125 (1959)
490. Rado, R.: Note on the transfinite case of Hall's theorem on representatives. J. Lond. Math. Soc. **42**, 321–324 (1967)
491. Ramsey, F.P.: On a problem of formal logic. Proc. Lond. Math. Soc. **30**, 264–286 (1930)
492. Randé, B.: Équations fonctionnelles de Mahler et applications aux suites p-régulières. Institut National de Recherche en Informatique et en Automatique (INRIA), Rocquencourt (1992). Thèse, Université de Bordeaux I, Talence, 1992
493. Rao, M.: Last cases of dejeans conjecture. Theor. Comput. Sci. **412**(27), 3010–3018 (2011)
494. Rao, M.: On some generalizations of abelian power avoidability. Theor. Comput. Sci. **601**, 39–46 (2015)
495. Rao, M., Rigo, M., Salimov, P.: Avoiding 2-binomial squares and cubes. Theor. Comput. Sci. **572**, 83–91 (2015)
496. Rao, M., Rosenfeld, M.: Avoidability of long k-abelian repetitions. Math. Comput. **85**(302), 3051–3060 (2016)
497. Rao, M., Rosenfeld, M.: Avoiding two consecutive blocks of same size and same sum over $\mathbb{Z}^2$. Manuscript (2016). ArXiv:1511.05875
498. Regev, A.: Existence of identities in $A \otimes B$. Isr. J. Math. **11**, 131–152 (1972)
499. Reiter, H.: Classical Harmonic Analysis and Locally Compact Groups. Clarendon Press, Oxford (1968)
500. Rényi, A.: Representations for real numbers and their ergodic properties. Acta Math. Acad. Sci. Hung. **8**, 477–493 (1957)
501. Reznick, B.: Some binary partition functions. In: Analytic Number Theory (Allerton Park, IL, 1989). Progress in Mathematics, vol. 85, pp. 451–477. Birkhäuser, Boston, MA (1990)
502. Rigo, M.: Construction of regular languages and recognizability of polynomials. Discret. Math. **254**, 485–496 (2002)
503. Rigo, M.: Formal Languages, Automata and Numeration Systems, Applications to Recognizability and Decidability, vol. 2. ISTE, Wiley (2014)
504. Rigo, M.: Formal Languages, Automata and Numeration Systems, Introduction to Combinatorics on Words, vol. 1. ISTE, Wiley (2014)

505. Rigo, M., Salimov, P.: Another Generalization of Abelian Equivalence: Binomial Complexity of Infinite Words, pp. 217–228. Springer, Berlin/Heidelberg (2013)
506. Rips, E.: Subgroups of small cancellation groups. Bull. Lond. Math. Soc. **14**, 45–47 (1982)
507. Robertson, N., Seymour, P.: Graph minors. v. excluding a planar graph. J. Comb. Theory Ser. B **41**(1), 92–114 (1986)
508. Robinson, R.M.: Undecidability and nonperiodicity for tilings of the plane. Invent. Math. **12**, 177–209 (1971)
509. Robinson, R.M.: Undecidable tiling problems in the hyperbolic plane. Invent. Math. **44**, 159–264 (1978)
510. Rosen, D.: A class of continued fractions associated with certain properly discontinuous groups. Duke Math. J. **21**, 549–563 (1954)
511. Rosenfeld, M.: Every binary pattern of length greater than 14 is Abelian-2-avoidable. In: MFCS 2016. LIPIcs, vol. 58, pp. 81:1–81:11 (2016)
512. Rosset, S.: A new proof of the Amitsur-Levitski identity. Isr. J. Math. **23**(2), 187–188 (1976)
513. Rota, G.C., Strang, G.: A note on the joint spectral radius. Nederl. Akad. Wetensch. Proc. Ser. A 63 = Indag. Math. **22**, 379–381 (1960)
514. Roth, K.F.: Rational approximations to algebraic numbers. Mathematika **2**, 1–20 (1955). corrigendum, 168
515. Rudin, W.: Functional Analysis, 2nd edn. McGraw-Hill, New York (1991)
516. Sablik, M., Fernique, T.: Local rules for computable planar tilings. In: Automata and Journées Automates Cellulaires, Corse, France (2012)
517. Sainte-Marie, C.F.: Question 48. L'intermédiaire des mathématiciens **1**, 107–110 (1894)
518. Sakarovitch, J.: Éléments de théorie des automates. Vuibert (2003). English corrected edition: Elements of Automata Theory, Cambridge University Press, 2009
519. Salo, V., Törmä, I.: Factor colorings of linearly recurrent words (2015). ArXiv:1504.0582
520. Salomaa, A., Soittola, M.: Automata-Theoretic Aspects of Formal Power Series. Springer, New York/Heidelberg (1978). Texts and Monographs in Computer Science
521. Savchuk, D., Vorobets, Y.: Automata generating free products of groups of order 2. J. Algebra **336-1**(1), 53–66 (2011)
522. Schaeffer, D.G.L., Shallit, J.: Subword complexity and k-synchronization. In: Developments in Language Theory. 17th International Conference, DLT 2013, Marne-la-Vallée, France, June 18–21, 2013. Proceedings, no. 7907 in Lecture Notes in Computer Science, pp. 252–263. Springer, Berlin (2013)
523. Schaeffer, L.: Deciding properties of automatic sequences. Master's Thesis, University of Waterloo (2013)
524. Schaeffer, L., Shallit, J.: The critical exponent is computable for automatic sequences. Int. J. Found. Comput. Sci. **23**(8), 1611–1626 (2012)
525. Scheerer, A.M.: Computable absolutely normal numbers and discrepancies. Math. Comput. (2017, to appear). ArXiv:1511.03582
526. Scheerer, A.M.: Normality in Pisot numeration systems. Ergodic Theory Dyn. Syst. **37**(2), 664–672 (2017)
527. Schiffer, J.: Discrepancy of normal numbers. Acta Arith. **47**(2), 175–186 (1986)
528. Schlickewei, H.P.: The $\mathfrak{p}$-adic Thue-Siegel-Roth-Schmidt theorem. Arch. Math. (Basel) **29**(3), 267–270 (1977)
529. Schmidt, W.: Über die Normalität von Zahlen zu verschiedenen Basen. Acta Arith. **7**, 299–309 (1961/1962)
530. Schmidt, W.: Irregularities of distribution. VII. Acta Arith. **21**, 4550 (1972)
531. Schnorr, C.P., Stimm, H.: Endliche automaten und zufallsfolgen. Acta Inform. **1**, 345–359 (1972)
532. Schützenberger, M.P.: Quelques problèmes combinatoires de la théorie des automates. Cours professé à l'Institut de Programmation (1999). J.-F. Perrot
533. Schweiger, F.: Ergodic Theory of Fibred Systems and Metric Number Theory. Oxford Science Publications/The Clarendon Press/Oxford University Press, New York (1995)

534. Segal, D.: Polycyclic Groups. Cambridge Tracts in Mathematics, vol. 82. Cambridge University Press, Cambridge (1983)
535. Selman, A.L.: Arithmetical reducibilities I. Math. Log. Q. **17**(1), 335–350 (1971)
536. Semenov, A.L.: Presburgerness of predicates regular in two number systems. Sibirskii Matematicheskii Zhurnal **18**, 403–418 (1977, in Russian). English translation in Sib. J. Math. **18**, 289–300 (1977)
537. Serre, J.P.: Trees. Springer, Berlin (1980)
538. Seward, B.: Burnside's problem, spanning trees and tilings. Geom. Topol. **18**(1), 179–210 (2014)
539. Shallit, J.: A Second Course in Formal Languages and Automata Theory. Cambridge University Press, Cambridge (2008)
540. Shallit, J.: Decidability and enumeration for automatic sequences: a survey. In: Computer Science—Theory and Applications. Lecture Notes in Computer Science, vol. 7913, pp. 49–63. Springer, Heidelberg (2013)
541. Shallit, J.: Enumeration and automatic sequences. Pure Math. Appl. (PU.M.A.) **25**(1), 96–106 (2015)
542. Shallit, J.O.: A generalization of automatic sequences. Theor. Comput. Sci. **61**, 1–16 (1988)
543. Shalom, Y.: The growth of linear groups. J. Algebra **199**(1), 169–174 (1998)
544. Shur, A.M.: Growth rates of complexity of power-free languages. Theor. Comput. Sci. **411**(34), 3209–3223 (2010)
545. Shur, A.M., Gorbunova, I.A.: On the growth rates of complexity of threshold languages. RAIRO Theor. Inform. Appl. **44**(1), 175–192 (2010)
546. Sidki, S.: Automorphisms of one-rooted trees: growth, circuit structure, and acyclicity. J. Math. Sci. **100**(1), 1925–1943 (2000)
547. Sierpiński, W.: Démonstration élémentaire du théorème de M. Borel sur les nombres absolument normaux et détermination effective d'un tel nombre. Bull. Soc. Math. France **45**, 132–144 (1917)
548. Sierpiński, W.: Sur un problème de la théorie des relations. Ann. Scuola Norm. Sup. Pisa **2**, 285–287 (1933)
549. Silva, P.V., Steinberg, B.: On a class of automata groups generalizing lamplighter groups. Int. J. Algebra Comput. **15**(5-6), 1213–1234 (2005)
550. Sipser, M.: Introduction to the Theory of Computation. International Thomson Publishing (1996)
551. Staiger, L.: Finite-state ω-languages. J. Comput. Syst. Sci. **27**(3), 434–448 (1983)
552. Stone, M.H.: Postulates for the barycentric calculus. Ann. Mat. Pura Appl. (4) **29**, 25–30 (1949)
553. Stoneham, R.: A general arithmetic construction of transcendental non-Liouville normal numbers from rational fractions. Acta Arith. **XVI**, 239–253 (1970). Errata to the paper in Acta Arith. XVII, 1971
554. Strogalov, A.S.: Regular languages with polynomial growth in the number of words. Diskret. Mat. **2**(3), 146–152 (1990)
555. Sudkamp, T.A.: Languages and Machines, An Introduction to the Theory of Computer Science. Addison-Wesley, Reading (1997)
556. Šunić, Z., Ventura, E.: The conjugacy problem in automaton groups is not solvable. J. Algebra **364**(0), 148–154 (2012)
557. Szwarc, R.: A short proof of the grigorchuk-cohen cogrowth theorem. Proc. Am. Math. Soc. **106**(3), 663–665 (1989-07)
558. Tamari, D.: A refined classification of semi-groups leading to generalised polynomial rings with a generalized degree concept. In: Proceedings of ICM, vol. 3, pp. 439–440, Amsterdam (1954)
559. Tarannikov, Y.: The minimal density of a letter in an infinite ternary square-free word is 0.2746 .... J. Integer Seq. **5**(2), Art. 02.2.2 (2002)
560. Tatch Moore, J.: Hindman's theorem, Ellis's lemma, and Thompson's group f. Zb. Rad. (Beogr.) **17**(25)(Selected topics in combinatorial analysis), 171–187 (2015)

561. Taylor, A.: A canonical partition relation for finite subsets of ω. J. Comb. Theory. Ser. A **21**, 137–146 (1976)
562. Thue, A.: Über unendliche Zeichenreihen. Norske vid. Selsk. Skr. Mat. Nat. Kl. **7**, 1–22 (1906). Reprinted in *Selected Mathematical Papers of Axel Thue*, T. Nagell, editor, Universitetsforlaget, Oslo, 1977, pp. 139–158
563. Thue, A.: Über die gegenseitige Lage gleicher Teile gewisser Zeichenreihen. Norske vid. Selsk. Skr. Mat. Nat. Kl. **1**, 1–67 (1912). Reprinted in *Selected Mathematical Papers of Axel Thue*, T. Nagell, editor, Universitetsforlaget, Oslo, 1977, pp. 413–478
564. Tits, J.: Free subgroups in linear groups. J. Algebra **20**(2), 250–270 (1972)
565. Tointon, M.C.H.: Characterizations of algebraic properties of groups in terms of harmonic functions. Groups Geom. Dyn. **10**(3), 1007–1049 (2016)
566. Töpfer, T.: Zero order estimates for functions satisfying generalized functional equations of Mahler type. Acta Arith. **85**(1), 1–12 (1998)
567. Törmä, I.: Quantifier extensions of multidimensional sofic shifts. Proc. Am. Math. Soc. **143**, 4775–4790 (2015)
568. Tunev, I.N., Shur, A.M.: On Two Stronger Versions of Dejean's Conjecture, pp. 800–812. Springer, Berlin/Heidelberg (2012)
569. Turing, A.: On computable numbers, with an application to the entscheidungsproblem. Proc. Lond. Math. Soc. **42**(2), 230265 (1936)
570. Turing, A.: A note on normal numbers. In: Collected Works of Alan M. Turing, Pure Mathematics, pp. 117–119. North Holland, Amsterdam (1992). Notes of editor, 263–265
571. Ugalde, E.: An alternative construction of normal numbers. J. Théorie Nombres Bordeaux **12**, 165–177 (2000)
572. Vaaler, J.D.: Some extremal functions in Fourier analysis. Bull. Am. Math. Soc. (N.S.) **12**(2), 183–216 (1985)
573. Vandehey, J.: A simpler normal number construction for simple Lüroth series. J. Integer Seq. **17**(6), Article 14.6.1, 18 (2014)
574. Vaughan-Lee, M.: The restricted Burnside problem. London Mathematical Society Monographs. New Series, vol. 8. Oxford University Press, Oxford (1993)
575. Volkmann, B.: Über Hausdorffsche Dimensionen von Mengen, die durch Zifferneigenschaften charakterisiert sind. VI. Math. Z. **68**, 439–449 (1958)
576. von Neumann, J.: Zur allgemeinen Theorie des Masses. Fundam. Math. **13**, 73–116 and 333 (1929). Collected works, vol. I, pages 599–643
577. von Neumann, J.: Einige sätze über messbare abbildungen. Ann. Math. (2) **33**(3), 574–586 (1932)
578. Wall, D.D.: Normal numbers. Ph.D. thesis, University of California, Berkeley, CA (1949)
579. Walters, P.: An Introduction to Ergodic Theory. Springer, New York (1982)
580. Wang, H.: Proving theorems by pattern recognition, II. Bell Syst. Tech. J. **40**(1), 1–41 (1961)
581. Weiss, B.: On the work of V. A. Rokhlin in ergodic theory. Ergodic Theory Dyn. Syst. **9**(4), 619–627 (1989)
582. Weiss, B.: Sofic groups and dynamical systems. Sankhyā Ser. A **62**(3), 350–359 (2000). Ergodic theory and harmonic analysis (Mumbai, 1999)
583. Weiss, B.: Monotileable amenable groups. In: Topology, Ergodic Theory, Real Algebraic Geometry. American Mathematical Society Translations Series 2, vol. 202, pp. 257–262. American Mathematical Society (2001)
584. Whyte, K.: Amenability, Bilipschitz equivalence, and the Von Neumann Conjecture. Duke Math. J. **99**(1), 93–112 (1999)
585. Willis, G.A.: Probability measures on groups and some related ideals in group algebras. J. Funct. Anal. **92**(1), 202–263 (1990)
586. Woess, W.: Graphs and groups with tree-like properties. J. Comb. Theory Ser. B **47**(3), 361–371 (1989)
587. Woess, W.: Random walks on infinite graphs and groups—a survey on selected topics. Bull. Lond. Math. Soc. **26**, 1–60 (1994)

588. Woess, W.: Random Walks on Infinite Graphs and Groups. Cambridge University Press, Cambridge (2000)
589. Wojcik, C., Zamboni, L.Q.: Monochromatic factorisations of words and periodicity (2017). Preprint
590. Yu, S.: Regular languages. In: Rozenberg, G., Salomaa, A. (eds.) Handbook of Formal Languages, vol. 1, pp. 41–110. Springer, Berlin (1997)
591. Zamboni, L.Q.: A note on coloring factors of words (2010). In: Oberwolfach Report 37/2010, Mini-workshop: Combinatorics on Words, August, 22–27
592. Zeckendorf, E.: Représentation des nombres naturels par une somme de nombres de Fibonacci ou de nombres Lucas. Bull. Soc. R. Liége **41**, 179–182 (1972)
593. Zel′manov, E.I.: Solution of the restricted Burnside problem for groups of odd exponent. Izv. Akad. Nauk SSSR Ser. Mat. **54**(1), 42–59, 221 (1990)
594. Zel′manov, E.I.: Solution of the restricted Burnside problem for 2-groups. Matematicheskiĭ Sbornik **182**(4), 568–592 (1991)
595. Zimin, A.: Blocking sets of terms. Math. USSR Sbornik **47**(2), 353–364 (1984)

Index

V. Berthé, M. Rigo (eds.), *Sequences, Groups, and Number Theory*,
Trends in Mathematics, https://doi.org/10.1007/978-3-319-69152-7

Q

R

The manufacturer's authorised representative in the EU is Springer Nature Customer Service Centre GmbH, Europaplatz 3, 69115 Heidelberg, Germany. If you have any concerns regarding our products, please contact ProductSafety@springernature.com

Printed and bound by CPI Group (UK) Ltd, Croydon, CR0 4YY
30/09/2026
02225643-0015